$168.95 per copy (in United States).
Price subject to change without prior notice.

0059

RSMeans

Square Foot Costs

30th Annual Edition

R.S. Means Company, Inc.
Construction Publishers & Consultants
63 Smiths Lane
Kingston, MA 02364-3008
(781) 422-5000

Copyright©2008 by R.S. Means Company, Inc.
All rights reserved.

Printed in the United States of America.
ISSN 0068-3531
ISBN 978-0-87629-209-9

Senior Editor
Barbara Balboni

Contributing Editors
Christopher Babbitt
Ted Baker
Robert A. Bastoni
John H. Chiang, PE
Gary W. Christensen
David G. Drain, PE
Cheryl Elsmore
Robert J. Kuchta
Robert C. McNichols
Melville J. Mossman, PE
Jeannene D. Murphy
Stephen C. Plotner
Eugene R. Spencer
Marshall J. Stetson
Phillip R. Waier, PE

Senior Vice President & General Manager
John Ware

Vice President of Operations
Dave Walsh

Vice President of Sales & Marketing
Sev Ritchie

Marketing Director
John M. Shea

Director of Product Development
Thomas J. Dion

Engineering Manager
Bob Mewis, CCC

Production Manager
Michael Kokernak

Production Coordinator
Jill Goodman

Technical Support
Jonathan Forgit
Mary Lou Geary
Roger Hancock
Gary L. Hoitt
Genevieve Medeiros
Debbie Panarelli
Paula Reale-Camelio
Kathryn S. Rodriguez
Sheryl A. Rose

Book & Cover Design
Norman R. Forgit

This book is printed on recycled paper (10% PCW cover, 20% PCW text) using soy-based printing ink. This book is recyclable.

Foreword

Our Mission

Since 1942, RSMeans has been actively engaged in construction cost publishing and consulting throughout North America.

Today, over 60 years after RSMeans began, our primary objective remains the same: to provide you, the construction and facilities professional, with the most current and comprehensive construction cost data possible.

Whether you are a contractor, an owner, an architect, an engineer, a facilities manager, or anyone else who needs a reliable construction cost estimate, you'll find this publication to be a highly useful and necessary tool.

With the constant flow of new construction methods and materials today, it's difficult to find the time to look at and evaluate all the different construction cost possibilities. In addition, because labor and material costs keep changing, last year's cost information is not a reliable basis for today's estimate or budget.

That's why so many construction professionals turn to RSMeans. We keep track of the costs for you, along with a wide range of other key information, from city cost indexes . . . to productivity rates . . . to crew composition . . . to contractor's overhead and profit rates.

RSMeans performs these functions by collecting data from all facets of the industry and organizing it in a format that is instantly accessible to you. From the preliminary budget to the detailed unit price estimate, you'll find the data in this book useful for all phases of construction cost determination.

The Staff, the Organization, and Our Services

When you purchase one of RSMeans' publications, you are, in effect, hiring the services of a full-time staff of construction and engineering professionals.

Our thoroughly experienced and highly qualified staff works daily at collecting, analyzing, and disseminating comprehensive cost information for your needs. These staff members have years of practical construction experience and engineering training prior to joining the firm. As a result, you can count on them not only for the cost figures, but also for additional background reference information that will help you create a realistic estimate.

The RSMeans organization is always prepared to help you solve construction problems through its four major divisions: Construction and Cost Data Publishing, Electronic Products and Services, Consulting and Business Solutions, and Professional Development Services.

Besides a full array of construction cost estimating books, RSMeans also publishes a number of other reference works for the construction industry. Subjects include construction estimating and project and business management; special topics such as HVAC, roofing, plumbing, and hazardous waste remediation; and a library of facility management references.

In addition, you can access all of our construction cost data electronically using *Means CostWorks®* CD or on the Web using Means CostWorks.com.

What's more, you can increase your knowledge and improve your construction estimating and management performance with an RSMeans Construction Seminar or In-House Training Program. These two-day seminar programs offer unparalleled opportunities for everyone in your organization to get updated on a wide variety of construction-related issues.

RSMeans is also a worldwide provider of construction cost management and analysis services for commercial and government owners.

In short, RSMeans can provide you with the tools and expertise for constructing accurate and dependable construction estimates and budgets in a variety of ways.

Robert Snow Means Established a Tradition of Quality That Continues Today

Robert Snow Means spent years building RSMeans, making certain he always delivered a quality product.

Today, at RSMeans, we do more than talk about the quality of our data and the usefulness of our books. We stand behind all of our data, from historical cost indexes to construction materials and techniques to current costs.

If you have any questions about our products or services, please call us toll-free at 1-800-334-3509. Our customer service representatives will be happy to assist you. You can also visit our Web site at www.rsmeans.com.

Table of Contents

Related RSMeans Products and Services

The Engineers at RSMeans, suggest the following products and services as companion information resources to *RSMeans Square Foot Costs:*

Construction Cost Data Books
Building Construction Cost Data 2009
Assemblies Cost Data 2009
Light Commercial Cost Data 2009
Facilities Construction Cost Data 2009

Reference Books
Building Security: Strategies & Costs
Designing & Building with the IBC, 2nd Ed.
Estimating Building Costs
Estimating Handbook, 2nd Ed.
How to Estimate with Means Data and CostWorks, 3rd Ed.
Project Scheduling and Management for Construction, 3rd Ed.
Square Foot & Assemblies Estimating Methods, 3rd Ed.

Seminars and In-House Training
Square Foot & Assemblies Estimating
Means CostWorks® Training
Means Data for Job Order Contracting (JOC)
Plan Reading & Material Takeoff
Scheduling & Project Management

RSMeans on the Internet
Visit RSMeans at **www.rsmeans.com.** The site contains useful interactive cost and reference material. Request or download **FREE** estimating software demos. Visit our bookstore for convenient ordering and to learn more about new publications and companion products.

RSMeans Electronic Data
Get the information found in RSMeans cost books electronically on *Means CostWorks®* CD or on the Web.

RSMeans Business Solutions
Engineers and Analysts offer research studies, benchmark analysis, predictive cost modeling, analytics, job order contracting, and real property management consultation, as well as custom-designed, web-based dashboards and calculators that apply RCD/RSMeans extensive databases. Clients include federal government agencies, architects, construction management firms, and institutional organizations such as school systems, health care facilities, associations, and corporations.

RSMeans for Job Order Contracting (JOC)
Best practice JOC tools for cost estimating and project management to help streamline delivery processes for renovation projects. Renovation is a $147 billion market in the U.S., and includes projects in school districts, municipalities, health care facilities, colleges and universities, and corporations.
- RSMeans Engineers consult in contracting methods and conduct JOC Facility Audits
- JOCWorks™ Software (Basic, Advanced, PRO levels)
- RSMeans Job Order Contracting Cost Data for the entire U.S.

Construction Costs for Software Applications
Over 25 unit price and assemblies cost databases are available through a number of leading estimating and facilities management software providers (listed below). For more information see the "Other RSMeans Products" pages at the back of this publication.

MeansData™ is also available to federal, state, and local government agencies as multi-year, multi-seat licenses.

- 4Clicks-Solutions, LLC
- ArenaSoft Estimating
- Beck Technology
- BSD – Building Systems Design, Inc.
- CMS – Construction Management Software
- Corecon Technologies, Inc.
- CorVet Systems
- Estimating Systems, Inc.
- Maximus Asset Solutions
- Sage Timberline Office
- US Cost, Inc.
- VFA – Vanderweil Facility Advisers
- WinEstimator, Inc.

How the Book Is Built: An Overview

A Powerful Construction Tool

You have in your hands one of the most powerful construction tools available today. A successful project is built on the foundation of an accurate and dependable estimate. This book will enable you to construct just such an estimate.

For the casual user the book is designed to be:

- quickly and easily understood so you can get right to your estimate.
- filled with valuable information so you can understand the necessary factors that go into the cost estimate.

For the regular user, the book is designed to be:

- a handy desk reference that can be quickly referred to for key costs.
- a comprehensive, fully reliable source of current construction costs for typical building structures when only minimum dimensions, specifications, and technical data are available. It is specifically useful in the conceptual or planning stage for preparing preliminary estimates which can be supported and developed into complete engineering estimates for construction.
- a source book for rapid estimating of replacement costs and preparing capital expenditure budgets.

To meet all of these requirements, we have organized the book into the following clearly defined sections.

How To Use the Book: The Details

This section contains an in-depth explanation of how the book is arranged and how you can use it to determine a reliable construction cost estimate. It includes information about how we develop our cost figures and how to completely prepare your estimate.

Residential Section

Model buildings for four classes of construction—economy, average, custom, and luxury—are developed and shown with complete costs per square foot.

Commercial, Industrial, Institutional Section

This section contains complete costs for 75 typical model buildings expressed as costs per square foot.

Assemblies Section

The cost data in this section has been organized in an "Assemblies" format. These assemblies are the functional elements of a building and are arranged according to the 7 divisions of the UNIFORMAT II classification system. The costs in this section include standard markups for the overhead and profit of the installing contractor. The assemblies costs are those used to calculate the total building costs in the Commercial/Industrial/Institutional section.

Reference Section

This section includes information on General Conditions, Location Factors, Historical Cost Indexes, a Glossary, and a listing of Abbreviations used in this book.

Location Factors: You can adjust the total project costs to over 900 locations throughout the U.S. and Canada by using the data in this section.

Historical Cost Indexes: These indexes provide you with data to adjust construction costs over time.

Glossary: A listing of common terms related to construction and their definitions is included.

Abbreviations: A listing of the abbreviations used throughout this book, along with the terms they represent, is included.

The Scope of This Book

This book is designed to be as comprehensive and as easy to use as possible. To that end we have made certain assumptions and limited its scope in two key ways:

1. We have established material prices based on a national average.
2. We have computed labor costs based on a 30-city national average of union wage rates.

All costs and adjustments in this manual were developed from the *RSMeans Building Construction Cost Data, Assemblies Cost Data, Residential Cost Data,* and *Light Commercial Cost Data.* Costs include labor, material, overhead and profit, and general conditions and architects' fees where specified. No allowance is made for sales tax, financing, premiums for material and labor, or unusual business and weather conditions.

What's Behind the Numbers? The Development of Cost Data

The staff at RSMeans continuously monitors developments in the construction industry in order to ensure reliable, thorough, and up-to-date cost information.

While overall construction costs may vary relative to general economic conditions, price fluctuations within the industry are dependent upon many factors. Individual price variations may, in fact, be opposite to overall economic trends. Therefore, costs are continually monitored and complete updates are published yearly. Also, new items are frequently added in response to changes in materials and methods.

Costs—$ (U.S.)

All costs represent U.S. national averages and are given in U.S. dollars. The RSMeans Location Factors can be used to adjust costs to a particular location. The Location Factors for Canada can be used to adjust U.S. national averages to local costs in Canadian dollars. No exchange rate conversion is necessary.

Factors Affecting Costs

Due to the almost limitless variation of building designs and combinations of construction methods and materials, costs in this book must be used with discretion. The editors have made every effort to develop costs for "typical" building types. By definition, a square foot estimate, due to the short time and limited amount of detail required, involves a relatively lower degree of accuracy than a detailed unit cost estimate, for example. The accuracy of an estimate will increase as the project is defined in more detailed components. The user should examine closely and compare the specific requirements of the project being estimated with the components included in the models in this book. Any differences should be accounted for in the estimate.

In many building projects, there may be factors that increase or decrease the cost beyond the range shown in this manual. Some of these factors are:

- Substitution of building materials or systems for those used in the model
- Custom-designed finishes, fixtures, and/or equipment
- Special structural qualities (allowance for earthquake, future expansion, high winds, long spans, and unusual shape)
- Special foundations and substructures to compensate for unusual soil conditions
- Inclusion of typical "owner-supplied" equipment in the building cost (special testing and monitoring equipment for hospitals, kitchen equipment for hotels, and offices that have cafeteria accommodations for employees, etc.)
- An abnormal shortage of labor or materials

- Isolated building site or rough terrain that would affect the transportation of personnel, material, or equipment
- Unusual climatic conditions during the construction process

We strongly urge that, for maximum accuracy, these factors be considered each time a structure is being evaluated.

If users require greater accuracy than this manual can provide, the editors recommend that the *RSMeans Building Construction Cost Data* or the *RSMeans Assemblies Cost Data* be consulted.

Final Checklist

Estimating can be a straightforward process provided you remember the basics. Here's a checklist of some of the steps you should remember to complete before finalizing your estimate.

Did you remember to . . .

- include the Location Factor for your city?
- mark up the entire estimate sufficiently for your purposes?
- include all components of your project in the final estimate?
- double check your figures for accuracy?
- call RSMeans if you have any questions about your estimate or the data you've found in our publications?

Remember, RSMeans stands behind its publications. If you have any questions about your estimate, about the costs you've used from our books, or even about the technical aspects of the job that may affect your estimate, feel free to call the RSMeans editors at 1-800-334-3509.

Residential Section

Table of Contents

Introduction to Residential Section

The residential section of this manual contains costs per square foot for four classes of construction in seven building types. Costs are listed for various exterior wall materials which are typical of the class and building type. There are cost tables for Wings and Ells with modification tables to adjust the base cost of each class of building. Non-standard items can easily be added to the standard structures.

Cost estimating for a residence is a three-step process:
(1) Identification
(2) Listing dimensions
(3) Calculations

Guidelines and a sample cost estimating procedure are shown on the following pages.

Identification

To properly identify a residential building, the class of construction, type, and exterior wall material must be determined. Page 6 has drawings and guidelines for determining the class of construction. There are also detailed specifications and additional drawings at the beginning of each set of tables to further aid in proper building class and type identification.

Sketches for eight types of residential buildings and their configurations are shown on pages 8 and 9. Definitions of living area are next to each sketch. Sketches and definitions of garage types are on page 10.

Living Area

Base cost tables are prepared as costs per square foot of living area. The living area of a residence is that area which is suitable and normally designed for full time living. It does not include basement recreation rooms or finished attics, although these areas are often considered full time living areas by the owners.

Living area is calculated from the exterior dimensions without the need to adjust for exterior wall thickness. When calculating the living area of a 1-1/2 story, two story, three story or tri-level residence, overhangs and other differences in size and shape between floors must be considered.

Only the floor area with a ceiling height of six feet or more in a 1-1/2 story residence is considered living area. In bi-levels and tri-levels, the areas that are below grade are considered living area, even when these areas may not be completely finished.

Base Tables and Modifications

Base cost tables show the base cost per square foot without a basement, with one full bath and one full kitchen for economy and average homes, and an additional half bath for custom and luxury models. Adjustments for finished and unfinished basements are part of the base cost tables. Adjustments for multi family residences, additional bathrooms, townhouses, alternate roofs, and air conditioning and heating systems are listed in modification tables below the base cost tables.

Costs for other modifications, including garages, breezeways and site improvements, are in the modification tables on pages 58 to 61.

Listing of Dimensions

To use this section of the manual, only the dimensions used to calculate the horizontal area of the building and additions and modifications are needed. The dimensions, normally the length and width, can come from drawings or field measurements. For ease in calculation, consider measuring in tenths of feet, i.e., 9 ft. 6 in. = 9.5 ft., 9 ft. 4 in. = 9.3 ft.

In all cases, make a sketch of the building. Any protrusions or other variations in shape should be noted on the sketch with dimensions.

Calculations

The calculations portion of the estimate is a two-step activity:
(1) The selection of appropriate costs from the tables
(2) Computations

Selection of Appropriate Costs

To select the appropriate cost from the base tables, the following information is needed:
(1) Class of construction
    (a) Economy
    (b) Average
    (c) Custom
    (d) Luxury
(2) Type of residence
    (a) 1 story
    (b) 1-1/2 story
    (c) 2 story
    (d) 3 story
    (e) Bi-level
    (f) Tri-level
(3) Occupancy
    (a) One family
    (b) Two family
    (c) Three family
(4) Building configuration
    (a) Detached
    (b) Town/Row house
    (c) Semi-detached
(5) Exterior wall construction
    (a) Wood frame
    (b) Brick veneer
    (c) Solid masonry
(6) Living areas

Modifications are classified by class, type, and size.

Computations

The computation process should take the following sequence:
(1) Multiply the base cost by the area.
(2) Add or subtract the modifications.
(3) Apply the location modifier.

When selecting costs, interpolate or use the cost that most nearly matches the structure under study. This applies to size, exterior wall construction, and class.

How to Use the Residential Square Foot Cost Pages

The following is a detailed explanation of a sample entry in the Residential Square Foot Cost Section. Each bold number below corresponds to the item being described on the facing page with the appropriate component of the sample entry following in parenthesis.

Prices listed are costs that include overhead and profit of the installing contractor. Total model costs include an additional mark-up for General Contractors' overhead and profit, and fees specific to class of construction.

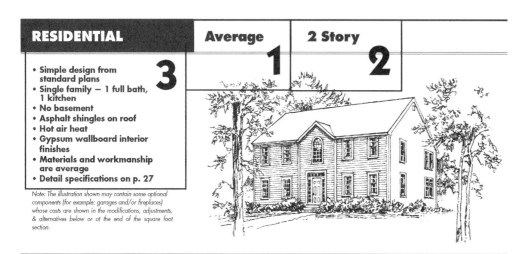

RESIDENTIAL | **Average** **1** | **2 Story** **2**

3
- Simple design from standard plans
- Single family — 1 full bath, 1 kitchen
- No basement
- Asphalt shingles on roof
- Hot air heat
- Gypsum wallboard interior finishes
- Materials and workmanship are average
- Detail specifications on p. 27

Note: The illustration shown may contain some optional components (for example: garages and/or fireplaces) whose costs are shown in the modifications, adjustments, & alternatives below or at the end of the square foot section.

Base cost per square foot of living area

5 Living Area

Exterior Wall **4**	1000	1200	1400	1600	1800	2000	2200	2600	3000	3400	3800
Wood Siding - Wood Frame	118.25	107.30	102.20	98.65	95.20	91.25	88.70	83.80	78.90	76.75	74.80
Brick Veneer - Wood Frame	128.75	117.05	111.35	107.35	103.35	99.15	**6**	90.55	85.20	82.70	80.50
Stucco on Wood Frame	115.20	104.45	99.65	96.20	92.80	89.00		81.80	77.05	75.00	73.20
Solid Masonry	143.40	130.60	124.00	119.45	114.70	110.10	106.60	99.95	93.85	91.00	88.35
Finished Basement, Add **7**	19.40	19.10	18.50	18.05	17.65	17.35	**8**	16.35	15.90	15.65	15.40
Unfinished Basement, Add	7.55	7.05	6.65	6.35	6.05	5.90		5.30	5.00	4.80	4.70

Modifications

Add to the total cost
Upgrade Kitchen Cabinets	$ + 3760
Solid Surface Countertops (Included)	
Full Bath - including plumbing, wall and floor finishes	+ 5522
Half Bath - including plumbing, wall and floor finishes **9**	+ 3416
One Car Attached Garage	+ 12,594
One Car Detached Garage	+ 16,766
Fireplace & Chimney	+ 5536

Adjustments

For multi family - add to total cost
Additional Kitchen	$ + 6433
Additional Bath	+ 5522
Additional Entry & Exit	+ 1622
Separate Heating	+ 1376
Separate Electric	+ 1870

For Townhouse/Rowhouse - Multiply cost per square foot by **10**
Inner Unit	+ .90
End Unit	+ .95

Alternatives

Add to or deduct from the cost per square foot of living area
Cedar Shake Roof	+ 2.25
Clay Tile Roof	+ 2.40
Slate Roof **11**	+ 3.35
Upgrade Walls to Skim Coat Plaster	+ .37
Upgrade Ceilings to Textured Finish	+ .47
Air Conditioning, in Heating Ductwork	+ 2.04
In Separate Ductwork	+ 5.12
Heating Systems, Hot Water	+ 1.62
Heat Pump	+ 2.50
Electric Heat	– .64
Not Heated	– 2.86

Additional upgrades or components
Kitchen Cabinets & Countertops	Page 58
Bathroom Vanities	59
Fireplaces & Chimneys	59
Windows, Skylights & Dormers	59
Appliances	60
Breezeways & Porches	60
Finished Attic	60
Garages **12**	61
Site Improvements	61
Wings & Ells	37

Important: See the Reference Section for Location Factors (to adjust for your city) and Estimating Forms

1 Class of Construction (Average)

The class of construction depends upon the design and specifications of the plan. The four classes are economy, average, custom, and luxury.

2 Type of Residence (2 Story)

The building type describes the number of stories or levels in the model. The seven building types are 1 story, 1-1/2 story, 2 story, 2-1/2 story, 3 story, bi-level, and tri-level.

3 Specification Highlights (Hot Air Heat)

These specifications include information concerning the components of the model, including the number of baths, roofing types, HVAC systems, and materials and workmanship. If the components listed are not appropriate, modifications can be made by consulting the information shown lower on the page or the Assemblies Section.

4 Exterior Wall System (Wood Siding - Wood Frame)

This section includes the types of exterior wall systems and the structural frame used. The exterior wall systems shown are typical of the class of construction and the building type shown.

5 Living Areas (2000 SF)

The living area is that area of the residence which is suitable and normally designed for full time living. It does not include basement recreation rooms or finished attics. Living area is calculated from the exterior dimensions without the need to adjust for exterior wall thickness. When calculating the living area of a 1-1/2 story, 2 story, 3 story or tri-level residence, overhangs and other differences in size and shape between floors must be considered. Only the floor area with a ceiling height of six feet or more in a 1-1/2 story residence is considered living area. In bi-levels and tri-levels, the areas that are below grade are considered living area, even when these areas may not be completely finished. A range of various living areas for the residential model are shown to aid in selection of values from the matrix.

6 Base Costs per Square Foot of Living Area ($91.25)

Base cost tables show the cost per square foot of living area without a basement, with one full bath and one full kitchen for economy and average homes, and an additional half bath for custom and luxury models. When selecting costs, interpolate or use the cost that most nearly matches the residence under consideration for size, exterior wall system, and class of construction. Prices listed are costs that include overhead and profit of the installing contractor, a general contractor markup, and an allowance for plans that vary by class of construction. For additional information on contractor overhead and architectural fees, see the Reference Section.

7 Basement Types (Finished)

The two types of basements are finished and unfinished. The specifications and components for both are shown on the Building Classes page in the Introduction to this section.

8 Additional Costs for Basements ($17.35)

These values indicate the additional cost per square foot of living area for either a finished or an unfinished basement.

9 Modifications and Adjustments (Upgrade Kitchen Cabinets $3760)

Modifications and Adjustments are costs added to or subtracted from the total cost of the residence. The total cost of the residence is equal to the cost per square foot of living area times the living area. Typical modifications and adjustments include kitchens, baths, garages and fireplaces.

10 Multiplier for Townhouse/ Rowhouse (Inner Unit .90)

The multipliers shown adjust the base costs per square foot of living area for the common wall condition encountered in townhouses or rowhouses.

11 Alternatives (Skim Coat Plaster $.37)

Alternatives are costs added to or subtracted from the base cost per square foot of living area. Typical alternatives include variations in kitchens, baths, roofing, and air conditioning and heating systems.

12 Additional Upgrades or Components (Wings & Ells page 37)

Costs for additional upgrades or components, including wings or ells, breezeways, porches, finished attics, and site improvements, are shown in other locations in the Residential Square Foot Cost Section.

Building Classes

Economy Class

An economy class residence is usually built from stock plans. The materials and workmanship are sufficient to satisfy building codes. Low construction cost is more important than distinctive features. The overall shape of the foundation and structure is seldom other than square or rectangular.

An unfinished basement includes 7' high 8" thick foundation wall composed of either concrete block or cast-in-place concrete.

Included in the finished basement cost are inexpensive paneling or drywall as the interior finish on the foundation walls, a low cost sponge backed carpeting adhered to the concrete floor, a drywall ceiling, and overhead lighting.

Custom Class

A custom class residence is usually built from plans and specifications with enough features to give the building a distinction of design. Materials and workmanship are generally above average with obvious attention given to construction details. Construction normally exceeds building code requirements.

An unfinished basement includes a 7'-6" high 10" thick cast-in-place concrete foundation wall or a 7'-6" high 12" thick concrete block foundation wall.

A finished basement includes painted drywall on insulated 2" x 4" wood furring as the interior finish to the concrete walls, a suspended ceiling, carpeting adhered to the concrete floor, overhead lighting and heating.

Average Class

An average class residence is a simple design and built from standard plans. Materials and workmanship are average, but often exceed minimum building codes. There are frequently special features that give the residence some distinctive characteristics.

An unfinished basement includes 7'-6" high 8" thick foundation wall composed of either cast-in-place concrete or concrete block.

Included in the finished basement are plywood paneling or drywall on furring that is fastened to the foundation walls, sponge backed carpeting adhered to the concrete floor, a suspended ceiling, overhead lighting and heating.

Luxury Class

A luxury class residence is built from an architect's plan for a specific owner. It is unique both in design and workmanship. There are many special features, and construction usually exceeds all building codes. It is obvious that primary attention is placed on the owner's comfort and pleasure. Construction is supervised by an architect.

An unfinished basement includes 8' high 12" thick foundation wall that is composed of cast-in-place concrete or concrete block.

A finished basement includes painted drywall on 2" x 4" wood furring as the interior finish, suspended ceiling, tackless carpet on wood subfloor with sleepers, overhead lighting and heating.

Configurations

Detached House

This category of residence is a free-standing separate building with or without an attached garage. It has four complete walls.

Semi-Detached House

This category of residence has two living units side-by-side. The common wall is a fireproof wall. Semi-detached residences can be treated as a row house with two end units. Semi-detached residences can be any of the building types.

Town/Row House

This category of residence has a number of attached units made up of inner units and end units. The units are joined by common walls. The inner units have only two exterior walls. The common walls are fireproof. The end units have three walls and a common wall. Town houses/row houses can be any of the building types.

Building Types

One Story

This is an example of a one-story dwelling. The living area of this type of residence is confined to the ground floor. The headroom in the attic is usually too low for use as a living area.

One-and-a-half Story

The living area in the upper level of this type of residence is 50% to 90% of the ground floor. This is made possible by a combination of this design's high-peaked roof and/or dormers. Only the upper level area with a ceiling height of 6 feet or more is considered living area. The living area of this residence is the sum of the ground floor area plus the area on the second level with a ceiling height of 6 feet or more.

One Story with Finished Attic

The main living area in this type of residence is the ground floor. The upper level or attic area has sufficient headroom for comfortable use as a living area. This is made possible by a high peaked roof. The living area in the attic is less than 50% of the ground floor. The living area of this type of residence is the ground floor area only. The finished attic is considered an adjustment.

Two Story

This type of residence has a second floor or upper level area which is equal or nearly equal to the ground floor area. The upper level of this type of residence can range from 90% to 110% of the ground floor area, depending on setbacks or overhangs. The living area is the sum of the ground floor area and the upper level floor area.

Two-and-one-half Story

This type of residence has two levels of equal or nearly equal area and a third level which has a living area that is 50% to 90% of the ground floor. This is made possible by a high peaked roof, extended wall heights and/or dormers. Only the upper level area with a ceiling height of 6 feet or more is considered living area. The living area of this residence is the sum of the ground floor area, the second floor area and the area on the third level with a ceiling height of 6 feet or more.

Bi-level

This type of residence has two living areas, one above the other. One area is about 4 feet below grade and the second is about 4 feet above grade. Both areas are equal in size. The lower level in this type of residence is originally designed and built to serve as a living area and not as a basement. Both levels have full ceiling heights. The living area is the sum of the lower level area and the upper level area.

Three Story

This type of residence has three levels which are equal or nearly equal. As in the 2 story residence, the second and third floor areas may vary slightly depending on setbacks or overhangs. The living area is the sum of the ground floor area and the two upper level floor areas.

Tri-level

This type of residence has three levels of living area: one at grade level, one about 4 feet below grade, and one about 4 feet above grade. All levels are originally designed to serve as living areas. All levels have full ceiling heights. The living area is a sum of the areas of each of the three levels.

Garage Types

Attached Garage

Shares a common wall with the dwelling. Access is typically through a door between dwelling and garage.

Basement Garage

Constructed under the roof of the dwelling but below the living area.

Built-In Garage

Constructed under the second floor living space and above basement level of dwelling. Reduces gross square feet of living area.

Detached Garage

Constructed apart from the main dwelling. Shares no common area or wall with the dwelling.

Building Components

1. Excavation
2. Sill Plate
3. Basement Window
4. Floor Joist
5. Shoe Plate
6. Studs
7. Drywall
8. Plate
9. Ceiling Joists
10. Rafters
11. Collar Ties
12. Ridge Rafter
13. Roof Sheathing
14. Roof Felt
15. Roof Shingles
16. Flashing
17. Flue Lining
18. Chimney
19. Roof Shingles
20. Gutter
21. Fascia
22. Downspout
23. Shutter
24. Window
25. Wall Shingles
26. Building Paper
27. Wall Sheathing
28. Fire Stop
29. Dampproofing
30. Foundation Wall
31. Backfill
32. Drainage Stone
33. Drainage Tile
34. Wall Footing
35. Gravel
36. Concrete Slab
37. Column Footing
38. Pipe Column
39. Expansion Joint
40. Girder
41. Sub-floor
42. Finish Floor
43. Attic Insulation
44. Soffit
45. Ceiling Strapping
46. Wall Insulation
47. Cross Bridging
48. Bulkhead Stairs

Exterior Wall Construction

Typical Frame Construction

Typical wood frame construction consists of wood studs with insulation between them. A typical exterior surface is made up of sheathing, building paper and exterior siding consisting of wood, vinyl, aluminum or stucco over the wood sheathing.

Brick Veneer

Typical brick veneer construction consists of wood studs with insulation between them. A typical exterior surface is sheathing, building paper and an exterior of brick tied to the sheathing, with metal strips.

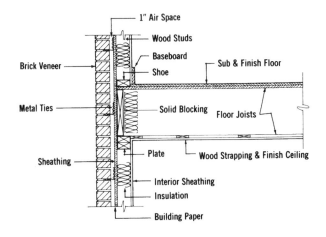

Stone

Typical solid masonry construction consists of a stone or block wall covered on the exterior with brick, stone or other masonry.

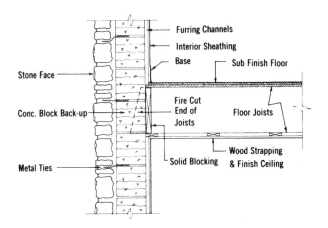

Residential Cost Estimate Worksheet

Worksheet Instructions

The residential cost estimate worksheet can be used as an outline for developing a residential construction or replacement cost. It is also useful for insurance appraisals. The design of the worksheet helps eliminate errors and omissions. To use the worksheet, follow the example below.

1. Fill out the owner's name, residence address, the estimator or appraiser's name, some type of project identifying number or code, and the date.

2. Determine from the plans, specifications, owner's description, photographs or any other means possible the class of construction. The models in this book use economy, average, custom and luxury as classes. Fill in the appropriate box.

3. Fill in the appropriate box for the residence type, configuration, occupancy, and exterior wall. If you require clarification, the pages preceding this worksheet describe each of these.

4. Next, the living area of the residence must be established. The heated or air conditioned space of the residence, not including the basement, should be measured. It is easiest to break the structure up into separate components as shown in the example. The main house (A), a one-and-one-half story wing (B), and a one story wing (C). The breezeway (D), garage (E), and open covered porch (F) will be treated differently. Data entry blocks for the living area are included on the worksheet for your use. Keep each level of each component separate, and fill out the blocks as shown.

5. By using the information on the worksheet, find the model, wing or ell in the following square foot cost pages that best matches the class, type, exterior finish and size of the residence being estimated. Use the *Modifications, Adjustments, and Alternatives* to determine the adjusted cost per square foot of living area for each component.

6. For each component, multiply the cost per square foot by the living area square footage. If the residence is a townhouse/rowhouse, a multiplier should be applied based upon the configuration.

7. The second page of the residential cost estimate worksheet has space for the additional components of a house. The cost for additional bathrooms, finished attic space, breezeways, porches, fireplaces, appliance or cabinet upgrades, and garages should be added on this page. The information for each of these components is found with the model being used, or in the *Modifications, Adjustments, and Alternatives* pages.

8. Add the total from page one of the estimate worksheet and the items listed on page two. The sum is the adjusted total building cost.

9. Depending on the use of the final estimated cost, one of the remaining two boxes should be filled out. Any additional items or exclusions should be added or subtracted at this time. The data contained in this book is a national average. Construction costs are different throughout the country. To allow for this difference, a location factor based upon the first three digits of the residence's zip code must be applied. The location factor is a multiplier that increases or decreases the adjusted total building cost. Find the appropriate location factor and calculate the local cost. If depreciation is a concern, a dollar figure should be subtracted at this point.

10. No residence will match a model exactly. Many differences will be found. At this level of estimating, a variation of plus or minus 10% should be expected.

Adjustments Instructions

No residence matches a model exactly in shape, material, or specifications.

The common differences are:

1. Two or more exterior wall systems:
   a. Partial basement
   b. Partly finished basement
2. Specifications or features that are between two classes
3. Crawl space instead of a basement

EXAMPLES

Below are quick examples. See pages 15-17 for complete examples of cost adjustments for these differences:

1. Residence "A" is an average one-story structure with 1,600 S.F. of living area and no basement. Three walls are wood siding on wood frame, and the fourth wall is brick veneer on wood frame. The brick veneer wall is 35% of the exterior wall area.

   Use page 28 to calculate the Base Cost per S.F. of Living Area.
   Wood Siding for 1,600 S.F.   = $93.55 per S.F.
   Brick Veneer for 1,600 S.F.   = $110.00 per S.F.
   .65 ($93.55) + .35 ($110.00) = $99.31 per S.F. of Living Area.

2. a. Residence "B" is the same as Residence "A"; However, it has an unfinished basement under 50% of the building. To adjust the $99.31 per S.F. of living area for this partial basement, use page 28.

   $99.31 + .50 ($9.50) = $104.06 per S.F. of Living Area.

   b. Residence "C" is the same as Residence "A"; However, it has a full basement under the entire building. 640 S.F. or 40% of the basement area is finished.

   Using Page 28:

   $99.31 + .40 ($27.65) + .60 ($9.50) = $116.07 per S.F. of Living Area.

3. When specifications or features of a building are between classes, estimate the percent deviation, and use two tables to calculate the cost per S.F.

   A two-story residence with wood siding and 1,800 S.F. of living area has features 30% better than Average, but 70% less than Custom.

   From pages 30 and 42:

   Custom 1,800 S.F. Base Cost   = $120.30 per S.F.
   Average 1,800 S.F. Base Cost   = $95.20 per S.F.
   DIFFERENCE   = $25.10 per S.F.

   Cost is $95.20 + .30 ($25.10) = $102.73 per S.F. of Living Area.

4. To add the cost of a crawl space, use the cost of an unfinished basement as a maximum. For specific costs of components to be added or deducted, such as vapor barrier, underdrain, and floor, see the "Assemblies" section, pages 229 to 450.

Model Residence Example

First Floor Plan

18'

12'   F

24'   E   D   C   A

8'

12'

16'

12'

28'

20'

24'   12'   12'   46'

B

18'

Second Floor Plan

46'

28'   A

20'

5'

5'

B

5'   13'

A = Main House
B = 1-1/2 Story Wing
C = 1 Story Wing
D = Breezeway
E = Garage
F = Open Covered Porch

15

RESIDENTIAL COST ESTIMATE

OWNERS NAME:	Albert Westenberg	APPRAISER:	Nicole Wojtowicz
RESIDENCE ADDRESS:	300 Sygiel Road	PROJECT:	# 55
CITY, STATE, ZIP CODE:	Three Rivers, MA 01080	DATE:	Jan. 1, 2009

CLASS OF CONSTRUCTION
- ☐ ECONOMY
- ☒ AVERAGE
- ☐ CUSTOM
- ☐ LUXURY

RESIDENCE TYPE
- ☐ 1 STORY
- ☐ 1 1/2 STORY
- ☒ 2 STORY
- ☐ 2 1/2 STORY
- ☐ 3 STORY
- ☐ BI-LEVEL
- ☐ TRI-LEVEL

CONFIGURATION
- ☒ DETACHED
- ☐ TOWN/ROW HOUSE
- ☐ SEMI-DETACHED

OCCUPANCY
- ☒ ONE FAMILY
- ☐ TWO FAMILY
- ☐ THREE FAMILY
- ☐ OTHER

EXTERIOR WALL SYSTEM
- ☒ WOOD SIDING - WOOD FRAME
- ☐ BRICK VENEER - WOOD FRAME
- ☐ STUCCO ON WOOD FRAME
- ☐ PAINTED CONCRETE BLOCK
- ☐ SOLID MASONRY (AVERAGE & CUSTOM)
- ☐ STONE VENEER - WOOD FRAME
- ☐ SOLID BRICK (LUXURY)
- ☐ SOLID STONE (LUXURY)

* LIVING AREA (Main Building)		
First Level	1288	S.F.
Second level	1288	S.F.
Third Level		S.F.
Total	2576	S.F.

* LIVING AREA (Wing or Ell)	(B)	
First Level	360	S.F.
Second level	310	S.F.
Third Level		S.F.
Total	670	S.F.

* LIVING AREA (Wing or Ell)	(C)	
First Level	192	S.F.
Second level		S.F.
Third Level		S.F.
Total	192	S.F.

* Basement Area is not part of living area.

MAIN BUILDING		COSTS PER S.F. LIVING AREA	
Cost per Square Foot of Living Area, from Page	42	$	83.80
Basement Addition: _____ % Finished, 100 % Unfinished		+	5.30
Roof Cover Adjustment: Cedar Shake Type, Page 42 (Add or Deduct)		( )	2.25
Central Air Conditioning: ☐ Separate Ducts ☒ Heating Ducts, Page 42		+	2.04
Heating System Adjustment: _____ Type, Page _____ (Add or Deduct)		( )	
Main Building: Adjusted Cost per S.F. of Living Area		$	93.39

MAIN BUILDING TOTAL COST	$ 93.39 /S.F.	x	2,576 S.F.	x	——————	=	$ 240,573
	Cost per S.F. Living Area		Living Area		Town/Row House Multiplier (Use 1 for Detached)		TOTAL COST

WING OR ELL (B) 1 - 1/2 STORY		COSTS PER S.F. LIVING AREA	
Cost per Square Foot of Living Area, from Page 56		$	78.30
Basement Addition: 100 % Finished, _____ % Unfinished		+	24.35
Roof Cover Adjustment: _____ Type, Page _____ (Add or Deduct)		( )	——
Central Air Conditioning: ☐ Separate Ducts ☒ Heating Ducts, Page 42		+	2.04
Heating System Adjustment: _____ Type, Page _____ (Add or Deduct)		( )	——
Wing or Ell (B): Adjusted Cost per S.F. of Living Area		$	104.69

WING OR ELL (B) TOTAL COST	$ 104.69 /S.F.	x	670 S.F.	=	$ 70,143
	Cost per S.F. Living Area		Living Area		TOTAL COST

WING OR ELL (C) 1 STORY		COSTS PER S.F. LIVING AREA	
Cost per Square Foot of Living Area, from Page # 56 (WOOD SIDING)		$	118.60
Basement Addition: _____ % Finished, _____ % Unfinished		+	——
Roof Cover Adjustment: _____ Type, Page _____ (Add or Deduct)		( )	——
Central Air Conditioning: ☐ Separate Ducts ☐ Heating Ducts, Page _____		+	——
Heating System Adjustment: _____ Type, Page _____ (Add or Deduct)		( )	——
Wing or Ell (C) Adjusted Cost per S.F. of Living Area		$	118.60

WING OR ELL (C) TOTAL COST	$ 118.60 /S.F.	x	192 S.F.	=	$ 22,772
	Cost per S.F. Living Area		Living Area		TOTAL COST

	TOTAL THIS PAGE	333,488

Page 1 of 2

16

RESIDENTIAL
COST ESTIMATE

Total Page 1						$	333,488
				QUANTITY	UNIT COST		
Additional Bathrooms:	**2** Full,	**1** Half	**2 @ 5522** **1 @ 3416**			+	14,460
Finished Attic:	**N/A** Ft. x	Ft.			S.F.		
Breezeway: ☑ Open ☐ ıclosed	**12** Ft. x	**12** Ft.		144	S.F. 23.52	+	3,387
Covered Porch: ☑ Open ☐	Enclosed **18** Ft. x	**12** Ft.		216	S.F. 34.85	+	7,528
Fireplace: ☑ Interior Chimney	☐ Exterior Chimney						
☑ No. of Flues (**2**)	☑ Additional Fireplaces	**1 - 2nd Story**				+	9,349
Appliances:						+	—
Kitchen Cabinets Adjustments:			(±)				—
☑ Garage ☐ Carport: **2**	Car(s) Description	**Wood, Attached**	(±)				21,144
Miscellaneous:						+	
			ADJUSTED TOTAL BUILDING COST			$	389,356

REPLACEMENT COST				INSURANCE COST		
ADJUSTED TOTAL BUILDING COST	$	389,356		ADJUSTED TOTAL BUILDING COST	$	
Site Improvements				Insurance Exclusions		
(A) Paving & Sidewalks	$			(A) Footings, sitework, Underground Piping	-$	
(B) Landscaping	$			(B) Architects Fees	-$	
(C) Fences	$			Total Building Cost Less Exclusion	$	
(D) Swimming Pools	$			Location Factor	x	
(E) Miscellaneous	$					
TOTAL	$	389,356		LOCAL INSURABLE REPLACEMENT COST	$	
Location Factor	x	1.040				
Location Replacement Cost	$	404,931				
Depreciation	-$	40,493				
LOCAL DEPRECIATED COST	$	364,438				

SKETCH AND ADDITIONAL CALCULATIONS

1 Story

© Home Planners, Inc.

1 - 1/2 Story

2 Story

Bi-Level

Tri-Level

©Design Basics, Inc.

	Components
1 Site Work	Site preparation for slab or excavation for lower level; 4' deep trench excavation for foundation wall.
2 Foundations	Continuous reinforced concrete footing, 8" deep x 18" wide; dampproofed and insulated 8" thick reinforced concrete block foundation wall, 4' deep; 4" concrete slab on 4" crushed stone base and polyethylene vapor barrier, trowel finish.
3 Framing	Exterior walls—2 x 4 wood studs, 16" O.C.; 1/2" insulation board sheathing; wood truss roof 24" O.C. or 2 x 6 rafters 16" O.C. with 1/2" plywood sheathing, 4 in 12 or 8 in 12 roof pitch.
4 Exterior Walls	Beveled wood siding and #15 felt building paper on insulated wood frame walls; sliding sash or double hung wood windows; 2 flush solid core wood exterior doors. **Alternates:** • Brick veneer on wood frame, has 4" veneer of common brick. • Stucco on wood frame has 1" thick colored stucco finish. • Painted concrete block has 8" concrete block sealed and painted on exterior and furring on the interior for the drywall.
5 Roofing	20 year asphalt shingles; #15 felt building paper; aluminum gutters, downspouts, drip edge and flashings.
6 Interiors	Walls and ceilings—1/2" taped and finished drywall, primed and painted with 2 coats; painted baseboard and trim; rubber backed carpeting 80%, asphalt tile 20%; hollow core wood interior doors.
7 Specialties	Economy grade kitchen cabinets—6 L.F. wall and base with plastic laminate counter top and kitchen sink; 30 gallon electric water heater.
8 Mechanical	1 lavatory, white, wall hung; 1 water closet, white; 1 bathtub, enameled steel, white; gas fired warm air heat.
9 Electrical	100 Amp. service; romex wiring; incandescent lighting fixtures, switches, receptacles.
10 Overhead and Profit	General Contractor overhead and profit.

Adjustments

Unfinished Basement:
7' high 8" concrete block or cast-in-place concrete walls.

Finished Basement:
Includes inexpensive paneling or drywall on foundation walls. Inexpensive sponge backed carpeting on concrete floor, drywall ceiling, and lighting.

- **Mass produced from stock plans**
- **Single family — 1 full bath, 1 kitchen**
- **No basement**
- **Asphalt shingles on roof**
- **Hot air heat**
- **Gypsum wallboard interior finishes**
- **Materials and workmanship are sufficient to meet codes**
- **Detail specifications on page 19**

Note: The illustration shown may contain some optional components (for example: garages and/or fireplaces) whose costs are shown in the modifications, adjustments, & alternatives below or at the end of the square foot section.

©Home Planners, Inc.

Base cost per square foot of living area

Exterior Wall	Living Area										
	600	800	1000	1200	1400	1600	1800	2000	2400	2800	3200
Wood Siding - Wood Frame	110.05	99.65	91.80	85.40	79.85	76.25	74.60	72.25	67.35	63.85	61.45
Brick Veneer - Wood Frame	115.80	104.75	96.45	89.60	83.60	79.85	78.05	75.45	70.35	66.60	64.00
Stucco on Wood Frame	104.40	94.60	87.25	81.30	76.05	72.80	71.25	69.00	64.45	61.15	58.95
Painted Concrete Block	109.15	98.90	91.05	84.80	79.25	75.75	74.05	71.70	66.90	63.40	61.05
Finished Basement, Add	27.45	25.90	24.70	23.65	22.85	22.30	22.00	21.50	20.90	20.45	20.00
Unfinished Basement, Add	12.15	10.85	9.95	9.10	8.45	8.00	7.75	7.40	6.90	6.50	6.20

Modifications

Add to the total cost

Upgrade Kitchen Cabinets	$ + 780
Solid Surface Countertops	+ 584
Full Bath - including plumbing, wall and floor finishes	+ 4400
Half Bath - including plumbing, wall and floor finishes	+ 2722
One Car Attached Garage	+ 11,165
One Car Detached Garage	+ 14,483
Fireplace & Chimney	+ 4965

Adjustments

For multi family - add to total cost

Additional Kitchen	$ + 3500
Additional Bath	+ 4400
Additional Entry & Exit	+ 1622
Separate Heating	+ 1376
Separate Electric	+ 1037

For Townhouse/Rowhouse -
Multiply cost per square foot by

Inner Unit	+ .95
End Unit	+ .97

Alternatives

Add to or deduct from the cost per square foot of living area

Composition Roll Roofing	– .75
Cedar Shake Roof	+ 4.75
Upgrade Walls and Ceilings to Skim Coat Plaster	+ .50
Upgrade Ceilings to Textured Finish	+ .47
Air Conditioning, in Heating Ductwork	+ 3.26
In Separate Ductwork	+ 6.21
Heating Systems, Hot Water	+ 1.70
Heat Pump	+ 2.06
Electric Heat	– 1.46
Not Heated	– 3.67

Additional upgrades or components

Kitchen Cabinets & Countertops	Page 58
Bathroom Vanities	59
Fireplaces & Chimneys	59
Windows, Skylights & Dormers	59
Appliances	60
Breezeways & Porches	60
Finished Attic	60
Garages	61
Site Improvements	61
Wings & Ells	25

- **Mass produced from stock plans**
- **Single family — 1 full bath, 1 kitchen**
- **No basement**
- **Asphalt shingles on roof**
- **Hot air heat**
- **Gypsum wallboard interior finishes**
- **Materials and workmanship are sufficient to meet codes**
- **Detail specifications on page 19**

Note: The illustration shown may contain some optional components (for example: garages and/or fireplaces) whose costs are shown in the modifications, adjustments, & alternatives below or at the end of the square foot section.

Base cost per square foot of living area

Exterior Wall	Living Area										
	600	800	1000	1200	1400	1600	1800	2000	2400	2800	3200
Wood Siding - Wood Frame	125.35	104.25	93.20	88.10	84.45	78.75	76.05	73.15	67.15	64.90	62.45
Brick Veneer - Wood Frame	133.40	110.10	98.55	93.20	89.25	83.15	80.25	77.10	70.60	68.20	65.50
Stucco on Wood Frame	117.50	98.55	87.95	83.15	79.70	74.45	72.00	69.30	63.75	61.65	59.50
Painted Concrete Block	124.15	103.35	92.35	87.35	83.70	78.10	75.45	72.55	66.60	64.35	62.00
Finished Basement, Add	21.15	18.00	17.10	16.50	16.05	15.45	15.10	14.75	14.10	13.80	13.45
Unfinished Basement, Add	10.70	8.20	7.55	7.05	6.70	6.20	5.95	5.65	5.15	4.90	4.65

Modifications

Add to the total cost

Upgrade Kitchen Cabinets	$ + 780
Solid Surface Countertops	+ 584
Full Bath - including plumbing, wall and floor finishes	+ 4400
Half Bath - including plumbing, wall and floor finishes	+ 2722
One Car Attached Garage	+ 11,165
One Car Detached Garage	+ 14,483
Fireplace & Chimney	+ 4965

Adjustments

For multi family - add to total cost

Additional Kitchen	$ + 3500
Additional Bath	+ 4400
Additional Entry & Exit	+ 1622
Separate Heating	+ 1376
Separate Electric	+ 1037

*For Townhouse/Rowhouse -
Multiply cost per square foot by*

Inner Unit	+ .95
End Unit	+ .97

Alternatives

Add to or deduct from the cost per square foot of living area

Composition Roll Roofing	– .50
Cedar Shake Roof	+ 3.45
Upgrade Walls and Ceilings to Skim Coat Plaster	+ .50
Upgrade Ceilings to Textured Finish	+ .47
Air Conditioning, in Heating Ductwork	+ 2.44
In Separate Ductwork	+ 5.46
Heating Systems, Hot Water	+ 1.62
Heat Pump	+ 2.27
Electric Heat	– 1.16
Not Heated	– 3.38

Additional upgrades or components

- Mass produced from stock plans
- Single family — 1 full bath, 1 kitchen
- No basement
- Asphalt shingles on roof
- Hot air heat
- Gypsum wallboard interior finishes
- Materials and workmanship are sufficient to meet codes
- Detail specifications on page 19

Note: The illustration shown may contain some optional components (for example: garages and/or fireplaces) whose costs are shown in the modifications, adjustments, & alternatives below or at the end of the square foot section.

Base cost per square foot of living area

Exterior Wall	Living Area										
	1000	1200	1400	1600	1800	2000	2200	2600	3000	3400	3800
Wood Siding - Wood Frame	99.00	89.55	85.20	82.15	79.20	75.70	73.40	69.15	64.85	63.00	61.35
Brick Veneer - Wood Frame	105.25	95.30	90.55	87.35	84.05	80.35	77.85	73.15	68.60	66.55	64.70
Stucco on Wood Frame	92.95	83.90	79.95	77.15	74.50	71.15	69.10	65.20	61.20	59.55	58.05
Painted Concrete Block	98.10	88.65	84.40	81.40	78.50	74.95	72.80	68.55	64.30	62.45	60.85
Finished Basement, Add	14.40	13.80	13.30	12.95	12.65	12.35	12.15	11.65	11.35	11.10	10.90
Unfinished Basement, Add	6.60	6.05	5.70	5.45	5.15	4.95	4.75	4.40	4.10	3.95	3.75

Modifications

Add to the total cost

Upgrade Kitchen Cabinets	$ + 780
Solid Surface Countertops	+ 584
Full Bath - including plumbing, wall and floor finishes	+ 4400
Half Bath - including plumbing, wall and floor finishes	+ 2722
One Car Attached Garage	+ 11,165
One Car Detached Garage	+ 14,483
Fireplace & Chimney	+ 5520

Adjustments

For multi family - add to total cost

Additional Kitchen	$ + 3500
Additional Bath	+ 4400
Additional Entry & Exit	+ 1622
Separate Heating	+ 1376
Separate Electric	+ 1037

For Townhouse/Rowhouse -
Multiply cost per square foot by

Inner Unit	+ .93
End Unit	+ .96

Alternatives

Add to or deduct from the cost per square foot of living area

Composition Roll Roofing	– .35
Cedar Shake Roof	+ 2.40
Upgrade Walls and Ceilings to Skim Coat Plaster	+ .51
Upgrade Ceilings to Textured Finish	+ .47
Air Conditioning, in Heating Ductwork	+ 1.97
In Separate Ductwork	+ 5
Heating Systems, Hot Water	+ 1.58
Heat Pump	+ 2.41
Electric Heat	– 1.01
Not Heated	– 3.19

Additional upgrades or components

Kitchen Cabinets & Countertops	Page 58
Bathroom Vanities	59
Fireplaces & Chimneys	59
Windows, Skylights & Dormers	59
Appliances	60
Breezeways & Porches	60
Finished Attic	60
Garages	61
Site Improvements	61
Wings & Ells	25

- **Mass produced from stock plans**
- **Single family — 1 full bath, 1 kitchen**
- **No basement**
- **Asphalt shingles on roof**
- **Hot air heat**
- **Gypsum wallboard interior finishes**
- **Materials and workmanship are sufficient to meet codes**
- **Detail specifications on page 19**

Note: The illustration shown may contain some optional components (for example: garages and/or fireplaces) whose costs are shown in the modifications, adjustments, & alternatives below or at the end of the square foot section.

Base cost per square foot of living area

Exterior Wall	Living Area										
	1000	1200	1400	1600	1800	2000	2200	2600	3000	3400	3800
Wood Siding - Wood Frame	91.90	82.90	78.95	76.25	73.70	70.35	68.30	64.50	60.60	58.95	57.50
Brick Veneer - Wood Frame	96.55	87.20	83.00	80.15	77.30	73.80	71.65	67.50	63.35	61.60	60.00
Stucco on Wood Frame	87.30	78.65	75.05	72.50	70.10	66.90	65.05	61.60	57.85	56.35	55.05
Painted Concrete Block	91.15	82.30	78.30	75.70	73.10	69.80	67.85	64.05	60.15	58.55	57.10
Finished Basement, Add	14.40	13.80	13.30	12.95	12.65	12.35	12.15	11.65	11.35	11.10	10.90
Unfinished Basement, Add	6.60	6.05	5.70	5.45	5.15	4.95	4.75	4.40	4.10	3.95	3.75

Modifications

Add to the total cost

Upgrade Kitchen Cabinets	$ + 780
Solid Surface Countertops	+ 584
Full Bath - including plumbing, wall and floor finishes	+ 4400
Half Bath - including plumbing, wall and floor finishes	+ 2722
One Car Attached Garage	+ 11,165
One Car Detached Garage	+ 14,483
Fireplace & Chimney	+ 4965

Adjustments

For multi family - add to total cost

Additional Kitchen	$ + 3500
Additional Bath	+ 4400
Additional Entry & Exit	+ 1622
Separate Heating	+ 1376
Separate Electric	+ 1037

For Townhouse/Rowhouse - Multiply cost per square foot by

Inner Unit	+ .94
End Unit	+ .97

Alternatives

Add to or deduct from the cost per square foot of living area

Composition Roll Roofing	– .35
Cedar Shake Roof	+ 2.40
Upgrade Walls and Ceilings to Skim Coat Plaster	+ .48
Upgrade Ceilings to Textured Finish	+ .47
Air Conditioning, in Heating Ductwork	+ 1.97
In Separate Ductwork	+ 5
Heating Systems, Hot Water	+ 1.58
Heat Pump	+ 2.41
Electric Heat	– 1.01
Not Heated	– 3.19

Additional upgrades or components

Kitchen Cabinets & Countertops	Page 58
Bathroom Vanities	59
Fireplaces & Chimneys	59
Windows, Skylights & Dormers	59
Appliances	60
Breezeways & Porches	60
Finished Attic	60
Garages	61
Site Improvements	61
Wings & Ells	25

- **Mass produced from stock plans**
- **Single family — 1 full bath, 1 kitchen**
- **No basement**
- **Asphalt shingles on roof**
- **Hot air heat**
- **Gypsum wallboard interior finishes**
- **Materials and workmanship are sufficient to meet codes**
- **Detail specifications on page 19**

Note: The illustration shown may contain some optional components (for example: garages and/or fireplaces) whose costs are shown in the modifications, adjustments, & alternatives below or at the end of the square foot section.

©Design Basics, Inc.

Base cost per square foot of living area

Exterior Wall	Living Area										
	1200	1500	1800	2000	2200	2400	2800	3200	3600	4000	4400
Wood Siding - Wood Frame	84.90	78.00	72.90	70.80	67.75	65.30	63.40	60.80	57.70	56.65	54.30
Brick Veneer - Wood Frame	89.15	81.90	76.35	74.15	70.85	68.30	66.30	63.45	60.20	59.05	56.55
Stucco on Wood Frame	80.65	74.20	69.45	67.55	64.60	62.35	60.65	58.15	55.30	54.30	52.10
Solid Masonry	84.25	77.40	72.35	70.35	67.25	64.85	63.00	60.40	57.30	56.30	53.90
Finished Basement, Add*	17.25	16.50	15.80	15.50	15.15	14.90	14.60	14.25	13.90	13.80	13.55
Unfinished Basement, Add*	7.20	6.60	6.05	5.85	5.55	5.35	5.10	4.85	4.60	4.45	4.30

*Basement under middle level only.

Modifications

Add to the total cost

Upgrade Kitchen Cabinets	$	+ 780
Solid Surface Countertops		+ 584
Full Bath - including plumbing, wall and floor finishes		+ 4400
Half Bath - including plumbing, wall and floor finishes		+ 2722
One Car Attached Garage		+ 11,165
One Car Detached Garage		+ 14,483
Fireplace & Chimney		+ 4965

Adjustments

For multi family - add to total cost

Additional Kitchen	$	+ 3500
Additional Bath		+ 4400
Additional Entry & Exit		+ 1622
Separate Heating		+ 1376
Separate Electric		+ 1037

For Townhouse/Rowhouse -
Multiply cost per square foot by

Inner Unit	+ .93
End Unit	+ .96

Alternatives

Add to or deduct from the cost per square foot of living area

Composition Roll Roofing	– .50
Cedar Shake Roof	+ 3.45
Upgrade Walls and Ceilings to Skim Coat Plaster	+ .44
Upgrade Ceilings to Textured Finish	+ .47
Air Conditioning, in Heating Ductwork	+ 1.69
In Separate Ductwork	+ 4.65
Heating Systems, Hot Water	+ 1.51
Heat Pump	+ 2.50
Electric Heat	– .88
Not Heated	– 3.09

Additional upgrades or components

Kitchen Cabinets & Countertops	Page 58
Bathroom Vanities	59
Fireplaces & Chimneys	59
Windows, Skylights & Dormers	59
Appliances	60
Breezeways & Porches	60
Finished Attic	60
Garages	61
Site Improvements	61
Wings & Ells	25

1 Story — Base cost per square foot of living area

Exterior Wall	Living Area							
	50	100	200	300	400	500	600	700
Wood Siding - Wood Frame	153.30	116.75	100.85	83.80	78.65	75.55	73.45	74.00
Brick Veneer - Wood Frame	167.80	127.10	109.50	89.60	83.85	80.40	78.05	78.45
Stucco on Wood Frame	139.10	106.60	92.40	78.15	73.55	70.80	68.95	69.65
Painted Concrete Block	151.10	115.20	99.55	82.95	77.85	74.80	72.80	73.30
Finished Basement, Add	41.80	34.10	30.95	25.60	24.55	23.90	23.50	23.15
Unfinished Basement, Add	23.00	17.10	14.65	10.50	9.65	9.20	8.85	8.60

1-1/2 Story — Base cost per square foot of living area

Exterior Wall	Living Area							
	100	200	300	400	500	600	700	800
Wood Siding - Wood Frame	120.95	97.20	82.65	73.90	69.50	67.35	64.70	63.90
Brick Veneer - Wood Frame	133.90	107.55	91.30	80.65	75.70	73.25	70.25	69.45
Stucco on Wood Frame	108.25	87.05	74.20	67.30	63.40	61.55	59.25	58.50
Painted Concrete Block	119.00	95.60	81.35	72.85	68.55	66.45	63.85	63.10
Finished Basement, Add	27.90	24.75	22.60	20.25	19.60	19.20	18.80	18.75
Unfinished Basement, Add	14.15	11.70	10.05	8.25	7.75	7.40	7.10	7.10

2 Story — Base cost per square foot of living area

Exterior Wall	Living Area							
	100	200	400	600	800	1000	1200	1400
Wood Siding - Wood Frame	123.30	91.55	77.70	64.00	59.50	56.80	54.95	55.70
Brick Veneer - Wood Frame	137.80	101.90	86.30	69.75	64.70	61.60	59.55	60.15
Stucco on Wood Frame	109.05	81.40	69.25	58.35	54.45	52.05	50.45	51.40
Painted Concrete Block	121.10	89.95	76.35	63.15	58.70	56.05	54.25	55.05
Finished Basement, Add	20.90	17.10	15.50	12.85	12.30	11.95	11.80	11.65
Unfinished Basement, Add	11.55	8.55	7.30	5.25	4.85	4.60	4.45	4.30

Base costs do not include bathroom or kitchen facilities. Use Modifications/Adjustments/Alternatives on pages 58-61 where appropriate.

Important: See the Reference Section for Location Factors (to adjust for your city) and Estimating Forms.

1 Story

1-1/2 Story

2 Story

2-1/2 Story

Bi-Level

Tri-Level

	Components
1 Site Work	Site preparation for slab or excavation for lower level; 4' deep trench excavation for foundation wall.
2 Foundations	Continuous reinforced concrete footing, 8" deep x 18" wide; dampproofed and insulated 8" thick reinforced concrete foundation wall, 4' deep; 4" concrete slab on 4" crushed stone base and polyethylene vapor barrier, trowel finish.
3 Framing	Exterior walls—2 x 4 wood studs, 16" O.C.; 1/2" plywood sheathing; 2 x 6 rafters 16" O.C. with 1/2" plywood sheathing, 4 in 12 or 8 in 12 roof pitch, 2 x 6 ceiling joists; 1/2" plywood subfloor on 1 x 2 wood sleepers 16" O.C.; 2 x 8 floor joists with 5/8" plywood subfloor on models with more than one level.
4 Exterior Walls	Beveled wood siding and #15 felt building paper on insulated wood frame walls; double hung windows; 3 flush solid core wood exterior doors with storms. **Alternates:** • Brick veneer on wood frame, has 4" veneer of common brick. • Stucco on wood frame has 1" thick colored stucco finish. • Solid masonry has a 6" concrete block load bearing wall with insulation and a brick or stone exterior. Also solid brick or stone walls.
5 Roofing	25 year asphalt shingles; #15 felt building paper; aluminum gutters, downspouts, drip edge and flashings.
6 Interiors	Walls and ceilings—1/2" taped and finished drywall, primed and painted with 2 coats; painted baseboard and trim, finished hardwood floor 40%, carpet with 1/2" underlayment 40%, vinyl tile with 1/2" underlayment 15%, ceramic tile with 1/2" underlayment 5%; hollow core and louvered interior doors.
7 Specialties	Average grade kitchen cabinets—14 L.F. wall and base with plastic laminate counter top and kitchen sink; 40 gallon electric water heater.
8 Mechanical	1 lavatory, white, wall hung; 1 water closet, white; 1 bathtub, enameled steel, white; gas fired warm air heat.
9 Electrical	100 Amp. service; romex wiring; incandescent lighting fixtures, switches, receptacles.
10 Overhead and Profit	General Contractor overhead and profit.

Adjustments

Unfinished Basement:
7'-6" high cast-in-place concrete walls 8" thick or 8" concrete block.

Finished Basement:
Includes plywood paneling or drywall on furring on foundation walls, sponge backed carpeting on concrete floor, suspended ceiling, lighting and heating.

- **Simple design from standard plans**
- **Single family — 1 full bath, 1 kitchen**
- **No basement**
- **Asphalt shingles on roof**
- **Hot air heat**
- **Gypsum wallboard interior finishes**
- **Materials and workmanship are average**
- **Detail specifications on p. 27**

Note: The illustration shown may contain some optional components (for example: garages and/or fireplaces) whose costs are shown in the modifications, adjustments, & alternatives below or at the end of the square foot section.

©Home Planners, Inc.

Base cost per square foot of living area

Exterior Wall	Living Area										
	600	800	1000	1200	1400	1600	1800	2000	2400	2800	3200
Wood Siding - Wood Frame	133.40	120.95	111.55	104.10	97.75	93.55	91.40	88.70	83.15	79.10	76.35
Brick Veneer - Wood Frame	153.55	140.00	129.85	121.60	114.60	110.00	107.65	104.60	98.55	94.15	91.05
Stucco on Wood Frame	141.00	128.80	119.70	112.40	106.30	102.15	100.10	97.50	92.10	88.15	85.50
Solid Masonry	167.10	152.10	140.80	131.40	123.60	118.40	115.80	112.20	105.50	100.60	97.05
Finished Basement, Add	33.50	32.30	30.80	29.40	28.30	27.65	27.25	26.60	25.80	25.10	24.55
Unfinished Basement, Add	13.75	12.45	11.50	10.60	9.90	9.50	9.25	8.85	8.40	7.95	7.65

Modifications

Add to the total cost

Upgrade Kitchen Cabinets	$ + 3760
Solid Surface Countertops (Included)	
Full Bath - including plumbing, wall and floor finishes	+ 5522
Half Bath - including plumbing, wall and floor finishes	+ 3416
One Car Attached Garage	+ 12,594
One Car Detached Garage	+ 16,766
Fireplace & Chimney	+ 4965

Adjustments

For multi family - add to total cost

Additional Kitchen	$ + 6433
Additional Bath	+ 5522
Additional Entry & Exit	+ 1622
Separate Heating	+ 1376
Separate Electric	+ 1870

For Townhouse/Rowhouse -
Multiply cost per square foot by

Inner Unit	+ .92
End Unit	+ .96

Alternatives

Add to or deduct from the cost per square foot of living area

Cedar Shake Roof	+ 4.45
Clay Tile Roof	+ 4.80
Slate Roof	+ 6.70
Upgrade Walls to Skim Coat Plaster	+ .31
Upgrade Ceilings to Textured Finish	+ .47
Air Conditioning, in Heating Ductwork	+ 3.37
In Separate Ductwork	+ 6.45
Heating Systems, Hot Water	+ 1.74
Heat Pump	+ 2.14
Electric Heat	– .81
Not Heated	– 3.06

Additional upgrades or components

Kitchen Cabinets & Countertops	Page 58
Bathroom Vanities	59
Fireplaces & Chimneys	59
Windows, Skylights & Dormers	59
Appliances	60
Breezeways & Porches	60
Finished Attic	60
Garages	61
Site Improvements	61
Wings & Ells	37

- **Simple design from standard plans**
- **Single family — 1 full bath, 1 kitchen**
- **No basement**
- **Asphalt shingles on roof**
- **Hot air heat**
- **Gypsum wallboard interior finishes**
- **Materials and workmanship are average**
- **Detail specifications on p. 27**

Note: The illustration shown may contain some optional components (for example: garages and/or fireplaces) whose costs are shown in the modifications, adjustments, & alternatives below or at the end of the square foot section.

©By Designer

Base cost per square foot of living area

Exterior Wall	Living Area										
	600	800	1000	1200	1400	1600	1800	2000	2400	2800	3200
Wood Siding - Wood Frame	149.05	124.95	112.10	106.10	101.75	95.30	92.10	88.75	82.00	79.35	76.55
Brick Veneer - Wood Frame	162.65	134.85	121.25	114.70	109.95	102.65	99.15	95.45	87.90	84.95	81.70
Stucco on Wood Frame	145.15	122.10	109.45	103.65	99.45	93.15	90.10	86.85	80.30	77.75	75.10
Solid Masonry	181.50	148.50	133.90	126.55	121.20	113.00	108.95	104.75	96.10	92.70	88.90
Finished Basement, Add	27.35	23.90	22.85	22.05	21.50	20.65	20.20	19.75	18.90	18.50	18.05
Unfinished Basement, Add	11.90	9.35	8.65	8.15	7.75	7.30	7.00	6.70	6.20	5.95	5.65

Modifications

Add to the total cost

Upgrade Kitchen Cabinets	$ + 3760
Solid Surface Countertops (Included)	
Full Bath - including plumbing, wall and floor finishes	+ 5522
Half Bath - including plumbing, wall and floor finishes	+ 3416
One Car Attached Garage	+ 12,594
One Car Detached Garage	+ 16,766
Fireplace & Chimney	+ 4965

Adjustments

For multi family - add to total cost

Additional Kitchen	$ + 6433
Additional Bath	+ 5522
Additional Entry & Exit	+ 1622
Separate Heating	+ 1376
Separate Electric	+ 1870

*For Townhouse/Rowhouse -
Multiply cost per square foot by*

Inner Unit	+ .92
End Unit	+ .96

Alternatives

Add to or deduct from the cost per square foot of living area

Cedar Shake Roof	+ 3.25
Clay Tile Roof	+ 3.50
Slate Roof	+ 4.85
Upgrade Walls to Skim Coat Plaster	+ .36
Upgrade Ceilings to Textured Finish	+ .47
Air Conditioning, in Heating Ductwork	+ 2.57
In Separate Ductwork	+ 5.64
Heating Systems, Hot Water	+ 1.65
Heat Pump	+ 2.38
Electric Heat	– .73
Not Heated	– 2.95

Additional upgrades or components

Kitchen Cabinets & Countertops	Page 58
Bathroom Vanities	59
Fireplaces & Chimneys	59
Windows, Skylights & Dormers	59
Appliances	60
Breezeways & Porches	60
Finished Attic	60
Garages	61
Site Improvements	61
Wings & Ells	37

Important: See the Reference Section for Location Factors (to adjust for your city) and Estimating Forms

- Simple design from standard plans
- Single family — 1 full bath, 1 kitchen
- No basement
- Asphalt shingles on roof
- Hot air heat
- Gypsum wallboard interior finishes
- Materials and workmanship are average
- Detail specifications on p. 27

Note: The illustration shown may contain some optional components (for example: garages and/or fireplaces) whose costs are shown in the modifications, adjustments, & alternatives below or at the end of the square foot section.

Base cost per square foot of living area

Exterior Wall	Living Area										
	1000	1200	1400	1600	1800	2000	2200	2600	3000	3400	3800
Wood Siding - Wood Frame	118.25	107.30	102.20	98.65	95.20	91.25	88.70	83.80	78.90	76.75	74.80
Brick Veneer - Wood Frame	128.75	117.05	111.35	107.35	103.35	99.15	96.20	90.55	85.20	82.70	80.50
Stucco on Wood Frame	115.20	104.45	99.65	96.20	92.80	89.00	86.55	81.80	77.05	75.00	73.20
Solid Masonry	143.40	130.60	124.00	119.45	114.70	110.10	106.60	99.95	93.85	91.00	88.35
Finished Basement, Add	19.40	19.10	18.50	18.05	17.65	17.35	16.95	16.35	15.90	15.65	15.40
Unfinished Basement, Add	7.55	7.05	6.65	6.35	6.05	5.90	5.65	5.30	5.00	4.80	4.70

Modifications

Add to the total cost

Upgrade Kitchen Cabinets	$ + 3760
Solid Surface Countertops (Included)	
Full Bath - including plumbing, wall and floor finishes	+ 5522
Half Bath - including plumbing, wall and floor finishes	+ 3416
One Car Attached Garage	+ 12,594
One Car Detached Garage	+ 16,766
Fireplace & Chimney	+ 5536

Adjustments

For multi family - add to total cost

Additional Kitchen	$ + 6433
Additional Bath	+ 5522
Additional Entry & Exit	+ 1622
Separate Heating	+ 1376
Separate Electric	+ 1870

For Townhouse/Rowhouse -
Multiply cost per square foot by

Inner Unit	+ .90
End Unit	+ .95

Alternatives

Add to or deduct from the cost per square foot of living area

Cedar Shake Roof	+ 2.25
Clay Tile Roof	+ 2.40
Slate Roof	+ 3.35
Upgrade Walls to Skim Coat Plaster	+ .37
Upgrade Ceilings to Textured Finish	+ .47
Air Conditioning, in Heating Ductwork	+ 2.04
In Separate Ductwork	+ 5.12
Heating Systems, Hot Water	+ 1.62
Heat Pump	+ 2.50
Electric Heat	– .64
Not Heated	– 2.86

Additional upgrades or components

Kitchen Cabinets & Countertops	Page 58
Bathroom Vanities	59
Fireplaces & Chimneys	59
Windows, Skylights & Dormers	59
Appliances	60
Breezeways & Porches	60
Finished Attic	60
Garages	61
Site Improvements	61
Wings & Ells	37

- **Simple design from standard plans**
- **Single family — 1 full bath, 1 kitchen**
- **No basement**
- **Asphalt shingles on roof**
- **Hot air heat**
- **Gypsum wallboard interior finishes**
- **Materials and workmanship are average**
- **Detail specifications on p. 27**

Note: The illustration shown may contain some optional components (for example: garages and/or fireplaces) whose costs are shown in the modifications, adjustments, & alternatives below or at the end of the square foot section.

Base cost per square foot of living area

Exterior Wall	Living Area										
	1200	1400	1600	1800	2000	2400	2800	3200	3600	4000	4400
Wood Siding - Wood Frame	117.35	110.30	100.85	99.05	95.60	90.15	85.80	81.30	79.10	74.95	73.70
Brick Veneer - Wood Frame	128.55	120.50	110.15	108.40	104.35	98.10	93.35	88.10	85.60	81.00	79.55
Stucco on Wood Frame	114.10	107.40	98.15	96.35	93.05	87.85	83.60	79.35	77.25	73.20	72.00
Solid Masonry	144.10	134.55	123.10	121.35	116.55	109.05	103.85	97.65	94.60	89.40	87.65
Finished Basement, Add	16.40	16.05	15.50	15.45	15.00	14.45	14.15	13.65	13.40	13.10	12.95
Unfinished Basement, Add	6.30	5.80	5.40	5.35	5.15	4.75	4.55	4.30	4.10	3.95	3.85

Modifications

Add to the total cost

Upgrade Kitchen Cabinets	$ + 3760
Solid Surface Countertops (Included)	
Full Bath - including plumbing, wall and floor finishes	+ 5522
Half Bath - including plumbing, wall and floor finishes	+ 3416
One Car Attached Garage	+ 12,594
One Car Detached Garage	+ 16,766
Fireplace & Chimney	+ 6293

Adjustments

For multi family - add to total cost

Additional Kitchen	$ + 6433
Additional Bath	+ 5522
Additional Entry & Exit	+ 1622
Separate Heating	+ 1376
Separate Electric	+ 1870

For Townhouse/Rowhouse -
Multiply cost per square foot by

Inner Unit	+ .90
End Unit	+ .95

Alternatives

Add to or deduct from the cost per square foot of living area

Cedar Shake Roof	+ 1.95
Clay Tile Roof	+ 2.10
Slate Roof	+ 2.90
Upgrade Walls to Skim Coat Plaster	+ .35
Upgrade Ceilings to Textured Finish	+ .47
Air Conditioning, in Heating Ductwork	+ 1.86
In Separate Ductwork	+ 4.94
Heating Systems, Hot Water	+ 1.46
Heat Pump	+ 2.57
Electric Heat	– 1.13
Not Heated	– 3.34

Additional upgrades or components

Kitchen Cabinets & Countertops	Page 58
Bathroom Vanities	59
Fireplaces & Chimneys	59
Windows, Skylights & Dormers	59
Appliances	60
Breezeways & Porches	60
Finished Attic	60
Garages	61
Site Improvements	61
Wings & Ells	37

- **Simple design from standard plans**
- **Single family — 1 full bath, 1 kitchen**
- **No basement**
- **Asphalt shingles on roof**
- **Hot air heat**
- **Gypsum wallboard interior finishes**
- **Materials and workmanship are average**
- **Detail specifications on p. 27**

Note: The illustration shown may contain some optional components (for example: garages and/or fireplaces) whose costs are shown in the modifications, adjustments, & alternatives below or at the end of the square foot section.

Base cost per square foot of living area

	Living Area										
Exterior Wall	1500	1800	2100	2500	3000	3500	4000	4500	5000	5500	6000
Wood Siding - Wood Frame	108.90	99.00	94.70	91.35	84.95	82.20	78.20	73.90	72.55	71.05	69.40
Brick Veneer - Wood Frame	119.45	108.70	103.80	100.00	92.85	89.65	85.00	80.20	78.60	76.90	74.90
Stucco on Wood Frame	105.85	96.15	92.10	88.85	82.65	80.05	76.25	72.15	70.80	69.35	67.85
Solid Masonry	134.05	122.25	116.45	112.15	103.80	100.05	94.45	88.90	87.00	85.00	82.50
Finished Basement, Add	14.10	14.05	13.60	13.35	12.80	12.55	12.15	11.85	11.75	11.60	11.40
Unfinished Basement, Add	5.15	4.75	4.55	4.35	4.05	3.85	3.65	3.45	3.35	3.25	3.20

Modifications

Add to the total cost

Upgrade Kitchen Cabinets	$ + 3760
Solid Surface Countertops (Included)	
Full Bath - including plumbing, wall and floor finishes	+ 5522
Half Bath - including plumbing, wall and floor finishes	+ 3416
One Car Attached Garage	+ 12,594
One Car Detached Garage	+ 16,766
Fireplace & Chimney	+ 6293

Adjustments

For multi family - add to total cost

Additional Kitchen	$ + 6433
Additional Bath	+ 5522
Additional Entry & Exit	+ 1622
Separate Heating	+ 1376
Separate Electric	+ 1870

For Townhouse/Rowhouse -
Multiply cost per square foot by

Inner Unit	+ .88
End Unit	+ .94

Alternatives

Add to or deduct from the cost per square foot of living area

Cedar Shake Roof	+ 1.50
Clay Tile Roof	+ 1.60
Slate Roof	+ 2.25
Upgrade Walls to Skim Coat Plaster	+ .37
Upgrade Ceilings to Textured Finish	+ .47
Air Conditioning, in Heating Ductwork	+ 1.86
In Separate Ductwork	+ 4.94
Heating Systems, Hot Water	+ 1.46
Heat Pump	+ 2.57
Electric Heat	− .88
Not Heated	− 3.10

Additional upgrades or components

Kitchen Cabinets & Countertops	Page 58
Bathroom Vanities	59
Fireplaces & Chimneys	59
Windows, Skylights & Dormers	59
Appliances	60
Breezeways & Porches	60
Finished Attic	60
Garages	61
Site Improvements	61
Wings & Ells	37

- **Simple design from standard plans**
- **Single family — 1 full bath, 1 kitchen**
- **No basement**
- **Asphalt shingles on roof**
- **Hot air heat**
- **Gypsum wallboard interior finishes**
- **Materials and workmanship are average**
- **Detail specifications on p. 27**

Note: The illustration shown may contain some optional components (for example: garages and/or fireplaces) whose costs are shown in the modifications, adjustments, & alternatives below or at the end of the square foot section.

Base cost per square foot of living area

Exterior Wall	Living Area										
	1000	1200	1400	1600	1800	2000	2200	2600	3000	3400	3800
Wood Siding - Wood Frame	110.40	100.05	95.45	92.20	89.05	85.35	83.05	78.70	74.25	72.30	70.60
Brick Veneer - Wood Frame	118.30	107.40	102.25	98.75	95.20	91.30	88.75	83.80	78.95	76.80	74.85
Stucco on Wood Frame	108.15	98.00	93.50	90.35	87.25	83.65	81.45	77.25	72.85	71.00	69.40
Solid Masonry	129.25	117.55	111.75	107.80	103.75	99.50	96.55	90.90	85.45	82.95	80.75
Finished Basement, Add	19.40	19.10	18.50	18.05	17.65	17.35	16.95	16.35	15.90	15.65	15.40
Unfinished Basement, Add	7.55	7.05	6.65	6.35	6.05	5.90	5.65	5.30	5.00	4.80	4.70

Modifications

Add to the total cost

Upgrade Kitchen Cabinets	$ + 3760
Solid Surface Countertops (Included)	
Full Bath - including plumbing, wall and floor finishes	+ 5522
Half Bath - including plumbing, wall and floor finishes	+ 3416
One Car Attached Garage	+ 12,594
One Car Detached Garage	+ 16,766
Fireplace & Chimney	+ 4965

Adjustments

For multi family - add to total cost

Additional Kitchen	$ + 6433
Additional Bath	+ 5522
Additional Entry & Exit	+ 1622
Separate Heating	+ 1376
Separate Electric	+ 1870

For Townhouse/Rowhouse -
Multiply cost per square foot by

Inner Unit	+ .91
End Unit	+ .96

Alternatives

Add to or deduct from the cost per square foot of living area

Cedar Shake Roof	+ 2.25
Clay Tile Roof	+ 2.40
Slate Roof	+ 3.35
Upgrade Walls to Skim Coat Plaster	+ .34
Upgrade Ceilings to Textured Finish	+ .47
Air Conditioning, in Heating Ductwork	+ 2.04
In Separate Ductwork	+ 5.12
Heating Systems, Hot Water	+ 1.62
Heat Pump	+ 2.50
Electric Heat	– .64
Not Heated	– 2.86

Additional upgrades or components

Kitchen Cabinets & Countertops	Page 58
Bathroom Vanities	59
Fireplaces & Chimneys	59
Windows, Skylights & Dormers	59
Appliances	60
Breezeways & Porches	60
Finished Attic	60
Garages	61
Site Improvements	61
Wings & Ells	37

Important: See the Reference Section for Location Factors (to adjust for your city) and Estimating Forms

- Simple design from standard plans
- Single family — 1 full bath, 1 kitchen
- No basement
- Asphalt shingles on roof
- Hot air heat
- Gypsum wallboard interior finishes
- Materials and workmanship are average
- Detail specifications on p. 27

Note: The illustration shown may contain some optional components (for example: garages and/or fireplaces) whose costs are shown in the modifications, adjustments, & alternatives below or at the end of the square foot section.

©Design Basics, Inc.

Base cost per square foot of living area

Exterior Wall	Living Area										
	1200	1500	1800	2100	2400	2700	3000	3400	3800	4200	4600
Wood Siding - Wood Frame	105.00	97.05	91.00	85.75	82.45	80.60	78.40	76.45	73.10	70.40	69.10
Brick Veneer - Wood Frame	112.25	103.60	97.00	91.20	87.50	85.45	83.00	80.90	77.25	74.30	72.90
Stucco on Wood Frame	102.85	95.15	89.30	84.20	81.00	79.20	77.10	75.20	71.95	69.35	68.05
Solid Masonry	122.30	112.75	105.20	98.65	94.55	92.20	89.35	87.05	82.95	79.65	78.05
Finished Basement, Add*	22.20	21.80	20.90	20.10	19.65	19.40	18.95	18.75	18.35	18.00	17.80
Unfinished Basement, Add*	8.35	7.70	7.15	6.70	6.40	6.25	5.95	5.85	5.60	5.40	5.30

*Basement under middle level only.

Modifications

Add to the total cost

Upgrade Kitchen Cabinets	$ + 3760
Solid Surface Countertops (Included)	
Full Bath - including plumbing, wall and floor finishes	+ 5522
Half Bath - including plumbing, wall and floor finishes	+ 3416
One Car Attached Garage	+ 12,594
One Car Detached Garage	+ 16,766
Fireplace & Chimney	+ 4965

Adjustments

For multi family - add to total cost

Additional Kitchen	$ + 6433
Additional Bath	+ 5522
Additional Entry & Exit	+ 1622
Separate Heating	+ 1376
Separate Electric	+ 1870

For Townhouse/Rowhouse - Multiply cost per square foot by

Inner Unit	+ .90
End Unit	+ .95

Alternatives

Add to or deduct from the cost per square foot of living area

Cedar Shake Roof	+ 3.25
Clay Tile Roof	+ 3.50
Slate Roof	+ 4.85
Upgrade Walls to Skim Coat Plaster	+ .30
Upgrade Ceilings to Textured Finish	+ .47
Air Conditioning, in Heating Ductwork	+ 1.72
In Separate Ductwork	+ 4.83
Heating Systems, Hot Water	+ 1.56
Heat Pump	+ 2.60
Electric Heat	– .52
Not Heated	– 2.76

Additional upgrades or components

Kitchen Cabinets & Countertops	Page 58
Bathroom Vanities	59
Fireplaces & Chimneys	59
Windows, Skylights & Dormers	59
Appliances	60
Breezeways & Porches	60
Finished Attic	60
Garages	61
Site Improvements	61
Wings & Ells	37

- **Post and beam frame**
- **Log exterior walls**
- **Simple design from standard plans**
- **Single family — 1 full bath, 1 kitchen**
- **No basement**
- **Asphalt shingles on roof**
- **Hot air heat**
- **Gypsum wallborad interior finishes**
- **Materials and workmanship are average**
- **Detail specification on page 27**

Note: The illustration shown may contain some optional components (for example: garages and/or fireplaces) whose costs are shown in the modifications, adjustments, & alternatives below or at the end of the square foot section.

Base cost per square foot of living area

Exterior Wall	Living Area										
	600	800	1000	1200	1400	1600	1800	2000	2400	2800	3200
6″ Log - Solid Wall	150.35	136.75	126.65	118.30	111.35	106.75	104.35	101.30	95.30	90.85	87.80
8″ Log - Solid Wall	152.65	138.85	128.50	120.00	112.85	108.15	105.75	102.60	96.45	91.95	88.85
Finished Basement, Add	33.50	32.30	30.80	29.40	28.30	27.65	27.25	26.60	25.80	25.10	24.55
Unfinished Basement, Add	13.75	12.45	11.50	10.60	9.90	9.50	9.25	8.85	8.40	7.95	7.65

Modifications

Add to the total cost

Upgrade Kitchen Cabinets	$ + 3760
Solid Surface Countertops (Included)	
Full Bath - including plumbing, wall and floor finishes	+ 5522
Half Bath - including plumbing, wall and floor finishes	+ 3416
One Car Attached Garage	+ 12,594
One Car Detached Garage	+ 16,766
Fireplace & Chimney	+ 4965

Adjustments

For multi family - add to total cost

Additional Kitchen	$ + 6433
Additional Bath	+ 5522
Additional Entry & Exit	+ 1622
Separate Heating	+ 1376
Separate Electric	+ 1870

For Townhouse/Rowhouse - Multiply cost per square foot by

Inner Unit	+ .92
End Unit	+ .96

Alternatives

Add to or deduct from the cost per square foot of living area

Cedar Shake Roof	+ 4.45
Air Conditioning, in Heating Ductwork	+ 3.37
In Separate Ductwork	+ 6.44
Heating Systems, Hot Water	+ 1.74
Heat Pump	+ 2.13
Electric Heat	– .84
Not Heated	– 3.06

Additional upgrades or components

Kitchen Cabinets & Countertops	Page 58
Bathroom Vanities	59
Fireplaces & Chimneys	59
Windows, Skylights & Dormers	59
Appliances	60
Breezeways & Porches	60
Finished Attic	60
Garages	61
Site Improvements	61
Wings & Ells	37

- Post and beam frame
- Log exterior walls
- Simple design from standard plans
- Single family — 1 full bath, 1 kitchen
- No basement
- Asphalt shingles on roof
- Hot air heat
- Gypsum wallboard interior finishes
- Materials and workmanship are average
- Detail specification on page 27

Note: The illustration shown may contain some optional components (for example: garages and/or fireplaces) whose costs are shown in the modifications, adjustments, & alternatives below or at the end of the square foot section.

Base cost per square foot of living area

Exterior Wall	Living Area										
	1000	1200	1400	1600	1800	2000	2200	2600	3000	3400	3800
6" Log-Solid	133.90	122.20	116.50	112.50	108.50	104.30	101.30	95.65	90.25	87.80	85.65
8" Log-Solid	118.25	107.30	102.20	98.65	95.20	91.25	88.70	83.80	78.90	76.75	74.80
Finished Basement, Add	19.40	19.10	18.50	18.05	17.65	17.35	16.95	16.35	15.90	15.65	15.40
Unfinished Basement, Add	7.55	7.05	6.65	6.35	6.05	5.90	5.65	5.30	5.00	4.80	4.70

Modifications

Add to the total cost

Upgrade Kitchen Cabinets	$ + 3760
Solid Surface Countertops (Included)	
Full Bath - including plumbing, wall and floor finishes	+ 5522
Half Bath - including plumbing, wall and floor finishes	+ 3416
One Car Attached Garage	+ 12,594
One Car Detached Garage	+ 16,766
Fireplace & Chimney	+ 5536

Adjustments

For multi family - add to total cost

Additional Kitchen	$ + 6433
Additional Bath	+ 5522
Additional Entry & Exit	+ 1622
Separate Heating	+ 1376
Separate Electric	+ 1870

For Townhouse/Rowhouse - Multiply cost per square foot by

Inner Unit	+ .92
End Unit	+ .96

Alternatives

Add to or deduct from the cost per square foot of living area

Cedar Shake Roof	+ 2.25
Air Conditioning, in Heating Ductwork	+ 2.04
In Separate Ductwork	+ 5.12
Heating Systems, Hot Water	+ 1.62
Heat Pump	+ 2.50
Electric Heat	– .64
Not Heated	– 2.86

Additional upgrades or components

Kitchen Cabinets & Countertops	Page 58
Bathroom Vanities	59
Fireplaces & Chimneys	59
Windows, Skylights & Dormers	59
Appliances	60
Breezeways & Porches	60
Finished Attic	60
Garages	61
Site Improvements	61
Wings & Ells	37

1 Story — Base cost per square foot of living area

Exterior Wall	Living Area							
	50	100	200	300	400	500	600	700
Wood Siding - Wood Frame	176.65	136.20	118.60	100.35	94.60	91.20	88.85	89.70
Brick Veneer - Wood Frame	190.80	143.35	122.85	99.70	93.00	89.00	86.25	86.90
Stucco on Wood Frame	169.30	130.95	114.20	97.30	91.85	88.60	86.40	87.35
Solid Masonry	235.30	178.15	153.55	123.65	115.60	110.75	109.10	109.05
Finished Basement, Add	53.30	44.20	39.75	32.45	30.95	30.10	29.45	29.10
Unfinished Basement, Add	24.95	18.85	16.30	12.10	11.20	10.70	10.40	10.15

1-1/2 Story — Base cost per square foot of living area

Exterior Wall	Living Area							
	100	200	300	400	500	600	700	800
Wood Siding - Wood Frame	142.40	114.75	98.30	88.70	83.65	81.30	78.30	77.40
Brick Veneer - Wood Frame	199.10	149.70	124.45	108.85	101.10	97.00	92.70	91.00
Stucco on Wood Frame	170.85	127.05	105.60	94.15	87.55	84.20	80.50	79.05
Solid Masonry	229.50	174.05	144.75	124.65	115.75	110.80	105.70	104.00
Finished Basement, Add	36.00	32.60	29.65	26.40	25.50	24.90	24.35	24.30
Unfinished Basement, Add	15.65	13.10	11.45	9.55	9.00	8.70	8.35	8.35

2 Story — Base cost per square foot of living area

Exterior Wall	Living Area							
	100	200	400	600	800	1000	1200	1400
Wood Siding - Wood Frame	142.10	106.75	91.30	76.55	71.50	68.45	66.45	67.50
Brick Veneer - Wood Frame	203.35	142.70	115.15	92.40	84.85	80.35	77.30	77.65
Stucco on Wood Frame	171.70	120.05	96.25	79.80	73.55	69.85	67.25	67.95
Solid Masonry	237.50	167.05	135.45	105.95	97.05	91.70	88.15	88.05
Finished Basement, Add	28.50	24.00	21.80	18.15	17.35	16.95	16.65	16.45
Unfinished Basement, Add	12.65	9.55	8.35	6.20	5.75	5.50	5.35	5.25

Base costs do not include bathroom or kitchen facilities. Use Modifications/Adjustments/Alternatives on pages 58-61 where appropriate.

Important: See the Reference Section for Location Factors (to adjust for your city) and Estimating Forms.

1 Story

1 - 1/2 Story

2 Story

2 - 1/2 Story

Bi-Level

Tri-Level

Components

1 Site Work

Site preparation for slab or excavation for lower level; 4' deep trench excavation for foundation wall.

2 Foundations

Continuous reinforced concrete footing, 8" deep x 18" wide; dampproofed and insulated 8" thick reinforced concrete foundation wall, 4' deep; 4" concrete slab on 4" crushed stone base and polyethylene vapor barrier, trowel finish.

3 Framing

Exterior walls—2 x 6 wood studs, 16" O.C.; 1/2" plywood sheathing; 2 x 8 rafters 16" O.C. with 1/2" plywood sheathing, 4 in 12, 6 in 12 or 8 in 12 roof pitch, 2 x 6 or 2 x 8 ceiling joists; 1/2" plywood subfloor on 1 x 3 wood sleepers 16" O.C.; 2 x 10 floor joists with 5/8" plywood subfloor on models with more than one level.

4 Exterior Walls

Horizontal beveled wood siding and #15 felt building paper on insulated wood frame walls; double hung windows; 3 flush solid core wood exterior doors with storms.

Alternates:
- Brick veneer on wood frame, has 4" veneer high quality face brick or select common brick.
- Stone veneer on wood frame has exterior veneer of field stone or 2" thick limestone.
- Solid masonry has an 8" concrete block wall with insulation and a brick or stone facing. It may be a solid brick or stone structure.

5 Roofing

30 year asphalt shingles; #15 felt building paper; aluminum gutters, downspouts and drip edge; copper flashings.

6 Interiors

Walls and ceilings—5/8" drywall, skim coat plaster, primed and painted with 2 coats; hardwood baseboard and trim; sanded and finished, hardwood floor 70%, ceramic tile with 1/2" underlayment 20%, vinyl tile with 1/2" underlayment 10%; wood panel interior doors, primed and painted with 2 coats.

7 Specialties

Custom grade kitchen cabinets—20 L.F. wall and base with plastic laminate counter top and kitchen sink; 4 L.F. bathroom vanity; 75 gallon electric water heater; medicine cabinet.

8 Mechanical

Gas fired warm air heat/air conditioning; one full bath including bathtub, corner shower, built in lavatory and water closet; one 1/2 bath including built in lavatory and water closet.

9 Electrical

100 Amp. service; romex wiring; incandescent lighting fixtures, switches, receptacles.

10 Overhead and Profit

General Contractor overhead and profit.

djustments

Finished Basement:
5' high cast-in-place concrete walls 10" thick or 12" concrete block.

Finished Basement:
Includes painted drywall on 2 x 4 wood furring with insulation, suspended ceiling, carpeting on concrete floor, heating and lighting.

- A distinct residence from designer's plans
- Single family — 1 full bath, 1 half bath, 1 kitchen
- No basement
- Asphalt shingles on roof
- Forced hot air heat/air conditioning
- Gypsum wallboard interior finishes
- Materials and workmanship are above average
- Detail specifications on page 39

Note: The illustration shown may contain some optional components (for example: garages and/or fireplaces) whose costs are shown in the modifications, adjustments, & alternatives below or at the end of the square foot section.

©Design Basics, Inc.

Base cost per square foot of living area

Exterior Wall	Living Area										
	800	1000	1200	1400	1600	1800	2000	2400	2800	3200	3600
Wood Siding - Wood Frame	165.30	150.00	137.85	128.05	121.35	117.75	113.40	105.25	99.30	95.15	90.85
Brick Veneer - Wood Frame	184.90	168.85	155.90	145.45	138.35	134.50	129.80	121.15	114.80	110.35	105.75
Stone Veneer - Wood Frame	187.80	171.45	158.25	147.60	140.35	136.45	131.60	122.75	116.35	111.80	107.05
Solid Masonry	193.50	176.60	162.85	151.85	144.30	140.20	135.15	126.05	119.40	114.60	109.65
Finished Basement, Add	51.05	50.90	48.80	47.10	45.90	45.30	44.35	43.10	42.15	41.20	40.50
Unfinished Basement, Add	21.35	20.15	19.05	18.15	17.60	17.20	16.75	16.05	15.60	15.15	14.70

Modifications

Add to the total cost

Upgrade Kitchen Cabinets	$ + 1102
Solid Surface Countertops (Included)	
Full Bath - including plumbing, wall and floor finishes	+ 6543
Half Bath - including plumbing, wall and floor finishes	+ 4048
Two Car Attached Garage	+ 24,917
Two Car Detached Garage	+ 28,578
Fireplace & Chimney	+ 5465

Adjustments

For multi family - add to total cost

Additional Kitchen	$ + 14,107
Additional Full Bath & Half Bath	+ 10,591
Additional Entry & Exit	+ 1622
Separate Heating & Air Conditioning	+ 6091
Separate Electric	+ 1870

For Townhouse/Rowhouse -
Multiply cost per square foot by

Inner Unit	+ .90
End Unit	+ .95

Alternatives

Add to or deduct from the cost per square foot of living area

Cedar Shake Roof	+ 3.95
Clay Tile Roof	+ 4.30
Slate Roof	+ 6.15
Upgrade Ceilings to Textured Finish	+ .47
Air Conditioning, in Heating Ductwork	Base System
Heating Systems, Hot Water	+ 1.78
Heat Pump	+ 2.12
Electric Heat	– 2.37
Not Heated	– 3.84

Additional upgrades or components

Kitchen Cabinets & Countertops	Page 58
Bathroom Vanities	59
Fireplaces & Chimneys	59
Windows, Skylights & Dormers	59
Appliances	60
Breezeways & Porches	60
Finished Attic	60
Garages	61
Site Improvements	61
Wings & Ells	47

- **A distinct residence from designer's plans**
- **Single family — 1 full bath, 1 half bath, 1 kitchen**
- **No basement**
- **Asphalt shingles on roof**
- **Forced hot air heat/air conditioning**
- **Gypsum wallboard interior finishes**
- **Materials and workmanship are above average**
- **Detail specifications on page 39**

Note: The illustration shown may contain some optional components (for example: garages and/or fireplaces) whose costs are shown in the modifications, adjustments, & alternatives below or at the end of the square foot section.

•Donald A. Gardner Architects, Inc.

Base cost per square foot of living area

Exterior Wall	Living Area										
	1000	1200	1400	1600	1800	2000	2400	2800	3200	3600	4000
Wood Siding - Wood Frame	148.70	138.50	131.35	122.50	117.55	112.60	103.25	99.00	95.10	92.10	87.75
Brick Veneer - Wood Frame	157.90	147.20	139.60	130.00	124.65	119.35	109.20	104.65	100.35	97.10	92.45
Stone Veneer - Wood Frame	161.00	150.05	142.30	132.45	127.00	121.55	111.20	106.50	102.05	98.80	94.00
Solid Masonry	166.90	155.60	147.55	137.30	131.55	125.90	114.95	110.10	105.35	102.00	96.95
Finished Basement, Add	33.80	34.10	33.20	31.85	31.10	30.40	29.10	28.50	27.80	27.40	26.80
Unfinished Basement, Add	14.35	13.75	13.25	12.60	12.25	11.85	11.15	10.85	10.50	10.30	9.95

Modifications

Add to the total cost

Upgrade Kitchen Cabinets	$ + 1102
Solid Surface Countertops (Included)	
Full Bath - including plumbing, wall and floor finishes	+ 6543
Half Bath - including plumbing, wall and floor finishes	+ 4048
Two Car Attached Garage	+ 24,917
Two Car Detached Garage	+ 28,578
Fireplace & Chimney	+ 5465

Adjustments

For multi family - add to total cost

Additional Kitchen	$ + 14,107
Additional Full Bath & Half Bath	+ 10,591
Additional Entry & Exit	+ 1622
Separate Heating & Air Conditioning	+ 6091
Separate Electric	+ 1870

For Townhouse/Rowhouse -
Multiply cost per square foot by

Inner Unit	+ .90
End Unit	+ .95

Alternatives

Add to or deduct from the cost per square foot of living area

Cedar Shake Roof	+ 2.85
Clay Tile Roof	+ 3.10
Slate Roof	+ 4.45
Upgrade Ceilings to Textured Finish	+ .47
Air Conditioning, in Heating Ductwork	Base System
Heating Systems, Hot Water	+ 1.69
Heat Pump	+ 2.23
Electric Heat	– 2.09
Not Heated	– 3.54

Additional upgrades or components

Kitchen Cabinets & Countertops	Page 58
Bathroom Vanities	59
Fireplaces & Chimneys	59
Windows, Skylights & Dormers	59
Appliances	60
Breezeways & Porches	60
Finished Attic	60
Garages	61
Site Improvements	61
Wings & Ells	47

- **A distinct residence from designer's plans**
- **Single family — 1 full bath, 1 half bath, 1 kitchen**
- **No basement**
- **Asphalt shingles on roof**
- **Forced hot air heat/air conditioning**
- **Gypsum wallboard interior finishes**
- **Materials and workmanship are above average**
- **Detail specifications on page 39**

Note: The illustration shown may contain some optional components (for example: garages and/or fireplaces) whose costs are shown in the modifications, adjustments, & alternatives below or at the end of the square foot section.

Base cost per square foot of living area

Exterior Wall	Living Area										
	1200	1400	1600	1800	2000	2400	2800	3200	3600	4000	4400
Wood Siding - Wood Frame	139.35	131.10	125.30	120.30	114.70	106.70	99.90	95.60	92.95	90.05	87.75
Brick Veneer - Wood Frame	149.20	140.35	134.10	128.60	122.70	113.90	106.50	101.75	98.80	95.65	93.15
Stone Veneer - Wood Frame	152.45	143.35	137.00	131.30	125.30	116.25	108.65	103.75	100.80	97.45	94.90
Solid Masonry	158.80	149.35	142.70	136.65	130.50	120.90	112.90	107.70	104.55	101.00	98.30
Finished Basement, Add	27.20	27.35	26.70	25.90	25.45	24.35	23.50	23.00	22.65	22.15	21.90
Unfinished Basement, Add	11.50	11.00	10.70	10.30	10.05	9.50	9.05	8.75	8.60	8.40	8.20

Modifications

Add to the total cost

Upgrade Kitchen Cabinets	$ + 1102
Solid Surface Countertops (Included)	
Full Bath - including plumbing, wall and floor finishes	+ 6543
Half Bath - including plumbing, wall and floor finishes	+ 4048
Two Car Attached Garage	+ 24,917
Two Car Detached Garage	+ 28,578
Fireplace & Chimney	+ 6167

Adjustments

For multi family - add to total cost

Additional Kitchen	$ + 14,107
Additional Full Bath & Half Bath	+ 10,591
Additional Entry & Exit	+ 1622
Separate Heating & Air Conditioning	+ 6091
Separate Electric	+ 1870

For Townhouse/Rowhouse -
Multiply cost per square foot by

Inner Unit	+ .87
End Unit	+ .93

Alternatives

Add to or deduct from the cost per square foot of living area

Cedar Shake Roof	+ 2
Clay Tile Roof	+ 2.15
Slate Roof	+ 3.10
Upgrade Ceilings to Textured Finish	+ .47
Air Conditioning, in Heating Ductwork	Base System
Heating Systems, Hot Water	+ 1.65
Heat Pump	+ 2.48
Electric Heat	– 2.09
Not Heated	– 3.35

Additional upgrades or components

Kitchen Cabinets & Countertops	Page 58
Bathroom Vanities	59
Fireplaces & Chimneys	59
Windows, Skylights & Dormers	59
Appliances	60
Breezeways & Porches	60
Finished Attic	60
Garages	61
Site Improvements	61
Wings & Ells	47

- **A distinct residence from designer's plans**
- **Single family — 1 full bath, 1 half bath, 1 kitchen**
- **No basement**
- **Asphalt shingles on roof**
- **Forced hot air heat/air conditioning**
- **Gypsum wallboard interior finishes**
- **Materials and workmanship are above average**
- **Detail specifications on page 39**

Note: The illustration shown may contain some optional components (for example: garages and/or fireplaces) whose costs are shown in the modifications, adjustments, & alternatives below or at the end of the square foot section.

Base cost per square foot of living area

Exterior Wall	Living Area										
	1500	1800	2100	2400	2800	3200	3600	4000	4500	5000	5500
Wood Siding - Wood Frame	136.90	123.40	115.55	110.75	104.65	98.80	95.55	90.40	87.70	85.30	82.85
Brick Veneer - Wood Frame	147.10	132.85	124.00	118.85	112.30	105.75	102.10	96.50	93.50	90.85	88.10
Stone Veneer - Wood Frame	150.50	135.95	126.80	121.40	114.80	108.05	104.30	98.50	95.40	92.60	89.85
Solid Masonry	157.10	142.00	132.20	126.60	119.70	112.50	108.50	102.45	99.15	96.20	93.20
Finished Basement, Add	21.55	21.55	20.45	19.95	19.50	18.70	18.30	17.85	17.50	17.20	16.95
Unfinished Basement, Add	9.25	8.75	8.20	7.90	7.70	7.30	7.10	6.85	6.65	6.50	6.35

Modifications

Add to the total cost

Upgrade Kitchen Cabinets	$ + 1102
Solid Surface Countertops (Included)	
Full Bath - including plumbing, wall and floor finishes	+ 6543
Half Bath - including plumbing, wall and floor finishes	+ 4048
Two Car Attached Garage	+ 24,917
Two Car Detached Garage	+ 28,578
Fireplace & Chimney	+ 6964

Adjustments

For multi family - add to total cost

Additional Kitchen	$ + 14,107
Additional Full Bath & Half Bath	+ 10,591
Additional Entry & Exit	+ 1622
Separate Heating & Air Conditioning	+ 6091
Separate Electric	+ 1870

For Townhouse/Rowhouse - Multiply cost per square foot by

Inner Unit	+ .87
End Unit	+ .94

Alternatives

Add to or deduct from the cost per square foot of living area

Cedar Shake Roof	+ 1.70
Clay Tile Roof	+ 1.85
Slate Roof	+ 2.70
Upgrade Ceilings to Textured Finish	+ .47
Air Conditioning, in Heating Ductwork	Base System
Heating Systems, Hot Water	+ 1.49
Heat Pump	+ 2.57
Electric Heat	– 3.67
Not Heated	– 3.35

Additional upgrades or components

Kitchen Cabinets & Countertops	Page 58
Bathroom Vanities	59
Fireplaces & Chimneys	59
Windows, Skylights & Dormers	59
Appliances	60
Breezeways & Porches	60
Finished Attic	60
Garages	61
Site Improvements	61
Wings & Ells	47

- **A distinct residence from designer's plans**
- **Single family — 1 full bath, 1 half bath, 1 kitchen**
- **No basement**
- **Asphalt shingles on roof**
- **Forced hot air heat/air conditioning**
- **Gypsum wallboard interior finishes**
- **Materials and workmanship are above average**
- **Detail specifications on page 39**

Note: The illustration shown may contain some optional components (for example: garages and/or fireplaces) whose costs are shown in the modifications, adjustments, & alternatives below or at the end of the square foot section.

Base cost per square foot of living area

Exterior Wall	Living Area										
	1500	1800	2100	2500	3000	3500	4000	4500	5000	5500	6000
Wood Siding - Wood Frame	136.20	123.00	116.60	111.20	102.90	98.80	93.70	88.35	86.50	84.45	82.35
Brick Veneer - Wood Frame	146.90	132.80	125.80	120.00	110.95	106.40	100.60	94.70	92.65	90.35	87.90
Stone Veneer - Wood Frame	150.35	136.05	128.85	122.90	113.50	108.85	102.85	96.75	94.65	92.30	89.70
Solid Masonry	157.20	142.40	134.75	128.60	118.70	113.75	107.30	100.85	98.55	96.10	93.25
Finished Basement, Add	18.95	18.95	18.30	17.75	17.00	16.55	15.95	15.45	15.25	15.10	14.80
Unfinished Basement, Add	8.10	7.70	7.35	7.05	6.70	6.50	6.15	5.95	5.80	5.70	5.60

Modifications

Add to the total cost

Upgrade Kitchen Cabinets	$ + 1102
Solid Surface Countertops (Included)	
Full Bath - including plumbing, wall and floor finishes	+ 6543
Half Bath - including plumbing, wall and floor finishes	+ 4048
Two Car Attached Garage	+ 24,917
Two Car Detached Garage	+ 28,578
Fireplace & Chimney	+ 6964

Adjustments

For multi family - add to total cost

Additional Kitchen	$ + 14,107
Additional Full Bath & Half Bath	+ 10,591
Additional Entry & Exit	+ 1622
Separate Heating & Air Conditioning	+ 6091
Separate Electric	+ 1870

For Townhouse/Rowhouse -
Multiply cost per square foot by

Inner Unit	+ .85
End Unit	+ .93

Alternatives

Add to or deduct from the cost per square foot of living area

Cedar Shake Roof	+ 1.30
Clay Tile Roof	+ 1.45
Slate Roof	+ 2.05
Upgrade Ceilings to Textured Finish	+ .47
Air Conditioning, in Heating Ductwork	Base System
Heating Systems, Hot Water	+ 1.49
Heat Pump	+ 2.57
Electric Heat	– 3.67
Not Heated	– 3.24

Additional upgrades or components

Kitchen Cabinets & Countertops	Page 58
Bathroom Vanities	59
Fireplaces & Chimneys	59
Windows, Skylights & Dormers	59
Appliances	60
Breezeways & Porches	60
Finished Attic	60
Garages	61
Site Improvements	61
Wings & Ells	47

- A distinct residence from designer's plans
- Single family — 1 full bath, 1 half bath, 1 kitchen
- No basement
- Asphalt shingles on roof
- Forced hot air heat/air conditioning
- Gypsum wallboard interior finishes
- Materials and workmanship are above average
- Detail specifications on page 39

Note: The illustration shown may contain some optional components (for example: garages and/or fireplaces) whose costs are shown in the modifications, adjustments, & alternatives below or at the end of the square foot section.

Base cost per square foot of living area

Exterior Wall	Living Area										
	1200	1400	1600	1800	2000	2400	2800	3200	3600	4000	4400
Wood Siding - Wood Frame	131.55	123.90	118.40	113.75	108.45	100.95	94.65	90.65	88.20	85.65	83.45
Brick Veneer - Wood Frame	138.95	130.85	125.00	119.95	114.40	106.40	99.60	95.25	92.60	89.85	87.55
Stone Veneer - Wood Frame	141.40	133.10	127.15	122.00	116.35	108.15	101.20	96.75	94.10	91.15	88.80
Solid Masonry	146.15	137.55	131.40	125.95	120.25	111.65	104.40	99.75	96.95	93.85	91.45
Finished Basement, Add	27.20	27.35	26.70	25.90	25.45	24.35	23.50	23.00	22.65	22.15	21.90
Unfinished Basement, Add	11.50	11.00	10.70	10.30	10.05	9.50	9.05	8.75	8.60	8.40	8.20

Modifications

Add to the total cost

Upgrade Kitchen Cabinets	$ + 1102
Solid Surface Countertops (Included)	
Full Bath - including plumbing, wall and floor finishes	+ 6543
Half Bath - including plumbing, wall and floor finishes	+ 4048
Two Car Attached Garage	+ 24,917
Two Car Detached Garage	+ 28,578
Fireplace & Chimney	+ 5465

Adjustments

For multi family - add to total cost

Additional Kitchen	$ + 14,107
Additional Full Bath & Half Bath	+ 10,591
Additional Entry & Exit	+ 1622
Separate Heating & Air Conditioning	+ 6091
Separate Electric	+ 1870

For Townhouse/Rowhouse -
Multiply cost per square foot by

Inner Unit	+ .89
End Unit	+ .95

Alternatives

Add to or deduct from the cost per square foot of living area

Cedar Shake Roof	+ 2
Clay Tile Roof	+ 2.15
Slate Roof	+ 3.10
Upgrade Ceilings to Textured Finish	+ .47
Air Conditioning, in Heating Ductwork	Base System
Heating Systems, Hot Water	+ 1.65
Heat Pump	+ 2.48
Electric Heat	– 2.09
Not Heated	– 3.24

Additional upgrades or components

Kitchen Cabinets & Countertops	Page 58
Bathroom Vanities	59
Fireplaces & Chimneys	59
Windows, Skylights & Dormers	59
Appliances	60
Breezeways & Porches	60
Finished Attic	60
Garages	61
Site Improvements	61
Wings & Ells	47

- **A distinct residence from designer's plans**
- **Single family — 1 full bath, 1 half bath, 1 kitchen**
- **No basement**
- **Asphalt shingles on roof**
- **Forced hot air heat/air conditioning**
- **Gypsum wallboard interior finishes**
- **Materials and workmanship are above average**
- **Detail specifications on page 39**

Note: The illustration shown may contain some optional components (for example: garages and/or fireplaces) whose costs are shown in the modifications, adjustments, & alternatives below or at the end of the square foot section.

©Design Basics, Inc.

Base cost per square foot of living area

Exterior Wall	Living Area										
	1200	1500	1800	2100	2400	2800	3200	3600	4000	4500	5000
Wood Siding - Wood Frame	135.65	123.50	114.30	106.65	101.60	97.95	93.55	88.95	86.95	82.60	80.20
Brick Veneer - Wood Frame	143.00	130.15	120.35	112.15	106.75	102.85	98.15	93.15	91.10	86.40	83.80
Stone Veneer - Wood Frame	145.40	132.35	122.30	113.95	108.40	104.45	99.60	94.55	92.40	87.65	84.90
Solid Masonry	150.10	136.65	126.15	117.45	111.75	107.65	102.55	97.30	95.10	90.05	87.25
Finished Basement, Add*	34.05	33.95	32.50	31.35	30.60	30.10	29.35	28.75	28.40	27.80	27.40
Unfinished Basement, Add*	14.25	13.40	12.65	12.10	11.70	11.45	11.05	10.70	10.55	10.20	10.00

*Basement under middle level only.

Modifications

Add to the total cost

Upgrade Kitchen Cabinets	$ + 1102
Solid Surface Countertops (Included)	
Full Bath - including plumbing, wall and floor finishes	+ 6543
Half Bath - including plumbing, wall and floor finishes	+ 4048
Two Car Attached Garage	+ 24,917
Two Car Detached Garage	+ 28,578
Fireplace & Chimney	+ 5465

Adjustments

For multi family - add to total cost

Additional Kitchen	$ + 14,107
Additional Full Bath & Half Bath	+ 10,591
Additional Entry & Exit	+ 1622
Separate Heating & Air Conditioning	+ 6091
Separate Electric	+ 1870

For Townhouse/Rowhouse - Multiply cost per square foot by

Inner Unit	+ .87
End Unit	+ .94

Alternatives

Add to or deduct from the cost per square foot of living area

Cedar Shake Roof	+ 2.85
Clay Tile Roof	+ 3.10
Slate Roof	+ 4.45
Upgrade Ceilings to Textured Finish	+ .47
Air Conditioning, in Heating Ductwork	Base System
Heating Systems, Hot Water	+ 1.60
Heat Pump	+ 2.59
Electric Heat	– 1.86
Not Heated	– 3.24

Additional upgrades or components

Kitchen Cabinets & Countertops	Page 58
Bathroom Vanities	59
Fireplaces & Chimneys	59
Windows, Skylights & Dormers	59
Appliances	60
Breezeways & Porches	60
Finished Attic	60
Garages	61
Site Improvements	61
Wings & Ells	47

1 Story — Base cost per square foot of living area

Exterior Wall	Living Area							
	50	100	200	300	400	500	600	700
Wood Siding - Wood Frame	215.50	168.20	147.65	125.50	118.75	114.75	112.00	112.85
Brick Veneer - Wood Frame	240.35	186.00	162.45	135.35	127.65	123.05	119.90	120.50
Stone Veneer - Wood Frame	248.55	191.80	167.35	138.60	130.55	125.80	122.50	122.95
Solid Masonry	264.55	203.30	176.90	144.95	136.30	131.15	127.60	127.85
Finished Basement, Add	80.85	69.20	62.90	52.20	50.15	48.85	48.00	47.40
Unfinished Basement, Add	59.45	40.90	32.30	24.75	22.70	21.50	20.70	20.15

1-1/2 Story — Base cost per square foot of living area

Exterior Wall	Living Area							
	100	200	300	400	500	600	700	800
Wood Siding - Wood Frame	172.75	142.40	123.85	112.60	107.05	104.40	100.85	99.85
Brick Veneer - Wood Frame	194.95	160.20	138.65	124.15	117.70	114.45	110.45	109.30
Stone Veneer - Wood Frame	202.25	166.00	143.50	128.00	121.20	117.70	113.50	112.35
Solid Masonry	216.55	177.45	153.05	135.40	128.05	124.20	119.65	118.45
Finished Basement, Add	54.15	50.30	46.05	41.40	40.10	39.25	38.50	38.35
Unfinished Basement, Add	35.45	26.85	22.80	19.40	18.15	17.35	16.70	16.45

2 Story — Base cost per square foot of living area

Exterior Wall	Living Area							
	100	200	400	600	800	1000	1200	1400
Wood Siding - Wood Frame	172.05	132.05	114.50	97.40	91.70	88.30	85.95	87.05
Brick Veneer - Wood Frame	196.90	149.85	129.30	107.25	100.55	96.60	93.85	94.70
Stone Veneer - Wood Frame	205.10	155.65	134.20	110.50	103.50	99.30	96.45	97.20
Solid Masonry	221.10	167.10	143.75	116.85	109.20	104.65	101.55	102.10
Finished Basement, Add	40.40	34.60	31.45	26.10	25.10	24.45	24.05	23.70
Unfinished Basement, Add	29.75	20.45	16.15	12.35	11.35	10.75	10.35	10.10

Base costs do not include bathroom or kitchen facilities. Use Modifications/Adjustments/Alternatives on pages 58-61 where appropriate.

1 Story

1 - 1/2 Story

2 Story

2 - 1/2 Story

Bi-Level

Tri-Level

	Components
1 Site Work	Site preparation for slab or excavation for lower level; 4' deep trench excavation for foundation wall.
2 Foundations	Continuous reinforced concrete footing, 8" deep x 18" wide; dampproofed and insulated 12" thick reinforced concrete foundation wall, 4' deep; 4" concrete slab on 4" crushed stone base and polyethylene vapor barrier, trowel finish.
3 Framing	Exterior walls—2 x 6 wood studs, 16" O.C.; 1/2" plywood sheathing; 2 x 8 rafters 16" O.C. with 1/2" plywood sheathing, 4 in 12, 6 in 12 or 8 in 12 roof pitch, 2 x 6 or 2 x 8 ceiling joists; 1/2" plywood subfloor on 1 x 3 wood sleepers 16" O.C.; 2 x 10 floor joists with 5/8" plywood subfloor on models with more than one level.
4 Exterior Walls	Face brick veneer and #15 felt building paper on insulated wood frame walls; double hung windows; 3 flush solid core wood exterior doors with storms. **Alternates:** • Wood siding on wood frame, has top quality cedar or redwood siding or hand split cedar shingles or shakes. • Solid brick may have solid brick exterior wall or brick on concrete block. • Solid stone concrete block with selected fieldstone or limestone exterior.
5 Roofing	Red cedar shingles; #15 felt building paper; aluminum gutters, downspouts and drip edge; copper flashings.
6 Interiors	Walls and ceilings—5/8" drywall, skim coat plaster, primed and painted with 2 coats; hardwood baseboard and trim; sanded and finished, hardwood floor 70%, ceramic tile with 1/2" underlayment 20%, vinyl tile with 1/2" underlayment 10%; wood panel interior doors, primed and painted with 2 coats.
7 Specialties	Luxury grade kitchen cabinets—25 L.F. wall and base with plastic laminate counter top and kitchen sink; 6 L.F. bathroom vanity; 75 gallon electric water heater; medicine cabinet.
8 Mechanical	Gas fired warm air heat/air conditioning; one full bath including bathtub, corner shower, built in lavatory and water closet; one 1/2 bath including built in lavatory and water closet.
9 Electrical	100 Amp. service; romex wiring; incandescent lighting fixtures, switches, receptacles.
10 Overhead and Profit	General Contractor overhead and profit.

Adjustments

Unfinished Basement:
8" high cast-in-place concrete walls 12" thick or 12" concrete block.

Finished Basement:
Includes painted drywall on 2 x 4 wood furring with insulation, suspended ceiling, carpeting on subfloor with sleepers, heating and lighting.

- **Unique residence built from an architect's plan**
- **Single family — 1 full bath, 1 half bath, 1 kitchen**
- **No basement**
- **Cedar shakes on roof**
- **Forced hot air heat/air conditioning**
- **Gypsum wallboard interior finishes**
- **Many special features**
- **Extraordinary materials and workmanship**
- **Detail specifications on page 49**

Note: The illustration shown may contain some optional components (for example: garages and/or fireplaces) whose costs are shown in the modifications, adjustments, & alternatives below or at the end of the square foot section.

©Home Planners, Inc.

Base cost per square foot of living area

Exterior Wall	Living Area										
	1000	1200	1400	1600	1800	2000	2400	2800	3200	3600	4000
Wood Siding - Wood Frame	189.45	174.90	163.35	155.35	150.85	145.65	136.05	129.15	124.30	119.25	115.20
Brick Veneer - Wood Frame	198.60	183.15	170.90	162.40	157.65	152.00	141.90	134.55	129.25	123.90	119.50
Solid Brick	215.60	198.50	184.85	175.45	170.25	163.90	152.70	144.50	138.50	132.55	127.55
Solid Stone	209.50	193.05	179.85	170.75	165.80	159.60	148.85	140.90	135.25	129.45	124.65
Finished Basement, Add	49.20	53.00	50.85	49.50	48.60	47.50	45.90	44.70	43.60	42.65	41.90
Unfinished Basement, Add	21.80	20.50	19.45	18.70	18.35	17.70	16.90	16.30	15.75	15.25	14.85

Modifications

Add to the total cost

Upgrade Kitchen Cabinets	$ + 1455
Solid Surface Countertops (Included)	
Full Bath - including plumbing, wall and floor finishes	+ 7590
Half Bath - including plumbing, wall and floor finishes	+ 4696
Two Car Attached Garage	+ 30,094
Two Car Detached Garage	+ 34,187
Fireplace & Chimney	+ 7722

Adjustments

For multi family - add to total cost

Additional Kitchen	$ + 18,062
Additional Full Bath & Half Bath	+ 12,286
Additional Entry & Exit	+ 2324
Separate Heating & Air Conditioning	+ 6091
Separate Electric	+ 1870

For Townhouse/Rowhouse -
Multiply cost per square foot by

Inner Unit	+ .90
End Unit	+ .95

Alternatives

Add to or deduct from the cost per square foot of living area

Heavyweight Asphalt Shingles	– 3.95
Clay Tile Roof	+ .35
Slate Roof	+ 2.20
Upgrade Ceilings to Textured Finish	+ .47
Air Conditioning, in Heating Ductwork	Base System
Heating Systems, Hot Water	+ 1.92
Heat Pump	+ 2.30
Electric Heat	– 2.09
Not Heated	– 4.19

Additional upgrades or components

Kitchen Cabinets & Countertops	Page 58
Bathroom Vanities	59
Fireplaces & Chimneys	59
Windows, Skylights & Dormers	59
Appliances	60
Breezeways & Porches	60
Finished Attic	60
Garages	61
Site Improvements	61
Wings & Ells	57

Important: See the Reference Section for Location Factors (to adjust for your city) and Estimating Forms

- **Unique residence built from an architect's plan**
- **Single family — 1 full bath, 1 half bath, 1 kitchen**
- **No basement**
- **Cedar shakes on roof**
- **Forced hot air heat/air conditioning**
- **Gypsum wallboard interior finishes**
- **Many special features**
- **Extraordinary materials and workmanship**
- **Detail specifications on page 49**

Note: The illustration shown may contain some optional components (for example: garages and/or fireplaces) whose costs are shown in the modifications, adjustments, & alternatives below or at the end of the square foot section.

©Larry E. Belk Designs

Base cost per square foot of living area

Exterior Wall	Living Area										
	1000	1200	1400	1600	1800	2000	2400	2800	3200	3600	4000
Wood Siding - Wood Frame	175.25	162.90	154.10	143.65	137.60	131.75	120.75	115.65	111.00	107.45	102.40
Brick Veneer - Wood Frame	185.85	172.85	163.50	152.25	145.75	139.50	127.60	122.10	116.95	113.20	107.75
Solid Brick	205.50	191.25	181.10	168.20	160.85	153.85	140.25	134.15	128.10	123.95	117.70
Solid Stone	198.50	184.65	174.85	162.45	155.45	148.70	135.75	129.80	124.10	120.15	114.10
Finished Basement, Add	34.60	37.55	36.40	34.80	33.85	33.05	31.40	30.60	29.70	29.30	28.55
Unfinished Basement, Add	15.75	15.00	14.45	13.65	13.20	12.75	11.95	11.55	11.10	10.85	10.50

Modifications

Add to the total cost

Upgrade Kitchen Cabinets	$ + 1455
Solid Surface Countertops (Included)	
Full Bath - including plumbing, wall and floor finishes	+ 7590
Half Bath - including plumbing, wall and floor finishes	+ 4696
Two Car Attached Garage	+ 30,094
Two Car Detached Garage	+ 34,187
Fireplace & Chimney	+ 7722

Adjustments

For multi family - add to total cost

Additional Kitchen	$ + 18,062
Additional Full Bath & Half Bath	+ 12,286
Additional Entry & Exit	+ 2324
Separate Heating & Air Conditioning	+ 6091
Separate Electric	+ 1870

For Townhouse/Rowhouse -
Multiply cost per square foot by

Inner Unit	+ .90
End Unit	+ .95

Alternatives

Add to or deduct from the cost per square foot of living area

Heavyweight Asphalt Shingles	– 2.85
Clay Tile Roof	+ .25
Slate Roof	+ 1.60
Upgrade Ceilings to Textured Finish	+ .47
Air Conditioning, in Heating Ductwork	Base System
Heating Systems, Hot Water	+ 1.83
Heat Pump	+ 2.55
Electric Heat	– 2.09
Not Heated	– 3.86

Additional upgrades or components

Kitchen Cabinets & Countertops	Page 58
Bathroom Vanities	59
Fireplaces & Chimneys	59
Windows, Skylights & Dormers	59
Appliances	60
Breezeways & Porches	60
Finished Attic	60
Garages	61
Site Improvements	61
Wings & Ells	57

- **Unique residence built from an architect's plan**
- **Single family — 1 full bath, 1 half bath, 1 kitchen**
- **No basement**
- **Cedar shakes on roof**
- **Forced hot air heat/air conditioning**
- **Gypsum wallboard interior finishes**
- **Many special features**
- **Extraordinary materials and workmanship**
- **Detail specifications on page 49**

Note: The illustration shown may contain some optional components (for example: garages and/or fireplaces) whose costs are shown in the modifications, adjustments, & alternatives below or at the end of the square foot section.

Base cost per square foot of living area

Exterior Wall	Living Area										
	1200	1400	1600	1800	2000	2400	2800	3200	3600	4000	4400
Wood Siding - Wood Frame	162.40	152.65	145.55	139.50	132.95	123.45	115.60	110.45	107.30	103.90	101.20
Brick Veneer - Wood Frame	173.75	163.20	155.65	149.05	142.15	131.70	123.15	117.50	114.05	110.30	107.35
Solid Brick	194.70	182.80	174.40	166.65	159.15	147.05	137.10	130.60	126.65	122.15	118.80
Solid Stone	187.20	175.80	167.70	160.30	153.05	141.60	132.10	125.90	122.20	117.85	114.70
Finished Basement, Add	27.80	30.15	29.35	28.40	27.85	26.50	25.50	24.75	24.35	23.75	23.45
Unfinished Basement, Add	12.65	12.05	11.65	11.20	10.90	10.25	9.70	9.35	9.15	8.90	8.70

Modifications

Add to the total cost

Upgrade Kitchen Cabinets	$ + 1455
Solid Surface Countertops (Included)	
Full Bath - including plumbing, wall and floor finishes	+ 7590
Half Bath - including plumbing, wall and floor finishes	+ 4696
Two Car Attached Garage	+ 30,094
Two Car Detached Garage	+ 34,187
Fireplace & Chimney	+ 8465

Adjustments

For multi family - add to total cost

Additional Kitchen	$ + 18,062
Additional Full Bath & Half Bath	+ 12,286
Additional Entry & Exit	+ 2324
Separate Heating & Air Conditioning	+ 6091
Separate Electric	+ 1870

For Townhouse/Rowhouse -
Multiply cost per square foot by

Inner Unit	+ .86
End Unit	+ .93

Alternatives

Add to or deduct from the cost per square foot of living area

Heavyweight Asphalt Shingles	– 2
Clay Tile Roof	+ .20
Slate Roof	+ 1.10
Upgrade Ceilings to Textured Finish	+ .47
Air Conditioning, in Heating Ductwork	Base System
Heating Systems, Hot Water	+ 1.78
Heat Pump	+ 2.69
Electric Heat	– 1.88
Not Heated	– 3.65

Additional upgrades or components

Kitchen Cabinets & Countertops	Page 58
Bathroom Vanities	59
Fireplaces & Chimneys	59
Windows, Skylights & Dormers	59
Appliances	60
Breezeways & Porches	60
Finished Attic	60
Garages	61
Site Improvements	61
Wings & Ells	57

- **Unique residence built from an architect's plan**
- **Single family — 1 full bath, 1 half bath, 1 kitchen**
- **No basement**
- **Cedar shakes on roof**
- **Forced hot air heat/air conditioning**
- **Gypsum wallboard interior finishes**
- **Many special features**
- **Extraordinary materials and workmanship**
- **Detail specifications on page 49**

Note: The illustration shown may contain some optional components (for example: garages and/or fireplaces) whose costs are shown in the modifications, adjustments, & alternatives below or at the end of the square foot section.

©Larry W. Garnett & Associates, Inc

Base cost per square foot of living area

Exterior Wall	Living Area										
	1500	1800	2100	2500	3000	3500	4000	4500	5000	5500	6000
Wood Siding - Wood Frame	157.80	142.20	132.85	126.10	116.65	109.85	103.40	100.15	97.30	94.45	91.45
Brick Veneer - Wood Frame	169.55	153.05	142.55	135.25	125.00	117.45	110.40	106.80	103.60	100.45	97.20
Solid Brick	191.40	173.15	160.50	152.25	140.40	131.45	123.45	119.10	115.35	111.65	107.90
Solid Stone	183.60	165.90	154.10	146.15	134.90	126.45	118.75	114.75	111.15	107.65	104.10
Finished Basement, Add	22.20	23.85	22.55	21.75	20.75	19.95	19.25	18.85	18.50	18.15	17.80
Unfinished Basement, Add	10.20	9.60	8.95	8.55	8.05	7.65	7.30	7.10	6.95	6.75	6.60

Modifications

Add to the total cost

Upgrade Kitchen Cabinets	$ + 1455
Solid Surface Countertops (Included)	
Full Bath - including plumbing, wall and floor finishes	+ 7590
Half Bath - including plumbing, wall and floor finishes	+ 4696
Two Car Attached Garage	+ 30,094
Two Car Detached Garage	+ 34,187
Fireplace & Chimney	+ 9263

Adjustments

For multi family - add to total cost

Additional Kitchen	$ + 18,062
Additional Full Bath & Half Bath	+ 12,286
Additional Entry & Exit	+ 2324
Separate Heating & Air Conditioning	+ 6091
Separate Electric	+ 1870

For Townhouse/Rowhouse -
Multiply cost per square foot by

Inner Unit	+ .86
End Unit	+ .93

Alternatives

Add to or deduct from the cost per square foot of living area

Heavyweight Asphalt Shingles	– 1.70
Clay Tile Roof	+ .15
Slate Roof	+ .95
Upgrade Ceilings to Textured Finish	+ .47
Air Conditioning, in Heating Ductwork	Base System
Heating Systems, Hot Water	+ 1.61
Heat Pump	+ 2.77
Electric Heat	– 3.70
Not Heated	– 3.65

Additional upgrades or components

Kitchen Cabinets & Countertops	Page 58
Bathroom Vanities	59
Fireplaces & Chimneys	59
Windows, Skylights & Dormers	59
Appliances	60
Breezeways & Porches	60
Finished Attic	60
Garages	61
Site Improvements	61
Wings & Ells	57

- **Unique residence built from an architect's plan**
- **Single family — 1 full bath, 1 half bath, 1 kitchen**
- **No basement**
- **Cedar shakes on roof**
- **Forced hot air heat/air conditioning**
- **Gypsum wallboard interior finishes**
- **Many special features**
- **Extraordinary materials and workmanship**
- **Detail specifications on page 49**

Note: The illustration shown may contain some optional components (for example: garages and/or fireplaces) whose costs are shown in the modifications, adjustments, & alternatives below or at the end of the square foot section.

Base cost per square foot of living area

Exterior Wall	Living Area										
	1500	1800	2100	2500	3000	3500	4000	4500	5000	5500	6000
Wood Siding - Wood Frame	156.65	141.30	133.60	127.25	117.55	112.60	106.60	100.45	98.20	95.80	93.30
Brick Veneer - Wood Frame	168.90	152.55	144.20	137.35	126.70	121.25	114.50	107.70	105.25	102.55	99.65
Solid Brick	191.65	173.60	163.80	156.10	143.75	137.35	129.20	121.25	118.40	115.15	111.50
Solid Stone	183.50	166.10	156.80	149.40	137.60	131.60	123.95	116.40	113.70	110.65	107.25
Finished Basement, Add	19.45	20.95	20.10	19.50	18.55	18.00	17.30	16.70	16.50	16.15	15.80
Unfinished Basement, Add	8.95	8.45	8.05	7.75	7.30	7.00	6.65	6.35	6.20	6.05	5.90

Modifications

Add to the total cost

Upgrade Kitchen Cabinets	$ + 1455
Solid Surface Countertops (Included)	
Full Bath - including plumbing, wall and floor finishes	+ 7590
Half Bath - including plumbing, wall and floor finishes	+ 4696
Two Car Attached Garage	+ 30,094
Two Car Detached Garage	+ 34,187
Fireplace & Chimney	+ 9263

Adjustments

For multi family - add to total cost

Additional Kitchen	$ + 18,062
Additional Full Bath & Half Bath	+ 12,286
Additional Entry & Exit	+ 2324
Separate Heating & Air Conditioning	+ 6091
Separate Electric	+ 1870

For Townhouse/Rowhouse -
Multiply cost per square foot by

Inner Unit	+ .84
End Unit	+ .92

Alternatives

Add to or deduct from the cost per square foot of living area

Heavyweight Asphalt Shingles	– 1.30
Clay Tile Roof	+ .10
Slate Roof	+ .75
Upgrade Ceilings to Textured Finish	+ .47
Air Conditioning, in Heating Ductwork	Base System
Heating Systems, Hot Water	+ 1.61
Heat Pump	+ 2.77
Electric Heat	– 3.70
Not Heated	– 3.54

Additional upgrades or components

Kitchen Cabinets & Countertops	Page 58
Bathroom Vanities	59
Fireplaces & Chimneys	59
Windows, Skylights & Dormers	59
Appliances	60
Breezeways & Porches	60
Finished Attic	60
Garages	61
Site Improvements	61
Wings & Ells	57

- **Unique residence built from an architect's plan**
- **Single family — 1 full bath, 1 half bath, 1 kitchen**
- **No basement**
- **Cedar shakes on roof**
- **Forced hot air heat/air conditioning**
- **Gypsum wallboard interior finishes**
- **Many special features**
- **Extraordinary materials and workmanship**
- **Detail specifications on page 49**

Note: The illustration shown may contain some optional components (for example: garages and/or fireplaces) whose costs are shown in the modifications, adjustments, & alternatives below or at the end of the square foot section.

Base cost per square foot of living area

Exterior Wall	Living Area										
	1200	1400	1600	1800	2000	2400	2800	3200	3600	4000	4400
Wood Siding - Wood Frame	153.35	144.20	137.50	131.90	125.70	116.90	109.55	104.75	101.85	98.85	96.25
Brick Veneer - Wood Frame	161.85	152.05	145.05	139.05	132.55	123.05	115.15	110.05	106.95	103.60	100.90
Solid Brick	177.65	166.80	159.15	152.30	145.35	134.55	125.65	119.85	116.35	112.45	109.50
Solid Stone	172.00	161.50	154.10	147.55	140.75	130.45	121.95	116.30	113.00	109.25	106.40
Finished Basement, Add	27.80	30.15	29.35	28.40	27.85	26.50	25.50	24.75	24.35	23.75	23.45
Unfinished Basement, Add	12.65	12.05	11.65	11.20	10.90	10.25	9.70	9.35	9.15	8.90	8.70

Modifications

Add to the total cost

Upgrade Kitchen Cabinets	$ + 1455
Solid Surface Countertops (Included)	
Full Bath - including plumbing, wall and floor finishes	+ 7590
Half Bath - including plumbing, wall and floor finishes	+ 4696
Two Car Attached Garage	+ 30,094
Two Car Detached Garage	+ 34,187
Fireplace & Chimney	+ 7722

Adjustments

For multi family - add to total cost

Additional Kitchen	$ + 18,062
Additional Full Bath & Half Bath	+ 12,286
Additional Entry & Exit	+ 2324
Separate Heating & Air Conditioning	+ 6091
Separate Electric	+ 1870

For Townhouse/Rowhouse -
Multiply cost per square foot by

Inner Unit	+ .89
End Unit	+ .94

Alternatives

Add to or deduct from the cost per square foot of living area

Heavyweight Asphalt Shingles	– 2
Clay Tile Roof	+ .20
Slate Roof	+ 1.10
Upgrade Ceilings to Textured Finish	+ .47
Air Conditioning, in Heating Ductwork	Base System
Heating Systems, Hot Water	+ 1.78
Heat Pump	+ 2.69
Electric Heat	– 1.88
Not Heated	– 3.65

Additional upgrades or components

Kitchen Cabinets & Countertops	Page 58
Bathroom Vanities	59
Fireplaces & Chimneys	59
Windows, Skylights & Dormers	59
Appliances	60
Breezeways & Porches	60
Finished Attic	60
Garages	61
Site Improvements	61
Wings & Ells	57

- **Unique residence built from an architect's plan**
- **Single family — 1 full bath, 1 half bath, 1 kitchen**
- **No basement**
- **Cedar shakes on roof**
- **Forced hot air heat/air conditioning**
- **Gypsum wallboard interior finishes**
- **Many special features**
- **Extraordinary materials and workmanship**
- **Detail specifications on page 49**

Note: The illustration shown may contain some optional components (for example: garages and/or fireplaces) whose costs are shown in the modifications, adjustments, & alternatives below or at the end of the square foot section.

©Home Planners, Inc.

Base cost per square foot of living area

Exterior Wall	Living Area										
	1500	1800	2100	2400	2800	3200	3600	4000	4500	5000	5500
Wood Siding - Wood Frame	144.80	133.90	124.90	118.80	114.50	109.35	104.00	101.60	96.55	93.65	90.60
Brick Veneer - Wood Frame	152.40	140.80	131.15	124.70	120.15	114.55	108.85	106.35	100.90	97.80	94.45
Solid Brick	166.60	153.60	142.80	135.60	130.60	124.25	117.90	115.05	108.85	105.35	101.65
Solid Stone	161.50	149.00	138.65	131.75	126.85	120.80	114.70	111.95	106.00	102.65	99.10
Finished Basement, Add*	32.80	35.25	33.80	32.90	32.30	31.35	30.60	30.20	29.35	28.85	28.40
Unfinished Basement, Add*	14.55	13.65	12.95	12.45	12.15	11.65	11.30	11.10	10.65	10.40	10.15

*Basement under middle level only.

Modifications

Add to the total cost

Upgrade Kitchen Cabinets	$ + 1455
Solid Surface Countertops (Included)	
Full Bath - including plumbing, wall and floor finishes	+ 7590
Half Bath - including plumbing, wall and floor finishes	+ 4696
Two Car Attached Garage	+ 30,094
Two Car Detached Garage	+ 34,187
Fireplace & Chimney	+ 7722

Adjustments

For multi family - add to total cost

Additional Kitchen	$ + 18,062
Additional Full Bath & Half Bath	+ 12,286
Additional Entry & Exit	+ 2324
Separate Heating & Air Conditioning	+ 6091
Separate Electric	+ 1870

For Townhouse/Rowhouse - Multiply cost per square foot by

Inner Unit	+ .86
End Unit	+ .93

Alternatives

Add to or deduct from the cost per square foot of living area

Heavyweight Asphalt Shingles	– 2.85
Clay Tile Roof	+ .25
Slate Roof	+ 1.60
Upgrade Ceilings to Textured Finish	+ .47
Air Conditioning, in Heating Ductwork	Base System
Heating Systems, Hot Water	+ 1.73
Heat Pump	+ 2.80
Electric Heat	– 1.66
Not Heated	– 3.54

Additional upgrades or components

Kitchen Cabinets & Countertops	Page 58
Bathroom Vanities	59
Fireplaces & Chimneys	59
Windows, Skylights & Dormers	59
Appliances	60
Breezeways & Porches	60
Finished Attic	60
Garages	61
Site Improvements	61
Wings & Ells	57

1 Story — Base cost per square foot of living area

Exterior Wall	Living Area							
	50	100	200	300	400	500	600	700
Wood Siding - Wood Frame	237.50	183.70	160.35	134.85	127.25	122.65	119.55	120.40
Brick Veneer - Wood Frame	266.00	204.05	177.35	146.20	137.45	132.15	128.60	129.15
Solid Brick	318.95	241.85	208.85	167.20	156.35	149.80	145.45	145.35
Solid Stone	299.95	228.35	197.55	159.70	149.55	143.50	139.40	139.55
Finished Basement, Add	87.35	79.35	71.20	57.75	55.05	53.50	52.30	51.60
Unfinished Basement, Add	45.50	35.05	30.75	23.45	22.05	21.15	20.55	20.15

1-1/2 Story — Base cost per square foot of living area

Exterior Wall	Living Area							
	100	200	300	400	500	600	700	800
Wood Siding - Wood Frame	189.45	154.85	133.80	120.85	114.45	111.30	107.40	106.30
Brick Veneer - Wood Frame	214.85	175.25	150.80	134.10	126.65	122.85	118.30	117.10
Solid Brick	262.15	213.00	182.30	158.60	149.40	144.30	138.55	137.25
Solid Stone	245.20	199.50	171.00	149.80	141.25	136.65	131.35	130.00
Finished Basement, Add	57.55	56.95	51.60	45.65	44.05	43.00	41.95	41.80
Unfinished Basement, Add	28.90	24.55	21.70	18.45	17.65	17.05	16.50	16.40

2 Story — Base cost per square foot of living area

Exterior Wall	Living Area							
	100	200	400	600	800	1000	1200	1400
Wood Siding - Wood Frame	186.50	140.85	120.90	101.05	94.60	90.65	88.00	89.15
Brick Veneer - Wood Frame	215.00	161.20	137.90	112.35	104.75	100.15	97.05	97.85
Solid Brick	267.90	199.00	169.40	133.40	123.65	117.80	113.90	114.05
Solid Stone	248.95	185.45	158.15	125.85	116.90	111.45	107.90	108.30
Finished Basement, Add	43.70	39.70	35.65	28.85	27.55	26.70	26.20	25.80
Unfinished Basement, Add	22.70	17.55	15.35	11.75	11.00	10.60	10.30	10.10

Base costs do not include bathroom or kitchen facilities. Use Modifications/Adjustments/Alternatives on pages 58-61 where appropriate.

Kitchen cabinets -
Base units, hardwood *(Cost per Unit)*

	Economy	Average	Custom	Luxury
24″ deep, 35″ high,				
One top drawer,				
One door below				
12″ wide	$ 210	$ 281	$ 373	$ 491
15″ wide	247	330	438	577
18″ wide	270	360	478	630
21″ wide	247	330	438	577
24″ wide	326	435	578	761
Four drawers				
12″ wide	296	395	525	691
15″ wide	322	430	571	752
18″ wide	341	455	605	796
24″ wide	363	485	645	848
Two top drawers,				
Two doors below				
27″ wide	333	445	591	778
30″ wide	382	510	678	892
33″ wide	337	450	598	787
36″ wide	348	465	618	813
42″ wide	375	500	665	875
48″ wide	401	535	711	936
Range or sink base				
(Cost per unit)				
Two doors below				
30″ wide	273	365	485	638
33″ wide	292	390	518	682
36″ wide	303	405	538	708
42″ wide	330	440	585	770
48″ wide	341	455	605	796
Corner Base Cabinet				
(Cost per unit)				
36″ wide	547	730	970	1277
Lazy Susan *(Cost per unit)*				
With revolving door	480	640	851	1120

Kitchen cabinets -
Wall cabinets, hardwood *(Cost per Unit)*

	Economy	Average	Custom	Luxury
12″ deep, 2 doors				
12″ high				
30″ wide	$ 186	$ 248	$ 329	$ 434
36″ wide	212	283	376	495
15″ high				
30″ wide	187	250	332	437
33″ wide	208	278	369	486
36″ wide	225	300	399	525
24″ high				
30″ wide	281	375	498	656
36″ wide	311	415	551	726
42″ wide	300	400	532	700
30″ high, 1 door				
12″ wide	184	246	327	430
15″ wide	200	267	355	467
18″ wide	220	294	391	514
24″ wide	251	335	445	586
30″ high, 2 doors				
27″ wide	270	360	478	630
30″ wide	292	390	518	682
36″ wide	326	435	578	761
42″ wide	348	465	618	813
48″ wide	360	480	638	840
Corner wall, 30″ high				
24″ wide	204	272	361	476
30″ wide	247	330	438	577
36″ wide	222	297	395	519
Broom closet				
84″ high, 24″ deep				
18″ wide	480	640	851	1120
Oven Cabinet				
84″ high, 24″ deep				
27″ wide	716	955	1270	1671

Kitchen countertops *(Cost per L.F.)*

	Economy	Average	Custom	Luxury
Solid Surface				
24″ wide, no backsplash	91	122	162	213
with backsplash	99	132	175	231
Stock plastic laminate, 24″ wide				
with backsplash	16	22	29	38
Custom plastic laminate, no splash				
7/8″ thick, alum. molding	25	34	45	60
1-1/4″ thick, no splash	34	46	61	80
Marble				
1/2″ - 3/4″ thick w/splash	46	62	82	108
Maple, laminated				
1-1/2″ thick w/splash	68	91	121	159
Stainless steel				
(per S.F.)	124	166	220	290
Cutting blocks, recessed				
16″ x 20″ x 1″ (each)	93	124	164	217

Vanity bases (Cost per Unit)

2 door, 30" high, 21" deep	Economy	Average	Custom	Luxury
24" wide	258	345	458	603
30" wide	300	400	532	700
36" wide	326	435	578	761
48" wide	401	535	711	936

Solid surface vanity tops (Cost Each)

Center bowl	Economy	Average	Custom	Luxury
22" x 25"	$ 260	$ 280	$ 303	$ 327
22" x 31"	297	320	346	374
22" x 37"	340	367	396	428
22" x 49"	420	453	489	529

Fireplaces & Chimneys (Cost per Unit)

	1-1/2 Story	2 Story	3 Story
Economy (prefab metal)			
Exterior chimney & 1 fireplace	$ 4995	$ 5520	$ 6054
Interior chimney & 1 fireplace	4787	5322	5568
Average (masonry)			
Exterior chimney & 1 fireplace	4964	5535	6292
Interior chimney & 1 fireplace	4650	5220	5684
For more than 1 flue, add	357	607	1018
For more than 1 fireplace, add	3519	3519	3519
Custom (masonry)			
Exterior chimney & 1 fireplace	5463	6166	6963
Interior chimney & 1 fireplace	5122	5798	6249
For more than 1 flue, add	428	742	1005
For more than 1 fireplace, add	3924	3924	3924
Luxury (masonry)			
Exterior chimney & 1 fireplace	7721	8464	9262
Interior chimney & 1 fireplace	7366	8054	8525
For more than 1 flue, add	637	1064	1485
For more than 1 fireplace, add	6086	6086	6086

Windows and Skylights (Cost Each)

	Economy	Average	Custom	Luxury
Fixed Picture Windows				
3'-6" x 4'-0"	$ 505	$ 546	$ 595	$ 637
4'-0" x 6'-0"	921	995	1075	1161
5'-0" x 6'-0"	1007	1087	1200	1269
6'-0" x 6'-0"	1028	1111	1200	1290
Bay/Bow Windows				
8'-0" x 5'-0"	1264	1365	1525	1593
10'-0" x 5'-0"	1478	1597	1625	1863
10'-0" x 6'-0"	2314	2500	2575	2916
12'-0" x 6'-0"	2936	3171	3350	3699
Palladian Windows				
3'-2" x 6'-4"		1782	2000	2079
4'-0" x 6'-0"		2106	2275	2457
5'-5" x 6'-10"		2523	2975	2943
8'-0" x 6'-0"		3032	3525	3537
Skylights				
46" x 21-1/2"	416	450	590	637
46" x 28"	453	490	625	675
57" x 44"	550	595	735	793

Dormers (Cost/S.F. of plan area)

	Economy	Average	Custom	Luxury
Framing and Roofing Only				
Gable dormer, 2" x 6" roof frame	$ 25	$ 28	$ 32	$ 52
2" x 8" roof frame	26	29	33	55
Shed dormer, 2" x 6" roof frame	15	17	21	34
2" x 8" roof frame	17	19	22	35
2" x 10" roof frame	18	21	23	36

Appliances   *(Cost per Unit)*

	Economy	Average	Custom	Luxury
Range				
30" free standing, 1 oven	$ 405	$ 1252	$ 1676	$ 2100
2 oven	1650	1750	1800	1850
30" built-in, 1 oven	830	1365	1632	1900
2 oven	1675	1912	2043	2175
21" free standing				
1 oven	400	450	475	500
Counter Top Ranges				
4 burner standard	335	557	668	780
As above with griddle	1925	2875	3350	3825
Microwave Oven	217	458	579	700
Compactor				
4 to 1 compaction	605	647	668	690
Deep Freeze				
15 to 23 C.F.	540	662	723	785
30 C.F.	955	1065	1120	1175
Dehumidifier, portable, auto.				
15 pint	165	189	202	214
30 pint	197	226	241	256
Washing Machine, automatic	485	1055	1340	1625
Water Heater				
Electric, glass lined				
30 gal.	570	695	757	820
80 gal.	1100	1387	1531	1675
Water Heater, Gas, glass lined				
30 gal.	790	970	1060	1150
50 gal.	1275	1562	1706	1850
Dishwasher, built-in				
2 cycles	360	482	543	605
4 or more cycles	470	580	990	1400
Dryer, automatic	500	1187	1531	1875
Garage Door Opener	495	597	648	700
Garbage Disposal	126	195	229	264
Heater, Electric, built-in				
1250 watt ceiling type	208	259	284	310
1250 watt wall type	260	282	293	305
Wall type w/blower				
1500 watt	260	299	318	338
3000 watt	455	523	557	591
Hood For Range, 2 speed				
30" wide	154	602	826	1050
42" wide	370	1210	1630	2050
Humidifier, portable				
7 gal. per day	176	202	215	228
15 gal. per day	212	243	259	275
Ice Maker, automatic				
13 lb. per day	995	1144	1218	1293
51 lb. per day	1500	1725	1837	1950
Refrigerator, no frost				
10-12 C.F.	545	607	638	670
14-16 C.F.	565	680	737	795
18-20 C.F.	705	977	1113	1250
21-29 C.F.	1025	2225	2825	3425
Sump Pump, 1/3 H.P.	272	386	443	500

Breezeway   *(Cost per S.F.)*

Class	Type	Area (S.F.)			
		50	100	150	200
Economy	Open	$ 22.73	$ 19.31	$ 16.21	$ 15.94
	Enclosed	109.01	84.19	69.91	61.24
Average	Open	30.49	26.89	23.52	21.37
	Enclosed	119.08	88.74	72.58	63.89
Custom	Open	42.27	37.21	32.37	29.63
	Enclosed	173.63	129.17	105.59	92.83
Luxury	Open	44.01	38.58	34.76	33.76
	Enclosed	177.22	131.43	106.32	95.18

Porches   *(Cost per S.F.)*

Class	Type	Area (S.F.)				
		25	50	100	200	300
Economy	Open	$ 65.15	$ 43.59	$ 34.07	$ 28.83	$ 24.60
	Enclosed	130.19	90.72	68.57	53.49	45.79
Average	Open	85.84	54.65	41.93	34.85	34.85
	Enclosed	155.77	105.54	79.83	61.81	52.37
Custom	Open	107.95	71.62	53.92	47.10	42.21
	Enclosed	214.63	146.64	111.58	86.90	74.86
Luxury	Open	116.52	76.16	56.26	50.55	45.07
	Enclosed	228.10	160.20	118.85	92.35	79.49

Finished attic   *(Cost per S.F.)*

Class	Area (S.F.)				
	400	500	600	800	1000
Economy	$ 17.38	$ 16.80	$ 16.10	$ 15.83	$ 15.24
Average	26.66	26.08	25.44	25.12	24.42
Custom	34.41	33.64	32.87	32.37	31.71
Luxury	43.74	42.68	41.68	40.69	40.02

Alarm system   *(Cost per System)*

	Burglar Alarm	Smoke Detector
Economy	$ 400	$ 69
Average	460	83
Custom	783	194
Luxury	1150	178

Sauna, prefabricated
(Cost per unit, including heater and controls—7' high)

Size	Cost
6' x 4'	$ 5225
6' x 5'	5825
6' x 6'	6200
6' x 9'	7600
8' x 10'	10,000
8' x 12'	11,700
10' x 12'	12,200

Garages *

(Costs include exterior wall systems comparable with the quality of the residence. Included in the cost is an allowance for one personnel door, manual overhead door(s) and electrical fixture.)

Class	Type									
	Detached			Attached			Built-in		Basement	
	One Car	Two Car	Three Car	One Car	Two Car	Three Car	One Car	Two Car	One Car	Two Car
Economy										
Wood	$14,483	$22,007	$29,531	$11,166	$19,141	$26,665	$-1575	$-3149	$1307	$1680
Masonry	19,665	28,492	37,318	14,408	23,686	32,513	-2211	-4421		
Average										
Wood	16,766	24,864	32,962	12,594	21,144	29,242	-1855	-3709	1498	2063
Masonry	21,019	30,186	39,354	15,255	24,874	34,042	-2377	-4008		
Custom										
Wood	18,734	28,578	38,422	14,451	24,917	34,761	-2985	-2712	2617	4295
Masonry	23,146	34,099	45,053	17,212	28,787	39,740	-3527	-3795		
Luxury										
Wood	21,971	34,187	46,403	17,256	30,094	42,310	-3127	-2995	3600	5811
Masonry	26,552	39,920	53,288	20,122	34,113	47,481	-3689	-4121		

*See the Introduction to this section for definitions of garage types.

Swimming pools (Cost per S.F.)

Residential	
Inground	$23.00—61.00
Deck equipment	1.30
Paint pool, preparation & 3 coats (epoxy)	4.34
Rubber base paint	3.72
Pool Cover	.86
Swimming Pool Heaters	
(not including wiring, external piping, base or pad)	
Gas	
155 MBH	$ 2500.00
190 MBH	2975.00
500 MBH	10,700.
Electric	
15 KW 7200 gallon pool	2625.00
24 KW 9600 gallon pool	3000.00
54 KW 24,000 gallon pool	4625.00

Wood and coal stoves

Wood Only	
Free Standing (minimum)	$ 1825
Fireplace Insert (minimum)	1839
Coal Only	
Free Standing	$ 2078
Fireplace Insert	2275
Wood and Coal	
Free Standing	$ 4275
Fireplace Insert	4381

Sidewalks (Cost per S.F.)

Concrete, 3000 psi with wire mesh	4" thick	$ 3.52
	5" thick	4.31
	6" thick	4.84
Precast concrete patio blocks (natural)	2" thick	10.45
Precast concrete patio blocks (colors)	2" thick	11.10
Flagstone, bluestone	1" thick	14.90
Flagstone, bluestone	1-1/2" thick	19.85
Slate (natural, irregular)	3/4" thick	15.00
Slate (random rectangular)	1/2" thick	23.00
Seeding		
Fine grading & seeding includes lime, fertilizer & seed per S.Y.		2.30
Lawn Sprinkler System	per S.F.	.91

Fencing (Cost per L.F.)

Chain Link, 4' high, galvanized	$ 16.75
Gate, 4' high (each)	155.00
Cedar Picket, 3' high, 2 rail	11.85
Gate (each)	175.00
3 Rail, 4' high	14.55
Gate (each)	186.00
Cedar Stockade, 3 Rail, 6' high	14.75
Gate (each)	185.00
Board & Battens, 2 sides 6' high, pine	21.00
6' high, cedar	29.50
No. 1 Cedar, basketweave, 6' high	16.70
Gate, 6' high (each)	217.00

Carport (Cost per S.F.)

Economy	$ 8.08
Average	12.26
Custom	18.78
Luxury	21.50

Insurance Exclusions

Insurance exclusions are a matter of policy coverage and are not standard or universal. When making an appraisal for insurance purposes, it is recommended that some time be taken in studying the insurance policy to determine the specific items to be excluded. Most homeowners insurance policies have, as part of their provisions, statements that read "the policy permits the insured to exclude from value, in determining whether the amount of the insurance equals or exceeds 80% of its replacement cost, such items as building excavation, basement footings, underground piping and wiring, piers and other supports which are below the undersurface of the lowest floor or where there is no basement below the ground." Loss to any of these items, however, is covered.

Costs for Excavation, Spread and Strip Footings and Underground Piping

This chart shows excluded items expressed as a percentage of total building cost.

Class	Number of Stories			
	1	1-1/2	2	3
Luxury	3.4%	2.7%	2.4%	1.9%
Custom	3.5%	2.7%	2.4%	1.9%
Average	5.0%	4.3%	3.8%	3.3%
Economy	5.0%	4.4%	4.1%	3.5%

Architect/Designer Fees

Architect/designer fees as presented in the following chart are typical ranges for 4 classes of residences. Factors affecting these ranges include economic conditions, size and scope of project, site selection, and standardization. Where superior quality and detail is required or where closer supervision is required, fees may run higher than listed. Lower quality or simplicity may dictate lower fees.

The editors have included architect/designer fees from the "mean" column in calculating costs.

Listed below are average costs expressed as a range and mean of total building cost for 4 classes of residences.

Class		Range	Mean
Luxury	— Architecturally designed and supervised	5%–15%	10%
Custom	— Architecturally modified designer plans	$2000–$3500	$2500
Average	— Designer plans	$800–$1500	$1100
Economy	— Stock plans	$100–$500	$300

Depreciation as generally defined is "the loss of value due to any cause." Specifically, depreciation can be broken down into three categories: **Physical Depreciation, Functional Obsolescence,** and **Economic Obsolescence.**

Physical Depreciation is the loss of value from the wearing out of the building's components. Such causes may be decay, dry rot, cracks, or structural defects.

Functional Obsolescence is the loss of value from features which render the residence less useful or desirable. Such features may be higher than normal ceilings, design and/or style changes, or outdated mechanical systems.

Economic Obsolescence is the loss of value from external features which render the residence less useful or desirable. These features may be changes in neighborhood socio-economic grouping, zoning changes, legislation, etc.

Depreciation as it applies to the residential portion of this manual deals with the observed physical condition of the residence being appraised. It is in essence *"the cost to cure."*

The ultimate method to arrive at *"the cost to cure"* is to analyze each component of the residence.

For example:

    Component, Roof Covering – Cost = $2400
    Average Life = 15 years
    Actual Life = 5 years
    Depreciation = 5 ÷ 15 = .33 or 33%

    $2400 × 33% = $792 = Amount of Depreciation

The following table, however, can be used as a guide in estimating the % depreciation using the actual age and general condition of the residence.

Depreciation Table — Residential

Age in Years	Good	Average	Poor
2	2%	3%	10%
5	4	6	20
10	7	10	25
15	10	15	30
20	15	20	35
25	18	25	40
30	24	30	45
35	28	35	50
40	32	40	55
45	36	45	60
50	40	50	65

Commercial/Industrial/ Institutional Section

Table of Contents

Introduction to the Commercial/ Industrial/Institutional Section

General

The Commercial/Industrial/Institutional section of this manual contains base building costs per square foot of floor area for 72 model buildings. Each model has a table of square foot costs for combinations of exterior wall and framing systems. This table is supplemented by a list of common additives and their unit costs. A breakdown of the component costs used to develop the base cost for the model is on the opposite page. The total cost derived from the base cost and additives must be modified using the appropriate location factor from the Reference Section.

This section may be used directly to estimate the construction cost of most types of buildings knowing only the floor area, exterior wall construction and framing systems. To adjust the base cost for components which are different than the model, use the tables from the Assemblies Section.

Building Identification & Model Selection

The building models in this section represent structures by use. Occupancy, however, does not necessarily identify the building, i.e., a restaurant could be a converted warehouse. In all instances, the building should be described and identified by its own physical characteristics. The model selection should also be guided by comparing specifications with the model. In the case of converted use, data from one model may be used to supplement data from another.

Adjustments

The base cost tables represent the base cost per square foot of floor area for buildings without a basement and without unusual special features. Basement costs and other common additives are listed below the base cost table. Cost adjustments can also be made to the model by using the tables from the Assemblies Section. This table is for example only.

Dimensions

All base cost tables are developed so that measurement can be readily made during the inspection process. Areas are calculated from exterior dimensions and story heights are measured from the top surface of one floor to the top surface of the floor above. Roof areas are measured by horizontal area covered and costs related to inclines are converted with appropriate factors. The precision of measurement is a matter of the user's choice and discretion. For ease in calculation, consideration should be given to measuring in tenths of a foot, i.e., 9 ft. 6 in. = 9.5 ft., 9 ft. 4 in. = 9.3 ft.

Floor Area

The term "Floor Area" as used in this section includes the sum of floor plate at grade level and above. This dimension is measured from the outside face of the foundation wall. Basement costs are calculated separately. The user must exercise his/her own judgment, where the lowest level floor is slightly below grade, whether to consider it at grade level or make the basement adjustment.

How to Use the Commercial/Industrial/Institutional Section

The following is a detailed explanation of a sample entry in the Commercial/Industrial/Institutional Square Foot Cost Section. Each bold number below corresponds to the item being described on the following page with the appropriate component or cost of the sample entry following in parenthesis.

Prices listed are costs that include overhead and profit of the installing contractor and additional mark-ups for General Conditions and Architects' Fees.

COMMERCIAL/INDUSTRIAL/INSTITUTIONAL	M.010 **1**	Apartment, 1-3 Story **2**

Costs per square foot of floor area

Exterior Wall	S.F. Area	8000	12000	15000	19000	22500	25000	29000	32000	36000
3	L.F. Perimeter	213	280	330	350	400	433	442	480	520
Face Brick with Concrete Block Back-up	Wood Joists	195.05	173.35	164.65	153.15	148.95	146.45	140.80	139.20	136.85
	Steel Joists	203.05	182.40	174.10	163.70	159.70	157.35	152.35	150.80	148.60
Stucco on Concrete Block	Wood Joists	171.20	151.70	143.85	134.80	131.00	128.85	124.55	123.10	121.15
	Steel Joists	189.35	169.70	161.85	152.65	148.80	146.65	142.30	140.90	138.85
Wood Siding	Wood Frame	171.30	151.80	143.95	134.85	131.05	128.90	124.55	123.10	121.20
Brick Veneer	Wood Frame	181.15	160.35	152.00	141.55	137.50	135.10	130.05	128.50	126.35
Perimeter Adj., Add or Deduct	Per 100 L.F.	20.80	13.90	11.10	8.80	7.40	6.70	5.75	5.20	4.65
Story Hgt. Adj., Add or Deduct **6**	Per 1 Ft.	3.85	3.40	3.20	2.70	2.60	2.60	2.30	2.20	2.20
	For Basement, add $33.00 per square foot of basement area									

4 **5** **7**

The above costs were calculated using the basic specifications shown on the facing page. These costs should be adjusted where necessary for design alternatives and owner's requirements. Reported completed project costs, for this type of structure, range from $55.10 to $204.90 per S.F. **8** **9**

Common additives

Description	Unit	$ Cost
Appliances		
Cooking range, 30" free standing		
1 oven	Each	430 - 2150
2 oven	Each	1675 - 1875
30" built-in		
1 oven	Each	855 - 2025
2 oven	Each	1725 - 1600
Counter top cook tops, 4 burner	Each	360 - 830
Microwave oven	Each	254 - 775
Combination range, refrig. & sink, 30" wide	Each	1550 - 4650
72" wide	Each	5025
Combination range, refrigerator, sink,		
microwave oven & icemaker	Each	5775
Compactor, residential, 4-1 compaction	Each	630 - 730
Dishwasher, built-in, 2 cycles	Each	515 - 910
4 cycles	Each	625 - 1700
Garbage disposer, sink type	Each	188 - 325
Hood for range, 2 speed, vented, 30" wide	Each	271 - 1250
42" wide	Each	490 - 2250
Refrigerator, no frost 10-12 C.F.	Each	570 - 705
18-20 C.F.	Each	735 - 1300

10

Description	Unit	$ Cost
Closed Circuit Surveillance, One station		
Camera and monitor	Each	1850
For additional camera stations, add	Each	1000
Elevators, Hydraulic passenger, 2 stops		
2000# capacity	Each	64,800
2500# capacity	Each	66,300
3500# capacity	Each	69,800
Additional stop, add	Each	7825
Emergency Lighting, 25 watt, battery operated		
Lead battery	Each	282
Nickel cadmium	Each	805
Laundry Equipment		
Dryer, gas, 16 lb. capacity	Each	885
30 lb. capacity	Each	3600
Washer, 4 cycle	Each	1075
Commercial	Each	1450
Smoke Detectors		
Ceiling type	Each	187
Duct type	Each	480

1 Model Number (M.010)

"M" distinguishes this section of the book and stands for model. The number designation is a sequential number.

2 Type of Building (Apartment, 1-3 Story)

There are 45 different types of commercial/industrial/ institutional buildings highlighted in this section.

3 Exterior Wall Construction and Building Framing Options (Face Brick with Concrete Block Back-up and Open Web Steel Bar Joists)

Three or more commonly used exterior walls, and in most cases, two typical building framing systems are presented for each type of building. The model selected should be based on the actual characteristics of the building being estimated.

4 Total Square Foot of Floor Area and Base Perimeter Used to Compute Base Costs (22,500 Square Feet and 400 Linear Feet)

Square foot of floor area is the total gross area of all floors at grade, and above, and does not include a basement. The perimeter in linear feet used for the base cost is for a generally rectangular economical building shape.

5 Cost per Square Foot of Floor Area ($159.70)

The highlighted cost is for a building of the selected exterior wall and framing system and floor area. Costs for buildings with floor areas other than those calculated may be interpolated between the costs shown.

6 Building Perimeter and Story Height Adjustments

Square foot costs for a building with a perimeter or floor to floor story height significantly different from the model used to calculate the base cost may be adjusted, add or deduct, to reflect the actual building geometry.

7 Cost per Square Foot of Floor Area for the Perimeter and/or Height Adjustment ($7.40 for Perimeter Difference and $2.60 for Story Height Difference)

Add (or deduct) $6.45 to the base square foot cost for each 100 feet of perimeter difference between the model and the actual building. Add (or deduct) $2.25 to the base square foot cost for each 1 foot of story height difference between the model and the actual building.

8 Optional Cost per Square Foot of Basement Floor Area ($33.00)

The cost of an unfinished basement for the building being estimated is $28.95 times the gross floor area of the basement.

9 Range of Cost per Square Foot of Floor Area for Similar Buildings ($55.10 to $204.90)

Many different buildings of the same type have been built using similar materials and systems. RSMeans historical cost data of actual construction projects indicates a range of $48.35 to $179.70 for this type of building.

10 Common Additives

Common components and/or systems used in this type of building are listed. These costs should be added to the total building cost. Additional selections may be found in the Assemblies Section.

The following is a detailed explanation of a specification and costs for a model building in the Commercial/Industrial/Institutional Square Foot Cost Section. Each bold number below corresponds to the item being described on the following page with the appropriate component of the sample entry following in parenthesis.

Prices listed are costs that include overhead and profit of the installing contractor.

Model costs calculated for a 3 story building with 10' story height and 22,500 square feet of floor area **1**

2 **Apartment, 1-3 Story**

				Unit	Unit Cost	Cost Per S.F.	% Of Sub-Total
A. SUBSTRUCTURE							
1010	Standard Foundations	Poured concrete; strip and spread footings		S.F. Ground	6.33	2.11	
1020	Special Foundations	N/A			–	–	
1030	Slab on Grade	4" reinforced concrete with vapor barrier and granular base		S.F. Slab	4.74	1.58	4.4%
2010	Basement Excavation	Site preparation for slab and trench for foundation wall and footing		S.F. Ground	.17	.06	
2020	Basement Walls	4' foundation wall		L.F. Wall	74	1.50	
B. SHELL							
B10 Superstructure							
1010	Floor Construction	Open web steel joists, slab form, concrete, interior steel columns		S.F. Floor	22.34	14.89	15.9%
1020	Roof Construction	Open web steel joists with rib metal deck, interior steel columns		S.F. Roof	11.64	3.88	
B20 Exterior Enclosure							
2010	Exterior Walls	Face brick with concrete block backup	88% of wall	S.F. Wall	23.99	11.26	
2020	Exterior Windows	Aluminum horizontal sliding	12% of wall	Each	492	2.10	11.5%
2030	Exterior Doors	Aluminum and glass		Each	1694	.30	
B30 Roofing							
3010	Roof Coverings	Built-up tar and gravel with flashing; perlite/EPS composite insulation		S.F. Roof **6**	5.79	1.93	1.6%
3020	Roof Openings	N/A		–	–	–	
C. INTERIORS							
1010	Partitions **3**	Gypsum board and sound deadening board on metal studs	10 S.F. of Floor/L.F. Partition **4**	S.F. Partition	6.37	5.66 **9**	
1020	Interior Doors	15% solid core wood, 85% hollow core wood	80 S.F. Floor/Door	Each	500	6.24	
1030	Fittings	Kitchen Cabinets		S.F. Floor	2.64	2.64	
2010	Stair Construction	Concrete filled metal pan	**5**	Flight	15,800 **7**	3.51 **8**	24.5%
3010	Wall Finishes	70% paint, 25% vinyl wall covering, 5% ceramic tile		S.F. Surface			
3020	Floor Finishes	60% carpet, 30% vinyl composition tile, 10% ceramic tile		S.F. Floor			
3030	Ceiling Finishes	Painted gypsum board on resilient channels		S.F. Ceiling			
D. SERVICES							
D10 Conveying							
1010	Elevators & Lifts	One hydraulic passenger elevator		Each	105,750	4.70	4.0%
1020	Escalators & Moving Walks	N/A		–	–	–	
D20 Plumbing							
2010	Plumbing Fixtures	Kitchen, bathroom and service fixtures, supply and drainage	1 Fixture/200 S.F. Floor	Each	2274	11.37	
2020	Domestic Water Distribution	Gas fired water heater		S.F. Floor	3.44	3.44	12.8%
2040	Rain Water Drainage	Roof drains		S.F. Roof	1.02	.34	
D30 HVAC							
3010	Energy Supply	Oil fired hot water, baseboard radiation		S.F. Floor	7.85	7.85	
3020	Heat Generating Systems	N/A		–	–	–	
3030	Cooling Generating Systems	Chilled water, air cooled condenser system		S.F. Floor	8.53	8.53	13.8%
3050	Terminal & Package Units	N/A		–	–	–	
3090	Other HVAC Sys. & Equipment	N/A		–	–	–	
D40 Fire Protection							
4010	Sprinklers	Wet pipe sprinkler system		S.F. Floor	2.73	2.73	2.3%
4020	Standpipes	N/A		–	–	–	
D50 Electrical							
5010	Electrical Service/Distribution	600 ampere service, panel board and feeders		S.F. Floor	2.64	2.64	
5020	Lighting & Branch Wiring	Incandescent fixtures, receptacles, switches, A.C. and misc. power		S.F. Floor	6.66	6.66	9.1%
5030	Communications & Security	Alarm systems, internet wiring, and emergency lighting		S.F. Floor	1.26	1.26	
5090	Other Electrical Systems	Emergency Generator, 11.5 kW		S.F. Floor	.19	.19	
E. EQUIPMENT & FURNISHINGS							
1010	Commercial Equipment	N/A		–	–		
1020	Institutional Equipment	N/A		–	–		0.0 %
1030	Vehicular Equipment	N/A		–	**10**		
1090	Other Equipment	N/A		–	–		
F. SPECIAL CONSTRUCTION							
1020	Integrated Construction	N/A		–	–	–	0.0 %
1040	Special Facilities	N/A		–	–	–	
G. BUILDING SITEWORK	**N/A**						
				Sub-Total		118.27	100%
	CONTRACTOR FEES (General Requirements: 10%, Overhead: 5%, Profit: 10%) **11**	**12**			25%	29.60	
	ARCHITECT FEES				8%	11.83	
				Total Building Cost		**159.70**	

**1 Building Description
(Model costs are calculated for a
three-story apartment building with
10' story height and 22,500 square
feet of floor area)**

The model highlighted is described in terms of building type,
number of stories, typical story height, and square footage.

**2 Type of Building
(Apartment, 1-3 Story)**

**3 Division C Interiors
(C1020 Interior Doors)**

System costs are presented in divisions according to the
7-division UNIFORMAT II classifications. Each of the
component systems are listed.

**4 Specification Highlights
(15% solid core wood;
85% hollow core wood)**

All systems in each subdivision are described with the
material and proportions used.

**5 Quality Criteria
(80 S.F. Floor/Door)**

The criteria used in determining quantities for the
calculations are shown.

6 Unit (Each)

The unit of measure shown in this column is the unit of
measure of the particular system shown that corresponds
to the unit cost.

7 Unit Cost ($500)

The cost per unit of measure of each system
subdivision.

8 Cost per Square Foot ($6.24)

The cost per square foot for each system is the unit cost of
the system times the total number of units divided by the
total square feet of building area.

9 % of Sub-Total (24.5%)

The percent of sub-total is the total cost per square foot of
all systems in the division divided by the sub-total cost per
square foot of the building.

10 Sub-Total ($118.27)

The sub-total is the total of all the system costs
per square foot.

**11 Project Fees
(Contractor Fees) (25%)
(Architect Fees) (8%)**

Contractor Fees to cover the general requirements,
overhead and profit of the General Contractor are added
as a percentage of the sub-total. Architect Fees, also as a
percentage of the sub-total, are also added. These values
vary with the building type.

12 Total Building Cost ($159.70)

The total building cost per square foot of building area is
the sum of the square foot costs of all the systems plus
the General Contractor's general requirements, overhead
and profit, and the Architect fees. The total building cost
is the amount which appears shaded in the Cost per
Square Foot of Floor Area table shown previously.

71

Examples

Example 1

This example illustrates the use of the base cost tables. The base cost is adjusted for different exterior wall systems, different story height and a partial basement.

CII APPRAISAL FIELD DATA FORM

COMMERCIAL
INDUSTRIAL
INSTITUTIONAL

Use sheets A & D for
App. Form 2
Use sheets A,B & C for
App. Form 3

CII APPRAISAL

1.	SUBJECT PROPERTY:	**Westernberg Holdings, Inc.**	5. DATE: **Jan. 1, 2009**
2.	BUILDING:	**# 623**	6. APPRAISER: **SJS**
3.	ADDRESS:	**55 Pine wood Rd., Bondsville, N.J. 07410**	
4.	BUILDING USE:	**Apartments**	7. YEAR BUILT: **1987**

North wall - Face brick w/concrete block back-up
East, West & South walls - Decorative concrete block

8.	EXTERIOR WALL CONSTRUCTION:					
9.	FRAME:	**Steel**				
10.	GROUND FLOOR AREA:	**60' x 191'**	**11,460**	S.F.	11. GROSS FLOOR AREA (EXCL. BASEMENT):	**80, 220 (7 Floors)** S.F.
12.	NUMBER OF STORIES:	**7**			13. STORY HEIGHT:	**11 ' - 4"**
14.	PERIMETER:	**502**		L.F.	15. BASEMENT AREA:	**4200 S.F.** S.F.
16.	GENERAL COMMENTS:					

Example 1 (continued)

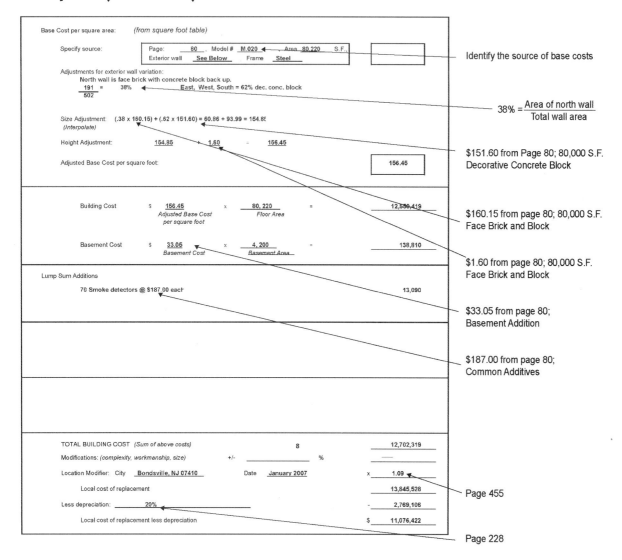

Base Cost per square area: *(from square foot table)*

Specify source:
Page: __80__ , Model # __M.020__ ◄———— Area __80,220__ S.F.
Exterior wall __See Below__ Frame __Steel__

————— Identify the source of base costs

Adjustments for exterior wall variation:
North wall is face brick with concrete block back up.

$$\frac{191}{502} = 38\%$$ ◄——— East, West, South = 62% dec. conc. block

$$38\% = \frac{\text{Area of north wall}}{\text{Total wall area}}$$

Size Adjustment: (.38 x 160.15) + (.62 x 151.60) = 60.86 + 93.99 = 154.85
(Interpolate)

Height Adjustment: __154.85__ + __1.60__ = __156.45__

$151.60 from Page 80; 80,000 S.F.
Decorative Concrete Block

Adjusted Base Cost per square foot: | 156.45 |

Building Cost $ __156.45__ x __80, 220__ = __12,550,419__
  Adjusted Base Cost per square foot *Floor Area*

$160.15 from page 80; 80,000 S.F.
Face Brick and Block

Basement Cost $ __33.05__ x __4, 200__ = __138,810__
  Basement Cost *Basement Area*

$1.60 from page 80; 80,000 S.F.
Face Brick and Block

Lump Sum Additions

70 Smoke detectors @ $187.00 each   13,090

$33.05 from page 80;
Basement Addition

$187.00 from page 80;
Common Additives

TOTAL BUILDING COST *(Sum of above costs)*   8   12,702,319

Modifications: *(complexity, workmanship, size)* +/- _____ %   ···········

Location Modifier: City __Bondsville, NJ 07410__ Date __January 2007__ x __1.09__ ◄

  Local cost of replacement   13,845,528

————— Page 455

Less depreciation: __20%__ _____   - __2,769,106__

  Local cost of replacement less depreciation $ __11,076,422__

————— Page 228

Example 2

This example shows how to modify a model building.

Model #M.020, 4–7 Story Apartment, page 80, matches the example quite closely. The model specifies a 6-story building with a 10′ story height.

The following adjustments must be made:
- add stone ashlar wall
- adjust model to a 5-story building
- change partitions
- change floor finish

CII APPRAISAL

		5. DATE:	**Jan. 2, 2009**	
1. SUBJECT PROPERTY:	**Westernberg Holdings, Inc.**	6. APPRAISER:	**SJS**	
2. BUILDING:	**# 623**			
3. ADDRESS:	**55 Pinewood Rd., Bondsville, N.J. 07410**			
4. BUILDING USE:	**Apartments**	7. YEAR BUILT:	**1989**	

```
              N
             170'



   W                              E
  64.7'



              S
```

North wall - 4" stone Ashlar w/8" block back up (split Face)

8. EXTERIOR WALL CONSTRUCTION: East, West & South walls - 12" Decorative concrete block

9. FRAME: **Steel**

10. GROUND FLOOR AREA: **11,000**   11. GROSS FLOOR AREA (EXCL. BASEMENT): **55,000** S.F.

12. NUMBER OF STORIES: **5**   13. STORY HEIGHT: **11 Ft.**

14. PERIMETER: **469.4**   15. BASEMENT AREA: None S.F.

16. GENERAL COMMENTS:

Two exterior wall types

Square foot area (excluding basement) from item 11 ___55,000___ S.F.
Perimeter from item 14 ___469.4___ Lin. Ft.
Item 17-Model square foot costs (from model sub-total) ___$123.97___

NO.	DESCRIPTION			UNIT	UNIT COST	NEW SF COST	MODEL SF COST	+/- CHANGE
A SUBSTRUCTURE								
A 1010	Standard Foundations	Bay size: *Same as model*		S.F. Gnd.	7.20	1.44	1.20	+ 0.24
A 1030	Slab on Grade	Material:	Thickness: *Same as model*	S.F. Slab	4.74	0.95	0.79	+ 0.16
A 2010	Basement Excavation	Depth:	Area: *11,000 S.F.*	S.F. Gnd.	0.25	0.05	0.04	+ 0.01
A 2020	Basement Walls			L.F. Walls	78.00	0.67	0.65	+ 0.02
B SHELL								
	B10 Superstructure							
B 1010	Floor Construction	Elevated floors: *Same as model*		S.F. Floor	28.37	22.70	23.64	- 0.94
B 1020	Roof Construction			S.F. Roof	10.86	2.17	1.81	+ 0.36
	B20 Exterior Closure							
B 2010	Exterior Walls	Material: *North wall- 4" stone ashlar on 8" CMU*	Thickness: *36% of wall*	S.F. Walls	23.99	9.79	10.66	- 0.87
		Material: *East, West, & South walls- 12" split rib block*	Thickness: *64% of wall*	S.F. Walls				
B 2020	Exterior Windows	Type:	% of wall	S.F. Wind.				
		Type:	14% of wall Each	S.F. Wind.	492	2.16	2.37	- 0.21
B 2030	Exterior Doors	Type: *Same as model*	Number: 4	Each	4280	0.31	0.28	+ 0.03
		Type:	Number:	Each				
	B30 Roofing							
B 3010	Roof Coverings	Material: *Same as model*		S.F. Roof	5.82	1.16	0.97	+ 0.19
B 3020	Roof Openings	N/A		S.F. Opng.				
C INTERIORS								
C 1010	Partitions: *High*	Material: *Gyp board on metal studs* Density: *8 S.F. Floor/L.F. Part.*		S.F. Part.	6.74	7.58	8.74	- 0.84
		Material:	Density:					
C 1020	Interior Doors	Type:	Number:	Each	601.00	7.51	7.51	—
C 1030	Fittings			S.F. Floor	3.17	3.17	3.17	—
C 2010	Stair Construction	*13 Flights*		Flight	12650	2.99	3.37	- 0.38
C 3010	Wall Finishes	Material: *Same as model* % of Wall	9'-0" *High*	S.F. Walls	1.43	3.22	2.85	+ 0.37
C 3020	Floor Finishes	Material: *Hardwood - 60% of floor*		S.F. Floor	5.07	8.39	5.07	+ 3.32
		Material: *Carpet - 30% of floor*						
		Material: *Tile -10% of floor*						
C 3030	Ceiling Finishes	Material: *Same as model*		S.F. Ceil.	3.71	3.71	3.71	—

From page 81, Model #M.020 Sub-total

$.25/S.F. ground area
—————————————
5 stories

$.25/S.F. ground area
—————————————
6 stories

Area of Elevated Floors
$28.37/S.F. x 4 x 11,000 S.F.
———————————————————————
55,000
Building Floor Area

10.86/S.F. roof
————————————
5 stories

Windows are 14% of wall area,
15 S.F. S.F./window

492 x .14 x 469.4 x 11 x 5 ÷ 55,000 = 2.16
————————————————————————————————
15

From Table C1010 124 page 337, C1010 126
on page 338 and C1010 128 page 339

$6.74/S.F. partition x 9 ft. high
——————————————————————
8 S.F. floor/L.F. partition

From Table C3020 410 page 351 & 352

Oak Strip, (max price)   @$9.50/S.F.
Carpet (26 oz.)   @$5.81/S.F.
Ceramic Tile   @$9.51/S.F.

.60 x 9.50 = 5.70
.30 x 5.81 = 1.74
.10 x 9.51 = .95
————————
8.39

North wall is 36% of wall area, e.g.   170 x 11
From Table B2010 128 2350 page 300   ————————
Stone ashiar @ $34.70/S.F. wall   469.4 x 11
East, west and south walls are 64% of wall area
From Table B2010 113 1530 page 294
Split ribbed block @$18.35/S.F. wall

New wall cost
(.36 x $34.70/S.F.) + (.64 x $18.35/S.F.) = $24.24/S.F. wall

Convert Cost per S.F. wall to cost per S.F. of floor
Windows are 14% of wall area

$9.79 S.F. floor = $24.24/S.F. wall x .86 x 469.4 L.F. x 11 ft. high x 5 stories
——
55,000 S.F. of floor

Example 2 (continued)

	NO.	SYSTEM/COMPONENT	DESCRIPTION	UNIT	UNIT COST	NEW S.F. COST	MODEL S.F. COST	+/- CHANGE
D		**SERVICES**						
		D10 Conveying						
D	1010	Elevators & Lifts	2 — Type: *Same as model* — Capacity: — Stops:	Each	224,100	8.15	7.47	+ 0.68
D	1020	Escalators & Moving Walks	Type:	Each				
		D20 Plumbing						
D	2010	Plumbing	*Kitchen, bath, + service fixtures* — *1 Fixture/215 S.F. Floor*	Each	2234	10.39	10.39	---
D	2020	Domestic Water Distribution		S.F. Floor	3.32	3.32	3.32	---
D	2040	Rain Water Drainage		S.F. Roof	1.14	0.23	0.19	+ 0.04
		D30 HVAC						
D	3010	Energy Supply		S.F. Floor	6.91	6.91	6.91	---
D	3020	Heat Generating Systems	Type: *Same as model*	S.F. Floor				
D	3030	Cooling Generating Systems	Type: *Same as model*	S.F. Floor	7.93	7.93	7.93	---
D	3090	Other HVAC Sys. & Equipment		Each				
		D40 Fire Protection						
D	4010	Sprinklers	*Same as model*	S.F. Floor	2.25	2.25	2.25	---
D	4020	Standpipes	*Same as model*	S.F. Floor	0.44	0.44	0.44	---
		D50 Electrical						
D	5010	Electrical Service/Distribution	*Same as model*	S.F. Floor	2.57	2.57	2.57	---
D	5020	Lighting & Branch Wiring		S.F. Floor	6.62	6.62	6.62	---
D	5030	Communications & Security		S.F. Floor	0.87	0.87	0.87	---
D	5090	Other Electrical Systems		S.F. Floor	0.18	0.18	0.18	---
E		**EQUIPMENT & FURNISHINGS**						
E	1010	Commercial Equipment		Each				
E	1020	Institutional Equipment		Each				
E	1030	Vehicular Equipment		Each				
E	1090	Other Equipment		Each				
F		**SPECIAL CONSTRUCTION**						
F	1020	Integrated Construction	N/A	S.F.				
F	1040	Special Facilities	N/A	S.F.				
G		**BUILDING SITEWORK**						

ITEM				Total Change	+ 3.86
17	Model 020 total	$ 123.97			
18	Adjusted S.F. cost $ 127.83	item 17 +/- changes			
19	Building area - from item 11	55,000	S.F. x adjusted S.F. cost..	$	7,030,650
20	Basement area - from item 15		S.F. x S.F. cost $	$	
21	Base building sub-total - item 20 + item 19			$	7,030,650
22	Miscellaneous addition (quality, etc.)			$	
23	Sub-total - item 22 + 21			$	7,030,650
24	General conditions -25 % of item 23			$	1,757,663
25	Sub-total - item 24 + item 23...			$	8,788,313
26	Architects fees 5.5 % of item 25			$	483,357
27	Sub-total - item 26 + item 27			$	9,271,670
28	Location modifier .... zip code 07410			x	1.09
29	Local replacement cost - item 28 x item 27			$	10,106,120
30	Depreciation 20 % of item 29			$	2,021,224
31	Depreciated local replacement cost - item 29 less item 30			$	8,084,896
32	Exclusions			$	
33	Net depreciated replacement cost - item 31 less item 32			$	8,084,896

Costs per square foot of floor area

Exterior Wall	S.F. Area	8000	12000	15000	19000	22500	25000	29000	32000	36000
	L.F. Perimeter	213	280	330	350	400	433	442	480	520
Face Brick with Concrete Block Back-up	Wood Joists	195.05	173.35	164.65	153.15	148.95	146.45	140.80	139.20	136.85
	Steel Joists	203.05	182.40	174.10	163.70	159.70	157.35	152.35	150.80	148.60
Stucco on Concrete Block	Wood Joists	171.20	151.70	143.85	134.80	131.00	128.80	124.55	123.10	121.15
	Steel Joists	189.35	169.70	161.85	152.65	148.80	146.65	142.30	140.90	138.85
Wood Siding	Wood Frame	171.30	151.80	143.95	134.85	131.05	128.90	124.55	123.10	121.20
Brick Veneer	Wood Frame	181.15	160.35	152.00	141.55	137.50	135.10	130.05	128.50	126.35
Perimeter Adj., Add or Deduct	Per 100 L.F.	20.80	13.90	11.10	8.80	7.40	6.70	5.75	5.20	4.65
Story Hgt. Adj., Add or Deduct	Per 1 Ft.	3.85	3.40	3.20	2.70	2.60	2.60	2.30	2.20	2.20
For Basement, add $33.00 per square foot of basement area										

The above costs were calculated using the basic specifications shown on the facing page. These costs should be adjusted where necessary for design alternatives and owner's requirements. Reported completed project costs, for this type of structure, range from $55.10 to $204.90 per S.F.

Common additives

Description	Unit	$ Cost	Description	Unit	$ Cost
Appliances			Closed Circuit Surveillance, One station		
Cooking range, 30" free standing			Camera and monitor	Each	1850
1 oven	Each	430 - 2150	For additional camera stations, add	Each	1000
2 oven	Each	1675 - 1875	Elevators, Hydraulic passenger, 2 stops		
30" built-in			2000# capacity	Each	64,800
1 oven	Each	855 - 2025	2500# capacity	Each	66,300
2 oven	Each	1725 - 1600	3500# capacity	Each	69,800
Counter top cook tops, 4 burner	Each	360 - 830	Additional stop, add	Each	7825
Microwave oven	Each	254 - 775	Emergency Lighting, 25 watt, battery operated		
Combination range, refrig. & sink, 30" wide	Each	1550 - 4650	Lead battery	Each	282
72" wide	Each	5025	Nickel cadmium	Each	805
Combination range, refrigerator, sink,			Laundry Equipment		
microwave oven & icemaker	Each	5775	Dryer, gas, 16 lb. capacity	Each	885
Compactor, residential, 4-1 compaction	Each	630 - 730	30 lb. capacity	Each	3600
Dishwasher, built-in, 2 cycles	Each	515 - 910	Washer, 4 cycle	Each	1075
4 cycles	Each	625 - 1700	Commercial	Each	1450
Garbage disposer, sink type	Each	188 - 325	Smoke Detectors		
Hood for range, 2 speed, vented, 30" wide	Each	271 - 1250	Ceiling type	Each	187
42" wide	Each	490 - 2250	Duct type	Each	480
Refrigerator, no frost 10-12 C.F.	Each	570 - 705			
18-20 C.F.	Each	735 - 1300			

Important: See the Reference Section for Location Factors

Model costs calculated for a 3 story building with 10' story height and 22,500 square feet of floor area

				Unit	Unit Cost	Cost Per S.F.	% Of Sub-Total
A. SUBSTRUCTURE							
1010	Standard Foundations	Poured concrete; strip and spread footings		S.F. Ground	6.33	2.11	
1020	Special Foundations	N/A		—	—	—	
1030	Slab on Grade	4" reinforced concrete with vapor barrier and granular base		S.F. Slab	4.74	1.58	4.4%
2010	Basement Excavation	Site preparation for slab and trench for foundation wall and footing		S.F. Ground	.17	.06	
2020	Basement Walls	4' foundation wall		L.F. Wall	74	1.50	
B. SHELL							
B10 Superstructure							
1010	Floor Construction	Open web steel joists, slab form, concrete, interior steel columns		S.F. Floor	22.34	14.89	15.9%
1020	Roof Construction	Open web steel joists with rib metal deck, interior steel columns		S.F. Roof	11.64	3.88	
B20 Exterior Enclosure							
2010	Exterior Walls	Face brick with concrete block backup	88% of wall	S.F. Wall	23.99	11.26	
2020	Exterior Windows	Aluminum horizontal sliding	12% of wall	Each	492	2.10	11.5%
2030	Exterior Doors	Aluminum and glass		Each	1694	.30	
B30 Roofing							
3010	Roof Coverings	Built-up tar and gravel with flashing; perlite/EPS composite insulation		S.F. Roof	5.79	1.93	1.6%
3020	Roof Openings	N/A		—	—	—	
C. INTERIORS							
1010	Partitions	Gypsum board and sound deadening board on metal studs	10 S.F. of Floor/L.F. Partition	S.F. Partition	6.37	5.66	
1020	Interior Doors	15% solid core wood, 85% hollow core wood	80 S.F. Floor/Door	Each	500	6.24	
1030	Fittings	Kitchen Cabinets		S.F. Floor	2.64	2.64	
2010	Stair Construction	Concrete filled metal pan		Flight	15,800	3.51	24.5%
3010	Wall Finishes	70% paint, 25% vinyl wall covering, 5% ceramic tile		S.F. Surface	1.29	2.30	
3020	Floor Finishes	60% carpet, 30% vinyl composition tile, 10% ceramic tile		S.F. Floor	5.07	5.07	
3030	Ceiling Finishes	Painted gypsum board on resilient channels		S.F. Ceiling	3.53	3.53	
D. SERVICES							
D10 Conveying							
1010	Elevators & Lifts	One hydraulic passenger elevator		Each	105,750	4.70	4.0%
1020	Escalators & Moving Walks	N/A		—	—	—	
D20 Plumbing							
2010	Plumbing Fixtures	Kitchen, bathroom and service fixtures, supply and drainage	1 Fixture/200 S.F. Floor	Each	2274	11.37	
2020	Domestic Water Distribution	Gas fired water heater		S.F. Floor	3.44	3.44	12.8%
2040	Rain Water Drainage	Roof drains		S.F. Roof	1.02	.34	
D30 HVAC							
3010	Energy Supply	Oil fired hot water, baseboard radiation		S.F. Floor	7.85	7.85	
3020	Heat Generating Systems	N/A		—	—	—	
3030	Cooling Generating Systems	Chilled water, air cooled condenser system		S.F. Floor	8.53	8.53	13.8%
3050	Terminal & Package Units	N/A		—	—	—	
3090	Other HVAC Sys. & Equipment	N/A		—	—	—	
D40 Fire Protection							
4010	Sprinklers	Wet pipe sprinkler system		S.F. Floor	2.73	2.73	2.3%
4020	Standpipes	N/A		—	—	—	
D50 Electrical							
5010	Electrical Service/Distribution	600 ampere service, panel board and feeders		S.F. Floor	2.64	2.64	
5020	Lighting & Branch Wiring	Incandescent fixtures, receptacles, switches, A.C. and misc. power		S.F. Floor	6.66	6.66	9.1%
5030	Communications & Security	Alarm systems, internet wiring, and emergency lighting		S.F. Floor	1.26	1.26	
5090	Other Electrical Systems	Emergency Generator, 11.5 kW		S.F. Floor	.19	.19	
E. EQUIPMENT & FURNISHINGS							
1010	Commercial Equipment	N/A		—	—	—	
1020	Institutional Equipment	N/A		—	—	—	0.0%
1030	Vehicular Equipment	N/A		—	—	—	
1090	Other Equipment	N/A		—	—	—	
F. SPECIAL CONSTRUCTION							
1020	Integrated Construction	N/A		—	—	—	0.0%
1040	Special Facilities	N/A		—	—	—	
G. BUILDING SITEWORK	**N/A**						

		Sub-Total	118.27	100%
CONTRACTOR FEES (General Requirements: 10%, Overhead: 5%, Profit: 10%)		25%	29.60	
ARCHITECT FEES		8%	11.83	
	Total Building Cost		**159.70**	

Costs per square foot of floor area

Exterior Wall	S.F. Area	40000	45000	50000	55000	60000	70000	80000	90000	100000
	L.F. Perimeter	366	400	433	466	500	510	530	550	594
Face Brick with Concrete Block Back-up	Steel Frame	170.25	168.80	167.55	166.65	165.80	162.40	160.15	158.40	157.60
	R/Conc. Frame	171.55	169.90	168.65	167.60	166.70	162.75	160.20	158.25	157.40
Decorative Concrete Block	Steel Frame	160.05	158.80	157.75	156.85	156.15	153.40	151.60	150.15	149.45
	R/Conc. Frame	157.60	156.25	155.25	154.35	153.60	150.75	148.95	147.50	146.80
Precast Concrete Panels	Steel Frame	169.20	167.60	166.40	165.35	164.50	160.65	158.20	156.25	155.40
	R/Conc. Frame	166.45	164.85	163.65	162.60	161.70	157.80	155.35	153.40	152.55
Perimeter Adj., Add or Deduct	Per 100 L.F.	6.80	6.00	5.45	4.90	4.55	3.80	3.35	2.95	2.65
Story Hgt. Adj., Add or Deduct	Per 1 Ft.	2.20	2.15	2.15	2.00	2.00	1.70	1.60	1.45	1.40

For Basement, add $33.05 per square foot of basement area

The above costs were calculated using the basic specifications shown on the facing page. These costs should be adjusted where necessary for design alternatives and owner's requirements. Reported completed project costs, for this type of structure, range from $66.00 to $187.60 per S.F.

Common additives

Description	Unit	$ Cost
Appliances		
Cooking range, 30" free standing		
1 oven	Each	430 - 2150
2 oven	Each	1675 - 1875
30" built-in		
1 oven	Each	855 - 2025
2 oven	Each	1725 - 1600
Counter top cook tops, 4 burner	Each	360 - 830
Microwave oven	Each	254 - 775
Combination range, refrig. & sink, 30" wide	Each	1550 - 4650
72" wide	Each	5025
Combination range, refrigerator, sink,		
microwave oven & icemaker	Each	5775
Compactor, residential, 4-1 compaction	Each	630 - 730
Dishwasher, built-in, 2 cycles	Each	515 - 910
4 cycles	Each	625 - 1700
Garbage disposer, sink type	Each	188 - 325
Hood for range, 2 speed, vented, 30" wide	Each	271 - 1250
42" wide	Each	490 - 2250
Refrigerator, no frost 10-12 C.F.	Each	570 - 705
18-20 C.F.	Each	735 - 1300

Description	Unit	$ Cost
Closed Circuit Surveillance, One station		
Camera and monitor	Each	1850
For additional camera stations, add	Each	1000
Elevators, Electric passenger, 5 stops		
2000# capacity	Each	158,700
3500# capacity	Each	167,200
5000# capacity	Each	170,700
Additional stop, add	Each	13,600
Emergency Lighting, 25 watt, battery operated		
Lead battery	Each	282
Nickel cadmium	Each	805
Laundry Equipment		
Dryer, gas, 16 lb. capacity	Each	885
30 lb. capacity	Each	3600
Washer, 4 cycle	Each	1075
Commercial	Each	1450
Smoke Detectors		
Ceiling type	Each	187
Duct type	Each	480

Important: See the Reference Section for Location Factors

Model costs calculated for a 6 story building with 10'-4" story height and 60,000 square feet of floor area

				Unit	Unit Cost	Cost Per S.F.	% Of Sub-Total
A.	**SUBSTRUCTURE**						
1010	Standard Foundations	Poured concrete; strip and spread footings and 4' foundation wall		S.F. Ground	7.20	1.20	
1020	Special Foundations	N/A		—	—	—	
1030	Slab on Grade	4" reinforced concrete with vapor barrier and granular base		S.F. Slab	4.74	.79	2.2%
2010	Basement Excavation	Site preparation for slab and trench for foundation wall and footing		S.F. Ground	.25	.04	
2020	Basement Walls	4' foundation wall		L.F. Wall	78	.65	
B.	**SHELL**						
	B10 Superstructure						
1010	Floor Construction	Open web steel joists, slab form, concrete, steel columns		S.F. Floor	28.37	23.64	20.5%
1020	Roof Construction	Open web steel joists with rib metal deck, steel columns		S.F. Roof	10.86	1.81	
	B20 Exterior Enclosure						
2010	Exterior Walls	Face brick with concrete block backup	86% of wall	S.F. Wall	23.99	10.66	
2020	Exterior Windows	Aluminum horizontal sliding	14% of wall	Each	492	2.37	10.7%
2030	Exterior Doors	Aluminum and glass		Each	4280	.28	
	B30 Roofing						
3010	Roof Coverings	Built-up tar and gravel with flashing; perlite/EPS composite insulation		S.F. Roof	5.82	.97	0.8%
3020	Roof Openings	N/A		—	—	—	
C.	**INTERIORS**						
1010	Partitions	Gypsum board and sound deadening board on metal studs	8 S.F. Floor/L.F. Partitions	S.F. Partition	6.74	6.74	
1020	Interior Doors	15% solid core wood, 85% hollow core wood	80 S.F Floor/Door	Each	601	7.51	
1030	Fittings	Kitchen cabinets		S.F. Floor	3.17	3.17	
2010	Stair Construction	Concrete filled metal pan		Flight	12,650	3.37	26.2%
3010	Wall Finishes	70% paint, 25% vinyl wall covering, 5% ceramic tile		S.F. Surface	1.43	2.85	
3020	Floor Finishes	60% carpet, 30% vinyl composition tile, 10% ceramic tile		S.F. Floor	5.07	5.07	
3030	Ceiling Finishes	Painted gypsum board on resilient channels		S.F. Ceiling	3.71	3.71	
D.	**SERVICES**						
	D10 Conveying						
1010	Elevators & Lifts	Two geared passenger elevators		Each	224,100	7.47	6.0%
1020	Escalators & Moving Walks	N/A		—	—	—	
	D20 Plumbing						
2010	Plumbing Fixtures	Kitchen, bathroom and service fixtures, supply and drainage	1 Fixture/215 S.F. Floor	Each	2234	10.39	
2020	Domestic Water Distribution	Gas fired water heater		S.F. Floor	3.32	3.32	11.2%
2040	Rain Water Drainage	Roof drains		S.F. Roof	1.14	.19	
	D30 HVAC						
3010	Energy Supply	Oil fired hot water, baseboard radiation		S.F. Floor	6.91	6.91	
3020	Heat Generating Systems	N/A		—	—	—	
3030	Cooling Generating Systems	Chilled water, air cooled condenser system		S.F. Floor	7.93	7.93	12.0%
3050	Terminal & Package Units	N/A		—	—	—	
3090	Other HVAC Sys. & Equipment	N/A		—	—	—	
	D40 Fire Protection						
4010	Sprinklers	Wet pipe sprinkler system		S.F. Floor	2.25	2.25	2.2%
4020	Standpipes	Standpipe		S.F. Floor	.44	.44	
	D50 Electrical						
5010	Electrical Service/Distribution	1600 ampere service, panel board and feeders		S.F. Floor	2.57	2.57	
5020	Lighting & Branch Wiring	Incandescent fixtures, receptacles, switches, A.C. and misc. power		S.F. Floor	6.62	6.62	8.3%
5030	Communications & Security	Alarm systems, internet wiring, emergency lighting, and intercom		S.F. Floor	.87	.87	
5090	Other Electrical Systems	Emergency generator, 11.5 kW		S.F. Floor	.18	.18	
E.	**EQUIPMENT & FURNISHINGS**						
1010	Commercial Equipment	N/A		—	—	—	
1020	Institutional Equipment	N/A		—	—	—	
1030	Vehicular Equipment	N/A		—	—	—	0.0 %
1090	Other Equipment	N/A		—	—	—	
F.	**SPECIAL CONSTRUCTION**						
1020	Integrated Construction	N/A		—	—	—	0.0 %
1040	Special Facilities	N/A		—	—	—	
G.	**BUILDING SITEWORK**	**N/A**					

		Sub-Total	123.97	100%
CONTRACTOR FEES (General Requirements: 10%, Overhead: 5%, Profit: 10%)		25%	30.98	
ARCHITECT FEES		7%	10.85	
	Total Building Cost		**165.80**	

Costs per square foot of floor area

Exterior Wall	S.F. Area	95000	112000	129000	145000	170000	200000	275000	400000	600000
	L.F. Perimeter	345	386	406	442	480	510	530	570	630
Ribbed Precast Concrete Panel	Steel Frame	220.95	216.90	212.35	210.20	206.55	202.70	194.70	188.20	183.40
	R/Conc. Frame	209.45	205.50	201.15	199.10	195.70	191.95	184.55	178.45	173.90
Face Brick with Concrete Block Back-up	Steel Frame	199.00	195.30	191.35	189.40	186.20	182.90	176.20	170.75	166.70
	R/Conc. Frame	204.90	201.20	197.20	195.30	192.15	188.80	182.10	176.65	172.65
Stucco on Concrete Block	Steel Frame	188.00	184.85	181.80	180.20	177.70	175.15	170.35	166.45	163.50
	R/Conc. Frame	193.90	190.75	187.70	186.10	183.60	181.10	176.30	172.35	169.40
Perimeter Adj., Add or Deduct	Per 100 L.F.	10.25	8.65	7.55	6.70	5.70	4.85	3.55	2.45	1.70
Story Hgt. Adj., Add or Deduct	Per 1 Ft.	3.25	3.05	2.75	2.70	2.55	2.25	1.70	1.30	0.95

For Basement, add $33.85 per square foot of basement area

The above costs were calculated using the basic specifications shown on the facing page. These costs should be adjusted where necessary for design alternatives and owner's requirements. Reported completed project costs, for this type of structure, range from $84.45 to $198.20 per S.F.

Common additives

Description	Unit	$ Cost
Appliances		
Cooking range, 30" free standing		
1 oven	Each	430 - 2150
2 oven	Each	1675 - 1875
30" built-in		
1 oven	Each	855 - 2025
2 oven	Each	1725 - 1600
Counter top cook tops, 4 burner	Each	360 - 830
Microwave oven	Each	254 - 775
Combination range, refrig. & sink, 30" wide	Each	1550 - 4650
72" wide	Each	5025
Combination range, refrigerator, sink,		
microwave oven & icemaker	Each	5775
Compactor, residential, 4-1 compaction	Each	630 - 730
Dishwasher, built-in, 2 cycles	Each	515 - 910
4 cycles	Each	625 - 1700
Garbage disposer, sink type	Each	188 - 325
Hood for range, 2 speed, vented, 30" wide	Each	271 - 1250
42" wide	Each	490 - 2250
Refrigerator, no frost 10-12 C.F.	Each	570 - 705
18-20 C.F.	Each	735 - 1300

Description	Unit	$ Cost
Closed Circuit Surveillance, One station		
Camera and monitor	Each	1850
For additional camera stations, add	Each	1000
Elevators, Electric passenger, 10 stops		
3000# capacity	Each	430,500
4000# capacity	Each	433,000
5000# capacity	Each	437,000
Additional stop, add	Each	13,600
Emergency Lighting, 25 watt, battery operated		
Lead battery	Each	282
Nickel cadmium	Each	805
Laundry Equipment		
Dryer, gas, 16 lb. capacity	Each	885
30 lb. capacity	Each	3600
Washer, 4 cycle	Each	1075
Commercial	Each	1450
Smoke Detectors		
Ceiling type	Each	187
Duct type	Each	480

Important: See the Reference Section for Location Factors

Model costs calculated for a 15 story building with 10'-6" story height and 145,000 square feet of floor area

				Unit	Unit Cost	Cost Per S.F.	% Of Sub-Total
A. SUBSTRUCTURE							
1010	Standard Foundations	CIP concrete pile caps		S.F. Ground	9.30	.62	
1020	Special Foundations	Steel H-piles, concrete grade beams		S.F. Ground	203	13.50	
1030	Slab on Grade	4" reinforced concrete with vapor barrier and granular base		S.F. Slab	4.74	.32	9.3%
2010	Basement Excavation	Site preparation for slab, piles and grade beam		S.F. Ground	.25	.02	
2020	Basement Walls	4' Foundation wall		L.F. Wall	74	.27	
B. SHELL							
	B10 Superstructure						
1010	Floor Construction	Open web steel joists, slab form, concrete, interior steel columns		S.F. Floor	25.32	23.63	15.2%
1020	Roof Construction	Open web steel joists with rib metal deck, interior steel columns		S.F. Roof	8.10	.54	
	B20 Exterior Enclosure						
2010	Exterior Walls	Ribbed precast concrete panel	87% of wall	S.F. Wall	38.35	16.02	
2020	Exterior Windows	Aluminum horizontal sliding	13% of wall	Each	492	2.05	12.8%
2030	Exterior Doors	Aluminum and glass		Each	2649	2.28	
	B30 Roofing						
3010	Roof Coverings	Built-up tar and gravel with flashing; perlite/EPS composite insulation		S.F. Roof	5.55	.37	0.2%
3020	Roof Openings	N/A		—	—	—	
C. INTERIORS							
1010	Partitions	Gypsum board on concrete block and metal studs	10 S.F. of Floor/L.F. Partition	S.F. Partition	13.27	13.27	
1020	Interior Doors	15% solid core wood, 85% hollow core wood	80 S.F. Floor/Door	Each	612	7.65	
1030	Fittings	Kitchen cabinets		S.F. Floor	3.27	3.27	
2010	Stair Construction	Concrete filled metal pan		Flight	12,650	3.75	24.9%
3010	Wall Finishes	70% paint, 25% vinyl wall covering, 5% ceramic tile		S.F. Surface	1.41	2.82	
3020	Floor Finishes	60% carpet, 30% vinyl composition tile, 10% ceramic tile		S.F. Floor	5.07	5.07	
3030	Ceiling Finishes	Painted gypsum board on resilient channels		S.F. Ceiling	3.71	3.71	
D. SERVICES							
	D10 Conveying						
1010	Elevators & Lifts	Four geared passenger elevators		Each	479,225	13.22	8.3%
1020	Escalators & Moving Walks	N/A		—	—	—	
	D20 Plumbing						
2010	Plumbing Fixtures	Kitchen, bathroom and service fixtures, supply and drainage	1 Fixture/210 S.F. Floor	Each	2499	11.90	
2020	Domestic Water Distribution	Gas fired water heater		S.F. Floor	3.80	3.80	10.0%
2040	Rain Water Drainage	Roof drains		S.F Roof	2.40	.16	
	D30 HVAC						
3010	Energy Supply	Oil fired hot water, baseboard radiation		S.F.Floor	6.91	6.91	
3020	Heat Generating Systems	N/A		—	—	—	
3030	Cooling Generating Systems	Chilled water, air cooled condenser system		S.F. Floor	7.93	7.93	9.4%
3050	Terminal & Package Units	N/A		—	—	—	
3090	Other HVAC Sys. & Equipment	N/A		—	—	—	
	D40 Fire Protection						
4010	Sprinklers	Wet pipe sprinkler system		S.F. Floor	2.18	2.18	2.1%
4020	Standpipes	Standpipe		S.F. Floor	1.15	1.15	
	D50 Electrical						
5010	Electrical Service/Distribution	4000 ampere service, panel board and feeders		S.F. Floor	2.18	2.18	
5020	Lighting & Branch Wiring	Incandescent fixtures, receptacles, switches, A.C. and misc. power		S.F. Floor	7.19	7.19	7.7%
5030	Communications & Security	Alarm systems, internet wiring, emergency lighting, antenna, intercom and security television		S.F. Floor	2.68	2.68	
5090	Other Electrical Systems	Emergency generator, 80 kW		S.F. Floor	.19	.19	
E. EQUIPMENT & FURNISHINGS							
1010	Commercial Equipment	N/A		—	—	—	
1020	Institutional Equipment	N/A		—	—	—	
1030	Vehicular Equipment	N/A		—	—	—	0.0 %
1090	Other Equipment	N/A		—	—	—	
F. SPECIAL CONSTRUCTION							
1020	Integrated Construction	N/A		—	—	—	0.0 %
1040	Special Facilities	N/A		—	—	—	
G. BUILDING SITEWORK	**N/A**						

		Sub-Total	158.65	100%
CONTRACTOR FEES (General Requirements: 10%, Overhead: 5%, Profit: 10%)		25%	39.65	
ARCHITECT FEES		6%	11.90	
	Total Building Cost		**210.20**	

Costs per square foot of floor area

Exterior Wall	S.F. Area	2000	3000	4000	5000	7500	10000	15000	25000	50000
	L.F. Perimeter	188	224	260	300	356	400	504	700	1200
Brick Veneer	Wood Frame	234.60	201.35	184.80	175.35	159.15	150.40	142.20	135.45	130.40
	Steel Frame	262.90	229.90	213.40	204.00	188.00	179.30	171.20	164.50	159.50
Brick Veneer on Block	Wood Truss	225.30	195.95	181.40	173.15	158.20	150.00	142.45	136.10	131.40
	Bearing Wall	273.15	238.50	221.20	211.35	194.25	184.95	176.35	169.10	163.80
Wood Siding	Wood Frame	219.95	189.50	174.35	165.60	151.15	143.40	136.15	130.10	125.65
Vinyl Siding	Wood Frame	218.40	188.50	173.55	164.95	150.90	143.50	136.45	130.60	126.25
Perimeter Adj., Add or Deduct	Per 100 L.F.	29.80	19.95	14.95	11.90	7.95	6.00	4.00	2.35	1.20
Story Hgt. Adj., Add or Deduct	Per 1 Ft.	3.70	2.95	2.55	2.35	1.85	1.60	1.35	1.05	0.95
For Basement, add $29.80 per square foot of basement area										

The above costs were calculated using the basic specifications shown on the facing page. These costs should be adjusted where necessary for design alternatives and owner's requirements. Reported completed project costs, for this type of structure, range from $82.95 to $208.05 per S.F.

Common additives

Description	Unit	$ Cost
Appliances		
Cooking range, 30" free standing		
1 oven	Each	430 - 2150
2 oven	Each	1675 - 1875
30" built-in		
1 oven	Each	855 - 2025
2 oven	Each	1725 - 1600
Counter top cook tops, 4 burner	Each	360 - 830
Microwave oven	Each	254 - 775
Combination range, refrig. & sink, 30" wide	Each	1550 - 4650
60" wide	Each	3800
72" wide	Each	5025
Combination range, refrigerator, sink,		
microwave oven & icemaker	Each	5775
Compactor, residential, 4-1 compaction	Each	630 - 730
Dishwasher, built-in, 2 cycles	Each	515 - 910
4 cycles	Each	625 - 1700
Garbage disposer, sink type	Each	188 - 325
Hood for range, 2 speed, vented, 30" wide	Each	271 - 1250
42" wide	Each	490 - 2250

Description	Unit	$ Cost
Appliances, cont.		
Refrigerator, no frost 10-12 C.F.	Each	570 - 705
14-16 C.F.	Each	590 - 840
18-20 C.F.	Each	735 - 1300
Laundry Equipment		
Dryer, gas, 16 lb. capacity	Each	885
30 lb. capacity	Each	3600
Washer, 4 cycle	Each	1075
Commercial	Each	1450
Sound System		
Amplifier, 250 watts	Each	2350
Speaker, ceiling or wall	Each	191
Trumpet	Each	365

Model costs calculated for a 1 story building with 10' story height and 10,000 square feet of floor area

				Unit	Unit Cost	Cost Per S.F.	% Of Sub-Total
A.	**SUBSTRUCTURE**						
1010	Standard Foundations	Poured concrete strip footings		S.F. Ground	2.06	2.06	
1020	Special Foundations	N/A		—	—	—	
1030	Slab on Grade	4" reinforced concrete with vapor barrier and granular base		S.F. Slab	4.74	4.74	9.6%
2010	Basement Excavation	Site preparation for slab and trench for foundation wall and footing		S.F. Ground	.46	.46	
2020	Basement Walls	4' foundation wall		L.F. Wall	78	3.12	
B.	**SHELL**						
	B10 Superstructure						
1010	Floor Construction	N/A		—	—	—	6.7 %
1020	Roof Construction	Plywood on wood trusses (pitched); fiberglass batt insulation		S.F. Roof	7.24	7.24	
	B20 Exterior Enclosure						
2010	Exterior Walls	Brick veneer on wood studs, insulated	80% of wall	S.F. Wall	24.22	7.75	
2020	Exterior Windows	Double hung wood	20% of wall	Each	588	2.94	11.0%
2030	Exterior Doors	Aluminum and glass; 18 ga. steel flush		Each	2108	1.26	
	B30 Roofing						
3010	Roof Coverings	Asphalt shingles with flashing (pitched); al. gutters & downspouts		S.F. Roof	2.29	2.29	2.1%
3020	Roof Openings	N/A		—	—	—	
C.	**INTERIORS**						
1010	Partitions	Gypsum board on wood studs	6 S.F. Floor/L.F. Partition	S.F. Partition	4.75	6.33	
1020	Interior Doors	Single leaf wood	80 S.F. Floor/Door	Each	565	6.65	
1030	Fittings	N/A		—	—	—	
2010	Stair Construction	N/A		—	—	—	29.9%
3010	Wall Finishes	Paint on drywall		S.F. Surface	.73	1.94	
3020	Floor Finishes	80% carpet, 20% ceramic tile		S.F. Floor	9.47	9.47	
3030	Ceiling Finishes	Gypsum board on wood furring, painted		S.F. Ceiling	7.98	7.98	
D.	**SERVICES**						
	D10 Conveying						
1010	Elevators & Lifts	N/A		—	—	—	0.0 %
1020	Escalators & Moving Walks	N/A		—	—	—	
	D20 Plumbing						
2010	Plumbing Fixtures	Kitchen, toilet and service fixtures, supply and drainage	1 Fixture/200 S.F. Floor	Each	1598	7.99	
2020	Domestic Water Distribution	Included with D2010		—	—	—	7.4%
2040	Rain Water Drainage	N/A		—	—	—	
	D30 HVAC						
3010	Energy Supply	N/A		—	—	—	
3020	Heat Generating Systems	N/A		—	—	—	
3030	Cooling Generating Systems	N/A		—	—	—	2.7 %
3050	Terminal & Package Units	Gas fired hot water boiler; terminal package heat/A/C units		S.F. Floor	2.94	2.94	
3090	Other HVAC Sys. & Equipment	N/A		—	—	—	
	D40 Fire Protection						
4010	Sprinklers	Wet pipe sprinkler system		S.F. Floor	2.77	2.77	2.6%
4020	Standpipes	N/A		—	—	—	
	D50 Electrical						
5010	Electrical Service/Distribution	600 ampere service, panel board, feeders and branch wiring		S.F. Floor	9.57	9.57	
5020	Lighting & Branch Wiring	Fluorescent & incandescent fixtures, receptacles, switches, A.C. and misc. power		S.F. Floor	10.55	10.55	26.8%
5030	Communications & Security	Alarm, fire detection, and communication system		S.F. Floor	8.75	8.75	
5090	Other Electrical Systems	Emergency generator, 7.5 kW		S.F. Floor	.19	.19	
E.	**EQUIPMENT & FURNISHINGS**						
1010	Commercial Equipment	Residential washers and dryers		S.F. Floor	.40	.40	
1020	Institutional Equipment	N/A		—	—	—	1.3%
1030	Vehicular Equipment	N/A		—	—	—	
1090	Other Equipment	Residential ranges and refrigerators		S.F. Floor	1	1	
F.	**SPECIAL CONSTRUCTION**						
1020	Integrated Construction	N/A		—	—	—	0.0 %
1040	Special Facilities	N/A		—	—	—	
G.	**BUILDING SITEWORK**	**N/A**					

		Sub-Total	108.39	100%
	CONTRACTOR FEES (General Requirements: 10%, Overhead: 5%, Profit: 10%)	25%	27.11	
	ARCHITECT FEES	11%	14.90	
	Total Building Cost		**150.40**	

Costs per square foot of floor area

Exterior Wall	S.F. Area	12000	15000	18000	21000	24000	27000	30000	33000	36000
	L.F. Perimeter	440	500	540	590	640	665	700	732	770
Face Brick with Concrete Block Back-up	Steel Frame	182.55	176.85	171.45	168.25	165.85	162.55	160.45	158.60	157.30
	Bearing Wall	176.40	171.25	166.40	163.50	161.35	158.40	156.50	154.80	153.70
Precast Concrete	Steel Frame	180.35	174.95	169.70	166.65	164.35	161.25	159.25	157.50	156.20
Decorative Concrete Block	Bearing Wall	158.65	155.00	151.60	149.55	148.00	145.95	144.65	143.50	142.65
Concrete Block	Steel Frame	170.50	167.55	164.95	163.30	162.10	160.55	159.55	158.70	158.05
	Bearing Wall	153.70	150.50	147.50	145.75	144.40	142.60	141.50	140.55	139.80
Perimeter Adj., Add or Deduct	Per 100 L.F.	12.60	10.10	8.40	7.20	6.25	5.60	5.05	4.60	4.20
Story Hgt. Adj., Add or Deduct	Per 1 Ft.	2.00	1.80	1.55	1.45	1.35	1.30	1.25	1.20	1.15
For Basement, add $27.05 per square foot of basement area										

The above costs were calculated using the basic specifications shown on the facing page. These costs should be adjusted where necessary for design alternatives and owner's requirements. Reported completed project costs, for this type of structure, range from $83.60 to $211.65 per S.F.

Common additives

Description	Unit	$ Cost
Closed Circuit Surveillance, One station		
Camera and monitor	Each	1850
For additional camera stations, add	Each	1000
Emergency Lighting, 25 watt, battery operated		
Lead battery	Each	282
Nickel cadmium	Each	805
Seating		
Auditorium chair, all veneer	Each	238
Veneer back, padded seat	Each	288
Upholstered, spring seat	Each	277
Classroom, movable chair & desk	Set	65 - 120
Lecture hall, pedestal type	Each	227 - 680
Smoke Detectors		
Ceiling type	Each	187
Duct type	Each	480
Sound System		
Amplifier, 250 watts	Each	2350
Speaker, ceiling or wall	Each	191
Trumpet	Each	365

Model costs calculated for a 1 story building with 24' story height and 24,000 square feet of floor area

Auditorium

			Unit	Unit Cost	Cost Per S.F.	% Of Sub-Total
A. SUBSTRUCTURE						
1010	Standard Foundations	Poured concrete; strip and spread footings	S.F. Ground	1.46	1.46	
1020	Special Foundations	N/A	—	—	—	
1030	Slab on Grade	6" reinforced concrete with vapor barrier and granular base	S.F. Slab	5.82	5.82	8.0%
2010	Basement Excavation	Site preparation for slab and trench for foundation wall and footing	S.F. Ground	.17	.17	
2020	Basement Walls	4' foundation wall	L.F. Wall	88	2.34	
B. SHELL						
	B10 Superstructure					
1010	Floor Construction	Open web steel joists, slab form, concrete (balcony)	S.F. Floor	26.88	3.36	11.9%
1020	Roof Construction	Metal deck on steel truss	S.F. Roof	11.29	11.29	
	B20 Exterior Enclosure					
2010	Exterior Walls	Precast concrete panel 80% of wall (adjusted for end walls)	S.F. Wall	34.47	17.65	
2020	Exterior Windows	Glass curtain wall 20% of wall	Each	39.10	5.01	19.9%
2030	Exterior Doors	Double aluminum and glass and hollow metal	Each	3573	1.79	
	B30 Roofing					
3010	Roof Coverings	Built-up tar and gravel with flashing; perlite/EPS composite insulation	S.F. Roof	5.33	5.33	4.5%
3020	Roof Openings	Roof hatches	S.F. Roof	.21	.21	
C. INTERIORS						
1010	Partitions	Concrete Block and toilet partitions 40 S.F. Floor/L.F. Partition	S.F. Partition	8.45	3.38	
1020	Interior Doors	Single leaf hollow metal 400 S.F. Floor/Door	Each	875	2.19	
1030	Fittings	N/A	—	—	—	
2010	Stair Construction	Concrete filled metal pan	Flight	18,950	2.37	21.5%
3010	Wall Finishes	70% paint, 30% epoxy coating	S.F. Surface	4.86	3.89	
3020	Floor Finishes	70% vinyl tile, 30% carpet	S.F. Floor	10.26	10.26	
3030	Ceiling Finishes	Fiberglass board, suspended	S.F. Ceiling	4.30	4.30	
D. SERVICES						
	D10 Conveying					
1010	Elevators & Lifts	One hydraulic passenger elevator	Each	87,600	3.65	3.0%
1020	Escalators & Moving Walks	N/A	—	—	—	
	D20 Plumbing					
2010	Plumbing Fixtures	Toilet and service fixtures, supply and drainage 1 Fixture/800 S.F. Floor	Each	3416	4.27	
2020	Domestic Water Distribution	Gas fired water heater	S.F. Floor	.37	.37	4.5%
2040	Rain Water Drainage	Roof drains	S.F. Roof	.84	.84	
	D30 HVAC					
3010	Energy Supply	N/A	—	—	—	
3020	Heat Generating Systems	Included in D3050	—	—	—	
3030	Cooling Generating Systems	N/A	—	—	—	9.9 %
3050	Terminal & Package Units	Single zone rooftop unit, gas heating, electric cooling	S.F. Floor	12.20	12.20	
3090	Other HVAC Sys. & Equipment	N/A	—	—	—	
	D40 Fire Protection					
4010	Sprinklers	Wet pipe sprinkler system	S.F. Floor	2.77	2.77	2.3%
4020	Standpipes	N/A	—	—	—	
	D50 Electrical					
5010	Electrical Service/Distribution	800 ampere service, panel board and feeders	S.F. Floor	2.45	2.45	
5020	Lighting & Branch Wiring	Fluorescent fixtures, receptacles, switches, A.C. and misc. power	S.F. Floor	11.12	11.12	14.6%
5030	Communications & Security	Alarm systems and emergency lighting, and public address system	S.F. Floor	3.16	3.16	
5090	Other Electrical Systems	Emergency generator, 100 kW	S.F. Floor	1.24	1.24	
E. EQUIPMENT & FURNISHINGS						
1010	Commercial Equipment	N/A	—	—	—	
1020	Institutional Equipment	N/A	—	—	—	
1030	Vehicular Equipment	N/A	—	—	—	0.0 %
1090	Other Equipment	N/A	—	—	—	
F. SPECIAL CONSTRUCTION						
1020	Integrated Construction	N/A	—	—	—	0.0 %
1040	Special Facilities	N/A	—	—	—	
G. BUILDING SITEWORK	**N/A**					

		Sub-Total	122.89	100%
CONTRACTOR FEES (General Requirements: 10%, Overhead: 5%, Profit: 10%)		25%	30.71	
ARCHITECT FEES		7%	10.75	

Total Building Cost	**164.35**

Costs per square foot of floor area

Exterior Wall	S.F. Area	2000	2700	3400	4100	4800	5500	6200	6900	7600
	L.F. Perimeter	180	208	236	256	280	303	317	337	357
Face Brick with Concrete Block Back-up	Steel Frame	243.05	227.70	218.70	211.00	206.35	202.65	198.40	195.85	193.80
	R/Conc. Frame	257.05	241.70	232.70	225.00	220.35	216.65	212.40	209.90	207.80
Precast Concrete Panel	Steel Frame	239.60	224.80	216.10	208.65	204.15	200.55	196.50	194.05	192.05
	R/Conc. Frame	253.60	238.80	230.15	222.65	218.15	214.55	210.50	208.00	206.10
Limestone with Concrete Block Back-up	Steel Frame	282.95	261.90	249.55	238.70	232.20	227.05	221.10	217.55	214.65
	R/Conc. Frame	296.95	275.90	263.55	252.70	246.20	241.05	235.10	231.50	228.65
Perimeter Adj., Add or Deduct	Per 100 L.F.	46.65	34.60	27.50	22.80	19.40	16.95	15.05	13.55	12.30
Story Hgt. Adj., Add or Deduct	Per 1 Ft.	4.35	3.80	3.40	3.05	2.85	2.70	2.55	2.40	2.30

For Basement, add $31.25 per square foot of basement area

The above costs were calculated using the basic specifications shown on the facing page. These costs should be adjusted where necessary for design alternatives and owner's requirements. Reported completed project costs, for this type of structure, range from $127.50 to $313.80 per S.F.

Common additives

Description	Unit	$ Cost
Bulletproof Teller Window, 44" x 60"	Each	4325
60" x 48"	Each	5300
Closed Circuit Surveillance, One station		
Camera and monitor	Each	1850
For additional camera stations, add	Each	1000
Counters, Complete	Station	5725
Door & Frame, 3' x 6'-8", bullet resistant steel with vision panel	Each	6150 - 7925
Drive-up Window, Drawer & micr., not incl. glass	Each	7425 - 12,300
Emergency Lighting, 25 watt, battery operated		
Lead battery	Each	282
Nickel cadmium	Each	805
Night Depository	Each	9250 - 13,900
Package Receiver, painted	Each	1825
stainless steel	Each	2825
Partitions, Bullet resistant to 8' high	L.F.	305 - 495
Pneumatic Tube Systems, 2 station	Each	31,400
With TV viewer	Each	58,000

Description	Unit	$ Cost
Service Windows, Pass thru, steel		
24" x 36"	Each	3275
48" x 48"	Each	3850
72" x 40"	Each	5375
Smoke Detectors		
Ceiling type	Each	187
Duct type	Each	480
Twenty-four Hour Teller		
Automatic deposit cash & memo	Each	51,500
Vault Front, Door & frame		
1 hour test, 32"x 78"	Opening	5175
2 hour test, 32" door	Opening	5600
40" door	Opening	6400
4 hour test, 32" door	Opening	6075
40" door	Opening	7125
Time lock, two movement, add	Each	1850

Model costs calculated for a 1 story building with 14' story height and 4,100 square feet of floor area

Bank

				Unit	Unit Cost	Cost Per S.F.	% Of Sub-Total
A. SUBSTRUCTURE							
1010	Standard Foundations	Poured concrete; strip and spread footings		S.F. Ground	4.18	4.18	
1020	Special Foundations	N/A		—	—	—	
1030	Slab on Grade	4" reinforced concrete with vapor barrier and granular base		S.F. Slab	4.74	4.74	9.3%
2010	Basement Excavation	Site preparation for slab and trench for foundation wall and footing		S.F. Ground	.26	.26	
2020	Basement Walls	4' Foundation wall		L.F. Wall	78	5.84	
B. SHELL							
	B10 Superstructure						
1010	Floor Construction	Cast-in-place columns		L.F. Column	120	5.83	13.4%
1020	Roof Construction	Cast-in-place concrete flat plate		S.F. Roof	15.90	15.90	
	B20 Exterior Enclosure						
2010	Exterior Walls	Face brick with concrete block backup	80% of wall	S.F. Wall	30.84	21.57	
2020	Exterior Windows	Horizontal aluminum sliding	20% of wall	Each	552	6.44	18.1%
2030	Exterior Doors	Double aluminum and glass and hollow metal		Each	2853	1.39	
	B30 Roofing						
3010	Roof Coverings	Built-up tar and gravel with flashing; perlite/EPS composite insulation		S.F. Roof	6.69	6.69	4.1%
3020	Roof Openings	N/A		—	—	—	
C. INTERIORS							
1010	Partitions	Gypsum board on metal studs	20 S.F. of Floor/L.F. Partition	S.F. Partition	9.64	4.82	
1020	Interior Doors	Single leaf hollow metal	200 S.F. Floor/Door	Each	875	4.38	
1030	Fittings	N/A		—	—	—	
2010	Stair Construction	N/A		—	—	—	13.7%
3010	Wall Finishes	50% vinyl wall covering, 50% paint		S.F. Surface	1.25	1.25	
3020	Floor Finishes	50% carpet tile, 40% vinyl composition tile, 10% quarry tile		S.F. Floor	5.42	5.42	
3030	Ceiling Finishes	Mineral fiber tile on concealed zee bars		S.F. Ceiling	6.38	6.38	
D. SERVICES							
	D10 Conveying						
1010	Elevators & Lifts	N/A		—	—	—	0.0%
1020	Escalators & Moving Walks	N/A		—	—	—	
	D20 Plumbing						
2010	Plumbing Fixtures	Toilet and service fixtures, supply and drainage	1 Fixture/580 S.F. Floor	Each	4048	6.98	
2020	Domestic Water Distribution	Gas fired water heater		S.F. Floor	.94	.94	5.5%
2040	Rain Water Drainage	Roof drains		S.F. Roof	.94	.94	
	D30 HVAC						
3010	Energy Supply	N/A		—	—	—	
3020	Heat Generating Systems	Included in D3050		—	—	—	
3030	Cooling Generating Systems	N/A		—	—	—	6.9%
3050	Terminal & Package Units	Single zone rooftop unit, gas heating, electric cooling		S.F. Floor	11.20	11.20	
3090	Other HVAC Sys. & Equipment	N/A		—	—	—	
	D40 Fire Protection						
4010	Sprinklers	Wet pipe sprinkler system		S.F. Floor	3.59	3.59	2.2%
4020	Standpipes	N/A		—	—	—	
	D50 Electrical						
5010	Electrical Service/Distribution	200 ampere service, panel board and feeders		S.F. Floor	2.40	2.40	
5020	Lighting & Branch Wiring	Fluorescent fixtures, receptacles, switches, A.C. and misc. power		S.F. Floor	10.06	10.06	12.8%
5030	Communications & Security	Alarm systems, internet and phone wiring, emergency lighting, and security television		S.F. Floor	7.79	7.79	
5090	Other Electrical Systems	Emergency generator, 15 kW, Uninterruptible power supply		S.F. Floor	.57	.57	
E. EQUIPMENT & FURNISHINGS							
1010	Commercial Equipment	Automatic teller, drive up window, night depository		S.F. Floor	7.44	7.44	
1020	Institutional Equipment	Closed circuit TV monitoring system		S.F. Floor	3.07	3.07	6.5%
1030	Vehicular Equipment	N/A		—	—	—	
1090	Other Equipment	N/A		—	—	—	
F. SPECIAL CONSTRUCTION							
1020	Integrated Construction	N/A		—	—	—	7.5%
1040	Special Facilities	Security vault door		S.F. Floor	12.09	12.09	
G. BUILDING SITEWORK	**N/A**						

		Sub-Total	162.16	100%
CONTRACTOR FEES (General Requirements: 10%, Overhead: 5%, Profit: 10%)		25%	40.54	
ARCHITECT FEES		11%	22.30	
	Total Building Cost		**225**	

Costs per square foot of floor area

Exterior Wall	S.F. Area	12000	14000	16000	18000	20000	22000	24000	26000	28000
	L.F. Perimeter	460	491	520	540	566	593	620	645	670
Concrete Block	Steel Roof Deck	160.25	157.20	154.75	152.70	151.10	149.90	148.85	147.95	147.20
Decorative Concrete Block	Steel Roof Deck	165.15	161.70	158.95	156.55	154.75	153.35	152.20	151.15	150.25
Face Brick on Concrete Block	Steel Roof Deck	173.05	168.90	165.65	162.75	160.55	158.85	157.50	156.20	155.15
Jumbo Brick on Concrete Block	Steel Roof Deck	168.75	165.00	161.95	159.35	157.40	155.80	154.60	153.45	152.50
Stucco on Concrete Block	Steel Roof Deck	163.35	160.00	157.35	155.10	153.35	152.05	150.95	149.95	149.10
Precast Concrete Panel	Steel Roof Deck	166.25	162.70	159.85	157.40	155.55	154.10	152.95	151.85	150.95
Perimeter Adj., Add or Deduct	Per 100 L.F.	4.80	4.05	3.55	3.10	2.85	2.60	2.40	2.20	2.05
Story Hgt. Adj., Add or Deduct	Per 1 Ft.	1.00	0.95	0.95	0.80	0.70	0.70	0.70	0.65	0.60
For Basement, add $30.85 per square foot of basement area										

The above costs were calculated using the basic specifications shown on the facing page. These costs should be adjusted where necessary for design alternatives and owner's requirements. Reported completed project costs, for this type of structure, range from $56.40 to $138.55 per S.F.

Common additives

Description	Unit	$ Cost
Bowling Alleys, Incl. alley, pinsetter,		
scorer, counter & misc. supplies, average	Lane	59,000
For automatic scorer, add (maximum)	Lane	10,900
Emergency Lighting, 25 watt, battery operated		
Lead battery	Each	282
Nickel cadmium	Each	805
Lockers, Steel, Single tier, 60" or 72"	Opening	191 - 310
2 tier, 60" or 72" total	Opening	107 - 141
5 tier, box lockers	Opening	65 - 83.50
Locker bench, lam., maple top only	L.F.	21
Pedestals, steel pipe	Each	63.50
Seating		
Auditorium chair, all veneer	Each	238
Veneer back, padded seat	Each	288
Upholstered, spring seat	Each	277
Sound System		
Amplifier, 250 watts	Each	2350
Speaker, ceiling or wall	Each	191
Trumpet	Each	365

Model costs calculated for a 1 story building with 14' story height and 20,000 square feet of floor area

Bowling Alley

				Unit	Unit Cost	Cost Per S.F.	% Of Sub-Total
A.	**SUBSTRUCTURE**						
1010	Standard Foundations	Poured concrete; strip and spread footings		S.F. Ground	1.25	1.25	
1020	Special Foundations	N/A		—	—	—	
1030	Slab on Grade	4" reinforced concrete with vapor barrier and granular base		S.F. Slab	4.74	4.74	7.7%
2010	Basement Excavation	Site preparation for slab and trench for foundation wall and footing		S.F. Ground	.46	.46	
2020	Basement Walls	4' Foundation wall		L.F. Wall	78	2.20	
B.	**SHELL**						
	B10	**Superstructure**					
1010	Floor Construction	N/A		—	—	—	9.8 %
1020	Roof Construction	Metal deck on open web steel joists with columns and beams		S.F. Roof	11.11	11.11	
	B20	**Exterior Enclosure**					
2010	Exterior Walls	Concrete block	90% of wall	S.F. Wall	11.05	3.94	
2020	Exterior Windows	Horizontal pivoted	10% of wall	Each	1840	3.04	6.6%
2030	Exterior Doors	Double aluminum and glass and hollow metal		Each	2089	.53	
	B30	**Roofing**					
3010	Roof Coverings	Built-up tar and gravel with flashing; perlite/EPS composite insulation		S.F. Roof	5.10	5.10	4.6%
3020	Roof Openings	Roof hatches		S.F. Roof	.14	.14	
C.	**INTERIORS**						
1010	Partitions	Concrete block	50 S.F. Floor/L.F. Partition	S.F. Partition	9.40	1.88	
1020	Interior Doors	Hollow metal single leaf	1000 S.F. Floor/Door	Each	875	.88	
1030	Fittings	N/A		—	—	—	
2010	Stair Construction	N/A		—	—	—	5.8%
3010	Wall Finishes	Paint and block filler		S.F. Surface	5.25	2.10	
3020	Floor Finishes	Vinyl tile	25% of floor	S.F. Floor	2.52	.63	
3030	Ceiling Finishes	Suspended fiberglass board	25% of area	S.F. Ceiling	4.30	1.08	
D.	**SERVICES**						
	D10	**Conveying**					
1010	Elevators & Lifts	N/A		—	—	—	0.0 %
1020	Escalators & Moving Walks	N/A		—	—	—	
	D20	**Plumbing**					
2010	Plumbing Fixtures	Toilet and service fixtures, supply and drainage	1 Fixture/2200 S.F. Floor	Each	2618	1.19	
2020	Domestic Water Distribution	Gas fired water heater		S.F. Floor	.18	.18	1.7%
2040	Rain Water Drainage	Roof drains		S.F. Roof	.55	.55	
	D30	**HVAC**					
3010	Energy Supply	N/A		—	—	—	
3020	Heat Generating Systems	Included in D3050		—	—	—	
3030	Cooling Generating Systems	N/A		—	—	—	12.3 %
3050	Terminal & Package Units	Single zone rooftop unit, gas heating, electric cooling		S.F. Floor	13.85	13.85	
3090	Other HVAC Sys. & Equipment	N/A		—	—	—	
	D40	**Fire Protection**					
4010	Sprinklers	Sprinklers, light hazard		S.F. Floor	2.33	2.33	2.1%
4020	Standpipes	N/A		—	—	—	
	D50	**Electrical**					
5010	Electrical Service/Distribution	800 ampere service, panel board and feeders		S.F. Floor	2.94	2.94	
5020	Lighting & Branch Wiring	High efficiency fluorescent fixtures, receptacles, switches, A.C. and misc. power		S.F. Floor	10.13	10.13	11.9%
5030	Communications & Security	Addressable alarm systems and emergency lighting		S.F. Floor	.31	.31	
5090	Other Electrical Systems	Emergency generator, 7.5 kW		S.F. Floor	.07	.07	
E.	**EQUIPMENT & FURNISHINGS**						
1010	Commercial Equipment	N/A		—	—	—	
1020	Institutional Equipment	N/A		—	—	—	37.5 %
1030	Vehicular Equipment	N/A		—	—	—	
1090	Other Equipment	Bowling Alley, ballrack, automatic scorers		S.F. Floor	42.33	42.33	
F.	**SPECIAL CONSTRUCTION**						
1020	Integrated Construction	N/A		—	—	—	0.0 %
1040	Special Facilities	N/A		—	—	—	
G.	**BUILDING SITEWORK**	**N/A**					

			Sub-Total	112.96	100%
CONTRACTOR FEES (General Requirements: 10%, Overhead: 5%, Profit: 10%)			25%	28.26	
ARCHITECT FEES			7%	9.88	

	Total Building Cost	**151.10**

Costs per square foot of floor area

Exterior Wall	S.F. Area	6000	8000	10000	12000	14000	16000	18000	20000	22000
	L.F. Perimeter	320	386	453	520	540	597	610	660	710
Face Brick with Concrete Block Back-up	Bearing Walls	152.60	146.50	142.95	140.55	136.10	134.70	131.65	130.70	129.85
	Steel Frame	156.05	150.05	146.50	144.10	139.80	138.40	135.35	134.45	133.65
Decorative Concrete Block	Bearing Walls	144.15	138.80	135.70	133.65	130.00	128.80	126.30	125.45	124.70
	Steel Frame	147.55	142.35	139.25	137.25	133.60	132.45	130.00	129.20	128.50
Precast Concrete Panels	Bearing Walls	147.95	142.30	139.00	136.75	132.75	131.45	128.70	127.80	127.10
	Steel Frame	151.40	145.80	142.55	140.35	136.45	135.15	132.40	131.55	130.80
Perimeter Adj., Add or Deduct	Per 100 L.F.	13.35	10.00	7.95	6.65	5.75	5.00	4.45	4.00	3.65
Story Hgt. Adj., Add or Deduct	Per 1 Ft.	2.20	1.95	1.80	1.80	1.60	1.55	1.45	1.35	1.35
Basement—Not Applicable										

The above costs were calculated using the basic specifications shown on the facing page. These costs should be adjusted where necessary for design alternatives and owner's requirements. Reported completed project costs, for this type of structure, range from $68.15 to $164.25 per S.F.

Common additives

Description	Unit	$ Cost
Directory Boards, Plastic, glass covered		
30" x 20"	Each	595
36" x 48"	Each	1450
Aluminum, 24" x 18"	Each	600
36" x 24"	Each	675
48" x 32"	Each	980
48" x 60"	Each	2025
Emergency Lighting, 25 watt, battery operated		
Lead battery	Each	282
Nickel cadmium	Each	805
Benches, Hardwood	L.F.	107 - 194
Ticket Printer	Each	7450
Turnstiles, One way		
4 arm, 46" dia., manual	Each	1050
Electric	Each	2225
High security, 3 arm		
65" dia., manual	Each	5250
Electric	Each	6450
Gate with horizontal bars		
65" dia., 7' high transit type	Each	5975

Important: See the Reference Section for Location Factors

Model costs calculated for a 1 story building with 14' story height and 12,000 square feet of floor area

				Unit	Unit Cost	Cost Per S.F.	% Of Sub-Total
A.	**SUBSTRUCTURE**						
	1010	Standard Foundations	Poured concrete; strip and spread footings	S.F. Ground	2.59	2.59	
	1020	Special Foundations	N/A	—	—	—	
	1030	Slab on Grade	6" reinforced concrete with vapor barrier and granular base	S.F. Slab	7.29	7.29	13.0%
	2010	Basement Excavation	Site preparation for slab and trench for foundation wall and footing	S.F. Ground	.26	.26	
	2020	Basement Walls	4' Foundation wall	L.F. Wall	80	3.47	
B.	**SHELL**						
	B10	**Superstructure**					
	1010	Floor Construction	N/A	—	—	—	7.7 %
	1020	Roof Construction	Metal deck on open web steel joists with columns and beams	S.F. Roof	8.11	8.11	
	B20	**Exterior Enclosure**					
	2010	Exterior Walls	Face brick with concrete block backup 70% of wall	S.F. Wall	30.85	13.10	
	2020	Exterior Windows	Store front 30% of wall	Each	28.75	5.24	19.0%
	2030	Exterior Doors	Double aluminum and glass	Each	4850	1.62	
	B30	**Roofing**					
	3010	Roof Coverings	Built-up tar and gravel with flashing; perlite/EPS composite insulation	S.F. Roof	5.96	5.96	5.7%
	3020	Roof Openings	N/A	—	—	—	
C.	**INTERIORS**						
	1010	Partitions	Lightweight concrete block 15 S.F. Floor/L.F. Partition	S.F. Partition	9.38	6.25	
	1020	Interior Doors	Hollow metal 150 S.F. Floor/Door	Each	875	5.83	
	1030	Fittings	N/A	—	—	—	
	2010	Stair Construction	N/A	—	—	—	28.3%
	3010	Wall Finishes	Glazed coating	S.F. Surface	1.76	2.34	
	3020	Floor Finishes	Quarry tile and vinyl composition tile	S.F. Floor	8.95	8.95	
	3030	Ceiling Finishes	Mineral fiber tile on concealed zee bars	S.F. Ceiling	6.38	6.38	
D.	**SERVICES**						
	D10	**Conveying**					
	1010	Elevators & Lifts	N/A	—	—	—	0.0 %
	1020	Escalators & Moving Walks	N/A	—	—	—	
	D20	**Plumbing**					
	2010	Plumbing Fixtures	Toilet and service fixtures, supply and drainage 1 Fixture/850 S.F. Floor	Each	2644	3.11	
	2020	Domestic Water Distribution	Electric water heater	S.F. Floor	.75	.75	4.3%
	2040	Rain Water Drainage	Roof drains	S.F. Roof	.61	.61	
	D30	**HVAC**					
	3010	Energy Supply	N/A	—	—	—	
	3020	Heat Generating Systems	Included in D3050	—	—	—	
	3030	Cooling Generating Systems	N/A	—	—	—	9.8 %
	3050	Terminal & Package Units	Single zone rooftop unit, gas heating, electric cooling	S.F. Floor	10.35	10.35	
	3090	Other HVAC Sys. & Equipment	N/A	—	—	—	
	D40	**Fire Protection**					
	4010	Sprinklers	Wet pipe sprinkler system	S.F. Floor	3.62	3.62	3.4%
	4020	Standpipes	N/A	—	—	—	
	D50	**Electrical**					
	5010	Electrical Service/Distribution	400 ampere service, panel board and feeders	S.F. Floor	1.68	1.68	
	5020	Lighting & Branch Wiring	Fluorescent fixtures, receptacles, switches, A.C. and misc. power	S.F. Floor	6.91	6.91	8.8%
	5030	Communications & Security	Alarm systems and emergency lighting	S.F. Floor	.26	.26	
	5090	Other Electrical Systems	Emergency generator, 7.5 kW	S.F. Floor	.40	.40	
E.	**EQUIPMENT & FURNISHINGS**						
	1010	Commercial Equipment	N/A	—	—	—	
	1020	Institutional Equipment	N/A	—	—	—	0.0 %
	1030	Vehicular Equipment	N/A	—	—	—	
	1090	Other Equipment	N/A	—	—	—	
F.	**SPECIAL CONSTRUCTION**						
	1020	Integrated Construction	N/A	—	—	—	0.0 %
	1040	Special Facilities	N/A	—	—	—	
G.	**BUILDING SITEWORK**	**N/A**					

			Sub-Total	105.08	100%
CONTRACTOR FEES (General Requirements: 10%, Overhead: 5%, Profit: 10%)			25%	26.28	
ARCHITECT FEES			7%	9.19	

	Total Building Cost	140.55

Costs per square foot of floor area

Exterior Wall	S.F. Area	600	800	1000	1200	1600	2000	2400	3000	4000
	L.F. Perimeter	100	114	128	139	164	189	214	250	314
Brick with Concrete Block Back-up	Steel Frame	341.15	327.60	319.35	312.55	305.00	300.50	297.50	294.25	291.50
	Bearing Walls	334.50	320.85	312.70	305.85	298.30	293.85	290.85	287.55	284.80
Concrete Block	Steel Frame	299.80	292.15	287.60	283.75	279.60	277.05	275.35	273.55	272.00
	Bearing Walls	293.10	285.50	280.95	277.10	272.90	270.35	268.70	266.85	265.35
Galvanized Steel Siding	Steel Frame	302.05	295.40	291.35	287.95	284.25	282.00	280.55	278.90	277.55
Metal Sandwich Panel	Steel Frame	299.30	290.85	285.85	281.65	277.00	274.15	272.35	270.35	268.65
Perimeter Adj., Add or Deduct	Per 100 L.F.	93.95	70.40	56.40	46.95	35.25	28.20	23.50	18.75	14.10
Story Hgt. Adj., Add or Deduct	Per 1 Ft.	5.75	4.80	4.45	3.95	3.50	3.25	3.10	2.80	2.70
Basement—Not Applicable										

The above costs were calculated using the basic specifications shown on the facing page. These costs should be adjusted where necessary for design alternatives and owner's requirements. Reported completed project costs, for this type of structure, range from $122.25 to $347.25 per S.F.

Common additives

Description	Unit	$ Cost
Air Compressors, Electric		
1-1/2 H.P., standard controls	Each	1325
Dual controls	Each	1700
5 H.P., 115/230 volt, standard control	Each	3450
Dual controls	Each	3600
Emergency Lighting, 25 watt, battery operated		
Lead battery	Each	282
Nickel cadmium	Each	805
Fence, Chain link, 6' high		
9 ga. wire, galvanized	L.F.	22.50
6 ga. wire	L.F.	32.50
Gate	Each	335
Product Dispenser with vapor recovery		
for 6 nozzles	Each	22,900
Laundry Equipment		
Dryer, gas, 16 lb. capacity	Each	885
30 lb. capacity	Each	3600
Washer, 4 cycle	Each	1075
Commercial	Each	1450

Description	Unit	$ Cost
Lockers, Steel, single tier, 60" or 72"	Opening	191 - 310
2 tier, 60" or 72" total	Opening	107 - 141
5 tier, box lockers	Opening	65 - 83.50
Locker bench, lam. maple top only	L.F.	21
Pedestals, steel pipe	Each	63.50
Paving, Bituminous		
Wearing course plus base course	Sq. Yard	10.50
Safe, Office type, 1 hour rating		
30" x 18" x 18"	Each	2400
60" x 36" x 18", double door	Each	8750
Sidewalks, Concrete 4" thick	S.F.	4.08
Yard Lighting,		
20' aluminum pole with 400 watt high pressure sodium fixture	Each	2825

Model costs calculated for a 1 story building with 12' story height and 800 square feet of floor area

Car Wash

				Unit	Unit Cost	Cost Per S.F.	% Of Sub-Total
A. SUBSTRUCTURE							
1010	Standard Foundations	Poured concrete; strip and spread footings		S.F. Ground	3.13	3.13	
1020	Special Foundations	N/A		—	—	—	
1030	Slab on Grade	5" reinforced concrete with vapor barrier and granular base		S.F. Slab	6.27	6.27	8.2%
2010	Basement Excavation	Site preparation for slab and trench for foundation wall and footing		S.F. Ground	.88	.88	
2020	Basement Walls	4' Foundation wall		L.F. Wall	69	9.84	
B. SHELL							
	B10 Superstructure						
1010	Floor Construction	N/A		—	—	—	4.2 %
1020	Roof Construction	Metal deck, open web steel joists, beams, columns		S.F. Roof	10.19	10.19	
	B20 Exterior Enclosure						
2010	Exterior Walls	Face brick with concrete block backup	70% of wall	S.F. Wall	30.85	36.93	
2020	Exterior Windows	Horizontal pivoted steel	5% of wall	Each	726	6.90	24.5%
2030	Exterior Doors	Steel overhead hollow metal		Each	3215	16.08	
	B30 Roofing						
3010	Roof Coverings	Built-up tar and gravel with flashing; perlite/EPS composite insulation		S.F. Roof	7.59	7.59	3.1%
3020	Roof Openings	N/A		—	—	—	
C. INTERIORS							
1010	Partitions	Concrete block	20 S.F. Floor/S.F. Partition	S.F. Partition	8.76	4.38	
1020	Interior Doors	Hollow metal	600 S.F. Floor/Door	Each	875	1.46	
1030	Fittings	N/A		—	—	—	
2010	Stair Construction	N/A		—	—	—	2.8%
3010	Wall Finishes	Paint		S.F. Surface	1.11	1.11	
3020	Floor Finishes	N/A		—	—	—	
3030	Ceiling Finishes	N/A		—	—	—	
D. SERVICES							
	D10 Conveying						
1010	Elevators & Lifts	N/A		—	—	—	0.0 %
1020	Escalators & Moving Walks	N/A		—	—	—	
	D20 Plumbing						
2010	Plumbing Fixtures	Toilet and service fixtures, supply and drainage	1 Fixture/160 S.F. Floor	Each	2890	18.06	
2020	Domestic Water Distribution	Gas fired water heater		S.F. Floor	34.20	34.20	22.8%
2040	Rain Water Drainage	Roof drains		S.F. Roof	3.57	3.57	
	D30 HVAC						
3010	Energy Supply	N/A		—	—	—	
3020	Heat Generating Systems	Unit heaters		Each	17.45	17.45	
3030	Cooling Generating Systems	N/A		—	—	—	7.1 %
3050	Terminal & Package Units	N/A		—	—	—	
3090	Other HVAC Sys. & Equipment	N/A		—	—	—	
	D40 Fire Protection						
4010	Sprinklers	N/A		—	—	—	0.0 %
4020	Standpipes	N/A		—	—	—	
	D50 Electrical						
5010	Electrical Service/Distribution	400 ampere service, panel board and feeders		S.F. Floor	22.10	22.10	
5020	Lighting & Branch Wiring	Fluorescent fixtures, receptacles, switches and misc. power		S.F. Floor	43.05	43.05	27.3%
5030	Communications & Security	N/A		—	—	—	
5090	Other Electrical Systems	Emergency generator, 7.5 kW		S.F. Floor	1.72	1.72	
E. EQUIPMENT & FURNISHINGS							
1010	Commercial Equipment	N/A		—	—	—	
1020	Institutional Equipment	N/A		—	—	—	0.0 %
1030	Vehicular Equipment	N/A		—	—	—	
1090	Other Equipment	N/A		—	—	—	
F. SPECIAL CONSTRUCTION							
1020	Integrated Construction	N/A		—	—	—	0.0 %
1040	Special Facilities	N/A		—	—	—	
G. BUILDING SITEWORK	**N/A**						

		Sub-Total	244.91	100%
CONTRACTOR FEES (General Requirements: 10%, Overhead: 5%, Profit: 10%)		25%	61.26	
ARCHITECT FEES		7%	21.43	
	Total Building Cost		**327.60**	

Costs per square foot of floor area

Exterior Wall	S.F. Area	2000	7000	12000	17000	22000	27000	32000	37000	42000
	L.F. Perimeter	180	340	470	540	640	740	762	793	860
Decorative Concrete Block	Wood Arch	237.90	177.95	164.90	155.15	151.55	149.25	144.70	141.60	140.35
	Steel Truss	241.60	181.70	168.60	158.95	155.30	152.95	148.45	145.35	144.05
Stone with Concrete Block Back-up	Wood Arch	278.60	199.95	182.65	169.55	164.70	161.65	155.45	151.30	149.60
	Steel Truss	276.30	197.60	180.25	167.20	162.35	159.25	153.10	149.00	147.30
Face Brick with Concrete Block Back-up	Wood Arch	267.65	194.00	177.85	165.70	161.15	158.30	152.55	148.70	147.15
	Steel Truss	265.30	191.70	175.50	163.35	158.80	155.95	150.25	146.40	144.80
Perimeter Adj., Add or Deduct	Per 100 L.F.	61.85	17.75	10.35	7.25	5.60	4.60	3.85	3.40	2.95
Story Hgt. Adj., Add or Deduct	Per 1 Ft.	3.60	1.95	1.60	1.30	1.15	1.05	0.95	0.90	0.85
For Basement, add $32.20 per square foot of basement area										

The above costs were calculated using the basic specifications shown on the facing page. These costs should be adjusted where necessary for design alternatives and owner's requirements. Reported completed project costs, for this type of structure, range from $78.05 to $319.15 per S.F.

Common additives

Description	Unit	$ Cost		Description	Unit	$ Cost
Altar, Wood, custom design, plain	Each	2450		Pews/Benches, Hardwood	L.F.	107 - 194
Deluxe	Each	12,700		Pulpits, Prefabricated, hardwood	Each	1600 - 9450
Granite or marble, average	Each	13,600		Railing, Hardwood	L.F.	203
Deluxe	Each	38,100		Steeples, translucent fiberglass		
Ark, Prefabricated, plain	Each	10,100		30" square, 15' high	Each	7675
Deluxe	Each	130,500		25' high	Each	8850
Baptistry, Fiberglass, incl. plumbing	Each	5475 - 8700		Painted fiberglass, 24" square, 14' high	Each	6675
Bells & Carillons, 48 bells	Each	982,000		28' high	Each	7125
24 bells	Each	462,000		Aluminum		
Confessional, Prefabricated wood				20' high, 3'- 6" base	Each	7800
Single, plain	Each	3500		35' high, 8'- 0" base	Each	30,200
Deluxe	Each	8600		60' high, 14'- 0" base	Each	65,000
Double, plain	Each	6600				
Deluxe	Each	19,200				
Emergency Lighting, 25 watt, battery operated						
Lead battery	Each	282				
Nickel cadmium	Each	805				
Lecterns, Wood, plain	Each	925				
Deluxe	Each	5800				

Important: See the Reference Section for Location Factors

Model costs calculated for a 1 story building with 24' story height and 17,000 square feet of floor area

				Unit	Unit Cost	Cost Per S.F.	% Of Sub-Total
A.	**SUBSTRUCTURE**						
1010	Standard Foundations	Poured concrete; strip and spread footings		S.F. Ground	3.93	3.93	
1020	Special Foundations	N/A		—	—	—	
1030	Slab on Grade	4" reinforced concrete with vapor barrier and granular base		S.F. Slab	4.74	4.74	10.5%
2010	Basement Excavation	Site preparation for slab and trench for foundation wall and footing		S.F. Ground	.17	.17	
2020	Basement Walls	4' Foundation wall		L.F. Wall	78	3.72	
B.	**SHELL**						
	B10 Superstructure						
1010	Floor Construction	N/A		—	—	—	15.9 %
1020	Roof Construction	Wood deck on laminated wood arches		S.F. Roof	17.84	19.03	
	B20 Exterior Enclosure						
2010	Exterior Walls	Face brick with concrete block backup	80% of wall (adjusted for end walls)	S.F. Wall	36.65	22.35	
2020	Exterior Windows	Aluminum, top hinged, in-swinging and curtain wall panels	20% of wall	S.F. Window	27.68	4.22	22.7%
2030	Exterior Doors	Double hollow metal swinging, single hollow metal		Each	1476	.52	
	B30 Roofing						
3010	Roof Coverings	Asphalt shingles with flashing; polystyrene insulation		S.F. Roof	5.08	5.08	4.3%
3020	Roof Openings	N/A		—	—	—	
C.	**INTERIORS**						
1010	Partitions	Plaster on metal studs	40 S.F. Floor/L.F. Partitions	S.F. Partition	14.05	8.43	
1020	Interior Doors	Hollow metal	400 S.F. Floor/Door	Each	875	2.19	
1030	Fittings	N/A		—	—	—	
2010	Stair Construction	N/A		—	—	—	17.1%
3010	Wall Finishes	Paint		S.F. Surface	1.03	1.23	
3020	Floor Finishes	Carpet		S.F. Floor	8.56	8.56	
3030	Ceiling Finishes	N/A		—	—	—	
D.	**SERVICES**						
	D10 Conveying						
1010	Elevators & Lifts	N/A		—	—	—	0.0 %
1020	Escalators & Moving Walks	N/A		—	—	—	
	D20 Plumbing						
2010	Plumbing Fixtures	Kitchen, toilet and service fixtures, supply and drainage	1 Fixture/2430 S.F. Floor	Each	3402	1.40	
2020	Domestic Water Distribution	Gas fired hot water heater		S.F. Floor	.24	.24	1.4%
2040	Rain Water Drainage	N/A		—	—	—	
	D30 HVAC						
3010	Energy Supply	Oil fired hot water, wall fin radiation		S.F. Floor	8.73	8.73	
3020	Heat Generating Systems	N/A		—	—	—	
3030	Cooling Generating Systems	N/A		—	—	—	16.0%
3050	Terminal & Package Units	Split systems with air cooled condensing units		S.F. Floor	10.40	10.40	
3090	Other HVAC Sys. & Equipment	N/A		—	—	—	
	D40 Fire Protection						
4010	Sprinklers	Wet pipe sprinkler system		S.F. Floor	2.77	2.77	2.3%
4020	Standpipes	N/A		—	—	—	
	D50 Electrical						
5010	Electrical Service/Distribution	400 ampere service, panel board and feeders		S.F. Floor	.78	.78	
5020	Lighting & Branch Wiring	Fluorescent fixtures, receptacles, switches, A.C. and misc. power		S.F. Floor	8.13	8.13	9.8%
5030	Communications & Security	Alarm systems, sound system and emergency lighting		S.F. Floor	2.70	2.70	
5090	Other Electrical Systems	Emergency generator, 7.5 kW		S.F. Floor	.11	.11	
E.	**EQUIPMENT & FURNISHINGS**						
1010	Commercial Equipment	N/A		—	—	—	
1020	Institutional Equipment	N/A		—	—	—	0.0 %
1030	Vehicular Equipment	N/A		—	—	—	
1090	Other Equipment	N/A		—	—	—	
F.	**SPECIAL CONSTRUCTION**						
1020	Integrated Construction	N/A		—	—	—	0.0 %
1040	Special Facilities	N/A		—	—	—	
G.	**BUILDING SITEWORK**	**N/A**					

			Sub-Total	119.43	**100%**
CONTRACTOR FEES (General Requirements: 10%, Overhead: 5%, Profit: 10%)		25%	29.85	
ARCHITECT FEES		11%	16.42	

Total Building Cost	**165.70**

Costs per square foot of floor area

Exterior Wall	S.F. Area	2000	4000	6000	8000	12000	15000	18000	20000	22000
	L.F. Perimeter	180	280	340	386	460	535	560	600	640
Stone Ashlar Veneer On Concrete Block	Wood Truss	249.40	212.15	**194.05**	183.65	171.90	168.30	163.50	162.10	160.95
	Steel Joists	248.50	211.25	193.15	182.75	170.95	167.35	162.55	161.20	160.00
Stucco on Concrete Block	Wood Truss	228.00	195.50	180.60	172.15	162.70	159.80	156.05	154.95	154.05
	Steel Joists	227.40	194.90	179.95	171.55	162.10	159.15	155.50	154.30	153.40
Brick Veneer	Wood Frame	239.45	204.25	187.55	178.00	167.30	163.95	159.70	158.40	157.35
Wood Shingles	Wood Frame	227.30	194.85	179.95	171.50	162.10	159.15	155.50	154.40	153.50
Perimeter Adj., Add or Deduct	Per 100 L.F.	42.20	21.05	14.05	10.45	6.95	5.55	4.70	4.20	3.85
Story Hgt. Adj., Add or Deduct	Per 1 Ft.	4.60	3.60	2.95	2.45	1.90	1.80	1.55	1.55	1.50
For Basement, add $28.25 per square foot of basement area										

The above costs were calculated using the basic specifications shown on the facing page. These costs should be adjusted where necessary for design alternatives and owner's requirements. Reported completed project costs, for this type of structure, range from $77.20 to $256.85 per S.F.

Common additives

Description	Unit	$ Cost
Bar, Front bar	L.F.	360
Back bar	L.F.	289
Booth, Upholstered, custom, straight	L.F.	202 - 375
"L" or "U" shaped	L.F.	210 - 355
Fireplaces, Brick not incl. chimney or foundation, 30" x 24" opening	Each	2800
Chimney, standard brick		
Single flue 16" x 20"	V.L.F.	91
20" x 20"	V.L.F.	107
2 flue, 20" x 32"	V.L.F.	163
Lockers, Steel, single tier, 60" or 72"	Opening	191 - 310
2 tier, 60" or 72" total	Opening	107 - 141
5 tier, box lockers	Opening	65 - 83.50
Locker bench, lam. maple top only	L.F.	21
Pedestals, steel pipe	Each	63.50
Refrigerators, Prefabricated, walk-in		
7'-6" High, 6' x 6'	S.F.	164
10' x 10'	S.F.	129
12' x 14'	S.F.	114
12' x 20'	S.F.	101

Description	Unit	$ Cost
Sauna, Prefabricated, complete, 6' x 4'	Each	5850
6' x 9'	Each	8525
8' x 8'	Each	10,100
10' x 12'	Each	14,000
Smoke Detectors		
Ceiling type	Each	187
Duct type	Each	480
Sound System		
Amplifier, 250 watts	Each	2350
Speaker, ceiling or wall	Each	191
Trumpet	Each	365
Steam Bath, Complete, to 140 C.F.	Each	2050
To 300 C.F.	Each	2300
To 800 C.F.	Each	6375
To 2500 C.F.	Each	9550
Swimming Pool Complete, gunite	S.F.	64 - 78.50
Tennis Court, Complete with fence		
Bituminous	Each	36,600 - 45,400
Clay	Each	38,500 - 46,700

Important: See the Reference Section for Location Factors

Model costs calculated for a 1 story building with 12' story height and 6,000 square feet of floor area

Club, Country

					Unit	Unit Cost	Cost Per S.F.	% Of Sub-Total

A. SUBSTRUCTURE

					Unit	Unit Cost	Cost Per S.F.	% Of Sub-Total
1010	Standard Foundations	Poured concrete; strip and spread footings			S.F. Ground	2.91	2.91	
1020	Special Foundations	N/A			—	—	—	
1030	Slab on Grade	4" reinforced concrete with vapor barrier and granular base			S.F. Slab	4.74	4.74	9.8%
2010	Basement Excavation	Site preparation for slab and trench for foundation wall and footing			S.F. Ground	.26	.26	
2020	Basement Walls	4' Foundation wall			L.F. Wall	78	5.74	

B. SHELL

B10 Superstructure

					Unit	Unit Cost	Cost Per S.F.	% Of Sub-Total
1010	Floor Construction	N/A			—	—	—	
1020	Roof Construction	Wood truss with plywood sheathing			S.F. Roof	6.61	6.61	4.7 %

B20 Exterior Enclosure

					Unit	Unit Cost	Cost Per S.F.	% Of Sub-Total
2010	Exterior Walls	Stone ashlar veneer on concrete block	65% of wall		S.F. Wall	34.71	15.34	
2020	Exterior Windows	Aluminum horizontal sliding	35% of wall		Each	492	7.80	18.3%
2030	Exterior Doors	Double aluminum and glass, hollow metal			Each	2428	2.44	

B30 Roofing

					Unit	Unit Cost	Cost Per S.F.	% Of Sub-Total
3010	Roof Coverings	Asphalt shingles			S.F. Roof	16.07	2.75	2.0%
3020	Roof Openings	N/A			—	—	—	

C. INTERIORS

					Unit	Unit Cost	Cost Per S.F.	% Of Sub-Total
1010	Partitions	Gypsum board on metal studs, load bearing	14 S.F. Floor/L.F. Partition		S.F. Partition	7.63	5.45	
1020	Interior Doors	Single leaf wood	140 S.F. Floor/Door		Each	565	4.03	
1030	Fittings	N/A			—	—	—	
2010	Stair Construction	N/A			—	—	—	17.2%
3010	Wall Finishes	40% vinyl wall covering, 40% paint, 20% ceramic tile			S.F. Surface	2.38	3.40	
3020	Floor Finishes	50% carpet, 30% hardwood tile, 20% ceramic tile			S.F. Floor	7.21	7.21	
3030	Ceiling Finishes	Gypsum plaster on wood furring			S.F. Ceiling	3.95	3.95	

D. SERVICES

D10 Conveying

					Unit	Unit Cost	Cost Per S.F.	% Of Sub-Total
1010	Elevators & Lifts	N/A			—	—	—	0.0 %
1020	Escalators & Moving Walks	N/A			—	—	—	

D20 Plumbing

					Unit	Unit Cost	Cost Per S.F.	% Of Sub-Total
2010	Plumbing Fixtures	Kitchen, toilet and service fixtures, supply and drainage	1 Fixture/125 S.F. Floor		Each	2399	19.19	
2020	Domestic Water Distribution	Gas fired hot water			S.F. Floor	6.27	6.27	18.2%
2040	Rain Water Drainage	N/A			—	—	—	

D30 HVAC

					Unit	Unit Cost	Cost Per S.F.	% Of Sub-Total
3010	Energy Supply	N/A			—	—	—	
3020	Heat Generating Systems	Included in D3050			—	—	—	
3030	Cooling Generating Systems	N/A			—	—	—	20.7 %
3050	Terminal & Package Units	Multizone rooftop unit, gas heating, electric cooling			S.F. Floor	28.90	28.90	
3090	Other HVAC Sys. & Equipment	N/A			—	—	—	

D40 Fire Protection

					Unit	Unit Cost	Cost Per S.F.	% Of Sub-Total
4010	Sprinklers	Wet pipe sprinkler system			S.F. Floor	3.59	3.59	2.6%
4020	Standpipes	N/A			—	—	—	

D50 Electrical

					Unit	Unit Cost	Cost Per S.F.	% Of Sub-Total
5010	Electrical Service/Distribution	400 ampere service, panel board and feeders			S.F. Floor	3.54	3.54	
5020	Lighting & Branch Wiring	Fluorescent fixtures, receptacles, switches, A.C. and misc. power			S.F. Floor	5.17	5.17	6.6%
5030	Communications & Security	Alarm systems and emergency lighting			S.F. Floor	.36	.36	
5090	Other Electrical Systems	Emergency generator, 11.5 kW			S.F. Floor	.22	.22	

E. EQUIPMENT & FURNISHINGS

					Unit	Unit Cost	Cost Per S.F.	% Of Sub-Total
1010	Commercial Equipment	N/A			—	—	—	
1020	Institutional Equipment	N/A			—	—	—	
1030	Vehicular Equipment	N/A			—	—	—	0.0 %
1090	Other Equipment	N/A			—	—	—	

F. SPECIAL CONSTRUCTION

					Unit	Unit Cost	Cost Per S.F.	% Of Sub-Total
1020	Integrated Construction	N/A			—	—	—	0.0 %
1040	Special Facilities	N/A			—	—	—	

G. BUILDING SITEWORK   N/A

Sub-Total		139.87	**100%**
CONTRACTOR FEES (General Requirements: 10%, Overhead: 5%, Profit: 10%)	25%	34.95	
ARCHITECT FEES	11%	19.23	
Total Building Cost		**194.05**	

Costs per square foot of floor area

Exterior Wall	S.F. Area	4000	8000	12000	17000	22000	27000	32000	37000	42000
	L.F. Perimeter	280	386	520	585	640	740	840	940	940
Stone Ashlar on Concrete Block	Steel Joists	184.70	157.05	150.00	140.10	134.25	132.25	130.75	129.75	126.65
	Wood Joists	177.95	151.10	144.25	134.65	129.00	127.05	125.65	124.60	121.65
Face Brick on Concrete Block	Steel Joists	181.90	155.10	148.25	138.75	133.10	131.10	129.70	128.75	125.75
	Wood Joists	175.85	149.70	143.05	133.75	128.30	126.35	124.95	124.05	121.15
Decorative Concrete Block	Steel Joists	173.05	149.00	142.75	134.35	129.45	127.65	126.40	125.50	122.90
	Wood Joists	166.15	142.75	136.70	128.55	123.75	122.05	120.80	119.95	117.50
Perimeter Adj., Add or Deduct	Per 100 L.F.	23.95	11.95	7.95	5.60	4.40	3.45	3.00	2.60	2.25
Story Hgt. Adj., Add or Deduct	Per 1 Ft.	3.85	2.65	2.35	1.95	1.60	1.45	1.45	1.40	1.25
For Basement, add $28.80 per square foot of basement area										

The above costs were calculated using the basic specifications shown on the facing page. These costs should be adjusted where necessary for design alternatives and owner's requirements. Reported completed project costs, for this type of structure, range from $78.95 to $226.85 per S.F.

Common additives

Description	Unit	$ Cost
Bar, Front bar	L.F.	360
Back bar	L.F.	289
Booth, Upholstered, custom, straight	L.F.	202 - 375
"L" or "U" shaped	L.F.	210 - 355
Emergency Lighting, 25 watt, battery operated		
Lead battery	Each	282
Nickel cadmium	Each	805
Flagpoles, Complete		
Aluminum, 20' high	Each	1650
40' High	Each	3475
70' High	Each	10,100
Fiberglass, 23' High	Each	1775
39'-5" High	Each	3325
59' High	Each	8225
Kitchen Equipment		
Broiler	Each	4025
Coffee urn, twin 6 gallon	Each	3075
Cooler, 6 ft. long, reach-in	Each	4925
Dishwasher, 10-12 racks per hr.	Each	4950
Food warmer, counter 1.2 kw	Each	735

Description	Unit	$ Cost
Kitchen Equipment, cont.		
Freezer, 44 C.F., reach-in	Each	3725
Ice cube maker, 50 lb. per day	Each	1750
Lockers, Steel, single tier, 60" or 72"	Opening	191 - 310
2 tier, 60" or 72" total	Opening	107 - 141
5 tier, box lockers	Opening	65 - 83.50
Locker bench, lam. maple top only	L.F.	21
Pedestals, steel pipe	Each	63.50
Refrigerators, Prefabricated, walk-in		
7'-6" High, 6' x 6'	S.F.	164
10' x 10'	S.F.	129
12' x 14'	S.F.	114
12' x 20'	S.F.	101
Smoke Detectors		
Ceiling type	Each	187
Duct type	Each	480
Sound System		
Amplifier, 250 watts	Each	2350
Speaker, ceiling or wall	Each	191
Trumpet	Each	365

Important: See the Reference Section for Location Factors

Model costs calculated for a 1 story building with 12' story height and 22,000 square feet of floor area

Club, Social

				Unit	Unit Cost	Cost Per S.F.	% Of Sub-Total
A. SUBSTRUCTURE							
1010	Standard Foundations	Poured concrete; strip and spread footings		S.F. Ground	1.83	1.83	
1020	Special Foundations	N/A		—	—	—	
1030	Slab on Grade	4" reinforced concrete with vapor barrier and granular base		S.F. Slab	4.74	4.74	10.3%
2010	Basement Excavation	Site preparation for slab and trench for foundation wall and footing		S.F. Ground	.17	.17	
2020	Basement Walls	4' Foundation wall		L.F. Wall	78	3.63	
B. SHELL							
	B10 Superstructure						
1010	Floor Construction	N/A		—	—	—	7.9 %
1020	Roof Construction	Metal deck on open web steel joists		S.F. Roof	7.97	7.97	
	B20 Exterior Enclosure						
2010	Exterior Walls	Stone, ashlar veneer on concrete block	65% of wall	S.F. Wall	34.68	7.87	
2020	Exterior Windows	Window wall	35% of wall	Each	44.15	5.39	13.9%
2030	Exterior Doors	Double aluminum and glass doors		Each	2428	.67	
	B30 Roofing						
3010	Roof Coverings	Built-up tar and gravel with flashing; perlite/EPS composite insulation		S.F. Roof	5.43	5.43	5.4%
3020	Roof Openings	N/A		—	—	—	
C. INTERIORS							
1010	Partitions	Lightweight concrete block	14 S.F. Floor/L.F. Partition	S.F. Partition	10.19	7.28	
1020	Interior Doors	Single leaf wood	140 S.F. Floor/Door	Each	565	4.03	
1030	Fittings	N/A		—	—	—	
2010	Stair Construction	N/A		—	—	—	26.6%
3010	Wall Finishes	65% paint, 25% vinyl wall covering, 10% ceramic tile		S.F. Surface	1.60	2.29	
3020	Floor Finishes	60% carpet, 35% hardwood, 15% ceramic tile		S.F. Floor	6.71	6.71	
3030	Ceiling Finishes	Mineral fiber tile on concealed zee bars		S.F. Ceiling	6.38	6.38	
D. SERVICES							
	D10 Conveying						
1010	Elevators & Lifts	N/A		—	—	—	0.0 %
1020	Escalators & Moving Walks	N/A		—	—	—	
	D20 Plumbing						
2010	Plumbing Fixtures	Kitchen, toilet and service fixtures, supply and drainage	1 Fixture/1050 S.F. Floor	Each	2720	2.59	
2020	Domestic Water Distribution	Gas fired hot water heater		S.F. Floor	.21	.21	3.3%
2040	Rain Water Drainage	Roof drains		S.F. Roof	.51	.51	
	D30 HVAC						
3010	Energy Supply	N/A		—	—	—	
3020	Heat Generating Systems	Included in D3050		—	—	—	
3030	Cooling Generating Systems	N/A		—	—	—	23.1 %
3050	Terminal & Package Units	Multizone rooftop unit, gas heating, electric cooling		S.F. Floor	23.20	23.20	
3090	Other HVAC Sys. & Equipment	N/A		—	—	—	
	D40 Fire Protection						
4010	Sprinklers	Wet pipe sprinkler system		S.F. Floor	2.77	2.77	2.8%
4020	Standpipes	N/A		—	—	—	
	D50 Electrical						
5010	Electrical Service/Distribution	400 ampere service, panel board and feeders		S.F. Floor	1.15	1.15	
5020	Lighting & Branch Wiring	Fluorescent fixtures, receptacles, switches, A.C. and misc. power		S.F. Floor	5.25	5.25	6.7%
5030	Communications & Security	Alarm systems and emergency lighting		S.F. Floor	.20	.20	
5090	Other Electrical Systems	Emergency generator, 11.5 kW		S.F. Floor	.12	.12	
E. EQUIPMENT & FURNISHINGS							
1010	Commercial Equipment	N/A		—	—	—	
1020	Institutional Equipment	N/A		—	—	—	0.0 %
1030	Vehicular Equipment	N/A		—	—	—	
1090	Other Equipment	N/A		—	—	—	
F. SPECIAL CONSTRUCTION							
1020	Integrated Construction	N/A		—	—	—	0.0 %
1040	Special Facilities	N/A		—	—	—	
G. BUILDING SITEWORK	**N/A**						

			Sub-Total	100.39	**100%**
CONTRACTOR FEES (General Requirements: 10%, Overhead: 5%, Profit: 10%)			25%	25.08	
ARCHITECT FEES			7%	8.78	
		Total Building Cost		**134.25**	

Costs per square foot of floor area

Exterior Wall	S.F. Area	15000	20000	28000	38000	50000	65000	85000	100000	150000
	L.F. Perimeter	350	400	480	550	630	660	750	825	1035
Face Brick with Concrete Block Back-up	Steel Frame	216.50	204.00	193.25	185.20	179.60	174.10	170.55	169.00	165.60
	Bearing Walls	214.30	200.70	188.95	179.95	173.65	167.30	163.30	161.50	157.60
Decorative Concrete Block	Steel Frame	208.90	197.45	187.65	180.45	175.45	170.75	167.70	166.25	163.30
	Bearing Walls	206.60	194.10	183.30	175.15	169.50	163.90	160.45	158.75	155.35
Stucco on Concrete Block	Steel Frame	205.30	194.05	184.40	177.35	172.45	167.85	164.85	163.45	160.55
	Bearing Walls	205.50	193.10	182.45	174.45	168.85	163.45	160.00	158.40	155.05
Perimeter Adj., Add or Deduct	Per 100 L.F.	8.95	6.70	4.80	3.45	2.70	2.05	1.60	1.30	0.85
Story Hgt. Adj., Add or Deduct	Per 1 Ft.	2.20	1.90	1.65	1.35	1.15	0.95	0.90	0.80	0.60
For Basement, add $33.55 per square foot of basement area										

The above costs were calculated using the basic specifications shown on the facing page. These costs should be adjusted where necessary for design alternatives and owner's requirements. Reported completed project costs, for this type of structure, range from $114.20 to $268.40 per S.F.

Common additives

Description	Unit	$ Cost
Carrels Hardwood	Each	660 - 990
Clock System		
20 Room	Each	16,000
50 Room	Each	39,100
Elevators, Hydraulic passenger, 2 stops		
1500# capacity	Each	62,800
2500# capacity	Each	66,300
3500# capacity	Each	69,800
Additional stop, add	Each	7825
Emergency Lighting, 25 watt, battery operated		
Lead battery	Each	282
Nickel cadmium	Each	805
Flagpoles, Complete		
Aluminum, 20' high	Each	1650
40' High	Each	3475
70' High	Each	10,100
Fiberglass, 23' High	Each	1775
39'-5" High	Each	3325
59' High	Each	8225

Description	Unit	$ Cost
Lockers, Steel, single tier, 60" or 72"	Opening	191 - 310
2 tier, 60" or 72" total	Opening	107 - 141
5 tier, box lockers	Opening	65 - 83.50
Locker bench, lam. maple top only	L.F.	21
Pedestals, steel pipe	Each	63.50
Seating		
Auditorium chair, all veneer	Each	238
Veneer back, padded seat	Each	288
Upholstered, spring seat	Each	277
Classroom, movable chair & desk	Set	65 - 120
Lecture hall, pedestal type	Each	227 - 680
Smoke Detectors		
Ceiling type	Each	187
Duct type	Each	480
Sound System		
Amplifier, 250 watts	Each	2350
Speaker, ceiling or wall	Each	191
Trumpet	Each	365
TV Antenna, Master system, 12 outlet 0	Outlet	315
30 outlet	Outlet	203
100 outlet	Outlet	194

Model costs calculated for a 2 story building with 12' story height and 50,000 square feet of floor area

				Unit	Unit Cost	Cost Per S.F.	% Of Sub-Total
A.	**SUBSTRUCTURE**						
1010	Standard Foundations	Poured concrete; strip and spread footings		S.F. Ground	1.32	.66	
1020	Special Foundations	N/A		—	—	—	
1030	Slab on Grade	4" reinforced concrete with vapor barrier and granular base		S.F. Slab	4.74	2.37	3.7%
2010	Basement Excavation	Site preparation for slab and trench for foundation wall and footing		S.F. Ground	.25	.13	
2020	Basement Walls	4' Foundation wall		L.F. Wall	74	1.50	
B.	**SHELL**						
	B10 Superstructure						
1010	Floor Construction	Open web steel joists, slab form, concrete		S.F. Floor	18.72	9.36	11.9%
1020	Roof Construction	Metal deck on open web steel joists, columns		S.F. Roof	11.48	5.74	
	B20 Exterior Enclosure						
2010	Exterior Walls	Decorative concrete block	65% of wall	S.F. Wall	15.06	2.96	
2020	Exterior Windows	Window wall	35% of wall	Each	37.60	3.98	5.9%
2030	Exterior Doors	Double glass and aluminum with transom		Each	4850	.58	
	B30 Roofing						
3010	Roof Coverings	Built-up tar and gravel with flashing; perlite/EPS composite insulation		S.F. Roof	5.28	2.64	2.1%
3020	Roof Openings	N/A		—	—	—	
C.	**INTERIORS**						
1010	Partitions	Concrete block	20 S.F. Floor/L.F. Partition	S.F. Partition	13.66	6.83	
1020	Interior Doors	Single leaf hollow metal	200 S.F. Floor/Door	Each	875	4.38	
1030	Fittings	Chalkboards, counters, cabinets		S.F. Floor	4.01	4.01	
2010	Stair Construction	Concrete filled metal pan		Flight	18,950	3.79	26.6%
3010	Wall Finishes	95% paint, 5% ceramic tile		S.F. Surface	3.56	3.56	
3020	Floor Finishes	70% vinyl composition tile, 25% carpet, 5% ceramic tile		S.F. Floor	4.72	4.72	
3030	Ceiling Finishes	Mineral fiber tile on concealed zee bars		S.F. Ceiling	6.38	6.38	
D.	**SERVICES**						
	D10 Conveying						
1010	Elevators & Lifts	Two hydraulic passenger elevators		Each	81,000	3.24	2.6%
1020	Escalators & Moving Walks	N/A		—	—	—	
	D20 Plumbing						
2010	Plumbing Fixtures	Toilet and service fixtures, supply and drainage	1 Fixture/455 S.F. Floor	Each	6056	13.31	
2020	Domestic Water Distribution	Oil fired hot water heater		S.F. Floor	1.58	1.58	12.3%
2040	Rain Water Drainage	Roof drains		S.F. Roof	1.42	.71	
	D30 HVAC						
3010	Energy Supply	N/A		—	—	—	
3020	Heat Generating Systems	Included in D3050		—	—	—	
3030	Cooling Generating Systems	N/A		—	—	—	14.4 %
3050	Terminal & Package Units	Multizone unit, gas heating, electric cooling		S.F. Floor	18.20	18.20	
3090	Other HVAC Sys. & Equipment	N/A		—	—	—	
	D40 Fire Protection						
4010	Sprinklers	Sprinklers, light hazard		S.F. Floor	2.04	2.04	1.6%
4020	Standpipes	N/A		—	—	—	
	D50 Electrical						
5010	Electrical Service/Distribution	2000 ampere service, panel board and feeders		S.F. Floor	4.80	4.80	
5020	Lighting & Branch Wiring	Fluorescent fixtures, receptacles, switches, A.C. and misc. power		S.F. Floor	11.70	11.70	19.0%
5030	Communications & Security	Alarm systems, internet wiring, communications systems and emergency lighting		S.F. Floor	6.88	6.88	
5090	Other Electrical Systems	Emergency generator, 100 kW		S.F. Floor	.67	.67	
E.	**EQUIPMENT & FURNISHINGS**						
1010	Commercial Equipment	N/A		—	—	—	
1020	Institutional Equipment	N/A		—	—	—	
1030	Vehicular Equipment	N/A		—	—	—	0.0 %
1090	Other Equipment	N/A		—	—	—	
F.	**SPECIAL CONSTRUCTION**						
1020	Integrated Construction	N/A		—	—	—	
1040	Special Facilities	N/A		—	—	—	0.0 %
G.	**BUILDING SITEWORK**	**N/A**					

		Sub-Total	126.72	100%
CONTRACTOR FEES (General Requirements: 10%, Overhead: 5%, Profit: 10%)		25%	31.69	
ARCHITECT FEES		7%	11.09	
Total Building Cost			**169.50**	

Costs per square foot of floor area

Exterior Wall	S.F. Area	10000	15000	25000	40000	55000	70000	80000	90000	100000
	L.F. Perimeter	260	320	400	476	575	628	684	721	772
Face Brick with Concrete Block Back-up	R/Conc. Frame	209.60	192.80	176.55	165.35	161.10	157.45	156.20	154.90	154.10
	Steel Frame	225.30	208.60	192.30	181.10	176.80	173.20	171.95	170.65	169.85
Decorative Concrete Block	R/Conc. Frame	194.80	180.85	167.80	159.15	155.70	153.00	152.00	151.05	150.40
	Steel Frame	209.50	195.60	182.55	173.90	170.45	167.75	166.75	165.80	165.15
Precast Concrete Panels	R/Conc. Frame	208.95	192.30	176.20	165.15	160.90	157.35	156.15	154.85	154.05
	Steel Frame	224.45	207.85	191.70	180.75	176.45	172.90	171.70	170.40	169.60
Perimeter Adj., Add or Deduct	Per 100 L.F.	17.90	11.95	7.15	4.55	3.20	2.50	2.25	2.00	1.80
Story Hgt. Adj., Add or Deduct	Per 1 Ft.	3.45	2.80	2.10	1.65	1.35	1.15	1.10	1.10	1.05
For Basement, add $32.65 per square foot of basement area										

The above costs were calculated using the basic specifications shown on the facing page. These costs should be adjusted where necessary for design alternatives and owner's requirements. Reported completed project costs, for this type of structure, range from $73.75 to $217.40 per S.F.

Common additives

Description	Unit	$ Cost
Carrels Hardwood	Each	660 - 990
Closed Circuit Surveillance, One station		
Camera and monitor	Each	1850
For additional camera stations, add	Each	1000
Elevators, Hydraulic passenger, 2 stops		
2000# capacity	Each	64,800
2500# capacity	Each	66,300
3500# capacity	Each	69,800
Additional stop, add	Each	7825
Emergency Lighting, 25 watt, battery operated		
Lead battery	Each	282
Nickel cadmium	Each	805
Furniture	Student	1900 - 3600
Intercom System, 25 station capacity		
Master station	Each	2650
Intercom outlets	Each	169
Handset	Each	470

Description	Unit	$ Cost
Kitchen Equipment		
Broiler	Each	4025
Coffee urn, twin 6 gallon	Each	3075
Cooler, 6 ft. long	Each	4925
Dishwasher, 10-12 racks per hr.	Each	4950
Food warmer	Each	735
Freezer, 44 C.F., reach-in	Each	3725
Ice cube maker, 50 lb. per day	Each	1750
Range with 1 oven	Each	2700
Laundry Equipment		
Dryer, gas, 16 lb. capacity	Each	885
30 lb. capacity	Each	3600
Washer, 4 cycle	Each	1075
Commercial	Each	1450
Smoke Detectors		
Ceiling type	Each	187
Duct type	Each	480
TV Antenna, Master system, 12 outlet	Outlet	315
30 outlet	Outlet	203
100 outlet	Outlet	194

Important: See the Reference Section for Location Factors

Model costs calculated for a 3 story building with 12' story height and 25,000 square feet of floor area

College, Dormitory, 2-3 Story

					Unit	Unit Cost	Cost Per S.F.	% Of Sub-Total
A. SUBSTRUCTURE								
1010	Standard Foundations	Poured concrete; strip and spread footings			S.F. Ground	6.03	2.01	
1020	Special Foundations	N/A			—	—	—	
1030	Slab on Grade	4" reinforced concrete with vapor barrier and granular base			S.F. Slab	4.74	1.58	3.7%
2010	Basement Excavation	Site preparation for slab and trench for foundation wall and footing			S.F. Ground	.17	.06	
2020	Basement Walls	4' Foundation wall			L.F. Wall	78	1.25	
B. SHELL								
	B10 Superstructure							
1010	Floor Construction	Concrete flat plate			S.F. Floor	24.84	16.56	16.0%
1020	Roof Construction	Concrete flat plate			S.F. Roof	13.50	4.50	
	B20 Exterior Enclosure							
2010	Exterior Walls	Face brick with concrete block backup	80% of wall		S.F. Wall	30.86	14.22	
2020	Exterior Windows	Aluminum horizontal sliding	20% of wall		Each	696	3.49	14.4%
2030	Exterior Doors	Double glass & aluminum doors			Each	5175	1.24	
	B30 Roofing							
3010	Roof Coverings	Single ply membrane, loose laid and ballasted; perlite/EPS composite insulation			S.F. Roof	5.07	1.69	1.3%
3020	Roof Openings	N/A			—	—	—	
C. INTERIORS								
1010	Partitions	Gypsum board on metal studs, concrete block	9 S.F. Floor/L.F. Partition		S.F. Partition	6.13	6.81	
1020	Interior Doors	Single leaf wood	90 S.F. Floor/Door		Each	565	6.28	
1030	Fittings	Closet shelving, mirrors, bathroom accessories			S.F. Floor	2.29	2.29	
2010	Stair Construction	Cast in place concrete			Flight	5075	2.64	23.8%
3010	Wall Finishes	95% paint, 5% ceramic tile			S.F. Surface	1.66	3.69	
3020	Floor Finishes	80% carpet, 10% vinyl composition tile, 10% ceramic tile			S.F. Floor	8.77	8.77	
3030	Ceiling Finishes	90% paint, 10% suspended fiberglass board			S.F. Ceiling	.92	.92	
D. SERVICES								
	D10 Conveying							
1010	Elevators & Lifts	One hydraulic passenger elevator			Each	111,500	4.46	3.4%
1020	Escalators & Moving Walks	N/A			—	—	—	
	D20 Plumbing							
2010	Plumbing Fixtures	Toilet and service fixtures, supply and drainage	1 Fixture/455 S.F. Floor		Each	4118	9.05	
2020	Domestic Water Distribution	Electric water heater			S.F. Floor	3.56	3.56	9.8%
2040	Rain Water Drainage	Roof drains			S.F. Roof	1.11	.37	
	D30 HVAC							
3010	Energy Supply	N/A			—	—	—	
3020	Heat Generating Systems	Included in D3050			—	—	—	
3030	Cooling Generating Systems	N/A			—	—	—	8.6 %
3050	Terminal & Package Units	Rooftop multizone unit system			S.F. Floor	11.40	11.40	
3090	Other HVAC Sys. & Equipment	N/A			—	—	—	
	D40 Fire Protection							
4010	Sprinklers	Wet pipe sprinkler system			S.F. Floor	2.35	2.35	1.8%
4020	Standpipes	N/A			—	—	—	
	D50 Electrical							
5010	Electrical Service/Distribution	800 ampere service, panel board and feeders			S.F. Floor	2.61	2.61	
5020	Lighting & Branch Wiring	Fluorescent fixtures, receptacles, switches, A.C. and misc. power			S.F. Floor	9.01	9.01	14.5%
5030	Communications & Security	Alarm systems, internet and phone wiring, communications systems and emergency lighting			S.F. Floor	7.46	7.46	
5090	Other Electrical Systems	Emergency generator, 7.5 kW			S.F. Floor	.07	.07	
E. EQUIPMENT & FURNISHINGS								
1010	Commercial Equipment	N/A			—	—	—	
1020	Institutional Equipment	N/A			—	—	—	
1030	Vehicular Equipment	N/A			—	—	—	2.8 %
2020	Other Equipment	Dormitory furniture			S.F. Floor	3.65	3.65	
F. SPECIAL CONSTRUCTION								
1020	Integrated Construction	N/A			—	—	—	0.0 %
1040	Special Facilities	N/A			—	—	—	
G. BUILDING SITEWORK	**N/A**							

			Sub-Total	131.99	100%
CONTRACTOR FEES (General Requirements: 10%, Overhead: 5%, Profit: 10%)			25%	33.01	
ARCHITECT FEES			7%	11.55	
		Total Building Cost		**176.55**	

Costs per square foot of floor area

Exterior Wall	S.F. Area	20000	35000	45000	65000	85000	110000	135000	160000	200000
	L.F. Perimeter	260	340	400	440	500	540	560	590	640
Face Brick with Concrete Block Back-up	R/Conc. Frame	206.20	188.70	184.05	174.75	170.70	166.50	163.30	161.45	159.35
	Steel Frame	223.15	205.65	201.05	191.75	187.70	183.50	180.30	178.45	176.35
Decorative Concrete Block	R/Conc. Frame	192.05	178.10	174.40	167.40	164.30	161.15	158.80	157.40	155.90
	Steel Frame	209.00	195.10	191.35	184.35	181.25	178.10	175.80	174.40	172.90
Precast Concrete Panels With Exposed Aggregate	R/Conc. Frame	205.30	188.05	183.45	174.25	170.25	166.15	163.05	161.20	159.20
	Steel Frame	222.15	204.90	200.35	191.20	187.20	183.10	179.95	178.10	176.10
Perimeter Adj., Add or Deduct	Per 100 L.F.	17.40	9.95	7.75	5.35	4.05	3.15	2.65	2.15	1.80
Story Hgt. Adj., Add or Deduct	Per 1 Ft.	3.50	2.60	2.40	1.80	1.60	1.35	1.15	1.00	0.85

For Basement, add $32.65 per square foot of basement area

The above costs were calculated using the basic specifications shown on the facing page. These costs should be adjusted where necessary for design alternatives and owner's requirements. Reported completed project costs, for this type of structure, range from $104.85 to $252.85 per S.F.

Common additives

Description	Unit	$ Cost	Description	Unit	$ Cost
Carrels Hardwood	Each	660 - 990	Kitchen Equipment		
Closed Circuit Surveillance, One station			Broiler	Each	4025
Camera and monitor	Each	1850	Coffee urn, twin, 6 gallon	Each	3075
For additional camera stations, add	Each	1000	Cooler, 6 ft. long	Each	4925
Elevators, Electric passenger, 5 stops			Dishwasher, 10-12 racks per hr.	Each	4950
2000# capacity	Each	158,700	Food warmer	Each	735
2500# capacity	Each	162,700	Freezer, 44 C.F., reach-in	Each	3725
3500# capacity	Each	163,700	Ice cube maker, 50 lb. per day	Each	1750
Additional stop, add	Each	13,600	Range with 1 oven	Each	2700
Emergency Lighting, 25 watt, battery operated			Laundry Equipment		
Lead battery	Each	282	Dryer, gas, 16 lb. capacity	Each	885
Nickel cadmium	Each	805	30 lb. capacity	Each	3600
Furniture	Student	1900 - 3600	Washer, 4 cycle	Each	1075
Intercom System, 25 station capacity			Commercial	Each	1450
Master station	Each	2650	Smoke Detectors		
Intercom outlets	Each	169	Ceiling type	Each	187
Handset	Each	470	Duct type	Each	480
			TV Antenna, Master system, 12 outlet	Outlet	315
			30 outlet	Outlet	203
			100 outlet	Outlet	194

Model costs calculated for a 6 story building with 12' story height and 85,000 square feet of floor area

				Unit	Unit Cost	Cost Per S.F.	% Of Sub-Total
A.	**SUBSTRUCTURE**						
1010	Standard Foundations	Poured concrete; strip and spread footings		S.F. Ground	9.06	1.51	
1020	Special Foundations	N/A		—	—	—	
1030	Slab on Grade	4" reinforced concrete with vapor barrier and granular base		S.F. Slab	4.74	.79	2.2%
2010	Basement Excavation	Site preparation for slab and trench for foundation wall and footing		S.F. Ground	.25	.04	
2020	Basement Walls	4' foundation wall		L.F. Wall	74	.64	
B.	**SHELL**						
	B10 Superstructure						
1010	Floor Construction	Concrete slab with metal deck and beams		S.F. Floor	31.60	26.33	22.3%
1020	Roof Construction	Concrete slab with metal deck and beams		S.F. Roof	23.40	3.90	
	B20 Exterior Enclosure						
2010	Exterior Walls	Decorative concrete block	80% of wall	S.F. Wall	16.70	5.66	
2020	Exterior Windows	Aluminum horizontal sliding	20% of wall	Each	492	2.78	6.4%
2030	Exterior Doors	Double glass & aluminum doors		Each	3214	.30	
	B30 Roofing						
3010	Roof Coverings	Single ply membrane, loose laid and ballasted; perlite/EPS composite insulation		S.F. Roof	4.74	.79	0.6%
3020	Roof Openings	N/A		—	—	—	
C.	**INTERIORS**						
1010	Partitions	Concrete block	9 S.F. Floor/L.F. Partition	S.F. Partition	9.38	10.42	
1020	Interior Doors	Single leaf wood	90 S.F. Floor/Door	Each	565	6.28	
1030	Fittings	Closet shelving, mirrors, bathroom accessories		S.F. Floor	1.87	1.87	
2010	Stair Construction	Concrete filled metal pan		Flight	18,950	4.02	27.5%
3010	Wall Finishes	95% paint, 5% ceramic tile		S.F. Surface	2.25	.5	
3020	Floor Finishes	80% carpet, 10% vinyl composition tile, 10% ceramic tile		S.F. Floor	8.77	8.77	
3030	Ceiling Finishes	Mineral fiber tile on concealed zee bars, paint		S.F. Ceiling	.92	.92	
D.	**SERVICES**						
	D10 Conveying						
1010	Elevators & Lifts	Four geared passenger elevators		Each	234,388	11.03	8.1%
1020	Escalators & Moving Walks	N/A		—	—	—	
	D20 Plumbing						
2010	Plumbing Fixtures	Toilet and service fixtures, supply and drainage	1 Fixture/390 S.F. Floor	Each	3338	8.56	
2020	Domestic Water Distribution	Oil fired water heater		S.F. Floor	.89	.89	7.2%
2040	Rain Water Drainage	Roof drains		S.F. Roof	1.80	.30	
	D30 HVAC						
3010	Energy Supply	Oil fired hot water, wall fin radiation		S.F. Floor	3.77	3.77	
3020	Heat Generating Systems	N/A		—	—	—	
3030	Cooling Generating Systems	Chilled water, air cooled condenser system		S.F. Floor	7.93	7.93	8.6%
3050	Terminal & Package Units	N/A		—	—	—	
3090	Other HVAC Sys. & Equipment	N/A		—	—	—	
	D40 Fire Protection						
4010	Sprinklers	Sprinklers, light hazard		S.F. Floor	2.25	2.25	1.7%
4020	Standpipes	N/A		—	—	—	
	D50 Electrical						
5010	Electrical Service/Distribution	1200 ampere service, panel board and feeders		S.F. Floor	1.21	1.21	
5020	Lighting & Branch Wiring	Fluorescent fixtures, receptacles, switches, A.C. and misc. power		S.F. Floor	8.88	8.88	13.1%
5030	Communications & Security	Alarm systems, internet and phone wiring, communications systems and emergency lighting		S.F. Floor	7.49	7.49	
5090	Other Electrical Systems	Emergency generator, 30 kW		S.F. Floor	.20	.20	
E.	**EQUIPMENT & FURNISHINGS**						
1010	Commercial Equipment	N/A		—	—	—	
1020	Institutional Equipment	N/A		—	—	—	
1030	Vehicular Equipment	N/A		—	—	—	2.2 %
2020	Other Equipment	Built-in dormitory furnishings		S.F. Floor	2.99	2.99	
F.	**SPECIAL CONSTRUCTION**						
1020	Integrated Construction	N/A		—	—	—	0.0 %
1040	Special Facilities	N/A		—	—	—	
G.	**BUILDING SITEWORK**	**N/A**					

			Sub-Total	135.52	**100%**
CONTRACTOR FEES (General Requirements: 10%, Overhead: 5%, Profit: 10%)			25%	33.87	
ARCHITECT FEES			7%	11.86	
		Total Building Cost		**181.25**	

Costs per square foot of floor area

Exterior Wall	S.F. Area	12000	20000	28000	37000	45000	57000	68000	80000	92000
	L.F. Perimeter	470	600	698	793	900	1060	1127	1200	1320
Face Brick with Concrete Brick Back-up	Steel Frame	269.45	221.50	199.45	185.55	178.55	171.60	165.95	161.65	159.00
	Bearing Walls	263.20	215.25	193.30	179.35	172.35	165.40	159.75	155.40	152.80
Decorative Concrete Block	Steel Frame	262.55	216.20	195.10	181.80	175.00	168.35	163.05	158.95	156.50
	Bearing Walls	256.55	210.25	189.10	175.85	169.05	162.40	157.10	153.00	150.45
Stucco on Concrete Block	Steel Frame	260.60	214.75	193.85	180.75	174.05	167.45	162.30	158.25	155.80
	Bearing Walls	254.65	208.75	187.90	174.75	168.05	161.50	156.30	152.30	149.80
Perimeter Adj., Add or Deduct	Per 100 L.F.	10.40	6.25	4.50	3.40	2.80	2.25	1.85	1.50	1.35
Story Hgt. Adj., Add or Deduct	Per 1 Ft.	1.80	1.45	1.20	1.05	0.95	0.90	0.80	0.65	0.65
For Basement, add $21.10 per square foot of basement area										

The above costs were calculated using the basic specifications shown on the facing page. These costs should be adjusted where necessary for design alternatives and owner's requirements. Reported completed project costs, for this type of structure, range from $151.00 to $282.00 per S.F.

Common additives

Description	Unit	$ Cost
Cabinets, Base, door units, metal	L.F.	256
Drawer units	L.F.	505
Tall storage cabinets, open	L.F.	480
With doors	L.F.	565
Wall, metal 12-1/2" deep, open	L.F.	192
With doors	L.F.	345
Carrels Hardwood	Each	660 - 990
Countertops, not incl. base cabinets, acid proof	S.F.	46.50 - 59.50
Stainless steel	S.F.	117
Fume Hood, Not incl. ductwork	L.F.	755 - 2575
Ductwork	Hood	5075 - 8375
Glassware Washer, Distilled water rinse	Each	6625 - 14,600
Seating		
Auditorium chair, all veneer	Each	238
Veneer back, padded seat	Each	288
Upholstered, spring seat	Each	277
Classroom, movable chair & desk	Set	65 - 120
Lecture hall, pedestal type	Each	227 - 680

Description	Unit	$ Cost
Safety Equipment, Eye wash, hand held	Each	445
Deluge shower	Each	800
Sink, One piece plastic		
Flask wash, freestanding	Each	2350
Tables, acid resist. top, drawers	L.F.	194
Titration Unit, Four 2000 ml reservoirs	Each	6050

Important: See the Reference Section for Location Factors

Model costs calculated for a 1 story building with 12' story height and 45,000 square feet of floor area

College, Laboratory

				Unit	Unit Cost	Cost Per S.F.	% Of Sub-Total
A. SUBSTRUCTURE							
1010	Standard Foundations	Poured concrete; strip and spread footings		S.F. Ground	3.16	3.16	
1020	Special Foundations	N/A		—	—	—	
1030	Slab on Grade	4" reinforced concrete with vapor barrier and granular base		S.F. Slab	4.74	4.74	11.3%
2010	Basement Excavation	Site preparation for slab and trench for foundation wall and footing		S.F. Ground	.17	.17	
2020	Basement Walls	4' foundation wall		L.F. Wall	74	6.06	
B. SHELL							
B10 Superstructure							
1010	Floor Construction	Metal deck on open web steel joists	(5680 S.F.)	S.F. Floor	18.69	2.43	6.7%
1020	Roof Construction	Metal deck on open web steel joists		S.F. Roof	5.93	5.93	
B20 Exterior Enclosure							
2010	Exterior Walls	Face brick with concrete block backup	75% of wall	S.F. Wall	30.83	5.55	
2020	Exterior Windows	Window wall	25% of wall	Each	37.60	2.26	7.7%
2030	Exterior Doors	Glass and metal doors and entrances with transom		Each	4222	1.87	
B30 Roofing							
3010	Roof Coverings	Built-up tar and gravel with flashing; perlite/EPS composite insulation		S.F. Roof	5.08	5.08	4.3%
3020	Roof Openings	Skylight		S.F. Roof	.32	.32	
C. INTERIORS							
1010	Partitions	Concrete block	10 S.F. Floor/L.F. Partition	S.F. Partition	10.90	10.90	
1020	Interior Doors	Single leaf-kalamein fire doors	820 S.F. Floor/Door	Each	945	1.15	
1030	Fittings	Lockers		S.F. Floor	.04	.04	
2010	Stair Construction	N/A		—	—	—	23.3%
3010	Wall Finishes	60% paint, 40% epoxy coating		S.F. Surface	2.67	5.33	
3020	Floor Finishes	60% epoxy, 20% carpet, 20% vinyl composition tile		S.F. Floor	5.45	5.45	
3030	Ceiling Finishes	Mineral fiber tile on concealed zee runners		S.F. Ceiling	6.38	6.38	
D. SERVICES							
D10 Conveying							
1010	Elevators & Lifts	N/A		—	—	—	0.0 %
1020	Escalators & Moving Walks	N/A		—	—	—	
D20 Plumbing							
2010	Plumbing Fixtures	Toilet and service fixtures, supply and drainage	1 Fixture/260 S.F. Floor	Each	4987	19.18	
2020	Domestic Water Distribution	Gas fired water heater		S.F. Floor	1.35	1.35	17.1%
2040	Rain Water Drainage	Roof drains		S.F. Roof	.85	.85	
D30 HVAC							
3010	Energy Supply	N/A		—	—	—	
3020	Heat Generating Systems	Included in D3050		—	—	—	
3030	Cooling Generating Systems	N/A		—	—	—	14.5 %
3050	Terminal & Package Units	Multizone unit, gas heating, electric cooling		S.F. Floor	18.20	18.20	
3090	Other HVAC Sys. & Equipment	N/A		—	—	—	
D40 Fire Protection							
4010	Sprinklers	Sprinklers, light hazard		S.F. Floor	2.33	2.33	1.9%
4020	Standpipes	N/A		—	—	—	
D50 Electrical							
5010	Electrical Service/Distribution	1000 ampere service, panel board and feeders		S.F. Floor	2.02	2.02	
5020	Lighting & Branch Wiring	Fluorescent fixtures, receptacles, switches, A.C. and misc. power		S.F. Floor	10.02	10.02	12.1%
5030	Communications & Security	Alarm systems, internet wiring, and emergency lighting		S.F. Floor	2.91	2.91	
5090	Other Electrical Systems	Emergency generator, 11.5 kW, Uninterruptible power supply		S.F. Floor	.20	.20	
E. EQUIPMENT & FURNISHINGS							
1010	Commercial Equipment	N/A		—	—	—	
1020	Institutional Equipment	Cabinets, fume hoods, lockers, glassware washer		S.F. Floor	1.44	1.44	1.1 %
1030	Vehicular Equipment	N/A		—	—	—	
1090	Other Equipment	N/A		—	—	—	
F. SPECIAL CONSTRUCTION							
1020	Integrated Construction	N/A		—	—	—	0.0 %
1040	Special Facilities	N/A		—	—	—	
G. BUILDING SITEWORK	**N/A**						

			Sub-Total	125.32	**100%**
CONTRACTOR FEES (General Requirements: 10%, Overhead: 5%, Profit: 10%)			25%	31.36	
ARCHITECT FEES			10%	15.67	
		Total Building Cost		**172.35**	

Costs per square foot of floor area

Exterior Wall	S.F. Area	15000	20000	25000	30000	35000	40000	45000	50000	55000
	L.F. Perimeter	354	425	457	513	568	583	629	644	683
Brick Face with Concrete Block Back-up	Steel Frame	175.50	170.30	165.10	162.65	160.85	158.20	157.05	155.20	154.45
	R/Conc. Frame	162.30	157.10	151.85	149.40	147.65	145.05	143.80	142.05	141.20
Precast Concrete Panel	Steel Frame	176.95	171.65	166.25	163.70	161.95	159.15	157.95	156.10	155.25
	R/Conc. Frame	164.15	158.70	153.30	150.75	148.85	146.10	144.90	143.00	142.15
Limestone Face Concrete Block Back-up	Steel Frame	191.75	184.95	177.65	174.35	172.05	168.20	166.65	164.05	162.95
	R/Conc. Frame	178.55	171.70	164.40	161.15	158.80	155.00	153.45	150.85	149.70
Perimeter Adj., Add or Deduct	Per 100 L.F.	8.95	6.70	5.35	4.40	3.80	3.30	3.00	2.75	2.45
Story Hgt. Adj., Add or Deduct	Per 1 Ft.	2.25	2.00	1.65	1.60	1.55	1.35	1.30	1.25	1.15
For Basement, add $35.00 per square foot of basement area										

The above costs were calculated using the basic specifications shown on the facing page. These costs should be adjusted where necessary for design alternatives and owner's requirements. Reported completed project costs, for this type of structure, range from $123.70 to $252.40 per S.F.

Common additives

Description	Unit	$ Cost
Carrels Hardwood	Each	660 - 990
Elevators, Hydraulic passenger, 2 stops		
2000# capacity	Each	64,800
2500# capacity	Each	66,300
3500# capacity	Each	69,800
Emergency Lighting, 25 watt, battery operated		
Lead battery	Each	282
Nickel cadmium	Each	805
Escalators, Metal		
32" wide, 10' story height	Each	143,000
20' story height	Each	172,000
48" wide, 10' Story height	Each	152,000
20' story height	Each	180,500
Glass		
32" wide, 10' story height	Each	137,000
20' story height	Each	165,000
48" wide, 10' story height	Each	48,300
20' story height	Each	175,000

Description	Unit	$ Cost
Lockers, Steel, Single tier, 60" or 72"	Opening	191 - 310
2 tier, 60" or 72" total	Opening	107 - 141
5 tier, box lockers	Opening	65 - 83.50
Locker bench, lam. maple top only	L.F.	21
Pedestals, steel pipe	Each	63.50
Sound System		
Amplifier, 250 watts	Each	2350
Speaker, ceiling or wall	Each	191
Trumpet	Each	365

Model costs calculated for a 2 story building with 12' story height and 25,000 square feet of floor area

College, Student

				Unit	Unit Cost	Cost Per S.F.	% Sub-Total
A. SUBSTRUCTURE							
1010	Standard Foundations	Poured concrete; strip and spread footings		S.F. Ground	4.54	2.27	
1020	Special Foundations	N/A		—	—	—	
1030	Slab on Grade	4" reinforced concrete with vapor barrier and granular base		S.F. Slab	4.74	2.37	5.4%
2010	Basement Excavation	Site preparation for slab and trench for foundation wall and footing		S.F. Ground	.17	.09	
2020	Basement Walls	4' foundation wall		L.F. Wall	78	1.43	
B. SHELL							
B10 Superstructure							
1010	Floor Construction	Concrete flat plate		S.F. Floor	27	13.50	18.6%
1020	Roof Construction	Concrete flat plate		S.F. Roof	15.22	7.61	
B20 Exterior Enclosure							
2010	Exterior Walls	Face brick with concrete block backup	75% of wall	S.F. Wall	30.85	10.15	
2020	Exterior Windows	Window wall	25% of wall	Each	37.60	4.14	12.9%
2030	Exterior Doors	Double aluminum and glass		Each	2385	.39	
B30 Roofing							
3010	Roof Coverings	Built-up tar and gravel with flashing; perlite/EPS composite insulation		S.F. Roof	5.74	2.87	2.6%
3020	Roof Openings	Roof hatches		S.F. Roof	.10	.05	
C. INTERIORS							
1010	Partitions	Gypsum board on metal studs	14 S.F. Floor/L.F. Partition	S.F. Partition	5.81	4.15	
1020	Interior Doors	Single leaf hollow metal	140 S.F. Floor/Door	Each	875	6.25	
1030	Fittings	N/A		—	—	—	
2010	Stair Construction	Cast in place concrete		Flight	7725	1.24	22.5%
3010	Wall Finishes	50% paint, 50% vinyl wall covering		S.F. Surface	2.10	3	
3020	Floor Finishes	50% carpet, 50% vinyl composition tile		S.F. Floor	6.57	6.57	
3030	Ceiling Finishes	Suspended fiberglass board		S.F. Ceiling	4.30	4.30	
D. SERVICES							
D10 Conveying							
1010	Elevators & Lifts	One hydraulic passenger elevator		Each	81,000	3.24	2.9%
1020	Escalators & Moving Walks	N/A		—	—	—	
D20 Plumbing							
2010	Plumbing Fixtures	Toilet and service fixtures, supply and drainage	1 Fixture/1040 S.F. Floor	Each	2153	2.07	
2020	Domestic Water Distribution	Gas fired water heater		S.F. Floor	.66	.66	2.7%
2040	Rain Water Drainage	Roof drains		S.F. Roof	.56	.28	
D30 HVAC							
3010	Energy Supply	N/A		—	—	—	
3020	Heat Generating Systems	Included in D3050		—	—	—	
3030	Cooling Generating Systems	N/A		—	—	—	16.0 %
3050	Terminal & Package Units	Multizone unit, gas heating, electric cooling		S.F. Floor	18.20	18.20	
3090	Other HVAC Sys. & Equipment	N/A		—	—	—	
D40 Fire Protection							
4010	Sprinklers	Wet pipe sprinkler system		S.F. Floor	2.46	2.46	2.2%
4020	Standpipes	N/A		—	—	—	
D50 Electrical							
5010	Electrical Service/Distribution	600 ampere service, panel board and feeders		S.F. Floor	2.26	2.26	
5020	Lighting & Branch Wiring	Fluorescent fixtures, receptacles, switches, A.C. and misc. power		S.F. Floor	11.60	11.60	14.3%
5030	Communications & Security	Alarm systems, internet wiring, communications systems and emergency lighting		S.F. Floor	2.24	2.24	
5090	Other Electrical Systems	Emergency generator, 11.5 kW		S.F. Floor	.15	.15	
E. EQUIPMENT & FURNISHINGS							
1010	Commercial Equipment	N/A		—	—	—	
1020	Institutional Equipment	N/A		—	—	—	
1030	Vehicular Equipment	N/A		—	—	—	0.0 %
1090	Other Equipment	N/A		—	—	—	
F. SPECIAL CONSTRUCTION							
1020	Integrated Construction	N/A		—	—	—	0.0 %
1040	Special Facilities	N/A		—	—	—	
G. BUILDING SITEWORK	**N/A**						

		Sub-Total	113.54	100%
CONTRACTOR FEES (General Requirements: 10%, Overhead: 5%, Profit: 10%)		25%	28.37	
ARCHITECT FEES		7%	9.94	

Total Building Cost	**151.85**

Costs per square foot of floor area

Exterior Wall	S.F. Area	4000	6000	8000	10000	12000	14000	16000	18000	20000
	L.F. Perimeter	260	340	420	453	460	510	560	610	600
Face Brick with Concrete Block Back-up	Bearing Walls	163.75	155.30	151.05	144.55	138.20	136.35	135.05	133.95	130.50
	Steel Frame	160.40	153.00	149.35	143.75	138.35	136.75	135.60	134.65	131.70
Decorative Concrete Block	Bearing Walls	144.00	137.40	134.10	129.05	124.20	122.75	121.70	120.85	118.25
	Steel Frame	147.00	141.30	138.50	134.35	130.35	129.20	128.35	127.60	125.50
Tilt Up Concrete Wall Panels	Bearing Walls	148.50	141.95	138.75	133.90	129.15	127.80	126.75	125.95	123.40
	Steel Frame	145.10	139.65	137.00	133.05	129.30	128.20	127.30	126.65	124.65
Perimeter Adj., Add or Deduct	Per 100 L.F.	21.80	14.50	10.90	8.65	7.25	6.25	5.40	4.80	4.35
Story Hgt. Adj., Add or Deduct	Per 1 Ft.	2.75	2.40	2.25	1.85	1.60	1.55	1.35	1.40	1.20
For Basement, add $30.65 per square foot of basement area										

The above costs were calculated using the basic specifications shown on the facing page. These costs should be adjusted where necessary for design alternatives and owner's requirements. Reported completed project costs, for this type of structure, range from $75.50 to $238.50 per S.F.

Common additives

Description	Unit	$ Cost
Bar, Front bar	L.F.	360
Back bar	L.F.	289
Booth, Upholstered, custom straight	L.F.	202 - 375
"L" or "U" shaped	L.F.	210 - 355
Bowling Alleys, incl. alley, pinsetter		
Scorer, counter & misc. supplies, average	Lane	59,000
For automatic scorer, add	Lane	10,900
Emergency Lighting, 25 watt, battery operated		
Lead battery	Each	282
Nickel cadmium	Each	805
Kitchen Equipment		
Broiler	Each	4025
Coffee urn, twin 6 gallon	Each	3075
Cooler, 6 ft. long	Each	4925
Dishwasher, 10-12 racks per hr.	Each	4950
Food warmer	Each	735
Freezer, 44 C.F., reach-in	Each	3725
Ice cube maker, 50 lb. per day	Each	1750
Range with 1 oven	Each	2700

Description	Unit	$ Cost
Movie Equipment		
Projector, 35mm	Each	12,200 - 16,700
Screen, wall or ceiling hung	S.F.	8.45 - 12.85
Partitions, Folding leaf, wood		
Acoustic type	S.F.	70.50 - 117
Seating		
Auditorium chair, all veneer	Each	238
Veneer back, padded seat	Each	288
Upholstered, spring seat	Each	277
Classroom, movable chair & desk	Set	65 - 120
Lecture hall, pedestal type	Each	227 - 680
Sound System		
Amplifier, 250 watts	Each	2350
Speaker, ceiling or wall	Each	191
Trumpet	Each	365
Stage Curtains, Medium weight	S.F.	10.35 - 36.50
Curtain Track, Light duty	L.F.	77.50
Swimming Pools, Complete, gunite	S.F.	64 - 78.50

Model costs calculated for a 1 story building with 12' story height and 10,000 square feet of floor area

					Unit	Unit Cost	Cost Per S.F.	% Of Sub-Total

A. SUBSTRUCTURE

1010	Standard Foundations	Poured concrete; strip and spread footings			S.F. Ground	2.33	2.33	
1020	Special Foundations	N/A			—	—	—	
1030	Slab on Grade	4" reinforced concrete with vapor barrier and granular base			S.F. Slab	4.74	4.74	14.2%
2010	Basement Excavation	Site preparation for slab and trench for foundation wall and footing			S.F. Ground	.26	.26	
2020	Basement Walls	4' foundation wall			L.F. Wall	74	7.73	

B. SHELL

B10 Superstructure

| 1010 | Floor Construction | N/A | | | — | — | — | |
| 1020 | Roof Construction | Metal deck on open web steel joists | | | S.F. Roof | 7.96 | 7.96 | 7.5 % |

B20 Exterior Enclosure

2010	Exterior Walls	Face brick with concrete block backup	80% of wall		S.F. Wall	30.84	13.41	
2020	Exterior Windows	Aluminum sliding	20% of wall		Each	582	1.98	15.3%
2030	Exterior Doors	Double aluminum and glass and hollow metal			Each	2216	.88	

B30 Roofing

| 3010 | Roof Coverings | Built-up tar and gravel with flashing; perlite/EPS composite insulation | | | S.F. Roof | 6.57 | 6.57 | 6.3% |
| 3020 | Roof Openings | Roof hatches | | | S.F. Roof | .09 | .09 | |

C. INTERIORS

1010	Partitions	Gypsum board on metal studs	14 S.F. Floor/L.F. Partition		S.F. Partition	7.80	5.57	
1020	Interior Doors	Single leaf hollow metal	140 S.F. Floor/Door		Each	875	6.25	
1030	Fittings	Toilet partitions, directory board, mailboxes			S.F. Floor	2.42	2.42	
2010	Stair Construction	N/A			—	—	—	27.6%
3010	Wall Finishes	Paint			S.F. Surface	1.58	2.26	
3020	Floor Finishes	50% carpet, 50% vinyl tile			S.F. Floor	6.37	6.37	
3030	Ceiling Finishes	Mineral fiber tile on concealed zee bars			S.F. Ceiling	6.38	6.38	

D. SERVICES

D10 Conveying

| 1010 | Elevators & Lifts | N/A | | | — | — | — | 0.0 % |
| 1020 | Escalators & Moving Walks | N/A | | | — | — | — | |

D20 Plumbing

2010	Plumbing Fixtures	Kitchen, toilet and service fixtures, supply and drainage	1 Fixture/910 S.F. Floor		Each	2584	2.84	
2020	Domestic Water Distribution	Electric water heater			S.F. Floor	6.29	6.29	9.0%
2040	Rain Water Drainage	Roof drains			S.F. Roof	.42	.42	

D30 HVAC

3010	Energy Supply	N/A			—	—	—	
3020	Heat Generating Systems	Included in D3050			—	—	—	
3030	Cooling Generating Systems	N/A			—	—	—	9.0 %
3050	Terminal & Package Units	Single zone rooftop unit, gas heating, electric cooling			S.F. Floor	9.52	9.52	
3090	Other HVAC Sys. & Equipment	N/A			—	—	—	

D40 Fire Protection

| 4010 | Sprinklers | Wet pipe sprinkler system | | | S.F. Floor | 2.77 | 2.77 | 2.6% |
| 4020 | Standpipes | N/A | | | — | — | — | |

D50 Electrical

5010	Electrical Service/Distribution	200 ampere service, panel board and feeders			S.F. Floor	1	1	
5020	Lighting & Branch Wiring	Incandescent fixtures, receptacles, switches, A.C. and misc. power			S.F. Floor	4.76	4.76	5.9%
5030	Communications & Security	Alarm systems and emergency lighting			S.F. Floor	.37	.37	
5090	Other Electrical Systems	Emergency generator, 15 kW			S.F. Floor	.16	.16	

E. EQUIPMENT & FURNISHINGS

1010	Commercial Equipment	Freezer, chest type			S.F. Floor	.77	.77	
1020	Institutional Equipment	N/A			—	—	—	
1030	Vehicular Equipment	N/A			—	—	—	2.6%
1090	Other Equipment	Kitchen equipment, directory board, mailboxes, built-in coat racks			S.F. Floor	1.98	1.98	

F. SPECIAL CONSTRUCTION

| 1020 | Integrated Construction | N/A | | | — | — | — | 0.0 % |
| 1040 | Special Facilities | N/A | | | — | — | — | |

G. BUILDING SITEWORK   N/A

				Sub-Total	106.08	**100%**

CONTRACTOR FEES (General Requirements: 10%, Overhead: 5%, Profit: 10%)	25%	26.54
ARCHITECT FEES	9%	11.93

Total Building Cost	**144.55**

Costs per square foot of floor area

Exterior Wall	S.F. Area	10000	12500	15000	17500	20000	22500	25000	30000	40000
	L.F. Perimeter	400	450	500	550	600	625	650	700	800
Brick Veneer / Brick Veneer	Steel Frame	285.30	282.00	279.75	278.10	276.90	275.05	273.55	271.30	268.60
	Bearing Wall	297.20	292.75	289.70	287.55	285.90	283.45	281.45	278.50	274.75
Tilt Up Concrete	Steel Frame	277.60	274.30	272.05	270.45	269.25	267.35	265.90	263.65	260.95
Concrete Panels	Bearing Wall	300.10	295.35	292.15	289.90	288.20	285.50	283.40	280.25	276.35
EIFS / EIFS	Steel Frame	282.15	279.40	277.55	276.20	275.15	273.60	272.35	270.50	268.20
	Bearing Wall	281.85	279.00	277.05	275.55	274.55	272.85	271.55	269.60	267.25
Perimeter Adj., Add or Deduct	Per 100 L.F.	8.30	6.55	5.45	4.75	4.10	3.65	3.35	2.80	2.00
Story Hgt. Adj., Add or Deduct	Per 1 Ft.	1.50	1.30	1.20	1.15	1.15	1.00	1.00	0.90	0.70

For Basement, add $33.65 per square foot of basement area

The above costs were calculated using the basic specifications shown on the facing page. These costs should be adjusted where necessary for design alternatives and owner's requirements. Reported completed project costs, for this type of structure, range from $155.80 to $316.15 per S.F.

Common additives

Description	Unit	$ Cost
Clock System		
20 room	Ea.	16,000
50 room	Ea.	39,100
Closed Circuit Surveillance, one station		
Camera and monitor	Total	1850
For additional camera stations, add	Ea.	1000
For zoom lens - remote control, add	Ea.	2925 - 9950
For automatic iris for low light, add	Ea.	2600
Directory Boards, plastic, glass covered		
30" x 20"	Ea.	595
36" x 48"	Ea.	1450
Aluminum, 24" x 18"	Ea.	600
36" x 24"	Ea.	675
48" x 32"	Ea.	980
48" x 60"	Ea.	2025
Emergency Lighting, 25 watt, battery operated		
Lead battery	Ea.	282
Nickel cadmium	Ea.	805

Description	Unit	$ Cost
Smoke Detectors		
Ceiling type	Ea.	187
Duct type	Ea.	480
Sound System		
Amplifier, 250 watts	Ea.	2350
Speakers, ceiling or wall	Ea.	191
Trumpets	Ea.	365

Model costs calculated for a 1 story building with 16'-6" story height and 22,500 square feet of floor area

Computer Data Center

				Unit	Unit Cost	Cost Per S.F.	% Of Sub-Total
A.	**SUBSTRUCTURE**						
1010	Standard Foundations	Poured concrete; strip and spread footings		S.F. Ground	1.65	1.65	
1020	Special Foundations	N/A		—	—	—	
1030	Slab on Grade	4" concrete with vapor barrier and granular base		S.F. Slab	4.74	4.74	4.3%
2010	Basement Excavation	Site preparation for slab and trench for foundation wall and footing		S.F. Ground	.17	.19	
2020	Basement Walls	4' foundation wall		L.F. Wall	78	2.17	
B.	**SHELL**						
	B10 Superstructure						
1010	Floor Construction	N/A		—	—	—	6.0 %
1020	Roof Construction	Steel joist, girders & deck on columns		S.F. Roof	12.30	12.30	
	B20 Exterior Enclosure						
2010	Exterior Walls	Brick veneer on metal studs		S.F. Wall	24.64	10.73	
2020	Exterior Windows	Aluminum sliding	5% of Wall	Each	552	.85	6.0%
2030	Exterior Doors	Aluminum and glass		Each	3233	.86	
	B30 Roofing						
3010	Roof Coverings	Single ply membrane, loose laid and ballasted; polyisocyanurate insulation		S.F. Roof	6.81	6.81	3.4%
3020	Roof Openings	Roof hatch		S.F. Roof	.09	.09	
C.	**INTERIORS**						
1010	Partitions	Gypsum board on metal studs	13 S.F. Floor/L.F. Partition	S.F. Partition	4.60	4.25	
1020	Interior Doors	Solid core wood, metal frame	370 S.F. Floor/Door	Each	602	1.62	
1030	Fittings	Toilet partitions		S.F. Floor	.39	.39	
2010	Stair Construction	N/A		—	—	—	10.6%
3010	Wall Finishes	90% paint, 10% ceramic tile		S.F. Surface	1.86	3.44	
3020	Floor Finishes	65% carpet tile, 10% ceramic tile, 10% quarry tile		S.F. Floor	5.79	5.79	
3030	Ceiling Finishes	Mineral fiber tile on concealed zee bars		S.F. Ceiling	6.38	6.38	
D.	**SERVICES**						
	D10 Conveying						
1010	Elevators & Lifts	N/A		—	—	—	0.0 %
1020	Escalators & Moving Walks	N/A		—	—	—	
	D20 Plumbing						
2010	Plumbing Fixtures	Toilet and service fixtures, supply and drainage	1 fixture/1180 S.F. Floor	Each	3764	3.19	
2020	Domestic Water Distribution	Gas fired water heater		S.F. Floor	.58	.58	2.3%
2040	Rain Water Drainage	Roof drains		S.F. Floor	.88	.88	
	D30 HVAC						
3010	Energy Supply	Hot water reheat		S.F. Floor	5.88	5.88	
3020	Heat Generating Systems	Boiler, oil fired hot water		Each	32,325	6.02	
3030	Cooling Generating Systems	Cooling tower		S.F. Floor	8.34	8.34	32.9%
3050	Terminal & Package Units	N/A		—	—	—	
3090	Other HVAC Sys. & Equipment	Heat exchanger, ductwork, air handling units, VAV terminals		S.F. Floor	47.46	47.46	
	D40 Fire Protection						
4010	Sprinklers	85% wet pipe sprinkler system, 15% preaction sprinkler system		S.F. Floor	3.32	3.32	1.9%
4020	Standpipes	Standpipes and hose systems		S.F. Floor	.50	.50	
	D50 Electrical						
5010	Electrical Service/Distribution	1200 ampere service, panel board and feeders		S.F. Floor	4.53	4.53	
5020	Lighting & Branch Wiring	High efficiency fluorescent fixtures, receptacles, A.C. and misc. power		S.F. Floor	15.41	15.41	31.7%
5030	Communications & Security	Addressable alarm & telephone systems, internet wiring and emergency lighting		S.F. Floor	30.73	30.73	
5090	Other Electrical Systems	Emergency generator, 350 kW; UPS system		S.F. Floor	14.58	14.58	
E.	**EQUIPMENT & FURNISHINGS**						
1010	Commercial Equipment	N/A		—	—	—	
1020	Institutional Equipment	N/A		—	—	—	0.0 %
1030	Vehicular Equipment	N/A		—	—	—	
1090	Other Equipment	N/A		—	—	—	
F.	**SPECIAL CONSTRUCTION & DEMOLITION**						
1020	Integrated Construction	Pedestal access floor	15% of Floor		1.95	1.95	0.9%
1040	Special Facilities	N/A		—	—	—	
G.	**BUILDING SITEWORK**	**N/A**					

			Sub-Total	205.63	**100%**
CONTRACTOR FEES (General Requirements: 10%, Overhead: 5%, Profit: 10%)			25%	51.43	
ARCHITECT FEES			7%	17.99	

Total Building Cost	275.05	

Costs per square foot of floor area

Exterior Wall	S.F. Area	16000	23000	30000	37000	44000	51000	58000	65000	72000
	L.F. Perimeter	597	763	821	968	954	1066	1090	1132	1220
Limestone with Concrete Block Back-up	R/Conc. Frame	207.00	198.50	189.20	186.60	179.95	178.35	175.20	173.15	172.20
	Steel Frame	213.10	204.60	195.20	192.65	186.00	184.45	181.20	179.20	178.25
Face Brick with Concrete Block Back-up	R/Conc. Frame	188.65	182.15	175.70	173.75	169.25	168.05	165.90	164.60	163.85
	Steel Frame	194.70	188.20	181.75	179.75	175.30	174.15	171.95	170.60	169.90
Stone with Concrete Block Back-up	R/Conc. Frame	190.55	183.90	177.05	175.05	170.35	169.10	166.85	165.45	164.70
	Steel Frame	196.75	190.00	183.25	181.15	176.50	175.25	173.00	171.55	170.85
Perimeter Adj., Add or Deduct	Per 100 L.F.	8.40	5.85	4.45	3.60	3.05	2.65	2.30	2.10	1.90
Story Hgt. Adj., Add or Deduct	Per 1 Ft.	3.00	2.70	2.20	2.10	1.75	1.70	1.55	1.35	1.45
For Basement, add $28.95 per square foot of basement area										

The above costs were calculated using the basic specifications shown on the facing page. These costs should be adjusted where necessary for design alternatives and owner's requirements. Reported completed project costs, for this type of structure, range from $132.90 to $247.60 per S.F.

Common additives

Description	Unit	$ Cost	Description	Unit	$ Cost
Benches, Hardwood	L.F.	107 - 194	Flagpoles, Complete		
Clock System			Aluminum, 20' high	Each	1650
20 room	Each	16,000	40' high	Each	3475
50 room	Each	39,100	70' high	Each	10,100
Closed Circuit Surveillance, One station			Fiberglass, 23' high	Each	1775
Camera and monitor	Each	1850	39'-5" high	Each	3325
For additional camera stations, add	Each	1000	59' high	Each	8225
Directory Boards, Plastic, glass covered			Intercom System, 25 station capacity		
30" x 20"	Each	595	Master station	Each	2650
36" x 48"	Each	1450	Intercom outlets	Each	169
Aluminum, 24" x 18"	Each	600	Handset	Each	470
36" x 24"	Each	675	Safe, Office type, 1 hour rating		
48" x 32"	Each	980	30" x 18" x 18"	Each	2400
48" x 60"	Each	2025	60" x 36" x 18", double door	Each	8750
Emergency Lighting, 25 watt, battery operated			Smoke Detectors		
Lead battery	Each	282	Ceiling type	Each	187
Nickel cadmium	Each	805	Duct type	Each	480

Model costs calculated for a 1 story building with 14' story height and 30,000 square feet of floor area

				Unit	Unit Cost	Cost Per S.F.	% Of Sub-Total
A. SUBSTRUCTURE							
1010	Standard Foundations	Poured concrete; strip and spread footings		S.F. Ground	1.72	1.72	
1020	Special Foundations	N/A		—	—	—	
1030	Slab on Grade	4" reinforced concrete with vapor barrier and granular base		S.F. Slab	4.74	4.74	6.3%
2010	Basement Excavation	Site preparation for slab and trench for foundation wall and footing		S.F. Ground	.17	.17	
2020	Basement Walls	4' foundation wall		L.F. Wall	78	2.28	
B. SHELL							
	B10 Superstructure						
1010	Floor Construction	Cast-in-place columns		L.F. Column	69	1.74	13.9%
1020	Roof Construction	Cast-in-place concrete waffle slab		S.F. Roof	17.95	17.95	
	B20 Exterior Enclosure						
2010	Exterior Walls	Limestone panels with concrete block backup	75% of wall	S.F. Wall	66	18.94	
2020	Exterior Windows	Aluminum with insulated glass	25% of wall	Each	696	2.90	15.8%
2030	Exterior Doors	Double wood		Each	2003	.47	
	B30 Roofing						
3010	Roof Coverings	Built-up tar and gravel with flashing; perlite/EPS composite insulation		S.F. Roof	5.09	5.09	3.7%
3020	Roof Openings	Roof hatches		S.F. Roof	.09	.09	
C. INTERIORS							
1010	Partitions	Plaster on metal studs	10 S.F. Floor/L.F. Partition	S.F. Partition	12.28	14.73	
1020	Interior Doors	Single leaf wood	100 S.F. Floor/Door	Each	565	5.65	
1030	Fittings	Toilet partitions		S.F. Floor	.73	.73	
2010	Stair Construction	N/A		—	—	—	34.3%
3010	Wall Finishes	70% paint, 20% wood paneling, 10% vinyl wall covering		S.F. Surface	2.31	5.55	
3020	Floor Finishes	60% hardwood, 20% carpet, 20% terrazzo		S.F. Floor	12.58	12.58	
3030	Ceiling Finishes	Gypsum plaster on metal lath, suspended		S.F. Ceiling	9.23	9.23	
D. SERVICES							
	D10 Conveying						
1010	Elevators & Lifts	N/A		—	—	—	0.0 %
1020	Escalators & Moving Walks	N/A		—	—	—	
	D20 Plumbing						
2010	Plumbing Fixtures	Toilet and service fixtures, supply and drainage	1 Fixture/1110 S.F. Floor	Each	2964	2.67	
2020	Domestic Water Distribution	Electric hot water heater		S.F. Floor	2.25	2.25	3.9%
2040	Rain Water Drainage	Roof drains		S.F. Roof	.58	.58	
	D30 HVAC						
3010	Energy Supply	N/A		—	—	—	
3020	Heat Generating Systems	Included in D3050		—	—	—	
3030	Cooling Generating Systems	N/A		—	—	—	12.9 %
3050	Terminal & Package Units	Multizone unit, gas heating, electric cooling		S.F. Floor	18.20	18.20	
3090	Other HVAC Sys. & Equipment	N/A		—	—	—	
	D40 Fire Protection						
4010	Sprinklers	Wet pipe sprinkler system		S.F. Floor	2.33	2.33	1.6%
4020	Standpipes	N/A		—	—	—	
	D50 Electrical						
5010	Electrical Service/Distribution	400 ampere service, panel board and feeders		S.F. Floor	.95	.95	
5020	Lighting & Branch Wiring	Fluorescent fixtures, receptacles, switches, A.C. and misc. power		S.F. Floor	9.03	9.03	7.7%
5030	Communications & Security	Alarm systems, internet wiring, and emergency lighting		S.F. Floor	.74	.74	
5090	Other Electrical Systems	Emergency generator, 11.5 kW		S.F. Floor	.13	.13	
E. EQUIPMENT & FURNISHINGS							
1010	Commercial Equipment	N/A		—	—	—	
1020	Institutional Equipment	N/A		—	—	—	0.0 %
1030	Vehicular Equipment	N/A		—	—	—	
1090	Other Equipment	N/A		—	—	—	
F. SPECIAL CONSTRUCTION							
1020	Integrated Construction	N/A		—	—	—	0.0 %
1040	Special Facilities	N/A		—	—	—	
G. BUILDING SITEWORK	**N/A**						
				Sub-Total		141.44	**100%**
	CONTRACTOR FEES (General Requirements: 10%, Overhead: 5%, Profit: 10%)				25%	35.38	
	ARCHITECT FEES				7%	12.38	
				Total Building Cost		**189.20**	

Costs per square foot of floor area

Exterior Wall	S.F. Area	30000	40000	45000	50000	60000	70000	80000	90000	100000
	L.F. Perimeter	410	493	535	533	600	666	733	800	795
Limestone with Concrete Block Back-up	R/Conc. Frame	233.00	221.90	218.25	212.05	207.20	203.60	201.05	199.05	194.70
	Steel Frame	241.50	230.35	226.70	220.50	215.70	212.10	209.50	207.45	203.20
Face Brick with Concrete Block Back-up	R/Conc. Frame	215.70	206.25	203.20	198.55	194.60	191.60	189.45	187.75	184.65
	Steel Frame	224.20	214.75	211.70	207.05	203.00	200.10	197.90	196.20	193.15
Stone with Concrete Block Back-up	R/Conc. Frame	217.65	208.00	204.90	200.05	196.00	192.90	190.70	189.00	185.75
	Steel Frame	226.05	216.50	213.35	208.55	204.40	201.40	199.20	197.45	194.20
Perimeter Adj., Add or Deduct	Per 100 L.F.	12.25	9.15	8.15	7.35	6.15	5.25	4.60	4.05	3.75
Story Hgt. Adj., Add or Deduct	Per 1 Ft.	3.95	3.55	3.45	3.10	2.90	2.80	2.65	2.55	2.30

For Basement, add $29.55 per square foot of basement area

The above costs were calculated using the basic specifications shown on the facing page. These costs should be adjusted where necessary for design alternatives and owner's requirements. Reported completed project costs, for this type of structure, range from $133.50 to $249.05 per S.F.

Common additives

Description	Unit	$ Cost
Benches, Hardwood	L.F.	107 - 194
Clock System		
20 room	Each	16,000
50 room	Each	39,100
Closed Circuit Surveillance, One station		
Camera and monitor	Each	1850
For additional camera stations, add	Each	1000
Directory Boards, Plastic, glass covered		
30" x 20"	Each	595
36" x 48"	Each	1450
Aluminum, 24" x 18"	Each	600
36" x 24"	Each	675
48" x 32"	Each	980
48" x 60"	Each	2025
Elevators, Hydraulic passenger, 2 stops		
1500# capacity	Each	62,800
2500# capacity	Each	66,300
3500# capacity	Each	69,800
Additional stop, add	Each	7825

Description	Unit	$ Cost
Emergency Lighting, 25 watt, battery operated		
Lead battery	Each	282
Nickel cadmium	Each	805
Flagpoles, Complete		
Aluminum, 20' high	Each	1650
40' high	Each	3475
70' high	Each	10,100
Fiberglass, 23' high	Each	1775
39'-5" high	Each	3325
59' high	Each	8225
Intercom System, 25 station capacity		
Master station	Each	2650
Intercom outlets	Each	169
Handset	Each	470
Safe, Office type, 1 hour rating		
30" x 18" x 18"	Each	2400
60" x 36" x 18", double door	Each	8750
Smoke Detectors		
Ceiling type	Each	187
Duct type	Each	480

Important: See the Reference Section for Location Factors

Model costs calculated for a 3 story building with 12' story height and 60,000 square feet of floor area

Courthouse, 2-3 Story

				Unit	Unit Cost	Cost Per S.F.	% Of Sub-Total
A.	**SUBSTRUCTURE**						
	1010	Standard Foundations	Poured concrete; strip and spread footings	S.F. Ground	3.96	1.32	
	1020	Special Foundations	N/A	—	—	—	
	1030	Slab on Grade	4" reinforced concrete with vapor barrier and granular base	S.F. Slab	4.74	1.58	2.5%
	2010	Basement Excavation	Site preparation for slab and trench for foundation wall and footing	S.F. Ground	.17	.06	
	2020	Basement Walls	4' foundation wall	L.F. Wall	74	.91	
B.	**SHELL**						
	B10	**Superstructure**					
	1010	Floor Construction	Concrete slab with metal deck and beams	S.F. Floor	32.46	21.64	18.9%
	1020	Roof Construction	Concrete slab with metal deck and beams	S.F. Roof	21.30	7.10	
	B20	**Exterior Enclosure**					
	2010	Exterior Walls	Face brick with concrete block backup — 75% of wall	S.F. Wall	33.15	8.95	
	2020	Exterior Windows	Horizontal pivoted steel — 25% of wall	Each	692	6.92	10.7%
	2030	Exterior Doors	Double aluminum and glass and hollow metal	Each	3087	.32	
	B30	**Roofing**					
	3010	Roof Coverings	Built-up tar and gravel with flashing; perlite/EPS composite insulation	S.F. Roof	5.46	1.82	1.2%
	3020	Roof Openings	N/A	—	—	—	
C.	**INTERIORS**						
	1010	Partitions	Plaster on metal studs — 10 S.F. Floor/L.F. Partition	S.F. Partition	12.41	12.41	
	1020	Interior Doors	Single leaf wood — 100 S.F. Floor/Door	Each	565	5.65	
	1030	Fittings	Toilet partitions	S.F. Floor	.37	.37	
	2010	Stair Construction	Concrete filled metal pan	Flight	22,125	3.68	32.0%
	3010	Wall Finishes	70% paint, 20% wood paneling, 10% vinyl wall covering	S.F. Surface	2.32	4.63	
	3020	Floor Finishes	60% hardwood, 20% terrazzo, 20% carpet	S.F. Floor	12.58	12.58	
	3030	Ceiling Finishes	Gypsum plaster on metal lath, suspended	S.F. Ceiling	9.23	9.23	
D.	**SERVICES**						
	D10	**Conveying**					
	1010	Elevators & Lifts	Five hydraulic passenger elevators	Each	168,240	14.02	9.2%
	1020	Escalators & Moving Walks	N/A	—	—	—	
	D20	**Plumbing**					
	2010	Plumbing Fixtures	Toilet and service fixtures, supply and drainage — 1 Fixture/665 S.F. Floor	Each	1769	2.66	
	2020	Domestic Water Distribution	Electric water heater	S.F. Floor	3.71	3.71	4.4%
	2040	Rain Water Drainage	Roof drains	S.F. Roof	.96	.32	
	D30	**HVAC**					
	3010	Energy Supply	N/A	—	—	—	
	3020	Heat Generating Systems	Included in D3050	—	—	—	
	3030	Cooling Generating Systems	N/A	—	—	—	12.0 %
	3050	Terminal & Package Units	Multizone unit, gas heating, electric cooling	S.F. Floor	18.20	18.20	
	3090	Other HVAC Sys. & Equipment	N/A	—	—	—	
	D40	**Fire Protection**					
	4010	Sprinklers	Wet pipe sprinkler system	S.F. Floor	2.35	2.35	1.5%
	4020	Standpipes	N/A	—	—	—	
	D50	**Electrical**					
	5010	Electrical Service/Distribution	800 ampere service, panel board and feeders	S.F. Floor	1.25	1.25	
	5020	Lighting & Branch Wiring	Fluorescent fixtures, receptacles, switches, A.C. and misc. power	S.F. Floor	9.17	9.17	7.5%
	5030	Communications & Security	Alarm systems, internet wiring, and emergency lighting	S.F. Floor	.74	.74	
	5090	Other Electrical Systems	Emergency generator, 15 kW	S.F. Floor	.21	.21	
E.	**EQUIPMENT & FURNISHINGS**						
	1010	Commercial Equipment	N/A	—	—	—	
	1020	Institutional Equipment	N/A	—	—	—	
	1030	Vehicular Equipment	N/A	—	—	—	0.0 %
	1090	Other Equipment	N/A	—	—	—	
F.	**SPECIAL CONSTRUCTION**						
	1020	Integrated Construction	N/A	—	—	—	
	1040	Special Facilities	N/A	—	—	—	0.0 %
G.	**BUILDING SITEWORK**	**N/A**					

		Sub-Total	151.80	100%
CONTRACTOR FEES (General Requirements: 10%, Overhead: 5%, Profit: 10%)			25%	37.92
ARCHITECT FEES			7%	13.28
Total Building Cost				**203**

Costs per square foot of floor area

Exterior Wall	S.F. Area	2000	5000	7000	10000	12000	15000	18000	21000	25000
	L.F. Perimeter	200	310	360	440	480	520	560	600	660
Tiltup Concrete Panel	Steel Joists	195.90	159.65	150.90	144.65	141.55	137.70	135.10	133.35	131.80
Decorative Concrete Block	Bearing Walls	188.00	152.95	144.50	138.45	135.50	131.85	129.35	127.65	126.10
Brick on Block	Steel Joists	236.30	193.40	182.50	174.80	170.95	166.05	162.75	160.40	158.40
Stucco on Concrete Block	Wood Truss	187.55	152.60	144.25	138.30	135.35	131.75	129.35	127.70	126.25
Brick Veneer	Wood Frame	200.20	162.10	152.75	146.05	142.70	138.50	135.70	133.75	132.05
Wood Siding	Wood Frame	181.15	151.35	144.25	139.10	136.65	133.60	131.50	130.05	128.80
Perimeter Adj., Add or Deduct	Per 100 L.F.	27.75	11.10	7.90	5.50	4.60	3.65	3.15	2.65	2.25
Story Hgt. Adj., Add or Deduct	Per 1 Ft.	2.90	1.75	1.45	1.25	1.10	1.00	0.95	0.75	0.70
For Basement, add $28.75 per square foot of basement area										

The above costs were calculated using the basic specifications shown on the facing page. These costs should be adjusted where necessary for design alternatives and owner's requirements. Reported completed project costs, for this type of structure, range from $57.55 to $209.80 per S.F.

Common additives

Description	Unit	$ Cost
Emergency Lighting, 25 watt, battery operated		
Lead battery	Each	282
Nickel cadmium	Each	805
Flagpoles, Complete		
Aluminum, 20' high	Each	1650
40' high	Each	3475
70' high	Each	10,100
Fiberglass, 23' high	Each	1775
39'-5" high	Each	3325
59' high	Each	8225
Gym Floor, Incl. sleepers and finish, maple	S.F.	12.90
Intercom System, 25 Station capacity		
Master station	Each	2650
Intercom outlets	Each	169
Handset	Each	470

Description	Unit	$ Cost
Lockers, Steel, single tier, 60" to 72"	Opening	191 - 310
2 tier, 60" to 72" total	Opening	107 - 141
5 tier, box lockers	Opening	65 - 83.50
Locker bench, lam. maple top only	L.F.	21
Pedestals, steel pipe	Each	63.50
Smoke Detectors		
Ceiling type	Each	187
Duct type	Each	480
Sound System		
Amplifier, 250 watts	Each	2350
Speaker, ceiling or wall	Each	191
Trumpet	Each	365

Model costs calculated for a 1 story building with 12' story height and 10,000 square feet of floor area

Day Care Center

				Unit	Unit Cost	Cost Per S.F.	% Of Sub-Total
A. SUBSTRUCTURE							
1010	Standard Foundations	Poured concrete; strip and spread footings		S.F. Ground	2.11	2.11	
1020	Special Foundations	N/A		—	—	—	
1030	Slab on Grade	4" concrete with vapor barrier and granular base		S.F. Slab	4.20	4.20	9.3%
2010	Basement Excavation	Site preparation for slab and trench for foundation wall and footing		S.F. Ground	.26	.26	
2020	Basement Walls	4' foundation wall		L.F. Wall	78	3.43	
B. SHELL							
	B10 Superstructure						
1010	Floor Construction	Wood beams on columns		S.F. Floor	.15	.15	7.8%
1020	Roof Construction	Wood trusses		S.F. Roof	8.23	8.23	
	B20 Exterior Enclosure						
2010	Exterior Walls	Brick veneer on wood studs	85% of wall	S.F. Wall	24.60	11.04	
2020	Exterior Windows	Window wall	15% of wall	Each	44.15	3.50	17.2%
2030	Exterior Doors	Aluminum and glass; steel		Each	2063	3.92	
	B30 Roofing						
3010	Roof Coverings	Asphalt shingles, 9" fiberglass batt insulation, gutters and downspouts		S.F. Roof	3.78	3.78	3.5%
3020	Roof Openings	N/A		—	—	—	
C. INTERIORS							
1010	Partitions	Gypsum board on wood studs	8 S.F. Floor/S.F. Partition	S.F. Partition	7.35	4.90	
1020	Interior Doors	Single leaf hollow metal	700 S.F. Floor/Door	Each	875	2.33	
1030	Fittings	Toilet partitions		S.F. Floor	.66	.66	
2010	Stair Construction	N/A		—	—	—	15.4%
3010	Wall Finishes	Paint		S.F. Surface	1.39	1.85	
3020	Floor Finishes	5% quarry tile, 95% vinyl composition tile		S.F. Floor	2.44	2.44	
3030	Ceiling Finishes	Fiberglass tile on tee grid		S.F. Ceiling	4.30	4.30	
D. SERVICES							
	D10 Conveying						
1010	Elevators & Lifts	N/A		—	—	—	0.0 %
1020	Escalators & Moving Walks	N/A		—	—	—	
	D20 Plumbing						
2010	Plumbing Fixtures	Toilet and service fixtures, supply and drainage	1 Fixture/455 S.F. Floor	Each	1501	8.43	
2020	Domestic Water Distribution	Electric water heater		S.F. Floor	2.22	2.22	9.9%
2040	Rain Water Drainage	N/A		—	—	—	
	D30 HVAC						
3010	Energy Supply	Oil fired hot water, wall fin radiation		S.F. Floor	9.61	9.61	
3020	Heat Generating Systems	N/A		—	—	—	
3030	Cooling Generating Systems	N/A		—	—	—	21.1%
3050	Terminal & Package Units	Split systems with air cooled condensing units		S.F. Floor	13.04	13.04	
3090	Other HVAC Sys. & Equipment	N/A		—	—	—	
	D40 Fire Protection						
4010	Sprinklers	Sprinkler, light hazard		S.F. Floor	2.77	2.77	2.6%
4020	Standpipes	N/A		—	—	—	
	D50 Electrical						
5010	Electrical Service/Distribution	200 ampere service, panel board and feeders		S.F. Floor	1	1	
5020	Lighting & Branch Wiring	Fluorescent fixtures, receptacles, switches, A.C. and misc. power		S.F. Floor	8.74	8.74	9.5%
5030	Communications & Security	Alarm systems and emergency lighting		S.F. Floor	.40	.40	
5090	Other Electrical Systems	Emergency generator, 15 kW		S.F. Floor	.08	.08	
E. EQUIPMENT & FURNISHINGS							
1010	Commercial Equipment	N/A		—	—	—	
1020	Institutional Equipment	Cabinets and countertop		S.F. Floor	3.79	3.79	3.5%
1030	Vehicular Equipment	N/A		—	—	—	
1090	Other Equipment	N/A		—	—	—	
F. SPECIAL CONSTRUCTION							
1020	Integrated Construction	N/A		—	—	—	0.0 %
1040	Special Facilities	N/A		—	—	—	
G. BUILDING SITEWORK	**N/A**						

		Sub-Total	107.18	100%
CONTRACTOR FEES (General Requirements: 10%, Overhead: 5%, Profit: 10%)		25%	26.81	
ARCHITECT FEES		9%	12.06	

Total Building Cost	**146.05**

Costs per square foot of floor area

Exterior Wall	S.F. Area	12000	18000	24000	30000	36000	42000	48000	54000	60000
	L.F. Perimeter	460	580	713	730	826	880	965	1006	1045
Concrete Block	Steel Frame	116.65	110.15	107.25	103.20	101.80	100.15	99.35	98.15	97.30
	Bearing Walls	113.50	107.05	104.10	100.05	98.65	97.05	96.25	95.05	94.25
Precast Concrete Panels	Steel Frame	122.90	115.40	112.05	107.15	105.55	103.55	102.65	101.25	100.15
Insulated Metal Panels	Steel Frame	119.15	112.25	109.20	104.80	103.30	101.55	100.65	99.45	98.45
Face Brick on Common Brick	Steel Frame	137.60	127.80	123.50	116.50	114.30	111.65	110.35	108.40	106.85
Tilt-up Concrete Panel	Steel Frame	121.05	113.90	110.65	105.95	104.40	102.60	101.65	100.30	99.35
Perimeter Adj., Add or Deduct	Per 100 L.F.	5.05	3.35	2.55	1.95	1.60	1.45	1.30	1.20	1.00
Story Hgt. Adj., Add or Deduct	Per 1 Ft.	0.65	0.60	0.55	0.40	0.40	0.40	0.35	0.40	0.35
For Basement, add $29.30 per square foot of basement area										

The above costs were calculated using the basic specifications shown on the facing page. These costs should be adjusted where necessary for design alternatives and owner's requirements. Reported completed project costs, for this type of structure, range from $42.75 to $165.40 per S.F.

Common additives

Description	Unit	$ Cost
Clock System		
20 room	Each	16,000
50 room	Each	39,100
Dock Bumpers, Rubber blocks		
4-1/2" thick, 10" high, 14" long	Each	69
24" long	Each	118
36" long	Each	120
12" high, 14" long	Each	120
24" long	Each	136
36" long	Each	156
6" thick, 10" high, 14" long	Each	118
24" long	Each	149
36" long	Each	189
20" high, 11" long	Each	185
Dock Boards, Heavy		
60" x 60" Aluminum, 5,000# cap.	Each	1250
9000# cap.	Each	1475
15,000# cap.	Each	1650

Description	Unit	$ Cost
Dock Levelers, Hinged 10 ton cap.		
6' x 8'	Each	6300
7' x 8'	Each	6800
Partitions, Woven wire, 10 ga., 1-1/2" mesh		
4' wide x 7' high	Each	185
8' high	Each	192
10' High	Each	241
Platform Lifter, Portable, 6'x 6'		
3000# cap.	Each	9150
4000# cap.	Each	11,300
Fixed, 6' x 8', 5000# cap.	Each	11,900

Model costs calculated for a 1 story building with 20' story height and 30,000 square feet of floor area

Factory, 1 Story

				Unit	Unit Cost	Cost Per S.F.	% Of Sub-Total
A. SUBSTRUCTURE							
1010	Standard Foundations	Poured concrete; strip and spread footings		S.F. Ground	1.42	1.42	
1020	Special Foundations	N/A		—	—	—	
1030	Slab on Grade	4" reinforced concrete with vapor barrier and granular base		S.F. Slab	6.27	6.27	13.1%
2010	Basement Excavation	Site preparation for slab and trench for foundation wall and footing		S.F. Ground	.17	.17	
2020	Basement Walls	4' foundation wall		L.F. Wall	78	2.28	
B. SHELL							
B10 Superstructure							
1010	Floor Construction	N/A		—	—	—	
1020	Roof Construction	Metal deck, open web steel joists, beams and columns		S.F. Roof	10.97	10.97	14.2 %
B20 Exterior Enclosure							
2010	Exterior Walls	Concrete block	75% of wall	S.F. Wall	7.10	2.59	
2020	Exterior Windows	Industrial horizontal pivoted steel	25% of wall	Each	807	3.07	8.7%
2030	Exterior Doors	Double aluminum and glass, hollow metal, steel overhead		Each	2137	1.06	
B30 Roofing							
3010	Roof Coverings	Built-up tar and gravel with flashing; perlite/EPS composite insulation		S.F. Roof	5.26	5.26	7.2%
3020	Roof Openings	Roof hatches		S.F. Roof	.32	.32	
C. INTERIORS							
1010	Partitions	Concrete block	60 S.F. Floor/L.F. Partition	S.F. Partition	7.90	1.58	
1020	Interior Doors	Single leaf hollow metal and fire doors	600 S.F. Floor/Door	Each	875	1.46	
1030	Fittings	Toilet partitions		S.F. Floor	1.47	1.47	
2010	Stair Construction	N/A		—	—	—	9.2%
3010	Wall Finishes	Paint		S.F. Surface	4.08	1.63	
3020	Floor Finishes	Vinyl composition tile	10% of floor	S.F. Floor	3.30	.33	
3030	Ceiling Finishes	Fiberglass board on exposed grid system	10% of area	S.F. Ceiling	6.38	.64	
D. SERVICES							
D10 Conveying							
1010	Elevators & Lifts	N/A		—	—	—	
1020	Escalators & Moving Walks	N/A		—	—	—	0.0 %
D20 Plumbing							
2010	Plumbing Fixtures	Toilet and service fixtures, supply and drainage	1 Fixture/1000 S.F. Floor	Each	3370	3.37	
2020	Domestic Water Distribution	Gas fired water heater		S.F. Floor	.45	.45	5.8%
2040	Rain Water Drainage	Roof drains		S.F. Roof	.65	.65	
D30 HVAC							
3010	Energy Supply	Oil fired hot water, unit heaters		S.F. Floor	8.20	8.20	
3020	Heat Generating Systems	N/A		—	—	—	
3030	Cooling Generating Systems	Chilled water, air cooled condenser system		S.F. Floor	10.29	10.29	24.0%
3050	Terminal & Package Units	N/A		—	—	—	
3090	Other HVAC Sys. & Equipment	N/A		—	—	—	
D40 Fire Protection							
4010	Sprinklers	Sprinklers, ordinary hazard		S.F. Floor	3.24	3.24	4.2%
4020	Standpipes	N/A		—	—	—	
D50 Electrical							
5010	Electrical Service/Distribution	600 ampere service, panel board and feeders		S.F. Floor	1.34	1.34	
5020	Lighting & Branch Wiring	High intensity discharge fixtures, receptacles, switches, A.C. and misc. power		S.F. Floor	8.66	8.66	13.5%
5030	Communications & Security	Alarm systems and emergency lighting		S.F. Floor	.41	.41	
5090	Other Electrical Systems	N/A		—	—	—	
E. EQUIPMENT & FURNISHINGS							
1010	Commercial Equipment	N/A		—	—	—	
1020	Institutional Equipment	N/A		—	—	—	
1030	Vehicular Equipment	N/A		—	—	—	0.0 %
1090	Other Equipment	N/A		—	—	—	
F. SPECIAL CONSTRUCTION							
1020	Integrated Construction	N/A		—	—	—	
1040	Special Facilities	N/A		—	—	—	0.0 %
G. BUILDING SITEWORK	**N/A**						

		Sub-Total	77.13	100%
CONTRACTOR FEES (General Requirements: 10%, Overhead: 5%, Profit: 10%)		25%	19.32	
ARCHITECT FEES		7%	6.75	
	Total Building Cost		**103.20**	

Costs per square foot of floor area

Exterior Wall	S.F. Area	20000	30000	40000	50000	60000	70000	80000	90000	100000
	L.F. Perimeter	362	410	493	576	600	628	660	700	744
Face Brick Common Brick Back-up	Steel Frame	158.75	144.75	139.95	137.05	132.75	129.75	127.55	126.15	125.10
	Concrete Frame	148.30	134.35	129.55	126.65	122.35	119.35	117.15	115.75	114.75
Face Brick Concrete Block Back-up	Steel Frame	156.00	142.80	138.20	135.50	131.40	128.60	126.55	125.25	124.25
	Concrete Frame	145.25	132.05	127.45	124.75	120.65	117.80	115.75	114.45	113.45
Stucco on Concrete Block	Steel Frame	145.05	134.50	130.80	128.50	125.35	123.15	121.55	120.55	119.75
	Concrete Frame	134.30	123.75	120.00	117.75	114.55	112.35	110.75	109.75	108.95
Perimeter Adj., Add or Deduct	Per 100 L.F.	12.45	8.30	6.25	5.00	4.15	3.55	3.15	2.80	2.50
Story Hgt. Adj., Add or Deduct	Per 1 Ft.	3.45	2.55	2.35	2.25	1.90	1.70	1.60	1.50	1.45
For Basement, add $31.95 per square foot of basement area										

The above costs were calculated using the basic specifications shown on the facing page. These costs should be adjusted where necessary for design alternatives and owner's requirements. Reported completed project costs, for this type of structure, range from $42.75 to $165.45 per S.F.

Common additives

Description	Unit	$ Cost
Clock System		
20 room	Each	16,000
50 room	Each	39,100
Dock Bumpers, Rubber blocks		
4-1/2" thick, 10" high, 14" long	Each	69
24" long	Each	118
36" long	Each	120
12" high, 14" long	Each	120
24" long	Each	136
36" long	Each	156
6" thick, 10" high, 14" long	Each	118
24" long	Each	149
36" long	Each	189
20" high, 11" long	Each	185
Dock Boards, Heavy		
60" x 60" Aluminum, 5,000# cap.	Each	1250
9000# cap.	Each	1475
15,000# cap.	Each	1650

Description	Unit	$ Cost
Dock Levelers, Hinged 10 ton cap.		
6' x 8'	Each	6300
7' x 8'	Each	6800
Elevator, Hydraulic freight, 2 stops		
3500# capacity	Each	149,600
4000# capacity	Each	158,600
Additional stop, add	Each	10,000
Partitions, Woven wire, 10 ga., 1-1/2" mesh		
4' Wide x 7" high	Each	185
8' High	Each	192
10' High	Each	241
Platform Lifter, Portable, 6' x 6'		
3000# cap.	Each	9150
4000# cap.	Each	11,300
Fixed, 6' x 8', 5000# cap.	Each	11,900

Important: See the Reference Section for Location Factors

Model costs calculated for a 3 story building with 12' story height and 90,000 square feet of floor area

Factory, 3 Story

				Unit	Unit Cost	Cost Per S.F.	% Of Sub-Total
A.	**SUBSTRUCTURE**						
1010	Standard Foundations	Poured concrete; strip and spread footings		S.F. Ground	6.54	2.18	
1020	Special Foundations	N/A		—	—	—	
1030	Slab on Grade	4" reinforced concrete with vapor barrier and granular base		S.F. Slab	9.58	3.19	6.9%
2010	Basement Excavation	Site preparation for slab and trench for foundation wall and footing		S.F. Ground	.17	.06	
2020	Basement Walls	4' foundation wall		L.F. Wall	78	.61	
B.	**SHELL**						
	B10 Superstructure						
1010	Floor Construction	Concrete flat slab		S.F. Floor	23.99	15.99	24.2%
1020	Roof Construction	Concrete flat slab		S.F. Roof	15.45	5.15	
	B20 Exterior Enclosure						
2010	Exterior Walls	Face brick with common brick backup	70% of wall	S.F. Wall	35.92	7.04	
2020	Exterior Windows	Industrial, horizontal pivoted steel	30% of wall	Each	1226	6.44	15.9%
2030	Exterior Doors	Double aluminum & glass, hollow metal, overhead doors		Each	2029	.39	
	B30 Roofing						
3010	Roof Coverings	Built-up tar and gravel with flashing; perlite/EPS composite insulation		S.F. Roof	5.10	1.70	2.1%
3020	Roof Openings	Roof hatches		S.F. Roof	.30	.10	
C.	**INTERIORS**						
1010	Partitions	Gypsum board on metal studs	50 S.F. Floor/L.F. Partition	S.F. Partition	3.90	.78	
1020	Interior Doors	Single leaf fire doors	500 S.F. Floor/Door	Each	945	1.89	
1030	Fittings	Toilet partitions		S.F. Floor	.73	.73	
2010	Stair Construction	Concrete		Flight	7725	.69	9.5%
3010	Wall Finishes	Paint		S.F. Surface	.73	.29	
3020	Floor Finishes	90% metallic hardener, 10% vinyl composition tile		S.F. Floor	3.51	3.51	
3030	Ceiling Finishes	Fiberglass board on exposed grid systems	10% of area	S.F. Ceiling	4.30	.43	
D.	**SERVICES**						
	D10 Conveying						
1010	Elevators & Lifts	Two hydraulic freight elevators		Each	230,400	5.12	5.9%
1020	Escalators & Moving Walks	N/A		—	—	—	
	D20 Plumbing						
2010	Plumbing Fixtures	Toilet and service fixtures, supply and drainage	1 Fixture/1345 S.F. Floor	Each	3806	2.83	
2020	Domestic Water Distribution	Gas fired water heater		S.F. Floor	.27	.27	3.9%
2040	Rain Water Drainage	Roof drains		S.F. Roof	.81	.27	
	D30 HVAC						
3010	Energy Supply	Oil fired hot water, unit heaters		S.F. Floor	3.94	3.94	
3020	Heat Generating Systems	N/A		—	—	—	
3030	Cooling Generating Systems	Chilled water, air cooled condenser system		S.F. Floor	10.29	10.29	16.3%
3050	Terminal & Package Units	N/A		—	—	—	
3090	Other HVAC Sys. & Equipment	N/A		—	—	—	
	D40 Fire Protection						
4010	Sprinklers	Wet pipe sprinkler system		S.F. Floor	2.86	2.86	3.3%
4020	Standpipes	N/A		—	—	—	
	D50 Electrical						
5010	Electrical Service/Distribution	800 ampere service, panel board and feeders		S.F. Floor	.83	.83	
5020	Lighting & Branch Wiring	High intensity discharge fixtures, switches, A.C. and misc. power		S.F. Floor	9.18	9.18	12.2%
5030	Communications & Security	Alarm systems and emergency lighting		S.F. Floor	.42	.42	
5090	Other Electrical Systems	Emergency generator, 30 kW		S.F. Floor	.19	.19	
E.	**EQUIPMENT & FURNISHINGS**						
1010	Commercial Equipment	N/A		—	—	—	
1020	Institutional Equipment	N/A		—	—	—	0.0%
1030	Vehicular Equipment	N/A		—	—	—	
1090	Other Equipment	N/A		—	—	—	
F.	**SPECIAL CONSTRUCTION**						
1020	Integrated Construction	N/A		—	—	—	0.0%
1040	Special Facilities	N/A		—	—	—	
G.	**BUILDING SITEWORK**	**N/A**					

		Sub-Total	87.37	100%
CONTRACTOR FEES (General Requirements: 10%, Overhead: 5%, Profit: 10%)		25%	21.83	
ARCHITECT FEES		6%	6.55	
	Total Building Cost		**115.75**	

Costs per square foot of floor area

Exterior Wall	S.F. Area	4000	4500	5000	5500	6000	6500	7000	7500	8000
	L.F. Perimeter	260	280	300	310	320	336	353	370	386
Face Brick Concrete Block Back-up	Steel Joists	166.25	163.15	160.50	157.00	154.05	152.30	150.85	149.70	148.55
	Bearing Walls	156.15	153.00	150.35	146.85	143.90	142.15	140.70	139.60	138.40
Decorative Concrete Block	Steel Joists	151.75	149.15	147.10	144.40	142.10	140.75	139.60	138.65	137.75
	Bearing Walls	141.85	139.30	137.15	134.50	132.20	130.85	129.70	128.80	127.85
Limestone with Concrete Block Back-up	Steel Joists	192.60	188.30	184.80	179.80	175.65	173.25	171.30	169.70	168.05
	Bearing Walls	182.65	178.40	174.95	169.95	165.75	163.35	161.40	159.75	158.20
Perimeter Adj., Add or Deduct	Per 100 L.F.	20.00	17.70	15.95	14.55	13.35	12.25	11.40	10.65	9.95
Story Hgt. Adj., Add or Deduct	Per 1 Ft.	2.65	2.45	2.40	2.20	2.15	2.05	2.05	2.00	1.90
For Basement, add $35.80 per square foot of basement area										

The above costs were calculated using the basic specifications shown on the facing page. These costs should be adjusted where necessary for design alternatives and owner's requirements. Reported completed project costs, for this type of structure, range from $69.15 to $204.65 per S.F.

Common additives

Description	Unit	$ Cost
Appliances		
Cooking range, 30" free standing		
1 oven	Each	430 - 2150
2 oven	Each	1675 - 1875
30" built-in		
1 oven	Each	855 - 2025
2 oven	Each	1725 - 1600
Counter top cook tops, 4 burner	Each	360 - 830
Microwave oven	Each	254 - 775
Combination range, refrig. & sink, 30" wide	Each	1550 - 4650
60" wide	Each	3800
72" wide	Each	5025
Combination range refrigerator, sink		
microwave oven & icemaker	Each	5775
Compactor, residential, 4-1 compaction	Each	630 - 730
Dishwasher, built-in, 2 cycles	Each	515 - 910
4 cycles	Each	625 - 1700
Garbage disposer, sink type	Each	188 - 325
Hood for range, 2 speed, vented, 30" wide	Each	271 - 1250
42" wide	Each	490 - 2250

Description	Unit	$ Cost
Appliances, cont.		
Refrigerator, no frost 10-12 C.F.	Each	570 - 705
14-16 C.F.	Each	590 - 840
18-20 C.F.	Each	735 - 1300
Lockers, Steel, single tier, 60" or 72"	Opening	191 - 310
2 tier, 60" or 72" total	Opening	107 - 141
5 tier, box lockers	Opening	65 - 83.50
Locker bench, lam. maple top only	L.F.	21
Pedestals, steel pipe	Each	63.50
Sound System		
Amplifier, 250 watts	Each	2350
Speaker, ceiling or wall	Each	191
Trumpet	Each	365

Important: See the Reference Section for Location Factors

Fire Station, 1 Story

Model costs calculated for a 1 story building with 14' story height and 6,000 square feet of floor area

				Unit	Unit Cost	Cost Per S.F.	% Of Sub-Total
A.	**SUBSTRUCTURE**						
1010	Standard Foundations	Poured concrete; strip and spread footings		S.F. Ground	3.37	3.37	
1020	Special Foundations	N/A		—	—	—	
1030	Slab on Grade	6" reinforced concrete with vapor barrier and granular base		S.F. Slab	6.27	6.27	12.5%
2010	Basement Excavation	Site preparation for slab and trench for foundation wall and footing		S.F. Ground	.46	.46	
2020	Basement Walls	4' foundation wall		L.F. Wall	78	4.16	
B.	**SHELL**						
	B10 Superstructure						
1010	Floor Construction	N/A		—	—	—	10.2 %
1020	Roof Construction	Metal deck, open web steel joists, beams on columns		S.F. Roof	11.69	11.69	
	B20 Exterior Enclosure						
2010	Exterior Walls	Face brick with concrete block backup	75% of wall	S.F. Wall	30.86	17.28	
2020	Exterior Windows	Aluminum insulated glass	10% of wall	Each	807	1.88	20.3%
2030	Exterior Doors	Single aluminum and glass, overhead, hollow metal	15% of wall	S.F. Door	35.63	3.99	
	B30 Roofing						
3010	Roof Coverings	Built-up tar and gravel with flashing; perlite/EPS composite insulation		S.F. Roof	6.34	6.34	5.7%
3020	Roof Openings	Skylights, roof hatches		S.F. Roof	.14	.14	
C.	**INTERIORS**						
1010	Partitions	Concrete block	17 S.F. Floor/L.F. Partition	S.F. Partition	8.43	4.96	
1020	Interior Doors	Single leaf hollow metal	500 S.F. Floor/Door	Each	875	1.75	
1030	Fittings	Toilet partitions		S.F. Floor	.73	.73	
2010	Stair Construction	N/A		—	—	—	13.3%
3010	Wall Finishes	Paint		S.F. Surface	1.82	2.14	
3020	Floor Finishes	50% vinyl tile, 50% paint		S.F. Floor	2.41	2.41	
3030	Ceiling Finishes	Fiberglass board on exposed grid, suspended	50% of area	S.F. Ceiling	6.38	3.19	
D.	**SERVICES**						
	D10 Conveying						
1010	Elevators & Lifts	N/A		—	—	—	0.0 %
1020	Escalators & Moving Walks	N/A		—	—	—	
	D20 Plumbing						
2010	Plumbing Fixtures	Kitchen, toilet and service fixtures, supply and drainage	1 Fixture/375 S.F. Floor	Each	2948	7.86	
2020	Domestic Water Distribution	Gast fired water heater		S.F. Floor	2.09	2.09	9.4%
2040	Rain Water Drainage	Roof drains		S.F. Roof	.82	.82	
	D30 HVAC						
3010	Energy Supply	N/A		—	—	—	
3020	Heat Generating Systems	Included in D3050		—	—	—	
3030	Cooling Generating Systems	N/A		—	—	—	18.4 %
3050	Terminal & Package Units	Rooftop multizone unit system		S.F. Floor	20.96	20.96	
3090	Other HVAC Sys. & Equipment	N/A		—	—	—	
	D40 Fire Protection						
4010	Sprinklers	Wet pipe sprinkler system		S.F. Floor	3.59	3.59	3.1%
4020	Standpipes	N/A		—	—	—	
	D50 Electrical						
5010	Electrical Service/Distribution	200 ampere service, panel board and feeders		S.F. Floor	1.39	1.39	
5020	Lighting & Branch Wiring	Fluorescent fixtures, receptacles, switches, A.C. and misc. power		S.F. Floor	6.23	6.23	7.0%
5030	Communications & Security	Alarm systems		S.F. Floor	.41	.41	
5090	Other Electrical Systems	N/A		—	—	—	
E.	**EQUIPMENT & FURNISHINGS**						
1010	Commercial Equipment	N/A		—	—	—	
1020	Institutional Equipment	N/A		—	—	—	0.0 %
1030	Vehicular Equipment	N/A		—	—	—	
1090	Other Equipment	N/A		—	—	—	
F.	**SPECIAL CONSTRUCTION**						
1020	Integrated Construction	N/A		—	—	—	0.0 %
1040	Special Facilities	N/A		—	—	—	
G.	**BUILDING SITEWORK**	**N/A**					

	Sub-Total	114.11	100%
CONTRACTOR FEES (General Requirements: 10%, Overhead: 5%, Profit: 10%)	25%	28.53	
ARCHITECT FEES	8%	11.41	
	Total Building Cost	**154.05**	

Costs per square foot of floor area

Exterior Wall	S.F. Area	6000	7000	8000	9000	10000	11000	12000	13000	14000
	L.F. Perimeter	220	240	260	280	286	303	320	336	353
Face Brick with Concrete Block Back-up	Steel Joists	175.25	169.40	164.90	161.50	156.85	154.35	152.35	150.50	149.00
	Precast Conc.	181.95	176.10	171.70	168.25	163.65	161.20	159.20	157.40	155.90
Decorative Concrete Block	Steel Joists	162.65	157.60	153.70	150.75	147.00	144.90	143.15	141.60	140.35
	Precast Conc.	169.30	164.35	160.45	157.50	153.80	151.70	150.00	148.50	147.20
Limestone with Concrete Block Back-up	Steel Joists	204.95	197.15	191.25	186.65	180.00	176.65	173.95	171.45	169.40
	Precast Conc.	211.40	203.65	197.75	193.25	186.60	183.30	180.60	178.10	176.15
Perimeter Adj., Add or Deduct	Per 100 L.F.	22.95	19.60	17.25	15.25	13.75	12.50	11.45	10.55	9.85
Story Hgt. Adj., Add or Deduct	Per 1 Ft.	3.00	2.75	2.65	2.50	2.30	2.30	2.15	2.05	2.05
For Basement, add $34.75 per square foot of basement area										

The above costs were calculated using the basic specifications shown on the facing page. These costs should be adjusted where necessary for design alternatives and owner's requirements. Reported completed project costs, for this type of structure, range from $69.00 to $204.65 per S.F.

Common additives

Description	Unit	$ Cost	Description	Unit	$ Cost
Appliances			Appliances, cont.		
Cooking range, 30" free standing			Refrigerator, no frost 10-12 C.F.	Each	570 - 705
1 oven	Each	430 - 2150	14-16 C.F.	Each	590 - 840
2 oven	Each	1675 - 1875	18-20 C.F.	Each	735 - 1300
30" built-in			Elevators, Hydraulic passenger, 2 stops		
1 oven	Each	855 - 2025	1500# capacity	Each	62,800
2 oven	Each	1725 - 1600	2500# capacity	Each	66,300
Counter top cook tops, 4 burner	Each	360 - 830	3500# capacity	Each	69,800
Microwave oven	Each	254 - 775	Lockers, Steel, single tier, 60" or 72"	Opening	191 - 310
Combination range, refrig. & sink, 30" wide	Each	1550 - 4650	2 tier, 60" or 72" total	Opening	107 - 141
60" wide	Each	3800	5 tier, box lockers	Opening	65 - 83.50
72" wide	Each	5025	Locker bench, lam. maple top only	L.F.	21
Combination range, refrigerator, sink,			Pedestals, steel pipe	Each	63.50
microwave oven & icemaker	Each	5775	Sound System		
Compactor, residential, 4-1 compaction	Each	630 - 730	Amplifier, 250 watts	Each	2350
Dishwasher, built-in, 2 cycles	Each	515 - 910	Speaker, ceiling or wall	Each	191
4 cycles	Each	625 - 1700	Trumpet	Each	365
Garbage diposer, sink type	Each	188 - 325			
Hood for range, 2 speed, vented, 30" wide	Each	271 - 1250			
42" wide	Each	490 - 2250			

Model costs calculated for a 2 story building with 14' story height and 10,000 square feet of floor area

				Unit	Unit Cost	Cost Per S.F.	% Of Sub-Total
A.	**SUBSTRUCTURE**						
1010	Standard Foundations	Poured concrete; strip and spread footings		S.F. Ground	3.16	1.58	
1020	Special Foundations	N/A		—	—	—	
1030	Slab on Grade	6" reinforced concrete with vapor barrier and granular base		S.F. Slab	6.27	3.14	6.8%
2010	Basement Excavation	Site preparation for slab and trench for foundation wall and footing		S.F. Ground	.46	.23	
2020	Basement Walls	4' foundation wall		L.F. Wall	88	2.50	
B.	**SHELL**						
	B10 Superstructure						
1010	Floor Construction	Open web steel joists, slab form, concrete		S.F. Floor	15.86	7.93	10.0%
1020	Roof Construction	Metal deck on open web steel joists		S.F. Roof	5.94	2.97	
	B20 Exterior Enclosure						
2010	Exterior Walls	Decorative concrete block	75% of wall	S.F. Wall	18.70	11.23	
2020	Exterior Windows	Aluminum insulated glass	10% of wall	Each	696	2.43	15.6%
2030	Exterior Doors	Single aluminum and glass, steel overhead, hollow metal	15% of wall	S.F. Door	27.72	3.33	
	B30 Roofing						
3010	Roof Coverings	Built-up tar and gravel with flashing; perlite/EPS composite insulation		S.F. Roof	6.50	3.25	3.0%
3020	Roof Openings	N/A		—	—	—	
C.	**INTERIORS**						
1010	Partitions	Concrete block	10 S.F. Floor/L.F. Partition	S.F. Partition	8.43	4.96	
1020	Interior Doors	Single leaf hollow metal	500 S.F. Floor/Door	Each	945	1.89	
1030	Fittings	Toilet partitions		S.F. Floor	.66	.66	
2010	Stair Construction	Concrete filled metal pan		Flight	22,125	4.43	18.5%
3010	Wall Finishes	Paint		S.F. Surface	3.08	3.62	
3020	Floor Finishes	50% vinyl tile, 50% paint		S.F. Floor	2.41	2.41	
3030	Ceiling Finishes	Fiberglass board on exposed grid, suspended	50% of area	S.F. Ceiling	4.30	2.15	
D.	**SERVICES**						
	D10 Conveying						
1010	Elevators & Lifts	One hydraulic passenger elevator		Each	80,900	8.09	7.4%
1020	Escalators & Moving Walks	N/A		—	—	—	
	D20 Plumbing						
2010	Plumbing Fixtures	Kitchen toilet and service fixtures, supply and drainage	1 Fixture/400 S.F. Floor	Each	2924	7.31	
2020	Domestic Water Distribution	Gas fired water heater		S.F. Floor	1.67	1.67	8.7%
2040	Rain Water Drainage	Roof drains		S.F. Roof	.92	.46	
	D30 HVAC						
3010	Energy Supply	N/A		—	—	—	
3020	Heat Generating Systems	Included in D3050		—	—	—	
3030	Cooling Generating Systems	N/A		—	—	—	19.3 %
3050	Terminal & Package Units	Rooftop multizone unit system		S.F. Floor	20.96	20.96	
3090	Other HVAC Sys. & Equipment	N/A		—	—	—	
	D40 Fire Protection						
4010	Sprinklers	Wet pipe sprinkler system		S.F. Floor	2.95	2.95	2.7%
4020	Standpipes	N/A		—	—	—	
	D50 Electrical						
5010	Electrical Service/Distribution	400 ampere service, panel board and feeders		S.F. Floor	1.64	1.64	
5020	Lighting & Branch Wiring	Fluorescent fixtures, receptacles, switches, A.C. and misc. power		S.F. Floor	6.44	6.44	8.0%
5030	Communications & Security	Alarm systems and emergency lighting		S.F. Floor	.42	.42	
5090	Other Electrical Systems	Emergency generator, 15 kW		S.F. Floor	.23	.23	
E.	**EQUIPMENT & FURNISHINGS**						
1010	Commercial Equipment	N/A		—	—	—	
1020	Institutional Equipment	N/A		—	—	—	0.0%
1030	Vehicular Equipment	N/A		—	—	—	
1090	Other Equipment	N/A		—	—	—	
F.	**SPECIAL CONSTRUCTION**						
1020	Integrated Construction	N/A		—	—	—	0.0 %
1040	Special Facilities	N/A		—	—	—	
G.	**BUILDING SITEWORK**	**N/A**					

		Sub-Total	108.88	**100%**
CONTRACTOR FEES (General Requirements: 10%, Overhead: 5%, Profit: 10%)		25%	27.23	
ARCHITECT FEES		8%	10.89	
		Total Building Cost	147	

Costs per square foot of floor area

Exterior Wall	S.F. Area	4000	5000	6000	8000	10000	12000	14000	16000	18000
	L.F. Perimeter	180	205	230	260	300	340	353	386	420
Cedar Beveled Siding	Wood Frame	170.75	161.25	154.85	145.35	140.20	136.90	133.20	131.30	129.75
Aluminum Siding	Wood Frame	169.75	160.35	153.90	144.40	139.25	135.90	132.25	130.35	128.80
Board and Batten	Wood Frame	166.70	158.55	153.05	144.70	140.20	137.35	133.95	132.30	131.00
Face Brick on Block	Bearing Wall	204.75	193.65	186.10	174.15	168.00	163.95	158.90	156.60	154.75
Stucco on Block	Bearing Wall	184.70	175.20	168.85	159.30	154.20	150.90	147.15	145.30	143.80
Decorative Block	Wood Joists	174.60	164.70	158.15	148.10	142.80	139.30	135.30	133.35	131.75
Perimeter Adj., Add or Deduct	Per 100 L.F.	16.00	12.80	10.70	7.95	6.45	5.35	4.60	3.95	3.55
Story Hgt. Adj., Add or Deduct	Per 1 Ft.	2.05	1.85	1.75	1.45	1.45	1.30	1.10	1.10	1.10
For Basement, add $22.10 per square foot of basement area										

The above costs were calculated using the basic specifications shown on the facing page. These costs should be adjusted where necessary for design alternatives and owner's requirements. Reported completed project costs, for this type of structure, range from $100.30 to $219.40 per S.F.

Common additives

Description	Unit	$ Cost
Appliances		
Cooking range, 30" free standing		
1 oven	Each	430 - 2150
2 oven	Each	1675 - 1875
30" built-in		
1 oven	Each	855 - 2025
2 oven	Each	1725 - 1600
Counter top cook tops, 4 burner	Each	360 - 830
Microwave oven	Each	254 - 775
Combination range, refrig. & sink, 30" wide	Each	1550 - 4650
60" wide	Each	3800
72" wide	Each	5025
Combination range, refrigerator, sink, microwave oven & icemaker	Each	5775
Compactor, residential, 4-1 compaction	Each	630 - 730
Dishwasher, built-in, 2 cycles	Each	515 - 910
4 cycles	Each	625 - 1700
Garbage disposer, sink type	Each	188 - 325
Hood for range, 2 speed, vented, 30" wide	Each	271 - 1250
42" wide	Each	490 - 2250

Description	Unit	$ Cost
Appliances, cont.		
Refrigerator, no frost 10-12 C.F.	Each	570 - 705
14-16 C.F.	Each	590 - 840
18-20 C.F.	Each	735 - 1300
Elevators, Hydraulic passenger, 2 stops		
1500# capacity	Each	62,800
2500# capacity	Each	66,300
3500# capacity	Each	69,800
Laundry Equipment		
Dryer, gas, 16 lb. capacity	Each	885
30 lb. capacity	Each	3600
Washer, 4 cycle	Each	1075
Commercial	Each	1450
Sound System		
Amplifier, 250 watts	Each	2350
Speaker, ceiling or wall	Each	191
Trumpet	Each	365

Important: See the Reference Section for Location Factors

Fraternity/Sorority House

Model costs calculated for a 2 story building with 10′ story height and 10,000 square feet of floor area

				Unit	Unit Cost	Cost Per S.F.	% Of Sub-Total
A. SUBSTRUCTURE							
1010	Standard Foundations	Poured concrete; strip and spread footings		S.F. Ground	3.08	1.54	
1020	Special Foundations	N/A		—	—	—	
1030	Slab on Grade	4″ reinforced concrete with vapor barrier and granular base		S.F. Slab	4.74	2.37	6.0%
2010	Basement Excavation	Site preparation for slab and trench for foundation wall and footing		S.F. Ground	.46	.23	
2020	Basement Walls	4′ foundation wall		L.F. Wall	69	2.08	
B. SHELL							
B10 Superstructure							
1010	Floor Construction	Plywood on wood joists		S.F. Floor	3.42	1.71	3.3%
1020	Roof Construction	Plywood on wood rafters (pitched)		S.F. Roof	3.07	1.72	
B20 Exterior Enclosure							
2010	Exterior Walls	Cedar bevel siding on wood studs, insulated	80% of wall	S.F. Wall	12.81	6.15	
2020	Exterior Windows	Double hung wood	20% of wall	Each	588	2.82	10.2%
2030	Exterior Doors	Solid core wood		Each	2127	1.50	
B30 Roofing							
3010	Roof Coverings	Asphalt shingles with flashing (pitched); rigid fiberglass insulation		S.F. Roof	4.14	2.07	2.0%
3020	Roof Openings	N/A		—	—	—	
C. INTERIORS							
1010	Partitions	Gypsum board on wood studs	25 S.F. Floor/L.F. Partition	S.F. Partition	7.47	2.39	
1020	Interior Doors	Single leaf wood	200 S.F. Floor/Door	Each	565	2.83	
1030	Fittings	N/A		—	—	—	
2010	Stair Construction	Wood		Flight	2174	.87	19.5%
3010	Wall Finishes	Paint		S.F. Surface	.73	.47	
3020	Floor Finishes	10% hardwood, 70% carpet, 20% ceramic tile		S.F. Floor	9.51	9.51	
3030	Ceiling Finishes	Gypsum board on wood furring		S.F. Ceiling	3.95	3.95	
D. SERVICES							
D10 Conveying							
1010	Elevators & Lifts	One hydraulic passenger elevator		Each	80,900	8.09	7.9%
1020	Escalators & Moving Walks	N/A		—	—	—	
D20 Plumbing							
2010	Plumbing Fixtures	Kitchen toilet and service fixtures, supply and drainage	1 Fixture/150 S.F. Floor	Each	962	6.41	
2020	Domestic Water Distribution	Gas fired water heater		S.F. Floor	1.90	1.90	8.1%
2040	Rain Water Drainage	N/A		—	—	—	
D30 HVAC							
3010	Energy Supply	Oil fired hot water, baseboard radiation		S.F. Floor	6.88	6.88	
3020	Heat Generating Systems	N/A		—	—	—	
3030	Cooling Generating Systems	N/A		—	—	—	14.0%
3050	Terminal & Package Units	Split system with air cooled condensing unit		S.F. Floor	7.49	7.49	
3090	Other HVAC Sys. & Equipment	N/A		—	—	—	
D40 Fire Protection							
4010	Sprinklers	Wet pipe sprinkler system		S.F. Floor	2.95	2.95	2.9%
4020	Standpipes	N/A		—	—	—	
D50 Electrical							
5010	Electrical Service/Distribution	600 ampere service, panel board and feeders		S.F. Floor	4.98	4.98	
5020	Lighting & Branch Wiring	Fluorescent fixtures, receptacles, switches, A.C. and misc. power		S.F. Floor	9.29	9.29	26.2%
5030	Communications & Security	Alarm, communication system, internet and phone wiring, and generator set		S.F. Floor	12.54	12.54	
5090	Other Electrical Systems	Emergency generator, 7.5 kW		S.F. Floor	.19	.19	
E. EQUIPMENT & FURNISHINGS							
1010	Commercial Equipment	N/A		—	—	—	
1020	Institutional Equipment	N/A		—	—	—	
1030	Vehicular Equipment	N/A		—	—	—	0.0%
1090	Other Equipment	N/A		—	—	—	
F. SPECIAL CONSTRUCTION							
1020	Integrated Construction	N/A		—	—	—	
1040	Special Facilities	N/A		—	—	—	0.0%
G. BUILDING SITEWORK	**N/A**						

		Sub-Total	102.93	100%
CONTRACTOR FEES (General Requirements: 10%, Overhead: 5%, Profit: 10%)		25%	25.69	
ARCHITECT FEES		9%	11.58	
	Total Building Cost		**140.20**	

131

Costs per square foot of floor area

Exterior Wall	S.F. Area	4000	6000	8000	10000	12000	14000	16000	18000	20000
	L.F. Perimeter	260	320	384	424	460	484	510	540	576
Vertical Redwood Siding	Wood Frame	159.15	141.30	132.70	126.35	121.95	118.50	115.95	114.05	112.65
Brick Veneer	Wood Frame	173.80	153.65	143.90	136.45	131.20	127.00	123.90	121.65	120.00
Aluminum Siding	Wood Frame	157.75	140.45	132.05	126.00	121.75	118.45	116.00	114.20	112.90
Brick on Block	Wood Truss	183.65	162.15	151.90	143.80	138.15	133.50	130.10	127.60	125.85
Limestone on Block	Wood Truss	214.10	187.10	174.20	163.55	155.95	149.50	144.85	141.50	139.20
Stucco on Block	Wood Truss	165.20	147.00	138.25	131.80	127.25	123.65	121.05	119.10	117.70
Perimeter Adj., Add or Deduct	Per 100 L.F.	11.30	7.55	5.60	4.50	3.80	3.15	2.75	2.50	2.25
Story Hgt. Adj., Add or Deduct	Per 1 Ft.	1.30	1.10	0.90	0.85	0.75	0.65	0.60	0.60	0.55
For Basement, add $29.50 per square foot of basement area										

The above costs were calculated using the basic specifications shown on the facing page. These costs should be adjusted where necessary for design alternatives and owner's requirements. Reported completed project costs, for this type of structure, range from $98.20 to $301.95 per S.F.

Common additives

Description	Unit	$ Cost
Autopsy Table, Standard	Each	9850
Deluxe	Each	19,400
Directory Boards, Plastic, glass covered		
30" x 20"	Each	595
36" x 48"	Each	1450
Aluminum, 24" x 18"	Each	600
36" x 24"	Each	675
48" x 32"	Each	980
48" x 60"	Each	2025
Emergency Lighting, 25 watt, battery operated		
Lead battery	Each	282
Nickel cadmium	Each	805
Mortuary Refrigerator, End operated		
Two capacity	Each	13,600
Six capacity	Each	24,500

Description	Unit	$ Cost
Planters, Precast concrete		
48" diam., 24" high	Each	755
7" diam., 36" high	Each	1250
Fiberglass, 36" diam., 24" high	Each	575
60" diam., 24" high	Each	955
Smoke Detectors		
Ceiling type	Each	187
Duct type	Each	480

Model costs calculated for a 1 story building with 12' story height and 10,000 square feet of floor area

				Unit	Unit Cost	Cost Per S.F.	% Of Sub-Total
A. SUBSTRUCTURE							
1010	Standard Foundations	Poured concrete; strip and spread footings		S.F. Ground	1.85	1.85	
1020	Special Foundations	N/A		—	—	—	
1030	Slab on Grade	4" reinforced concrete with vapor barrier and granular base		S.F. Slab	4.74	4.74	10.9%
2010	Basement Excavation	Site preparation for slab and trench for foundation wall and footing		S.F. Ground	.26	.26	
2020	Basement Walls	4' foundation wall		L.F. Wall	69	3.22	
B. SHELL							
B10 Superstructure							
1010	Floor Construction	N/A		—	—	—	4.7 %
1020	Roof Construction	Plywood on wood truss		S.F. Roof	4.36	4.36	
B20 Exterior Enclosure							
2010	Exterior Walls	1" x 4" vertical T & G redwood siding on wood studs	90% of wall	S.F. Wall	10.42	4.77	
2020	Exterior Windows	Double hung wood	10% of wall	Each	471	2	8.6%
2030	Exterior Doors	Wood swinging double doors, single leaf hollow metal		Each	2091	1.25	
B30 Roofing							
3010	Roof Coverings	Single ply membrane, fully adhered; polyisocyanurate sheets		S.F. Roof	5.95	5.95	6.4%
3020	Roof Openings	N/A		—	—	—	
C. INTERIORS							
1010	Partitions	Gypsum board on wood studs with sound deadening board	15 S.F. Floor/L.F. Partition	S.F. Partition	8.79	4.69	
1020	Interior Doors	Single leaf wood	150 S.F. Floor/Door	Each	565	3.76	
1030	Fittings	N/A		—	—	—	
2010	Stair Construction	N/A		—	—	—	28.9%
3010	Wall Finishes	50% wallpaper, 25% wood paneling, 25% paint		S.F. Surface	3.28	3.50	
3020	Floor Finishes	70% carpet, 30% ceramic tile		S.F. Floor	10.60	10.60	
3030	Ceiling Finishes	Fiberglass board on exposed grid, suspended		S.F. Ceiling	4.30	4.30	
D. SERVICES							
D10 Conveying							
1010	Elevators & Lifts	N/A		—	—	—	0.0 %
1020	Escalators & Moving Walks	N/A		—	—	—	
D20 Plumbing							
2010	Plumbing Fixtures	Toilet and service fixtures, supply and drainage	1 Fixture/770 S.F. Floor	Each	2087	2.71	
2020	Domestic Water Distribution	Electric water heater		S.F. Floor	8.88	8.88	13.0%
2040	Rain Water Drainage	Roof drain		S.F. Roof	.47	.47	
D30 HVAC							
3010	Energy Supply	N/A		—	—	—	
3020	Heat Generating Systems	Included in D3030		—	—	—	
3030	Cooling Generating Systems	Multizone rooftop unit, gas heating, electric cooling		S.F. Floor	15.50	15.50	16.7 %
3050	Terminal & Package Units	N/A		—	—	—	
3090	Other HVAC Sys. & Equipment	N/A		—	—	—	
D40 Fire Protection							
4010	Sprinklers	Wet pipe sprinkler system		S.F. Floor	2.77	2.77	3.0%
4020	Standpipes	N/A		—	—	—	
D50 Electrical							
5010	Electrical Service/Distribution	400 ampere service, panel board and feeders		S.F. Floor	1.72	1.72	
5020	Lighting & Branch Wiring	Fluorescent fixtures, receptacles, switches, A.C. and misc. power		S.F. Floor	5.15	5.15	7.7%
5030	Communications & Security	Alarm systems and emergency lighting		S.F. Floor	.22	.22	
5090	Other Electrical Systems	Emergency generator, 15 kW		S.F. Floor	.08	.08	
E. EQUIPMENT & FURNISHINGS							
1010	Commercial Equipment	N/A		—	—	—	
1020	Institutional Equipment	N/A		—	—	—	
1030	Vehicular Equipment	N/A		—	—	—	0.0 %
1090	Other Equipment	N/A		—	—	—	
F. SPECIAL CONSTRUCTION							
1020	Integrated Construction	N/A		—	—	—	
1040	Special Facilities	N/A		—	—	—	0.0 %
G. BUILDING SITEWORK	**N/A**						

		Sub-Total	92.75	100%
CONTRACTOR FEES (General Requirements: 10%, Overhead: 5%, Profit: 10%)		25%	23.17	
ARCHITECT FEES		9%	10.43	

	Total Building Cost	**126.35**

Costs per square foot of floor area

Exterior Wall	S.F. Area	12000	14000	16000	19000	21000	23000	26000	28000	30000
	L.F. Perimeter	440	474	510	556	583	607	648	670	695
E.I.F.S. on Concrete Block	Steel Frame	121.15	118.30	116.35	113.70	112.25	111.00	109.50	108.65	107.85
Tilt-up Concrete Wall	Steel Frame	118.05	115.45	113.60	111.25	109.90	108.70	107.40	106.60	105.85
Face Brick with Concrete Block Back-up	Bearing Walls	120.65	117.35	115.05	111.95	110.30	108.75	107.00	105.95	105.05
	Steel Frame	127.20	123.85	121.50	118.45	116.75	115.20	113.50	112.40	111.50
Stucco on Concrete Block	Bearing Walls	111.20	108.60	106.75	104.35	103.05	101.85	100.50	99.75	98.95
	Steel Frame	118.20	115.55	113.75	111.30	110.00	108.85	107.45	106.65	105.90
Perimeter Adj., Add or Deduct	Per 100 L.F.	7.05	6.10	5.25	4.50	4.10	3.65	3.30	3.00	2.85
Story Hgt. Adj., Add or Deduct	Per 1 Ft.	1.65	1.55	1.45	1.30	1.35	1.20	1.20	1.05	1.05
For Basement, add $33.50 per square foot of basement area										

The above costs were calculated using the basic specifications shown on the facing page. These costs should be adjusted where necessary for design alternatives and owner's requirements. Reported completed project costs, for this type of structure, range from $52.25 to $152.00 per S.F.

Common additives

Description	Unit	$ Cost
Emergency Lighting, 25 watt, battery operated		
Lead battery	Each	282
Nickel cadmium	Each	805
Smoke Detectors		
Ceiling type	Each	187
Duct type	Each	480
Sound System		
Amplifier, 250 watts	Each	2350
Speaker, ceiling or wall	Each	191
Trumpet	Each	365

Model costs calculated for a 1 story building with 14' story height and 21,000 square feet of floor area

Garage, Auto Sales

				Unit	Unit Cost	Cost Per S.F.	% Of Sub-Total
A. SUBSTRUCTURE							
1010	Standard Foundations	Poured concrete; strip and spread footings		S.F. Ground	1.45	1.45	
1020	Special Foundations	N/A		—	—	—	
1030	Slab on Grade	4" reinforced concrete with vapor barrier and granular base		S.F. Slab	7.29	7.29	13.3%
2010	Basement Excavation	Site preparation for slab and trench for foundation wall and footing		S.F. Ground	.26	.26	
2020	Basement Walls	4' foundation wall		L.F. Wall	78	2.17	
B. SHELL							
	B10 Superstructure						
1010	Floor Construction	N/A		—	—	—	14.2 %
1020	Roof Construction	Metal deck, open web steel joists, beams, columns		S.F. Roof	11.92	11.92	
	B20 Exterior Enclosure						
2010	Exterior Walls	E.I.F.S.	70% of wall	S.F. Wall	27.75	7.55	
2020	Exterior Windows	Window wall	30% of wall	Each	50	5.85	19.8%
2030	Exterior Doors	Double aluminum and glass, hollow metal, steel overhead		Each	3788	3.25	
	B30 Roofing						
3010	Roof Coverings	Built-up tar and gravel with flashing; perlite/EPS composite insulation		S.F. Roof	5.25	5.25	6.3%
3020	Roof Openings	N/A		—	—	—	
C. INTERIORS							
1010	Partitions	Gypsum board on metal studs	28 S.F. Floor/L.F. Partition	S.F. Partition	4.15	1.78	
1020	Interior Doors	Hollow metal	280 S.F. Floor/Door	Each	875	3.13	
1030	Fittings	N/A		—	—	—	
2010	Stair Construction	N/A		—	—	—	12.0%
3010	Wall Finishes	Paint		S.F. Surface	.74	.63	
3020	Floor Finishes	50% vinyl tile, 50% paint		S.F. Floor	2.41	2.41	
3030	Ceiling Finishes	Fiberglass board on exposed grid, suspended	50% of area	S.F. Ceiling	4.30	2.15	
D. SERVICES							
	D10 Conveying						
1010	Elevators & Lifts	N/A		—	—	—	0.0 %
1020	Escalators & Moving Walks	N/A		—	—	—	
	D20 Plumbing						
2010	Plumbing Fixtures	Toilet and service fixtures, supply and drainage	1 Fixture/1500 S.F. Floor	Each	2610	1.74	
2020	Domestic Water Distribution	Gas fired water heater		S.F. Floor	1.49	1.49	6.4%
2040	Rain Water Drainage	Roof drains		S.F. Roof	2.13	2.13	
	D30 HVAC						
3010	Energy Supply	N/A		—	—	—	
3020	Heat Generating Systems	N/A		—	—	—	
3030	Cooling Generating Systems	N/A		—	—	—	10.2 %
3050	Terminal & Package Units	Single zone rooftop unit, gas heating, electric cooling		S.F. Floor	8.21	8.21	
3090	Other HVAC Sys. & Equipment	Underfloor garage exhaust system		S.F. Floor	.32	.32	
	D40 Fire Protection						
4010	Sprinklers	Wet pipe sprinkler system		S.F. Floor	3.62	3.62	4.7%
4020	Standpipes	Standpipe		S.F. Floor	.34	.34	
	D50 Electrical						
5010	Electrical Service/Distribution	200 ampere service, panel board and feeders		S.F. Floor	.42	.42	
5020	Lighting & Branch Wiring	T-8 fluorescent fixtures, receptacles, switches, A.C. and misc. power		S.F. Floor	6.28	6.28	11.7%
5030	Communications & Security	Addressable alarm systems, partial internet wiring and emergency lighting		S.F. Floor	3.02	3.02	
5090	Other Electrical Systems	Emergency generator, 7.5 kW		S.F. Floor	.07	.07	
E. EQUIPMENT & FURNISHINGS							
1010	Commercial Equipment	N/A		—	—	—	
1020	Institutional Equipment	N/A		—	—	—	1.4 %
1030	Vehicular Equipment	Hoists, compressor, fuel pump		S.F. Floor	1.21	1.21	
1090	Other Equipment	N/A		—	—	—	
F. SPECIAL CONSTRUCTION							
1020	Integrated Construction	N/A		—	—	—	0.0 %
1040	Special Facilities	N/A		—	—	—	
G. BUILDING SITEWORK	**N/A**						

		Sub-Total	83.94	100%
	CONTRACTOR FEES (General Requirements: 10%, Overhead: 5%, Profit: 10%)		25%	20.96
	ARCHITECT FEES		7%	7.35

Total Building Cost	**112.25**

Costs per square foot of floor area

Exterior Wall	S.F. Area	85000	115000	145000	175000	205000	235000	265000	295000	325000
	L.F. Perimeter	529	638	723	823	923	951	1037	1057	1132
Face Brick with Concrete Block Back-up	Steel Frame	72.00	70.55	69.50	68.90	68.55	67.80	67.50	67.05	66.85
	R/Conc. Frame	53.00	51.55	50.50	49.90	49.55	48.80	48.50	48.05	47.85
Precast Concrete Precast Concrete	Steel Frame	74.90	73.20	72.00	71.25	70.80	69.95	69.55	69.00	68.80
	R/Conc. Frame	55.25	53.55	52.30	51.60	51.15	50.25	49.90	49.35	49.05
Reinforced Concrete Reinforced Concrete	Steel Frame	71.30	70.00	69.15	68.60	68.30	67.70	67.45	67.10	66.90
	R/Conc. Frame	51.25	50.00	49.10	48.60	48.25	47.65	47.40	47.05	46.85
Perimeter Adj., Add or Deduct	Per 100 L.F.	1.55	1.15	0.90	0.75	0.60	0.55	0.50	0.40	0.40
Story Hgt. Adj., Add or Deduct	Per 1 Ft.	0.55	0.45	0.45	0.40	0.35	0.35	0.30	0.30	0.30
Basement—Not Applicable										

The above costs were calculated using the basic specifications shown on the facing page. These costs should be adjusted where necessary for design alternatives and owner's requirements. Reported completed project costs, for this type of structure, range from $ 32.05 to $ 124.05 per S.F.

Common additives

Description	Unit	$ Cost
Elevators, Electric passenger, 5 stops		
2000# capacity	Each	158,700
3500# capacity	Each	167,200
5000# capacity	Each	170,700
Barrier gate w/programmable controller	Each	4000
Booth for attendant, average	Each	12,400
Fee computer	Each	15,000
Ticket spitter with time/date stamp	Each	7450
Mag strip encoding	Each	21,000
Collection station, pay on foot	Each	126,000
Parking control software	Each	25,200 - 108,500
Painting, Parking stalls	Stall	13.25
Parking Barriers		
Timber with saddles, 4" x 4"	L.F.	7.10
Precast concrete, 6" x 10" x 6'	Each	81
Traffic Signs, directional, 12" x 18", high density	Each	84.50

Important: See the Reference Section for Location Factors

Model costs calculated for a 5 story building with 10' story height and 145,000 square feet of floor area

Garage, Parking

				Unit	Unit Cost	Cost Per S.F.	% Of Sub-Total
A. SUBSTRUCTURE							
1010	Standard Foundations	Poured concrete; strip and spread footings		S.F. Ground	6.90	1.38	
1020	Special Foundations	N/A		—	—	—	
1030	Slab on Grade	6" reinforced concrete with vapor barrier and granular base		S.F. Slab	6.14	1.22	10.5%
2010	Basement Excavation	Site preparation for slab and trench for foundation wall and footing		S.F. Ground	.17	.03	
2020	Basement Walls	4' foundation wall		L.F. Wall	69	1.38	
B. SHELL							
	B10 Superstructure						
1010	Floor Construction	Double tee precast concrete slab, precast concrete columns		S.F. Floor	22.35	17.88	46.9%
1020	Roof Construction	N/A		—	—	—	
	B20 Exterior Enclosure						
2010	Exterior Walls	Face brick with concrete block backup	40% of story height	S.F. Wall	30.89	3.08	
2020	Exterior Windows	N/A		—	—	—	8.1%
2030	Exterior Doors	N/A		—	—	—	
	B30 Roofing						
3010	Roof Coverings	N/A		—	—	—	0.0%
3020	Roof Openings	N/A		—	—	—	
C. INTERIORS							
1010	Partitions	Concrete block		S.F. Partition	29.38	1.13	
1020	Interior Doors	Hollow metal		Each	17,500	.12	
1030	Fittings	N/A		—	—	—	
2010	Stair Construction	Concrete		Flight	4025	.28	4.3%
3010	Wall Finishes	Paint		S.F. Surface	1.56	.12	
3020	Floor Finishes	N/A		—	—	—	
3030	Ceiling Finishes	N/A		—	—	—	
D. SERVICES							
	D10 Conveying						
1010	Elevators & Lifts	Two hydraulic passenger elevators		Each	164,575	2.27	6.0%
1020	Escalators & Moving Walks	N/A		—	—	—	
	D20 Plumbing						
2010	Plumbing Fixtures	Toilet and service fixtures, supply and drainage	1 Fixture/18,125 S.F. Floor	Each	725	.04	
2020	Domestic Water Distribution	Electric water heater		S.F. Floor	.07	.07	3.6%
2040	Rain Water Drainage	Roof drains		S.F. Roof	6.40	1.28	
	D30 HVAC						
3010	Energy Supply	N/A		—	—	—	
3020	Heat Generating Systems	N/A		—	—	—	
3030	Cooling Generating Systems	N/A		—	—	—	0.0%
3050	Terminal & Package Units	N/A		—	—	—	
3090	Other HVAC Sys. & Equipment	N/A		—	—	—	
	D40 Fire Protection						
4010	Sprinklers	Dry pipe sprinkler system		S.F. Floor	3.35	3.35	9.0%
4020	Standpipes	Standpipes and hose systems		S.F. Floor	.08	.08	
	D50 Electrical						
5010	Electrical Service/Distribution	400 ampere service, panel board and feeders		S.F. Floor	.24	.24	
5020	Lighting & Branch Wiring	T-8 fluorescent fixtures, receptacles, switches and misc. power		S.F. Floor	2.85	2.85	8.5%
5030	Communications & Security	Addressable alarm systems and emergency lighting		S.F. Floor	.10	.10	
5090	Other Electrical Systems	Emergency generator, 7.5 kW		S.F. Floor	.06	.06	
E. EQUIPMENT & FURNISHINGS							
1010	Commercial Equipment	N/A		—	—	—	
1020	Institutional Equipment	N/A		—	—	—	
1030	Vehicular Equipment	Ticket dispensers, booths, automatic gates		S.F. Floor	1.16	1.16	3.0%
1090	Other Equipment	N/A		—	—	—	
F. SPECIAL CONSTRUCTION							
1020	Integrated Construction	N/A		—	—	—	0.0%
1040	Special Facilities	N/A		—	—	—	
G. BUILDING SITEWORK	**N/A**						

			Sub-Total	38.12	100%
CONTRACTOR FEES (General Requirements: 10%, Overhead: 5%, Profit: 10%)			25%	9.52	
ARCHITECT FEES			6%	2.86	
		Total Building Cost		**50.50**	

Costs per square foot of floor area

Exterior Wall	S.F. Area	20000	30000	40000	50000	75000	100000	125000	150000	175000
	L.F. Perimeter	400	500	600	650	775	900	1000	1100	1185
Reinforced Concrete	R/Conc. Frame	93.55	87.50	84.45	81.60	77.70	75.80	74.35	73.50	72.75
Perimeter Adj., Add or Deduct	Per 100 L.F.	5.40	3.60	2.65	2.15	1.45	1.10	0.90	0.70	0.65
Story Hgt. Adj., Add or Deduct	Per 1 Ft.	2.05	1.65	1.55	1.30	1.05	0.85	0.90	0.75	0.70
Basement—Not Applicable										

The above costs were calculated using the basic specifications shown on the facing page. These costs should be adjusted where necessary for design alternatives and owner's requirements. Reported completed project costs, for this type of structure, range from $46.60 to $111.05 per S.F.

Common additives

Description	Unit	$ Cost
Elevators, Hydraulic passenger, 2 stops		
1500# capacity	Each	62,800
2500# capacity	Each	66,300
3500# capacity	Each	69,800
Barrier gate w/programmable controller	Each	4000
Booth for attendant, average	Each	12,400
Fee computer	Each	15,000
Ticket spitter with time/date stamp	Each	7450
Mag strip encoding	Each	21,000
Collection station, pay on foot	Each	126,000
Parking control software	Each	25,200 - 108,500
Painting, Parking stalls	Stall	13.25
Parking Barriers		
Timber with saddles, 4" x 4"	L.F.	7.10
Precast concrete, 6" x 10" x 6'	Each	81
Traffic Signs, directional, 12" x 18"	Each	84.50

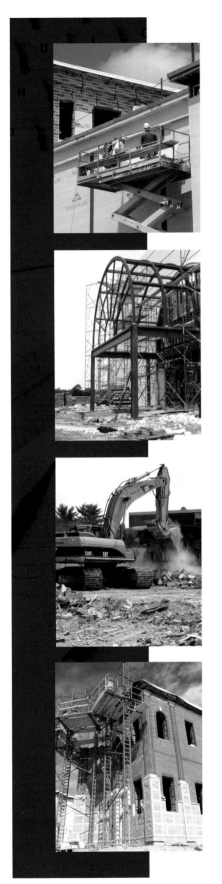

RSMeans Business Solutions

Leaders in Cost Engineering for the Design, Construction, and Facilities Management Industry...

Market Analysis and Third Party Validation – RSMeans conducts research studies from numerous perspectives, including escalation costs in local markets, cost estimates for legal damages, and trend research for industry associations, as well as product adoption and productivity studies for building product manufacturers. Means is often called upon to produce "third party validation" construction cost estimates for owners, architects, and boards of directors.

Predictive Cost Models for Facility Owners – From early planning, cost modeling is driven by close attention to standards, codes, methods, and business processes that impact cost and operational effectiveness. Predictive cost models offer cost control and standardization for facility owners, particularly in repetitive building environments. RSMeans custom models are based upon actual architectural plans and historical budgets, with cost data and location factors adjusted to regions anywhere in North America.

Custom Cost Databases and Web-Based Dashboards – Custom applications of the RSMeans cost databases use the building blocks of construction cost engineering, including a standard code of accounts, activity descriptions, material, labor, and equipment costs. RSMeans has in-house technical expertise for rapid deployment of cost databases, online cost calculators, and web-based analytic dashboards.

Job Order Contracting and FacilitiesWorks™ – A delivery method for construction renovation to reduce cycle times and overhead costs. RSMeans offers consulting in JOC management practices, FacilitiesWorks™ software for renovation cost estimating and project management, and on-site training programs. JOC is available to government agencies, colleges and universities, school districts, municipalities, and health care systems.

Life Cycle Cost Analysis and Sustainability – Building upon its extensive facility cost databases, RSMeans offers life cycle cost studies and sustainability models to analyze total cost of ownership (TCO) for high performance buildings. Analyses include capital costs, life cycle and operations costs, as well as studies specific to integrated building automation, energy efficiency and security systems. Cost savings realized in operations, maintenance and replacement are analyzed enterprise-wide.

CORPORATE • HEALTH CARE • EDUCATION • GOVERNMENT • MUNICIPAL • RETAIL • ASSOCIATIONS • BUILDING PRODUCT MANUFACTURERS • LEGAL

For more information contact RSMeans Business Solutions at 781-422-5101 or visit our website at www.rsmeans.com or email consulting@rsmeans.com

RSMeans

Model costs calculated for a 2 story building with 10' story height and 100,000 square feet of floor area

Garage, Underground Parking

			Unit	Unit Cost	Cost Per S.F.	% Of Sub-Total
A. SUBSTRUCTURE						
1010	Standard Foundations	Poured concrete; strip and spread footings and waterproofing	S.F. Ground	7.86	3.93	
1020	Special Foundations	N/A	—	—	—	
1030	Slab on Grade	5" reinforced concrete with vapor barrier and granular base	S.F. Slab	6.27	3.14	20.4%
2010	Basement Excavation	Excavation 24' deep	S.F. Ground	8.80	4.40	
2020	Basement Walls	N/A	—	—	—	
B. SHELL						
	B10 Superstructure					
1010	Floor Construction	Cast-in-place concrete beam and slab, concrete columns	S.F. Floor	28.40	14.20	48.7%
1020	Roof Construction	Cast-in-place concrete beam and slab, concrete columns	S.F. Roof	26.32	13.16	
	B20 Exterior Enclosure					
2010	Exterior Walls	Cast-in place concrete	S.F. Wall	20.44	3.68	
2020	Exterior Windows	N/A	—	—	—	6.9%
2030	Exterior Doors	Steel overhead, hollow metal	Each	3965	.17	
	B30 Roofing					
3010	Roof Coverings	Neoprene membrane traffic deck	S.F. Roof	4.22	2.11	3.8%
3020	Roof Openings	N/A	—	—	—	
C. INTERIORS						
1010	Partitions	Concrete block	S.F. Partition	38.48	.74	
1020	Interior Doors	Hollow metal	Each	7000	.07	
1030	Fittings	N/A	—	—	—	
2010	Stair Construction	Concrete	Flight	6400	.32	2.2%
3010	Wall Finishes	Paint	S.F. Surface	2.34	.09	
3020	Floor Finishes	N/A	—	—	—	
3030	Ceiling Finishes	N/A	—	—	—	
D. SERVICES						
	D10 Conveying					
1010	Elevators & Lifts	Two hydraulic passenger elevators	Each	81,000	1.62	2.9%
1020	Escalators & Moving Walks	N/A	—	—	—	
	D20 Plumbing					
2010	Plumbing Fixtures	Drainage in parking areas, toilets, & service fixtures 1 Fixture/5000 S.F. Floor	Each	.04	.04	
2020	Domestic Water Distribution	Electric water heater	S.F. Floor	.10	.10	2.0%
2040	Rain Water Drainage	Roof drains	S.F. Roof	2	1	
	D30 HVAC					
3010	Energy Supply	N/A	—	—	—	
3020	Heat Generating Systems	N/A	—	—	—	
3030	Cooling Generating Systems	N/A	—	—	—	0.3 %
3050	Terminal & Package Units	Exhaust fans	S.F. Floor	.15	.15	
3090	Other HVAC Sys. & Equipment	N/A	—	—	—	
	D40 Fire Protection					
4010	Sprinklers	Dry pipe sprinkler system	S.F. Floor	3.33	3.33	6.2%
4020	Standpipes	Dry standpipe system, class 3	S.F. Floor	.13	.13	
	D50 Electrical					
5010	Electrical Service/Distribution	200 ampere service, panel board and feeders	S.F. Floor	.13	.13	
5020	Lighting & Branch Wiring	T-8 fluorescent fixtures, receptacles, switches and misc. power	S.F. Floor	3.06	3.06	6.0%
5030	Communications & Security	Addressable alarm systems and emergency lighting	S.F. Floor	.14	.14	
5090	Other Electrical Systems	Emergency generator, 11.5 kW	S.F. Floor	.06	.06	
E. EQUIPMENT & FURNISHINGS						
1010	Commercial Equipment	N/A	—	—	—	
1020	Institutional Equipment	N/A	—	—	—	
1030	Vehicular Equipment	Ticket dispensers, booths, automatic gates	S.F. Floor	.36	.36	0.6 %
1090	Other Equipment	N/A	—	—	—	
F. SPECIAL CONSTRUCTION						
1020	Integrated Construction	N/A	—	—	—	0.0%
1040	Special Facilities	N/A	—	—	—	
G. BUILDING SITEWORK	**N/A**					

			Sub-Total	56.13	100%
CONTRACTOR FEES (General Requirements: 10%, Overhead: 5%, Profit: 10%)			25%	14.06	
ARCHITECT FEES			8%	5.61	
		Total Building Cost		**75.80**	

Costs per square foot of floor area

Exterior Wall	S.F. Area	2000	4000	6000	8000	10000	12000	14000	16000	18000
	L.F. Perimeter	180	260	340	420	500	580	586	600	610
Concrete Block	Wood Joists	140.85	121.55	115.10	111.80	109.85	108.60	105.00	102.70	100.75
	Steel Joists	144.45	125.70	119.45	116.30	114.40	113.20	109.65	107.35	105.35
Poured Concrete Poured Concrete	Wood Joists	152.30	130.05	122.65	118.90	116.65	115.20	110.75	107.85	105.35
	Steel Joists	156.95	134.70	127.30	123.55	121.35	119.85	115.40	112.45	109.95
Stucco	Wood Frame	147.20	124.50	116.90	113.20	110.90	109.35	104.75	101.80	99.20
Insulated Metal Panels	Steel Frame	143.35	126.20	120.45	117.55	115.80	114.70	111.50	109.55	107.75
Perimeter Adj., Add or Deduct	Per 100 L.F.	24.50	12.25	8.10	6.15	4.90	4.10	3.60	3.05	2.70
Story Hgt. Adj., Add or Deduct	Per 1 Ft.	1.80	1.25	1.05	1.05	1.10	1.00	0.85	0.75	0.65
For Basement, add $31.50 per square foot of basement area										

The above costs were calculated using the basic specifications shown on the facing page. These costs should be adjusted where necessary for design alternatives and owner's requirements. Reported completed project costs, for this type of structure, range from $66.85 to $200.80 per S.F.

Common additives

Description	Unit	$ Cost
Air Compressors		
Electric 1-1/2 H.P., standard controls	Each	1325
Dual controls	Each	1700
5 H.P. 115/230 Volt, standard controls	Each	3450
Dual controls	Each	3600
Product Dispenser		
with vapor recovery for 6 nozzles	Each	22,900
Hoists, Single post		
8000# cap., swivel arm	Each	8950
Two post, adjustable frames, 11,000# cap.	Each	12,600
24,000# cap.	Each	18,700
7500# Frame support	Each	10,600
Four post, roll on ramp	Each	9825
Lockers, Steel, single tier, 60" or 72"	Opening	191 - 310
2 tier, 60" or 72" total	Opening	107 - 141
5 tier, box lockers	Opening	65 - 83.50
Locker bench, lam. maple top only	L.F.	21
Pedestals, steel pipe	Each	63.50
Lube Equipment		
3 reel type, with pumps, no piping	Each	11,500
Spray Painting Booth, 26' long, complete	Each	20,400

Important: See the Reference Section for Location Factors

Model costs calculated for a 1 story building with 14' story height and 10,000 square feet of floor area

				Unit	Unit Cost	Cost Per S.F.	% Of Sub-Total
A. SUBSTRUCTURE							
1010	Standard Foundations	Poured concrete; strip and spread footings		S.F. Ground	2.27	2.27	
1020	Special Foundations	N/A		—	—	—	
1030	Slab on Grade	6" reinforced concrete with vapor barrier and granular base		S.F. Slab	7.29	7.29	16.2%
2010	Basement Excavation	Site preparation for slab and trench for foundation wall and footing		S.F. Ground	.26	.26	
2020	Basement Walls	4' foundation wall		L.F. Wall	78	3.91	
B. SHELL							
B10 Superstructure							
1010	Floor Construction	N/A		—	—	—	7.8 %
1020	Roof Construction	Metal deck on open web steel joists		S.F. Roof	6.61	6.61	
B20 Exterior Enclosure							
2010	Exterior Walls	Concrete block	80% of wall	S.F. Wall	11.05	6.19	
2020	Exterior Windows	Hopper type commercial steel	5% of wall	Each	492	1.15	10.9%
2030	Exterior Doors	Steel overhead and hollow metal	15% of wall	S.F. Door	18.10	1.90	
B30 Roofing							
3010	Roof Coverings	Built-up tar and gravel; perlite/EPS composite insulation		S.F. Roof	6.01	6.01	7.1%
3020	Roof Openings	Skylight		S.F. Roof	.02	.02	
C. INTERIORS							
1010	Partitions	Concrete block	50 S.F. Floor/L.F. Partition	S.F. Partition	7	1.40	
1020	Interior Doors	Single leaf hollow metal	3000 S.F. Floor/Door	Each	875	.30	
1030	Fittings	Toilet partitions		S.F. Floor	.22	.22	
2010	Stair Construction	N/A		—	—	—	7.2%
3010	Wall Finishes	Paint		S.F. Surface	6.68	2.67	
3020	Floor Finishes	90% metallic floor hardener, 10% vinyl composition tile		S.F. Floor	1.17	1.17	
3030	Ceiling Finishes	Gypsum board on wood joists in office and washrooms	10% of area	S.F. Ceiling	3.20	.32	
D. SERVICES							
D10 Conveying							
1010	Elevators & Lifts	N/A		—	—	—	0.0 %
1020	Escalators & Moving Walks	N/A		—	—	—	
D20 Plumbing							
2010	Plumbing Fixtures	Toilet and service fixtures, supply and drainage	1 Fixture/500 S.F. Floor	Each	1275	2.55	
2020	Domestic Water Distribution	Gas fired water heater		S.F. Floor	.42	.42	5.9%
2040	Rain Water Drainage	Roof drains		S.F. Roof	2.03	2.03	
D30 HVAC							
3010	Energy Supply	N/A		—	—	—	
3020	Heat Generating Systems	N/A		—	—	—	
3030	Cooling Generating Systems	N/A		—	—	—	10.5 %
3050	Terminal & Package Units	Single zone AC unit		S.F. Floor	8.40	8.40	
3090	Other HVAC Sys. & Equipment	Garage exhaust system		S.F. Floor	.50	.50	
D40 Fire Protection							
4010	Sprinklers	Sprinklers, ordinary hazard		S.F. Floor	3.62	3.62	5.1%
4020	Standpipes	Standpipe		S.F. Floor	.72	.72	
D50 Electrical							
5010	Electrical Service/Distribution	200 ampere service, panel board and feeders		S.F. Floor	.51	.51	
5020	Lighting & Branch Wiring	T-8 fluorescent fixtures, receptacles, switches, A.C. and misc. power		S.F. Floor	7.64	7.64	13.5%
5030	Communications & Security	Addressable alarm systems, partial internet wiring and emergency lighting		S.F. Floor	3.17	3.17	
5090	Other Electrical Systems	Emergency generator, 15 kW		S.F. Floor	.08	.08	
E. EQUIPMENT & FURNISHINGS							
1010	Commercial Equipment	N/A		—	—	—	
1020	Institutional Equipment	N/A		—	—	—	15.8 %
1030	Vehicular Equipment	Hoists		S.F. Floor	13.42	13.42	
1090	Other Equipment	N/A		—	—	—	
F. SPECIAL CONSTRUCTION							
1020	Integrated Construction	N/A		—	—	—	0.0 %
1040	Special Facilities	N/A		—	—	—	
G. BUILDING SITEWORK	**N/A**						

		Sub-Total	84.75	100%
CONTRACTOR FEES (General Requirements: 10%, Overhead: 5%, Profit: 10%)			25%	21.17	
ARCHITECT FEES			8%	8.48	

Total Building Cost	**114.40**

Costs per square foot of floor area

Exterior Wall	S.F. Area	600	800	1000	1200	1400	1600	1800	2000	2200
	L.F. Perimeter	100	120	126	140	153	160	170	180	190
Face Brick with Concrete Block Back-up	Wood Truss	253.05	235.85	216.15	207.45	200.75	193.25	188.55	184.80	181.70
	Steel Joists	247.70	230.50	210.80	202.15	195.40	187.90	183.25	179.45	176.35
Enameled Sandwich Panel	Steel Frame	227.75	212.60	196.05	188.55	182.80	176.45	172.45	169.25	166.65
Tile on Concrete Block	Steel Joists	263.75	245.10	223.35	213.85	206.55	198.20	193.05	188.90	185.50
Aluminum Siding	Wood Frame	215.75	202.25	187.95	181.35	176.30	170.85	167.45	164.65	162.35
Wood Siding	Wood Frame	218.20	204.45	189.80	183.05	177.95	172.35	168.80	165.95	163.60
Perimeter Adj., Add or Deduct	Per 100 L.F.	111.30	83.45	66.80	55.65	47.75	41.75	37.15	33.40	30.30
Story Hgt. Adj., Add or Deduct	Per 1 Ft.	7.25	6.45	5.45	5.10	4.80	4.35	4.05	3.90	3.75
Basement—Not Applicable										

The above costs were calculated using the basic specifications shown on the facing page. These costs should be adjusted where necessary for design alternatives and owner's requirements. Reported completed project costs, for this type of structure, range from $48.15 to $243.25 per S.F.

Common additives

Description	Unit	$ Cost
Air Compressors		
Electric 1-1/2 H.P., standard controls	Each	1325
Dual controls	Each	1700
5 H.P. 115/230 volt, standard controls	Each	3450
Dual controls	Each	3600
Product Dispenser		
with vapor recovery for 6 nozzles	Each	22,900
Hoists, Single post		
8000# cap. swivel arm	Each	8950
Two post, adjustable frames, 11,000# cap.	Each	12,600
24,000# cap.	Each	18,700
7500# cap.	Each	10,600
Four post, roll on ramp	Each	9825
Lockers, Steel, single tier, 60" or 72"	Opening	191 - 310
2 tier, 60" or 72" total	Opening	107 - 141
5 tier, box lockers	Each	65 - 83.50
Locker bench, lam. maple top only	L.F.	21
Pedestals, steel pipe	Each	63.50
Lube Equipment		
3 reel type, with pumps, no piping	Each	11,500

Important: See the Reference Section for Location Factors

Garage, Service Station

			Unit	Unit Cost	Cost Per S.F.	% Of Sub-Total
A.	**SUBSTRUCTURE**					
1010	Standard Foundations	Poured concrete; strip and spread footings	S.F. Ground	4.32	4.32	
1020	Special Foundations	N/A	—	—	—	
1030	Slab on Grade	6" reinforced concrete with vapor barrier and granular base	S.F. Slab	7.29	7.29	13.5%
2010	Basement Excavation	Site preparation for slab and trench for foundation wall and footing	S.F. Ground	.88	.88	
2020	Basement Walls	4' foundation wall	L.F. Wall	69	7.54	
B.	**SHELL**					
	B10 Superstructure					
1010	Floor Construction	N/A	—	—	—	4.4 %
1020	Roof Construction	Plywood on wood trusses	S.F. Roof	6.61	6.61	
	B20 Exterior Enclosure					
2010	Exterior Walls	Face brick with concrete block backup 60% of wall	S.F. Wall	30.86	24.28	
2020	Exterior Windows	Store front and metal top hinged outswinging 20% of wall	Each	37.60	9.86	30.0%
2030	Exterior Doors	Steel overhead, aluminum & glass and hollow metal 20% of wall	Each	39.73	10.42	
	B30 Roofing					
3010	Roof Coverings	Asphalt shingles with flashing; perlite/EPS composite insulation	S.F. Roof	2.94	2.94	2.0%
3020	Roof Openings	N/A	—	—	—	
C.	**INTERIORS**					
1010	Partitions	Concrete block 25 S.F. Floor/L.F. Partition	S.F. Partition	6.97	2.23	
1020	Interior Doors	Single leaf hollow metal 700 S.F. Floor/Door	Each	875	1.25	
1030	Fittings	Toilet partitions	S.F. Floor	3.14	3.14	
2010	Stair Construction	N/A	—	—	—	8.3%
3010	Wall Finishes	Paint	S.F. Surface	4.98	3.19	
3020	Floor Finishes	Vinyl composition tile 35% of floor area	S.F. Floor	3.23	1.13	
3030	Ceiling Finishes	Painted gypsum board on furring in sales area & washrooms 35% of floor area	S.F. Ceiling	3.95	1.38	
D.	**SERVICES**					
	D10 Conveying					
1010	Elevators & Lifts	N/A	—	—	—	0.0 %
1020	Escalators & Moving Walks	N/A	—	—	—	
	D20 Plumbing					
2010	Plumbing Fixtures	Toilet and service fixtures, supply and drainage 1 Fixture/235 S.F. Floor	Each	2345	9.98	
2020	Domestic Water Distribution	Gas fired water heater	S.F. Floor	2.98	2.98	8.7%
2040	Rain Water Drainage	N/A	—	—	—	
	D30 HVAC					
3010	Energy Supply	N/A	—	—	—	
3020	Heat Generating Systems	N/A	—	—	—	
3030	Cooling Generating Systems	N/A	—	—	—	14.6 %
3050	Terminal & Package Units	Single zone AC unit	S.F. Floor	18.50	18.50	
3090	Other HVAC Sys. & Equipment	Underfloor exhaust system	S.F. Floor	3.16	3.16	
	D40 Fire Protection					
4010	Sprinklers	Wet pipe sprinkler system	S.F. Floor	7.25	7.25	7.1%
4020	Standpipes	Standpipe	S.F. Floor	3.30	3.30	
	D50 Electrical					
5010	Electrical Service/Distribution	200 ampere service, panel board and feeders	S.F. Floor	4.59	4.59	
5020	Lighting & Branch Wiring	High efficiency fluorescent fixtures, receptacles, switches, A.C. and misc. power	S.F. Floor	5.72	5.72	11.5%
5030	Communications & Security	Addressable alarm systems and emergency lighting	S.F. Floor	6.19	6.19	
5090	Other Electrical Systems	Emergency generator	S.F. Floor	.59	.59	
E.	**EQUIPMENT & FURNISHINGS**					
1010	Commercial Equipment	N/A	—	—	—	
1020	Institutional Equipment	N/A	—	—	—	0.0 %
1030	Vehicular Equipment	N/A	—	—	—	
1090	Other Equipment	N/A	—	—	—	
F.	**SPECIAL CONSTRUCTION**					
1020	Integrated Construction	N/A	—	—	—	0.0 %
1040	Special Facilities	N/A	—	—	—	
G.	**BUILDING SITEWORK** N/A					

		Sub-Total	148.72	100%
CONTRACTOR FEES (General Requirements: 10%, Overhead: 5%, Profit: 10%)		25%	37.16	
ARCHITECT FEES		8%	14.87	
	Total Building Cost		**200.75**	

Costs per square foot of floor area

Exterior Wall	S.F. Area	12000	16000	20000	25000	30000	35000	40000	45000	50000
	L.F. Perimeter	440	520	600	700	708	780	841	910	979
Reinforced Concrete Block	Lam. Wood Arches	154.80	147.95	143.85	140.55	136.00	134.15	132.50	131.45	130.60
	Rigid Steel Frame	164.60	157.75	153.70	150.35	145.80	143.95	142.30	141.25	140.40
Face Brick with Concrete Block Back-up	Lam. Wood Arches	180.75	171.00	165.15	160.40	152.65	149.90	147.40	145.75	144.50
	Rigid Steel Frame	190.55	180.80	174.95	170.20	162.50	159.75	157.20	155.55	154.30
Metal Sandwich Panels	Lam. Wood Arches	150.60	144.25	140.45	137.40	133.25	131.60	130.10	129.10	128.35
	Rigid Steel Frame	160.40	154.05	150.30	147.20	143.10	141.40	139.90	138.95	138.20
Perimeter Adj., Add or Deduct	Per 100 L.F.	6.50	4.90	3.95	3.15	2.60	2.20	1.95	1.75	1.55
Story Hgt. Adj., Add or Deduct	Per 1 Ft.	0.90	0.80	0.80	0.70	0.55	0.55	0.55	0.50	0.50
Basement—Not Applicable										

The above costs were calculated using the basic specifications shown on the facing page. These costs should be adjusted where necessary for design alternatives and owner's requirements. Reported completed project costs, for this type of structure, range from $73.05 to $218.00 per S.F.

Common additives

Description	Unit	$ Cost
Bleachers, Telescoping, manual		
To 15 tier	Seat	115 - 160
16-20 tier	Seat	235 - 288
21-30 tier	Seat	249 - 300
For power operation, add	Seat	45.50 - 71.50
Gym Divider Curtain, Mesh top		
Manual roll-up	S.F.	11.60
Gym Mats		
2" naugahyde covered	S.F.	3.95
2" nylon	S.F.	6.85
1-1/2" wall pads	S.F.	9.05
1" wrestling mats	S.F.	6.05
Scoreboard		
Basketball, one side	Each	3175 - 11,200
Basketball Backstop		
Wall mtd., 6' extended, fixed	Each	2125 - 2725
Swing up, wall mtd.	Each	2275 - 6775

Description	Unit	$ Cost
Lockers, Steel, single tier, 60" or 72"	Opening	191 - 310
2 tier, 60" or 72" total	Opening	107 - 141
5 tier, box lockers	Opening	65 - 83.50
Locker bench, lam. maple top only	L.F.	21
Pedestals, steel pipe	Each	63.50
Sound System		
Amplifier, 250 watts	Each	2350
Speaker, ceiling or wall	Each	191
Trumpet	Each	365
Emergency Lighting, 25 watt, battery operated		
Lead battery	Each	282
Nickel cadmium	Each	805

Important: See the Reference Section for Location Factors

Model costs calculated for a 1 story building with 25' story height and 20,000 square feet of floor area

Gymnasium

				Unit	Unit Cost	Cost Per S.F.	% Of Sub-Total
A.	**SUBSTRUCTURE**						
1010	Standard Foundations	Poured concrete; strip and spread footings		S.F. Ground	1.75	1.75	
1020	Special Foundations	N/A		—	—	—	
1030	Slab on Grade	4" reinforced concrete with vapor barrier and granular base		S.F. Slab	4.74	4.74	8.4%
2010	Basement Excavation	Site preparation for slab and trench for foundation wall and footing		S.F. Ground	.17	.17	
2020	Basement Walls	4' foundation wall		L.F. Wall	69	2.42	
B.	**SHELL**						
	B10 Superstructure						
1010	Floor Construction	N/A		—	—	—	16.4 %
1020	Roof Construction	Wood deck on laminated wood arches		S.F. Roof	17.66	17.66	
	B20 Exterior Enclosure						
2010	Exterior Walls	Reinforced concrete block (end walls included)	90% of wall	S.F. Wall	12.10	8.17	
2020	Exterior Windows	Metal horizontal pivoted	10% of wall	Each	497	3.73	11.6%
2030	Exterior Doors	Aluminum and glass, hollow metal, steel overhead		Each	1724	.53	
	B30 Roofing						
3010	Roof Coverings	EPDM, 60 mils, fully adhered; polyisocyanurate insulation		S.F. Roof	4.86	4.86	4.5%
3020	Roof Openings	N/A		—			
C.	**INTERIORS**						
1010	Partitions	Concrete block	50 S.F. Floor/L.F. Partition	S.F. Partition	8.45	1.69	
1020	Interior Doors	Single leaf hollow metal	500 S.F. Floor/Door	Each	875	1.75	
1030	Fittings	Toilet partitions		S.F. Floor	.44	.44	
2010	Stair Construction	N/A		—	—	—	21.1%
3010	Wall Finishes	50% paint, 50% ceramic tile		S.F. Surface	8.95	3.58	
3020	Floor Finishes	90% hardwood, 10% ceramic tile		S.F. Floor	14.25	14.25	
3030	Ceiling Finishes	Mineral fiber tile on concealed zee bars	15% of area	S.F. Ceiling	6.38	.96	
D.	**SERVICES**						
	D10 Conveying						
1010	Elevators & Lifts	N/A		—	—	—	0.0 %
1020	Escalators & Moving Walks	N/A		—	—	—	
	D20 Plumbing						
2010	Plumbing Fixtures	Toilet and service fixtures, supply and drainage	1 Fixture/515 S.F. Floor	Each	2838	5.51	
2020	Domestic Water Distribution	Electric water heater		S.F. Floor	4.95	4.95	9.7%
2040	Rain Water Drainage	N/A		—	—	—	
	D30 HVAC						
3010	Energy Supply	N/A		—	—	—	
3020	Heat Generating Systems	Included in D3050		—	—	—	
3030	Cooling Generating Systems	N/A		—	—	—	9.6 %
3050	Terminal & Package Units	Single zone rooftop unit, gas heating, electric cooling		S.F. Floor	10.35	10.35	
3090	Other HVAC Sys. & Equipment	N/A		—	—	—	
	D40 Fire Protection						
4010	Sprinklers	Wet pipe sprinkler system		S.F. Floor	2.77	2.77	2.6%
4020	Standpipes	N/A		—	—	—	
	D50 Electrical						
5010	Electrical Service/Distribution	400 ampere service, panel board and feeders		S.F. Floor	1.02	1.02	
5020	Lighting & Branch Wiring	Fluorescent fixtures, receptacles, switches, A.C. and misc. power		S.F. Floor	8.49	8.49	10.7%
5030	Communications & Security	Alarm systems, sound system and emergency lighting		S.F. Floor	1.82	1.82	
5090	Other Electrical Systems	Emergency generator, 7.5 kW		S.F. Floor	.19	.19	
E.	**EQUIPMENT & FURNISHINGS**						
1010	Commercial Equipment	N/A		—	—	—	
1020	Institutional Equipment	N/A		—	—	—	5.4 %
1030	Vehicular Equipment	N/A		—	—	—	
1090	Other Equipment	Bleachers, sauna, weight room		S.F. Floor	5.77	5.77	
F.	**SPECIAL CONSTRUCTION**						
1020	Integrated Construction	N/A		—	—	—	0.0 %
1040	Special Facilities	N/A		—	—	—	
G.	**BUILDING SITEWORK**	**N/A**					

		Sub-Total	107.57	**100%**
CONTRACTOR FEES (General Requirements: 10%, Overhead: 5%, Profit: 10%)			25%	26.87
ARCHITECT FEES			7%	9.41
		Total Building Cost	**143.85**	

Costs per square foot of floor area

Exterior Wall	S.F. Area	5000	10000	15000	20000	30000	40000	50000	75000	100000
	L.F. Perimeter	300	410	500	580	710	830	930	1150	1300
Concrete Block Reinforced	Steel Frame	130.55	115.00	108.75	105.20	100.85	98.45	96.70	94.00	92.10
	Bearing Walls	138.75	122.90	116.60	113.00	108.50	106.15	104.35	101.60	99.70
Precast Concrete	Steel Frame	150.05	128.35	119.65	114.65	108.55	105.25	102.75	99.00	96.30
	Bearing Walls	157.60	135.85	127.15	122.15	116.05	112.75	110.25	106.50	103.80
Galv. Steel Siding	Steel Frame	131.20	115.45	109.10	105.50	101.05	98.65	96.85	94.15	92.20
Metal Sandwich Panel	Steel Frame	136.90	119.35	112.30	108.30	103.35	100.65	98.70	95.65	93.50
Perimeter Adj., Add or Deduct	Per 100 L.F.	16.35	8.20	5.50	4.10	2.75	2.05	1.60	1.10	0.85
Story Hgt. Adj., Add or Deduct	Per 1 Ft.	1.60	1.10	0.90	0.80	0.65	0.60	0.55	0.45	0.40
Basement—Not Applicable										

The above costs were calculated using the basic specifications shown on the facing page. These costs should be adjusted where necessary for design alternatives and owner's requirements. Reported completed project costs, for this type of structure, range from $41.95 to $191.20 per S.F.

Common additives

Description	Unit	$ Cost
Closed Circuit Surveillance, One station		
Camera and monitor	Each	1850
For additional camera stations, add	Each	1000
Emergency Lighting, 25 watt, battery operated		
Lead battery	Each	282
Nickel cadmium	Each	805
Lockers, Steel, single tier, 60″ or 72″	Opening	191 - 310
2 tier, 60″ or 72″ total	Opening	107 - 141
5 tier, box lockers	Opening	65 - 83.50
Locker bench, lam. maple top only	L.F.	21
Pedestals, steel pipe	Each	63.50
Safe, Office type, 1 hour rating		
30″ x 18″ x 18″	Each	2400
60″ x 36″ x 18″, double door	Each	8750
Sound System		
Amplifier, 250 watts	Each	2350
Speaker, ceiling or wall	Each	191
Trumpet	Each	365

Important: See the Reference Section for Location Factors

Model costs calculated for a 1 story building with 24' story height and 20,000 square feet of floor area

Hangar, Aircraft

				Unit	Unit Cost	Cost Per S.F.	% Of Sub-Total
A. SUBSTRUCTURE							
1010	Standard Foundations	Poured concrete; strip and spread footings and		S.F. Ground	1.94	1.94	
1020	Special Foundations	N/A		—	—	—	
1030	Slab on Grade	6" reinforced concrete with vapor barrier and granular base		S.F. Slab	7.29	7.29	14.9%
2010	Basement Excavation	Site preparation for slab and trench for foundation wall and footing		S.F. Ground	.26	.26	
2020	Basement Walls	4' foundation wall		L.F. Wall	78	2.27	
B. SHELL							
	B10 Superstructure						
1010	Floor Construction	N/A		—	—	—	15.6 %
1020	Roof Construction	Metal deck, open web steel joists, beams, columns		S.F. Roof	12.30	12.30	
	B20 Exterior Enclosure						
2010	Exterior Walls	Galvanized steel siding		S.F. Wall	10.34	3.60	
2020	Exterior Windows	Industrial horizontal pivoted steel	20% of wall	Each	1840	10.67	28.9%
2030	Exterior Doors	Steel overhead and sliding	30% of wall	Each	40.85	8.53	
	B30 Roofing						
3010	Roof Coverings	Elastomeric membrane; fiberboard insulation		S.F. Roof	4.09	4.09	5.4%
3020	Roof Openings	Roof hatches		S.F. Roof	.13	.13	
C. INTERIORS							
1010	Partitions	Concrete block	200 S.F. Floor/L.F. Partition	S.F. Partition	8.80	.44	
1020	Interior Doors	Single leaf hollow metal	5000 S.F. Floor/Door	Each	875	.18	
1030	Fittings	Toilet partitions		S.F. Floor	.24	.24	
2010	Stair Construction	N/A		—	—	—	1.2%
3010	Wall Finishes	Paint		S.F. Surface	1.10	.11	
3020	Floor Finishes	N/A		—	—	—	
3030	Ceiling Finishes	N/A		—	—	—	
D. SERVICES							
	D10 Conveying						
1010	Elevators & Lifts	N/A		—	—	—	0.0 %
1020	Escalators & Moving Walks	N/A		—	—	—	
	D20 Plumbing						
2010	Plumbing Fixtures	Toilet and service fixtures, supply and drainage	1 Fixture/1000 S.F. Floor	Each	2810	2.81	
2020	Domestic Water Distribution	Gas fired water heater		S.F. Floor	.62	.62	4.3%
2040	Rain Water Drainage	N/A		—	—	—	
	D30 HVAC						
3010	Energy Supply	N/A		—	—	—	
3020	Heat Generating Systems	Unit heaters		Each	6.48	6.48	
3030	Cooling Generating Systems	N/A		—	—	—	8.5 %
3050	Terminal & Package Units	Exhaust fan		S.F. Floor	.24	.24	
3090	Other HVAC Sys. & Equipment	N/A		—	—	—	
	D40 Fire Protection						
4010	Sprinklers	Sprinklers, extra hazard		S.F. Floor	5.43	5.43	6.9%
4020	Standpipes	N/A		—	—	—	
	D50 Electrical						
5010	Electrical Service/Distribution	200 ampere service, panel board and feeders		S.F. Floor	.87	.87	
5020	Lighting & Branch Wiring	High intensity discharge fixtures, receptacles, switches and misc. power		S.F. Floor	9.84	9.84	14.3%
5030	Communications & Security	Alarm systems and emergency lighting		S.F. Floor	.46	.46	
5090	Other Electrical Systems	Emergency generator, 7.5 kW		S.F. Floor	.07	.07	
E. EQUIPMENT & FURNISHINGS							
1010	Commercial Equipment	N/A		—	—	—	
1020	Institutional Equipment	N/A		—	—	—	
1030	Vehicular Equipment	N/A		—	—	—	0.0 %
1090	Other Equipment	N/A		—	—	—	
F. SPECIAL CONSTRUCTION							
1020	Integrated Construction	N/A		—	—	—	0.0 %
1040	Special Facilities	N/A		—	—	—	
G. BUILDING SITEWORK	**N/A**						

	Sub-Total	78.87	**100%**
CONTRACTOR FEES (General Requirements: 10%, Overhead: 5%, Profit: 10%)	25%	19.73	
ARCHITECT FEES	7%	6.90	
	Total Building Cost	105.50	

Costs per square foot of floor area

Exterior Wall	S.F. Area	25000	40000	55000	70000	85000	100000	115000	130000	145000
	L.F. Perimeter	388	520	566	666	766	866	878	962	1045
Face Brick with Structural Facing Tile	Steel Frame	305.15	296.65	289.20	286.70	285.05	283.95	281.30	280.60	280.05
	R/Conc. Frame	310.00	301.45	294.00	291.50	289.90	288.75	286.15	285.45	284.85
Face Brick with Concrete Block Back-up	Steel Frame	296.65	289.50	283.55	281.50	280.15	279.20	277.15	276.55	276.10
	R/Conc. Frame	294.55	287.45	281.45	279.40	278.05	277.10	275.10	274.50	274.00
Precast Concrete Panels	Steel Frame	304.10	295.80	288.50	286.05	284.50	283.35	280.85	280.10	279.55
	R/Conc. Frame	308.95	300.60	293.30	290.85	289.30	288.15	285.70	284.90	284.35
Perimeter Adj., Add or Deduct	Per 100 L.F.	9.20	5.75	4.15	3.30	2.75	2.25	2.00	1.80	1.60
Story Hgt. Adj., Add or Deduct	Per 1 Ft.	2.65	2.20	1.75	1.65	1.60	1.50	1.35	1.30	1.25
For Basement, add $32.90 per square foot of basement area										

The above costs were calculated using the basic specifications shown on the facing page. These costs should be adjusted where necessary for design alternatives and owner's requirements. Reported completed project costs, for this type of structure, range from $161.20 to $407.15 per S.F.

Common additives

Description	Unit	$ Cost
Cabinet, Base, door units, metal	L.F.	256
Drawer units	L.F.	505
Tall storage cabinets, 7' high, open	L.F.	480
With doors	L.F.	565
Wall, metal 12-1/2" deep, open	L.F.	192
With doors	L.F.	345
Closed Circuit TV (Patient monitoring)		
One station camera & monitor	Each	1850
For additional camera, add	Each	1000
For automatic iris for low light, add	Each	2600
Hubbard Tank, with accessories		
Stainless steel, 125 GPM 45 psi	Each	27,600
For electric hoist, add	Each	3000
Mortuary Refrigerator, End operated		
2 capacity	Each	13,600
6 capacity	Each	24,500

Description	Unit	$ Cost
Nurses Call Station		
Single bedside call station	Each	310
Ceiling speaker station	Each	143
Emergency call station	Each	192
Pillow speaker	Each	296
Double bedside call station	Each	385
Duty station	Each	325
Standard call button	Each	169
Master control station for 20 stations	Each	6025
Sound System		
Amplifier, 250 watts	Each	2350
Speaker, ceiling or wall	Each	191
Trumpet	Each	365
Station, Dietary with ice	Each	16,800
Sterilizers		
Single door, steam	Each	166,000
Double door, steam	Each	213,500
Portable, countertop, steam	Each	3975 - 6225
Gas	Each	41,200
Automatic washer/sterilizer	Each	57,000

Model costs calculated for a 3 story building with 12' story height and 55,000 square feet of floor area

					Unit	Unit Cost	Cost Per S.F.	% Of Sub-Total
A.	**SUBSTRUCTURE**							
1010	Standard Foundations	Poured concrete; strip and spread footings			S.F. Ground	5.49	1.83	
1020	Special Foundations	N/A			—	—	—	
1030	Slab on Grade	4" reinforced concrete with vapor barrier and granular base			S.F. Slab	4.74	1.58	2.0%
2010	Basement Excavation	Site preparation for slab and trench for foundation wall and footing			S.F. Ground	.17	.06	
2020	Basement Walls	4' foundation wall			L.F. Wall	78	.80	
B.	**SHELL**							
	B10 Superstructure							
1010	Floor Construction	Cast-in-place concrete beam and slab			S.F. Floor	23.76	15.84	10.3%
1020	Roof Construction	Cast-in-place concrete beam and slab			S.F. Roof	19.11	6.37	
	B20 Exterior Enclosure							
2010	Exterior Walls	Face brick and structural facing tile	85% of wall		S.F. Wall	44.01	13.86	
2020	Exterior Windows	Aluminum sliding	15% of wall		Each	696	1.68	7.3%
2030	Exterior Doors	Double aluminum and glass and sliding doors			Each	2746	.30	
	B30 Roofing							
3010	Roof Coverings	Built-up tar and gravel with flashing; perlite/EPS composite insulation			S.F. Roof	5.46	1.82	0.8%
3020	Roof Openings	Roof hatches			S.F. Roof	.03	.01	
C.	**INTERIORS**							
1010	Partitions	Concrete block, gypsum board on metal studs	9 S.F. Floor/L.F. Partition		S.F. Partition	8.35	9.28	
1020	Interior Doors	Single leaf hollow metal	90 S.F. Floor/Door		Each	904	10.03	
1030	Fittings	Hospital curtains			S.F. Floor	.87	.87	
2010	Stair Construction	Concrete filled metal pan			Flight	5975	1.42	20.8%
3010	Wall Finishes	40% vinyl wall covering, 35% ceramic tile, 25% epoxy coating			S.F. Surface	3.37	7.49	
3020	Floor Finishes	60% vinyl tile, 20% ceramic, 20% terrazzo			S.F. Floor	9.42	9.42	
3030	Ceiling Finishes	Plaster on suspended metal lath			S.F. Ceiling	6.38	6.38	
D.	**SERVICES**							
	D10 Conveying							
1010	Elevators & Lifts	Two hydraulic hospital elevators			Each	121,550	4.42	2.0%
1020	Escalators & Moving Walks	N/A			—	—	—	
	D20 Plumbing							
2010	Plumbing Fixtures	Medical, patient & specialty fixtures, supply and drainage	1 Fixture/224 S.F. Floor		Each	5600	25	
2020	Domestic Water Distribution	Electric water heater			S.F. Floor	11.12	11.12	16.9%
2040	Rain Water Drainage	Roof drains			S.F. Roof	1.35	.45	
	D30 HVAC							
3010	Energy Supply	Conditioned air with hot water reheat system			S.F. Floor	4.60	4.60	
3020	Heat Generating Systems	Boiler			Each	95,050	2.51	
3030	Cooling Generating Systems	Conditioned air, chillers, cooling towers			S.F. Floor	5.76	5.76	18.2%
3050	Terminal & Package Units	N/A			—	—	—	
3090	Other HVAC Sys. & Equipment	Conditioned air, ductwork, surgery air curtain			S.F. Floor	26.47	26.47	
	D40 Fire Protection							
4010	Sprinklers	Wet pipe sprinkler system			S.F. Floor	2.43	2.43	1.4%
4020	Standpipes	Standpipe			S.F. Floor	.63	.63	
	D50 Electrical							
5010	Electrical Service/Distribution	1200 ampere service, panel board and feeders			S.F. Floor	1.86	1.86	
5020	Lighting & Branch Wiring	High efficiency hospital grade light fixtures, receptacles, switches, A.C. and misc. power			S.F. Floor	15.73	15.73	
5030	Communications & Security	Addressable alarm systems, internet wiring, communications systems, emergency lighting			S.F. Floor	4.89	4.89	12.4%
5090	Other Electrical Systems	Emergency generator, 200 kW with fuel tank, Uninterruptible power supply			S.F. Floor	4.23	4.23	
E.	**EQUIPMENT & FURNISHINGS**							
1010	Commercial Equipment	N/A			—	—	—	
1020	Institutional Equipment	Medical gas system, kitchen equip, lab equip., curtain partitions			S.F. Floor	10.91	10.91	7.7 %
1030	Vehicular Equipment	N/A			—	—	—	
2020	Other Equipment	Patient wall systems			S.F. Floor	5.73	5.73	
F.	**SPECIAL CONSTRUCTION**							
1020	Integrated Construction	N/A			—	—	—	0.0 %
1040	Special Facilities	N/A			—	—	—	
G.	**BUILDING SITEWORK**	**N/A**						

			Sub-Total	215.78	100%
	CONTRACTOR FEES (General Requirements: 10%, Overhead: 5%, Profit: 10%)			25%	53.94
	ARCHITECT FEES			9%	24.28
			Total Building Cost	294	

Costs per square foot of floor area

Exterior Wall	S.F. Area	100000	125000	150000	175000	200000	225000	250000	275000	300000
	L.F. Perimeter	594	705	816	783	866	950	1033	1116	1200
Face Brick with Structural Facing Tile	Steel Frame	280.75	277.00	274.45	268.85	267.35	266.20	265.20	264.40	263.85
	R/Conc. Frame	281.35	277.50	274.85	269.25	267.70	266.50	265.50	264.70	264.10
Face Brick with Concrete Block Back-up	Steel Frame	275.25	271.65	269.25	264.45	263.05	262.00	261.10	260.30	259.75
	R/Conc. Frame	276.00	272.40	269.95	265.15	263.75	262.70	261.80	261.05	260.50
Precast Concrete Panels With Exposed Aggregate	Steel Frame	276.05	272.40	269.95	265.00	263.65	262.55	261.60	260.85	260.30
	R/Conc. Frame	276.80	273.15	270.70	265.75	264.35	263.25	262.30	261.60	261.00
Perimeter Adj., Add or Deduct	Per 100 L.F.	4.65	3.70	3.10	2.60	2.30	2.00	1.85	1.75	1.50
Story Hgt. Adj., Add or Deduct	Per 1 Ft.	2.10	1.90	1.90	1.55	1.45	1.40	1.45	1.45	1.35
For Basement, add $33.95 per square foot of basement area										

The above costs were calculated using the basic specifications shown on the facing page. These costs should be adjusted where necessary for design alternatives and owner's requirements. Reported completed project costs, for this type of structure, range from $164.90 to $402.10 per S.F.

Common additives

Description	Unit	$ Cost
Cabinets, Base, door units, metal	L.F.	256
Drawer units	L.F.	505
Tall storage cabinets, 7' high, open	L.F.	480
With doors	L.F.	565
Wall, metal 12-1/2" deep, open	L.F.	192
With doors	L.F.	345
Closed Circuit TV (Patient monitoring)		
One station camera & monitor	Each	1850
For additional camera add	Each	1000
For automatic iris for low light add	Each	2600
Hubbard Tank, with accessories		
Stainless steel, 125 GPM 45 psi	Each	27,600
For electric hoist, add	Each	3000
Mortuary Refrigerator, End operated		
2 capacity	Each	13,600
6 capacity	Each	24,500

Description	Unit	$ Cost
Nurses Call Station		
Single bedside call station	Each	310
Ceiling speaker station	Each	143
Emergency call station	Each	192
Pillow speaker	Each	296
Double bedside call station	Each	385
Duty station	Each	325
Standard call button	Each	169
Master control station for 20 stations	Each	6025
Sound System		
Amplifier, 250 watts	Each	2350
Speaker, ceiling or wall	Each	191
Trumpet	Each	365
Station, Dietary with ice	Each	16,800
Sterilizers		
Single door, steam	Each	166,000
Double door, steam	Each	213,500
Portable, counter top, steam	Each	3975 - 6225
Gas	Each	41,200
Automatic washer/sterilizer	Each	57,000

Model costs calculated for a 6 story building with 12′ story height and 200,000 square feet of floor area

				Unit	Unit Cost	Cost Per S.F.	% Of Sub-Total
A.	**SUBSTRUCTURE**						
1010	Standard Foundations	Poured concrete; strip and spread footings		S.F. Ground	14.46	2.41	
1020	Special Foundations	N/A		—	—	—	
1030	Slab on Grade	4″ reinforced concrete with vapor barrier and granular base		S.F. Slab	7.29	1.22	2.0%
2010	Basement Excavation	Site preparation for slab and trench for foundation wall and footing		S.F. Ground	.17	.03	
2020	Basement Walls	4′ foundation wall		L.F. Wall	78	.34	
B.	**SHELL**						
	B10 Superstructure						
1010	Floor Construction	Concrete slab with metal deck and beams, steel columns		S.F. Floor	23.48	19.57	10.9%
1020	Roof Construction	Metal deck, open web steel joists, beams, interior columns		S.F. Roof	10.44	1.74	
	B20 Exterior Enclosure						
2010	Exterior Walls	Face brick and structural facing tile	70% of wall	S.F. Wall	43.99	9.60	
2020	Exterior Windows	Aluminum sliding	30% of wall	Each	552	3.45	7.0%
2030	Exterior Doors	Double aluminum and glass and sliding doors		Each	5115	.72	
	B30 Roofing						
3010	Roof Coverings	Built-up tar and gravel with flashing; perlite/EPS composite insulation		S.F. Roof	7.14	1.19	0.6%
3020	Roof Openings	Roof hatches		S.F. Roof	.18	.03	
C.	**INTERIORS**						
1010	Partitions	Gypsum board on metal studs with sound deadening board	9 S.F. Floor/L.F. Partition	S.F. Partition	7.71	8.57	
1020	Interior Doors	Single leaf hollow metal	90 S.F. Floor/Door	Each	904	10.03	
1030	Fittings	Hospital curtains		S.F. Floor	.95	.95	
2010	Stair Construction	Concrete filled metal pan		Flight	12,650	1.64	23.6%
3010	Wall Finishes	40% vinyl wall covering, 35% ceramic tile, 25% epoxy coating		S.F. Surface	3.36	7.47	
3020	Floor Finishes	60% vinyl tile, 20% ceramic, 20% terrazzo		S.F. Floor	10.20	10.20	
3030	Ceiling Finishes	Plaster on suspended metal lath		S.F. Ceiling	7.44	7.44	
D.	**SERVICES**						
	D10 Conveying						
1010	Elevators & Lifts	Six geared hospital elevators		Each	215,333	6.46	3.3%
1020	Escalators & Moving Walks	N/A		—	—	—	
	D20 Plumbing						
2010	Plumbing Fixtures	Kitchen, toilet and service fixtures, supply and drainage	1 Fixture/416 S.F. Floor	Each	4489	10.79	
2020	Domestic Water Distribution	Electric water heater		S.F. Floor	6.81	6.81	9.2%
2040	Rain Water Drainage	Roof drains		S.F. Floor	3.06	.51	
	D30 HVAC						
3010	Energy Supply	Oil fired hot water, wall fin radiation		S.F. Floor	4.01	4.01	
3020	Heat Generating Systems	Hot water boilers, steam boiler for services		Each	30,475	.38	
3030	Cooling Generating Systems	Chilled water units		S.F. Floor	2.70	2.70	17.8%
3050	Terminal & Package Units	N/A		—	—	—	
3090	Other HVAC Sys. & Equipment	Conditioned air with reheat, operating room air curtains		S.F. Floor	27.78	27.78	
	D40 Fire Protection						
4010	Sprinklers	Wet pipe sprinkler system		S.F. Floor	2.29	2.29	1.4%
4020	Standpipes	Standpipe		S.F. Floor	.49	.49	
	D50 Electrical						
5010	Electrical Service/Distribution	4000 ampere service, panel board and feeders		S.F. Floor	4.17	4.17	
5020	Lighting & Branch Wiring	High efficiency hospital grade light fixtures, receptacles, switches, A.C. and misc. power		S.F. Floor	18.32	18.32	14.7%
5030	Communications & Security	Addressable alarm systems, internet wiring, communications system, emergency lighting		S.F. Floor	2.20	2.20	
5090	Other Electrical Systems	Emergency generator, 800 kW with fuel tank, uninterruptible power supply		S.F. Floor	4.17	4.17	
E.	**EQUIPMENT & FURNISHINGS**						
1010	Commercial Equipment	N/A		—	—	—	
1020	Institutional Equipment	Medical gases, curtain partitions		S.F. Floor	14.75	14.75	9.4%
1030	Vehicular Equipment	N/A		—	—	—	
2020	Other Equipment	Patient wall systems		S.F. Floor	3.78	3.78	
F.	**SPECIAL CONSTRUCTION**						
1020	Integrated Construction	N/A		—	—	—	0.0%
1040	Special Facilities	N/A		—	—	—	
G.	**BUILDING SITEWORK**	**N/A**					
				Sub-Total		196.21	100%
	CONTRACTOR FEES (General Requirements: 10%, Overhead: 5%, Profit: 10%)				25%	49.07	
	ARCHITECT FEES				9%	22.07	
				Total Building Cost		**267.35**	

151

Costs per square foot of floor area

Exterior Wall	S.F. Area	35000	55000	75000	95000	115000	135000	155000	175000	195000
	L.F. Perimeter	314	401	497	555	639	722	754	783	850
Face Brick with Concrete Block Back-up	Steel Frame	182.95	173.80	169.95	166.40	164.80	**163.70**	161.70	160.15	159.60
	R/Conc. Frame	180.95	171.85	167.95	164.40	162.85	161.75	159.75	158.20	157.60
Glass and Metal Curtain Walls	Steel Frame	177.25	168.95	165.40	162.35	160.90	159.90	158.20	156.85	156.35
	R/Conc. Frame	175.50	167.25	163.70	160.60	159.15	158.10	156.50	155.15	154.60
Precast Concrete Panels	Steel Frame	189.05	178.55	174.15	170.00	168.15	166.90	164.55	162.80	162.10
	R/Conc. Frame	188.20	177.55	173.10	168.85	167.00	165.70	163.35	161.50	160.85
Perimeter Adj., Add or Deduct	Per 100 L.F.	9.50	6.05	4.40	3.50	2.85	2.45	2.15	1.95	1.75
Story Hgt. Adj., Add or Deduct	Per 1 Ft.	2.65	2.15	1.95	1.70	1.60	1.60	1.45	1.30	1.25
For Basement, add $32.30 per square foot of basement area										

The above costs were calculated using the basic specifications shown on the facing page. These costs should be adjusted where necessary for design alternatives and owner's requirements. Reported completed project costs, for this type of structure, range from $108.75 to $208.75 per S.F.

Common additives

Description	Unit	$ Cost	Description	Unit	$ Cost
Bar, Front bar	L.F.	360	Laundry Equipment		
Back bar	L.F.	289	Folders, blankets & sheets, king size	Each	66,500
Booth, Upholstered, custom, straight	L.F.	202 - 375	Ironers, 110" single roll	Each	35,500
"L" or "U" shaped	L.F.	210 - 355	Combination washer extractor 50#	Each	12,200
Closed Circuit Surveillance, One station			125#	Each	32,800
Camera and monitor	Each	1850	Sauna, Prefabricated, complete		
For additional camera stations, add	Each	1000	6' x 4'	Each	5850
Directory Boards, Plastic, glass covered			6' x 6'	Each	6950
30" x 20"	Each	595	6' x 9'	Each	8525
36" x 48"	Each	1450	8' x 8'	Each	10,100
Aluminum, 24" x 18"	Each	600	10' x 12'	Each	14,000
48" x 32"	Each	980	Smoke Detectors		
48" x 60"	Each	2025	Ceiling type	Each	187
Elevators, Electric passenger, 5 stops			Duct type	Each	480
3500# capacity	Each	167,200	Sound System		
5000# capacity	Each	170,700	Amplifier, 250 watts	Each	2350
Additional stop, add	Each	13,600	Speaker, ceiling or wall	Each	191
Emergency Lighting, 25 watt, battery operated			Trumpet	Each	365
Lead battery	Each	282	TV Antenna, Master system, 12 outlet	Outlet	315
Nickel cadmium	Each	805	30 outlet	Outlet	203
			100 outlet	Outlet	194

Model costs calculated for a 6 story building with 10′ story height and 135,000 square feet of floor area

				Unit	Unit Cost	Cost Per S.F.	% Of Sub-Total
A.	**SUBSTRUCTURE**						
1010	Standard Foundations	Poured concrete; strip and spread footings		S.F. Ground	6.24	1.04	
1020	Special Foundations	N/A		—	—	—	
1030	Slab on Grade	4″ reinforced concrete with vapor barrier and granular base		S.F. Slab	4.74	.79	2.2%
2010	Basement Excavation	Site preparation for slab and trench for foundation wall and footing		S.F. Ground	.17	.03	
2020	Basement Walls	4′ foundation wall		L.F. Wall	78	.84	
B.	**SHELL**						
	B10 Superstructure						
1010	Floor Construction	Concrete slab with metal deck, beams, columns		S.F. Floor	23.24	19.37	17.3%
1020	Roof Construction	Metal deck, open web steel joists, beams, columns		S.F. Roof	12.12	2.02	
	B20 Exterior Enclosure						
2010	Exterior Walls	Face brick with concrete block backup	80% of wall	S.F. Wall	30.85	7.92	
2020	Exterior Windows	Window wall	20% of wall	Each	46.80	3	9.0%
2030	Exterior Doors	Glass amd metal doors and entrances		Each	3107	.24	
	B30 Roofing						
3010	Roof Coverings	Built-up tar and gravel with flashing; perlite/EPS composite insulation		S.F. Roof	5.70	.95	0.8%
3020	Roof Openings	Roof hatches		S.F. Roof	.06	.01	
C.	**INTERIORS**						
1010	Partitions	Gypsum board and sound deadening board, steel studs	9 S.F. Floor/L.F. Partition	S.F. Partition	7.02	6.24	
1020	Interior Doors	Single leaf hollow metal	90 S.F. Floor/Door	Each	875	9.73	
1030	Fittings	N/A		—	—	—	
2010	Stair Construction	Concrete filled metal pan		Flight	15,800	1.64	24.3%
3010	Wall Finishes	20% paint, 75% vinyl cover, 5% ceramic tile		S.F. Surface	1.80	3.20	
3020	Floor Finishes	80% carpet, 10% vinyl composition tile, 10% ceramic tile		S.F. Floor	5.12	5.12	
3030	Ceiling Finishes	Mineral fiber tile applied with adhesive		S.F. Ceiling	4.10	4.10	
D.	**SERVICES**						
	D10 Conveying						
1010	Elevators & Lifts	Four geared passenger elevators		Each	224,100	6.64	5.4%
1020	Escalators & Moving Walks	N/A		—	—	—	
	D20 Plumbing						
2010	Plumbing Fixtures	Kitchen, toilet and service fixtures, supply and drainage	1 Fixture/155 S.F. Floor	Each	2807	18.11	
2020	Domestic Water Distribution	Oil fired water heater		S.F. Floor	.47	.47	15.2%
2040	Rain Water Drainage	Roof drains		S.F. Roof	1.20	.20	
	D30 HVAC						
3010	Energy Supply	Gas fired hot water, wall fin radiation		S.F. Floor	4.15	4.15	
3020	Heat Generating Systems	N/A		—	—	—	
3030	Cooling Generating Systems	Chilled water, fan coil units		S.F. Floor	11.20	11.20	12.4%
3050	Terminal & Package Units	N/A		—	—	—	
3090	Other HVAC Sys. & Equipment	N/A		—	—	—	
	D40 Fire Protection						
4010	Sprinklers	Sprinkler system, light hazard		S.F. Floor	2.31	2.31	2.2%
4020	Standpipes	Standpipes and hose systems		S.F. Floor	.43	.43	
	D50 Electrical						
5010	Electrical Service/Distribution	2000 ampere service, panel board and feeders		S.F. Floor	1.40	1.40	
5020	Lighting & Branch Wiring	High efficiency fluorescent fixtures, receptacles, switches, A.C. and misc. power		S.F. Floor	8.08	8.08	11.2%
5030	Communications & Security	Addressable alarm systems, internet wiring, communications systems and emergency lighting		S.F. Floor	3.81	3.81	
5090	Other Electrical Systems	Emergency generator, 250 kW		S.F. Floor	.51	.51	
E.	**EQUIPMENT & FURNISHINGS**						
1010	Commercial Equipment	N/A		—	—	—	
1020	Institutional Equipment	N/A		—	—	—	
1030	Vehicular Equipment	N/A		—	—	—	0.0 %
1090	Other Equipment	N/A		—	—	—	
F.	**SPECIAL CONSTRUCTION**						
1020	Integrated Construction	N/A		—	—	—	0.0 %
1040	Special Facilities	N/A		—	—	—	
G.	**BUILDING SITEWORK**	**N/A**					

				Sub-Total	123.55	**100%**
	CONTRACTOR FEES (General Requirements: 10%, Overhead: 5%, Profit: 10%)			25%	30.88	
	ARCHITECT FEES			6%	9.27	
				Total Building Cost	**163.70**	

Costs per square foot of floor area

Exterior Wall	S.F. Area	140000	243000	346000	450000	552000	655000	760000	860000	965000
	L.F. Perimeter	403	587	672	800	936	1073	1213	1195	1312
Face Brick with Concrete Block Back-up	Steel Frame	171.10	164.35	159.40	157.45	156.35	155.70	155.15	153.35	152.90
	R/Conc. Frame	172.25	165.55	160.55	158.60	157.55	156.80	156.30	154.50	154.05
Face Brick Veneer On Steel Studs	Steel Frame	168.75	162.45	157.85	156.05	155.00	154.35	153.90	152.20	151.80
	R/Conc. Frame	172.60	166.25	161.70	159.90	158.85	158.20	157.70	156.05	155.65
Glass and Metal Curtain Walls	Steel Frame	200.70	186.85	179.45	176.05	174.00	172.65	171.65	169.80	169.05
	R/Conc. Frame	195.15	181.45	174.15	170.75	168.75	167.40	166.40	164.60	163.90
Perimeter Adj., Add or Deduct	Per 100 L.F.	5.60	3.35	2.25	1.75	1.45	1.20	1.05	0.90	0.85
Story Hgt. Adj., Add or Deduct	Per 1 Ft.	2.15	1.90	1.45	1.35	1.30	1.15	1.20	1.00	1.05

For Basement, add $33.55 per square foot of basement area

The above costs were calculated using the basic specifications shown on the facing page. These costs should be adjusted where necessary for design alternatives and owner's requirements. Reported completed project costs, for this type of structure, range from $120.70 to $211.05 per S.F.

Common additives

Description	Unit	$ Cost
Bar, Front bar	L.F.	360
Back bar	L.F.	289
Booth, Upholstered, custom, straight	L.F.	202 - 375
"L" or "U" shaped	L.F.	210 - 355
Closed Circuit Surveillance, One station		
Camera and monitor	Each	1850
For additional camera stations, add	Each	1000
Directory Boards, Plastic, glass covered		
30" x 20"	Each	595
36" x 48"	Each	1450
Aluminum, 24" x 18"	Each	600
48" x 32"	Each	980
48" x 60"	Each	2025
Elevators, Electric passenger, 10 stops		
3500# capacity	Each	430,500
5000# capacity	Each	437,000
Additional stop, add	Each	13,600
Emergency Lighting, 25 watt, battery operated		
Lead battery	Each	282
Nickel cadmium	Each	805

Description	Unit	$ Cost
Laundry Equipment		
Folders, blankets & sheets, king size	Each	66,500
Ironers, 110" single roll	Each	35,500
Combination washer & extractor 50#	Each	12,200
125#	Each	32,800
Sauna, Prefabricated, complete		
6' x 4'	Each	5850
6' x 6'	Each	6950
6' x 9'	Each	8525
8' x 8'	Each	10,100
10' x 12'	Each	14,000
Smoke Detectors		
Ceiling type	Each	187
Duct type	Each	480
Sound System		
Amplifier, 250 watts	Each	2350
Speaker, ceiling or wall	Each	191
Trumpet	Each	365
TV Antenna, Master system, 12 outlet	Outlet	315
30 outlet	Outlet	203
100 outlet	Outlet	194

Model costs calculated for a 15 story building with 10' story height and 450,000 square feet of floor area

				Unit	Unit Cost	Cost Per S.F.	% Of Sub-Total
A.	**SUBSTRUCTURE**						
1010	Standard Foundations	CIP concrete pile caps		S.F. Ground	5.85	.39	
1020	Special Foundations	Steel H-piles, concrete grade beams		S.F. Ground	113	7.55	
1030	Slab on Grade	4" reinforced concrete with vapor barrier and granular base		S.F. Slab	4.74	.32	6.4%
2010	Basement Excavation	Site preparation for slab, piles and grade beams		S.F. Ground	.17	.01	
2020	Basement Walls	4' foundation wall		L.F. Wall	78	.18	
B.	**SHELL**						
	B10 Superstructure						
1010	Floor Construction	Open web steel joists, slab form, concrete, columns		S.F. Floor	23.18	21.63	16.8%
1020	Roof Construction	Metal deck, open web steel joists, beams, columns		S.F. Roof	10.65	.71	
	B20 Exterior Enclosure						
2010	Exterior Walls	N/A		—	—	—	
2020	Exterior Windows	Glass and metal curtain walls	100% of wall	Each	25	6.67	5.2 %
2030	Exterior Doors	Glass and metal doors and entrances		Each	2798	.20	
	B30 Roofing						
3010	Roof Coverings	Built-up tar and gravel with flashing; perlite/EPS composite insulation		S.F. Roof	5.40	.36	0.3%
3020	Roof Openings	N/A		—	—	—	
C.	**INTERIORS**						
1010	Partitions	Gypsum board and sound deadening board, steel studs	9 S.F. Floor/L.F. Partition	S.F. Partition	6.82	6.06	
1020	Interior Doors	Single leaf hollow metal	90 S.F. Floor/Door	Each	875	9.73	
1030	Fittings	N/A		—	—	—	
2010	Stair Construction	Concrete filled metal pan		Flight	15,800	3.19	23.3%
3010	Wall Finishes	20% paint, 75% vinyl cover, 5% ceramic tile		S.F. Surface	1.79	3.19	
3020	Floor Finishes	80% carpet tile, 10% vinyl composition tile, 10% ceramic tile		S.F. Floor	5.12	5.12	
3030	Ceiling Finishes	Gypsum board on resilient channel		S.F. Ceiling	3.71	3.71	
D.	**SERVICES**						
	D10 Conveying						
1010	Elevators & Lifts	One geared freight, six geared passenger elevators		Each	489,750	6.53	4.9%
1020	Escalators & Moving Walks	N/A		—	—	—	
	D20 Plumbing						
2010	Plumbing Fixtures	Kitchen, toilet and service fixtures, supply and drainage	1 Fixture/165 S.F. Floor	Each	3214	19.48	
2020	Domestic Water Distribution	Electric water heater		S.F. Floor	4.50	4.50	18.1%
2040	Rain Water Drainage	Roof drains		S.F. Roof	1.95	.13	
	D30 HVAC						
3010	Energy Supply	Gas fired hot water, wall fin radiation		S.F.Floor	2.17	2.17	
3020	Heat Generating Systems	N/A		—	—	—	
3030	Cooling Generating Systems	Chilled water, fan coil units		S.F. Floor	11.20	11.20	10.1%
3050	Terminal & Package Units	N/A		—	—	—	
3090	Other HVAC Sys. & Equipment	N/A		—	—	—	
	D40 Fire Protection						
4010	Sprinklers	Sprinkler system, light hazard		S.F. Floor	3.29	3.29	4.7%
4020	Standpipes	Standpipes, fire pump and hose systems		S.F. Floor	2.90	2.90	
	D50 Electrical						
5010	Electrical Service/Distribution	6000 ampere service, panel board and feeders		S.F. Floor	1.52	1.52	
5020	Lighting & Branch Wiring	High efficiency fluorescent fixtures, receptacles, switches, A.C. and misc. power		S.F. Floor	7.96	7.96	
5030	Communications & Security	Addressable alarm systems, internet wiring, communications systems and emergency lighting		S.F. Floor	3.79	3.79	10.3%
5090	Other Electrical Systems	Emergency generator, 500 kW		S.F. Floor	.38	.38	
E.	**EQUIPMENT & FURNISHINGS**						
1010	Commercial Equipment	N/A		—	—	—	
1020	Institutional Equipment	N/A		—	—	—	
1030	Vehicular Equipment	N/A		—	—	—	0.0 %
1090	Other Equipment	N/A		—	—	—	
F.	**SPECIAL CONSTRUCTION**						
1020	Integrated Construction	N/A		—	—	—	0.0 %
1040	Special Facilities	N/A		—	—	—	
G.	**BUILDING SITEWORK**	**N/A**					

	Sub-Total	132.87	**100%**
CONTRACTOR FEES (General Requirements: 10%, Overhead: 5%, Profit: 10%)	25%	33.21	
ARCHITECT FEES	6%	9.97	
Total Building Cost		**176.05**	

Costs per square foot of floor area

Exterior Wall	S.F. Area	5500	12000	20000	30000	40000	60000	80000	100000	145000
	L.F. Perimeter	310	330	400	490	530	590	680	760	920
Face Brick with Concrete Block Back-up	Steel Frame	443.00	324.30	290.15	273.30	261.65	248.95	243.65	240.20	235.50
	R/Conc. Frame	439.65	320.90	286.70	269.85	258.15	245.50	240.15	236.75	232.00
Stucco on Concrete Block	Steel Frame	410.10	308.40	278.65	263.90	254.05	243.40	238.85	235.90	232.00
	R/Conc. Frame	406.45	304.70	274.95	260.15	250.35	239.70	235.15	232.25	228.25
Reinforced Concrete	Steel Frame	419.10	312.80	281.85	266.50	256.15	245.00	240.20	237.15	233.00
	R/Conc. Frame	415.40	309.10	278.10	262.85	252.45	241.25	236.50	233.40	229.30
Perimeter Adj., Add or Deduct	Per 100 L.F.	48.00	22.00	13.25	8.75	6.55	4.40	3.25	2.60	1.85
Story Hgt. Adj., Add or Deduct	Per 1 Ft.	11.40	5.55	4.05	3.30	2.65	1.95	1.70	1.55	1.25

For Basement, add $46.40 per square foot of basement area

The above costs were calculated using the basic specifications shown on the facing page. These costs should be adjusted where necessary for design alternatives and owner's requirements. Reported completed project costs, for this type of structure, range from $175.65 to $466.50 per S.F.

Common additives

Description	Unit	$ Cost
Clock System		
20 room	Each	16,000
50 room	Each	39,100
Closed Circuit Surveillance, One station		
Camera and monitor	Each	1850
For additional camera stations, add	Each	1000
Elevators, Hydraulic passenger, 2 stops		
1500# capacity	Each	62,800
2500# capacity	Each	66,300
3500# capacity	Each	69,800
Additional stop, add	Each	7825
Emergency Generators, Complete system, gas		
15 KW	Each	16,500
70 KW	Each	31,400
85 KW	Each	34,900
115 KW	Each	67,500
170 KW	Each	88,000
Diesel, 50 KW	Each	30,200
100 KW	Each	43,300
150 KW	Each	52,000
350 KW	Each	82,500

Description	Unit	$ Cost
Emergency Lighting, 25 watt, battery operated		
Nickel cadmium	Each	805
Flagpoles, Complete		
Aluminum, 20' high	Each	1650
40' High	Each	3475
70' High	Each	10,100
Fiberglass, 23' High	Each	1775
39'-5" High	Each	3325
59' High	Each	8225
Laundry Equipment		
Folders, blankets, & sheets	Each	66,500
Ironers, 110" single roll	Each	35,500
Combination washer & extractor, 50#	Each	12,200
125#	Each	32,800
Safe, Office type, 1 hour rating		
30" x 18" x 18"	Each	2400
60" x 36" x 18", double door	Each	8750
Sound System		
Amplifier, 250 watts	Each	2350
Speaker, ceiling or wall	Each	191
Trumpet	Each	365

Important: See the Reference Section for Location Factors

Model costs calculated for a 3 story building with 12' story height and 40,000 square feet of floor area

Jail

				Unit	Unit Cost	Cost Per S.F.	% Of Sub-Total
A.	**SUBSTRUCTURE**						
1010	Standard Foundations	Poured concrete; strip and spread footings		S.F. Ground	2.16	.72	
1020	Special Foundations	N/A		—	—	—	
1030	Slab on Grade	4" reinforced concrete with vapor barrier and granular base		S.F. Slab	4.74	1.58	1.7%
2010	Basement Excavation	Site preparation for slab and trench for foundation wall and footing		S.F. Ground	.26	.09	
2020	Basement Walls	4' foundation wall		L.F. Wall	69	.92	
B.	**SHELL**						
	B10 Superstructure						
1010	Floor Construction	Concrete slab with metal deck and beams		S.F. Floor	31.70	21.13	15.0%
1020	Roof Construction	Concrete slab with metal deck and beams		S.F. Roof	24.54	8.18	
	B20 Exterior Enclosure						
2010	Exterior Walls	Face brick with reinforced concrete block backup	85% of wall	S.F. Wall	30.85	12.51	
2020	Exterior Windows	Bullet resisting and metal horizontal pivoted	15% of wall	Each	146	10.48	11.8%
2030	Exterior Doors	Metal		Each	5450	.13	
	B30 Roofing						
3010	Roof Coverings	Built-up tar and gravel with flashing; perlite/EPS composite insulation		S.F. Roof	5.64	1.88	1.0%
3020	Roof Openings	Roof hatches		S.F. Roof	.06	.02	
C.	**INTERIORS**						
1010	Partitions	Concrete block	45 S.F. of Floor/L.F. Partition	S.F. Partition	9.38	2.50	
1020	Interior Doors	Hollow metal fire doors	930 S.F. Floor/Door	Each	875	.94	
1030	Fittings	N/A		—	—	—	
2010	Stair Construction	Concrete filled metal pan		Flight	18,950	2.37	6.5%
3010	Wall Finishes	Paint		S.F. Surface	3.68	1.96	
3020	Floor Finishes	70% vinyl composition tile, 20% carpet, 10% ceramic tile		S.F. Floor	3.08	3.08	
3030	Ceiling Finishes	Mineral tile on zee runners	30% of area	S.F. Ceiling	6.38	1.92	
D.	**SERVICES**						
	D10 Conveying						
1010	Elevators & Lifts	One hydraulic passenger elevator		Each	160,800	4.02	2.1%
1020	Escalators & Moving Walks	N/A		—	—	—	
	D20 Plumbing						
2010	Plumbing Fixtures	Toilet and service fixtures, supply and drainage	1 Fixture/78 S.F. Floor	Each	2276	29.18	
2020	Domestic Water Distribution	Electric water heater		S.F. Floor	5.67	5.67	18.1%
2040	Rain Water Drainage	Roof drains		S.F. Roof	1.71	.57	
	D30 HVAC						
3010	Energy Supply	N/A		—	—	—	
3020	Heat Generating Systems	Included in D3050		—	—	—	
3030	Cooling Generating Systems	N/A		—	—	—	7.3 %
3050	Terminal & Package Units	Rooftop multizone unit systems		S.F. Floor	14.25	14.25	
3090	Other HVAC Sys. & Equipment	N/A		—	—	—	
	D40 Fire Protection						
4010	Sprinklers	N/A		—	—	—	0.2 %
4020	Standpipes	Standpipes and hose systems		S.F. Floor	.48	.48	
	D50 Electrical						
5010	Electrical Service/Distribution	600 ampere service, panel board and feeders		S.F. Floor	1.26	1.26	
5020	Lighting & Branch Wiring	Fluorescent fixtures, receptacles, switches, A.C. and misc. power		S.F. Floor	9.19	9.19	7.5%
5030	Communications & Security	Alarm systems, intercom, and emergency lighting		S.F. Floor	3.52	3.52	
5090	Other Electrical Systems	Emergency generator, 80 kW		S.F. Floor	.68	.68	
E.	**EQUIPMENT & FURNISHINGS**						
1010	Commercial Equipment	N/A		—	—	—	
1020	Institutional Equipment	Prefabricated cells, visitor cubicles		S.F. Floor	56	56.37	28.8 %
1030	Vehicular Equipment	N/A		—	—	—	
1090	Other Equipment	N/A		—	—	—	
F.	**SPECIAL CONSTRUCTION**						
1020	Integrated Construction	N/A		—	—	—	0.0 %
1040	Special Facilities	N/A		—	—	—	
G.	**BUILDING SITEWORK**	**N/A**					

		Sub-Total	195.60	**100%**
CONTRACTOR FEES (General Requirements: 10%, Overhead: 5%, Profit: 10%)		25%	48.93	
ARCHITECT FEES		7%	17.12	
		Total Building Cost	261.65	

Costs per square foot of floor area

Exterior Wall	S.F. Area	1000	2000	3000	4000	5000	10000	15000	20000	25000
	L.F. Perimeter	126	179	219	253	283	400	490	568	632
Decorative Concrete Block	Steel Frame	249.60	214.05	199.85	191.95	186.80	174.70	169.65	166.90	164.90
	Bearing Walls	248.50	212.90	198.70	190.75	185.65	173.45	168.40	165.55	163.65
Face Brick with Concrete Block Back-up	Steel Frame	280.05	236.50	218.75	208.70	202.15	186.40	179.80	176.05	173.40
	Bearing Walls	277.30	233.30	215.35	205.20	198.50	182.60	175.85	172.05	169.40
Metal Sandwich Panel	Steel Frame	250.75	215.85	202.00	194.30	189.25	177.45	172.60	169.85	167.95
Precast Concrete Panel	Bearing Walls	251.70	215.15	200.55	192.40	187.05	174.45	169.25	166.30	164.30
Perimeter Adj., Add or Deduct	Per 100 L.F.	53.75	26.90	17.95	13.50	10.75	5.40	3.65	2.65	2.10
Story Hgt. Adj., Add or Deduct	Per 1 Ft.	3.60	2.55	2.15	1.85	1.60	1.15	0.95	0.75	0.75
For Basement, add $30.75 per square foot of basement area										

The above costs were calculated using the basic specifications shown on the facing page. These costs should be adjusted where necessary for design alternatives and owner's requirements. Reported completed project costs, for this type of structure, range from $97.80 to $239.05 per S.F.

Common additives

Description	Unit	$ Cost
Closed Circuit Surveillance, One station		
Camera and monitor	Each	1850
For additional camera stations, add	Each	1000
Emergency Lighting, 25 watt, battery operated		
Lead battery	Each	282
Nickel cadmium	Each	805
Laundry Equipment		
Dryers, coin operated 30 lb.	Each	3600
Double stacked	Each	7200
50 lb.	Each	3875
Dry cleaner 20 lb.	Each	42,300
30 lb.	Each	60,000
Washers, coin operated	Each	1450
Washer/extractor 20 lb.	Each	6475
30 lb.	Each	10,400
50 lb.	Each	12,200
75 lb.	Each	23,400
Smoke Detectors		
Ceiling type	Each	187
Duct type	Each	480

Important: See the Reference Section for Location Factors

Model costs calculated for a 1 story building with 12' story height and 3,000 square feet of floor area

Laundromat

			Unit	Unit Cost	Cost Per S.F.	% Of Sub-Total
A. SUBSTRUCTURE						
1010	Standard Foundations	Poured concrete; strip and spread footings	S.F. Ground	3.34	3.34	
1020	Special Foundations	N/A	—	—	—	
1030	Slab on Grade	5" reinforced concrete with vapor barrier and granular base	S.F. Slab	5.15	5.15	9.4%
2010	Basement Excavation	Site preparation for slab and trench for foundation wall and footing	S.F. Ground	.46	.46	
2020	Basement Walls	4' Foundation wall	L.F. Wall	69	5.03	
B. SHELL						
	B10 Superstructure					
1010	Floor Construction	N/A	—	—	—	7.8 %
1020	Roof Construction	Metal deck, open web steel joists, beams, columns	S.F. Roof	11.64	11.64	
	B20 Exterior Enclosure					
2010	Exterior Walls	Decorative concrete block 90% of wall	S.F. Wall	15.35	12.10	
2020	Exterior Windows	Store front 10% of wall	Each	39.65	3.47	11.3%
2030	Exterior Doors	Double aluminum and glass	Each	4125	1.38	
	B30 Roofing					
3010	Roof Coverings	Built-up tar and gravel with flashing; perlite/EPS composite insulation	S.F. Roof	8.75	8.75	5.9%
3020	Roof Openings	N/A	—	—	—	
C. INTERIORS						
1010	Partitions	Gypsum board on metal studs 60 S.F. Floor/L.F. Partition	S.F. Partition	23.40	3.90	
1020	Interior Doors	Single leaf wood 750 S.F. Floor/Door	Each	565	.76	
1030	Fittings	N/A	—	—	—	
2010	Stair Construction	N/A	—	—	—	6.7%
3010	Wall Finishes	Paint	S.F. Surface	.75	.25	
3020	Floor Finishes	Vinyl composition tile	S.F. Floor	1.85	1.85	
3030	Ceiling Finishes	Fiberglass board on exposed grid system	S.F. Ceiling	3.20	3.20	
D. SERVICES						
	D10 Conveying					
1010	Elevators & Lifts	N/A	—	—	—	0.0 %
1020	Escalators & Moving Walks	N/A	—	—	—	
	D20 Plumbing					
2010	Plumbing Fixtures	Toilet and service fixtures, supply and drainage 1 Fixture/600 S.F. Floor	Each	4758	7.93	
2020	Domestic Water Distribution	Gas fired hot water heater	S.F. Floor	40.41	40.41	33.0%
2040	Rain Water Drainage	Roof drains	S.F. Roof	.94	.94	
	D30 HVAC					
3010	Energy Supply	N/A	—	—	—	
3020	Heat Generating Systems	Included in D3050	—	—	—	
3030	Cooling Generating Systems	N/A	—	—	—	5.7 %
3050	Terminal & Package Units	Rooftop single zone unit systems	S.F. Floor	8.59	8.59	
3090	Other HVAC Sys. & Equipment	N/A	—	—	—	
	D40 Fire Protection					
4010	Sprinklers	Sprinkler, ordinary hazard	S.F. Floor	3.92	3.92	2.6%
4020	Standpipes	N/A	—	—	—	
	D50 Electrical					
5010	Electrical Service/Distribution	200 ampere service, panel board and feeders	S.F. Floor	3.56	3.56	
5020	Lighting & Branch Wiring	Fluorescent fixtures, receptacles, switches, A.C. and misc. power	S.F. Floor	22.33	22.33	17.6%
5030	Communications & Security	Alarm systems and emergency lighting	S.F. Floor			
5090	Other Electrical Systems	Emergency generator, 7.5 kW	S.F. Floor	.46	.46	
E. EQUIPMENT & FURNISHINGS						
1010	Commercial Equipment	N/A	—	—	—	
1020	Institutional Equipment	N/A	—	—	—	0.0 %
1030	Vehicular Equipment	N/A	—	—	—	
1090	Other Equipment	N/A	—	—	—	
F. SPECIAL CONSTRUCTION						
1020	Integrated Construction	N/A	—	—	—	0.0 %
1040	Special Facilities	N/A	—	—	—	
G. BUILDING SITEWORK	**N/A**					

		Sub-Total	149.42	100%
CONTRACTOR FEES (General Requirements: 10%, Overhead: 5%, Profit: 10%)		25%	37.36	
ARCHITECT FEES		7%	13.07	

Total Building Cost	**199.85**

Costs per square foot of floor area

Exterior Wall	S.F. Area	7000	10000	13000	16000	19000	22000	25000	28000	31000
	L.F. Perimeter	240	300	336	386	411	435	472	510	524
Face Brick with Concrete Block Back-up	R/Conc. Frame	176.25	166.30	157.95	154.05	149.35	145.85	144.05	142.65	140.30
	Steel Frame	182.90	172.90	164.50	160.70	155.90	152.45	150.65	149.20	146.90
Limestone with Concrete Block	R/Conc. Frame	213.80	199.15	186.20	180.45	172.95	167.50	164.70	162.50	158.75
	Steel Frame	220.35	205.75	192.80	187.10	179.60	174.10	171.30	169.15	165.40
Precast Concrete Panels	R/Conc. Frame	173.95	164.25	156.15	152.40	147.85	144.50	142.75	141.35	139.15
	Steel Frame	180.55	170.90	162.80	159.05	154.50	151.10	149.40	148.00	145.75
Perimeter Adj., Add or Deduct	Per 100 L.F.	22.95	16.00	12.30	10.00	8.40	7.30	6.40	5.70	5.10
Story Hgt. Adj., Add or Deduct	Per 1 Ft.	3.45	3.05	2.60	2.45	2.15	2.00	1.90	1.80	1.70
For Basement, add $44.00 per square foot of basement area										

The above costs were calculated using the basic specifications shown on the facing page. These costs should be adjusted where necessary for design alternatives and owner's requirements. Reported completed project costs, for this type of structure, range from $93.95 to $241.15 per S.F.

Common additives

Description	Unit	$ Cost
Carrels Hardwood	Each	660 - 990
Closed Circuit Surveillance, One station		
Camera and monitor	Each	1850
For additional camera stations, add	Each	1000
Elevators, Hydraulic passenger, 2 stops		
1500# capacity	Each	62,800
2500# capacity	Each	66,300
3500# capacity	Each	69,800
Emergency Lighting, 25 watt, battery operated		
Lead battery	Each	282
Nickel cadmium	Each	805
Flagpoles, Complete		
Aluminum, 20' high	Each	1650
40' high	Each	3475
70' high	Each	10,100
Fiberglass, 23' high	Each	1775
39'-5" high	Each	3325
59' high	Each	8225

Description	Unit	$ Cost
Library Furnishings		
Bookshelf, 90" high, 10" shelf double face	L.F.	173
single face	L.F.	166
Charging desk, built-in with counter		
Plastic laminated top	L.F.	355
Reading table, laminated		
top 60" x 36"	Each	710

Model costs calculated for a 2 story building with 14' story height and 22,000 square feet of floor area

Library

				Unit	Unit Cost	Cost Per S.F.	% Of Sub-Total
A. SUBSTRUCTURE							
1010	Standard Foundations	Poured concrete; strip and spread footings		S.F. Ground	3.84	1.92	
1020	Special Foundations	N/A		—	—	—	
1030	Slab on Grade	4" reinforced concrete with vapor barrier and granular base		S.F. Slab	4.74	2.37	5.5%
2010	Basement Excavation	Site preparation for slab and trench for foundation wall and footing		S.F. Ground	.26	.13	
2020	Basement Walls	4' foundation wall		L.F. Wall	78	1.54	
B. SHELL							
	B10 Superstructure						
1010	Floor Construction	Concrete waffle slab		S.F. Floor	25.08	12.54	19.6%
1020	Roof Construction	Concrete waffle slab		S.F. Roof	17.36	8.68	
	B20 Exterior Enclosure						
2010	Exterior Walls	Face brick with concrete block backup	90% of wall	S.F. Wall	32.01	15.95	
2020	Exterior Windows	Window wall	10% of wall	Each	48.55	2.69	17.7%
2030	Exterior Doors	Double aluminum and glass, single leaf hollow metal		Each	4850	.45	
	B30 Roofing						
3010	Roof Coverings	Single ply membrane, EPDM, fully adhered; perlite/EPS composite insulation		S.F. Roof	4.84	2.42	2.3%
3020	Roof Openings	Roof hatches		S.F. Roof	.08	.04	
C. INTERIORS							
1010	Partitions	Gypsum board on metal studs	30 S.F. Floor/L.F. Partition	S.F. Partition	10.88	4.35	
1020	Interior Doors	Single leaf wood	300 S.F. Floor/Door	Each	565	1.89	
1030	Fittings	N/A		—	—	—	
2010	Stair Construction	Concrete filled metal pan		Flight	9075	.83	16.7%
3010	Wall Finishes	Paint		S.F. Surface	.74	.59	
3020	Floor Finishes	50% carpet, 50% vinyl tile		S.F. Floor	4.02	4.02	
3030	Ceiling Finishes	Mineral fiber on concealed zee bars		S.F. Ceiling	6.38	6.38	
D. SERVICES							
	D10 Conveying						
1010	Elevators & Lifts	One hydraulic passenger elevator		Each	84,260	3.83	3.5%
1020	Escalators & Moving Walks	N/A		—	—	—	
	D20 Plumbing						
2010	Plumbing Fixtures	Toilet and service fixtures, supply and drainage	1 Fixture/1835 S.F. Floor	Each	2551	1.39	
2020	Domestic Water Distribution	Gas fired water heater		S.F. Floor	.90	.90	2.5%
2040	Rain Water Drainage	Roof drains		S.F. Roof	.74	.37	
	D30 HVAC						
3010	Energy Supply	N/A		—	—	—	
3020	Heat Generating Systems	Included in D3050		—	—	—	
3030	Cooling Generating Systems	N/A		—	—	—	17.9 %
3050	Terminal & Package Units	Multizone unit, gas heating, electric cooling		S.F. Floor	19.30	19.30	
3090	Other HVAC Sys. & Equipment	N/A		—	—	—	
	D40 Fire Protection						
4010	Sprinklers	Wet pipe sprinkler system		S.F. Floor	2.46	2.46	2.3%
4020	Standpipes	N/A		—	—	—	
	D50 Electrical						
5010	Electrical Service/Distribution	400 ampere service, panel board and feeders		S.F. Floor	1.07	1.07	
5020	Lighting & Branch Wiring	Fluorescent fixtures, receptacles, switches, A.C. and misc. power		S.F. Floor	10.40	10.40	12.0%
5030	Communications & Security	Alarm systems, internet wiring, and emergency lighting		S.F. Floor	1.30	1.30	
5090	Other Electrical Systems	Emergency generator, 7.5 kW, Uninterruptible power supply		S.F. Floor	.23	.23	
E. EQUIPMENT & FURNISHINGS							
1010	Commercial Equipment	N/A		—	—	—	
1020	Institutional Equipment	N/A		—	—	—	
1030	Vehicular Equipment	N/A		—	—	—	0.0 %
1090	Other Equipment	N/A		—	—	—	
F. SPECIAL CONSTRUCTION							
1020	Integrated Construction	N/A		—	—	—	0.0 %
1040	Special Facilities	N/A		—	—	—	
G. BUILDING SITEWORK	**N/A**						

			Sub-Total	108.04	100%
CONTRACTOR FEES (General Requirements: 10%, Overhead: 5%, Profit: 10%)			25%	27.01	
ARCHITECT FEES			8%	10.80	

Total Building Cost	**145.85**

161

Costs per square foot of floor area

Exterior Wall	S.F. Area	4000	5500	7000	8500	10000	11500	13000	14500	16000
	L.F. Perimeter	280	320	380	440	453	503	510	522	560
Face Brick with Concrete Block Back-up	Steel Joists	185.30	175.40	171.60	169.20	164.40	162.90	159.60	157.25	156.40
	Wood Truss	190.10	180.20	176.40	173.95	169.15	167.65	164.40	162.00	161.15
Stucco on Concrete Block	Steel Joists	172.40	164.70	161.65	159.65	156.05	154.90	152.35	150.60	149.95
	Wood Truss	177.15	169.45	166.40	164.45	160.80	159.65	157.15	155.40	154.65
Brick Veneer	Wood Truss	185.10	176.10	172.60	170.30	165.95	164.55	161.65	159.45	158.70
Wood Siding	Wood Frame	174.05	166.90	164.00	162.15	158.80	157.70	155.40	153.80	153.15
Perimeter Adj., Add or Deduct	Per 100 L.F.	16.65	12.10	9.60	7.85	6.65	5.80	5.15	4.65	4.20
Story Hgt. Adj., Add or Deduct	Per 1 Ft.	3.40	2.85	2.65	2.50	2.20	2.15	1.95	1.75	1.70
For Basement, add $28.65 per square foot of basement area										

The above costs were calculated using the basic specifications shown on the facing page. These costs should be adjusted where necessary for design alternatives and owner's requirements. Reported completed project costs, for this type of structure, range from $79.20 to $203.85 per S.F.

Common additives

Description	Unit	$ Cost	Description	Unit	$ Cost
Cabinets, Hospital, base			Directory Boards, Plastic, glass covered		
Laminated plastic	L.F.	355	30" x 20"	Each	595
Stainless steel	L.F.	575	36" x 48"	Each	1450
Counter top, laminated plastic	L.F.	69.50	Aluminum, 24" x 18"	Each	600
Stainless steel	L.F.	169	36" x 24"	Each	675
For drop-in sink, add	Each	950	48" x 32"	Each	980
Nurses station, door type			48" x 60"	Each	2025
Laminated plastic	L.F.	400	Heat Therapy Unit		
Enameled steel	L.F.	385	Humidified, 26" x 78" x 28"	Each	3550
Stainless steel	L.F.	670	Smoke Detectors		
Wall cabinets, laminated plastic	L.F.	259	Ceiling type	Each	187
Enameled steel	L.F.	305	Duct type	Each	480
Stainless steel	L.F.	540	Tables, Examining, vinyl top		
			with base cabinets	Each	1700 - 4550
			Utensil Washer, Sanitizer	Each	11,700
			X-Ray, Mobile	Each	14,000 - 79,000

Important: See the Reference Section for Location Factors

Model costs calculated for a 1 story building with 10′ story height and 7,000 square feet of floor area

				Unit	Unit Cost	Cost Per S.F.	% Of Sub-Total
A.	**SUBSTRUCTURE**						
1010	Standard Foundations	Poured concrete; strip and spread footings		S.F. Ground	2.14	2.14	
1020	Special Foundations	N/A		—	—	—	
1030	Slab on Grade	4″ reinforced concrete with vapor barrier and granular base		S.F. Slab	4.74	4.74	8.8%
2010	Basement Excavation	Site preparation for slab and trench for foundation wall and footing		S.F. Ground	.26	.26	
2020	Basement Walls	4′ foundation wall		L.F. Wall	78	4.23	
B.	**SHELL**						
	B10 Superstructure						
1010	Floor Construction	N/A		—	—	—	5.1%
1020	Roof Construction	Plywood on wood trusses		S.F. Roof	6.61	6.61	
	B20 Exterior Enclosure						
2010	Exterior Walls	Face brick with concrete block backup	70% of wall	S.F. Wall	32	12.16	
2020	Exterior Windows	Wood double hung	30% of wall	Each	588	5.64	15.7%
2030	Exterior Doors	Aluminum and glass doors and entrance with transoms		Each	1476	2.53	
	B30 Roofing						
3010	Roof Coverings	Asphalt shingles with flashing (Pitched); rigid fiber glass insulation, gutters		S.F. Roof	3.79	3.79	2.9%
3020	Roof Openings	N/A		—	—	—	
C.	**INTERIORS**						
1010	Partitions	Gypsum bd. & sound deadening bd. on wood studs w/insul.	6 S.F. Floor/L.F. Partition	S.F. Partition	8.51	11.35	
1020	Interior Doors	Single leaf wood	60 S.F. Floor/Door	Each	565	9.41	
1030	Fittings	N/A		—	—	—	
2010	Stair Construction	N/A		—	—	—	27.9%
3010	Wall Finishes	50% paint, 50% vinyl wall covering		S.F. Surface	1.24	3.30	
3020	Floor Finishes	50% carpet, 50% vinyl composition tile		S.F. Floor	6.57	6.57	
3030	Ceiling Finishes	Mineral fiber tile on concealed zee bars		S.F. Ceiling	5.49	5.49	
D.	**SERVICES**						
	D10 Conveying						
1010	Elevators & Lifts	N/A		—	—	—	0.0 %
1020	Escalators & Moving Walks	N/A		—	—	—	
	D20 Plumbing						
2010	Plumbing Fixtures	Toilet and service fixtures, supply and drainage	1 Fixture/195 S.F. Floor	Each	3173	16.27	
2020	Domestic Water Distribution	Gas fired water heater		S.F. Floor	1.64	1.64	13.8%
2040	Rain Water Drainage	N/A		—	—	—	
	D30 HVAC						
3010	Energy Supply	N/A		—	—	—	
3020	Heat Generating Systems	Included in D3050		—	—	—	
3030	Cooling Generating Systems	N/A		—	—	—	10.4 %
3050	Terminal & Package Units	Multizone unit, gas heating, electric cooling		S.F. Floor	13.50	13.50	
3090	Other HVAC Sys. & Equipment	N/A		—	—	—	
	D40 Fire Protection						
4010	Sprinklers	Wet pipe sprinkler system		S.F. Floor	3.59	3.59	2.8%
4020	Standpipes	N/A		—	—	—	
	D50 Electrical						
5010	Electrical Service/Distribution	200 ampere service, panel board and feeders		S.F. Floor	1.45	1.45	
5020	Lighting & Branch Wiring	Fluorescent fixtures, receptacles, switches, A.C. and misc. power		S.F. Floor	7.01	7.01	12.5%
5030	Communications & Security	Alarm systems, internet and phone wiring, intercom system, and emergency lighting		S.F. Floor	6.95	6.95	
5090	Other Electrical Systems	Emergency generator, 7.5 kW		S.F. Floor	.83	.83	
E.	**EQUIPMENT & FURNISHINGS**						
1010	Commercial Equipment	N/A		—	—	—	
1020	Institutional Equipment	N/A		—	—	—	0.0 %
1030	Vehicular Equipment	N/A		—	—	—	
1090	Other Equipment	N/A		—	—	—	
F.	**SPECIAL CONSTRUCTION**						
1020	Integrated Construction	N/A		—	—	—	0.0 %
1040	Special Facilities	N/A		—	—	—	
G.	**BUILDING SITEWORK**	**N/A**					

			Sub-Total	129.46	100%
CONTRACTOR FEES (General Requirements: 10%, Overhead: 5%, Profit: 10%)			25%	32.38	
ARCHITECT FEES			9%	14.56	

Total Building Cost	**176.40**

163

Costs per square foot of floor area

Exterior Wall	S.F. Area	4000	5500	7000	8500	10000	11500	13000	14500	16000
	L.F. Perimeter	180	210	240	270	286	311	336	361	386
Face Brick with Concrete Block Back-up	Steel Joists	216.40	205.55	199.40	195.30	190.95	188.55	186.75	185.25	184.15
	Wood Joists	216.45	205.60	199.40	195.35	190.95	188.60	186.75	185.30	184.15
Stucco on Concrete Block	Steel Joists	204.70	195.65	190.45	187.05	183.40	181.55	180.00	178.75	177.80
	Wood Joists	204.05	195.10	189.95	186.65	183.10	181.20	179.70	178.50	177.50
Brick Veneer	Wood Frame	211.05	201.05	195.30	191.60	187.55	185.40	183.70	182.35	181.25
Wood Siding	Wood Frame	199.50	191.30	186.50	183.50	180.25	178.45	177.10	176.00	175.15
Perimeter Adj., Add or Deduct	Per 100 L.F.	28.75	20.90	16.40	13.60	11.50	10.00	8.90	8.00	7.15
Story Hgt. Adj., Add or Deduct	Per 1 Ft.	4.20	3.60	3.15	3.00	2.70	2.55	2.35	2.35	2.25
For Basement, add $31.75 per square foot of basement area										

The above costs were calculated using the basic specifications shown on the facing page. These costs should be adjusted where necessary for design alternatives and owner's requirements. Reported completed project costs, for this type of structure, range from $81.25 to $277.65 per S.F.

Common additives

Description	Unit	$ Cost
Cabinets, Hospital, base		
Laminated plastic	L.F.	355
Stainless steel	L.F.	575
Counter top, laminated plastic	L.F.	69.50
Stainless steel	L.F.	169
For drop-in sink, add	Each	950
Nurses station, door type		
Laminated plastic	L.F.	400
Enameled steel	L.F.	385
Stainless steel	L.F.	670
Wall cabinets, laminated plastic	L.F.	259
Enameled steel	L.F.	305
Stainless steel	L.F.	540
Elevators, Hydraulic passenger, 2 stops		
1500# capacity	Each	62,800
2500# capacity	Each	66,300
3500# capacity	Each	69,800

Description	Unit	$ Cost
Directory Boards, Plastic, glass covered		
30" x 20"	Each	595
36" x 48"	Each	1450
Aluminum, 24" x 18"	Each	600
36" x 24"	Each	675
48" x 32"	Each	980
48" x 60"	Each	2025
Emergency Lighting, 25 watt, battery operated		
Lead battery	Each	282
Nickel cadmium	Each	805
Heat Therapy Unit		
Humidified, 26" x 78" x 28"	Each	3550
Smoke Detectors		
Ceiling type	Each	187
Duct type	Each	480
Tables, Examining, vinyl top		
with base cabinets	Each	1700 - 4550
Utensil Washer, Sanitizer	Each	11,700
X-Ray, Mobile	Each	14,000 - 79,000

Model costs calculated for a 2 story building with 10′ story height and 7,000 square feet of floor area

				Unit	Unit Cost	Cost Per S.F.	% Of Sub-Total
A.	**SUBSTRUCTURE**						
1010	Standard Foundations	Poured concrete; strip and spread footings		S.F. Ground	3.42	1.71	
1020	Special Foundations	N/A		—	—	—	
1030	Slab on Grade	4″ reinforced concrete with vapor barrier and granular base		S.F. Slab	4.74	2.37	4.9%
2010	Basement Excavation	Site preparation for slab and trench for foundation wall and footing		S.F. Ground	.17	.09	
2020	Basement Walls	4′ foundation wall		L.F. Wall	78	2.67	
B.	**SHELL**						
	B10 Superstructure						
1010	Floor Construction	Open web steel joists, slab form, concrete, columns		S.F. Floor	15.72	7.86	8.1%
1020	Roof Construction	Metal deck, open web steel joists, beams, columns		S.F. Roof	7	3.50	
	B20 Exterior Enclosure						
2010	Exterior Walls	Stucco on concrete block		S.F. Wall	17.50	8.40	
2020	Exterior Windows	Outward projecting metal	30% of wall	Each	492	6.75	11.8%
2030	Exterior Doors	Aluminum and glass doors with transoms		Each	4850	1.38	
	B30 Roofing						
3010	Roof Coverings	Built-up tar and gravel with flashing; perlite/EPS composite insulation		S.F. Roof	6.66	3.33	2.5%
3020	Roof Openings	Roof hatches		S.F. Roof	.24	.12	
C.	**INTERIORS**						
1010	Partitions	Gypsum bd. & acous. insul. on metal studs	6 S.F. Floor/L.F. Partition	S.F. Partition	7.29	9.72	
1020	Interior Doors	Single leaf wood	60 S.F. Floor/Door	Each	565	9.41	
1030	Fittings	N/A		—	—	—	
2010	Stair Construction	Concrete filled metal pan		Flight	15,800	4.51	27.9%
3010	Wall Finishes	45% paint, 50% vinyl wall coating, 5% ceramic tile		S.F. Surface	1.24	3.30	
3020	Floor Finishes	50% carpet, 50% vinyl composition tile		S.F. Floor	6.57	6.57	
3030	Ceiling Finishes	Mineral fiber tile on concealed zee bars		S.F. Ceiling	5.49	5.49	
D.	**SERVICES**						
	D10 Conveying						
1010	Elevators & Lifts	One hydraulic hospital elevator		Each	103,460	14.78	10.6%
1020	Escalators & Moving Walks	N/A		—	—	—	
	D20 Plumbing						
2010	Plumbing Fixtures	Toilet and service fixtures, supply and drainage	1 Fixture/160 S.F. Floor	Each	1934	12.09	
2020	Domestic Water Distribution	Gas fired water heater		S.F. Floor	1.84	1.84	10.1%
2040	Rain Water Drainage	Roof drains		S.F. Roof	.48	.24	
	D30 HVAC						
3010	Energy Supply	N/A		—	—	—	
3020	Heat Generating Systems	Included in D3050		—	—	—	
3030	Cooling Generating Systems	N/A		—	—	—	9.7 %
3050	Terminal & Package Units	Multizone unit, gas heating, electric cooling		S.F. Floor	13.50	13.50	
3090	Other HVAC Sys. & Equipment	N/A		—	—	—	
	D40 Fire Protection						
4010	Sprinklers	Wet pipe sprinkler system		S.F. Floor	2.95	2.95	2.1%
4020	Standpipes	N/A		—	—	—	
	D50 Electrical						
5010	Electrical Service/Distribution	400 ampere service, panel board and feeders		S.F. Floor	2.40	2.40	
5020	Lighting & Branch Wiring	Fluorescent fixtures, receptacles, switches, A.C. and misc. power		S.F. Floor	7.01	7.01	12.3%
5030	Communications & Security	Alarm system, internet and phone wiring, intercom system, and emergency lighting		S.F. Floor	6.95	6.95	
5090	Other Electrical Systems	Emergency generator, 7.5 kW		S.F. Floor	.83	.83	
E.	**EQUIPMENT & FURNISHINGS**						
1010	Commercial Equipment	N/A		—	—	—	
1020	Institutional Equipment	N/A		—	—	—	
1030	Vehicular Equipment	N/A		—	—	—	0.0 %
1090	Other Equipment	N/A		—	—	—	
F.	**SPECIAL CONSTRUCTION**						
1020	Integrated Construction	N/A		—	—	—	0.0 %
1040	Special Facilities	N/A		—	—	—	
G.	**BUILDING SITEWORK**	**N/A**					

		Sub-Total	139.77	100%
	CONTRACTOR FEES (General Requirements: 10%, Overhead: 5%, Profit: 10%)	25%	34.96	
	ARCHITECT FEES	9%	15.72	
	Total Building Cost		**190.45**	

Costs per square foot of floor area

Exterior Wall	S.F. Area	2000	3000	4000	6000	8000	10000	12000	14000	16000
	L.F. Perimeter	240	260	280	380	480	560	580	660	740
Brick Veneer	Wood Frame	181.45	157.55	145.65	139.45	136.30	133.30	128.45	127.40	126.60
Aluminum Siding	Wood Frame	162.10	143.55	134.30	129.20	126.60	124.25	120.60	119.75	119.15
Wood Siding	Wood Frame	162.40	143.80	134.45	129.35	126.70	124.35	120.70	119.90	119.25
Wood Shingles	Wood Frame	167.10	147.20	137.25	131.85	129.10	126.55	122.60	121.70	121.05
Precast Concrete Block	Wood Truss	161.25	142.95	133.85	128.75	126.20	123.85	120.25	119.45	118.85
Brick on Concrete Block	Wood Truss	188.70	162.80	149.80	143.25	139.90	136.65	131.30	130.20	129.40
Perimeter Adj., Add or Deduct	Per 100 L.F.	28.75	19.15	14.30	9.55	7.20	5.75	4.75	4.05	3.65
Story Hgt. Adj., Add or Deduct	Per 1 Ft.	4.70	3.40	2.70	2.45	2.30	2.20	1.85	1.80	1.80
For Basement, add $21.85 per square foot of basement area										

The above costs were calculated using the basic specifications shown on the facing page. These costs should be adjusted where necessary for design alternatives and owner's requirements. Reported completed project costs, for this type of structure, range from $65.60 to $310.55 per S.F.

Common additives

Description	Unit	$ Cost
Closed Circuit Surveillance, One station		
Camera and monitor	Each	1850
For additional camera stations, add	Each	1000
Emergency Lighting, 25 watt, battery operated		
Lead battery	Each	282
Nickel cadmium	Each	805
Laundry Equipment		
Dryer, gas, 16 lb. capacity	Each	885
30 lb. capacity	Each	3600
Washer, 4 cycle	Each	1075
Commercial	Each	1450
Sauna, Prefabricated, complete		
6' x 4'	Each	5850
6' x 6'	Each	6950
6' x 9'	Each	8525
8' x 8'	Each	10,100
8' x 10'	Each	11,300
10' x 12'	Each	14,000
Smoke Detectors		
Ceiling type	Each	187
Duct type	Each	480

Description	Unit	$ Cost
Swimming Pools, Complete, gunite	S.F.	64 - 78.50
TV Antenna, Master system, 12 outlet	Outlet	315
30 outlet	Outlet	203
100 outlet	Outlet	194

Model costs calculated for a 1 story building with 9' story height and 8,000 square feet of floor area

				Unit	Unit Cost	Cost Per S.F.	% Of Sub-Total
A.	**SUBSTRUCTURE**						
1010	Standard Foundations	Poured concrete; strip and spread footings		S.F. Ground	3.25	3.25	
1020	Special Foundations	N/A		—	—	—	
1030	Slab on Grade	4" reinforced concrete with vapor barrier and granular base		S.F. Slab	4.74	4.74	14.0%
2010	Basement Excavation	Site preparation for slab and trench for foundation wall and footing		S.F. Ground	.26	.26	
2020	Basement Walls	4' foundation wall		L.F. Wall	74	6.04	
B.	**SHELL**						
	B10 Superstructure						
1010	Floor Construction	N/A		—	—	—	6.5 %
1020	Roof Construction	Plywood on wood trusses		S.F. Roof	6.61	6.61	
	B20 Exterior Enclosure						
2010	Exterior Walls	Face brick on wood studs with sheathing, insulation and paper	80% of wall	S.F. Wall	24.58	10.62	
2020	Exterior Windows	Wood double hung	20% of wall	Each	471	3.64	18.0%
2030	Exterior Doors	Wood solid core		Each	1476	4.06	
	B30 Roofing						
3010	Roof Coverings	Asphalt shingles with flashing (pitched); rigid fiberglass insulation		S.F. Roof	3.62	3.62	3.6%
3020	Roof Openings	N/A		—	—	—	
C.	**INTERIORS**						
1010	Partitions	Gypsum bd. and sound deadening bd. on wood studs	9 S.F. Floor/L.F. Partition	S.F. Partition	9.07	8.06	
1020	Interior Doors	Single leaf hollow core wood	300 S.F. Floor/Door	Each	488	1.62	
1030	Fittings	N/A		—	—	—	
2010	Stair Construction	N/A		—	—	—	25.0%
3010	Wall Finishes	90% paint, 10% ceramic tile		S.F. Surface	1.36	2.41	
3020	Floor Finishes	85% carpet, 15% ceramic tile		S.F. Floor	9.45	9.45	
3030	Ceiling Finishes	Painted gypsum board on furring		S.F. Ceiling	3.95	3.95	
D.	**SERVICES**						
	D10 Conveying						
1010	Elevators & Lifts	N/A		—	—	—	0.0 %
1020	Escalators & Moving Walks	N/A		—	—	—	
	D20 Plumbing						
2010	Plumbing Fixtures	Toilet and service fixtures, supply and drainage	1 Fixture/90 S.F. Floor	Each	1592	17.69	
2020	Domestic Water Distribution	Gas fired water heater		S.F. Floor	1.36	1.36	18.7%
2040	Rain Water Drainage	N/A		—	—	—	
	D30 HVAC						
3010	Energy Supply	N/A		—	—	—	
3020	Heat Generating Systems	Included in D3050		—	—	—	
3030	Cooling Generating Systems	N/A		—	—	—	2.7 %
3050	Terminal & Package Units	Through the wall electric heating and cooling units		S.F. Floor	2.76	2.76	
3090	Other HVAC Sys. & Equipment	N/A		—	—	—	
	D40 Fire Protection						
4010	Sprinklers	Wet pipe sprinkler system		S.F. Floor	3.59	3.59	3.5%
4020	Standpipes	N/A		—	—	—	
	D50 Electrical						
5010	Electrical Service/Distribution	200 ampere service, panel board and feeders		S.F. Floor	1.29	1.29	
5020	Lighting & Branch Wiring	Fluorescent fixtures, receptacles, switches and misc. power		S.F. Floor	6.47	6.47	8.0%
5030	Communications & Security	Alarm systems		S.F. Floor	.40	.40	
5090	Other Electrical Systems	N/A		—	—	—	
E.	**EQUIPMENT & FURNISHINGS**						
1010	Commercial Equipment	N/A		—	—	—	
1020	Institutional Equipment	N/A		—	—	—	
1030	Vehicular Equipment	N/A		—	—	—	0.0 %
1090	Other Equipment	N/A		—	—	—	
F.	**SPECIAL CONSTRUCTION**						
1020	Integrated Construction	N/A		—	—	—	0.0 %
1040	Special Facilities	N/A		—	—	—	
G.	**BUILDING SITEWORK**	**N/A**					

			Sub-Total	101.89	**100%**
CONTRACTOR FEES (General Requirements: 10%, Overhead: 5%, Profit: 10%)			25%	25.49	
ARCHITECT FEES			7%	8.92	

Total Building Cost	**136.30**

Costs per square foot of floor area

Exterior Wall	S.F. Area	25000	37000	49000	61000	73000	81000	88000	96000	104000
	L.F. Perimeter	433	593	606	720	835	911	978	1054	1074
Decorative Concrete Block	Wood Joists	161.35	156.70	151.00	149.40	148.35	147.80	147.40	147.05	146.15
	Precast Conc.	172.70	168.10	162.35	160.75	159.75	159.15	158.80	158.40	157.50
Stucco on Concrete Block	Wood Joists	160.55	156.00	150.20	148.70	147.65	147.05	146.70	146.30	145.40
	Precast Conc.	172.55	168.00	162.25	160.70	159.65	159.05	158.70	158.30	157.40
Wood Siding	Wood Frame	157.70	153.30	148.20	146.75	145.75	145.15	144.85	144.45	143.70
Brick Veneer	Wood Frame	164.85	159.90	153.30	151.55	150.45	149.85	149.45	149.05	148.00
Perimeter Adj., Add or Deduct	Per 100 L.F.	4.60	3.20	2.35	1.90	1.55	1.40	1.35	1.20	1.05
Story Hgt. Adj., Add or Deduct	Per 1 Ft.	1.60	1.50	1.15	1.05	1.05	1.00	1.05	0.95	0.95

For Basement, add $28.70 per square foot of basement area

The above costs were calculated using the basic specifications shown on the facing page. These costs should be adjusted where necessary for design alternatives and owner's requirements. Reported completed project costs, for this type of structure, range from $60.40 to $310.55 per S.F.

Common additives

Description	Unit	$ Cost	Description	Unit	$ Cost
Closed Circuit Surveillance, One station			Sauna, Prefabricated, complete		
Camera and monitor	Each	1850	6' x 4'	Each	5850
For additional camera station, add	Each	1000	6' x 6'	Each	6950
Elevators, Hydraulic passenger, 2 stops			6' x 9'	Each	8525
1500# capacity	Each	62,800	8' x 8'	Each	10,100
2500# capacity	Each	66,300	8' x 10'	Each	11,300
3500# capacity	Each	69,800	10' x 12'	Each	14,000
Additional stop, add	Each	7825	Smoke Detectors		
Emergency Lighting, 25 watt, battery operated			Ceiling type	Each	187
Lead battery	Each	282	Duct type	Each	480
Nickel cadmium	Each	805	Swimming Pools, Complete, gunite	S.F.	64 - 78.50
Laundry Equipment			TV Antenna, Master system, 12 outlet	Outlet	315
Dryer, gas, 16 lb. capacity	Each	885	30 outlet	Outlet	203
30 lb. capacity	Each	3600	100 outlet	Outlet	194
Washer, 4 cycle	Each	1075			
Commercial	Each	1450			

Model costs calculated for a 3 story building with 9' story height and 49,000 square feet of floor area

Motel, 2-3 Story

				Unit	Unit Cost	Cost Per S.F.	% Of Sub-Total
A.	**SUBSTRUCTURE**						
1010	Standard Foundations	Poured concrete; strip and spread footings		S.F. Ground	1.47	.49	
1020	Special Foundations	N/A		—	—	—	
1030	Slab on Grade	4" reinforced concrete with vapor barrier and granular base		S.F. Slab	4.74	1.58	3.3%
2010	Basement Excavation	Site preparation for slab and trench for foundation wall and footing		S.F. Ground	.17	.06	
2020	Basement Walls	4' foundation wall		L.F. Wall	78	1.93	
B.	**SHELL**						
	B10 Superstructure						
1010	Floor Construction	Precast concrete plank		S.F. Floor	10.41	6.94	8.4%
1020	Roof Construction	Precast concrete plank		S.F. Roof	10.02	3.34	
	B20 Exterior Enclosure						
2010	Exterior Walls	Decorative concrete block	85% of wall	S.F. Wall	16.70	4.74	
2020	Exterior Windows	Aluminum sliding	15% of wall	Each	552	1.84	9.9%
2030	Exterior Doors	Aluminum and glass doors and entrance with transom		Each	1667	5.50	
	B30 Roofing						
3010	Roof Coverings	Built-up tar and gravel with flashing; perlite/EPS composite insulation		S.F. Roof	5.97	1.99	1.7%
3020	Roof Openings	Roof hatches		S.F. Roof	.12	.04	
C.	**INTERIORS**						
1010	Partitions	Concrete block	7 S.F. Floor/L.F. Partition	S.F. Partition	17.27	19.74	
1020	Interior Doors	Wood hollow core	70 S.F. Floor/Door	Each	488	6.97	
1030	Fittings	N/A		—	—	—	
2010	Stair Construction	Concrete filled metal pan		Flight	15,800	3.87	38.2%
3010	Wall Finishes	90% paint, 10% ceramic tile		S.F. Surface	1.35	3.09	
3020	Floor Finishes	85% carpet, 5% vinyl composition tile, 10% ceramic tile		S.F. Floor	9.45	9.45	
3030	Ceiling Finishes	Textured finish		S.F. Ceiling	3.64	3.64	
D.	**SERVICES**						
	D10 Conveying						
1010	Elevators & Lifts	Two hydraulic passenger elevators		Each	104,860	4.28	3.5%
1020	Escalators & Moving Walks	N/A		—	—	—	
	D20 Plumbing						
2010	Plumbing Fixtures	Toilet and service fixtures, supply and drainage	1 Fixture/180 S.F. Floor	Each	4581	25.45	
2020	Domestic Water Distribution	Gas fired water heater		S.F. Floor	.94	.94	22.0%
2040	Rain Water Drainage	Roof drains		S.F. Roof	1.53	.51	
	D30 HVAC						
3010	Energy Supply	N/A		—	—	—	
3020	Heat Generating Systems	Included in D3050		—	—	—	
3030	Cooling Generating Systems	N/A		—	—	—	3.5 %
3050	Terminal & Package Units	Through the wall electric heating and cooling units		S.F. Floor	4.30	4.30	
3090	Other HVAC Sys. & Equipment	N/A		—	—	—	
	D40 Fire Protection						
4010	Sprinklers	Sprinklers, light hazard		S.F. Floor	1.91	1.91	1.6%
4020	Standpipes	N/A		—	—	—	
	D50 Electrical						
5010	Electrical Service/Distribution	800 ampere service, panel board and feeders		S.F. Floor	2.05	2.05	
5020	Lighting & Branch Wiring	Fluorescent fixtures, receptacles, switches and misc. power		S.F. Floor	7.14	7.14	8.1%
5030	Communications & Security	Alarm systems and emergency lighting		S.F. Floor	.62	.62	
5090	Other Electrical Systems	Emergency generator, 7.5 kW		S.F. Floor	.12	.12	
E.	**EQUIPMENT & FURNISHINGS**						
1010	Commercial Equipment	N/A		—	—	—	
1020	Institutional Equipment	N/A		—	—	—	
1030	Vehicular Equipment	N/A		—	—	—	0.0 %
1090	Other Equipment	N/A		—	—	—	
F.	**SPECIAL CONSTRUCTION**						
1020	Integrated Construction	N/A		—	—	—	0.0 %
1040	Special Facilities	N/A		—	—	—	
G.	**BUILDING SITEWORK**	**N/A**					

				Sub-Total	122.53	**100%**
	CONTRACTOR FEES (General Requirements: 10%, Overhead: 5%, Profit: 10%)		25%	30.63	
	ARCHITECT FEES		6%	9.19	

	Total Building Cost	**162.35**

Costs per square foot of floor area

Exterior Wall	S.F. Area	9000	10000	12000	13000	14000	15000	16000	18000	20000
	L.F. Perimeter	385	410	440	460	480	500	510	547	583
Decorative Concrete Block	Steel Joists	157.45	153.55	146.30	143.80	141.65	139.80	137.70	134.90	132.65
Painted Concrete Block	Steel Joists	148.70	145.15	138.75	136.55	134.60	133.00	131.20	128.75	126.75
Face Brick on Conc. Block	Steel Joists	171.35	166.85	158.20	155.30	152.80	150.65	148.00	144.85	142.15
Precast Concrete Panels	Steel Joists	173.50	168.95	160.05	157.10	154.55	152.35	149.70	146.35	143.65
Tilt-up Panels	Steel Joists	159.25	155.30	147.85	145.30	143.10	141.25	139.05	136.25	133.95
Metal Sandwich Panels	Steel Joists	153.85	150.10	143.20	140.85	138.80	137.00	135.05	132.35	130.25
Perimeter Adj., Add or Deduct	Per 100 L.F.	8.60	7.75	6.45	6.00	5.55	5.15	4.85	4.35	3.95
Story Hgt. Adj., Add or Deduct	Per 1 Ft.	1.20	1.15	1.05	1.00	0.95	0.95	0.90	0.90	0.85
Basement—Not Applicable										

The above costs were calculated using the basic specifications shown on the facing page. These costs should be adjusted where necessary for design alternatives and owner's requirements. Reported completed project costs, for this type of structure, range from $ 77.30 to $ 202.15 per S.F.

Common additives

Description	Unit	$ Cost
Emergency Lighting, 25 watt, battery operated		
Lead battery	Each	282
Nickel cadmium	Each	805
Seating		
Auditorium chair, all veneer	Each	238
Veneer back, padded seat	Each	288
Upholstered, spring seat	Each	277
Classroom, movable chair & desk	Set	65 - 120
Lecture hall, pedestal type	Each	227 - 680
Smoke Detectors		
Ceiling type	Each	187
Duct type	Each	480
Sound System		
Amplifier, 250 watts	Each	2350
Speaker, ceiling or wall	Each	191
Trumpet	Each	365

Model costs calculated for a 1 story building with 20' story height and 12,000 square feet of floor area

Movie Theater

				Unit	Unit Cost	Cost Per S.F.	% Of Sub-Total
A. SUBSTRUCTURE							
1010	Standard Foundations	Poured concrete; strip and spread footings		S.F. Ground	1.45	1.45	
1020	Special Foundations	N/A		—	—	—	
1030	Slab on Grade	4" reinforced concrete with vapor barrier and granular base		S.F. Slab	4.74	4.74	11.2%
2010	Basement Excavation	Site preparation for slab and trench for foundation wall and footing		S.F. Ground	3.17	3.17	
2020	Basement Walls	4' foundation wall		L.F. Wall	78	2.86	
B. SHELL							
B10 Superstructure							
1010	Floor Construction	Open web steel joists, slab form, concrete	mezzanine 2250 S.F.	S.F. Floor	12.41	2.32	14.4%
1020	Roof Construction	Metal deck on open web steel joists		S.F. Roof	13.42	13.42	
B20 Exterior Enclosure							
2010	Exterior Walls	Decorative concrete block	100% of wall	S.F. Wall	18.70	13.71	
2020	Exterior Windows	N/A		—	—	—	13.5%
2030	Exterior Doors	Sliding mallfront aluminum and glass and hollow metal		Each	2125	1.06	
B30 Roofing							
3010	Roof Coverings	Built-up tar and gravel with flashing; perlite/EPS composite insulation		S.F. Roof	5.56	5.56	5.1%
3020	Roof Openings	N/A		—	—	—	
C. INTERIORS							
1010	Partitions	Concrete block	40 S.F. Floor/L.F. Partition	S.F. Partition	8.45	3.38	
1020	Interior Doors	Single leaf hollow metal	705 S.F. Floor/Door	Each	875	1.24	
1030	Fittings	Toilet partitions		S.F. Floor	.87	.87	
2010	Stair Construction	Concrete filled metal pan		Flight	18,950	3.16	21.5%
3010	Wall Finishes	Paint		S.F. Surface	4.13	3.30	
3020	Floor Finishes	Carpet 50%, ceramic tile 5%	50% of area	S.F. Floor	10.44	5.22	
3030	Ceiling Finishes	Mineral fiber tile on concealed zee runners suspended		S.F. Ceiling	6.38	6.38	
D. SERVICES							
D10 Conveying							
1010	Elevators & Lifts	N/A		—	—	—	0.0 %
1020	Escalators & Moving Walks	N/A		—	—	—	
D20 Plumbing							
2010	Plumbing Fixtures	Toilet and service fixtures, supply and drainage	1 Fixture/500 S.F. Floor	Each	1880	3.76	
2020	Domestic Water Distribution	Gas fired water heater		S.F. Floor	.24	.24	4.0%
2040	Rain Water Drainage	Roof drains		S.F. Roof	.40	.40	
D30 HVAC							
3010	Energy Supply	N/A		—	—	—	
3020	Heat Generating Systems	Included in D3050		—	—	—	
3030	Cooling Generating Systems	N/A		—	—	—	8.7 %
3050	Terminal & Package Units	Single zone rooftop unit, gas heating, electric cooling		S.F. Floor	9.52	9.52	
3090	Other HVAC Sys. & Equipment	N/A		—	—	—	
D40 Fire Protection							
4010	Sprinklers	Wet pipe sprinkler system		S.F. Floor	3.34	3.34	3.1%
4020	Standpipes	N/A		—	—	—	
D50 Electrical							
5010	Electrical Service/Distribution	400 ampere service, panel board and feeders		S.F. Floor	1.77	1.77	
5020	Lighting & Branch Wiring	Fluorescent fixtures, receptacles, switches, A.C. and misc. power		S.F. Floor	4.60	4.60	6.4%
5030	Communications & Security	Alarm systems, sound system and emergency lighting		S.F. Floor	.51	.51	
5090	Other Electrical Systems	Emergency generator, 7.5 kW		S.F. Floor	.16	.16	
E. EQUIPMENT & FURNISHINGS							
1010	Commercial Equipment	N/A		—	—	—	
1020	Institutional Equipment	Projection equipment, screen		S.F. Floor	8.88	8.88	12.1 %
1030	Vehicular Equipment	N/A		—	—	—	
2010	Other Equipment	Movie theater seating		S.F. Floor	4.34	4.34	
F. SPECIAL CONSTRUCTION							
1020	Integrated Construction	N/A		—	—	—	0.0 %
1040	Special Facilities	N/A		—	—	—	
G. BUILDING SITEWORK	**N/A**						

		Sub-Total	109.36	**100%**
CONTRACTOR FEES (General Requirements: 10%, Overhead: 5%, Profit: 10%)			25%	27.37
ARCHITECT FEES			7%	9.57
		Total Building Cost	**146.30**	

171

Costs per square foot of floor area

Exterior Wall	S.F. Area	10000	15000	20000	25000	30000	35000	40000	45000	50000
	L.F. Perimeter	286	370	453	457	513	568	624	680	735
Precast Concrete Panels	Bearing Walls	174.45	164.55	159.55	152.20	149.65	147.85	146.50	145.45	144.60
	Steel Frame	186.90	177.00	172.05	164.65	162.15	160.30	158.95	157.90	157.05
Face Brick with Concrete Block Back-up	Bearing Walls	170.60	161.15	156.45	149.65	147.35	145.65	144.40	143.40	142.60
	Steel Joists	182.75	173.40	168.70	161.95	159.55	157.90	156.65	155.60	154.80
Stucco on Concrete Block	Bearing Walls	160.05	152.15	148.15	143.00	141.05	139.70	138.65	137.85	137.20
	Steel Joists	172.30	164.40	160.35	155.25	153.30	151.90	150.85	150.05	149.45
Perimeter Adj., Add or Deduct	Per 100 L.F.	13.80	9.25	6.95	5.55	4.65	3.95	3.45	3.05	2.75
Story Hgt. Adj., Add or Deduct	Per 1 Ft.	3.25	2.85	2.60	2.10	1.95	1.85	1.75	1.75	1.65
For Basement, add $29.80 per square foot of basement area										

The above costs were calculated using the basic specifications shown on the facing page. These costs should be adjusted where necessary for design alternatives and owner's requirements. Reported completed project costs, for this type of structure, range from $82.95 to $208.00 per S.F.

Common additives

Description	Unit	$ Cost
Beds, Manual	Each	800 - 2750
Elevators, Hydraulic passenger, 2 stops		
1500# capacity	Each	62,800
2500# capacity	Each	66,300
3500# capacity	Each	69,800
Emergency Lighting, 25 watt, battery operated		
Lead battery	Each	282
Nickel cadmium	Each	805
Intercom System, 25 station capacity		
Master station	Each	2650
Intercom outlets	Each	169
Handset	Each	470
Kitchen Equipment		
Broiler	Each	4025
Coffee urn, twin 6 gallon	Each	3075
Cooler, 6 ft. long	Each	4925
Dishwasher, 10-12 racks per hr.	Each	4950
Food warmer	Each	735
Freezer, 44 C.F., reach-in	Each	3725

Description	Unit	$ Cost
Kitchen Equipment, cont.		
Ice cube maker, 50 lb. per day	Each	1750
Range with 1 oven	Each	2700
Laundry Equipment		
Dryer, gas, 16 lb. capacity	Each	885
30 lb. capacity	Each	3600
Washer, 4 cycle	Each	1075
Commercial	Each	1450
Nurses Call System		
Single bedside call station	Each	310
Pillow speaker	Each	296
Refrigerator, Prefabricated, walk-in		
7'-6" high, 6' x 6'	S.F.	164
10' x 10'	S.F.	129
12' x 14'	S.F.	114
12' x 20'	S.F.	101
TV Antenna, Master system, 12 outlet	Outlet	315
30 outlet	Outlet	203
100 outlet	Outlet	194
Whirlpool Bath, Mobile, 18" x 24" x 60"	Each	4725
X-Ray, Mobile	Each	14,000 - 79,000

Important: See the Reference Section for Location Factors

Model costs calculated for a 2 story building with 10' story height and 25,000 square feet of floor area

Nursing Home

				Unit	Unit Cost	Cost Per S.F.	% Of Sub-Total
A. SUBSTRUCTURE							
1010	Standard Foundations	Poured concrete; strip and spread footings		S.F. Ground	1.44	.72	
1020	Special Foundations	N/A		—	—	—	
1030	Slab on Grade	4" reinforced concrete with vapor barrier and granular base		S.F. Slab	4.74	2.37	4.4%
2010	Basement Excavation	Site preparation for slab and trench for foundation wall and footing		S.F. Ground	.26	.13	
2020	Basement Walls	4' foundation wall		L.F. Wall	69	1.64	
B. SHELL							
	B10 Superstructure						
1010	Floor Construction	Pre-cast double tees with concrete topping		S.F. Floor	12.48	6.24	10.5%
1020	Roof Construction	Pre-cast double tees		S.F. Roof	10.50	5.25	
	B20 Exterior Enclosure						
2010	Exterior Walls	Precast concrete panels		S.F. Wall	37.78	11.74	
2020	Exterior Windows	Wood double hung	15% of wall	Each	552	2.02	12.9%
2030	Exterior Doors	Double aluminum & glass doors, single leaf hollow metal		Each	2151	.42	
	B30 Roofing						
3010	Roof Coverings	Built-up tar and gravel with flashing; perlite/EPS composite insulation		S.F. Roof	6.02	3.01	2.8%
3020	Roof Openings	Roof hatches		S.F. Roof	.08	.04	
C. INTERIORS							
1010	Partitions	Gypsum board on metal studs	8 S.F. Floor/L.F. Partition	S.F. Partition	6.99	6.99	
1020	Interior Doors	Single leaf wood	80 S.F. Floor/Door	Each	565	7.06	
1030	Fittings	N/A		—	—	—	
2010	Stair Construction	Concrete stair		Flight	6400	1.54	23.6%
3010	Wall Finishes	50% vinyl wall coverings, 45% paint, 5% ceramic tile		S.F. Surface	1.54	3.08	
3020	Floor Finishes	95% vinyl tile, 5% ceramic tile		S.F. Floor	3.56	3.56	
3030	Ceiling Finishes	Painted gypsum board		S.F. Ceiling	3.71	3.71	
D. SERVICES							
	D10 Conveying						
1010	Elevators & Lifts	One hydraulic hospital elevator		Each	103,500	4.14	3.8%
1020	Escalators & Moving Walks	N/A		—	—	—	
	D20 Plumbing						
2010	Plumbing Fixtures	Kitchen, toilet and service fixtures, supply and drainage	1 Fixture/230 S.F. Floor	Each	2788	12.12	
2020	Domestic Water Distribution	Oil fired water heater		S.F. Floor	1.26	1.26	12.5%
2040	Rain Water Drainage	Roof drains		S.F. Roof	.60	.30	
	D30 HVAC						
3010	Energy Supply	Oil fired hot water, wall fin radiation		S.F. Floor	7.50	7.50	
3020	Heat Generating Systems	N/A		—	—	—	
3030	Cooling Generating Systems	N/A		—	—	—	13.7%
3050	Terminal & Package Units	Split systems with air cooled condensing units		S.F. Floor	7.55	7.55	
3090	Other HVAC Sys. & Equipment	N/A		—	—	—	
	D40 Fire Protection						
4010	Sprinklers	Sprinkler, light hazard		S.F. Floor	4.91	4.91	4.5%
4020	Standpipes	N/A		—	—	—	
	D50 Electrical						
5010	Electrical Service/Distribution	800 ampere service, panel board and feeders		S.F. Floor	2.09	2.09	
5020	Lighting & Branch Wiring	Incandescent fixtures, receptacles, switches, A.C. and misc. power		S.F. Floor	9.13	9.13	11.3%
5030	Communications & Security	Alarm systems and emergency lighting		S.F. Floor	.68	.68	
5090	Other Electrical Systems	Emergency generator, 15 kW		S.F. Floor	.49	.49	
E. EQUIPMENT & FURNISHINGS							
1010	Commercial Equipment	N/A		—	—	—	
1020	Institutional Equipment	N/A		—	—	—	
1030	Vehicular Equipment	N/A		—	—	—	0.0%
1090	Other Equipment	N/A		—	—	—	
F. SPECIAL CONSTRUCTION							
1020	Integrated Construction	N/A		—	—	—	
1040	Special Facilities	N/A		—	—	—	0.0%
G. BUILDING SITEWORK	**N/A**						

		Sub-Total	109.69	100%
CONTRACTOR FEES (General Requirements: 10%, Overhead: 5%, Profit: 10%)		25%	27.43	
ARCHITECT FEES		11%	15.08	
	Total Building Cost		**152.20**	

Costs per square foot of floor area

Exterior Wall	S.F. Area	2000	3000	5000	7000	9000	12000	15000	20000	25000
	L.F. Perimeter	220	260	320	360	420	480	520	640	700
Wood Siding	Wood Truss	196.90	176.55	158.45	149.40	145.35	140.70	137.25	135.10	132.70
Brick Veneer	Wood Truss	247.65	221.40	197.95	185.95	180.80	174.60	170.05	167.20	164.00
Brick on Block	Wood Truss	191.75	164.00	139.00	126.10	120.70	114.00	109.00	106.05	102.50
	Steel Roof Deck	241.90	212.95	186.75	173.30	167.60	160.65	155.40	152.30	148.55
EIFS on Metal Studs	Steel Roof Deck	227.15	202.55	180.50	169.30	164.50	158.75	154.45	151.80	148.75
Tiltup Concrete Panel	Steel Roof Deck	227.15	202.25	179.90	168.55	163.70	157.85	153.55	150.85	147.85
Perimeter Adj., Add or Deduct	Per 100 L.F.	22.80	15.20	9.15	6.50	5.05	3.80	3.05	2.30	1.80
Story Hgt. Adj., Add or Deduct	Per 1 Ft.	2.65	2.05	1.55	1.25	1.20	0.95	0.90	0.80	0.70

For Basement, add $30.20 per square foot of basement area

The above costs were calculated using the basic specifications shown on the facing page. These costs should be adjusted where necessary for design alternatives and owner's requirements. Reported completed project costs, for this type of structure, range from $63.35 to $244.95 per S.F.

Common additives

Description	Unit	$ Cost
Clock System		
20 room	Each	16,000
50 room	Each	39,100
Closed Circuit Surveillance, One station		
Camera and monitor	Each	1850
For additional camera stations, add	Each	1000
Directory Boards, Plastic, glass covered		
30" x 20"	Each	595
36" x 48"	Each	1450
Aluminum, 24" x 18"	Each	600
36" x 24"	Each	675
48" x 32"	Each	980
48" x 60"	Each	2025
Emergency Lighting, 25 watt, battery operated		
Lead battery	Each	282
Nickel cadmium	Each	805

Description	Unit	$ Cost
Smoke Detectors		
Ceiling type	Each	187
Duct type	Each	480
Sound System		
Amplifier, 250 watts	Each	2350
Speaker, ceiling or wall	Each	191
Trumpet	Each	365
TV Antenna, Master system, 12 outlet	Outlet	315
30 outlet	Outlet	203
100 outlet	Outlet	194

Important: See the Reference Section for Location Factors

Model costs calculated for a 1 story building with 12' story height and 7,000 square feet of floor area

				Unit	Unit Cost	Cost Per S.F.	% Of Sub-Total
A.	**SUBSTRUCTURE**						
1010	Standard Foundations	Poured concrete; strip and spread footings		S.F. Ground	2.76	2.76	
1020	Special Foundations	N/A		—	—	—	
1030	Slab on Grade	4" reinforced concrete with vapor barrier and granular base		S.F. Slab	4.74	4.74	9.4%
2010	Basement Excavation	Site preparation for slab and trench for foundation wall and footing		S.F. Ground	.26	.26	
2020	Basement Walls	4' foundation wall		L.F. Wall	147	4.20	
B.	**SHELL**						
	B10 Superstructure						
1010	Floor Construction	N/A		—	—	—	9.3 %
1020	Roof Construction	Steel joists, girders & deck on columns		S.F. Roof	11.71	11.71	
	B20 Exterior Enclosure						
2010	Exterior Walls	E.I.F.S. on metal studs		S.F. Wall	18.70	9.23	
2020	Exterior Windows	Aluminum outward projecting	20% of wall	Each	696	3.73	12.3%
2030	Exterior Doors	Aluminum and glass, hollow metal		Each	2987	2.55	
	B30 Roofing						
3010	Roof Coverings	Single ply membrane, loose laid and ballasted		S.F. Roof	8.50	8.50	6.8%
3020	Roof Openings	Roof hatch		S.F. Roof	.16	.16	
C.	**INTERIORS**						
1010	Partitions	Gypsum board on metal studs	20 S.F. Floor/L.F. Partition	S.F. Partition	8.20	4.10	
1020	Interior Doors	Single leaf hollow metal	200 S.F. Floor/Door	Each	875	4.38	
1030	Fittings	Toilet partitions		S.F. Floor	.52	.52	
2010	Stair Construction	N/A		—	—	—	19.2%
3010	Wall Finishes	60% vinyl wall covering, 40% paint		S.F. Surface	1.33	1.33	
3020	Floor Finishes	60% carpet, 30% vinyl composition tile, 10% ceramic tile		S.F. Floor	7.62	7.62	
3030	Ceiling Finishes	Mineral fiber tile on concealed zee bars		S.F. Ceiling	6.38	6.38	
D.	**SERVICES**						
	D10 Conveying						
1010	Elevators & Lifts	N/A		—	—	—	0.0 %
1020	Escalators & Moving Walks	N/A		—	—	—	
	D20 Plumbing						
2010	Plumbing Fixtures	Toilet and service fixtures, supply and drainage	1 Fixture/1320 S.F. Floor	Each	6349	4.81	
2020	Domestic Water Distribution	Gas fired water heater		S.F. Floor	1.11	1.11	5.2%
2040	Rain Water Drainage	Roof drains		S.F. Roof	.68	.68	
	D30 HVAC						
3010	Energy Supply	N/A		—	—	—	
3020	Heat Generating Systems	Included in D3050		—	—	—	
3030	Cooling Generating Systems	N/A		—	—	—	15.0 %
3050	Terminal & Package Units	Multizone unit gas heating, electric cooling		S.F. Floor	19.05	19.05	
3090	Other HVAC Sys. & Equipment	N/A		—	—	—	
	D40 Fire Protection						
4010	Sprinklers	Sprinkler system, light hazard		S.F. Floor	2.77	2.77	3.1%
4020	Standpipes	Standpipes and hose systems		S.F. Floor	1.13	1.13	
	D50 Electrical						
5010	Electrical Service/Distribution	400 ampere service, panel board and feeders		S.F. Floor	5.85	5.85	
5020	Lighting & Branch Wiring	High efficiency fluorescent fixtures, receptacles, switches, A.C. and misc. power		S.F. Floor	11.68	11.68	19.6%
5030	Communications & Security	Addressable alarm systems, internet and phone wiring, and emergency lighting		S.F. Floor	7.14	7.14	
5090	Other Electrical Systems	Emergency generator, 7.5 kW		S.F. Floor	.20	.20	
E.	**EQUIPMENT & FURNISHINGS**						
1010	Commercial Equipment	N/A		—	—	—	
1020	Institutional Equipment	N/A		—	—	—	0.0 %
1030	Vehicular Equipment	N/A		—	—	—	
1090	Other Equipment	N/A		—	—	—	
F.	**SPECIAL CONSTRUCTION**						
1020	Integrated Construction	N/A		—	—	—	0.0 %
1040	Special Facilities	N/A		—	—	—	
G.	**BUILDING SITEWORK**	**N/A**					

		Sub-Total	126.59	**100%**
	CONTRACTOR FEES (General Requirements: 10%, Overhead: 5%, Profit: 10%)		25%	31.63
	ARCHITECT FEES		7%	11.08
		Total Building Cost	**169.30**	

Costs per square foot of floor area

Exterior Wall	S.F. Area	5000	8000	12000	16000	20000	35000	50000	65000	80000
	L.F. Perimeter	220	260	310	330	360	440	490	548	580
Face Brick with Concrete Block Back-up	Wood Joists	235.30	204.45	186.75	174.45	168.05	155.30	149.05	145.90	143.35
	Steel Joists	243.30	212.45	194.70	182.45	176.00	163.20	157.05	153.90	151.35
Glass and Metal Curtain Wall	Steel Frame	286.85	246.25	222.90	206.20	197.50	180.15	171.60	167.25	163.75
	R/Conc. Frame	279.85	239.75	216.70	200.15	191.55	174.30	165.85	161.55	158.05
Wood Siding	Wood Frame	188.10	165.95	153.35	145.00	140.50	131.85	127.80	125.65	124.00
Brick Veneer	Wood Frame	210.80	182.75	166.65	155.60	149.80	138.40	132.90	130.00	127.80
Perimeter Adj., Add or Deduct	Per 100 L.F.	36.85	23.05	15.40	11.55	9.20	5.25	3.70	2.85	2.30
Story Hgt. Adj., Add or Deduct	Per 1 Ft.	6.00	4.40	3.55	2.80	2.45	1.70	1.35	1.15	1.00
For Basement, add $33.65 per square foot of basement area										

The above costs were calculated using the basic specifications shown on the facing page. These costs should be adjusted where necessary for design alternatives and owner's requirements. Reported completed project costs, for this type of structure, range from $66.30 to $256.80 per S.F.

Common additives

Description	Unit	$ Cost
Clock System		
20 room	Each	16,000
50 room	Each	39,100
Closed Circuit Surveillance, One station		
Camera and monitor	Each	1850
For additional camera stations, add	Each	1000
Directory Boards, Plastic, glass covered		
30" x 20"	Each	595
36" x 48"	Each	1450
Aluminum, 24" x 18"	Each	600
36" x 24"	Each	675
48" x 32"	Each	980
48" x 60"	Each	2025
Elevators, Hydraulic passenger, 2 stops		
1500# capacity	Each	62,800
2500# capacity	Each	66,300
3500# capacity	Each	69,800
Additional stop, add	Each	7825
Emergency Lighting, 25 watt, battery operated		
Lead battery	Each	282
Nickel cadmium	Each	805

Description	Unit	$ Cost
Smoke Detectors		
Ceiling type	Each	187
Duct type	Each	480
Sound System		
Amplifier, 250 watts	Each	2350
Speaker, ceiling or wall	Each	191
Trumpet	Each	365
TV Antenna, Master system, 12 outlet	Outlet	315
30 outlet	Outlet	203
100 outlet	Outlet	194

Important: See the Reference Section for Location Factors

Model costs calculated for a 3 story building with 12' story height and 20,000 square feet of floor area

				Unit	Unit Cost	Cost Per S.F.	% Of Sub-Total
A.	**SUBSTRUCTURE**						
1010	Standard Foundations	Poured concrete; strip and spread footings		S.F. Ground	7.35	2.45	
1020	Special Foundations	N/A		—	—	—	
1030	Slab on Grade	4" reinforced concrete with vapor barrier and granular base		S.F. Slab	4.74	1.58	4.4%
2010	Basement Excavation	Site preparation for slab and trench for foundation wall and footing		S.F. Ground	.17	.06	
2020	Basement Walls	4' foundation wall		L.F. Wall	74	1.64	
B.	**SHELL**						
	B10 Superstructure						
1010	Floor Construction	Open web steel joists, slab form, concrete, columns		S.F. Floor	19.79	13.19	12.2%
1020	Roof Construction	Metal deck, open web steel joists, columns		S.F. Roof	8.43	2.81	
	B20 Exterior Enclosure						
2010	Exterior Walls	Face brick with concrete block backup	80% of wall	S.F. Wall	30.84	15.99	
2020	Exterior Windows	Aluminum outward projecting	20% of wall	Each	696	3.93	15.8%
2030	Exterior Doors	Aluminum and glass, hollow metal		Each	2987	.90	
	B30 Roofing						
3010	Roof Coverings	Built-up tar and gravel with flashing; perlite/EPS composite		S.F. Roof	6.33	2.11	1.6%
3020	Roof Openings	N/A		—	—	—	
C.	**INTERIORS**						
1010	Partitions	Gypsum board on metal studs	20 S.F. Floor/L.F. Partition	S.F. Partition	9.43	3.77	
1020	Interior Doors	Single leaf hollow metal	200 S.F. Floor/Door	Each	875	4.38	
1030	Fittings	Toilet partitions		S.F. Floor	1.10	1.10	
2010	Stair Construction	Concrete filled metal pan		Flight	15,800	5.53	22.7%
3010	Wall Finishes	60% vinyl wall covering, 40% paint		S.F. Surface	1.34	1.07	
3020	Floor Finishes	60% carpet, 30% vinyl composition tile, 10% ceramic tile		S.F. Floor	7.62	7.62	
3030	Ceiling Finishes	Mineral fiber tile on concealed zee bars		S.F. Ceiling	6.38	6.38	
D.	**SERVICES**						
	D10 Conveying						
1010	Elevators & Lifts	Two hydraulic passenger elevators		Each	117,500	11.75	8.9%
1020	Escalators & Moving Walks	N/A		—	—	—	
	D20 Plumbing						
2010	Plumbing Fixtures	Toilet and service fixtures, supply and drainage	1 Fixture/1320 S.F. Floor	Each	3775	2.86	
2020	Domestic Water Distribution	Gas fired water heater		S.F. Floor	.38	.38	2.8%
2040	Rain Water Drainage	Roof drains		S.F. Roof	1.53	.51	
	D30 HVAC						
3010	Energy Supply	N/A		—	—	—	
3020	Heat Generating Systems	Included in D3050		—	—	—	
3030	Cooling Generating Systems	N/A		—	—	—	11.8 %
3050	Terminal & Package Units	Multizone unit gas heating, electric cooling		S.F. Floor	15.50	15.50	
3090	Other HVAC Sys. & Equipment	N/A		—	—	—	
	D40 Fire Protection						
4010	Sprinklers	Wet pipe sprinkler system		S.F. Floor	2.96	2.96	2.8%
4020	Standpipes	Standpipes and hose systems		S.F. Floor	.72	.72	
	D50 Electrical						
5010	Electrical Service/Distribution	1000 ampere service, panel board and feeders		S.F. Floor	4.55	4.55	
5020	Lighting & Branch Wiring	High efficiency fluorescent fixtures, receptacles, switches, A.C. and misc. power		S.F. Floor	11.20	11.20	17.0%
5030	Communications & Security	Addressable alarm systems, internet and phone wiring, and emergency lighting		S.F. Floor	6.42	6.42	
5090	Other Electrical Systems	Emergency generator, 7.5 kW, uninterruptible power supply		S.F. Floor	.22	.22	
E.	**EQUIPMENT & FURNISHINGS**						
1010	Commercial Equipment	N/A		—	—	—	
1020	Institutional Equipment	N/A		—	—	—	0.0 %
1030	Vehicular Equipment	N/A		—	—	—	
1090	Other Equipment	N/A		—	—	—	
F.	**SPECIAL CONSTRUCTION**						
1020	Integrated Construction	N/A		—	—	—	0.0 %
1040	Special Facilities	N/A		—	—	—	
G.	**BUILDING SITEWORK**	**N/A**					

			Sub-Total	131.58	100%
CONTRACTOR FEES (General Requirements: 10%, Overhead: 5%, Profit: 10%)			25%	32.91	
ARCHITECT FEES			7%	11.51	

	Total Building Cost	**176**

Costs per square foot of floor area

Exterior Wall	S.F. Area	20000	40000	60000	80000	100000	150000	200000	250000	300000
	L.F. Perimeter	260	360	400	420	460	520	600	640	700
Precast Concrete Panel	Steel Frame	219.75	191.10	176.00	167.10	162.85	155.70	152.70	150.00	148.55
	R/Conc. Frame	215.70	186.70	171.55	162.60	158.30	151.10	148.10	145.40	143.95
Face Brick with Concrete Block Back-up	Steel Frame	212.30	185.15	171.35	163.35	159.45	152.90	150.15	147.80	146.45
	R/Conc. Frame	205.90	179.90	166.50	158.65	154.85	148.55	145.85	143.50	142.20
Limestone Panel Concrete Block Back-up	Steel Frame	256.90	216.70	194.95	182.05	175.95	165.60	161.20	157.30	155.20
	R/Conc. Frame	252.25	212.05	190.25	177.40	171.25	160.90	156.55	152.65	150.55
Perimeter Adj., Add or Deduct	Per 100 L.F.	27.40	13.65	9.10	6.80	5.50	3.65	2.75	2.25	1.90
Story Hgt. Adj., Add or Deduct	Per 1 Ft.	5.70	3.90	2.90	2.30	2.05	1.50	1.35	1.15	1.05
For Basement, add $36.40 per square foot of basement area										

The above costs were calculated using the basic specifications shown on the facing page. These costs should be adjusted where necessary for design alternatives and owner's requirements. Reported completed project costs, for this type of structure, range from $74.60 to $219.35 per S.F.

Common additives

Description	Unit	$ Cost
Clock System		
20 room	Each	16,000
50 room	Each	39,100
Closed Circuit Surveillance, One station		
Camera and monitor	Each	1850
For additional camera stations, add	Each	1000
Directory Boards, Plastic, glass covered		
30" x 20"	Each	595
36" x 48"	Each	1450
Aluminum, 24" x 18"	Each	600
36" x 24"	Each	675
48" x 32"	Each	980
48" x 60"	Each	2025
Elevators, Electric passenger, 5 stops		
2000# capacity	Each	158,700
3500# capacity	Each	167,200
5000# capacity	Each	170,700
Additional stop, add	Each	13,600
Emergency Lighting, 25 watt, battery operated		
Lead battery	Each	282
Nickel cadmium	Each	805

Description	Unit	$ Cost
Intercom System, 25 station capacity		
Master station	Each	2650
Intercom outlets	Each	169
Handset	Each	470
Smoke Detectors		
Ceiling type	Each	187
Duct type	Each	480
Sound System		
Amplifier, 250 watts	Each	2350
Speaker, ceiling or wall	Each	191
Trumpet	Each	365
TV Antenna, Master system, 12 oulet	Outlet	315
30 outlet	Outlet	203
100 outlet	Outlet	194

Important: See the Reference Section for Location Factors

Model costs calculated for a 8 story building with 12' story height and 80,000 square feet of floor area

				Unit	Unit Cost	Cost Per S.F.	% Of Sub-Total
A.	**SUBSTRUCTURE**						
1010	Standard Foundations	Poured concrete; strip and spread footings		S.F. Ground	12.08	1.51	
1020	Special Foundations	N/A		—	—	—	
1030	Slab on Grade	4" reinforced concrete with vapor barrier and granular base		S.F. Slab	4.74	.59	2.1%
2010	Basement Excavation	Site preparation for slab and trench for foundation wall and footing		S.F. Ground	.26	.03	
2020	Basement Walls	4' foundation wall		L.F. Wall	78	.53	
B.	**SHELL**						
	B10 Superstructure						
1010	Floor Construction	Concrete slab with metal deck and beams		S.F. Floor	26.25	22.97	19.0%
1020	Roof Construction	Metal deck, open web steel joists, interior columns		S.F. Roof	8.08	1.01	
	B20 Exterior Enclosure						
2010	Exterior Walls	Precast concrete panels	80% of wall	S.F. Wall	38.24	15.42	
2020	Exterior Windows	Vertical pivoted steel	20% of wall	Each	552	3.71	15.3%
2030	Exterior Doors	Double aluminum and glass doors and entrance with transoms		Each	3542	.22	
	B30 Roofing						
3010	Roof Coverings	Built-up tar and gravel with flashing; perlite/EPS composite insulation		S.F. Roof	5.52	.69	0.5%
3020	Roof Openings	N/A		—	—	—	
C.	**INTERIORS**						
1010	Partitions	Gypsum board on metal studs	30 S.F. Floor/L.F. Partition	S.F. Partition	9.09	3.03	
1020	Interior Doors	Single leaf hollow metal	400 S.F. Floor/Door	Each	875	2.19	
1030	Fittings	Toilet Partitions		S.F. Floor	.73	.73	
2010	Stair Construction	Concrete filled metal pan		Flight	15,800	3.36	19.2%
3010	Wall Finishes	60% vinyl wall covering, 40% paint		S.F. Surface	1.35	.90	
3020	Floor Finishes	60% carpet, 30% vinyl composition tile, 10% ceramic tile		S.F. Floor	7.62	7.62	
3030	Ceiling Finishes	Mineral fiber tile on concealed zee bars		S.F. Ceiling	6.38	6.38	
D.	**SERVICES**						
	D10 Conveying						
1010	Elevators & Lifts	Four geared passenger elevators		Each	292,600	14.63	11.6%
1020	Escalators & Moving Walks	N/A		—	—	—	
	D20 Plumbing						
2010	Plumbing Fixtures	Toilet and service fixtures, supply and drainage	1 Fixture/1370 S.F. Floor	Each	2781	2.03	
2020	Domestic Water Distribution	Gas fired water heater		S.F. Floor	.42	.42	2.1%
2040	Rain Water Drainage	Roof drains		S.F. Roof	1.84	.23	
	D30 HVAC						
3010	Energy Supply	N/A		—	—	—	
3020	Heat Generating Systems	Included in D3050		—	—	—	
3030	Cooling Generating Systems	N/A		—	—	—	12.3 %
3050	Terminal & Package Units	Multizone unit gas heating, electric cooling		S.F. Floor	15.50	15.50	
3090	Other HVAC Sys. & Equipment	N/A		—	—	—	
	D40 Fire Protection						
4010	Sprinklers	Wet pipe sprinkler system		S.F. Floor	2.33	2.33	2.7%
4020	Standpipes	Standpipes and hose systems		S.F. Floor	1.07	1.07	
	D50 Electrical						
5010	Electrical Service/Distribution	1600 ampere service, panel board and feeders		S.F. Floor	1.86	1.86	
5020	Lighting & Branch Wiring	High efficiency fluorescent fixtures, receptacles, switches, A.C. and misc. power		S.F. Floor	11.11	11.11	15.1%
5030	Communications & Security	Addressable alarm systems, internet and phone wiring, emergency lighting		S.F. Floor	5.05	5.05	
5090	Other Electrical Systems	Emergency generator, 100 kW, uninterruptible power supply		S.F. Floor	1.02	1.02	
E.	**EQUIPMENT & FURNISHINGS**						
1010	Commercial Equipment	N/A		—	—	—	
1020	Institutional Equipment	N/A		—	—	—	0.0 %
1030	Vehicular Equipment	N/A		—	—	—	
1090	Other Equipment	N/A		—	—	—	
F.	**SPECIAL CONSTRUCTION**						
1020	Integrated Construction	N/A		—	—	—	0.0 %
1040	Special Facilities	N/A		—	—	—	
G.	**BUILDING SITEWORK**	**N/A**					

			Sub-Total	126.14	100%
CONTRACTOR FEES (General Requirements: 10%, Overhead: 5%, Profit: 10%)			25%	31.50	
ARCHITECT FEES			6%	9.46	
			Total Building Cost	**167.10**	

Costs per square foot of floor area

Exterior Wall	S.F. Area	120000	145000	170000	200000	230000	260000	400000	600000	800000
	L.F. Perimeter	420	450	470	490	510	530	600	730	820
Double Glazed Heat Absorbing Tinted Plate Glass Panels	Steel Frame	173.85	167.95	163.20	158.85	155.65	153.20	145.95	142.00	139.50
	R/Conc. Frame	161.85	156.15	151.55	147.35	144.25	141.85	134.75	130.95	128.55
Face Brick with Concrete Block Back-up	Steel Frame	168.15	162.95	158.65	154.85	152.05	149.85	143.50	140.05	137.85
	R/Conc. Frame	177.25	172.25	168.15	164.45	161.70	159.60	153.45	150.10	147.95
Precast Concrete Panel With Exposed Aggregate	Steel Frame	169.95	164.55	160.15	156.15	153.20	150.95	144.30	140.65	138.35
	R/Conc. Frame	157.90	152.75	148.45	144.60	141.75	139.55	133.15	129.60	127.40
Perimeter Adj., Add or Deduct	Per 100 L.F.	8.35	6.95	5.85	5.00	4.40	3.85	2.50	1.70	1.30
Story Hgt. Adj., Add or Deduct	Per 1 Ft.	3.30	2.95	2.60	2.30	2.15	1.95	1.40	1.20	0.95
For Basement, add $36.40 per square foot of basement area										

The above costs were calculated using the basic specifications shown on the facing page. These costs should be adjusted where necessary for design alternatives and owner's requirements. Reported completed project costs, for this type of structure, range from $93.60 to $228.35 per S.F.

Common additives

Description	Unit	$ Cost
Clock System		
20 room	Each	16,000
50 room	Each	39,100
Directory Boards, Plastic, glass covered		
30" x 20"	Each	595
36" x 48"	Each	1450
Aluminum, 24" x 18"	Each	600
36" x 24"	Each	675
48" x 32"	Each	980
48" x 60"	Each	2025
Elevators, Electric passenger, 10 stops		
3000# capacity	Each	430,500
4000# capacity	Each	433,000
5000# capacity	Each	437,000
Additional stop, add	Each	13,600
Emergency Lighting, 25 watt, battery operated		
Lead battery	Each	282
Nickel cadmium	Each	805

Description	Unit	$ Cost
Escalators, Metal		
32" wide, 10' story height	Each	143,000
20' story height	Each	172,000
48" wide, 10' story height	Each	152,000
20' story height	Each	180,500
Glass		
32" wide, 10' story height	Each	137,000
20' story height	Each	165,000
48" wide, 10' story height	Each	48,300
20' story height	Each	175,000
Smoke Detectors		
Ceiling type	Each	187
Duct type	Each	480
Sound System		
Amplifier, 250 watts	Each	2350
Speaker, ceiling or wall	Each	191
Trumpet	Each	365
TV Antenna, Master system, 12 oulet	Outlet	315
30 oulet	Outlet	203
100 outlet	Outlet	194

Important: See the Reference Section for Location Factors

Model costs calculated for a 16 story building with 10′ story height and 260,000 square feet of floor area

				Unit	Unit Cost	Cost Per S.F.	% Of Sub-Total
A. SUBSTRUCTURE							
1010	Standard Foundations	CIP concrete pile caps		S.F. Ground	9.92	.62	
1020	Special Foundations	Steel H-piles, concrete grade beams		S.F. Ground	62	3.86	
1030	Slab on Grade	4″ reinforced concrete with vapor barrier and granular base		S.F. Slab	4.74	.30	4.5%
2010	Basement Excavation	Site preparation for slab, piles and grade beams		S.F. Ground	.26	.02	
2020	Basement Walls	4′ foundation wall		L.F. Wall	78	.38	
B. SHELL							
B10 Superstructure							
1010	Floor Construction	Concrete slab, metal deck, beams		S.F. Floor	29.90	28.03	24.8%
1020	Roof Construction	Metal deck, open web steel joists, beams, columns		S.F. Roof	9.76	.61	
B20 Exterior Enclosure							
2010	Exterior Walls	N/A		—	—	—	
2020	Exterior Windows	Double glazed heat absorbing, tinted plate glass wall panels	100% of wall	Each	41.45	13.52	12.2 %
2030	Exterior Doors	Double aluminum & glass doors		Each	5571	.60	
B30 Roofing							
3010	Roof Coverings	Single ply membrane, fully adhered; perlite/EPS composite insulation		S.F. Roof	5.60	.35	0.3%
3020	Roof Openings	N/A		—	—	—	
C. INTERIORS							
1010	Partitions	Gypsum board on metal studs	30 S.F. Floor/L.F. Partition	S.F. Partition	10.20	2.72	
1020	Interior Doors	Single leaf hollow metal	400 S.F. Floor/Door	Each	875	2.19	
1030	Fittings	Toilet partitions		S.F. Floor	.42	.42	
2010	Stair Construction	Concrete filled metal pan		Flight	18,950	2.55	17.1%
3010	Wall Finishes	60% vinyl wall covering, 40% paint		S.F. Surface	1.33	.71	
3020	Floor Finishes	60% carpet tile, 30% vinyl composition tile, 10% ceramic tile		S.F. Floor	4.81	4.81	
3030	Ceiling Finishes	Mineral fiber tile on concealed zee bars		S.F. Ceiling	6.38	6.38	
D. SERVICES							
D10 Conveying							
1010	Elevators & Lifts	Four geared passenger elevators		Each	479,050	7.37	6.4%
1020	Escalators & Moving Walks	N/A		—	—	—	
D20 Plumbing							
2010	Plumbing Fixtures	Toilet and service fixtures, supply and drainage	1 Fixture/1345 S.F. Floor	Each	4022	2.99	
2020	Domestic Water Distribution	Oil fired water heater		S.F. Floor	.25	.25	2.9%
2040	Rain Water Drainage	Roof drains		S.F. Roof	2.08	.13	
D30 HVAC							
3010	Energy Supply	N/A		—	—	—	
3020	Heat Generating Systems	Boiler, heat exchanger and fans		Each	388,485	2.04	
3030	Cooling Generating Systems	Chilled water, fan coil units		S.F. Floor	13.59	13.59	13.5 %
3050	Terminal & Package Units	N/A		—	—	—	
3090	Other HVAC Sys. & Equipment	N/A		—	—	—	
D40 Fire Protection							
4010	Sprinklers	Sprinkler system, light hazard		S.F. Floor	2.25	2.25	2.4%
4020	Standpipes	Standpipes and hose systems		S.F. Floor	.51	.51	
D50 Electrical							
5010	Electrical Service/Distribution	2400 ampere service, panel board and feeders		S.F. Floor	1.10	1.10	
5020	Lighting & Branch Wiring	High efficiency fluorescent fixtures, receptacles, switches, A.C. and misc. power		S.F. Floor	10.99	10.99	15.9%
5030	Communications & Security	Addressable alarm systems, internet and phone wiring, emergency lighting		S.F. Floor	5.81	5.81	
5090	Other Electrical Systems	Emergency generator, 200 kW, uninterruptible power supply		S.F. Floor	.51	.51	
E. EQUIPMENT & FURNISHINGS							
1010	Commercial Equipment	N/A		—	—	—	
1020	Institutional Equipment	N/A		—	—	—	
1030	Vehicular Equipment	N/A		—	—	—	0.0 %
1090	Other Equipment	N/A		—	—	—	
F. SPECIAL CONSTRUCTION							
1020	Integrated Construction	N/A		—	—	—	0.0 %
1040	Special Facilities	N/A		—	—	—	
G. BUILDING SITEWORK	**N/A**						
				Sub-Total		115.61	**100%**
	CONTRACTOR FEES (General Requirements: 10%, Overhead: 5%, Profit: 10%)				25%	28.92	
	ARCHITECT FEES				6%	8.67	
				Total Building Cost		**153.20**	

Costs per square foot of floor area

Exterior Wall	S.F. Area	5000	7500	10000	12500	15000	17500	20000	22500	25000
	L.F. Perimeter	284	350	410	450	490	530	570	600	632
Brick Veneer	Wood Truss	396.40	383.15	375.95	370.00	366.10	363.35	361.25	359.15	357.60
Brick Veneer	Steel Frame	404.35	391.65	384.75	379.15	375.35	372.75	370.80	368.75	367.35
Decorative Concrete Block	Steel Frame	403.05	390.60	383.75	378.30	374.65	372.05	370.15	368.20	366.80
Decorative Concrete Block	Bearing Wall	392.25	384.00	379.55	376.15	373.80	372.25	371.05	369.85	369.00
Tiltup Concrete Panel	Steel Frame	400.65	388.65	382.15	376.85	373.35	370.90	369.00	367.20	365.80
EIFS on Concrete Block	Steel Frame	402.10	388.05	380.55	374.75	370.90	368.20	366.20	364.25	362.80
Perimeter Adj., Add or Deduct	Per 100 L.F.	19.30	12.80	9.60	7.75	6.50	5.50	4.80	4.30	3.85
Story Hgt. Adj., Add or Deduct	Per 1 Ft.	2.90	2.45	2.10	1.80	1.65	1.55	1.45	1.40	1.35
For Basement, add $ 28.65 per square foot of basement area										

The above costs were calculated using the basic specifications shown on the facing page. These costs should be adjusted where necessary for design alternatives and owner's requirements. Reported completed project costs, for this type of structure, range from $ 163.90 to $ 849.80 per S.F.

Common additives

Description	Unit	$ Cost
Cabinets, base, door units, metal	L.F.	256
Drawer units	L.F.	505
Tall storage cabinets, open, 7' high	L.F.	480
With glazed doors	L.F.	565
Wall cabinets, metal, 12-1/2" deep, open	L.F.	192
With doors	L.F.	345
Counter top, laminated plastic, no backsplash	L.F.	69.50
Stainless steel counter top	L.F.	169
For drop-in stainless 43" x 21" sink, add	Ea.	950
Nurse Call Station		
Single bedside call station	Ea.	310
Ceiling speaker station	Ea.	143
Emergency call station	Ea.	192
Pillow speaker	Ea.	296
Standard call button	Ea.	169
Master control station for 20 stations	Total	6025

Description	Unit	$ Cost
Directory Boards, Plastic, glass covered		
Plastic, glass covered, 30" x 20"	Ea.	595
36" x 48"	Ea.	1450
Building directory, alum., black felt panels, 1 door, 24" x 18"	Ea.	600
36" x 24"	Ea.	675
48" x 32"	Ea.	980
48" x 60"	Ea.	2025
Tables, Examining, vinyl top		
with base cabinets	Ea.	1700
Utensil washer-sanitizer	Ea.	11,700
X-ray, mobile, minimum	Ea.	14,000

Model costs calculated for a 1 story building with 15'-4" story height and 12,500 square feet of floor area

Outpatient Surgery Center

			Unit	Unit Cost	Cost Per S.F.	% Of Sub-Total
A. SUBSTRUCTURE						
1010	Standard Foundations	Poured concrete; strip and spread footings	S.F. Ground	1.97	1.97	
1020	Special Foundations	N/A	—	—	—	
1030	Slab on Grade	4" reinforced concrete with vapor barrier and granular base	S.F. Slab	5.85	5.85	3.9%
2010	Basement Excavation	Site preparation for slab and trench for foundation wall and footing	S.F. Ground	.26	.26	
2020	Basement Walls	4' Foundation wall	L.F. Wall	78	2.81	
B. SHELL						
	B10 Superstructure					
1010	Floor Construction	N/A	—	—	—	3.2 %
1020	Roof Construction	Steel Joists and girders on columns	S.F. Roof	8.87	8.87	
	B20 Exterior Enclosure					
2010	Exterior Walls	Brick veneer 85% of wall	S.F. Wall	20.85	9.78	
2020	Exterior Windows	Aluminum frame, fixed 15% of wall	Eacj	740	4.08	6.7%
2030	Exterior Doors	Aluminum and glass entry; steel flush doors	Each	5962	4.77	
	B30 Roofing					
3010	Roof Coverings	Single-ply membrane, loose laid and ballasted; polyisocyanurate insulation	S.F. Roof	7.30	7.30	2.7%
3020	Roof Openings	Roof hatches	S.F. Roof	.09	.09	
C. INTERIORS						
1010	Partitions	Gypsum board on metal studs 7 S.F. Floor/L.F. Partition	S.F. Partition	7.54	10.77	
1020	Interior Doors	Single and double leaf fire doors 200 S.F. Floor/Door	Each	1062	5.31	
1030	Fittings	Stainless steel toilet partitions, steel lockers	S.F. Floor	1.53	1.53	
2010	Stair Construction	N/A	—	—	—	12.8%
3010	Wall Finishes	80% Paint, 10% vinyl wall covering, 10% ceramic tile	S.F. Surface	1.81	5.17	
3020	Floor Finishes	10% Carpet, 80% vinyl sheet, 10% ceramic tile	S.F. Floor	8.93	8.93	
3030	Ceiling Finishes	Mineral fiber tile and plastic coated tile on suspension system	S.F. Ceiling	3.68	3.68	
D. SERVICES						
	D10 Conveying					
1010	Elevators & Lifts	N/A	—	—	—	0.0 %
1020	Escalators & Moving Walks	N/A	—	—	—	
	D20 Plumbing					
2010	Plumbing Fixtures	Medical, patient & specialty fixtures, supply and drainage 1 fixture/350 S.F. Floor	Each	2713	7.75	
2020	Domestic Water Distribution	Electric water heater	S.F. Floor	1.50	1.50	4.0%
2040	Rain Water Drainage	Roof drains	S.F. Roof	1.77	1.77	
	D30 HVAC					
3010	Energy Supply	Hot water reheat for surgery	S.F. Floor	10.68	10.68	
3020	Heat Generating Systems	Boiler	Each	44,925	4.72	
3030	Cooling Generating Systems	N/A	—	—	—	34.9%
3050	Terminal & Package Units	N/A	—	—	—	
3090	Other HVAC Sys. & Equipment	Rooftop AC, CAV, roof vents, heat exchanger, surgical air curtains	S.F. Floor	81	81.06	
	D40 Fire Protection					
4010	Sprinklers	Wet pipe sprinkler system	S.F. Floor	2.77	2.77	1.5%
4020	Standpipes	Standpipe	S.F. Floor	1.31	1.31	
	D50 Electrical					
5010	Electrical Service/Distribution	1200 ampere service, panel board and feeders	S.F. Floor	8.16	8.16	
5020	Lighting & Branch Wiring	T-8 fluorescent light fixtures, receptacles, switches, A.C. and misc. power	S.F. Floor	15.78	15.78	14.7%
5030	Communications & Security	Addressable alarm system, internet wiring, communications system, emergency lighting	S.F. Floor	9.39	9.39	
5090	Other Electrical Systems	Emergency generator, 300 kW with fuel tank	S.F. Floor	7.22	7.22	
E. EQUIPMENT & FURNISHINGS						
1010	Commercial Equipment	N/A	—	—	—	
1020	Institutional Equipment	Medical gas system, cabinetry, scrub sink, sterilizer	S.F. Floor	42.40	42.40	15.7 %
1030	Vehicular Equipment	N/A	—	—	—	
1090	Other Equipment	Ice maker, washing machine	S.F. Floor	.91	.91	
F. SPECIAL CONSTRUCTION & DEMOLITION						
1020	Integrated Construction	N/A	—	—	—	0.0 %
1040	Special Facilities	N/A	—	—	—	
G. BUILDING SITEWORK	**N/A**					

				Sub-Total	276.59	100%
	CONTRACTOR FEES (General Requirements: 10%, Overhead: 5%, Profit: 10%)			25%	69.14	
	ARCHITECT FEES			9%	31.12	

	Total Building Cost	**376.85**

Costs per square foot of floor area

Exterior Wall	S.F. Area	7000	9000	11000	13000	15000	17000	19000	21000	23000
	L.F. Perimeter	240	280	303	325	354	372	397	422	447
Limestone with Concrete Block Back-up	Bearing Walls	262.20	244.90	230.05	219.55	213.05	206.40	202.10	198.60	195.75
	R/Conc. Frame	274.55	257.85	243.75	233.75	227.50	221.30	217.20	213.85	211.15
Face Brick with Concrete Block Back-up	Bearing Walls	223.35	209.55	198.70	191.10	186.10	181.45	178.25	175.65	173.55
	R/Conc. Frame	243.15	229.30	218.50	210.85	205.90	201.25	198.05	195.40	193.35
Decorative Concrete Block	Bearing Walls	212.45	199.65	189.95	183.10	178.65	174.50	171.60	169.30	167.35
	R/Conc. Frame	233.10	220.30	210.60	203.75	199.30	195.15	192.25	189.90	188.00
Perimeter Adj., Add or Deduct	Per 100 L.F.	35.85	27.95	22.85	19.30	16.75	14.80	13.25	11.95	11.00
Story Hgt. Adj., Add or Deduct	Per 1 Ft.	6.40	5.80	5.10	4.65	4.40	4.15	3.90	3.75	3.65
For Basement, add $26.45 per square foot of basement area										

The above costs were calculated using the basic specifications shown on the facing page. These costs should be adjusted where necessary for design alternatives and owner's requirements. Reported completed project costs, for this type of structure, range from $107.50 to $280.50 per S.F.

Common additives

Description	Unit	$ Cost
Cells Prefabricated, 5'-6' wide,		
7'-8' high, 7'-8' deep	Each	12,400
Elevators, Hydraulic passenger, 2 stops		
1500# capacity	Each	62,800
2500# capacity	Each	66,300
3500# capacity	Each	69,800
Emergency Lighting, 25 watt, battery operated		
Lead battery	Each	282
Nickel cadmium	Each	805
Flagpoles, Complete		
Aluminum, 20' high	Each	1650
40' high	Each	3475
70' high	Each	10,100
Fiberglass, 23' high	Each	1775
39'-5" high	Each	3325
59' high	Each	8225

Description	Unit	$ Cost
Lockers, Steel, Single tier, 60" to 72"	Opening	191 - 310
2 tier, 60" or 72" total	Opening	107 - 141
5 tier, box lockers	Opening	65 - 83.50
Locker bench, lam. maple top only	L.F.	21
Pedestals, steel pipe	Each	63.50
Safe, Office type, 1 hour rating		
30" x 18" x 18"	Each	2400
60" x 36" x 18", double door	Each	8750
Shooting Range, Incl. bullet traps,		
target provisions, and contols,		
not incl. structural shell	Each	38,300
Smoke Detectors		
Ceiling type	Each	187
Duct type	Each	480
Sound System		
Amplifier, 250 watts	Each	2350
Speaker, ceiling or wall	Each	191
Trumpet	Each	365

Important: See the Reference Section for Location Factors

Model costs calculated for a 2 story building with 12' story height and 11,000 square feet of floor area

				Unit	Unit Cost	Cost Per S.F.	% Of Sub-Total
A. SUBSTRUCTURE							
1010	Standard Foundations	Poured concrete; strip and spread footings		S.F. Ground	2.90	1.45	
1020	Special Foundations	N/A		—	—	—	
1030	Slab on Grade	4" reinforced concrete with vapor barrier and granular base		S.F. Slab	4.74	2.37	4.0%
2010	Basement Excavation	Site preparation for slab and trench for foundation wall and footing		S.F. Ground	.26	.13	
2020	Basement Walls	4' foundation wall		L.F. Wall	78	2.86	
B. SHELL							
	B10 Superstructure						
1010	Floor Construction	Open web steel joists, slab form, concrete		S.F. Floor	13.28	6.64	5.6%
1020	Roof Construction	Metal deck on open web steel joists		S.F. Roof	5.68	2.84	
	B20 Exterior Enclosure						
2010	Exterior Walls	Limestone with concrete block backup	80% of wall	S.F. Wall	66	34.85	
2020	Exterior Windows	Metal horizontal sliding	20% of wall	Each	1041	9.18	27.1%
2030	Exterior Doors	Hollow metal		Each	2360	1.72	
	B30 Roofing						
3010	Roof Coverings	Built-up tar and gravel with flashing; perlite/EPS composite insulation		S.F. Roof	6.44	3.22	1.9%
3020	Roof Openings	N/A		—	—	—	
C. INTERIORS							
1010	Partitions	Concrete block	20 S.F. Floor/L.F. Partition	S.F. Partition	8.44	4.22	
1020	Interior Doors	Single leaf kalamein fire door	200 S.F. Floor/Door	Each	875	4.38	
1030	Fittings	Toilet partitions		S.F. Floor	1.15	1.15	
2010	Stair Construction	Concrete filled metal pan		Flight	18,950	3.45	16.1%
3010	Wall Finishes	90% paint, 10% ceramic tile		S.F. Surface	3.34	3.34	
3020	Floor Finishes	70% vinyl composition tile, 20% carpet tile, 10% ceramic tile		S.F. Floor	4.18	4.18	
3030	Ceiling Finishes	Mineral fiber tile on concealed zee bars		S.F. Ceiling	6.38	6.38	
D. SERVICES							
	D10 Conveying						
1010	Elevators & Lifts	One hydraulic passenger elevator		Each	80,850	7.35	4.4%
1020	Escalators & Moving Walks	N/A		—	—	—	
	D20 Plumbing						
2010	Plumbing Fixtures	Toilet and service fixtures, supply and drainage	1 Fixture/580 S.F. Floor	Each	3840	6.62	
2020	Domestic Water Distribution	Oil fired water heater		S.F. Floor	2.70	2.70	6.4%
2040	Rain Water Drainage	Roof drains		S.F. Roof	2.94	1.47	
	D30 HVAC						
3010	Energy Supply	N/A		—	—	—	
3020	Heat Generating Systems	N/A		—	—	—	
3030	Cooling Generating Systems	N/A		—	—	—	11.3 %
3050	Terminal & Package Units	Multizone HVAC air cooled system		S.F. Floor	19.05	19.05	
3090	Other HVAC Sys. & Equipment	N/A		—	—	—	
	D40 Fire Protection						
4010	Sprinklers	Wet pipe sprinkler system		S.F. Floor	2.95	2.95	2.2%
4020	Standpipes	Standpipe		S.F. Floor	.79	.79	
	D50 Electrical						
5010	Electrical Service/Distribution	400 ampere service, panel board and feeders		S.F. Floor	1.84	1.84	
5020	Lighting & Branch Wiring	T-8 fluorescent fixtures, receptacles, switches, A.C. and misc. power		S.F. Floor	11.35	11.35	11.8%
5030	Communications & Security	Addressable alarm systems, internet wiring, intercom and emergency lighting		S.F. Floor	6.54	6.54	
5090	Other Electrical Systems	Emergency generator, 15 kW		S.F. Floor	.21	.21	
E. EQUIPMENT & FURNISHINGS							
1010	Commercial Equipment	N/A		—	—	—	
1020	Institutional Equipment	Lockers, detention rooms, cells, gasoline dispensers		S.F. Floor	13.52	13.52	9.2 %
1030	Vehicular Equipment	Gasoline dispenser system		S.F. Floor	2.08	2.08	
1090	Other Equipment	N/A		—	—	—	
F. SPECIAL CONSTRUCTION							
1020	Integrated Construction	N/A		—	—	—	0.0 %
1040	Special Facilities	N/A		—	—	—	
G. BUILDING SITEWORK	**N/A**						

			Sub-Total	168.83	**100%**
CONTRACTOR FEES (General Requirements: 10%, Overhead: 5%, Profit: 10%)			25%	42.23	
ARCHITECT FEES			9%	18.99	

Total Building Cost	**230.05**

Costs per square foot of floor area

Exterior Wall	S.F. Area	5000	7000	9000	11000	13000	15000	17000	19000	21000
	L.F. Perimeter	300	380	420	468	486	513	540	580	620
Face Brick with Concrete Block Back-up	Steel Frame	161.45	151.95	142.85	137.60	132.05	128.45	125.65	124.10	122.85
	Bearing Walls	158.70	149.20	140.05	134.90	129.30	125.70	122.85	121.35	120.05
Limestone with Concrete Block Back-up	Steel Frame	193.55	181.00	167.80	160.40	152.05	146.70	142.60	140.40	138.60
	Bearing Walls	190.60	178.05	164.80	157.40	149.10	143.80	139.70	137.45	135.70
Decorative Concrete Block	Steel Frame	150.30	141.90	134.15	129.75	125.15	122.10	119.75	118.50	117.35
	Bearing Walls	147.60	139.15	131.45	126.95	122.35	119.35	117.00	115.70	114.60
Perimeter Adj., Add or Deduct	Per 100 L.F.	17.45	12.50	9.65	7.90	6.75	5.85	5.15	4.60	4.10
Story Hgt. Adj., Add or Deduct	Per 1 Ft.	2.70	2.45	2.10	1.90	1.70	1.50	1.45	1.35	1.30
For Basement, add $27.90 per square foot of basement area										

The above costs were calculated using the basic specifications shown on the facing page. These costs should be adjusted where necessary for design alternatives and owner's requirements. Reported completed project costs, for this type of structure, range from $89.20 to $231.50 per S.F.

Common additives

Description	Unit	$ Cost
Closed Circuit Surveillance, One station		
Camera and monitor	Each	1850
For additional camera stations, add	Each	1000
Emergency Lighting, 25 watt, battery operated		
Lead battery	Each	282
Nickel cadmium	Each	805
Flagpoles, Complete		
Aluminum, 20' high	Each	1650
40' high	Each	3475
70' high	Each	10,100
Fiberglass, 23' high	Each	1775
39'-5" high	Each	3325
59' high	Each	8225

Description	Unit	$ Cost
Mail Boxes, Horizontal, key lock, 15" x 6" x 5"	Each	53
Double 15" x 12" x 5"	Each	98
Quadruple 15" x 12" x 10"	Each	125
Vertical, 6" x 5" x 15", aluminum	Each	47.50
Bronze	Each	58.50
Steel, enameled	Each	47.50
Scales, Dial type, 5 ton cap.		
8' x 6' platform	Each	11,000
9' x 7' platform	Each	14,100
Smoke Detectors		
Ceiling type	Each	187
Duct type	Each	480

Model costs calculated for a 1 story building with 14' story height and 13,000 square feet of floor area

				Unit	Unit Cost	Cost Per S.F.	% Of Sub-Total
A.	**SUBSTRUCTURE**						
1010	Standard Foundations	Poured concrete; strip and spread footings		S.F. Ground	2.02	2.02	
1020	Special Foundations	N/A		—	—	—	
1030	Slab on Grade	4" reinforced concrete with vapor barrier and granular base		S.F. Slab	4.74	4.74	11.2%
2010	Basement Excavation	Site preparation for slab and trench for foundation wall and footing		S.F. Ground	.26	.26	
2020	Basement Walls	4' foundation wall		L.F. Wall	78	3.79	
B.	**SHELL**						
	B10	**Superstructure**					
1010	Floor Construction	Steel column fireproofing		L.F. Column	30.76	.09	11.0%
1020	Roof Construction	Metal deck, open web steel joists, columns		S.F. Roof	10.57	10.57	
	B20	**Exterior Enclosure**					
2010	Exterior Walls	Face brick with concrete block backup	80% of wall	S.F. Wall	30.86	12.92	
2020	Exterior Windows	Double strength window glass	20% of wall	Each	696	3.17	17.7%
2030	Exterior Doors	Double aluminum & glass, single aluminum, hollow metal, steel overhead		Each	2283	1.06	
	B30	**Roofing**					
3010	Roof Coverings	Built-up tar and gravel with flashing; perlite/EPS composite insulation		S.F. Roof	5.74	5.74	5.9%
3020	Roof Openings	N/A		—	—	—	
C.	**INTERIORS**						
1010	Partitions	Concrete block	15 S.F. Floor/L.F. Partition	S.F. Partition	8.44	6.75	
1020	Interior Doors	Single leaf hollow metal	150 S.F. Floor/Door	Each	875	5.83	
1030	Fittings	Toilet partitions, cabinets, shelving, lockers		S.F. Floor	1.58	1.58	
2010	Stair Construction	N/A		—	—	—	22.1%
3010	Wall Finishes	Paint		S.F. Surface	2.03	3.25	
3020	Floor Finishes	50% vinyl tile, 50% paint		S.F. Floor	2.41	2.41	
3030	Ceiling Finishes	Mineral fiber tile on concealed zee bars	25% of area	S.F. Ceiling	6.38	1.60	
D.	**SERVICES**						
	D10	**Conveying**					
1010	Elevators & Lifts	N/A		—	—	—	0.0 %
1020	Escalators & Moving Walks	N/A		—	—	—	
	D20	**Plumbing**					
2010	Plumbing Fixtures	Toilet and service fixtures, supply and drainage	1 Fixture/1180 S.F. Floor	Each	1994	1.69	
2020	Domestic Water Distribution	Gas fired water heater		S.F. Floor	.32	.32	3.2%
2040	Rain Water Drainage	Roof drains		S.F. Roof	1.06	1.06	
	D30	**HVAC**					
3010	Energy Supply	N/A		—	—	—	
3020	Heat Generating Systems	Included in D3050		—	—	—	
3030	Cooling Generating Systems	N/A		—	—	—	8.2 %
3050	Terminal & Package Units	Single zone, gas heating, electric cooling		S.F. Floor	7.98	7.98	
3090	Other HVAC Sys. & Equipment	N/A		—	—	—	
	D40	**Fire Protection**					
4010	Sprinklers	Wet pipe sprinkler system		S.F. Floor	3.62	3.62	4.5%
4020	Standpipes	Standpipe		S.F. Floor	.71	.71	
	D50	**Electrical**					
5010	Electrical Service/Distribution	600 ampere service, panel board and feeders		S.F. Floor	3.10	3.10	
5020	Lighting & Branch Wiring	T-8 fluorescent fixtures, receptacles, switches, A.C. and misc. power		S.F. Floor	8.39	8.39	16.3%
5030	Communications & Security	Addressable alarm systems, internet wiring and emergency lighting		S.F. Floor	4.14	4.14	
5090	Other Electrical Systems	Emergency generator, 15 kW		S.F. Floor	.13	.13	
E.	**EQUIPMENT & FURNISHINGS**						
1010	Commercial Equipment	N/A		—	—	—	
1020	Institutional Equipment	N/A		—	—	—	0.0 %
1030	Vehicular Equipment	N/A		—	—	—	
1090	Other Equipment	N/A		—	—	—	
F.	**SPECIAL CONSTRUCTION**						
1020	Integrated Construction	N/A		—	—	—	0.0 %
1040	Special Facilities	N/A		—	—	—	
G.	**BUILDING SITEWORK**	**N/A**					

		Sub-Total		96.92	**100%**
CONTRACTOR FEES (General Requirements: 10%, Overhead: 5%, Profit: 10%)			25%	24.23	
ARCHITECT FEES			9%	10.90	
		Total Building Cost		**132.05**	

Costs per square foot of floor area

Exterior Wall	S.F. Area	5000	10000	15000	21000	25000	30000	40000	50000	60000
	L.F. Perimeter	287	400	500	600	660	700	834	900	1000
Face Brick with Concrete Block Back-up	Steel Frame	252.95	208.70	192.70	182.15	177.55	171.75	166.50	161.35	158.80
	Bearing Walls	240.20	201.20	186.95	177.35	173.15	167.70	162.95	158.00	155.60
Concrete Block	Steel Frame	218.30	184.15	172.00	164.30	161.00	157.10	153.40	149.95	148.25
Brick Veneer	Steel Frame	245.35	203.00	187.75	177.80	173.45	168.10	163.25	158.45	156.10
Galvanized Steel Siding	Steel Frame	208.25	176.90	165.95	158.95	156.05	152.55	149.30	146.35	144.75
Metal Sandwich Panel	Steel Frame	211.50	179.35	168.10	160.95	157.90	154.30	150.95	147.80	146.25
Perimeter Adj., Add or Deduct	Per 100 L.F.	29.80	15.00	9.95	7.10	5.95	4.95	3.75	2.95	2.50
Story Hgt. Adj., Add or Deduct	Per 1 Ft.	6.05	4.30	3.55	3.05	2.75	2.45	2.25	1.90	1.75
For Basement, add $26.85 per square foot of basement area										

The above costs were calculated using the basic specifications shown on the facing page. These costs should be adjusted where necessary for design alternatives and owner's requirements. Reported completed project costs, for this type of structure, range from $84.50 to $263.50 per S.F.

Common additives

Description	Unit	$ Cost	Description	Unit	$ Cost
Bar, Front Bar	L.F.	360	Lockers, Steel, single tier, 60" or 72"	Opening	191 - 310
Back Bar	L.F.	289	2 tier, 60" or 72" total	Opening	107 - 141
Booth, Upholstered, custom straight	L.F.	202 - 375	5 tier, box lockers	Opening	65 - 83.50
"L" or "U" shaped	L.F.	210 - 355	Locker bench, lam. maple top only	L.F.	21
Bleachers, Telescoping, manual			Pedestals, steel pipe	Each	63.50
To 15 tier	Seat	115 - 160	Sauna, Prefabricated, complete		
21-30 tier	Seat	249 - 300	6' x 4'	Each	5850
Courts			6' x 9'	Each	8525
Ceiling	Court	8900	8' x 8'	Each	10,100
Floor	Court	15,100	8' x 10'	Each	11,300
Walls	Court	25,600	10' x 12'	Each	14,000
Emergency Lighting, 25 watt, battery operated			Sound System		
Lead battery	Each	282	Amplifier, 250 watts	Each	2350
Nickel cadmium	Each	805	Speaker, ceiling or wall	Each	191
Kitchen Equipment			Trumpet	Each	365
Broiler	Each	4025	Steam Bath, Complete, to 140 C.F.	Each	2050
Cooler, 6 ft. long, reach-in	Each	4925	To 300 C.F.	Each	2300
Dishwasher, 10-12 racks per hr.	Each	4950	To 800 C.F.	Each	6375
Food warmer, counter 1.2 KW	Each	735	To 2500 C.F.	Each	9550
Freezer, reach-in, 44 C.F.	Each	3725			
Ice cube maker, 50 lb. per day	Each	1750			

Important: See the Reference Section for Location Factors

Racquetball Court

Model costs calculated for a 2 story building with 12' story height and 30,000 square feet of floor area

				Unit	Unit Cost	Cost Per S.F.	% Of Sub-Total
A.	**SUBSTRUCTURE**						
1010	Standard Foundations	Poured concrete; strip and spread footings		S.F. Ground	3.46	1.73	
1020	Special Foundations	N/A		—	—	—	
1030	Slab on Grade	5" reinforced concrete with vapor barrier and granular base		S.F. Slab	6.27	4.08	6.1%
2010	Basement Excavation	Site preparation for slab and trench for foundation wall and footing		S.F. Ground	.26	.17	
2020	Basement Walls	4' foundation wall		L.F. Wall	78	1.82	
B.	**SHELL**						
	B10 Superstructure						
1010	Floor Construction	Open web steel joists, slab form, concrete, columns	50% of area	S.F. Floor	20.38	10.19	12.5%
1020	Roof Construction	Metal deck on open web steel joists, columns		S.F. Roof	11.54	5.77	
	B20 Exterior Enclosure						
2010	Exterior Walls	Face brick with concrete block backup	95% of wall	S.F. Wall	31.99	17.02	
2020	Exterior Windows	Storefront	5% of wall	S.F. Window	93	2.59	15.7%
2030	Exterior Doors	Aluminum and glass and hollow metal		Each	2704	.36	
	B30 Roofing						
3010	Roof Coverings	Built-up tar and gravel with flashing; perlite/EPS composite insulation		S.F. Roof	7.16	3.58	2.9%
3020	Roof Openings	Roof hatches		S.F. Roof	.12	.06	
C.	**INTERIORS**						
1010	Partitions	Concrete block, gypsum board on metal studs	25 S.F. Floor/L.F. Partition	S.F. Partition	13.08	5.23	
1020	Interior Doors	Single leaf hollow metal	810 S.F. Floor/Door	Each	875	1.08	
1030	Fittings	Toilet partitions		S.F. Floor	.59	.59	
2010	Stair Construction	Concrete filled metal pan		Flight	15,800	1.58	12.7%
3010	Wall Finishes	Paint		S.F. Surface	1.21	.97	
3020	Floor Finishes	80% carpet tile, 20% ceramic tile	50% of floor area	S.F. Floor	5.74	2.87	
3030	Ceiling Finishes	Mineral fiber tile on concealed zee bars	60% of area	S.F. Ceiling	6.38	3.83	
D.	**SERVICES**						
	D10 Conveying						
1010	Elevators & Lifts	N/A		—	—	—	0.0 %
1020	Escalators & Moving Walks	N/A		—	—	—	
	D20 Plumbing						
2010	Plumbing Fixtures	Kitchen, bathroom and service fixtures, supply and drainage	1 Fixture/1000 S.F. Floor	Each	2970	2.97	
2020	Domestic Water Distribution	Gas fired water heater		S.F. Floor	2.63	2.63	4.7%
2040	Rain Water Drainage	Roof drains		S.F. Roof	.80	.40	
	D30 HVAC						
3010	Energy Supply	N/A		—	—	—	
3020	Heat Generating Systems	Included in D3050		—	—	—	
3030	Cooling Generating Systems	N/A		—	—	—	20.6 %
3050	Terminal & Package Units	Multizone unit, gas heating, electric cooling		S.F. Floor	26.20	26.20	
3090	Other HVAC Sys. & Equipment	N/A		—	—	—	
	D40 Fire Protection						
4010	Sprinklers	Sprinklers, light hazard		S.F. Floor	.42	.42	0.3%
4020	Standpipes	N/A		—	—	—	
	D50 Electrical						
5010	Electrical Service/Distribution	400 ampere service, panel board and feeders		S.F. Floor	.84	.84	
5020	Lighting & Branch Wiring	Fluorescent and high intensity discharge fixtures, receptacles, switches, A.C. and misc. power		S.F. Floor	11.91	11.91	10.3%
5030	Communications & Security	Alarm systems and emergency lighting		S.F. Floor	.28	.28	
5090	Other Electrical Systems	Emergency generator, 15 kW		S.F. Floor	.06	.06	
E.	**EQUIPMENT & FURNISHINGS**						
1010	Commercial Equipment	N/A		—	—	—	
1020	Institutional Equipment	N/A		—	—	—	14.1 %
1030	Vehicular Equipment	N/A		—	—	—	
1090	Other Equipment	Courts, sauna baths		S.F. Floor	17.97	17.97	
F.	**SPECIAL CONSTRUCTION**						
1020	Integrated Construction	N/A		—	—	—	0.0 %
1040	Special Facilities	N/A		—	—	—	
G.	**BUILDING SITEWORK**	**N/A**					

		Sub-Total	127.20	100%
CONTRACTOR FEES (General Requirements: 10%, Overhead: 5%, Profit: 10%)		25%	31.83	
ARCHITECT FEES		8%	12.72	
	Total Building Cost		**171.75**	

Costs per square foot of floor area

Exterior Wall	S.F. Area	5000	6000	7000	8000	9000	10000	11000	12000	13000
	L.F. Perimeter	286	320	353	386	397	425	454	460	486
Face Brick with Concrete Block Back-up	Steel Joists	175.55	171.40	168.40	166.15	162.45	160.80	159.60	157.00	156.00
	Wood Joists	173.15	168.85	165.70	163.35	159.40	157.75	156.45	153.70	152.65
Stucco on Concrete Block	Steel Joists	176.10	172.55	169.85	167.85	164.60	163.20	162.15	159.85	159.00
	Wood Joists	161.75	158.20	155.65	153.70	150.60	149.25	148.15	146.00	145.20
Limestone with Concrete Block Back-up	Steel Joists	214.75	208.50	203.80	200.35	194.25	191.75	189.80	185.55	184.05
	Wood Joists	201.00	194.90	190.25	186.85	180.90	178.45	176.55	172.35	170.85
Perimeter Adj., Add or Deduct	Per 100 L.F.	15.95	13.35	11.40	10.00	8.95	8.05	7.20	6.65	6.25
Story Hgt. Adj., Add or Deduct	Per 1 Ft.	2.75	2.65	2.45	2.30	2.05	2.05	2.00	1.85	1.85
For Basement, add $28.25 per square foot of basement area										

The above costs were calculated using the basic specifications shown on the facing page. These costs should be adjusted where necessary for design alternatives and owner's requirements. Reported completed project costs, for this type of structure, range from $73.80 to $228.50 per S.F.

Common additives

Description	Unit	$ Cost
Carrels Hardwood	Each	660 - 990
Emergency Lighting, 25 watt, battery operated		
Lead battery	Each	282
Nickel cadmium	Each	805
Flagpoles, Complete		
Aluminum, 20' high	Each	1650
40' high	Each	3475
70' high	Each	10,100
Fiberglass, 23' high	Each	1775
39'-5" high	Each	3325
59' high	Each	8225
Gym Floor, Incl. sleepers and finish, maple	S.F.	12.90
Intercom System, 25 Station capacity		
Master station	Each	2650
Intercom outlets	Each	169
Handset	Each	470

Description	Unit	$ Cost
Lockers, Steel, single tier, 60" to 72"	Opening	191 - 310
2 tier, 60" to 72" total	Opening	107 - 141
5 tier, box lockers	Opening	65 - 83.50
Locker bench, lam. maple top only	L.F.	21
Pedestals, steel pipe	Each	63.50
Seating		
Auditorium chair, all veneer	Each	238
Veneer back, padded seat	Each	288
Upholstered, spring seat	Each	277
Classroom, movable chair & desk	Set	65 - 120
Lecture hall, pedestal type	Each	227 - 680
Smoke Detectors		
Ceiling type	Each	187
Duct type	Each	480
Sound System		
Amplifier, 250 watts	Each	2350
Speaker, ceiling or wall	Each	191
Trumpet	Each	365
Swimming Pools, Complete, gunite	S.F.	64 - 78.50

Model costs calculated for a 1 story building with 12' story height and 10,000 square feet of floor area

				Unit	Unit Cost	Cost Per S.F.	% Of Sub-Total
A. SUBSTRUCTURE							
1010	Standard Foundations	Poured concrete; strip and spread footings		S.F. Ground	2.43	2.43	
1020	Special Foundations	N/A		—	—	—	
1030	Slab on Grade	4" reinforced concrete with vapor barrier and granular base		S.F. Slab	4.74	4.74	9.1%
2010	Basement Excavation	Site preparation for slab and trench for foundation wall and footing		S.F. Ground	.26	.26	
2020	Basement Walls	4' foundation wall		L.F. Wall	78	3.31	
B. SHELL							
	B10 Superstructure						
1010	Floor Construction	N/A		—	—	—	9.0 %
1020	Roof Construction	Metal deck, open web steel joists, beams, interior columns		S.F. Roof	10.57	10.57	
	B20 Exterior Enclosure						
2010	Exterior Walls	Face brick with concrete block backup	85% of wall	S.F. Wall	30.84	13.37	
2020	Exterior Windows	Window wall	15% of wall	Each	44.15	3.38	16.1%
2030	Exterior Doors	Double aluminum and glass		Each	5638	2.26	
	B30 Roofing						
3010	Roof Coverings	Built-up tar and gravel with flashing; perlite/EPS composite insulation		S.F. Roof	5.75	5.75	4.9%
3020	Roof Openings	N/A		—	—	—	
C. INTERIORS							
1010	Partitions	Concrete block	8 S.F. Floor/S.F. Partition	S.F. Partition	8.45	10.56	
1020	Interior Doors	Single leaf hollow metal	700 S.F. Floor/Door	Each	875	1.25	
1030	Fittings	Toilet partitions		S.F. Floor	1.76	1.76	
2010	Stair Construction	N/A		—	—	—	26.6%
3010	Wall Finishes	Paint		S.F. Surface	2.04	5.11	
3020	Floor Finishes	50% vinyl tile, 50% carpet		S.F. Floor	6.37	6.37	
3030	Ceiling Finishes	Mineral fiber tile on concealed zee bars		S.F. Ceiling	6.38	6.38	
D. SERVICES							
	D10 Conveying						
1010	Elevators & Lifts	N/A		—	—	—	0.0 %
1020	Escalators & Moving Walks	N/A		—	—	—	
	D20 Plumbing						
2010	Plumbing Fixtures	Toilet and service fixtures, supply and drainage	1 Fixture/455 S.F. Floor	Each	2789	6.13	
2020	Domestic Water Distribution	Gas fired water heater		S.F. Floor	1.21	1.21	6.9%
2040	Rain Water Drainage	Roof drains		S.F. Roof	.86	.86	
	D30 HVAC						
3010	Energy Supply	N/A		—	—	—	
3020	Heat Generating Systems	N/A		—	—	—	
3030	Cooling Generating Systems	N/A		—	—	—	18.8 %
3050	Terminal & Package Units	Multizone HVAC air cooled system		S.F. Floor	22.15	22.15	
3090	Other HVAC Sys. & Equipment	N/A		—	—	—	
	D40 Fire Protection						
4010	Sprinklers	Sprinkler, light hazard		S.F. Floor	2.77	2.77	2.7%
4020	Standpipes	Standpipe		S.F. Floor	.46	.46	
	D50 Electrical						
5010	Electrical Service/Distribution	200 ampere service, panel board and feeders		S.F. Floor	1	1	
5020	Lighting & Branch Wiring	T-8 fluorescent fixtures, receptacles, switches, A.C. and misc. power		S.F. Floor	2.70	2.70	5.9%
5030	Communications & Security	Addressable alarm systems, partial internet wiring and emergency lighting		S.F. Floor	3.17	3.17	
5090	Other Electrical Systems	Emergency generator, 15 kW		S.F. Floor	.08	.08	
E. EQUIPMENT & FURNISHINGS							
1010	Commercial Equipment	N/A		—	—	—	
1020	Institutional Equipment	N/A		—	—	—	
1030	Vehicular Equipment	N/A		—	—	—	0.0 %
1090	Other Equipment	N/A		—	—	—	
F. SPECIAL CONSTRUCTION							
1020	Integrated Construction	N/A		—	—	—	0.0 %
1040	Special Facilities	N/A		—	—	—	
G. BUILDING SITEWORK	**N/A**						

			Sub-Total	118.03	100%
CONTRACTOR FEES (General Requirements: 10%, Overhead: 5%, Profit: 10%)			25%	29.49	
ARCHITECT FEES			9%	13.28	

Total Building Cost	**160.80**

Costs per square foot of floor area

Exterior Wall	S.F. Area	2000	2800	3500	4200	5000	5800	6500	7200	8000
	L.F. Perimeter	180	212	240	268	300	314	336	344	368
Wood Siding	Wood Frame	219.10	207.30	201.40	197.50	194.40	190.45	188.60	186.10	184.70
Brick Veneer	Wood Frame	232.95	219.00	212.05	207.35	203.65	198.80	196.60	193.45	191.80
Face Brick with Concrete Block Back-up	Wood Joists	232.70	217.75	210.25	205.25	201.25	196.00	193.65	190.25	188.50
	Steel Joists	235.45	218.75	210.55	204.95	200.55	194.65	191.95	188.10	186.15
Stucco on Concrete Block	Wood Joists	224.75	212.05	205.75	201.50	198.15	193.85	191.90	189.10	187.65
	Steel Joists	212.45	199.75	193.50	189.20	185.90	181.60	179.60	176.85	175.35
Perimeter Adj., Add or Deduct	Per 100 L.F.	26.55	19.00	15.30	12.65	10.65	9.15	8.15	7.40	6.75
Story Hgt. Adj., Add or Deduct	Per 1 Ft.	2.65	2.20	2.10	1.90	1.85	1.65	1.55	1.45	1.40
For Basement, add $32.65 per square foot of basement area										

The above costs were calculated using the basic specifications shown on the facing page. These costs should be adjusted where necessary for design alternatives and owner's requirements. Reported completed project costs, for this type of structure, range from $114.40 to $266.60 per S.F.

Common additives

Description	Unit	$ Cost
Bar, Front Bar	L.F.	360
Back bar	L.F.	289
Booth, Upholstered, custom straight	L.F.	202 - 375
"L" or "U" shaped	L.F.	210 - 355
Cupola, Stock unit, redwood		
30" square, 37" high, aluminum roof	Each	680
Copper roof	Each	340
Fiberglass, 5'-0" base, 63" high	Each	3500 - 4350
6'-0" base, 63" high	Each	4850 - 7025
Decorative Wood Beams, Non load bearing		
Rough sawn, 4" x 6"	L.F.	8.30
4" x 8"	L.F.	9.45
4" x 10"	L.F.	10.65
4" x 12"	L.F.	11.95
8" x 8"	L.F.	15.85
Emergency Lighting, 25 watt, battery operated		
Lead battery	Each	282
Nickel cadmium	Each	805

Description	Unit	$ Cost
Fireplace, Brick, not incl. chimney or foundation		
30" x 29" opening	Each	2800
Chimney, standard brick		
Single flue, 16" x 20"	V.L.F.	91
20" x 20"	V.L.F.	107
2 Flue, 20" x 24"	V.L.F.	133
20" x 32"	V.L.F.	163
Kitchen Equipment		
Broiler	Each	4025
Coffee urn, twin 6 gallon	Each	3075
Cooler, 6 ft. long	Each	4925
Dishwasher, 10-12 racks per hr.	Each	4950
Food warmer, counter, 1.2 KW	Each	735
Freezer, 44 C.F., reach-in	Each	3725
Ice cube maker, 50 lb. per day	Each	1750
Range with 1 oven	Each	2700
Refrigerators, Prefabricated, walk-in		
7'-6" high, 6' x 6'	S.F.	164
10' x 10'	S.F.	129
12' x 14'	S.F.	114
12' x 20'	S.F.	101

Model costs calculated for a 1 story building with 12' story height and 5,000 square feet of floor area

Restaurant

				Unit	Unit Cost	Cost Per S.F.	% Of Sub-Total

A. SUBSTRUCTURE

				Unit	Unit Cost	Cost Per S.F.	% Of Sub-Total
1010	Standard Foundations	Poured concrete; strip and spread footings		S.F. Ground	2.64	2.64	
1020	Special Foundations	N/A		—	—	—	
1030	Slab on Grade	4" reinforced concrete with vapor barrier and granular base		S.F. Slab	4.74	4.74	8.6%
2010	Basement Excavation	Site preparation for slab and trench for foundation wall and footing		S.F. Ground	.46	.46	
2020	Basement Walls	4' foundation wall		L.F. Wall	78	4.68	

B. SHELL

B10 Superstructure

				Unit	Unit Cost	Cost Per S.F.	% Of Sub-Total
1010	Floor Construction	Wood columns		S.F. Floor	.39	.98	5.8%
1020	Roof Construction	Plywood on wood truss (pitched)		S.F. Roof	6.60	7.39	

B20 Exterior Enclosure

				Unit	Unit Cost	Cost Per S.F.	% Of Sub-Total
2010	Exterior Walls	Cedar siding on wood studs with insulation	70% of wall	S.F. Wall	10.83	5.46	
2020	Exterior Windows	Storefront windows	30% of wall	Each	40	8.63	12.9%
2030	Exterior Doors	Aluminum and glass doors and entrance with transom		Each	4641	4.65	

B30 Roofing

				Unit	Unit Cost	Cost Per S.F.	% Of Sub-Total
3010	Roof Coverings	Cedar shingles with flashing (pitched); rigid fiberglass insulation		S.F. Roof	18.91	8.71	6.0%
3020	Roof Openings	Skylights		S.F. Roof	.08	.08	

C. INTERIORS

				Unit	Unit Cost	Cost Per S.F.	% Of Sub-Total
1010	Partitions	Gypsum board on wood studs	25 S.F. Floor/L.F. Partition	S.F. Partition	9.40	3.76	
1020	Interior Doors	Hollow core wood	250 S.F. Floor/Door	Each	488	1.95	
1030	Fittings	Toilet partitions		S.F. Floor	.73	.73	
2010	Stair Construction	N/A		—	—	—	15.9%
3010	Wall Finishes	75% paint, 25% ceramic tile		S.F. Surface	2.29	1.83	
3020	Floor Finishes	65% carpet, 35% quarry tile		S.F. Floor	8.48	8.48	
3030	Ceiling Finishes	Mineral fiber tile on concealed zee bars		S.F. Ceiling	6.38	6.38	

D. SERVICES

D10 Conveying

				Unit	Unit Cost	Cost Per S.F.	% Of Sub-Total
1010	Elevators & Lifts	N/A		—	—	—	0.0 %
1020	Escalators & Moving Walks	N/A		—	—	—	

D20 Plumbing

				Unit	Unit Cost	Cost Per S.F.	% Of Sub-Total
2010	Plumbing Fixtures	Kitchen, bathroom and service fixtures, supply and drainage	1 Fixture/355 S.F. Floor	Each	3010	8.48	
2020	Domestic Water Distribution	Gas fired water heater		S.F. Floor	6.25	6.25	11.1%
2040	Rain Water Drainage	Roof drains		S.F. Roof	1.42	1.42	

D30 HVAC

				Unit	Unit Cost	Cost Per S.F.	% Of Sub-Total
3010	Energy Supply	N/A		—	—	—	
3020	Heat Generating Systems	Included in D3050		—	—	—	
3030	Cooling Generating Systems	N/A		—	—	—	23.0 %
3050	Terminal & Package Units	Multizone unit, gas heating, electric cooling; kitchen ventilation		S.F. Floor	33.43	33.43	
3090	Other HVAC Sys. & Equipment	N/A		—	—	—	

D40 Fire Protection

				Unit	Unit Cost	Cost Per S.F.	% Of Sub-Total
4010	Sprinklers	Sprinklers, light hazard and ordinary hazard kitchen		S.F. Floor	6.12	6.12	5.3%
4020	Standpipes	Standpipe			1.59	1.59	

D50 Electrical

				Unit	Unit Cost	Cost Per S.F.	% Of Sub-Total
5010	Electrical Service/Distribution	400 ampere service, panel board and feeders		S.F. Floor	4.25	4.25	
5020	Lighting & Branch Wiring	T-8 fluorescent fixtures, receptacles, switches, A.C. and misc. power		S.F. Floor	9.14	9.14	11.3%
5030	Communications & Security	Addressable alarm systems and emergency lighting		S.F. Floor	2.76	2.76	
5090	Other Electrical Systems	Emergency generator, 15 kW		S.F. Floor	.33	.33	

E. EQUIPMENT & FURNISHINGS

				Unit	Unit Cost	Cost Per S.F.	% Of Sub-Total
1010	Commercial Equipment	N/A		—	—	—	
1020	Institutional Equipment	N/A		—	—	—	0.0 %
1030	Vehicular Equipment	N/A		—	—	—	
1090	Other Equipment	N/A		—	—	—	

F. SPECIAL CONSTRUCTION

				Unit	Unit Cost	Cost Per S.F.	% Of Sub-Total
1020	Integrated Construction	N/A		—	—	—	0.0 %
1040	Special Facilities	N/A		—	—	—	

G. BUILDING SITEWORK   N/A

					Sub-Total	145.32	100%
	CONTRACTOR FEES (General Requirements: 10%, Overhead: 5%, Profit: 10%)				25%	36.36	
	ARCHITECT FEES				7%	12.72	

				Total Building Cost		**194.40**	

Costs per square foot of floor area

Exterior Wall	S.F. Area	2000	2800	3500	4000	5000	5800	6500	7200	8000
	L.F. Perimeter	180	212	240	260	300	314	336	344	368
Face Brick with Concrete Block Back-up	Bearing Walls	217.55	207.60	202.60	200.10	196.60	192.50	190.80	188.05	186.80
	Steel Frame	221.65	211.65	206.70	204.20	200.65	196.60	194.85	192.15	190.90
Concrete Block With Stucco	Bearing Walls	205.65	197.45	193.40	191.30	188.45	185.15	183.70	181.45	180.45
	Steel Frame	210.25	202.25	198.25	196.25	193.50	190.20	188.80	186.60	185.60
Wood Siding	Wood Frame	196.55	189.40	185.85	184.05	181.60	178.65	177.40	175.45	174.55
Brick Veneer	Steel Frame	216.80	207.80	203.25	200.95	197.80	194.10	192.50	190.05	188.90
Perimeter Adj., Add or Deduct	Per 100 L.F.	34.95	24.90	19.95	17.45	14.00	12.10	10.70	9.70	8.80
Story Hgt. Adj., Add or Deduct	Per 1 Ft.	4.60	3.75	3.50	3.30	3.00	2.80	2.60	2.45	2.35
Basement—Not Applicable										

The above costs were calculated using the basic specifications shown on the facing page. These costs should be adjusted where necessary for design alternatives and owner's requirements. Reported completed project costs, for this type of structure, range from $ 114.15 to $ 222.55 per S.F.

Common additives

Description	Unit	$ Cost	Description	Unit	$ Cost
Bar, Front Bar	L.F.	360	Refrigerators, Prefabricated, walk-in		
Back bar	L.F.	289	7'-6" High, 6' x 6'	S.F.	164
Booth, Upholstered, custom straight	L.F.	202 - 375	10' x 10'	S.F.	129
"L" or "U" shaped	L.F.	210 - 355	12' x 14'	S.F.	114
Drive-up Window	Each	7425 - 12,300	12' x 20'	S.F.	101
Emergency Lighting, 25 watt, battery operated			Serving		
Lead battery	Each	282	Counter top (Stainless steel)	L.F.	169
Nickel cadmium	Each	805	Base cabinets	L.F.	355 - 575
Kitchen Equipment			Sound System		
Broiler	Each	4025	Amplifier, 250 watts	Each	2350
Coffee urn, twin 6 gallon	Each	3075	Speaker, ceiling or wall	Each	191
Cooler, 6 ft. long	Each	4925	Trumpet	Each	365
Dishwasher, 10-12 racks per hr.	Each	4950	Storage		
Food warmer, counter, 1.2 KW	Each	735	Shelving	S.F.	9.40
Freezer, 44 C.F., reach-in	Each	3725	Washing		
Ice cube maker, 50 lb. per day	Each	1750	Stainless steel counter	L.F.	169
Range with 1 oven	Each	2700			

Model costs calculated for a 1 story building with 10' story height and 4,000 square feet of floor area

Restaurant, Fast Food

				Unit	Unit Cost	Cost Per S.F.	% Of Sub-Total
A.	**SUBSTRUCTURE**						
1010	Standard Foundations	Poured concrete; strip and spread footings		S.F. Ground	2.57	2.57	
1020	Special Foundations	N/A		—	—	—	
1030	Slab on Grade	4" reinforced concrete with vapor barrier and granular base		S.F. Slab	4.74	4.74	8.7%
2010	Basement Excavation	Site preparation for slab and trench for foundation wall and footing		S.F. Ground	.46	.46	
2020	Basement Walls	4' foundation wall		L.F. Wall	78	5.07	
B.	**SHELL**						
	B10 Superstructure						
1010	Floor Construction	N/A		—	—	—	6.0 %
1020	Roof Construction	Metal deck on open web steel joists		S.F. Roof	8.90	8.90	
	B20 Exterior Enclosure						
2010	Exterior Walls	Face brick with concrete block backup	70% of wall	S.F. Wall	30.86	14.04	
2020	Exterior Windows	Window wall	30% of wall	Each	45.75	8.92	20.2%
2030	Exterior Doors	Aluminum and glass		Each	3478	6.96	
	B30 Roofing						
3010	Roof Coverings	Built-up tar and gravel with flashing; perlite/EPS composite insulation		S.F. Roof	6.10	6.10	4.4%
3020	Roof Openings	Roof hatches		S.F. Roof	.47	.47	
C.	**INTERIORS**						
1010	Partitions	Gypsum board on metal studs	25 S.F. Floor/L.F. Partition	S.F. Partition	5.78	2.08	
1020	Interior Doors	Hollow core wood	1000 S.F. Floor/Door	Each	488	.49	
1030	Fittings	N/A		—	—	—	
2010	Stair Construction	N/A		—	—	—	17.6%
3010	Wall Finishes	Paint		S.F. Surface	2.47	1.78	
3020	Floor Finishes	Quarry tile		S.F. Floor	15.36	15.36	
3030	Ceiling Finishes	Mineral fiber tile on concealed zee bars		S.F. Ceiling	6.38	6.38	
D.	**SERVICES**						
	D10 Conveying						
1010	Elevators & Lifts	N/A		—	—	—	0.0 %
1020	Escalators & Moving Walks	N/A		—	—	—	
	D20 Plumbing						
2010	Plumbing Fixtures	Kitchen, bathroom and service fixtures, supply and drainage	1 Fixture/400 S.F. Floor	Each	3092	7.73	
2020	Domestic Water Distribution	Gas fired water heater		S.F. Floor	4.52	4.52	9.4%
2040	Rain Water Drainage	Roof drains		S.F. Roof	1.65	1.65	
	D30 HVAC						
3010	Energy Supply	N/A		—	—	—	
3020	Heat Generating Systems	Included in D3050		—	—	—	
3030	Cooling Generating Systems	N/A		—	—	—	12.2 %
3050	Terminal & Package Units	Multizone unit, gas heating, electric cooling; kitchen ventilation		S.F. Floor	18.15	18.15	
3090	Other HVAC Sys. & Equipment	N/A		—	—	—	
	D40 Fire Protection						
4010	Sprinklers	Sprinklers, light hazard		S.F. Floor	6.12	6.12	5.2%
4020	Standpipes	Standpipe		S.F. Floor	1.66	1.66	
	D50 Electrical						
5010	Electrical Service/Distribution	400 ampere service, panel board and feeders		S.F. Floor	4.30	4.30	
5020	Lighting & Branch Wiring	T-8 fluorescent fixtures, receptacles, switches, A.C. and misc. power		S.F. Floor	9.89	9.89	12.2%
5030	Communications & Security	Addressable alarm systems and emergency lighting		S.F. Floor	3.44	3.44	
5090	Other Electrical Systems	Emergency generator, 7.5 kW		S.F. Floor	.38	.38	
E.	**EQUIPMENT & FURNISHINGS**						
1010	Commercial Equipment	N/A		—	—	—	
1020	Institutional Equipment	N/A		—	—	—	4.1 %
1030	Vehicular Equipment	N/A		—	—	—	
1090	Other Equipment	Walk-in refrigerator		S.F. Floor	6.06	6.06	
F.	**SPECIAL CONSTRUCTION**						
1020	Integrated Construction	N/A		—	—	—	0.0 %
1040	Special Facilities	N/A		—	—	—	
G.	**BUILDING SITEWORK**	**N/A**					

	Sub-Total	148.22	**100%**
CONTRACTOR FEES (General Requirements: 10%, Overhead: 5%, Profit: 10%)	25%	37.06	
ARCHITECT FEES	8%	14.82	
	Total Building Cost	**200.10**	

Costs per square foot of floor area

Exterior Wall	S.F. Area	10000	15000	20000	25000	30000	35000	40000	45000	50000
	L.F. Perimeter	450	500	600	700	740	822	890	920	966
Face Brick with Concrete Block Back-up	Steel Frame	201.50	180.05	172.70	168.25	162.65	160.20	157.95	155.00	153.15
	Lam. Wood Truss	195.65	175.00	167.85	163.55	158.15	155.80	153.65	150.90	149.00
Concrete Block	Steel Frame	199.80	186.15	181.05	177.95	174.50	172.90	171.45	169.80	168.60
	Lam. Wood Truss	171.40	157.75	152.65	149.50	146.10	144.50	143.10	141.35	140.20
Galvanized Steel Siding	Steel Frame	189.50	175.05	169.70	166.45	162.85	161.10	159.60	157.80	156.55
Metal Sandwich Panel	Steel Joists	199.10	182.20	176.05	172.45	168.10	166.10	164.35	162.15	160.70
Perimeter Adj., Add or Deduct	Per 100 L.F.	13.35	8.90	6.65	5.35	4.45	3.80	3.30	3.00	2.70
Story Hgt. Adj., Add or Deduct	Per 1 Ft.	2.05	1.55	1.40	1.30	1.10	1.10	1.00	1.00	0.85
Basement—Not Applicable										

The above costs were calculated using the basic specifications shown on the facing page. These costs should be adjusted where necessary for design alternatives and owner's requirements. Reported completed project costs, for this type of structure, range from $67.55 to $210.10 per S.F.

Common additives

Description	Unit	$ Cost
Bar, Front Bar	L.F.	360
Back bar	L.F.	289
Booth, Upholstered, custom straight	L.F.	202 - 375
"L" or "U" shaped	L.F.	210 - 355
Bleachers, Telescoping, manual		
To 15 tier	Seat	115 - 160
16-20 tier	Seat	235 - 288
21-30 tier	Seat	249 - 300
For power operation, add	Seat	45.50 - 71.50
Emergency Lighting, 25 watt, battery operated		
Lead battery	Each	282
Nickel cadmium	Each	805
Lockers, Steel, single tier, 60" or 72"	Opening	191 - 310
2 tier, 60" or 72" total	Opening	107 - 141
5 tier, box lockers	Opening	65 - 83.50
Locker bench, lam. maple top only	L.F.	21
Pedestals, steel pipe	Each	63.50

Description	Unit	$ Cost
Rink		
Dasher boards & top guard	Each	176,500
Mats, rubber	S.F.	20.50
Score Board	Each	12,300 - 16,300

Important: See the Reference Section for Location Factors

Model costs calculated for a 1 story building with 24' story height and 30,000 square feet of floor area

Rink, Hockey/Indoor Soccer

				Unit	Unit Cost	Cost Per S.F.	% Of Sub-Total
A.	**SUBSTRUCTURE**						
1010	Standard Foundations	Poured concrete; strip and spread footings		S.F. Ground	1.75	1.75	
1020	Special Foundations	N/A		—	—	—	
1030	Slab on Grade	6" reinforced concrete with vapor barrier and granular base		S.F. Slab	5.82	5.82	7.8%
2010	Basement Excavation	Site preparation for slab and trench for foundation wall and footing		S.F. Ground	.17	.17	
2020	Basement Walls	4' foundation wall		L.F. Wall	74	2.49	
B.	**SHELL**						
	B10 Superstructure						
1010	Floor Construction	Wide flange beams and columns		S.F. Floor	31.07	21.75	31.6%
1020	Roof Construction	Metal deck on steel joist		S.F. Roof	19.53	19.53	
	B20 Exterior Enclosure						
2010	Exterior Walls	Concrete block	95% of wall	S.F. Wall	11.20	6.30	
2020	Exterior Windows	Store front	5% of wall	Each	43.50	1.29	6.4%
2030	Exterior Doors	Aluminum and glass, hollow metal, overhead		Each	2846	.76	
	B30 Roofing						
3010	Roof Coverings	Elastomeric neoprene membrane with flashing; perlite/EPS composite insulation		S.F. Roof	4.37	4.37	3.5%
3020	Roof Openings	Roof hatches		S.F. Roof	.23	.23	
C.	**INTERIORS**						
1010	Partitions	Concrete block	140 S.F. Floor/L.F. Partition	S.F. Partition	8.40	.72	
1020	Interior Doors	Hollow metal	2500 S.F. Floor/Door	Each	875	.35	
1030	Fittings	N/A		—	—	—	
2010	Stair Construction	N/A		—	—	—	5.5%
3010	Wall Finishes	Paint		S.F. Surface	10.73	1.84	
3020	Floor Finishes	80% rubber mat, 20% paint	50% of floor area	S.F. Floor	7.34	3.67	
3030	Ceiling Finishes	Mineral fiber tile on concealed zee bar	10% of area	S.F. Ceiling	6.38	.64	
D.	**SERVICES**						
	D10 Conveying						
1010	Elevators & Lifts	N/A		—	—	—	0.0%
1020	Escalators & Moving Walks	N/A		—	—	—	
	D20 Plumbing						
2010	Plumbing Fixtures	Toilet and service fixtures, supply and drainage	1 Fixture/1070 S.F. Floor	Each	3167	2.96	
2020	Domestic Water Distribution	Oil fired water heater		S.F. Floor	4.38	4.38	5.6%
2040	Rain Water Drainage	N/A		—	—	—	
	D30 HVAC						
3010	Energy Supply	Oil fired hot water, unit heaters	10% of area	S.F. Floor	.74	.74	
3020	Heat Generating Systems	N/A		—	—	—	
3030	Cooling Generating Systems	N/A		—	—	—	10.1%
3050	Terminal & Package Units	Single zone, electric cooling	90% of area	S.F. Floor	12.47	12.47	
3090	Other HVAC Sys. & Equipment	N/A		—	—	—	
	D40 Fire Protection						
4010	Sprinklers	N/A		—	—	—	0.0%
4020	Standpipes	N/A		—	—	—	
	D50 Electrical						
5010	Electrical Service/Distribution	600 ampere service, panel board and feeders		S.F. Floor	1.89	1.89	
5020	Lighting & Branch Wiring	High intensity discharge and fluorescent fixtures, receptacles, switches, A.C. and misc. power		S.F. Floor	7.40	7.40	8.1%
5030	Communications & Security	Alarm systems, emergency lighting and public address		S.F. Floor	1.04	1.04	
5090	Other Electrical Systems	Emergency generator		S.F. Floor	.19	.19	
E.	**EQUIPMENT & FURNISHINGS**						
1010	Commercial Equipment	N/A		—	—	—	
1020	Institutional Equipment	N/A		—	—	—	0.0%
1030	Vehicular Equipment	N/A		—	—	—	
1090	Other Equipment	N/A		—	—	—	
F.	**SPECIAL CONSTRUCTION**						
1020	Integrated Construction	N/A		—	—	—	21.3%
1040	Special Facilities	Dasher boards and rink (including ice making system)		S.F. Floor	27.74	27.74	
G.	**BUILDING SITEWORK**	**N/A**					

		Sub-Total	130.49	100%
CONTRACTOR FEES (General Requirements: 10%, Overhead: 5%, Profit: 10%)		25%	32.59	
ARCHITECT FEES		7%	11.42	
	Total Building Cost		**174.50**	

Costs per square foot of floor area

Exterior Wall	S.F. Area	25000	30000	35000	40000	45000	50000	55000	60000	65000
	L.F. Perimeter	900	1050	1200	1350	1510	1650	1800	1970	2100
Face Brick with Concrete Block Back-up	Steel Frame	171.80	170.15	168.90	168.05	167.60	166.80	166.35	166.35	165.65
	Bearing Walls	163.25	161.60	160.40	159.55	159.10	158.30	157.85	157.85	157.15
Stucco on Concrete Block	Steel Frame	164.50	163.00	161.95	161.25	160.80	160.10	159.75	159.70	159.10
	Bearing Walls	155.95	154.50	153.45	152.70	152.25	151.60	151.25	151.25	150.65
Decorative Concrete Block	Steel Frame	164.45	162.95	161.85	161.10	160.65	159.95	159.55	159.55	158.95
	Bearing Walls	160.40	158.95	157.80	157.05	156.60	155.95	155.55	155.55	154.90
Perimeter Adj., Add or Deduct	Per 100 L.F.	4.30	3.55	3.05	2.70	2.40	2.20	1.95	1.80	1.65
Story Hgt. Adj., Add or Deduct	Per 1 Ft.	1.55	1.50	1.50	1.45	1.45	1.45	1.45	1.50	1.45

For Basement, add $24.20 per square foot of basement area

The above costs were calculated using the basic specifications shown on the facing page. These costs should be adjusted where necessary for design alternatives and owner's requirements. Reported completed project costs, for this type of structure, range from $78.90 to $200.65 per S.F.

Common additives

Description	Unit	$ Cost
Bleachers, Telescoping, manual		
To 15 tier	Seat	115 - 160
16-20 tier	Seat	235 - 288
21-30 tier	Seat	249 - 300
For power operation, add	Seat	45.50 - 71.50
Carrels Hardwood	Each	660 - 990
Clock System		
20 room	Each	16,000
50 room	Each	39,100
Emergency Lighting, 25 watt, battery operated		
Lead battery	Each	282
Nickel cadmium	Each	805
Flagpoles, Complete		
Aluminum, 20' high	Each	1650
40' high	Each	3475
Fiberglass, 23' high	Each	1775
39'-5" high	Each	3325
Kitchen Equipment		
Broiler	Each	4025
Cooler, 6 ft. long, reach-in	Each	4925

Description	Unit	$ Cost
Kitchen Equipment, cont.		
Dishwasher, 10-12 racks per hr.	Each	4950
Food warmer, counter, 1.2 KW	Each	735
Freezer, 44 C.F., reach-in	Each	3725
Ice cube maker, 50 lb. per day	Each	1750
Range with 1 oven	Each	2700
Lockers, Steel, single tier, 60" to 72"	Opening	191 - 310
2 tier, 60" to 72" total	Opening	107 - 141
5 tier, box lockers	Opening	65 - 83.50
Locker bench, lam. maple top only	L.F.	21
Pedestals, steel pipe	Each	63.50
Seating		
Auditorium chair, all veneer	Each	238
Veneer back, padded seat	Each	288
Upholstered, spring seat	Each	277
Classroom, movable chair & desk	Set	65 - 120
Lecture hall, pedestal type	Each	227 - 680
Sound System		
Amplifier, 250 watts	Each	2350
Speaker, ceiling or wall	Each	191
Trumpet	Each	365

Model costs calculated for a 1 story building with 15' story height and 45,000 square feet of floor area

School, Elementary

				Unit	Unit Cost	Cost Per S.F.	% Of Sub-Total
A.	**SUBSTRUCTURE**						
1010	Standard Foundations	Poured concrete; strip and spread footings		S.F. Ground	5.03	5.03	
1020	Special Foundations	N/A		—	—	—	
1030	Slab on Grade	4" reinforced concrete with vapor barrier and granular base		S.F. Slab	4.74	4.74	12.1%
2010	Basement Excavation	Site preparation for slab and trench for foundation wall and footing		S.F. Ground	.17	.17	
2020	Basement Walls	4' foundation wall		L.F. Wall	78	4.45	
B.	**SHELL**						
	B10 Superstructure						
1010	Floor Construction	N/A		—	—	—	4.4 %
1020	Roof Construction	Metal deck on open web steel joists		S.F. Roof	5.19	5.19	
	B20 Exterior Enclosure						
2010	Exterior Walls	Face brick with concrete block backup	70% of wall	S.F. Wall	30.85	10.87	
2020	Exterior Windows	Steel outward projecting	25% of wall	Each	696	4.57	13.5%
2030	Exterior Doors	Metal and glass	5% of wall	Each	3215	.57	
	B30 Roofing						
3010	Roof Coverings	Single-ply membrane with flashing; polyisocyanurate insulation		S.F. Roof	7.78	7.78	6.5%
3020	Roof Openings	N/A		—	—	—	
C.	**INTERIORS**						
1010	Partitions	Concrete block	20 S.F. Floor/L.F. Partition	S.F. Partition	8.76	4.38	
1020	Interior Doors	Single leaf kalamein fire doors	700 S.F. Floor/Door	Each	875	1.25	
1030	Fittings	Toilet partitions		S.F. Floor	2.03	2.03	
2010	Stair Construction	N/A		—	—	—	21.0%
3010	Wall Finishes	75% paint, 15% glazed coating, 10% ceramic tile		S.F. Surface	4.08	4.08	
3020	Floor Finishes	65% vinyl composition tile, 25% carpet, 10% terrazzo		S.F. Floor	6.87	6.87	
3030	Ceiling Finishes	Mineral fiber tile on concealed zee bars		S.F. Ceiling	6.38	6.38	
D.	**SERVICES**						
	D10 Conveying						
1010	Elevators & Lifts	N/A		—	—	—	0.0 %
1020	Escalators & Moving Walks	N/A		—	—	—	
	D20 Plumbing						
2010	Plumbing Fixtures	Kitchen, bathroom and service fixtures, supply and drainage	1 Fixture/625 S.F. Floor	Each	6394	10.23	
2020	Domestic Water Distribution	Gas fired water heater		S.F. Floor	.48	.48	9.8%
2040	Rain Water Drainage	Roof drains		S.F. Roof	1	1	
	D30 HVAC						
3010	Energy Supply	Oil fired hot water, wall fin radiation		S.F. Floor	8.73	8.73	
3020	Heat Generating Systems	N/A		—	—	—	
3030	Cooling Generating Systems	N/A		—	—	—	17.9%
3050	Terminal & Package Units	Split systems with air cooled condensing units		S.F. Floor	12.60	12.60	
3090	Other HVAC Sys. & Equipment	N/A		—	—	—	
	D40 Fire Protection						
4010	Sprinklers	Sprinklers, light hazard		S.F. Floor	2.33	2.33	2.2%
4020	Standpipes	Standpipe		S.F. Floor	.30	.30	
	D50 Electrical						
5010	Electrical Service/Distribution	800 ampere service, panel board and feeders		S.F. Floor	1.41	1.41	
5020	Lighting & Branch Wiring	High efficiency fluorescent fixtures, receptacles, switches, A.C. and misc. power		S.F. Floor	9.91	9.91	12.4%
5030	Communications & Security	Addressable alarm systems, internet wiring, communications systems and emergency lighting		S.F. Floor	3.37	3.37	
5090	Other Electrical Systems	Emergency generator, 15 kW		S.F. Floor	.08	.08	
E.	**EQUIPMENT & FURNISHINGS**						
1010	Commercial Equipment	N/A		—	—	—	
1020	Institutional Equipment	Chalkboards		S.F. Floor	.13	.13	0.1%
1030	Vehicular Equipment	N/A		—	—	—	
1090	Other Equipment	N/A		—	—	—	
F.	**SPECIAL CONSTRUCTION**						
1020	Integrated Construction	N/A		—	—	—	0.0 %
1040	Special Facilities	N/A		—	—	—	
G.	**BUILDING SITEWORK**	**N/A**					

		Sub-Total	118.93	**100%**
CONTRACTOR FEES (General Requirements: 10%, Overhead: 5%, Profit: 10%)		25%	29.76	
ARCHITECT FEES		7%	10.41	
		Total Building Cost	**159.10**	

Costs per square foot of floor area

Exterior Wall	S.F. Area	50000	70000	90000	110000	130000	150000	170000	190000	210000
	L.F. Perimeter	850	1140	1420	1700	1980	2280	2560	2840	3120
Face Brick with Concrete Block Back-up	Steel Frame	166.05	162.30	160.10	158.65	157.65	157.15	156.55	156.10	155.75
	R/Conc. Frame	167.85	163.40	160.85	159.15	157.95	157.35	156.70	156.10	155.70
Decorative Concrete Block	Steel Frame	161.20	157.00	154.60	153.00	151.90	151.35	150.70	150.25	149.85
	R/Conc. Frame	160.00	155.80	153.45	151.80	150.75	150.15	149.55	149.00	148.65
Limestone with Concrete Block Back-up	Steel Frame	185.30	180.15	177.05	175.00	173.55	172.95	172.15	171.45	170.90
	R/Conc. Frame	189.65	184.45	181.35	179.30	177.85	177.25	176.45	175.75	175.25
Perimeter Adj., Add or Deduct	Per 100 L.F.	3.90	2.80	2.15	1.70	1.45	1.25	1.15	1.00	0.90
Story Hgt. Adj., Add or Deduct	Per 1 Ft.	1.95	1.95	1.80	1.70	1.75	1.75	1.75	1.75	1.65

For Basement, add $31.60 per square foot of basement area

The above costs were calculated using the basic specifications shown on the facing page. These costs should be adjusted where necessary for design alternatives and owner's requirements. Reported completed project costs, for this type of structure, range from $91.70 to $215.95 per S.F.

Common additives

Description	Unit	$ Cost
Bleachers, Telescoping, manual		
To 15 tier	Seat	115 - 160
16-20 tier	Seat	235 - 288
21-30 tier	Seat	249 - 300
For power operation, add	Seat	45.50 - 71.50
Carrels Hardwood	Each	660 - 990
Clock System		
20 room	Each	16,000
50 room	Each	39,100
Elevators, Hydraulic passenger, 2 stops		
1500# capacity	Each	62,800
2500# capacity	Each	66,300
Emergency Lighting, 25 watt, battery operated		
Lead battery	Each	282
Nickel cadmium	Each	805
Flagpoles, Complete		
Aluminum, 20' high	Each	1650
40' high	Each	3475
Fiberglass, 23' high	Each	1775
39'-5" high	Each	3325

Description	Unit	$ Cost
Kitchen Equipment		
Broiler	Each	4025
Cooler, 6 ft. long, reach-in	Each	4925
Dishwasher, 10-12 racks per hr.	Each	4950
Food warmer, counter, 1.2 KW	Each	735
Freezer, 44 C.F., reach-in	Each	3725
Lockers, Steel, single tier, 60" or 72"	Opening	191 - 310
2 tier, 60" or 72" total	Opening	107 - 141
5 tier, box lockers	Opening	65 - 83.50
Locker bench, lam. maple top only	L.F.	21
Pedestals, steel pipe	Each	63.50
Seating		
Auditorium chair, all veneer	Each	238
Veneer back, padded seat	Each	288
Upholstered, spring seat	Each	277
Classroom, movable chair & desk	Set	65 - 120
Lecture hall, pedestal type	Each	227 - 680
Sound System		
Amplifier, 250 watts	Each	2350
Speaker, ceiling or wall	Each	191
Trumpet	Each	365

Important: See the Reference Section for Location Factors

Model costs calculated for a 2 story building with 15' story height and 130,000 square feet of floor area

				Unit	Unit Cost	Cost Per S.F.	% Of Sub-Total
A. SUBSTRUCTURE							
1010	Standard Foundations	Poured concrete; strip and spread footings		S.F. Ground	2.64	1.32	
1020	Special Foundations	N/A		—	—	—	
1030	Slab on Grade	4" reinforced concrete with vapor barrier and granular base		S.F. Slab	4.74	2.37	4.2%
2010	Basement Excavation	Site preparation for slab and trench for foundation wall and footing		S.F. Ground	.17	.09	
2020	Basement Walls	4' foundation wall		L.F. Wall	78	1.19	
B. SHELL							
	B10 Superstructure						
1010	Floor Construction	Concrete slab without drop panel, concrete columns		S.F. Floor	21.80	10.90	15.7%
1020	Roof Construction	Concrete slab without drop panel		S.F. Roof	15.22	7.61	
	B20 Exterior Enclosure						
2010	Exterior Walls	Face brick with concrete block backup	75% of wall	S.F. Wall	30.84	10.57	
2020	Exterior Windows	Window wall	25% of wall	Each	67	7.64	15.8%
2030	Exterior Doors	Metal and glass		Each	1899	.42	
	B30 Roofing						
3010	Roof Coverings	Single-ply membrane and standing seam metal; polyisocyanurate insulation		S.F. Roof	11.40	5.70	4.9%
3020	Roof Openings	Roof hatches		S.F. Roof	.10	.05	
C. INTERIORS							
1010	Partitions	Concrete block	25 S.F. Floor/L.F. Partition	S.F. Partition	11.35	5.45	
1020	Interior Doors	Single leaf kalamein fire doors	700 S.F. Floor/Door	Each	875	1.25	
1030	Fittings	Toilet partitions, chalkboards		S.F. Floor	1.44	1.44	
2010	Stair Construction	Concrete filled metal pan		Flight	15,800	.73	21.0%
3010	Wall Finishes	75% paint, 15% glazed coating, 10% ceramic tile		S.F. Surface	3.10	2.98	
3020	Floor Finishes	70% vinyl composition tile, 20% carpet, 10% terrazzo		S.F. Floor	6.56	6.56	
3030	Ceiling Finishes	Mineral fiber tile on concealed zee bars		S.F. Ceiling	6.38	6.38	
D. SERVICES							
	D10 Conveying						
1010	Elevators & Lifts	One hydraulic passenger elevator		Each	80,600	.62	0.5%
1020	Escalators & Moving Walks	N/A		—	—	—	
	D20 Plumbing						
2010	Plumbing Fixtures	Kitchen, bathroom and service fixtures, supply and drainage	1 Fixture/860 S.F. Floor	Each	3956	4.60	
2020	Domestic Water Distribution	Gas fired water heater		S.F. Floor	.96	.96	5.2%
2040	Rain Water Drainage	Roof drains		S.F. Roof	1.22	.61	
	D30 HVAC						
3010	Energy Supply	Oil fired hot water, wall fin radiation		S.F. Floor	4.33	4.33	
3020	Heat Generating Systems	N/A		—	—	—	
3030	Cooling Generating Systems	Chilled water, cooling tower systems		S.F. Floor	14.78	14.78	16.2%
3050	Terminal & Package Units	N/A		—	—	—	
3090	Other HVAC Sys. & Equipment	N/A		—	—	—	
	D40 Fire Protection						
4010	Sprinklers	Sprinklers, light hazard		S.F. Floor	2.04	2.04	1.7%
4020	Standpipes	N/A		—	—	—	
	D50 Electrical						
5010	Electrical Service/Distribution	2000 ampere service, panel board and feeders		S.F. Floor	1.37	1.37	
5020	Lighting & Branch Wiring	Fluorescent fixtures, receptacles, switches, A.C. and misc. power		S.F. Floor	9.67	9.67	12.8%
5030	Communications & Security	Alarm systems, internet wiring, communications systems and emergency lighting		S.F. Floor	3.57	3.57	
5090	Other Electrical Systems	Emergency generator, 250 kW		S.F. Floor	.52	.52	
E. EQUIPMENT & FURNISHINGS							
1010	Commercial Equipment	N/A		—	—	—	
1020	Institutional Equipment	Laboratory casework and counters		S.F. Floor	1.59	1.59	2.0%
1030	Vehicular Equipment	N/A		—	—	—	
1090	Other Equipment	Built-in athletic equipment, bleachers and scoreboard		S.F. Floor	.77	.77	
F. SPECIAL CONSTRUCTION							
1020	Integrated Construction	N/A		—	—	—	0.0%
1040	Special Facilities	N/A		—	—	—	
G. BUILDING SITEWORK	**N/A**						

		Sub-Total	118.08	100%
	CONTRACTOR FEES (General Requirements: 10%, Overhead: 5%, Profit: 10%)	25%	29.54	
	ARCHITECT FEES	7%	10.33	
	Total Building Cost		**157.95**	

Costs per square foot of floor area

Exterior Wall	S.F. Area	50000	65000	80000	95000	110000	125000	140000	155000	170000
	L.F. Perimeter	850	1060	1280	1490	1700	1920	2140	2340	2560
Face Brick with Concrete Block Back-up	Steel Frame	166.40	164.65	163.75	162.90	162.30	162.05	161.75	161.40	161.25
	Bearing Walls	158.15	156.35	155.45	154.60	154.00	153.75	153.50	153.10	152.95
Concrete Block Stucco Face	Steel Frame	160.45	158.95	158.15	157.50	157.05	156.75	156.55	156.25	156.10
	Bearing Walls	150.60	149.10	148.30	147.65	147.20	146.90	146.70	146.40	146.25
Decorative Concrete Block	Steel Frame	160.45	158.95	158.10	157.40	156.95	156.65	156.40	156.10	155.95
	Bearing Walls	151.55	150.00	149.15	148.50	148.00	147.75	147.45	147.15	147.00
Perimeter Adj., Add or Deduct	Per 100 L.F.	3.45	2.65	2.10	1.85	1.60	1.35	1.30	1.10	0.95
Story Hgt. Adj., Add or Deduct	Per 1 Ft.	1.75	1.70	1.55	1.60	1.60	1.55	1.60	1.55	1.50

For Basement, add $32.90 per square foot of basement area

The above costs were calculated using the basic specifications shown on the facing page. These costs should be adjusted where necessary for design alternatives and owner's requirements. Reported completed project costs, for this type of structure, range from $86.55 to $203.40 per S.F.

Common additives

Description	Unit	$ Cost
Bleachers, Telescoping, manual		
To 15 tier	Seat	115 - 160
16-20 tier	Seat	235 - 288
21-30 tier	Seat	249 - 300
For power operation, add	Seat	45.50 - 71.50
Carrels Hardwood	Each	660 - 990
Clock System		
20 room	Each	16,000
50 room	Each	39,100
Elevators, Hydraulic passenger, 2 stops		
1500# capacity	Each	62,800
2500# capacity	Each	66,300
Emergency Lighting, 25 watt, battery operated		
Lead battery	Each	282
Nickel cadmium	Each	805
Flagpoles, Complete		
Aluminum, 20' high	Each	1650
40' high	Each	3475
Fiberglass, 23' high	Each	1775
39'-5" high	Each	3325

Description	Unit	$ Cost
Kitchen Equipment		
Broiler	Each	4025
Cooler, 6 ft. long, reach-in	Each	4925
Dishwasher, 10-12 racks per hr.	Each	4950
Food warmer, counter, 1.2 KW	Each	735
Freezer, 44 C.F., reach-in	Each	3725
Lockers, Steel, single tier, 60" to 72"	Opening	191 - 310
2 tier, 60" to 72" total	Opening	107 - 141
5 tier, box lockers	Opening	65 - 83.50
Locker bench, lam. maple top only	L.F.	21
Pedestals, steel pipe	Each	63.50
Seating		
Auditorium chair, all veneer	Each	238
Veneer back, padded seat	Each	288
Upholstered, spring seat	Each	277
Classroom, movable chair & desk	Set	65 - 120
Lecture hall, pedestal type	Each	227 - 680
Sound System		
Amplifier, 250 watts	Each	2350
Speaker, ceiling or wall	Each	191
Trumpet	Each	365

Model costs calculated for a 2 story building with 15' story height and 110,000 square feet of floor area

School, Jr High, 2-3 Story

			Unit	Unit Cost	Cost Per S.F.	% Of Sub-Total
A. SUBSTRUCTURE						
1010	Standard Foundations	Poured concrete; strip and spread footings	S.F. Ground	2.44	1.22	
1020	Special Foundations	N/A	—	—	—	
1030	Slab on Grade	4" reinforced concrete with vapor barrier and granular base	S.F. Slab	4.74	2.37	4.0%
2010	Basement Excavation	Site preparation for slab and trench for foundation wall and footing	S.F. Ground	.17	.09	
2020	Basement Walls	4' foundation wall	L.F. Wall	78	1.21	
B. SHELL						
	B10 Superstructure					
1010	Floor Construction	Open web steel joists, slab form, concrete, columns	S.F. Floor	30.90	15.45	18.3%
1020	Roof Construction	Metal deck, open web steel joists, columns	S.F. Roof	13.48	6.74	
	B20 Exterior Enclosure					
2010	Exterior Walls	Face brick with concrete block backup 75% of wall	S.F. Wall	30.86	10.73	
2020	Exterior Windows	Window wall 25% of wall	Each	51	5.88	14.1%
2030	Exterior Doors	Double aluminum & glass	Each	2041	.53	
	B30 Roofing					
3010	Roof Coverings	Single-ply membrane and standing seam metal; polyisocyanurate insulation	S.F. Roof	11.54	5.77	4.8%
3020	Roof Openings	Roof hatches	S.F. Roof	.04	.02	
C. INTERIORS						
1010	Partitions	Concrete block 20 S.F. Floor/L.F. Partition	S.F. Partition	8.43	5.06	
1020	Interior Doors	Single leaf kalamein fire doors 750 S.F. Floor/Door	Each	875	1.17	
1030	Fittings	Toilet partitions, chalkboards	S.F. Floor	1.50	1.50	
2010	Stair Construction	Concrete filled metal pan	Flight	15,800	.86	23.6%
3010	Wall Finishes	50% paint, 40% glazed coatings, 10% ceramic tile	S.F. Surface	3.72	4.46	
3020	Floor Finishes	50% vinyl composition tile, 30% carpet, 20% terrrazzo	S.F. Floor	9.26	9.26	
3030	Ceiling Finishes	Mineral fiberboard on concealed zee bars	S.F. Ceiling	6.38	6.38	
D. SERVICES						
	D10 Conveying					
1010	Elevators & Lifts	One hydraulic passenger elevator	Each	81,400	.74	0.6%
1020	Escalators & Moving Walks	N/A	—	—	—	
	D20 Plumbing					
2010	Plumbing Fixtures	Kitchen, toilet and service fixtures, supply and drainage 1 Fixture/1170 S.F. Floor	Each	4996	4.27	
2020	Domestic Water Distribution	Gas fired water heater	S.F. Floor	.49	.49	4.2%
2040	Rain Water Drainage	Roof drains	S.F. Roof	.78	.39	
	D30 HVAC					
3010	Energy Supply	N/A	—	—	—	
3020	Heat Generating Systems	Included in D3050	—	—	—	
3030	Cooling Generating Systems	N/A	—	—	—	15.0 %
3050	Terminal & Package Units	Multizone unit, gas heating, electric cooling	S.F. Floor	18.20	18.20	
3090	Other HVAC Sys. & Equipment	N/A	—	—	—	
	D40 Fire Protection					
4010	Sprinklers	Sprinklers, light hazard 10% of area	S.F. Floor	.42	.42	0.3%
4020	Standpipes	N/A	—	—	—	
	D50 Electrical					
5010	Electrical Service/Distribution	1600 ampere service, panel board and feeders	S.F. Floor	1.36	1.36	
5020	Lighting & Branch Wiring	Fluorescent fixtures, receptacles, switches, A.C. and misc. power	S.F. Floor	9.76	9.76	12.6%
5030	Communications & Security	Alarm systems, internet wiring, communications systems and emergency lighting	S.F. Floor	3.83	3.83	
5090	Other Electrical Systems	Emergency generator, 100 kW	S.F. Floor	.39	.39	
E. EQUIPMENT & FURNISHINGS						
1010	Commercial Equipment	N/A	—	—	—	
1020	Institutional Equipment	Laboratory casework and counters	S.F. Floor	1.86	1.86	2.3 %
1030	Vehicular Equipment	N/A	—	—	—	
1090	Other Equipment	Bleachers, basketball backstops	S.F. Floor	.96	.96	
F. SPECIAL CONSTRUCTION						
1020	Integrated Construction	N/A	—	—	—	0.0 %
1040	Special Facilities	N/A	—	—	—	
G. BUILDING SITEWORK	**N/A**					

			Sub-Total	121.37	100%
CONTRACTOR FEES (General Requirements: 10%, Overhead: 5%, Profit: 10%)			25%	30.31	
ARCHITECT FEES			7%	10.62	
		Total Building Cost		**162.30**	

203

Costs per square foot of floor area

Exterior Wall	S.F. Area	20000	30000	40000	50000	60000	70000	80000	90000	100000
	L.F. Perimeter	440	590	740	900	1050	1200	1360	1510	1660
Face Brick with Concrete Block Back-up	Steel Frame	167.75	162.50	159.90	158.70	157.55	156.80	156.40	155.90	155.55
	Bearing Walls	162.30	157.05	154.55	153.30	152.25	151.45	151.05	150.60	150.25
Decorative Concrete Block	Steel Frame	154.80	150.95	149.05	148.10	147.25	146.75	146.40	146.05	145.80
	Bearing Walls	148.85	144.90	143.05	142.10	141.30	140.70	140.40	140.05	139.80
Steel Siding on Steel Studs	Steel Frame	149.40	146.05	144.45	143.60	143.00	142.50	142.15	141.90	141.65
Metal Sandwich Panel	Steel Frame	149.50	146.20	144.60	143.75	143.10	142.60	142.30	142.05	141.80
Perimeter Adj., Add or Deduct	Per 100 L.F.	8.85	5.85	4.40	3.45	3.00	2.50	2.20	1.95	1.80
Story Hgt. Adj., Add or Deduct	Per 1 Ft.	2.15	1.90	1.80	1.70	1.75	1.65	1.60	1.65	1.60
For Basement, add $31.90 per square foot of basement area										

The above costs were calculated using the basic specifications shown on the facing page. These costs should be adjusted where necessary for design alternatives and owner's requirements. Reported completed project costs, for this type of structure, range from $77.70 to $221.95 per S.F.

Common additives

Description	Unit	$ Cost		Description	Unit	$ Cost
Carrels Hardwood	Each	660 - 990		Flagpoles, Complete		
Clock System				Aluminum, 20' high	Each	1650
20 room	Each	16,000		40' high	Each	3475
50 room	Each	39,100		Fiberglass, 23' high	Each	1775
Directory Boards, Plastic, glass covered				39'-5" high	Each	3325
30" x 20"	Each	595		Seating		
36" x 48"	Each	1450		Auditorium chair, all veneer	Each	238
Aluminum, 24" x 18"	Each	600		Veneer back, padded seat	Each	288
36" x 24"	Each	675		Upholstered, spring seat	Each	277
48" x 32"	Each	980		Classroom, movable chair & desk	Set	65 - 120
48" x 60"	Each	2025		Lecture hall, pedestal type	Each	227 - 680
Elevators, Hydraulic passenger, 2 stops				Shops & Workroom:		
1500# capacity	Each	62,800		Benches, metal	Each	605
2500# capacity	Each	66,300		Parts bins 6'-3" high, 3' wide, 12" deep, 72 bins	Each	550
3500# capacity	Each	69,800		Shelving, metal 1' x 3'	S.F.	10.85
Emergency Lighting, 25 watt, battery operated				Wide span 6' wide x 24" deep	S.F.	9.40
Lead battery	Each	282		Sound System		
Nickel cadmium	Each	805		Amplifier, 250 watts	Each	2350
				Speaker, ceiling or wall	Each	191
				Trumpet	Each	365

Important: See the Reference Section for Location Factors

Model costs calculated for a 2 story building with 16' story height and 40,000 square feet of floor area

School, Vocational

				Unit	Unit Cost	Cost Per S.F.	% Of Sub-Total

A. SUBSTRUCTURE

1010	Standard Foundations	Poured concrete; strip and spread footings		S.F. Ground	2.74	1.37	
1020	Special Foundations	N/A		—	—	—	
1030	Slab on Grade	5" reinforced concrete with vapor barrier and granular base		S.F. Slab	11.80	5.91	7.6%
2010	Basement Excavation	Site preparation for slab and trench for foundation wall and footing		S.F. Ground	.17	.09	
2020	Basement Walls	4' foundation wall		L.F. Wall	78	1.45	

B. SHELL

B10 Superstructure

1010	Floor Construction	Open web steel joists, slab form, concrete, beams, columns		S.F. Floor	20.46	10.23	12.1%
1020	Roof Construction	Metal deck, open web steel joists, beams, columns		S.F. Roof	7.50	3.75	

B20 Exterior Enclosure

2010	Exterior Walls	Face brick with concrete block backup	85% of wall	S.F. Wall	30.86	15.53	
2020	Exterior Windows	Tubular aluminum framing with insulated glass	15% of wall	Each	51	4.51	17.8%
2030	Exterior Doors	Metal and glass doors		Each	2530	.51	

B30 Roofing

3010	Roof Coverings	Single-ply membrane with polyisocyanurate insulation		S.F. Roof	5.72	2.86	2.6%
3020	Roof Openings	Roof hatches		S.F. Roof	.28	.14	

C. INTERIORS

1010	Partitions	Concrete block	20 S.F. Floor/L.F. Partition	S.F. Partition	8.43	5.06	
1020	Interior Doors	Single leaf kalamein fire doors	600 S.F. Floor/Door	Each	875	1.46	
1030	Fittings	Toilet partitions, chalkboards		S.F. Floor	1.75	1.75	
2010	Stair Construction	Concrete filled metal pan		Flight	15,800	1.58	21.8%
3010	Wall Finishes	50% paint, 40% glazed coating, 10% ceramic tile		S.F. Surface	4.08	4.90	
3020	Floor Finishes	70% vinyl composition tile, 20% carpet, 10% terrazzo		S.F. Floor	6.56	6.56	
3030	Ceiling Finishes	Mineral fiber tile on concealed zee bars	60% of area	S.F. Ceiling	6.38	3.83	

D. SERVICES

D10 Conveying

1010	Elevators & Lifts	One hydraulic passenger elevator		Each	81,200	2.03	1.8%
1020	Escalators & Moving Walks	N/A		—	—	—	

D20 Plumbing

2010	Plumbing Fixtures	Toilet and service fixtures, supply and drainage	1 Fixture/700 S.F. Floor	Each	3234	4.62	
2020	Domestic Water Distribution	Gas fired water heater		S.F. Floor	.78	.78	5.1%
2040	Rain Water Drainage	Roof drains		S.F. Roof	.96	.48	

D30 HVAC

3010	Energy Supply	Oil fired hot water, wall fin radiation		S.F. Floor	8.23	8.23	
3020	Heat Generating Systems	N/A		—	—	—	
3030	Cooling Generating Systems	Chilled water, cooling tower systems		S.F. Floor	12.55	12.55	18.0%
3050	Terminal & Package Units	N/A		—	—	—	
3090	Other HVAC Sys. & Equipment	N/A		—	—	—	

D40 Fire Protection

4010	Sprinklers	Sprinklers, light hazard		S.F. Floor	1.30	1.30	1.1%
4020	Standpipes	N/A		—	—	—	

D50 Electrical

5010	Electrical Service/Distribution	800 ampere service, panel board and feeders		S.F. Floor	1.60	1.60	
5020	Lighting & Branch Wiring	Fluorescent fixtures, receptacles, switches, A.C. and misc. power		S.F. Floor	9.79	9.79	12.1%
5030	Communications & Security	Alarm systems, internet wiring, communications systems and emergency lighting		S.F. Floor	2.45	2.45	
5090	Other Electrical Systems	Emergency generator, 11.5 kW		S.F. Floor	.11	.11	

E. EQUIPMENT & FURNISHINGS

1010	Commercial Equipment	N/A		—	—	—	
1020	Institutional Equipment	Stainless steel countertops		S.F. Floor	.12	.12	0.1%
1030	Vehicular Equipment	N/A		—	—	—	
1090	Other Equipment	N/A		—	—	—	

F. SPECIAL CONSTRUCTION

1020	Integrated Construction	N/A		—	—	—	0.0%
1040	Special Facilities	N/A		—	—	—	

G. BUILDING SITEWORK   N/A

Sub-Total		115.55	**100%**

CONTRACTOR FEES (General Requirements: 10%, Overhead: 5%, Profit: 10%)	25%	28.89	
ARCHITECT FEES	7%	10.11	

Total Building Cost	**154.55**	

Costs per square foot of floor area

Exterior Wall	S.F. Area	1000	2000	3000	4000	6000	8000	10000	12000	15000
	L.F. Perimeter	126	179	219	253	310	358	400	438	490
Wood Siding	Wood Frame	123.30	110.60	104.80	101.45	97.35	94.95	93.30	92.05	90.75
Face Brick Veneer	Wood Frame	152.25	130.40	120.55	114.75	107.75	103.65	100.80	98.75	96.40
Stucco on Concrete Block	Steel Frame	146.95	129.20	121.20	116.50	110.85	107.50	105.20	103.50	101.65
	Bearing Walls	142.35	124.65	116.60	111.90	106.30	102.95	100.70	98.95	97.10
Metal Sandwich Panel	Steel Frame	161.60	139.85	130.00	124.20	117.25	113.15	110.35	108.25	105.95
Precast Concrete	Steel Frame	187.55	157.85	144.45	136.45	127.00	121.40	117.55	114.70	111.55
Perimeter Adj., Add or Deduct	Per 100 L.F.	34.90	17.40	11.60	8.65	5.85	4.35	3.50	2.90	2.30
Story Hgt. Adj., Add or Deduct	Per 1 Ft.	2.40	1.65	1.35	1.10	0.95	0.80	0.70	0.70	0.60

For Basement, add $24.10 per square foot of basement area

The above costs were calculated using the basic specifications shown on the facing page. These costs should be adjusted where necessary for design alternatives and owner's requirements. Reported completed project costs, for this type of structure, range from $64.35 to $223.05 per S.F.

Common additives

Description	Unit	$ Cost
Check Out Counter		
Single belt	Each	2800
Double belt	Each	4675
Emergency Lighting, 25 watt, battery operated		
Lead battery	Each	282
Nickel cadmium	Each	805
Refrigerators, Prefabricated, walk-in		
7'-6" high, 6' x 6'	S.F.	164
10' x 10'	S.F.	129
12' x 14'	S.F.	114
12' x 20'	S.F.	101
Refrigerated Food Cases		
Dairy, multi deck, 12' long	Each	11,300
Delicatessen case, single deck, 12' long	Each	7625
Multi deck, 18 S.F. shelf display	Each	8525
Freezer, self-contained chest type, 30 C.F.	Each	7650
Glass door upright, 78 C.F.	Each	10,500

Description	Unit	$ Cost
Refrigerated Food Cases, cont.		
Frozen food, chest type, 12' long	Each	7700
Glass door reach-in, 5 door	Each	14,500
Island case 12' long, single deck	Each	8675
Multi deck	Each	18,100
Meat cases, 12' long, single deck	Each	6400
Multi deck	Each	10,700
Produce, 12' long single deck	Each	8350
Multi deck	Each	9125
Safe, Office type, 1 hour rating		
30" x 18" x 18"	Each	2400
60" x 36" x 18", double door	Each	8750
Smoke Detectors		
Ceiling type	Each	187
Duct type	Each	480
Sound System		
Amplifier, 250 watts	Each	2350
Speaker, ceiling or wall	Each	191
Trumpet	Each	365

Important: See the Reference Section for Location Factors

Model costs calculated for a 1 story building with 12' story height and 4,000 square feet of floor area

				Unit	Unit Cost	Cost Per S.F.	% Of Sub-Total
A.	**SUBSTRUCTURE**						
1010	Standard Foundations	Poured concrete; strip and spread footings		S.F. Ground	1.64	1.64	
1020	Special Foundations	N/A		—	—	—	
1030	Slab on Grade	4" reinforced concrete with vapor barrier and granular base		S.F. Slab	4.74	4.74	14.8%
2010	Basement Excavation	Site preparation for slab and trench for foundation wall and footing		S.F. Ground	.46	.46	
2020	Basement Walls	4' foundation wall		L.F. Wall	69	4.36	
B.	**SHELL**						
	B10 Superstructure						
1010	Floor Construction	N/A		—	—	—	
1020	Roof Construction	Wood truss with plywood sheathing		S.F. Roof	6.61	6.61	8.7 %
	B20 Exterior Enclosure						
2010	Exterior Walls	Wood siding on wood studs, insulated	80% of wall	S.F. Wall	8.68	5.27	
2020	Exterior Windows	Storefront	20% of wall	Each	41.50	4.18	15.2%
2030	Exterior Doors	Double aluminum and glass, solid core wood		Each	2776	2.09	
	B30 Roofing						
3010	Roof Coverings	Asphalt shingles; rigid fiberglass insulation		S.F. Roof	3.94	3.94	5.2%
3020	Roof Openings	N/A		—	—	—	
C.	**INTERIORS**						
1010	Partitions	Gypsum board on wood studs	60 S.F. Floor/L.F. Partition	S.F. Partition	12.12	2.02	
1020	Interior Doors	Single leaf wood, hollow metal	1300 S.F. Floor/Door	Each	996	.76	
1030	Fittings	N/A		—	—	—	
2010	Stair Construction	N/A		—	—	—	15.6%
3010	Wall Finishes	Paint		S.F. Surface	1.05	.35	
3020	Floor Finishes	Vinyl composition tile		S.F. Floor	3.24	3.24	
3030	Ceiling Finishes	Mineral fiber tile on wood furring		S.F. Ceiling	5.49	5.49	
D.	**SERVICES**						
	D10 Conveying						
1010	Elevators & Lifts	N/A		—	—	—	0.0 %
1020	Escalators & Moving Walks	N/A		—	—	—	
	D20 Plumbing						
2010	Plumbing Fixtures	Toilet and service fixtures, supply and drainage	1 Fixture/1000 S.F. Floor	Each	2100	2.10	
2020	Domestic Water Distribution	Gas fired water heater		S.F. Floor	1.12	1.12	4.2%
2040	Rain Water Drainage	N/A		—	—	—	
	D30 HVAC						
3010	Energy Supply	N/A		—	—	—	
3020	Heat Generating Systems	Included in D3050		—	—	—	
3030	Cooling Generating Systems	N/A		—	—	—	9.5 %
3050	Terminal & Package Units	Single zone rooftop unit, gas heating, electric cooling		S.F. Floor	7.17	7.17	
3090	Other HVAC Sys. & Equipment	N/A		—	—	—	
	D40 Fire Protection						
4010	Sprinklers	Sprinkler, ordinary hazard		S.F. Floor	3.59	3.59	4.7%
4020	Standpipes	N/A		—	—	—	
	D50 Electrical						
5010	Electrical Service/Distribution	200 ampere service, panel board and feeders		S.F. Floor	2.67	2.67	
5020	Lighting & Branch Wiring	Fluorescent fixtures, receptacles, switches, A.C. and misc. power		S.F. Floor	9.83	9.83	22.0%
5030	Communications & Security	Alarm systems and emergency lighting		S.F. Floor	3.84	3.84	
5090	Other Electrical Systems	Emergency generator, 7.5 kW		S.F. Floor	.34	.34	
E.	**EQUIPMENT & FURNISHINGS**						
1010	Commercial Equipment	N/A		—	—	—	
1020	Institutional Equipment	N/A		—	—	—	
1030	Vehicular Equipment	N/A		—	—	—	0.0 %
1090	Other Equipment	N/A		—	—	—	
F.	**SPECIAL CONSTRUCTION**						
1020	Integrated Construction	N/A		—	—	—	0.0 %
1040	Special Facilities	N/A		—	—	—	
G.	**BUILDING SITEWORK**	**N/A**					

				Sub-Total	75.81	**100%**
CONTRACTOR FEES (General Requirements: 10%, Overhead: 5%, Profit: 10%)				25%	19.01	
ARCHITECT FEES				7%	6.63	

	Total Building Cost	**101.45**

Costs per square foot of floor area

Exterior Wall	S.F. Area	50000	65000	80000	95000	110000	125000	140000	155000	170000
	L.F. Perimeter	920	1065	1167	1303	1333	1433	1533	1633	1733
Face Brick with Concrete Block Back-up	R/Conc. Frame	116.40	113.90	111.80	110.75	109.05	108.35	107.70	107.25	106.80
	Steel Frame	103.15	100.65	98.65	97.50	95.90	95.20	94.50	94.05	93.60
Decorative Concrete Block	R/Conc. Frame	110.70	108.80	107.35	106.50	105.35	104.80	104.30	104.00	103.65
	Steel Joists	98.50	96.50	94.90	94.10	92.80	92.25	91.75	91.35	91.05
Precast Concrete Panels	R/Conc. Frame	113.25	111.10	109.40	108.50	107.20	106.55	106.00	105.65	105.25
	Steel Joists	99.60	97.45	95.70	94.80	93.45	92.85	92.30	91.95	91.60
Perimeter Adj., Add or Deduct	Per 100 L.F.	1.80	1.40	1.15	0.95	0.85	0.70	0.70	0.60	0.55
Story Hgt. Adj., Add or Deduct	Per 1 Ft.	0.95	0.75	0.75	0.70	0.60	0.55	0.55	0.50	0.55
For Basement, add $23.70 per square foot of basement area										

The above costs were calculated using the basic specifications shown on the facing page. These costs should be adjusted where necessary for design alternatives and owner's requirements. Reported completed project costs, for this type of structure, range from $55.50 to $138.15 per S.F.

Common additives

Description	Unit	$ Cost
Closed Circuit Surveillance, One station		
Camera and monitor	Each	1850
For additional camera stations, add	Each	1000
Directory Boards, Plastic, glass covered		
30" x 20"	Each	595
36" x 48"	Each	1450
Aluminum, 24" x 18"	Each	600
36" x 24"	Each	675
48" x 32"	Each	980
48" x 60"	Each	2025
Emergency Lighting, 25 watt, battery operated		
Lead battery	Each	282
Nickel cadmium	Each	805
Safe, Office type, 1 hour rating		
30" x 18" x 18"	Each	2400
60" x 36" x 18", double door	Each	8750
Sound System		
Amplifier, 250 watts	Each	2350
Speaker, ceiling or wall	Each	191
Trumpet	Each	365

Model costs calculated for a 1 story building with 14' story height and 110,000 square feet of floor area

Store, Department, 1 Story

				Unit	Unit Cost	Cost Per S.F.	% Of Sub-Total
A. SUBSTRUCTURE							
1010	Standard Foundations	Poured concrete; strip and spread footings		S.F. Ground	.90	.90	
1020	Special Foundations	N/A		—	—	—	
1030	Slab on Grade	4" reinforced concrete with vapor barrier and granular base		S.F. Slab	4.74	4.74	8.3%
2010	Basement Excavation	Site preparation for slab and trench for foundation wall and footing		S.F. Ground	.25	.25	
2020	Basement Walls	4' foundation wall		L.F. Wall	78	.95	
B. SHELL							
B10 Superstructure							
1010	Floor Construction	Cast-in-place concrete columns		L.F. Column	69	.88	27.3%
1020	Roof Construction	Precast concrete beam and plank, concrete columns		S.F. Roof	21.55	21.55	
B20 Exterior Enclosure							
2010	Exterior Walls	Face brick with concrete block backup	90% of wall	S.F. Wall	32.03	4.89	
2020	Exterior Windows	Storefront	10% of wall	Each	62	1.05	7.7%
2030	Exterior Doors	Sliding electric operated entrance, hollow metal		Each	6845	.37	
B30 Roofing							
3010	Roof Coverings	Built-up tar and gravel with flashing; perlite/EPS composite insulation		S.F. Roof	4.74	4.74	5.8%
3020	Roof Openings	Roof hatches		S.F. Roof	.05	.05	
C. INTERIORS							
1010	Partitions	Gypsum board on metal studs	60 S.F. Floor/L.F. Partition	S.F. Partition	9.54	1.59	
1020	Interior Doors	Single leaf hollow metal	600 S.F. Floor/Door	Each	875	1.46	
1030	Fittings	N/A		—	—	—	
2010	Stair Construction	N/A		—	—	—	24.0%
3010	Wall Finishes	Paint		S.F. Surface	.75	.25	
3020	Floor Finishes	50% ceramic tile, 50% carpet tile		S.F. Floor	12.37	12.37	
3030	Ceiling Finishes	Mineral fiber board on exposed grid system, suspended		S.F. Ceiling	4.10	4.10	
D. SERVICES							
D10 Conveying							
1010	Elevators & Lifts	N/A		—	—	—	0.0 %
1020	Escalators & Moving Walks	N/A		—	—	—	
D20 Plumbing							
2010	Plumbing Fixtures	Toilet and service fixtures, supply and drainage	1 Fixture/4075 S.F. Floor	Each	4727	1.16	
2020	Domestic Water Distribution	Gas fired water heater		S.F. Floor	.29	.29	2.5%
2040	Rain Water Drainage	Roof drains		S.F. Roof	.61	.61	
D30 HVAC							
3010	Energy Supply	N/A		—	—	—	
3020	Heat Generating Systems	Included in D3050		—	—	—	
3030	Cooling Generating Systems	N/A		—	—	—	8.9 %
3050	Terminal & Package Units	Single zone unit gas heating, electric cooling		S.F. Floor	7.35	7.35	
3090	Other HVAC Sys. & Equipment	N/A		—	—	—	
D40 Fire Protection							
4010	Sprinklers	Sprinklers, light hazard		S.F. Floor	2.33	2.33	3.1%
4020	Standpipes	Standpipe		S.F. Floor	.20	.20	
D50 Electrical							
5010	Electrical Service/Distribution	1200 ampere service, panel board and feeders		S.F. Floor	.92	.92	
5020	Lighting & Branch Wiring	High efficiency fluorescent fixtures, receptacles, switches, A.C. and misc. power		S.F. Floor	8.12	8.12	12.4%
5030	Communications & Security	Addressable alarm systems, internet wiring and emergency lighting		S.F. Floor	1.15	1.15	
5090	Other Electrical Systems	Emergency generator, 7.5 kW		S.F. Floor	.04	.04	
E. EQUIPMENT & FURNISHINGS							
1010	Commercial Equipment	N/A		—	—	—	
1020	Institutional Equipment	N/A		—	—	—	
1030	Vehicular Equipment	N/A		—	—	—	0.0 %
1090	Other Equipment	N/A		—	—	—	
F. SPECIAL CONSTRUCTION							
1020	Integrated Construction	N/A		—	—	—	0.0 %
1040	Special Facilities	N/A		—	—	—	
G. BUILDING SITEWORK	**N/A**						

Sub-Total		82.31	**100%**
CONTRACTOR FEES (General Requirements: 10%, Overhead: 5%, Profit: 10%)		25%	20.57
ARCHITECT FEES		6%	6.17
Total Building Cost		**109.05**	

Costs per square foot of floor area

Exterior Wall	S.F. Area	50000	65000	80000	95000	110000	125000	140000	155000	170000
	L.F. Perimeter	533	593	670	715	778	840	871	923	976
Face Brick with Concrete Block Back-up	Steel Frame	150.50	144.45	141.20	138.10	136.25	134.95	133.20	132.25	131.45
	R/Conc. Frame	147.60	141.50	138.30	135.20	133.35	132.00	130.30	129.35	128.55
Face Brick on Steel Studs	Steel Frame	146.00	140.60	137.65	134.95	133.25	132.10	130.65	129.75	129.00
	R/Conc. Frame	145.30	139.90	136.95	134.25	132.60	131.40	129.95	129.05	128.35
Precast Concrete Panels Exposed Aggregate	Steel Frame	150.45	144.45	141.25	138.15	136.35	134.95	133.30	132.30	131.50
	R/Conc. Frame	149.20	143.15	139.95	136.85	135.00	133.70	132.00	131.05	130.25
Perimeter Adj., Add or Deduct	Per 100 L.F.	4.95	3.80	3.15	2.65	2.30	1.90	1.85	1.65	1.45
Story Hgt. Adj., Add or Deduct	Per 1 Ft.	1.55	1.30	1.20	1.10	1.05	0.95	0.95	0.85	0.85

For Basement, add $40.20 per square foot of basement area

The above costs were calculated using the basic specifications shown on the facing page. These costs should be adjusted where necessary for design alternatives and owner's requirements. Reported completed project costs, for this type of structure, range from $57.35 to $153.55 per S.F.

Common additives

Description	Unit	$ Cost
Closed Circuit Surveillance, One station		
Camera and monitor	Each	1850
For additional camera stations, add	Each	1000
Directory Boards, Plastic, glass covered		
30" x 20"	Each	595
36" x 48"	Each	1450
Aluminum, 24" x 18"	Each	600
36" x 24"	Each	675
48" x 32"	Each	980
48" x 60"	Each	2025
Elevators, Hydraulic passenger, 2 stops		
1500# capacity	Each	62,800
2500# capacity	Each	66,300
3500# capacity	Each	69,800
Additional stop, add	Each	7825
Emergency Lighting, 25 watt, battery operated		
Lead battery	Each	282
Nickel cadmium	Each	805

Description	Unit	$ Cost
Escalators, Metal		
32" wide, 10' story height	Each	143,000
20' story height	Each	172,000
48" wide, 10' story height	Each	152,000
20' story height	Each	180,500
Glass		
32" wide, 10' story height	Each	137,000
20' story height	Each	165,000
48" wide, 10' story height	Each	48,300
20' story height	Each	175,000
Safe, Office type, 1 hour rating		
30" x 18" x 18"	Each	2400
60" x 36" x 18", double door	Each	8750
Sound System		
Amplifier, 250 watts	Each	2350
Speaker, ceiling or wall	Each	191
Trumpet	Each	365

Model costs calculated for a 3 story building with 16' story height and 95,000 square feet of floor area

Store, Department, 3 Story

				Unit	Unit Cost	Cost Per S.F.	% Of Sub-Total
A. SUBSTRUCTURE							
1010	Standard Foundations	Poured concrete; strip and spread footings		S.F. Ground	3.66	1.22	
1020	Special Foundations	N/A		—	—	—	
1030	Slab on Grade	4" reinforced concrete with vapor barrier and granular base		S.F. Slab	4.74	1.58	3.3%
2010	Basement Excavation	Site preparation for slab and trench for foundation wall and footing		S.F. Ground	.17	.06	
2020	Basement Walls	4' foundation wall		L.F. Wall	74	.62	
B. SHELL							
	B10 Superstructure						
1010	Floor Construction	Concrete slab with metal deck and beams, steel columns		S.F. Floor	31.95	21.30	24.3%
1020	Roof Construction	Metal deck, open web steel joists, beams, columns		S.F. Roof	11.94	3.98	
	B20 Exterior Enclosure						
2010	Exterior Walls	Face brick with concrete block backup	90% of wall	S.F. Wall	31.99	10.40	
2020	Exterior Windows	Storefront	10% of wall	Each	43.70	1.58	13.0%
2030	Exterior Doors	Revolving and sliding panel, mall-front		Each	10,787	1.59	
	B30 Roofing						
3010	Roof Coverings	Built-up tar and gravel with flashing; perlite/EPS composite insulation		S.F. Roof	5.16	1.72	1.7%
3020	Roof Openings	Roof hatches		S.F. Roof	.18	.06	
C. INTERIORS							
1010	Partitions	Gypsum board on metal studs	60 S.F. Floor/L.F. Partition	S.F. Partition	5.82	.97	
1020	Interior Doors	Single leaf hollow metal	600 S.F. Floor/Door	Each	875	1.46	
1030	Fittings	N/A		—	—	—	
2010	Stair Construction	Concrete filled metal pan		Flight	22,125	2.33	24.3%
3010	Wall Finishes	70% paint, 20% vinyl wall covering, 10% ceramic tile		S.F. Surface	4.29	1.43	
3020	Floor Finishes	50% carpet tile, 40% marble tile, 10% terrazzo		S.F. Floor	12.77	12.77	
3030	Ceiling Finishes	Mineral fiber tile on concealed zee bars		S.F. Ceiling	6.38	6.38	
D. SERVICES							
	D10 Conveying						
1010	Elevators & Lifts	One hydraulic passenger, one hydraulic freight		Each	323,950	3.41	9.3%
1020	Escalators & Moving Walks	Four escalators		Each	149,150	6.28	
	D20 Plumbing						
2010	Plumbing Fixtures	Toilet and service fixtures, supply and drainage	1 Fixture/2570 S.F. Floor	Each	3290	1.28	
2020	Domestic Water Distribution	Gas fired water heater		S.F. Floor	.33	.33	1.9%
2040	Rain Water Drainage	Roof drains		S.F. Roof	1.23	.41	
	D30 HVAC						
3010	Energy Supply	N/A		—	—	—	
3020	Heat Generating Systems	Included in D3050		—	—	—	
3030	Cooling Generating Systems	N/A		—	—	—	7.1%
3050	Terminal & Package Units	Multizone rooftop unit, gas heating, electric cooling		S.F. Floor	7.35	7.35	
3090	Other HVAC Sys. & Equipment	N/A		—	—	—	
	D40 Fire Protection						
4010	Sprinklers	Sprinklers, light hazard		S.F. Floor	1.96	1.96	3.1%
4020	Standpipes	Standpipes		S.F. Floor	1.28	1.28	
	D50 Electrical						
5010	Electrical Service/Distribution	1200 ampere service, panel board and feeders		S.F. Floor	1.46	1.46	
5020	Lighting & Branch Wiring	High efficiency fluorescent fixtures, receptacles, switches, A.C. and misc. power		S.F. Floor	8.49	8.49	12.0%
5030	Communications & Security	Addressable alarm systems, internet wiring and emergency lighting		S.F. Floor	2.21	2.21	
5090	Other Electrical Systems	Emergency generator, 50 kW		S.F. Floor	.32	.32	
E. EQUIPMENT & FURNISHINGS							
1010	Commercial Equipment	N/A		—	—	—	
1020	Institutional Equipment	N/A		—	—	—	0.0%
1030	Vehicular Equipment	N/A		—	—	—	
1090	Other Equipment	N/A		—	—	—	
F. SPECIAL CONSTRUCTION							
1020	Integrated Construction	N/A		—	—	—	0.0%
1040	Special Facilities	N/A		—	—	—	
G. BUILDING SITEWORK	**N/A**						

		Sub-Total	104.23	100%
CONTRACTOR FEES (General Requirements: 10%, Overhead: 5%, Profit: 10%)		25%	26.05	
ARCHITECT FEES		6%	7.82	
	Total Building Cost		**138.10**	

Costs per square foot of floor area

Exterior Wall	S.F. Area	4000	6000	8000	10000	12000	15000	18000	20000	22000
	L.F. Perimeter	260	340	360	410	440	490	540	565	594
Split Face Concrete Block	Steel Joists	135.20	122.95	112.15	107.55	103.45	99.50	97.00	95.40	94.25
Stucco on Concrete Block	Steel Joists	131.35	119.60	109.45	105.10	101.25	97.50	95.15	93.65	92.60
Painted Concrete Block	Steel Joists	125.65	114.20	104.50	100.25	96.55	92.95	90.70	89.30	88.20
Face Brick on Concrete Block	Steel Joists	152.00	137.60	123.75	118.15	112.90	107.90	104.70	102.65	101.10
Painted Reinforced Concrete	Steel Joists	142.00	128.85	116.85	111.85	107.25	102.90	100.05	98.30	97.00
Tilt-up Concrete Panels	Steel Joists	134.15	122.05	111.45	106.90	102.85	99.00	96.45	94.95	93.70
Perimeter Adj., Add or Deduct	Per 100 L.F.	15.40	10.25	7.75	6.15	5.20	4.20	3.40	3.05	2.85
Story Hgt. Adj., Add or Deduct	Per 1 Ft.	1.90	1.70	1.30	1.20	1.10	1.00	0.85	0.85	0.80
For Basement, add $33.35 per square foot of basement area										

The above costs were calculated using the basic specifications shown on the facing page. These costs should be adjusted where necessary for design alternatives and owner's requirements. Reported completed project costs, for this type of structure, range from $56.00 to $194.65 per S.F.

Common additives

Description	Unit	$ Cost
Emergency Lighting, 25 watt, battery operated		
Lead battery	Each	282
Nickel cadmium	Each	805
Safe, Office type, 1 hour rating		
30" x 18" x 18"	Each	2400
60" x 36" x 18", double door	Each	8750
Smoke Detectors		
Ceiling type	Each	187
Duct type	Each	480
Sound System		
Amplifier, 250 watts	Each	2350
Speaker, ceiling or wall	Each	191
Trumpet	Each	365

Model costs calculated for a 1 story building with 14' story height and 8,000 square feet of floor area

				Unit	Unit Cost	Cost Per S.F.	% Of Sub-Total
A. SUBSTRUCTURE							
1010	Standard Foundations	Poured concrete; strip and spread footings		S.F. Ground	1.91	1.91	
1020	Special Foundations	N/A		—	—	—	
1030	Slab on Grade	4" reinforced concrete with vapor barrier and granular base		S.F. Slab	4.74	4.74	12.5%
2010	Basement Excavation	Site preparation for slab and trench for foundation wall and footing		S.F. Ground	.26	.26	
2020	Basement Walls	4' foundation wall		L.F. Wall	78	3.51	
B. SHELL							
B10 Superstructure							
1010	Floor Construction	N/A		—	—	—	10.4 %
1020	Roof Construction	Metal deck, open web steel joists, beams, interior columns		S.F. Roof	8.61	8.61	
B20 Exterior Enclosure							
2010	Exterior Walls	Decorative concrete block	90% of wall	S.F. Wall	16.70	9.47	
2020	Exterior Windows	Storefront windows	10% of wall	Each	43.70	2.75	15.3%
2030	Exterior Doors	Sliding entrance door and hollow metal service doors		Each	1983	.49	
B30 Roofing							
3010	Roof Coverings	Built-up tar and gravel with flashing; perlite/EPS composite insulation		S.F. Roof	5.84	5.84	7.1%
3020	Roof Openings	Roof hatches		S.F. Roof	.10	.10	
C. INTERIORS							
1010	Partitions	Gypsum board on metal studs	60 S.F. Floor/L.F. Partition	S.F. Partition	5.82	.97	
1020	Interior Doors	Single leaf hollow metal	600 S.F. Floor/Door	Each	875	1.46	
1030	Fittings	N/A		—	—	—	
2010	Stair Construction	N/A		—	—	—	16.7%
3010	Wall Finishes	Paint		S.F. Surface	5.52	1.84	
3020	Floor Finishes	Vinyl tile		S.F. Floor	3.24	3.24	
3030	Ceiling Finishes	Mineral fiber tile on concealed zee bars		S.F. Ceiling	6.38	6.38	
D. SERVICES							
D10 Conveying							
1010	Elevators & Lifts	N/A		—	—	—	0.0 %
1020	Escalators & Moving Walks	N/A		—	—	—	
D20 Plumbing							
2010	Plumbing Fixtures	Toilet and service fixtures, supply and drainage	1 Fixture/890 S.F. Floor	Each	2421	2.72	
2020	Domestic Water Distribution	Gas fired water heater		S.F. Floor	3.90	3.90	9.3%
2040	Rain Water Drainage	Roof drains		S.F. Roof	1.12	1.12	
D30 HVAC							
3010	Energy Supply	N/A		—	—	—	
3020	Heat Generating Systems	Included in D3050		—	—	—	
3030	Cooling Generating Systems	N/A		—	—	—	8.8 %
3050	Terminal & Package Units	Single zone unit, gas heating, electric cooling		S.F. Floor	7.35	7.35	
3090	Other HVAC Sys. & Equipment	N/A		—	—	—	
D40 Fire Protection							
4010	Sprinklers	Wet pipe sprinkler system		S.F. Floor	3.62	3.62	4.4%
4020	Standpipes	N/A		—	—	—	
D50 Electrical							
5010	Electrical Service/Distribution	400 ampere service, panel board and feeders		S.F. Floor	2.53	2.53	
5020	Lighting & Branch Wiring	Fluorescent fixtures, receptacles, switches, A.C. and misc. power		S.F. Floor	9.67	9.67	15.4%
5030	Communications & Security	Alarm systems and emergency lighting		S.F. Floor	.40	.40	
5090	Other Electrical Systems	Emergency generator, 15 kW		S.F. Floor	.21	.21	
E. EQUIPMENT & FURNISHINGS							
1010	Commercial Equipment	N/A		—	—	—	
1020	Institutional Equipment	N/A		—	—	—	
1030	Vehicular Equipment	N/A		—	—	—	0.0 %
1090	Other Equipment	N/A		—	—	—	
F. SPECIAL CONSTRUCTION							
1020	Integrated Construction	N/A		—	—	—	0.0 %
1040	Special Facilities	N/A		—	—	—	
G. BUILDING SITEWORK	**N/A**						

			Sub-Total	83.09	100%
CONTRACTOR FEES (General Requirements: 10%, Overhead: 5%, Profit: 10%)			25%	20.75	
ARCHITECT FEES			8%	8.31	

Total Building Cost	**112.15**

Costs per square foot of floor area

Exterior Wall	S.F. Area	12000	16000	20000	26000	32000	38000	44000	52000	60000
	L.F. Perimeter	450	510	570	650	716	780	840	920	1020
Face Brick with Concrete Block Back-up	Steel Frame	139.60	128.85	122.40	116.05	111.50	108.35	106.00	103.65	102.35
	Bearing Walls	135.60	124.80	118.40	111.95	107.45	104.30	101.95	99.60	98.30
Stucco on Concrete Block	Steel Frame	128.55	119.45	113.95	108.60	104.95	102.30	100.35	98.40	97.35
	Bearing Walls	129.40	120.35	114.85	109.50	105.75	103.15	101.20	99.30	98.20
Precast Concrete Panels	Steel Frame	141.65	130.55	123.90	117.35	112.70	109.45	107.00	104.60	103.25
Metal Sandwich Panels	Steel Frame	121.45	113.45	108.60	103.90	100.70	98.40	96.75	95.05	94.10
Perimeter Adj., Add or Deduct	Per 100 L.F.	9.50	7.10	5.70	4.35	3.60	3.00	2.60	2.20	1.90
Story Hgt. Adj., Add or Deduct	Per 1 Ft.	1.95	1.65	1.45	1.25	1.20	1.10	0.95	0.95	0.85
For Basement, add $24.25 per square foot of basement area										

The above costs were calculated using the basic specifications shown on the facing page. These costs should be adjusted where necessary for design alternatives and owner's requirements. Reported completed project costs, for this type of structure, range from $65.40 to $187.30 per S.F.

Common additives

Description	Unit	$ Cost
Check Out Counter		
Single belt	Each	2800
Double belt	Each	4675
Scanner, registers, guns & memory 2 lanes	Each	16,700
10 lanes	Each	158,500
Power take away	Each	5800
Emergency Lighting, 25 watt, battery operated		
Lead battery	Each	282
Nickel cadmium	Each	805
Refrigerators, Prefabricated, walk-in		
7'-6" High, 6' x 6'	S.F.	164
10' x 10'	S.F.	129
12' x 14'	S.F.	114
12' x 20'	S.F.	101
Refrigerated Food Cases		
Dairy, multi deck, 12' long	Each	11,300
Delicatessen case, single deck, 12' long	Each	7625
Multi deck, 18 S.F. shelf display	Each	8525
Freezer, self-contained chest type, 30 C.F.	Each	7650
Glass door upright, 78 C.F.	Each	10,500

Description	Unit	$ Cost
Refrigerated Food Cases, cont.		
Frozen food, chest type, 12' long	Each	7700
Glass door reach in, 5 door	Each	14,500
Island case 12' long, single deck	Each	8675
Multi deck	Each	18,100
Meat case 12' long, single deck	Each	6400
Multi deck	Each	10,700
Produce, 12' long single deck	Each	8350
Multi deck	Each	9125
Safe, Office type, 1 hour rating		
30" x 18" x 18"	Each	2400
60" x 36" x 18", double door	Each	8750
Smoke Detectors		
Ceiling type	Each	187
Duct type	Each	480
Sound System		
Amplifier, 250 watts	Each	2350
Speaker, ceiling or wall	Each	191
Trumpet	Each	365

Important: See the Reference Section for Location Factors

Model costs calculated for a 1 story building with 18' story height and 44,000 square feet of floor area

Supermarket

				Unit	Unit Cost	Cost Per S.F.	% Of Sub-Total
A. SUBSTRUCTURE							
1010	Standard Foundations	Poured concrete; strip and spread footings		S.F. Ground	1.12	1.12	
1020	Special Foundations	N/A		—	—	—	
1030	Slab on Grade	4" reinforced concrete with vapor barrier and granular base		S.F. Slab	4.74	4.74	9.9%
2010	Basement Excavation	Site preparation for slab and trench for foundation wall and footing		S.F. Ground	.17	.17	
2020	Basement Walls	4' foundation wall		L.F. Wall	78	1.49	
B. SHELL							
B10 Superstructure							
1010	Floor Construction	N/A		—	—	—	12.2 %
1020	Roof Construction	Metal deck, open web steel joists, beams, interior columns		S.F. Roof	9.28	9.28	
B20 Exterior Enclosure							
2010	Exterior Walls	Face brick with concrete block backup	85% of wall	S.F. Wall	30.85	9.01	
2020	Exterior Windows	Storefront windows	15% of wall	Each	69	3.58	18.1%
2030	Exterior Doors	Sliding entrance doors with electrical operator, hollow metal		Each	4839	1.21	
B30 Roofing							
3010	Roof Coverings	Built-up tar and gravel with flashing; perlite/EPS composite insulation		S.F. Roof	4.98	4.98	6.7%
3020	Roof Openings	Roof hatches		S.F. Roof	.10	.10	
C. INTERIORS							
1010	Partitions	50% concrete block, 50% gypsum board on metal studs	50 S.F. Floor/L.F. Partition	S.F. Partition	7.08	1.70	
1020	Interior Doors	Single leaf hollow metal	2000 S.F. Floor/Door	Each	875	.44	
1030	Fittings	N/A		—	—	—	
2010	Stair Construction	N/A		—	—	—	18.2%
3010	Wall Finishes	Paint		S.F. Surface	4.46	2.14	
3020	Floor Finishes	Vinyl composition tile		S.F. Floor	3.24	3.24	
3030	Ceiling Finishes	Mineral fiber tile on concealed zee bars		S.F. Ceiling	6.38	6.38	
D. SERVICES							
D10 Conveying							
1010	Elevators & Lifts	N/A		—	—	—	0.0 %
1020	Escalators & Moving Walks	N/A		—	—	—	
D20 Plumbing							
2010	Plumbing Fixtures	Toilet and service fixtures, supply and drainage	1 Fixture/1820 S.F. Floor	Each	1638	.90	
2020	Domestic Water Distribution	Gas fired water heater		S.F. Floor	.31	.31	2.6%
2040	Rain Water Drainage	Roof drains		S.F. Roof	.75	.75	
D30 HVAC							
3010	Energy Supply	N/A		—	—	—	
3020	Heat Generating Systems	Included in D3050		—	—	—	
3030	Cooling Generating Systems	N/A		—	—	—	6.2 %
3050	Terminal & Package Units	Single zone rooftop unit, gas heating, electric cooling		S.F. Floor	4.73	4.73	
3090	Other HVAC Sys. & Equipment	N/A		—	—	—	
D40 Fire Protection							
4010	Sprinklers	Sprinklers, light hazard		S.F. Floor	2.77	2.77	4.2%
4020	Standpipes	Standpipe		S.F. Floor	.42	.42	
D50 Electrical							
5010	Electrical Service/Distribution	1600 ampere service, panel board and feeders		S.F. Floor	3.43	3.43	
5020	Lighting & Branch Wiring	High efficiency fluorescent fixtures, receptacles, switches, A.C. and misc. power		S.F. Floor	11.34	11.34	22.0%
5030	Communications & Security	Addressable alarm systems, partial internet wiring and emergency lighting		S.F. Floor	1.95	1.95	
5090	Other Electrical Systems	Emergency generator, 15 kW		S.F. Floor	.04	.04	
E. EQUIPMENT & FURNISHINGS							
1010	Commercial Equipment	N/A		—	—	—	
1020	Institutional Equipment	N/A		—	—	—	0.0 %
1030	Vehicular Equipment	N/A		—	—	—	
1090	Other Equipment	N/A		—	—	—	
F. SPECIAL CONSTRUCTION							
1020	Integrated Construction	N/A		—	—	—	0.0 %
1040	Special Facilities	N/A		—	—	—	
G. BUILDING SITEWORK	**N/A**						

		Sub-Total	76.22	100%
CONTRACTOR FEES (General Requirements: 10%, Overhead: 5%, Profit: 10%)		25%	19.06	
ARCHITECT FEES		7%	6.67	
	Total Building Cost		**101.95**	

Costs per square foot of floor area

Exterior Wall	S.F. Area	10000	16000	20000	22000	24000	26000	28000	30000	32000
	L.F. Perimeter	420	510	600	640	660	684	706	740	737
Face Brick with Concrete Block Back-up	Wood Truss	234.05	213.15	208.70	206.65	203.65	201.25	199.10	198.00	195.00
	Precast Conc.	271.15	246.20	241.10	238.70	235.05	232.20	229.65	228.30	224.65
Metal Sandwich Panel	Wood Truss	229.90	214.10	210.65	209.05	206.80	205.05	203.45	202.65	200.45
Precast Concrete Panel	Precast Conc.	241.00	223.20	219.40	217.65	215.10	213.05	211.30	210.40	207.85
Painted Concrete Block	Wood Frame	228.90	213.75	210.35	208.85	206.70	205.00	203.50	202.70	200.60
	Precast Conc.	247.20	228.10	224.05	222.15	219.40	217.25	215.30	214.25	211.55
Perimeter Adj., Add or Deduct	Per 100 L.F.	16.70	10.40	8.35	7.60	6.95	6.45	6.00	5.50	5.25
Story Hgt. Adj., Add or Deduct	Per 1 Ft.	2.10	1.50	1.50	1.45	1.35	1.35	1.25	1.20	1.15
For Basement, add $35.95 per square foot of basement area										

The above costs were calculated using the basic specifications shown on the facing page. These costs should be adjusted where necessary for design alternatives and owner's requirements. Reported completed project costs, for this type of structure, range from $103.30 to $303.70 per S.F.

Common additives

Description	Unit	$ Cost
Bleachers, Telescoping, manual		
To 15 tier	Seat	115 - 160
16-20 tier	Seat	235 - 288
21-30 tier	Seat	249 - 300
For power operation, add	Seat	45.50 - 71.50
Emergency Lighting, 25 watt, battery operated		
Lead battery	Each	282
Nickel cadmium	Each	805
Lockers, Steel, single tier, 60″ or 72″	Opening	191 - 310
2 tier, 60″ or 72″ total	Opening	107 - 141
5 tier, box lockers	Opening	65 - 83.50
Locker bench, lam. maple top only	L.F.	21
Pedestal, steel pipe	Each	63.50
Pool Equipment		
Diving stand, 3 meter	Each	12,800
1 meter	Each	6400
Diving board, 16′ aluminum	Each	3750
Fiberglass	Each	3050
Lifeguard chair, fixed	Each	2875
Portable	Each	2800
Lights, underwater, 12 volt, 300 watt	Each	830

Description	Unit	$ Cost
Sauna, Prefabricated, complete		
6′ x 4′	Each	5850
6′ x 6′	Each	6950
6′ x 9′	Each	8525
8′ x 8′	Each	10,100
8′ x 10′	Each	11,300
10′ x 12′	Each	14,000
Sound System		
Amplifier, 250 watts	Each	2350
Speaker, ceiling or wall	Each	191
Trumpet	Each	365
Steam Bath, Complete, to 140 C.F.	Each	2050
To 300 C.F.	Each	2300
To 800 C.F.	Each	6375
To 2500 C.F.	Each	9550

Model costs calculated for a 1 story building with 24' story height and 20,000 square feet of floor area

Swimming Pool, Enclosed

					Unit	Unit Cost	Cost Per S.F.	% Of Sub-Total
A.	**SUBSTRUCTURE**							
1010	Standard Foundations	Poured concrete; strip and spread footings			S.F. Ground	1.61	1.61	
1020	Special Foundations	N/A			—	—	—	
1030	Slab on Grade	4" reinforced concrete with vapor barrier and granular base			S.F. Slab	4.74	4.74	15.9%
2010	Basement Excavation	Site preparation for slab and trench for foundation wall and footing			S.F. Ground	8.80	8.80	
2020	Basement Walls	4' foundation wall			L.F. Wall	78	9.39	
B.	**SHELL**							
	B10 Superstructure							
1010	Floor Construction	N/A			—	—	—	10.8 %
1020	Roof Construction	Wood deck on laminated wood truss			S.F. Roof	16.63	16.63	
	B20 Exterior Enclosure							
2010	Exterior Walls	Face brick with concrete block backup		80% of wall	S.F. Wall	32	18.43	
2020	Exterior Windows	Outward projecting steel		20% of wall	Each	43.70	6.30	16.4%
2030	Exterior Doors	Double aluminum and glass, hollow metal			Each	2234	.56	
	B30 Roofing							
3010	Roof Coverings	Asphalt shingles with flashing; perlite/EPS composite insulation			S.F. Roof	1.90	1.90	1.2%
3020	Roof Openings	N/A			—	—	—	
C.	**INTERIORS**							
1010	Partitions	Concrete block	100 S.F. Floor/L.F. Partition		S.F. Partition	8.40	.84	
1020	Interior Doors	Single leaf hollow metal	1000 S.F. Floor/Door		Each	875	.88	
1030	Fittings	Toilet partitions			S.F. Floor	.18	.18	
2010	Stair Construction	N/A			—	—	—	15.8%
3010	Wall Finishes	Acrylic glazed coating			S.F. Surface	10.10	2.02	
3020	Floor Finishes	70% terrazzo, 30% ceramic tile			S.F. Floor	19.63	19.63	
3030	Ceiling Finishes	Mineral fiber tile on concealed zee bars		20% of area	S.F. Ceiling	4.10	.82	
D.	**SERVICES**							
	D10 Conveying							
1010	Elevators & Lifts	N/A			—	—	—	0.0 %
1020	Escalators & Moving Walks	N/A			—	—	—	
	D20 Plumbing							
2010	Plumbing Fixtures	Toilet and service fixtures, supply and drainage	1 Fixture/210 S.F. Floor		Each	2266	10.79	
2020	Domestic Water Distribution	Gas fired water heater			S.F. Floor	2.90	2.90	8.9%
2040	Rain Water Drainage	N/A			—	—	—	
	D30 HVAC							
3010	Energy Supply	Terminal unit heaters	10% of area		S.F. Floor	.74	.74	
3020	Heat Generating Systems	N/A			—	—	—	
3030	Cooling Generating Systems	N/A			—	—	—	8.5%
3050	Terminal & Package Units	Single zone unit gas heating, electric cooling			S.F. Floor	12.47	12.47	
3090	Other HVAC Sys. & Equipment	N/A			—	—	—	
	D40 Fire Protection							
4010	Sprinklers	Sprinklers, light hazard			S.F. Floor	1.39	1.39	0.9%
4020	Standpipes	N/A			—	—	—	
	D50 Electrical							
5010	Electrical Service/Distribution	400 ampere service, panel board and feeders			S.F. Floor	1.07	1.07	
5020	Lighting & Branch Wiring	Fluorescent fixtures, receptacles, switches, A.C. and misc. power			S.F. Floor	8.75	8.75	6.6%
5030	Communications & Security	Alarm systems and emergency lighting			S.F. Floor	.32	.32	
5090	Other Electrical Systems	Emergency generator, 15 kW			S.F. Floor	.08	.08	
E.	**EQUIPMENT & FURNISHINGS**							
1010	Commercial Equipment	N/A			—	—	—	
1020	Institutional Equipment	N/A			—	—	—	15.1 %
1030	Vehicular Equipment	N/A			—	—	—	
1090	Other Equipment	Swimming pool			S.F. Floor	23.36	23.36	
F.	**SPECIAL CONSTRUCTION**							
1020	Integrated Construction	N/A			—	—	—	0.0 %
1040	Special Facilities	N/A			—	—	—	
G.	**BUILDING SITEWORK**	**N/A**						

		Sub-Total	154.60	100%
CONTRACTOR FEES (General Requirements: 10%, Overhead: 5%, Profit: 10%)			25%	38.64
ARCHITECT FEES			8%	15.46
		Total Building Cost	**208.70**	

Costs per square foot of floor area

Exterior Wall	S.F. Area	2000	3000	4000	5000	6000	7000	8000	9000	10000
	L.F. Perimeter	180	220	260	286	320	353	368	397	425
Face Brick with Concrete Block Back-up	Steel Frame	200.20	178.95	168.35	159.35	154.55	151.05	146.35	144.15	142.20
	Bearing Walls	195.15	173.90	163.30	154.30	149.55	146.05	141.30	139.10	137.20
Limestone with Concrete Block Back-up	Steel Frame	242.20	213.20	198.70	186.05	179.45	174.60	167.80	164.75	162.05
	Bearing Walls	237.20	208.20	193.70	181.00	174.45	169.55	162.75	159.75	157.05
Decorative Concrete Block	Steel Frame	185.65	167.05	157.85	150.10	145.95	142.95	138.90	137.05	135.35
	Bearing Walls	180.65	162.05	152.75	145.05	140.90	137.90	133.80	131.95	130.30
Perimeter Adj., Add or Deduct	Per 100 L.F.	46.25	30.85	23.10	18.45	15.50	13.20	11.55	10.30	9.30
Story Hgt. Adj., Add or Deduct	Per 1 Ft.	5.40	4.35	3.85	3.40	3.20	3.00	2.70	2.65	2.55

For Basement, add $34.85 per square foot of basement area

The above costs were calculated using the basic specifications shown on the facing page. These costs should be adjusted where necessary for design alternatives and owner's requirements. Reported completed project costs, for this type of structure, range from $91.30 to $329.45 per S.F.

Common additives

Description	Unit	$ Cost
Emergency Lighting, 25 watt, battery operated		
Lead battery	Each	282
Nickel cadmium	Each	805
Smoke Detectors		
Ceiling type	Each	187
Duct type	Each	480
Emergency Generators, complete system, gas		
15 kw	Each	16,500
85 kw	Each	34,900
170 kw	Each	88,000
Diesel, 50 kw	Each	30,200
150 kw	Each	52,000
350 kw	Each	82,500

Model costs calculated for a 1 story building with 12' story height and 5,000 square feet of floor area

				Unit	Unit Cost	Cost Per S.F.	% Of Sub-Total
A. SUBSTRUCTURE							
1010	Standard Foundations	Poured concrete; strip and spread footings		S.F. Ground	2.76	2.76	
1020	Special Foundations	N/A		—	—	—	
1030	Slab on Grade	4" reinforced concrete with vapor barrier and granular base		S.F. Slab	4.74	4.74	10.8%
2010	Basement Excavation	Site preparation for slab and trench for foundation wall and footing		S.F. Ground	.46	.46	
2020	Basement Walls	4' foundation wall		L.F. Wall	78	4.46	
B. SHELL							
B10 Superstructure							
1010	Floor Construction	N/A		—	—	—	10.1 %
1020	Roof Construction	Metal deck, open web steel joists, beams, columns		S.F. Roof	11.64	11.64	
B20 Exterior Enclosure							
2010	Exterior Walls	Face brick with concrete block backup	80% of wall	S.F. Wall	30.85	16.94	
2020	Exterior Windows	Outward projecting steel	20% of wall	Each	1605	11.02	25.5%
2030	Exterior Doors	Single aluminum glass with transom		Each	3378	1.36	
B30 Roofing							
3010	Roof Coverings	Built-up tar and gravel with flashing; perlite/EPS composite insulation		S.F. Roof	6.24	6.24	5.4%
3020	Roof Openings						
C. INTERIORS							
1010	Partitions	Double layer gypsum board on metal studs	15 S.F. Floor/L.F. Partition	S.F. Partition	5.81	3.87	
1020	Interior Doors	Single leaf hollow metal	150 S.F. Floor/Door	Each	875	5.83	
1030	Fittings	Toilet partitions		S.F. Floor	1.76	1.76	
2010	Stair Construction	N/A		—	—	—	23.9%
3010	Wall Finishes	Paint		S.F. Surface	1.88	2.51	
3020	Floor Finishes	90% carpet, 10% terrazzo		S.F. Floor	9.20	9.20	
3030	Ceiling Finishes	Fiberglass board on exposed grid system, suspended		S.F. Ceiling	4.30	4.30	
D. SERVICES							
D10 Conveying							
1010	Elevators & Lifts	N/A		—	—	—	0.0 %
1020	Escalators & Moving Walks	N/A		—	—	—	
D20 Plumbing							
2010	Plumbing Fixtures	Kitchen, toilet and service fixtures, supply and drainage	1 Fixture/715 S.F. Floor	Each	2567	3.59	
2020	Domestic Water Distribution	Gas fired water heater		S.F. Floor	1.43	1.43	6.1%
2040	Rain Water Drainage	Roof drains		S.F. Roof	1.95	1.95	
D30 HVAC							
3010	Energy Supply	N/A		—	—	—	
3020	Heat Generating Systems	Included in D3050		—	—	—	
3030	Cooling Generating Systems	N/A		—	—	—	7.0 %
3050	Terminal & Package Units	Single zone unit, gas heating, electric cooling		S.F. Floor	8	8	
3090	Other HVAC Sys. & Equipment	N/A		—	—	—	
D40 Fire Protection							
4010	Sprinklers	Wet pipe sprinkler system		S.F. Floor	3.92	3.92	3.4%
4020	Standpipes	N/A		—	—	—	
D50 Electrical							
5010	Electrical Service/Distribution	200 ampere service, panel board and feeders		S.F. Floor	1.68	1.68	
5020	Lighting & Branch Wiring	Fluorescent fixtures, receptacles, switches, A.C. and misc. power		S.F. Floor	5.73	5.73	7.7%
5030	Communications & Security	Alarm systems and emergency lighting		S.F. Floor	1.29	1.29	
5090	Other Electrical Systems	Emergency generator, 15 kW		S.F. Floor	.16	.16	
E. EQUIPMENT & FURNISHINGS							
1010	Commercial Equipment	N/A		—	—	—	
1020	Institutional Equipment	N/A		—	—	—	
1030	Vehicular Equipment	N/A		—	—	—	0.0 %
1090	Other Equipment	N/A		—	—	—	
F. SPECIAL CONSTRUCTION							
1020	Integrated Construction	N/A		—	—	—	0.0 %
1040	Special Facilities	N/A		—	—	—	
G. BUILDING SITEWORK	**N/A**						

			Sub-Total	114.84	100%
CONTRACTOR FEES (General Requirements: 10%, Overhead: 5%, Profit: 10%)			25%	28.72	
ARCHITECT FEES			11%	15.79	

Total Building Cost	**159.35**

Costs per square foot of floor area

Exterior Wall	S.F. Area	5000	6500	8000	9500	11000	14000	17500	21000	24000
	L.F. Perimeter	300	360	386	396	435	510	550	620	680
Face Brick with Concrete Block Back-up	Steel Joists	151.55	145.40	138.45	132.30	129.90	126.45	121.85	119.85	118.65
	Wood Joists	149.75	143.55	136.30	130.05	127.55	124.00	119.25	117.25	116.00
Stone with Concrete Block Back-up	Steel Joists	154.20	147.90	140.60	134.20	131.65	128.05	123.25	121.15	119.90
	Wood Joists	152.40	145.95	138.45	131.90	129.30	125.65	120.65	118.50	117.25
Brick Veneer	Wood Frame	141.10	135.35	128.85	123.25	121.05	117.80	113.60	111.70	110.65
E.I.F.S.	Wood Frame	129.35	124.30	118.95	114.40	112.50	109.65	106.20	104.65	103.70
Perimeter Adj., Add or Deduct	Per 100 L.F.	15.20	11.65	9.50	8.05	6.90	5.40	4.35	3.65	3.15
Story Hgt. Adj., Add or Deduct	Per 1 Ft.	2.80	2.55	2.20	1.90	1.80	1.65	1.50	1.35	1.30
For Basement, add $28.70 per square foot of basement area										

The above costs were calculated using the basic specifications shown on the facing page. These costs should be adjusted where necessary for design alternatives and owner's requirements. Reported completed project costs, for this type of structure, range from $86.30 to $227.00 per S.F.

Common additives

Description	Unit	$ Cost
Directory Boards, Plastic, glass covered		
30" x 20"	Each	595
36" x 48"	Each	1450
Aluminum, 24" x 18"	Each	600
36" x 24"	Each	675
48" x 32"	Each	980
48" x 60"	Each	2025
Emergency Lighting, 25 watt, battery operated		
Lead battery	Each	282
Nickel cadmium	Each	805
Flagpoles, Complete		
Aluminum, 20' high	Each	1650
40' high	Each	3475
70' high	Each	10,100
Fiberglass, 23' high	Each	1775
39'-5" high	Each	3325
59' high	Each	8225
Safe, Office type, 1 hour rating		
30" x 18" x 18"	Each	2400
60" x 36" x 18", double door	Each	8750

Description	Unit	$ Cost
Smoke Detectors		
Ceiling type	Each	187
Duct type	Each	480
Vault Front, Door & frame		
1 Hour test, 32" x 78"	Opening	5175
2 Hour test, 32" door	Opening	5600
40" door	Opening	6400
4 Hour test, 32" door	Opening	6075
40" door	Opening	7125
Time lock movement; two movement	Each	1850

Model costs calculated for a 1 story building with 12' story height and 11,000 square feet of floor area

				Unit	Unit Cost	Cost Per S.F.	% Of Sub-Total
A. SUBSTRUCTURE							
1010	Standard Foundations	Poured concrete; strip and spread footings		S.F. Ground	1.71	1.71	
1020	Special Foundations	N/A		—	—	—	
1030	Slab on Grade	4" reinforced concrete with vapor barrier and granular base		S.F. Slab	4.74	4.74	10.3%
2010	Basement Excavation	Site preparation for slab and trench for foundation wall and footing		S.F. Ground	.26	.26	
2020	Basement Walls	4' foundation wall		L.F. Wall	78	3.09	
B. SHELL							
B10 Superstructure							
1010	Floor Construction	N/A		—	—	—	9.3 %
1020	Roof Construction	Metal deck, open web steel joists, beams, interior columns		S.F. Roof	8.90	8.90	
B20 Exterior Enclosure							
2010	Exterior Walls	Face brick with concrete block backup	70% of wall	S.F. Wall	30.83	10.24	
2020	Exterior Windows	Metal outward projecting	30% of wall	Each	696	4.31	16.2%
2030	Exterior Doors	Metal and glass with transom		Each	2398	.88	
B30 Roofing							
3010	Roof Coverings	Built-up tar and gravel with flashing; perlite/EPS composite insulation		S.F. Roof	5.82	5.82	6.1%
3020	Roof Openings	Roof hatches		S.F. Roof			
C. INTERIORS							
1010	Partitions	Gypsum board on metal studs	20 S.F. Floor/L.F. Partition	S.F. Partition	8.50	4.25	
1020	Interior Doors	Wood solid core	200 S.F. Floor/Door	Each	565	2.83	
1030	Fittings	Toilet partitions		S.F. Floor	.50	.50	
2010	Stair Construction	N/A		—	—	—	23.8%
3010	Wall Finishes	90% paint, 10% ceramic tile		S.F. Surface	1.34	1.34	
3020	Floor Finishes	70% carpet tile, 15% terrazzo, 15% vinyl composition tile		S.F. Floor	7.42	7.42	
3030	Ceiling Finishes	Mineral fiber tile on concealed zee bars		S.F. Ceiling	6.38	6.38	
D. SERVICES							
D10 Conveying							
1010	Elevators & Lifts	N/A		—	—	—	0.0 %
1020	Escalators & Moving Walks	N/A		—	—	—	
D20 Plumbing							
2010	Plumbing Fixtures	Kitchen, toilet and service fixtures, supply and drainage	1 Fixture/500 S.F. Floor	Each	2965	5.93	
2020	Domestic Water Distribution	Gas fired water heater		S.F. Floor	1.16	1.16	8.2%
2040	Rain Water Drainage	Roof drains		S.F. Roof	.74	.74	
D30 HVAC							
3010	Energy Supply	N/A		—	—	—	
3020	Heat Generating Systems	Included in D3050		—	—	—	
3030	Cooling Generating Systems	N/A		—	—	—	8.4 %
3050	Terminal & Package Units	Multizone unit, gas heating, electric cooling		S.F. Floor	7.98	7.98	
3090	Other HVAC Sys. & Equipment	N/A		—	—	—	
D40 Fire Protection							
4010	Sprinklers	Wet pipe sprinkler system		S.F. Floor	2.77	2.77	2.9%
4020	Standpipes	N/A		—	—	—	
D50 Electrical							
5010	Electrical Service/Distribution	400 ampere service, panel board and feeders		S.F. Floor	1.84	1.84	
5020	Lighting & Branch Wiring	Fluorescent fixtures, receptacles, switches, A.C. and misc. power		S.F. Floor	9.63	9.63	14.8%
5030	Communications & Security	Alarm systems, internet wiring, and emergency lighting		S.F. Floor	2.48	2.48	
5090	Other Electrical Systems	Emergency generator, 15 kW		S.F. Floor	.15	.15	
E. EQUIPMENT & FURNISHINGS							
1010	Commercial Equipment	N/A		—	—	—	
1020	Institutional Equipment	N/A		—	—	—	0.0 %
1030	Vehicular Equipment	N/A		—	—	—	
1090	Other Equipment	N/A		—	—	—	
F. SPECIAL CONSTRUCTION							
1020	Integrated Construction	N/A		—	—	—	0.0 %
1040	Special Facilities	N/A		—	—	—	
G. BUILDING SITEWORK	**N/A**						
				Sub-Total		95.35	**100%**
	CONTRACTOR FEES (General Requirements: 10%, Overhead: 5%, Profit: 10%)				25%	23.82	
	ARCHITECT FEES				9%	10.73	
				Total Building Cost		**129.90**	

Costs per square foot of floor area

Exterior Wall	S.F. Area	8000	10000	12000	15000	18000	24000	28000	35000	40000
	L.F. Perimeter	206	233	260	300	320	360	393	451	493
Face Brick with Concrete Block Back-up	Steel Frame	198.70	187.30	179.75	172.10	165.00	156.10	152.65	148.55	146.55
	R/Conc. Frame	203.70	192.35	184.80	177.10	170.00	161.10	157.65	153.60	151.60
Stone with Concrete Block Back-up	Steel Frame	188.85	179.10	172.60	166.05	159.75	151.95	148.95	145.35	143.65
	R/Conc. Frame	207.10	195.40	187.60	179.75	172.35	163.10	159.50	155.30	153.20
Limestone with Concrete Block Back-up	Steel Frame	229.70	215.35	205.80	196.20	186.40	174.10	169.50	164.05	161.40
	R/Conc. Frame	234.70	220.40	210.85	201.20	191.40	179.15	174.55	169.10	166.45
Perimeter Adj., Add or Deduct	Per 100 L.F.	23.15	18.55	15.50	12.35	10.25	7.75	6.60	5.30	4.70
Story Hgt. Adj., Add or Deduct	Per 1 Ft.	3.50	3.25	2.95	2.75	2.45	2.05	1.90	1.80	1.75
For Basement, add $27.95 per square foot of basement area										

The above costs were calculated using the basic specifications shown on the facing page. These costs should be adjusted where necessary for design alternatives and owner's requirements. Reported completed project costs, for this type of structure, range from $77.35 to $227.00 per S.F.

Common additives

Description	Unit	$ Cost
Directory Boards, Plastic, glass covered		
30" x 20"	Each	595
36" x 48"	Each	1450
Aluminum, 24" x 18"	Each	600
36" x 24"	Each	675
48" x 32"	Each	980
48" x 60"	Each	2025
Elevators, Hydraulic passenger, 2 stops		
1500# capacity	Each	62,800
2500# capacity	Each	66,300
3500# capacity	Each	69,800
Additional stop, add	Each	7825
Emergency Lighting, 25 watt, battery operated		
Lead battery	Each	282
Nickel cadmium	Each	805

Description	Unit	$ Cost
Flagpoles, Complete		
Aluminum, 20' high	Each	1650
40' high	Each	3475
70' high	Each	10,100
Fiberglass, 23' high	Each	1775
39'-5" high	Each	3325
59' high	Each	8225
Safe, Office type, 1 hour rating		
30" x 18" x 18"	Each	2400
60" x 36" x 18", double door	Each	8750
Smoke Detectors		
Ceiling type	Each	187
Duct type	Each	480
Vault Front, Door & frame		
1 Hour test, 32" x 78"	Opening	5175
2 Hour test, 32" door	Opening	5600
40" door	Opening	6400
4 Hour test, 32" door	Opening	6075
40" door	Opening	7125
Time lock movement; two movement	Each	1850

Model costs calculated for a 3 story building with 12' story height and 18,000 square feet of floor area

				Unit	Unit Cost	Cost Per S.F.	% Of Sub-Total
A. SUBSTRUCTURE							
1010	Standard Foundations	Poured concrete; strip and spread footings		S.F. Ground	4.02	1.34	
1020	Special Foundations	N/A		—	—	—	
1030	Slab on Grade	4" reinforced concrete with vapor barrier and granular base		S.F. Slab	4.74	1.58	3.9%
2010	Basement Excavation	Site preparation for slab and trench for foundation wall and footing		S.F. Ground	.17	.06	
2020	Basement Walls	4' foundation wall		L.F. Wall	74	1.54	
B. SHELL							
	B10 Superstructure						
1010	Floor Construction	Open web steel joists, slab form, concrete, wide flange steel columns		S.F. Floor	27.87	18.58	18.6%
1020	Roof Construction	Metal deck, open web steel joists, beams, interior columns		S.F. Roof	9.78	3.26	
	B20 Exterior Enclosure						
2010	Exterior Walls	Stone with concrete block backup	70% of wall	S.F. Wall	34.69	15.54	
2020	Exterior Windows	Metal outward projecting	10% of wall	Each	1025	2.37	15.8%
2030	Exterior Doors	Metal and glass with transoms		Each	2234	.62	
	B30 Roofing						
3010	Roof Coverings	Built-up tar and gravel with flashing; perlite/EPS composite insulation		S.F. Roof	5.79	1.93	1.6%
3020	Roof Openings	N/A		—	—	—	
C. INTERIORS							
1010	Partitions	Gypsum board on metal studs	20 S.F. Floor/L.F. Partition	S.F. Partition	9.44	4.72	
1020	Interior Doors	Wood solid core	200 S.F. Floor/Door	Each	565	2.83	
1030	Fittings	Toilet partitions		S.F. Floor	.30	.30	
2010	Stair Construction	Concrete filled metal pan		Flight	18,950	6.31	25.0%
3010	Wall Finishes	90% paint, 10% ceramic tile		S.F. Surface	1.34	1.34	
3020	Floor Finishes	70% carpet tile, 15% terrazzo, 15% vinyl composition tile		S.F. Floor	7.42	7.42	
3030	Ceiling Finishes	Mineral fiber tile on concealed zee bars		S.F. Ceiling	6.38	6.38	
D. SERVICES							
	D10 Conveying						
1010	Elevators & Lifts	Two hydraulic elevators		Each	117,540	13.06	11.1%
1020	Escalators & Moving Walks	N/A		—	—	—	
	D20 Plumbing						
2010	Plumbing Fixtures	Toilet and service fixtures, supply and drainage	1 Fixture/1385 S.F. Floor	Each	2341	1.69	
2020	Domestic Water Distribution	Gas fired water heater		S.F. Floor	.40	.40	2.1%
2040	Rain Water Drainage	Roof drains		S.F. Roof	1.17	.39	
	D30 HVAC						
3010	Energy Supply	N/A		—	—	—	
3020	Heat Generating Systems	Included in D3050		—	—	—	
3030	Cooling Generating Systems	N/A		—	—	—	6.8 %
3050	Terminal & Package Units	Multizone unit, gas heating, electric cooling		S.F. Floor	7.98	7.98	
3090	Other HVAC Sys. & Equipment	N/A		—	—	—	
	D40 Fire Protection						
4010	Sprinklers	Sprinklers, light hazard		S.F. Floor	2.58	2.58	2.2%
4020	Standpipes	N/A		—	—	—	
	D50 Electrical						
5010	Electrical Service/Distribution	400 ampere service, panel board and feeders		S.F. Floor	1.59	1.59	
5020	Lighting & Branch Wiring	Fluorescent fixtures, receptacles, switches, A.C. and misc. power		S.F. Floor	10.79	10.79	12.8%
5030	Communications & Security	Alarm systems, internet wiring, and emergency lighting		S.F. Floor	2.48	2.48	
5090	Other Electrical Systems	Emergency generator, 15 kW		S.F. Floor	.18	.18	
E. EQUIPMENT & FURNISHINGS							
1010	Commercial Equipment	N/A		—	—	—	
1020	Institutional Equipment	N/A		—	—	—	0.0 %
1030	Vehicular Equipment	N/A		—	—	—	
1090	Other Equipment	N/A		—	—	—	
F. SPECIAL CONSTRUCTION							
1020	Integrated Construction	N/A		—	—	—	0.0 %
1040	Special Facilities	N/A		—	—	—	
G. BUILDING SITEWORK	**N/A**						

		Sub-Total	117.26	100%
CONTRACTOR FEES (General Requirements: 10%, Overhead: 5%, Profit: 10%)		25%	29.30	
ARCHITECT FEES		9%	13.19	
	Total Building Cost		**159.75**	

Costs per square foot of floor area

Exterior Wall	S.F. Area	10000	15000	20000	25000	30000	35000	40000	50000	60000
	L.F. Perimeter	410	500	600	640	700	766	833	966	1000
Tiltup Concrete Panels	Steel Frame	114.90	104.90	100.30	95.60	92.95	91.25	89.95	88.25	85.65
Brick with Block Back-up	Bearing Walls	129.50	115.80	109.60	102.85	99.10	96.75	95.00	92.50	88.70
Concrete Block	Steel Frame	110.00	102.60	99.20	95.60	93.65	92.35	91.45	90.10	88.15
Concrete Block	Bearing Walls	106.00	98.50	95.05	91.40	89.40	88.10	87.15	85.80	83.80
Galvanized Steel Siding	Steel Frame	118.95	110.90	107.25	103.30	101.10	99.75	98.65	97.25	95.00
Metal Sandwich Panels	Steel Frame	119.60	110.45	106.30	101.65	99.10	97.50	96.35	94.65	92.05
Perimeter Adj., Add or Deduct	Per 100 L.F.	8.15	5.45	4.05	3.25	2.75	2.30	2.10	1.65	1.40
Story Hgt. Adj., Add or Deduct	Per 1 Ft.	1.05	0.80	0.75	0.60	0.55	0.55	0.55	0.45	0.45
For Basement, add $26.85 per square foot of basement area										

The above costs were calculated using the basic specifications shown on the facing page. These costs should be adjusted where necessary for design alternatives and owner's requirements. Reported completed project costs, for this type of structure, range from $34.35 to $137.20 per S.F.

Common additives

Description	Unit	$ Cost
Dock Leveler, 10 ton cap.		
6' x 8'	Each	6300
7' x 8'	Each	6800
Emergency Lighting, 25 watt, battery operated		
Lead battery	Each	282
Nickel cadmium	Each	805
Fence, Chain link, 6' high		
9 ga. wire	L.F.	22.50
6 ga. wire	L.F.	32.50
Gate	Each	335
Flagpoles, Complete		
Aluminum, 20' high	Each	1650
40' high	Each	3475
70' high	Each	10,100
Fiberglass, 23' high	Each	1775
39'-5" high	Each	3325
59' high	Each	8225
Paving, Bituminous		
Wearing course plus base course	S.Y.	10.50
Sidewalks, Concrete 4" thick	S.F.	4.08

Description	Unit	$ Cost
Sound System		
Amplifier, 250 watts	Each	2350
Speaker, ceiling or wall	Each	191
Trumpet	Each	365
Yard Lighting, 20' aluminum pole with 400 watt high pressure sodium fixture.	Each	2825

Important: See the Reference Section for Location Factors

Model costs calculated for a 1 story building with 24' story height and 30,000 square feet of floor area

Warehouse

				Unit	Unit Cost	Cost Per S.F.	% Of Sub-Total
A. SUBSTRUCTURE							
1010	Standard Foundations	Poured concrete; strip and spread footings		S.F. Ground	1.44	1.44	
1020	Special Foundations	N/A		—	—	—	
1030	Slab on Grade	5" reinforced concrete with vapor barrier and granular base		S.F. Slab	11.80	11.80	23.6%
2010	Basement Excavation	Site preparation for slab and trench for foundation wall and footing		S.F. Ground	.17	.17	
2020	Basement Walls	4' foundation wall		L.F. Wall	78	2.36	
B. SHELL							
	B10 Superstructure						
1010	Floor Construction	Mezzanine: open web steel joists, slab form, concrete beams, columns	10% of area	S.F. Floor	23.40	2.34	16.1%
1020	Roof Construction	Metal deck, open web steel joists, beams, columns		S.F. Roof	8.41	8.41	
	B20 Exterior Enclosure						
2010	Exterior Walls	Concrete block	95% of wall	S.F. Wall	11.05	5.88	
2020	Exterior Windows	N/A		—	—	—	10.3%
2030	Exterior Doors	Steel overhead, hollow metal	5% of wall	Each	2780	1.02	
	B30 Roofing						
3010	Roof Coverings	Built-up tar and gravel with flashing; perlite/EPS composite insulation		S.F. Roof	5.10	5.10	8.1%
3020	Roof Openings	Roof hatches and skylight		S.F. Roof	.30	.30	
C. INTERIORS							
1010	Partitions	Concrete block (office and washrooms)	100 S.F. Floor/L.F. Partition	S.F. Partition	8.50	.68	
1020	Interior Doors	Single leaf hollow metal	5000 S.F. Floor/Door	Each	875	.18	
1030	Fittings	N/A		—	—	—	
2010	Stair Construction	Steel gate with rails		Flight	14,925	1	9.5%
3010	Wall Finishes	Paint		S.F. Surface	12	1.92	
3020	Floor Finishes	90% hardener, 10% vinyl composition tile		S.F. Floor	1.92	1.92	
3030	Ceiling Finishes	Suspended mineral tile on zee channels in office area	10% of area	S.F. Ceiling	6.38	.64	
D. SERVICES							
	D10 Conveying						
1010	Elevators & Lifts	N/A		—	—	—	0.0%
1020	Escalators & Moving Walks	N/A		—	—	—	
	D20 Plumbing						
2010	Plumbing Fixtures	Toilet and service fixtures, supply and drainage	1 Fixture/2500 S.F. Floor	Each	2950	1.18	
2020	Domestic Water Distribution	Gas fired water heater		S.F. Floor	.26	.26	4.0%
2040	Rain Water Drainage	Roof drains		S.F. Roof	1.23	1.23	
	D30 HVAC						
3010	Energy Supply	N/A		—	—	—	
3020	Heat Generating Systems	Ventilation with heat		Each	68,100	2.45	
3030	Cooling Generating Systems	N/A		—	—	—	4.9%
3050	Terminal & Package Units	Single zone unit gas, heating, electric cooling	10% of area	S.F. Floor	.82	.82	
3090	Other HVAC Sys. & Equipment	N/A		—	—	—	
	D40 Fire Protection						
4010	Sprinklers	Sprinklers, ordinary hazard		S.F. Floor	3.21	3.21	7.4%
4020	Standpipes	Standpipe		S.F. Floor	1.74	1.74	
	D50 Electrical						
5010	Electrical Service/Distribution	200 ampere service, panel board and feeders		S.F. Floor	.46	.46	
5020	Lighting & Branch Wiring	High efficiency fluorescent fixtures, receptacles, switches, A.C. and misc. power		S.F. Floor	5.67	5.67	12.6%
5030	Communications & Security	Addressable alarm systems		S.F. Floor	2.27	2.27	
5090	Other Electrical Systems	N/A		—	—	—	
E. EQUIPMENT & FURNISHINGS							
1010	Commercial Equipment	N/A		—	—	—	
1020	Institutional Equipment	N/A		—	—	—	3.6%
1030	Vehicular Equipment	Dock boards, dock levelers		S.F. Floor	2.39	2.39	
1090	Other Equipment	N/A		—	—	—	
F. SPECIAL CONSTRUCTION							
1020	Integrated Construction	N/A		—	—	—	0.0%
1040	Special Facilities	N/A		—	—	—	
G. BUILDING SITEWORK	**N/A**						

			Sub-Total	66.84	100%
CONTRACTOR FEES (General Requirements: 10%, Overhead: 5%, Profit: 10%)			25%	16.71	
ARCHITECT FEES			7%	5.85	
			Total Building Cost	**89.40**	

Costs per square foot of floor area

Exterior Wall	S.F. Area	2000	3000	5000	8000	12000	20000	30000	50000	100000
	L.F. Perimeter	180	220	300	420	580	900	1300	2100	4100
Concrete Block	Steel Frame	171.55	150.00	132.85	123.05	117.70	113.35	111.25	109.50	108.25
	R/Conc. Frame	149.85	136.75	126.35	120.45	117.20	114.60	113.25	112.20	111.40
Metal Sandwich Panel	Steel Frame	147.40	133.00	121.55	115.10	111.55	108.65	107.25	106.15	105.20
Tilt-up Concrete Panel	R/Conc. Frame	160.95	146.95	135.80	129.50	126.00	123.25	121.80	120.70	119.85
Precast Concrete Panel	Steel Frame	149.90	134.95	123.05	116.30	112.55	109.60	108.10	106.95	106.00
	R/Conc. Frame	170.45	154.50	141.75	134.55	130.60	127.40	125.80	124.55	123.55
Perimeter Adj., Add or Deduct	Per 100 L.F.	26.10	17.45	10.45	6.55	4.35	2.70	1.70	1.05	0.50
Story Hgt. Adj., Add or Deduct	Per 1 Ft.	2.65	2.10	1.65	1.55	1.45	1.35	1.20	1.25	1.15
Basement—Not Applicable										

The above costs were calculated using the basic specifications shown on the facing page. These costs should be adjusted where necessary for design alternatives and owner's requirements. Reported completed project costs, for this type of structure, range from $32.90 to $188.75 per S.F.

Common additives

Description	Unit	$ Cost
Dock Leveler, 10 ton cap.		
6' x 8'	Each	6300
7' x 8'	Each	6800
Emergency Lighting, 25 watt, battery operated		
Lead battery	Each	282
Nickel cadmium	Each	805
Fence, Chain link, 6' high		
9 ga. wire	L.F.	22.50
6 ga. wire	L.F.	32.50
Gate	Each	335
Flagpoles, Complete		
Aluminum, 20' high	Each	1650
40' high	Each	3475
70' high	Each	10,100
Fiberglass, 23' high	Each	1775
39'-5" high	Each	3325
59' high	Each	8225
Paving, Bituminous		
Wearing course plus base course	S.Y.	10.50
Sidewalks, Concrete 4" thick	S.F.	4.08

Description	Unit	$ Cost
Sound System		
Amplifier, 250 watts	Each	2350
Speaker, ceiling or wall	Each	191
Trumpet	Each	365
Yard Lighting, 20' aluminum pole with 400 watt high pressure sodium fixture	Each	2825

Model costs calculated for a 1 story building with 12' story height and 20,000 square feet of floor area

				Unit	Unit Cost	Cost Per S.F.	% Of Sub-Total
A.	**SUBSTRUCTURE**						
1010	Standard Foundations	Poured concrete; strip and spread footings		S.F. Ground	1.93	1.93	
1020	Special Foundations	N/A		—	—	—	
1030	Slab on Grade	4" reinforced concrete with vapor barrier and granular base		S.F. Slab	9.58	9.58	17.5%
2010	Basement Excavation	Site preparation for slab and trench for foundation wall and footing		S.F. Ground	.26	.26	
2020	Basement Walls	4' foundation wall		L.F. Wall	69	3.10	
B.	**SHELL**						
	B10 Superstructure						
1010	Floor Construction	N/A		—	—	—	12.5 %
1020	Roof Construction	Metal deck, open web steel joists, beams, columns		S.F. Roof	10.57	10.57	
	B20 Exterior Enclosure						
2010	Exterior Walls	Concrete block	70% of wall	S.F. Wall	12.46	4.71	
2020	Exterior Windows	Aluminum projecting	5% of wall	Each	779	7.01	17.6%
2030	Exterior Doors	Steel overhead, hollow metal	25% of wall	Each	1446	3.19	
	B30 Roofing						
3010	Roof Coverings	Built-up tar and gravel with flashing; perlite/EPS composite insulation		S.F. Roof	4.61	4.61	5.4%
3020	Roof Openings	N/A		—	—	—	
C.	**INTERIORS**						
1010	Partitions	Concrete block, gypsum board on metal studs	10.65 S.F. Floor/L.F. Partition	S.F. Partition	8.27	14.18	
1020	Interior Doors	Single leaf hollow metal	300 S.F. Floor/Door	Each	1440	4.80	
1030	Fittings	N/A		—	—	—	
2010	Stair Construction	N/A		—	—	—	22.4%
3010	Wall Finishes	N/A		—	—	—	
3020	Floor Finishes	N/A		—	—	—	
3030	Ceiling Finishes	N/A		—	—	—	
D.	**SERVICES**						
	D10 Conveying						
1010	Elevators & Lifts	N/A		—	—	—	0.0 %
1020	Escalators & Moving Walks	N/A		—	—	—	
	D20 Plumbing						
2010	Plumbing Fixtures	Toilet and service fixtures, supply and drainage	1 Fixture/5000 S.F. Floor	Each	2100	.42	
2020	Domestic Water Distribution	Gas fired water heater		S.F. Floor	.20	.20	1.9%
2040	Rain Water Drainage	Roof drains		S.F. Roof	1.03	1.03	
	D30 HVAC						
3010	Energy Supply	N/A		—	—	—	
3020	Heat Generating Systems	Oil fired hot water, unit heaters		Each	7.98	7.98	
3030	Cooling Generating Systems	N/A		—	—	—	9.4 %
3050	Terminal & Package Units	N/A		—	—	—	
3090	Other HVAC Sys. & Equipment	N/A		—	—	—	
	D40 Fire Protection						
4010	Sprinklers	Wet pipe sprinkler system		S.F. Floor	3.62	3.62	4.3%
4020	Standpipes	N/A		—	—	—	
	D50 Electrical						
5010	Electrical Service/Distribution	200 ampere service, panel board and feeders		S.F. Floor	1.66	1.66	
5020	Lighting & Branch Wiring	Fluorescent fixtures, receptacles, switches and misc. power		S.F. Floor	5.42	5.42	8.9%
5030	Communications & Security	Alarm systems		S.F. Floor	.39	.39	
5090	Other Electrical Systems	Emergency generator, 7.5 kW		S.F. Floor	.10	.10	
E.	**EQUIPMENT & FURNISHINGS**						
1010	Commercial Equipment	N/A		—	—	—	
1020	Institutional Equipment	N/A		—	—	—	0.0 %
1030	Vehicular Equipment	N/A		—	—	—	
1090	Other Equipment	N/A		—	—	—	
F.	**SPECIAL CONSTRUCTION**						
1020	Integrated Construction	N/A		—	—	—	0.0 %
1040	Special Facilities	N/A		—	—	—	
G.	**BUILDING SITEWORK**	**N/A**					

		Sub-Total	84.76	100%
CONTRACTOR FEES (General Requirements: 10%, Overhead: 5%, Profit: 10%)		25%	21.17	
ARCHITECT FEES		7%	7.42	

	Total Building Cost	**113.35**

Deterioration occurs within the structure itself and is determined by the observation of both materials and equipment.

Curable deterioration can be remedied either by maintenance, repair or replacement, within prudent economic limits.

Incurable deterioration that has progressed to the point of actually affecting the structural integrity of the structure, making repair or replacement not economically feasible.

Actual Versus Observed Age

The observed age of a structure refers to the age that the structure appears to be. Periodic maintenance, remodeling, and renovation all tend to reduce the amount of deterioration that has taken place, thereby decreasing the observed age. Actual age, on the other hand, relates solely to the year that the structure was built.

The Depreciation Table shown here relates to the observed age.

Obsolescence arises from conditions either occurring within the structure (functional) or caused by factors outside the limits of the structure (economic).

Functional obsolescence is any inadequacy caused by outmoded design, dated construction materials, or oversized or undersized areas, all of which cause excessive operation costs.

Incurable is so costly as not to economically justify the capital expenditure required to correct the deficiency.

Economic obsolescence is caused by factors outside the limits of the structure. The prime causes of economic obsolescence are:

- zoning and environmental laws
- government legislation
- negative neighborhood influences
- business climate
- proximity to transportation facilities

Depreciation Table
Commercial/Industrial/Institutional

Building Material			
Observed Age (Years)	Frame	Masonry On Wood	Masonry On Masonry Or Steel
1	1%	0%	0%
2	2	1	0
3	3	2	1
4	4	3	2
5	6	5	3
10	20	15	8
15	25	20	15
20	30	25	20
25	35	30	25
30	40	35	30
35	45	40	35
40	50	45	40
45	55	50	45
50	60	55	50
55	65	60	55
60	70	65	60

Assemblies Section

Table of Contents

Table of Contents

Introduction to the Assemblies Section

This section of the manual contains the technical description and installed cost of all components used in the commercial/industrial/institutional section. In most cases, there is an illustration of the component to aid in identification.

All Costs Include:

Materials purchased in lot sizes typical of normal construction projects with discounts appropriate to established contractors.

Installation work performed by union labor working under standard conditions at a normal pace.

Installing Contractor's Overhead & Profit

Standard installing contractor's overhead & profit are included in the assemblies costs.

These assemblies costs contain no provisions for General Contractor's overhead and profit or Architect's fees, nor do they include premiums for labor or materials, or savings which may be realized under certain economic situations.

The assemblies tables are arranged by physical size (dimensions) of the component wherever possible.

How to Use the Assemblies Cost Tables

The following is a detailed explanation of a sample Assemblies Cost Table. Most Assembly Tables are separated into two parts: 1) an illustration of the system to be estimated with descriptive notes; and 2) the costs for similar systems with dimensional and/or size variations. Next to each bold number below is the item being described with the appropriate component of the sample entry following in parenthesis. In most cases, if the work is to be subcontracted, the general contractor will need to add an additional markup (RSMeans suggests using 10%) to the "Total" figures.

System/Line Numbers (A1030 120 2240)

Each Assemblies Cost Line has been assigned a unique identification number based on the UNIFORMAT II classification sstem.

UNIFORMAT II Major Group

A1030, 120, 2240,

UNIFORMAT II Level 3

Means Major Classification

Means Individual Line Number

A10 Foundations

A1030 Slab on Grade

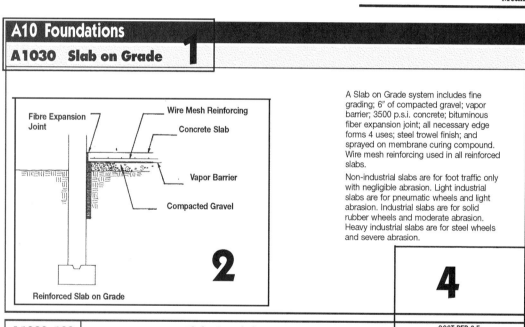

Fibre Expansion Joint

Wire Mesh Reinforcing

Concrete Slab

Vapor Barrier

Compacted Gravel

Reinforced Slab on Grade

A Slab on Grade system includes fine grading; 6" of compacted gravel; vapor barrier; 3500 p.s.i. concrete; bituminous fiber expansion joint; all necessary edge forms 4 uses; steel trowel finish; and sprayed on membrane curing compound. Wire mesh reinforcing used in all reinforced slabs.

Non-industrial slabs are for foot traffic only with negligible abrasion. Light industrial slabs are for pneumatic wheels and light abrasion. Industrial slabs are for solid rubber wheels and moderate abrasion. Heavy industrial slabs are for steel wheels and severe abrasion.

A1030 120	Plain & Reinforced	COST PER S.F.		
		MAT.	INST.	TOTAL
2220	Slab on grade, 4" thick, non industrial, non reinforced	1.88	2.32	4.20
2240	Reinforced	2.08	2.66	4.74
2260	Light industrial, non reinforced	2.46	2.85	5.31
2280	Reinforced	2.66	3.19	5.85
2300	Industrial, non reinforced	3.08	6	9.08
2320	Reinforced	3.28	6.30	9.58
3340	5" thick, non industrial, non reinforced	2.22	2.39	4.61
3360	Reinforced	2.42	2.73	5.15
3380	Light industrial, non reinforced	2.81	2.92	5.73
3400	Reinforced	3.01	3.26	6.27
3420	Heavy industrial, non reinforced	4.05	7.15	11.20
3440	Reinforced	4.25	7.55	11.80
4460	6" thick, non industrial, non reinforced	2.68	2.34	5.02
4480	Reinforced	3.03	2.79	5.82
4500	Light industrial, non reinforced	3.27	2.87	6.14
4520	Reinforced	3.82	3.47	7.29
4540	Heavy industrial, non reinforced	4.53	7.25	11.78
4560	Reinforced	4.88	7.70	12.58
5580	7" thick, non industrial, non reinforced	3.03	2.42	5.45
5600	Reinforced	3.46	2.90	6.36

2 Illustration

At the top of most assembly pages is an illustration, a brief description, and the design criteria used to develop the cost.

3 System Description

The components of a typical system are listed in the description to show what has been included in the development of the total system price. The rest of the table contains prices for other similar systems with dimensional and/or size variations.

4 Unit of Measure for Each System (S.F.)

Costs shown in the three right hand columns have been adjusted by the component quantity and unit of measure for the entire system. In this example, "Cost per S.F." is the unit of measure for this system or "assembly."

5 Materials (2.08)

This column contains the Materials Cost of each component. These cost figures are bare costs plus 10% for profit.

6 Installation (2.66)

Installation includes labor and equipment plus the installing contractor's overhead and profit. Equipment costs are the bare rental costs plus 10% for profit. The labor overhead and profit is defined on the inside back cover of this book.

7 Total (4.74)

The figure in this column is the sum of the material and installation costs.

Material Cost	+	Installation Cost	=	Total
$2.08	+	$2.66	=	$4.74

Example

Assemblies costs can be used to calculate the cost for each component of a building and, accumulated with appropriate markups, produce a cost for the complete structure. Components from a model building with similar characteristics in conjunction with Assemblies components may be used to calculate costs for a complete structure.

Example:

Outline Specifications

General	Building size—60' x 100', 8 suspended floors, 12' floor to floor, 4' high parapet above roof, full basement 11'-8" floor to floor, bay size 25' x 30', ceiling heights - 9' in office area and 8' in core area.
A Substructure	Concrete, spread and strip footings, concrete walls waterproofed, 4" slab on grade.
B Shell	
B10 Superstructure	Columns, wide flange; 3 hr. fire rated; Floors, composite steel frame & deck with concrete slab; Roof, steel beams, open web joists and deck.
B20 Exterior Enclosure	Walls; North, East & West, brick & lightweight concrete block with 2" cavity insulation, 25% window; South, 8" lightweight concrete block insulated, 10% window. Doors, aluminum & glass. Windows, aluminum, 3'-0" x 5'-4", insulating glass.
B30 Roofing	Tar & gravel, 2" rigid insulation, R 12.5.
C Interiors	Core—6" lightweight concrete block, full height with plaster on exposed faces to ceiling height.
	Corridors—1st & 2nd floor; 3-5/8" steel studs with F.R. gypsum board, full height.
	Exterior Wall—plaster, ceiling height.
	Fittings—Toilet accessories, directory boards.
	Doors—hollow metal.
	Wall Finishes—lobby, mahogany paneling on furring, remainder plaster & gypsum board, paint.
	Floor Finishes—1st floor lobby, corridors & toilet rooms, terrazzo; remainder concrete, tenant developed. 2nd thru 8th, toilet rooms, ceramic tile; office & corridor, carpet.
	Ceiling Finishes—24" x 48" fiberglass board on Tee grid.
D Services	
D10 Conveying	2-2500 lb. capacity, 200 F.P.M. geared elevators, 9 stops.
D20 Plumbing	Fixtures, see sketch.
	Roof Drains—2-4" C.I. pipe.
D30 HVAC	Heating—fin tube radiation, forced hot water.
	Air Conditioning—chilled water with water cooled condenser.
D40 Fire Protection	Fire Protection—4" standpipe, 9 hose cabinets.
D50 Electrical	Lighting, 1st thru 8th, 15 fluorescent fixtures/1000 S.F., 3 Watts/S.F.
	Basement, 10 fluorescent fixtures/1000 S.F., 2 Watts/S.F.
	Receptacles, 1st thru 8th, 16.5/1000 S.F., 2 Watts/S.F.
	Basement, 10 receptacles/1000 S.F., 1.2 Watts/S.F.
	Air Conditioning, 4 Watts/S.F.
	Miscellaneous Connections, 1.2 Watts/S.F.
	Elevator Power, 2-10 H.P., 230 volt motors.
	Wall Switches, 2/1000 S.F.
	Service, panel board & feeder, 2000 Amp.
	Fire Detection System, pull stations, signals, smoke and heat detectors.
	Emergency Lighting Generator, 30 KW.
E Equipment & Furnishings	NA.
F Special Construction	NA.
G Building Sitework	NA.

Front Elevation

Basement Plan

Ground Floor Plan

Typical Floor Plan

Example

If spread footing & column sizes are unknown, develop approximate loads as follows. Enter tables with these loads to determine costs.

Approximate loads/S.F. for roof & floors.
Roof. Assume 40 psf superimposed load.
Steel joists, beams & deck.
Table B1020 112—Line 3900

Superimposed Load Ranges

Apartments & Residential Structures	65	to	75 psf
Assembly Areas & Retail Stores	110	to	125 psf
Commercial & Manufacturing	150	to	250 psf
Offices	75	to	100 psf

B1020 112	Steel Joists, Beams, & Deck on Columns							
	BAY SIZE (FT.)	SUPERIMPOSED LOAD (P.S.F.)	DEPTH (IN.)	TOTAL LOAD (P.S.F.)	COLUMN ADD	COST PER S.F.		
						MAT.	INST.	TOTAL
3500	25x30	20	22	40		6.65	1.49	8.14
3600					columns	1.36	.27	1.63
3900		40	25	60		8	1.76	9.76
4000					columns	1.63	.32	1.95

B1010 256	Composite Beams, Deck & Slab							
	BAY SIZE (FT.)	SUPERIMPOSED LOAD (P.S.F.)	SLAB THICKNESS (IN.)	TOTAL DEPTH (FT.-IN.)	TOTAL LOAD (P.S.F.)	COST PER S.F.		
						MAT.	INST.	TOTAL
3400	25x30	40	5-1/2	1 - 11-1/2	83	14	5.20	19.20
3600		75	5-1/2	1 - 11-1/2	119	15.10	5.25	20.35
3900		125	5-1/2	1 - 11-1/2	170	17.60	5.95	23.55
4000		200	6-1/4	2 - 6-1/4	252	22	6.80	28.80

Floors–Total load, 119 psf.
Interior foundation load.
Roof

[(25' x 30' x 60 psf) + 8 floors x (25' x 30' x 119 psf)] x 1/1000 lb./Kip =	759 Kips
Approximate Footing Loads, Interior footing =	759 Kips
Exterior footing (1/2 bay) 759 k x .6 =	455 Kips
Corner footing (1/4 bay) 759 k x .45 =	342 Kips
[Factors to convert Interior load to Exterior & Corner loads]	
Approximate average Column load 759 k/2 =	379 Kips

PRELIMINARY ESTIMATE

PROJECT	**Office Building**	TOTAL SITE AREA		
BUILDING TYPE		OWNER		
LOCATION		ARCHITECT		
DATE OF CONSTRUCTION		ESTIMATED CONSTRUCTION PERIOD		

BRIEF DESCRIPTION **Building size: 60' x 100' - 8 structural floors - 12' floor to floor**

4' high parapet above roof, full basement 11' - 8" floor to floor, bay size

25' x 30', ceiling heights - 9' in office area + 8' in core area

TYPE OF PLAN		TYPE OF CONSTRUCTION		
QUALITY		BUILDING CAPACITY		

Floor			**Wall Area**				
Below Grade Levels			Foundation Walls	L.F.		Ht.	S.F.
Area		S.F.	Frost Walls	L.F.		Ht.	S.F.
Area		S.F.	Exterior Closure		Total		S.F.
Total Area		S.F.	Comment				
Ground Floor			Fenestration		%		S.F.
Area		S.F.			%		S.F.
Area		S.F.	Exterior Wall		%		S.F.
Total Area		S.F.			%		S.F.
Supported Levels			**Site Work**				
Area			Parking		S.F. (For		Cars)
Area		S.F.	Access Roads		L.F. (X		Ft. Wide)
Area		S.F.	Sidewalk		L.F. (X		Ft. Wide)
Area		S.F.	Landscaping		S.F. (		% Unbuilt Site)
Area		S.F.	**Building Codes**				
Total Area		S.F.	City		Country		
Miscellaneous			National		Other		
Area		S.F.	**Loading**				
Area		S.F.	Roof	psf	Ground Floor		psf
Area		S.F.	Supported Floors	psf	Corridor		psf
Area		S.F.	Balcony	psf	Partition, allow		psf
Total Area		S.F.	Miscellaneous				psf
Net Finished Area		S.F.	Live Load Reduction				
Net Floor Area		S.F.	Wind				
Gross Floor Area	**54,000**	S.F.	Earthquake		Zone		
Roof			Comment				
Total Area		S.F.	Soil Type				
Comments			Bearing Capacity				K.S.F.
			Frost Depth				Ft.
Volume			**Frame**				
Depth of Floor System			Type		Bay Spacing		
Minimum		In.	Foundation				
Maximum		In.	Special				
Foundation Wall Height		Ft.	Substructure				
Floor to Floor Height		Ft.	Comment				
Floor to Ceiling Height		Ft.	Superstructure, Vertical				
Subgrade Volume		C.F.	Fireproofing		☐ Columns		Hrs.
Above Grade Volume		C.F.	☐ Girders	Hrs.	☐ Beams		Hrs.
Total Building Volume	**648,000**	C.F.	☐ Floor	Hrs.	☐ None		

NUMBER			QTY.	UNIT	TOTAL COST UNIT	TOTAL COST TOTAL	COST PER S.F.
A	**SUBSTRUCTURE**						
A1010 210 7900	Corner Footings 8'-6" SQ. x 27"		4	Ea.	1,865.00	7,460	
8010	Exterior	9' -6" SQ. x 30"	8		2,525.00	20,200	
8300	Interior	12" SQ.	3		4,650.00	13,950	
A1010 110 2700	Strip	2' Wide x 1' Thick					
	320 L.F. [(4 x 8.5) + 8 x 9.5)] =		320	L.F.	39.50	12,640	
A2020 110 7262	Foundation Wall 12' High, 1' Thick		320		241.00	77,120	
A1010 320 2800	Foundation Waterproofing		320		19.27	6,166	
A1030 120 2240	4", non-industrial, reinf. slab on grade		6000	S.F.	4.74	28,440	
A2010 110 3440	Building Excavation + Backfill		6000	S.F.	5.87	35,220	
3500	(Interpolated ; 12' Between						
4620	8' and 16' ; 6,000 Between						
4680	4,000 and 10,000 S.F.						
	Total					201,196	3.73
B	**SHELL**						
B10	**Superstructure**						
B1010 208 5800	Columns: Load 379K Use 400K						
	Exterior 12 x 96' + Interior 3 x 108'		1476	V.L.F.	168.15	248,189	
B1010 720 3650	Column Fireproofing - Interior 4 sides						
3700	Exterior 1/2 (Interpolated for 12")		900	V.L.F	31.35	28,215	
B1010 256 3600	Floors: Composite steel + lt. wt. conc.		48000	S.F.	20.35	976,800	
B1020 112 3900	Roof: Open web joists, beams + decks		6000	S.F.	9.76	58,560	
	Total					1,311,764	24.29
B20	**Exterior Enclosure**						
B2010 134 1200	4" Brick + 6" Block - Insulated						
	75% NE + W Walls 220' x 100' x .75		16500	S.F.	29.70	490,050	
B2010 110 3410	8" Block - Insulated 90%						
	100' x 100' x .9		9000	S.F.	11.35	102,150	
B2030 110 6950	Double Aluminum + Glass Door		1	Pr.	4,850.00	4,850	
6300	Single		2	Ea.	2,425.00	4,850	
B2020 106 8800	Aluminum windows - Insul. glass						
	(5,500 S.F. + 1,000 S.F.)/(3 x 5.33)		406	Ea.	862.00	349,972	
B2010 105 8480	Precast concrete coping		320	L.F.	33.50	10,720	
	Total					962,592	17.83
B30	**Roofing**						
B3010 320 4020	Insulation: Perlite/Polyisocyanurate		6000	S.F.	2.22	13,320	
B3010 105 1400	Roof: 4 Ply T & G composite		6000		2.52	15,120	
B3010 430 0300	Flashing: Aluminum - fabric backed		640		2.95	1,888	
B3020 210 0500	Roof hatch		1	Ea.	1,151.00	1,151	
	Total					31,479	0.58

ASSEMBLY NUMBER		QTY.	UNIT	TOTAL COST		COST PER S.F.
				UNIT	TOTAL	
C	INTERIORS					
C1010 104 5500	Core partitions: 6" Lt. Wt. concrete					
	block 288 L.F. x 11.5' x 8 Floors	26,449	S.F.	8.44	223,230	
C1010 144 0920	Plaster: [(196 L.F. x 8') + (144 L.F. x 9')] 8					
	+ Ext. Wall [(320 x 9 x 8 Floors) - 6500]	39,452	S.F.	2.82	111,255	
C1010 126 5400	Corridor partitions 1st + 2nd floors					
	steel studs + F.R. gypsum board	2,530	S.F.	3.93	9,943	
C1020 102 2600	Interior doors	66	Ea.	875.00	57,750	
	Wall finishes - Lobby: Mahogany w/furring					
C1010 128 0652	Furring: 150 L.F. x 9' + Ht.	1,350	S.F.	1.38	1,863	
C1030 710 0120	Towel Dispenser	16	Ea.	77.00	1,232	
C1030 710 0140	Grab Bar	32	Ea.	53.00	1,696	
C1030 710 0160	Mirror	31	Ea.	274.00	8,494	
C1030 710 0180	Toilet Tissue Dispenser	31	Ea.	38.00	1,178	
C1030 510 0140	Directory Board	8	Ea.	288.00	2,304	
C2010 110 0780	Stairs: Steel w/conc. fill	18	Flt.	22,125.00	398,250	
C3010 230 1662	Mahogany paneling	1,350	S.F.	5.24	7,074	
C3010 230 0080	Paint, primer + 2 coats finish, on plaster or gyp. board					
	39,452 + [(220' x 9' x 2) - 1350] + 72	42,134	S.F.	1.03	43,398	
	Floor Finishes					
C3020 410 1100	1st floor lobby + terrazzo	2,175	S.F.	17.56	38,193	
	Remainder: Concrete - Tenant finished					
C3020 410 0060	2nd - 8th: Carpet - 4725 S.F. x 7	33,075	S.F.	4.17	137,923	
C3020 410 1720	Ceramic Tile - 300 S.F. x 7	2,100	S.F.	9.51	19,971	
C3030 240 2780	Suspended Ceiling Tiles- 5200 S.F. x 8	41,600	S.F.	2.02	84,032	
C3030 240 3260	Suspended Ceiling Grid- 5200 S.F. x 8	41,600	S.F.	1.31	54,496	
C1030 110 0700	Toilet partitions	31	Ea.	742.00	23,002	
0760	Handicapped addition	16	Ea.	300.00	4,800	
	Total				1,230,083	22.78
D	SERVICES					
D10	Conveying					
	2 Elevators - 2500# 200'/min. Geared					
D1010 140 1600	5 Floors $162,700					
1800	15 Floors $ 410,000					
	Diff. $247,300/10 = 24,730/Flr.					
	$162,700 + (4 x 24,730) = $261,620/Ea.	2	Ea.	261,620.00	523,240	
	Total				523,240	9.69
D20	Plumbing					
D2010 310 1600	Plumbing - Lavatories	31	Ea.	1,045.00	32,395	
D2010 440 4340	-Service Sink	8	Ea.	2,825.00	22,600	
D2010 210 2000	-Urinals	8	Ea.	1,210.00	9,680	

ASSEMBLY NUMBER		QTY.	UNIT	TOTAL COST		COST PER S.F.
				UNIT	TOTAL	
D2010 110 2080	-Water Closets	31	Ea.	1,880.00	58,280	
	Water Control, Waste Vent Piping	45%			55,330	
D2040 210 4200	- Roof Drains, 4" C.I.	2	Ea.	1,420.00	2,840	
4240	- Pipe, 9 Flr. x 12' x 12' Ea.	216	L.F.	31.80	6,869	
	Total				187,994	3.48
D30	HVAC					
D3010 520 2000	10,000 S.F. @ $8.73/S.F. Interpolate					
2040	10,000 S.F. @ $3.77/S.F.	48,000	S.F.	6.31	302,880	
	Cooling - Chilled Water, Air Cool, Cond.					
D3030 115 4000	40,000 S.F. @11.55/S.F. Interpolate	48,000	S.F.	12.11	581,280	
4040	60,000 S.F. @12.35/S.F.					
	Total				884,160	16.37
D40	Fire Protection					
D4020 310 0560	Wet Stand Pipe: 4" x 10' 1st floor	12/10	Ea.	6,500.00	7,800	
0580	Additional Floors: (12'/10 x 8)	9.6	Ea.	1,760.00	16,896	
D4020 410 8400	Cabinet Assembly	9	Ea.	990.00	8,910	
	Total				33,606	0.62
D50	Electrical					
D5020 210 0280	Office Lighting, 15/1000 S.F. - 3 Watts/S.F.	48,000	S.F.	7.28	349,440	
0240	Basement Lighting, 10/1000 S.F. - 2 Watts/S.F.	6,000	S.F.	4.83	28,980	
D5020 110 0640	Office Receptacles 16.5/1000 S.F. - 2 Watts/S.F.	48,000	S.F.	4.33	207,840	
0560	Basement Receptacles 10/1000 S.F. - 1.2 Watts/S.F.	6,000	S.F.	3.29	19,740	
D5020 140 0280	Central A.C. - 4 Watts/S.F.	48,000	S.F.	0.58	27,840	
D5020 135 0320	Misc. Connections - 1.2 Watts/S.F.	48,000	S.F.	0.30	14,400	
D5020 145 0680	Elevator Motor Power - 10 H.P.	2	Ea.	2,950.00	5,900	
D5020 130 0280	Wall Switches - 2/1000 S.F.	54,000	S.F.	0.43	23,220	
D5010 120 0560	2000 Amp Service	1	Ea.	42,500.00	42,500	
D5010 240 0400	Switchgear	1	Ea.	53,200.00	53,200	
D5010 230 0560	Feeder	50	L.F.	614.00	30,700	
D5030 910 0400	Fire Detection System - 50 Detectors	1	Ea.	30,400.00	30,400	
D5090 210 0320	Emergency Generator - 30 kW	30	kW	559.00	16,770	
	Total				850,930	15.76
E	EQUIPMENT & FURNISHINGS					
G	BUILDING SITEWORK					
	Miscellaneous					

Preliminary Estimate Cost Summary

PROJECT: Office Building	TOTAL AREA	54,000 S.F.	SHEET NO.
LOCATION	TOTAL VOLUME	648,000 C.F.	ESTIMATE NO.
ARCHITECT	COST PER S.F.		DATE
OWNER	COST PER C.F.		NO OF STORIES
QUANTITIES BY	EXTENSIONS BY		CHECKED BY

DIV	DESCRIPTION	SUBTOTAL COST	COST/S.F.		PERCENTAGE
A	SUBSTRUCTURE	201,196	$	3.73	
B10	SHELL: SUPERSTRUCTURE	1,311,764	$	24.29	
B20	SHELL: EXTERIOR ENCLOSURE	962,592	$	17.83	
B30	SHELL: ROOFING	31,479	$	0.58	
C	INTERIORS	1,230,083	$	22.78	
D10	SERVICES: CONVEYING	523,240	$	9.69	
D20	SERVICES: PLUMBING	187,994	$	3.48	
D30	SERVICES: HVAC	884,160	$	16.37	
D40	SERVICES: FIRE PROTECTION	33,606	$	0.62	
D50	SERVICES: ELECTRICAL	850,930	$	15.76	
E	EQUIPMENT & FURNISHINGS				
F	SPECIAL CONSTRUCTION & DEMO.				
G	BUILDING SITEWORK				
	BUILDING SUBTOTAL	$ 6,217,044			$ 6,217,044

Sales Tax % x Subtotal /2	N/A		$ -
General Conditions 10 % x Subtotal			$ 621,704
		Subtotal "A"	$ 6,838,748
Overhead 5 % x Subtotal "A"			$ 341,937
		Subtotal "B"	$ 7,180,686
Profit 10% x Subtotal "B"			$ 718,069
		Subtotal "C"	$ 7,898,754
Location Factor % x Subtotal "C"	N/A	Localized Cost	
Architects Fee 6.5% x Localized Cost =			$ 513,419
Contingency x Localized Cost =	N/A		$ -
		Project Total Cost	$ 8,412,173
Square Foot Cost $8,412,173/ 54,000 S.F. =		S.F. Cost	$ 155.78
Cubic Foot Cost $8,412,173/ 648,000 C.F. =		C.F. Cost	$ 12.98

241

A1010 Standard Foundations

The Strip Footing System includes: excavation; hand trim; all forms needed for footing placement; forms for 2″ x 6″ keyway (four uses); dowels; and 3,000 p.s.i. concrete.

The footing size required varies for different soils. Soil bearing capacities are listed for 3 KSF and 6 KSF. Depths of the system range from 8″ to 24″. Widths range from 16″ to 96″. Smaller strip footings may not require reinforcement.

A1010 110	Strip Footings	COST PER L.F.		
		MAT.	INST.	TOTAL
2100	Strip footing, load 2.6 KLF, soil capacity 3 KSF, 16″ wide x 8″ deep plain	9.65	12.30	21.95
2300	Load 3.9 KLF, soil capacity, 3 KSF, 24″ wide x 8″ deep, plain	11.55	13.50	25.05
2500	Load 5.1 KLF, soil capacity 3 KSF, 24″ wide x 12″ deep, reinf.	19	20.50	39.50
2700	Load 11.1 KLF, soil capacity 6 KSF, 24″ wide x 12″ deep, reinf.	19	20.50	39.50
2900	Load 6.8 KLF, soil capacity 3 KSF, 32″ wide x 12″ deep, reinf.	23	22.50	45.50
3100	Load 14.8 KLF, soil capacity 6 KSF, 32″ wide x 12″ deep, reinf.	23	22.50	45.50
3300	Load 9.3 KLF, soil capacity 3 KSF, 40″ wide x 12″ deep, reinf.	26.50	24	50.50
3500	Load 18.4 KLF, soil capacity 6 KSF, 40″ wide x 12″ deep, reinf.	27	24.50	51.50
4500	Load 10 KLF, soil capacity 3 KSF, 48″ wide x 16″ deep, reinf.	36	30.50	66.50
4700	Load 22 KLF, soil capacity 6 KSF, 48″ wide, 16″ deep, reinf.	37.50	31	68.50
5700	Load 15 KLF, soil capacity 3 KSF, 72″ wide x 20″ deep, reinf.	61.50	43	104.50
5900	Load 33 KLF, soil capacity 6 KSF, 72″ wide x 20″ deep, reinf.	66.50	46.50	113

A10 Foundations

A1010 Standard Foundations

Waterproofing

Spread Footings

Strip Footing

A1010 210	Spread Footings	COST EACH		
		MAT.	INST.	TOTAL
7090	Spread footings, 3000 psi concrete, chute delivered			
7100	Load 25K, soil capacity 3 KSF, 3'-0" sq. x 12" deep	60	107	167
7150	Load 50K, soil capacity 3 KSF, 4'-6" sq. x 12" deep	131	183	314
7200	Load 50K, soil capacity 6 KSF, 3'-0" sq. x 12" deep	60	107	167
7250	Load 75K, soil capacity 3 KSF, 5'-6" sq. x 13" deep	211	258	469
7300	Load 75K, soil capacity 6 KSF, 4'-0" sq. x 12" deep	106	157	263
7350	Load 100K, soil capacity 3 KSF, 6'-0" sq. x 14" deep	268	310	578
7410	Load 100K, soil capacity 6 KSF, 4'-6" sq. x 15" deep	162	215	377
7450	Load 125K, soil capacity 3 KSF, 7'-0" sq. x 17" deep	425	445	870
7500	Load 125K, soil capacity 6 KSF, 5'-0" sq. x 16" deep	210	258	468
7550	Load 150K, soil capacity 3 KSF 7'-6" sq. x 18" deep	515	515	1,030
7610	Load 150K, soil capacity 6 KSF, 5'-6" sq. x 18" deep	280	325	605
7650	Load 200K, soil capacity 3 KSF, 8'-6" sq. x 20" deep	740	685	1,425
7700	Load 200K, soil capacity 6 KSF, 6'-0" sq. x 20" deep	370	400	770
7750	Load 300K, soil capacity 3 KSF, 10'-6" sq. x 25" deep	1,350	1,125	2,475
7810	Load 300K, soil capacity 6 KSF, 7'-6" sq. x 25" deep	705	675	1,380
7850	Load 400K, soil capacity 3 KSF, 12'-6" sq. x 28" deep	2,175	1,675	3,850
7900	Load 400K, soil capacity 6 KSF, 8'-6" sq. x 27" deep	985	880	1,865
8010	Load 500K, soil capacity 6 KSF, 9'-6" sq. x 30" deep	1,375	1,150	2,525
8100	Load 600K, soil capacity 6 KSF, 10'-6" sq. x 33" deep	1,850	1,475	3,325
8200	Load 700K, soil capacity 6 KSF, 11'-6" sq. x 36" deep	2,375	1,800	4,175
8300	Load 800K, soil capacity 6 KSF, 12'-0" sq. x 37" deep	2,650	2,000	4,650
8400	Load 900K, soil capacity 6 KSF, 13'-0" sq. x 39" deep	3,275	2,375	5,650
8500	Load 1000K, soil capacity 6 KSF, 13'-6" sq. x 41" deep	3,725	2,625	6,350

A1010 320	Foundation Dampproofing	COST PER L.F.		
		MAT.	INST.	TOTAL
1000	Foundation dampproofing, bituminous, 1 coat, 4' high	.72	3.93	4.65
1400	8' high	1.44	7.85	9.29
1800	12' high	2.16	12.20	14.36
2000	2 coats, 4' high	1.44	4.85	6.29
2400	8' high	2.88	9.70	12.58
2800	12' high	4.32	14.95	19.27
3000	Asphalt with fibers, 1/16" thick, 4' high	1.64	4.85	6.49
3400	8' high	3.28	9.70	12.98
3800	12' high	4.92	14.95	19.87
4000	1/8" thick, 4' high	2.88	5.75	8.63

A1010 Standard Foundations

A1010 320	Foundation Dampproofing	COST PER L.F.		
		MAT.	INST.	TOTAL
4400	8' high	5.75	11.55	17.30
4800	12' high	8.65	17.70	26.35
5000	Asphalt coated board and mastic, 1/4" thick, 4' high	3.48	5.30	8.78
5400	8' high	6.95	10.60	17.55
5800	12' high	10.45	16.25	26.70
6000	1/2" thick, 4' high	5.20	7.35	12.55
6400	8' high	10.45	14.70	25.15
6800	12' high	15.65	22.50	38.15
7000	Cementitious coating, on walls, 1/8" thick coating, 4' high	7.10	7.50	14.60
7400	8' high	14.25	14.95	29.20
7800	12' high	21.50	22.50	44
8000	Cementitious/metallic slurry, 2 coat, 1/4" thick, 2' high	26.50	181	207.50
8400	4' high	53	360	413
8800	6' high	79.50	545	624.50

A1030 Slab on Grade

Fibre Expansion Joint

Wire Mesh Reinforcing

Concrete Slab

Vapor Barrier

Compacted Gravel

Reinforced Slab on Grade

A Slab on Grade system includes fine grading; 6" of compacted gravel; vapor barrier; 3500 p.s.i. concrete; bituminous fiber expansion joint; all necessary edge forms 4 uses; steel trowel finish; and sprayed on membrane curing compound. Wire mesh reinforcing used in all reinforced slabs.

Non-industrial slabs are for foot traffic only with negligible abrasion. Light industrial slabs are for pneumatic wheels and light abrasion. Industrial slabs are for solid rubber wheels and moderate abrasion. Heavy industrial slabs are for steel wheels and severe abrasion.

A1030 120	Plain & Reinforced	COST PER S.F.		
		MAT.	**INST.**	**TOTAL**
2220	Slab on grade, 4" thick, non industrial, non reinforced	1.88	2.32	4.20
2240	Reinforced	2.08	2.66	4.74
2260	Light industrial, non reinforced	2.46	2.85	5.31
2280	Reinforced	2.66	3.19	5.85
2300	Industrial, non reinforced	3.08	6	9.08
2320	Reinforced	3.28	6.30	9.58
3340	5" thick, non industrial, non reinforced	2.22	2.39	4.61
3360	Reinforced	2.42	2.73	5.15
3380	Light industrial, non reinforced	2.81	2.92	5.73
3400	Reinforced	3.01	3.26	6.27
3420	Heavy industrial, non reinforced	4.05	7.15	11.20
3440	Reinforced	4.25	7.55	11.80
4460	6" thick, non industrial, non reinforced	2.68	2.34	5.02
4480	Reinforced	3.03	2.79	5.82
4500	Light industrial, non reinforced	3.27	2.87	6.14
4520	Reinforced	3.82	3.47	7.29
4540	Heavy industrial, non reinforced	4.53	7.25	11.78
4560	Reinforced	4.88	7.70	12.58
5580	7" thick, non industrial, non reinforced	3.03	2.42	5.45
5600	Reinforced	3.46	2.90	6.36
5620	Light industrial, non reinforced	3.63	2.95	6.58
5640	Reinforced	4.06	3.43	7.49
5660	Heavy industrial, non reinforced	4.89	7.15	12.04
5680	Reinforced	5.25	7.60	12.85
6700	8" thick, non industrial, non reinforced	3.37	2.47	5.84
6720	Reinforced	3.73	2.87	6.60
6740	Light industrial, non reinforced	3.98	3	6.98
6760	Reinforced	4.34	3.40	7.74
6780	Heavy industrial, non reinforced	5.25	7.25	12.50
6800	Reinforced	5.75	7.65	13.40

A2010 Basement Excavation

Sand & Gravel 1:1

Clay 1/2:1

Line of Excavation

In general, the following items are accounted for in the table below.

Costs:
1) Excavation for building or other structure to depth and extent indicated.
2) Backfill compacted in place
3) Haul of excavated waste
4) Replacement of unsuitable backfill material with bank run gravel.

A2010 110	Building Excavation & Backfill	COST PER S.F.		
		MAT.	INST.	TOTAL
2220	Excav & fill, 1000 S.F., 4' sand, gravel, or common earth, on site storage		.88	.88
2240	Off site storage		1.31	1.31
2260	Clay excavation, bank run gravel borrow for backfill	3	1.97	4.97
2280	8' deep, sand, gravel, or common earth, on site storage		5.25	5.25
2300	Off site storage		11.30	11.30
2320	Clay excavation, bank run gravel borrow for backfill	12.20	9.25	21.45
2340	16' deep, sand, gravel, or common earth, on site storage		14.35	14.35
2350	Off site storage		28	28
2360	Clay excavation, bank run gravel borrow for backfill	33.50	24	57.50
3380	4000 S.F., 4' deep, sand, gravel, or common earth, on site storage		.46	.46
3400	Off site storage		.90	.90
3420	Clay excavation, bank run gravel borrow for backfill	1.53	1	2.53
3440	8' deep, sand, gravel, or common earth, on site storage		3.66	3.66
3460	Off site storage		6.35	6.35
3480	Clay excavation, bank run gravel borrow for backfill	5.55	5.55	11.10
3500	16' deep, sand, gravel, or common earth, on site storage		8.80	8.80
3520	Off site storage		17.25	17.25
3540	Clay, excavation, bank run gravel borrow for backfill	14.85	13.20	28.05
4560	10,000 S.F., 4' deep, sand gravel, or common earth, on site storage		.26	.26
4580	Off site storage		.52	.52
4600	Clay excavation, bank run gravel borrow for backfill	.93	.61	1.54
4620	8' deep, sand, gravel, or common earth, on site storage		3.17	3.17
4640	Off site storage		4.79	4.79
4660	Clay excavation, bank run gravel borrow for backfill	3.36	4.31	7.67
4680	16' deep, sand, gravel, or common earth, on site storage		7.15	7.15
4700	Off site storage		12.10	12.10
4720	Clay excavation, bank run gravel borrow for backfill	8.90	9.80	18.70
5740	30,000 S.F., 4' deep, sand, gravel, or common earth, on site storage		.17	.17
5760	Off site storage		.33	.33
5780	Clay excavation, bank run gravel borrow for backfill	.54	.35	.89
5860	16' deep, sand, gravel, or common earth, on site storage		6.10	6.10
5880	Off site storage		8.80	8.80
5900	Clay excavation, bank run gravel borrow for backfill	4.95	7.60	12.55
6910	100,000 S.F., 4' deep, sand, gravel, or common earth, on site storage		.07	.07
6940	8' deep, sand, gravel, or common earth, on site storage		2.63	2.63
6970	16' deep, sand, gravel, or common earth, on site storage		5.50	5.50
6980	Off site storage		6.95	6.95
6990	Clay excavation, bank run gravel borrow for backfill	2.67	6.35	9.02

A20 Basement Construction

A2020 Basement Walls

The Foundation Bearing Wall System includes: forms up to 16' high (four uses); 3,000 p.s.i. concrete placed and vibrated; and form removal with breaking form ties and patching walls. The wall systems list walls from 6" to 16" thick and are designed with minimum reinforcement.

Excavation and backfill are not included.

Please see the reference section for further design and cost information.

A2020 110					Walls, Cast in Place			
	WALL HEIGHT (FT.)	PLACING METHOD	CONCRETE (C.Y. per L.F.)	REINFORCING (LBS. per L.F.)	WALL THICKNESS (IN.)	COST PER L.F.		
						MAT.	INST.	TOTAL
1500	4'	direct chute	.074	3.3	6	18.30	45	63.30
1520			.099	4.8	8	22.50	46.50	69
1540			.123	6.0	10	26	47.50	73.50
1561			.148	7.2	12	29.50	48.50	78
1580			.173	8.1	14	33	49	82
1600			.197	9.44	16	37	50.50	87.50
3000	6'	direct chute	.111	4.95	6	27.50	68	95.50
3020			.149	7.20	8	33.50	70	103.50
3040			.184	9.00	10	39	71	110
3061			.222	10.8	12	44.50	72.50	117
5000	8'	direct chute	.148	6.6	6	36.50	90.50	127
5020			.199	9.6	8	44.50	93	137.50
5040			.250	12	10	52.50	95	147.50
5061			.296	14.39	12	59.50	97	156.50
6020	10'	direct chute	.248	12	8	56	116	172
6040			.307	14.99	10	65	118	183
6061			.370	17.99	12	74	121	195
7220	12'	pumped	.298	14.39	8	67	144	211
7240			.369	17.99	10	78	147	225
7262			.444	21.59	12	89	152	241
9220	16'	pumped	.397	19.19	8	89.50	192	281.50
9240			.492	23.99	10	104	196	300
9260			.593	28.79	12	119	202	321

B1010   Floor Construction

General: It is desirable for purposes of consistency and simplicity to maintain constant column sizes throughout building height. To do this, concrete strength may be varied (higher strength concrete at lower stories and lower strength concrete at upper stories), as well as varying the amount of reinforcing.

The table provides probably minimum column sizes with related costs and weight per lineal foot of story height.

B1010 203		C.I.P. Column, Square Tied						
	LOAD (KIPS)	STORY HEIGHT (FT.)	COLUMN SIZE (IN.)	COLUMN WEIGHT (P.L.F.)	CONCRETE STRENGTH (PSI)	COST PER V.L.F.		
						MAT.	INST.	TOTAL
0640	100	10	10	96	4000	12.80	42	54.80
0680		12	10	97	4000	13.10	42.50	55.60
0720		14	12	142	4000	17.30	51.50	68.80
0840	200	10	12	140	4000	19.45	53.50	72.95
0860		12	12	142	4000	19.90	53.50	73.40
0900		14	14	196	4000	22.50	59.50	82
0920	300	10	14	192	4000	24.50	61	85.50
0960		12	14	194	4000	25	61.50	86.50
0980		14	16	253	4000	28.50	68	96.50
1020	400	10	16	248	4000	31	70	101
1060		12	16	251	4000	31.50	71	102.50
1080		14	16	253	4000	32.50	71.50	104
1200	500	10	18	315	4000	35.50	78.50	114
1250		12	20	394	4000	43	89	132
1300		14	20	397	4000	44.50	90.50	135
1350	600	10	20	388	4000	45.50	91.50	137
1400		12	20	394	4000	47	92.50	139.50
1600		14	20	397	4000	48	93.50	141.50
3400	900	10	24	560	4000	62	113	175
3800		12	24	567	4000	64	115	179
4000		14	24	571	4000	65.50	116	181.50
7300	300	10	14	192	6000	25	60.50	85.50
7500		12	14	194	6000	25.50	61	86.50
7600		14	14	196	6000	26	61.50	87.50
8000	500	10	16	248	6000	32.50	70	102.50
8050		12	16	251	6000	33	71	104
8100		14	16	253	6000	34	71.50	105.50
8200	600	10	18	315	6000	37	79	116
8300		12	18	319	6000	38	80	118
8400		14	18	321	6000	39	80.50	119.50
8800	800	10	20	388	6000	45.50	89	134.50
8900		12	20	394	6000	47	90.50	137.50
9000		14	20	397	6000	48	91.50	139.50

B10 Superstructure

B1010 Floor Construction

B1010 203				C.I.P. Column, Square Tied					
	LOAD (KIPS)	STORY HEIGHT (FT.)	COLUMN SIZE (IN.)	COLUMN WEIGHT (P.L.F.)	CONCRETE STRENGTH (PSI)	COST PER V.L.F.			
						MAT.	INST.	TOTAL	
9100	900	10	20	388	6000	50.50	93.50	144	
9300		12	20	394	6000	51.50	94.50	146	
9600		14	20	397	6000	53	95.50	148.50	

B1010 204				C.I.P. Column, Square Tied-Minimum Reinforcing					
	LOAD (KIPS)	STORY HEIGHT (FT.)	COLUMN SIZE (IN.)	COLUMN WEIGHT (P.L.F.)	CONCRETE STRENGTH (PSI)	COST PER V.L.F.			
						MAT.	INST.	TOTAL	
9913	150	10-14	12	135	4000	15.10	49.50	64.60	
9918	300	10-14	16	240	4000	24.50	65	89.50	
9924	500	10-14	20	375	4000	37	84	121	
9930	700	10-14	24	540	4000	51.50	104	155.50	
9936	1000	10-14	28	740	4000	68.50	127	195.50	
9942	1400	10-14	32	965	4000	85	143	228	
9948	1800	10-14	36	1220	4000	107	167	274	

B1010 Floor Construction

Concentric Load

Eccentric Load

General: Data presented here is for plant produced members transported 50 miles to 100 miles to the site and erected.

Design and pricing assumptions:
Normal wt. concrete, f'c = 5 KSI

Main reinforcement, fy = 60 KSI
Ties, fy = 40 KSI

Minimum design eccentricity, 0.1t.

Concrete encased structural steel haunches are assumed where practical; otherwise galvanized rebar haunches are assumed.

Base plates are integral with columns.

Foundation anchor bolts, nuts and washers are included in price.

B1010 206			Tied, Concentric Loaded Precast Concrete Columns					
	LOAD (KIPS)	STORY HEIGHT (FT.)	COLUMN SIZE (IN.)	COLUMN WEIGHT (P.L.F.)	LOAD LEVELS	COST PER V.L.F.		
						MAT.	INST.	TOTAL
0560	100	10	12x12	164	2	123	11.25	134.25
0570		12	12x12	162	2	117	9.40	126.40
0580		14	12x12	161	2	117	9.40	126.40
0590	150	10	12x12	166	3	117	9.40	126.40
0600		12	12x12	169	3	116	8.45	124.45
0610		14	12x12	162	3	116	8.45	124.45
0620	200	10	12x12	168	4	117	10.35	127.35
0630		12	12x12	170	4	117	9.40	126.40
0640		14	12x12	220	4	116	9.40	125.40

B1010 207			Tied, Eccentric Loaded Precast Concrete Columns					
	LOAD (KIPS)	STORY HEIGHT (FT.)	COLUMN SIZE (IN.)	COLUMN WEIGHT (P.L.F.)	LOAD LEVELS	COST PER V.L.F.		
						MAT.	INST.	TOTAL
1130	100	10	12x12	161	2	115	11.25	126.25
1140		12	12x12	159	2	114	9.40	123.40
1150		14	12x12	159	2	114	9.40	123.40
1390	600	10	18x18	385	4	194	10.35	204.35
1400		12	18x18	380	4	194	9.40	203.40
1410		14	18x18	375	4	193	9.40	202.40
1480	800	10	20x20	490	4	194	10.35	204.35
1490		12	20x20	480	4	195	9.40	204.40
1500		14	20x20	475	4	195	9.40	204.40

B10 Superstructure

B1010 Floor Construction

(A) Wide Flange

(B) Pipe

(C) Pipe, Concrete Filled

(G) Rectangular Tube, Concrete Filled

B1010 208				Steel Columns				
	LOAD (KIPS)	UNSUPPORTED HEIGHT (FT.)	WEIGHT (P.L.F.)	SIZE (IN.)	TYPE	COST PER V.L.F.		
						MAT.	INST.	TOTAL
1000	25	10	13	4	A	26	10.15	36.15
1020			7.58	3	B	15.15	10.15	25.30
1040			15	3-1/2	C	18.40	10.15	28.55
1120			20	4x3	G	21.50	10.15	31.65
1200		16	16	5	A	29.50	7.60	37.10
1220			10.79	4	B	19.95	7.60	27.55
1240			36	5-1/2	C	27.50	7.60	35.10
1320			64	8x6	G	42.50	7.60	50.10
1600	50	10	16	5	A	32	10.15	42.15
1620			14.62	5	B	29	10.15	39.15
1640			24	4-1/2	C	22	10.15	32.15
1720			28	6x3	G	28	10.15	38.15
1800		16	24	8	A	44.50	7.60	52.10
1840			36	5-1/2	C	27.50	7.60	35.10
1920			64	8x6	G	42.50	7.60	50.10
2000		20	28	8	A	49	7.60	56.60
2040			49	6-5/8	C	34	7.60	41.60
2120			64	8x6	G	40.50	7.60	48.10
2200	75	10	20	6	A	40	10.15	50.15
2240			36	4-1/2	C	55	10.15	65.15
2320			35	6x4	G	32	10.15	42.15
2400		16	31	8	A	57.50	7.60	65.10
2440			49	6-5/8	C	36	7.60	43.60
2520			64	8x6	G	42.50	7.60	50.10
2600		20	31	8	A	54	7.60	61.60
2640			81	8-5/8	C	51.50	7.60	59.10
2720			64	8x6	G	40.50	7.60	48.10
2800	100	10	24	8	A	48	10.15	58.15
2840			35	4-1/2	C	55	10.15	65.15
2920			46	8x4	G	39	10.15	49.15

B1010 Floor Construction

B1010 208		Steel Columns						
	LOAD (KIPS)	UNSUPPORTED HEIGHT (FT.)	WEIGHT (P.L.F.)	SIZE (IN.)	TYPE	COST PER V.L.F.		
						MAT.	INST.	TOTAL
3000	100	16	31	8	A	57.50	7.60	65.10
3040			56	6-5/8	C	53.50	7.60	61.10
3120			64	8x6	G	42.50	7.60	50.10
3200		20	40	8	A	70	7.60	77.60
3240			81	8-5/8	C	51.50	7.60	59.10
3320			70	8x6	G	58	7.60	65.60
3400	125	10	31	8	A	62	10.15	72.15
3440			81	8	C	58.50	10.15	68.65
3520			64	8x6	G	46	10.15	56.15
3600	125	16	40	8	A	74	7.60	81.60
3640			81	8	C	54.50	7.60	62.10
3720			64	8x6	G	42.50	7.60	50.10
3800		20	48	8	A	84	7.60	91.60
3840			81	8	C	51.50	7.60	59.10
3920			60	8x6	G	58	7.60	65.60
4000	150	10	35	8	A	70	10.15	80.15
4040			81	8-5/8	C	58.50	10.15	68.65
4120			64	8x6	G	46	10.15	56.15
4200		16	45	10	A	83	7.60	90.60
4240			81	8-5/8	C	54.50	7.60	62.10
4320			70	8x6	G	61.50	7.60	69.10
4400		20	49	10	A	85.50	7.60	93.10
4440			123	10-3/4	C	73.50	7.60	81.10
4520			86	10x6	G	57	7.60	64.60
4600	200	10	45	10	A	90	10.15	100.15
4640			81	8-5/8	C	58.50	10.15	68.65
4720			70	8x6	G	66	10.15	76.15
4800		16	49	10	A	90.50	7.60	98.10
4840			123	10-3/4	C	77.50	7.60	85.10
4920			85	10x6	G	71	7.60	78.60
5200	300	10	61	14	A	122	10.15	132.15
5240			169	12-3/4	C	103	10.15	113.15
5320			86	10x6	G	99	10.15	109.15
5400		16	72	12	A	133	7.60	140.60
5440			169	12-3/4	C	95	7.60	102.60
5600		20	79	12	A	138	7.60	145.60
5640			169	12-3/4	C	90	7.60	97.60
5800	400	10	79	12	A	158	10.15	168.15
5840			178	12-3/4	C	134	10.15	144.15
6000		16	87	12	A	161	7.60	168.60
6040			178	12-3/4	C	124	7.60	131.60
6400	500	10	99	14	A	198	10.15	208.15
6600		16	109	14	A	201	7.60	208.60
6800		20	120	12	A	210	7.60	217.60
7000	600	10	120	12	A	240	10.15	250.15
7200		16	132	14	A	244	7.60	251.60
7400		20	132	14	A	231	7.60	238.60
7600	700	10	136	12	A	272	10.15	282.15
7800		16	145	14	A	268	7.60	275.60
8000		20	145	14	A	254	7.60	261.60
8200	800	10	145	14	A	289	10.15	299.15
8300		16	159	14	A	294	7.60	301.60
8400		20	176	14	A	310	7.60	317.60

B10 Superstructure

B1010 Floor Construction

| **Steel Columns**

	LOAD (KIPS)	UNSUPPORTED HEIGHT (FT.)	WEIGHT (P.L.F.)	SIZE (IN.)	TYPE	COST PER V.L.F.		
						MAT.	INST.	TOTAL
8800	900	10	159	14	A	315	10.15	325.15
8900		16	176	14	A	325	7.60	332.60
9000		20	193	14	A	340	7.60	347.60
9100	1000	10	176	14	A	350	10.15	360.15
9200		16	193	14	A	355	7.60	362.60
9300		20	211	14	A	370	7.60	377.60

B1010 Floor Construction

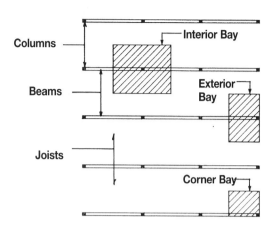

Description: Table below lists costs per S.F. of bay size for wood columns of various sizes and unsupported heights and the maximum allowable total load per S.F. per bay size.

B1010 210	Wood Columns						
	NOMINAL COLUMN SIZE (IN.)	BAY SIZE (FT.)	UNSUPPORTED HEIGHT (FT.)	MATERIAL (BF per M.S.F.)	TOTAL LOAD (P.S.F.)	COST PER S.F.	
						MAT.	INST.	TOTAL
1000	4 x 4	10 x 8	8	133	100	.19	.22	.41
1050			10	167	60	.24	.28	.52
1200		10 x 10	8	106	80	.15	.17	.32
1250			10	133	50	.19	.22	.41
1400		10 x 15	8	71	50	.10	.12	.22
1450			10	88	30	.13	.15	.28
1600		15 x 15	8	47	30	.07	.08	.15
1650			10	59	15	.09	.10	.19
2000	6 x 6	10 x 15	8	160	230	.28	.24	.52
2050			10	200	210	.36	.31	.67
2200		15 x 15	8	107	150	.19	.16	.35
2250			10	133	140	.24	.20	.44
2400		15 x 20	8	80	110	.14	.12	.26
2450			10	100	100	.18	.15	.33
2600		20 x 20	8	60	80	.11	.09	.20
2650			10	75	70	.13	.11	.24
2800		20 x 25	8	48	60	.09	.07	.16
2850			10	60	50	.11	.09	.20
3400	8 x 8	20 x 20	8	107	160	.24	.15	.39
3450			10	133	160	.30	.19	.49
3600		20 x 25	8	85	130	.12	.14	.26
3650			10	107	130	.16	.18	.34
3800		25 x 25	8	68	100	.17	.09	.26
3850			10	85	100	.21	.11	.32
4200	10 x 10	20 x 25	8	133	210	.33	.18	.51
4250			10	167	210	.42	.22	.64
4400		25 x 25	8	107	160	.27	.14	.41
4450			10	133	160	.33	.18	.51
4700	12 x 12	20 x 25	8	192	310	.49	.24	.73
4750			10	240	310	.61	.30	.91
4900		25 x 25	8	154	240	.39	.19	.58
4950			10	192	240	.49	.24	.73

B10 Superstructure

B1010 Floor Construction

"T" Shaped
Precast Beams

"L" Shaped
Precast Beams

B1010 214			"T" Shaped Precast Beams					
	SPAN (FT.)	SUPERIMPOSED LOAD (K.L.F.)	SIZE W X D (IN.)	BEAM WEIGHT (P.L.F.)	TOTAL LOAD (K.L.F.)	COST PER L.F.		
						MAT.	INST.	TOTAL
2300	15	2.8	12x16	260	3.06	166	20	186
2500		8.37	12x28	515	8.89	204	19.75	223.75
8900	45	3.34	12x60	1165	4.51	345	18.80	363.80
9900		6.5	24x60	1915	8.42	420	18.80	438.80

B1010 215			"L" Shaped Precast Beams					
	SPAN (FT.)	SUPERIMPOSED LOAD (K.L.F.)	SIZE W X D (IN.)	BEAM WEIGHT (P.L.F.)	TOTAL LOAD (K.L.F.)	COST PER L.F.		
						MAT.	INST.	TOTAL
2250	15	2.58	12x16	230	2.81	122	20	142
2400		5.92	12x24	370	6.29	144	18.80	162.80
4000	25	2.64	12x28	435	3.08	157	11.25	168.25
4450		6.44	18x36	790	7.23	203	14.10	217.10
5300	30	2.80	12x36	565	3.37	184	11.25	195.25
6400		8.66	24x44	1245	9.90	272	20.50	292.50

B10 Superstructure

B1010 Floor Construction

Cast in Place Floor Slab, One Way
General: Solid concrete slabs of uniform depth reinforced for flexure in one direction and for temperature and shrinkage in the other direction.

Cast in Place Beam & Slab, One Way
General: Solid concrete one way slab cast monolithically with reinforced concrete beams and girders.

B1010 217 — Cast in Place Slabs, One Way

	SLAB DESIGN & SPAN (FT.)	SUPERIMPOSED LOAD (P.S.F.)	THICKNESS (IN.)	TOTAL LOAD (P.S.F.)		COST PER S.F.		
						MAT.	INST.	TOTAL
2500	Single 8	40	4	90		5.50	7.15	12.65
2600		75	4	125		5.50	7.80	13.30
2700		125	4-1/2	181		5.70	7.85	13.55
2800		200	5	262		6.35	8.10	14.45
3000	Single 10	40	4	90		5.75	7.70	13.45
3100		75	4	125		5.75	7.70	13.45
3200		125	5	188		6.30	7.90	14.20
3300		200	7-1/2	293		8.15	8.90	17.05
3500	Single 15	40	5-1/2	90		6.75	7.85	14.60
3600		75	6-1/2	156		7.45	8.25	15.70
3700		125	7-1/2	219		8	8.45	16.45
3800		200	8-1/2	306		8.60	8.65	17.25
4000	Single 20	40	7-1/2	115		7.95	8.30	16.25
4100		75	9	200		8.55	8.25	16.80
4200		125	10	250		9.45	8.60	18.05
4300		200	10	324		10.30	9	19.30

B1010 219 — Cast in Place Beam & Slab, One Way

	BAY SIZE (FT.)	SUPERIMPOSED LOAD (P.S.F.)	MINIMUM COL. SIZE (IN.)	SLAB THICKNESS (IN.)	TOTAL LOAD (P.S.F.)	COST PER S.F.		
						MAT.	INST.	TOTAL
5000	20x25	40	12	5-1/2	121	6.15	9.90	16.05
5200		125	16	5-1/2	215	7.40	11.35	18.75
5500	25x25	40	12	6	129	6.45	9.70	16.15
5700		125	18	6	227	8.55	11.95	20.50
7500	30x35	40	16	8	158	8.10	11	19.10
7600		75	18	8	196	8.80	11.35	20.15
7700		125	22	8	254	10.05	12.60	22.65
7800		200	26	8	332	11.15	13.05	24.20
8000	35x35	40	16	9	169	9.20	11.35	20.55
8200		75	20	9	213	10.05	12.35	22.40
8400		125	24	9	272	11.25	12.85	24.10
8600		200	26	9	355	12.55	13.75	26.30

B10 Superstructure

B1010 Floor Construction

General: Solid concrete two way slab cast monolithically with reinforced concrete support beams and girders.

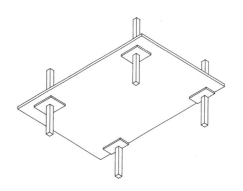

General: Flat Slab: Solid uniform depth concrete two way slabs with drop panels at columns and no column capitals.

B1010 220		Cast in Place Beam & Slab, Two Way						
	BAY SIZE (FT.)	SUPERIMPOSED LOAD (P.S.F.)	MINIMUM COL. SIZE (IN.)	SLAB THICKNESS (IN.)	TOTAL LOAD (P.S.F.)	COST PER S.F.		
						MAT.	INST.	TOTAL
4000	20 x 25	40	12	7	141	7.10	9.95	17.05
4300		75	14	7	181	8.40	10.90	19.30
4500		125	16	7	236	8.50	11.20	19.70
5100	25 x 25	40	12	7-1/2	149	7.40	10.10	17.50
5200		75	16	7-1/2	185	8.25	10.80	19.05
5300		125	18	7-1/2	250	8.90	11.75	20.65
7600	30 x 35	40	16	10	188	10.05	11.45	21.50
7700		75	18	10	225	10.80	11.90	22.70
8000		125	22	10	282	12.15	12.80	24.95
8500	35 x 35	40	16	10-1/2	193	10.90	11.70	22.60
8600		75	20	10-1/2	233	11.55	12.15	23.70
9000		125	24	10-1/2	287	13.05	13.05	26.10

B1010 222		Cast in Place Flat Slab with Drop Panels						
	BAY SIZE (FT.)	SUPERIMPOSED LOAD (P.S.F.)	MINIMUM COL. SIZE (IN.)	SLAB & DROP (IN.)	TOTAL LOAD (P.S.F.)	COST PER S.F.		
						MAT.	INST.	TOTAL
1960	20 x 20	40	12	7 - 3	132	6	7.95	13.95
1980		75	16	7 - 4	168	6.50	8.15	14.65
2000		125	18	7 - 6	221	7.30	8.40	15.70
3200	25 x 25	40	12	8-1/2 - 5-1/2	154	7.15	8.30	15.45
4000		125	20	8-1/2 - 8-1/2	243	8.50	8.95	17.45
4400		200	24	9 - 8-1/2	329	8.95	9.15	18.10
5000	25 x 30	40	14	9-1/2 - 7	168	7.85	8.65	16.50
5200		75	18	9-1/2 - 7	203	8.60	9	17.60
5600		125	22	9-1/2 - 8	256	9.15	9.20	18.35
6400	30 x 30	40	14	10-1/2 - 7-1/2	182	8.55	8.90	17.45
6600		75	18	10-1/2 - 7-1/2	217	9.35	9.25	18.60
6800		125	22	10-1/2 - 9	269	9.90	9.45	19.35
7400	30 x 35	40	16	11-1/2 - 9	196	9.40	9.25	18.65
7900		75	20	11-1/2 - 9	231	10.25	9.65	19.90
8000		125	24	11-1/2 - 11	284	10.80	9.85	20.65
9000	35 x 35	40	16	12 - 9	202	9.65	9.30	18.95
9400		75	20	12 - 11	240	10.65	9.75	20.40
9600		125	24	12 - 11	290	11.10	9.95	21.05

B10 Superstructure

B1010 Floor Construction

Multispan Joist Slab

General: Flat Plates: Solid uniform depth concrete two way slab without drops or interior beams. Primary design limit is shear at columns.

General: Combination of thin concrete slab and monolithic ribs at uniform spacing to reduce dead weight and increase rigidity.

B1010 223	Cast in Place Flat Plate							
	BAY SIZE (FT.)	SUPERIMPOSED LOAD (P.S.F.)	MINIMUM COL. SIZE (IN.)	SLAB THICKNESS (IN.)	TOTAL LOAD (P.S.F.)	COST PER S.F.		
						MAT.	INST.	TOTAL
3000	15 x 20	40	14	7	127	5.85	7.65	13.50
3400		75	16	7-1/2	169	6.30	7.80	14.10
3600		125	22	8-1/2	231	6.95	8	14.95
3800		175	24	8-1/2	281	7.05	8	15.05
4200	20 x 20	40	16	7	127	5.85	7.65	13.50
4400		75	20	7-1/2	175	6.40	7.80	14.20
4600		125	24	8-1/2	231	7	8	15
5000		175	24	8-1/2	281	7.05	8.05	15.10
5600	20 x 25	40	18	8-1/2	146	6.90	8	14.90
6000		75	20	9	188	7.15	8.05	15.20
6400		125	26	9-1/2	244	7.85	8.35	16.20
6600		175	30	10	300	8.15	8.45	16.60
7000	25 x 25	40	20	9	152	7.15	8.05	15.20
7400		75	24	9-1/2	194	7.60	8.30	15.90
7600		125	30	10	250	8.20	8.50	16.70

B1010 226	Cast in Place Multispan Joist Slab							
	BAY SIZE (FT.)	SUPERIMPOSED LOAD (P.S.F.)	MINIMUM COL. SIZE (IN.)	RIB DEPTH (IN.)	TOTAL LOAD (P.S.F.)	COST PER S.F.		
						MAT.	INST.	TOTAL
2000	15 x 15	40	12	8	115	6.65	9.40	16.05
2100		75	12	8	150	6.70	9.40	16.10
2200		125	12	8	200	6.90	9.50	16.40
2300		200	14	8	275	7.15	9.90	17.05
2600	15 x 20	40	12	8	115	6.80	9.40	16.20
2800		75	12	8	150	7	9.90	16.90
3000		125	14	8	200	7.30	10.05	17.35
3300		200	16	8	275	7.75	10.30	18.05
3600	20 x 20	40	12	10	120	7	9.25	16.25
3900		75	14	10	155	7.35	9.80	17.15
4000		125	16	10	205	7.45	10	17.45
4100		200	18	10	280	7.90	10.45	18.35
6200	30 x 30	40	14	14	131	7.85	9.70	17.55
6400		75	18	14	166	8.15	10	18.15
6600		125	20	14	216	8.75	10.55	19.30
6700		200	24	16	297	9.45	10.95	20.40

B10 Superstructure

B1010 Floor Construction

Waffle Slab

B1010 227	Cast in Place Waffle Slab						

	BAY SIZE (FT.)	SUPERIMPOSED LOAD (P.S.F.)	MINIMUM COL. SIZE (IN.)	RIB DEPTH (IN.)	TOTAL LOAD (P.S.F.)	COST PER S.F.		
						MAT.	INST.	TOTAL
3900	20 x 20	40	12	8	144	6.65	10.05	16.70
4000		75	12	8	179	6.85	10.15	17
4100		125	16	8	229	7.15	10.25	17.40
4200		200	18	8	304	7.90	10.60	18.50
4400	20 x 25	40	12	8	146	6.85	10.10	16.95
4500		75	14	8	181	7.20	10.25	17.45
4600		125	16	8	231	7.45	10.35	17.80
4700		200	18	8	306	8.10	10.65	18.75
4900	25 x 25	40	12	10	150	7.15	10.20	17.35
5000		75	16	10	185	7.55	10.40	17.95
5300		125	18	10	235	7.95	10.60	18.55
5500		200	20	10	310	8.40	10.75	19.15
5700	25 x 30	40	14	10	154	7.35	10.30	17.65
5800		75	16	10	189	7.75	10.50	18.25
5900		125	18	10	239	8.15	10.65	18.80
6000		200	20	12	329	9.35	11.10	20.45
6400	30 x 30	40	14	12	169	8	10.50	18.50
6500		75	18	12	204	8.35	10.65	19
6600		125	20	12	254	8.60	10.75	19.35
6700		200	24	12	329	9.95	11.35	21.30
6900	30 x 35	40	16	12	169	8.25	10.60	18.85
7000		75	18	12	204	8.25	10.60	18.85
7100		125	22	12	254	9.05	10.95	20
7200		200	26	14	334	10.55	11.60	22.15
7400	35 x 35	40	16	14	174	8.90	10.85	19.75
7500		75	20	14	209	9.30	11.05	20.35
7600		125	24	14	259	9.65	11.20	20.85
7700		200	26	16	346	10.85	11.75	22.60
8000	35 x 40	40	18	14	176	9.35	11.05	20.40
8300		75	22	14	211	9.90	11.30	21.20
8500		125	26	16	271	10.50	11.55	22.05
8750		200	30	20	372	11.95	12.15	24.10
9200	40 x 40	40	18	14	176	9.90	11.30	21.20
9400		75	24	14	211	10.55	11.60	22.15
9500		125	26	16	271	10.75	11.65	22.40
9700	40 x 45	40	20	16	186	10.30	11.45	21.75
9800		75	24	16	221	10.95	11.75	22.70
9900		125	28	16	271	11.30	11.90	23.20

B10 Superstructure

B1010 Floor Construction

Precast Plank with No Topping

Precast Plank with 2″ Concrete Topping

B1010 229			Precast Plank with No Topping					
	SPAN (FT.)	SUPERIMPOSED LOAD (P.S.F.)	TOTAL DEPTH (IN.)	DEAD LOAD (P.S.F.)	TOTAL LOAD (P.S.F.)	COST PER S.F.		
						MAT.	INST.	TOTAL
0720	10	40	4	50	90	6.25	3.14	9.39
0750		75	6	50	125	7.35	2.68	10.03
0770		100	6	50	150	7.35	2.68	10.03
0800	15	40	6	50	90	7.35	2.68	10.03
0820		75	6	50	125	7.35	2.68	10.03
0850		100	6	50	150	7.35	2.68	10.03
0950	25	40	6	50	90	7.35	2.68	10.03
0970		75	8	55	130	8.05	2.35	10.40
1000		100	8	55	155	8.05	2.35	10.40
1200	30	40	8	55	95	8.05	2.35	10.40
1300		75	8	55	130	8.05	2.35	10.40
1400		100	10	70	170	8.50	2.09	10.59
1500	40	40	10	70	110	8.50	2.09	10.59
1600		75	12	70	145	9.15	1.88	11.03
1700	45	40	12	70	110	9.15	1.88	11.03

B1010 230			Precast Plank with 2″ Concrete Topping					
	SPAN (FT.)	SUPERIMPOSED LOAD (P.S.F.)	TOTAL DEPTH (IN.)	DEAD LOAD (P.S.F.)	TOTAL LOAD (P.S.F.)	COST PER S.F.		
						MAT.	INST.	TOTAL
2000	10	40	6	75	115	7.25	5.20	12.45
2100		75	8	75	150	8.35	4.74	13.09
2200		100	8	75	175	8.35	4.74	13.09
2500	15	40	8	75	115	8.35	4.74	13.09
2600		75	8	75	150	8.35	4.74	13.09
2700		100	8	75	175	8.35	4.74	13.09
3100	25	40	8	75	115	8.35	4.74	13.09
3200		75	8	75	150	8.35	4.74	13.09
3300		100	10	80	180	9.05	4.41	13.46
3400	30	40	10	80	120	9.05	4.41	13.46
3500		75	10	80	155	9.05	4.41	13.46
3600		100	10	80	180	9.05	4.41	13.46
4000	40	40	12	95	135	9.50	4.15	13.65
4500		75	14	95	170	10.15	3.94	14.09
5000	45	40	14	95	135	10.15	3.94	14.09

B1010 Floor Construction

Precast Double "T" Beams with No Topping

Precast Double "T" Beams with 2" Topping

Most widely used for moderate span floors and moderate and long span roofs. At shorter spans, they tend to be competitive with hollow core slabs. They are also used as wall panels.

B1010 234			Precast Double "T" Beams with No Topping					
	SPAN (FT.)	SUPERIMPOSED LOAD (P.S.F.)	DBL. "T" SIZE D (IN.) W (FT.)	CONCRETE "T" TYPE	TOTAL LOAD (P.S.F.)	COST PER S.F.		
						MAT.	INST.	TOTAL
4300	50	30	20x8	Lt. Wt.	66	8.60	1.44	10.04
4400		40	20x8	Lt. Wt.	76	9.20	1.30	10.50
4500		50	20x8	Lt. Wt.	86	9.30	1.56	10.86
4600		75	20x8	Lt. Wt.	111	9.40	1.78	11.18
5600	70	30	32x10	Lt. Wt.	78	10.05	.99	11.04
5750		40	32x10	Lt. Wt.	88	10.15	1.19	11.34
5900		50	32x10	Lt. Wt.	98	10.20	1.38	11.58
6000		75	32x10	Lt. Wt.	123	10.30	1.67	11.97
6100		100	32x10	Lt. Wt.	148	10.55	2.26	12.81
6200	80	30	32x10	Lt. Wt.	78	10.20	1.38	11.58
6300		40	32x10	Lt. Wt.	88	10.30	1.67	11.97
6400		50	32x10	Lt. Wt.	98	10.45	1.96	12.41

B1010 235			Precast Double "T" Beams With 2" Topping					
	SPAN (FT.)	SUPERIMPOSED LOAD (P.S.F.)	DBL. "T" SIZE D (IN.) W (FT.)	CONCRETE "T" TYPE	TOTAL LOAD (P.S.F.)	COST PER S.F.		
						MAT.	INST.	TOTAL
7100	40	30	18x8	Reg. Wt.	120	7.85	2.84	10.69
7200		40	20x8	Reg. Wt.	130	7.90	2.70	10.60
7300		50	20x8	Reg. Wt.	140	8	2.96	10.96
7400		75	20x8	Reg. Wt.	165	8.05	3.07	11.12
7500		100	20x8	Reg. Wt.	190	8.15	3.45	11.60
7550	50	30	24x8	Reg. Wt.	120	8.45	2.84	11.29
7600		40	24x8	Reg. Wt.	130	8.50	2.93	11.43
7750		50	24x8	Reg. Wt.	140	8.50	2.95	11.45
7800		75	24x8	Reg. Wt.	165	8.60	3.32	11.92
7900		100	32x10	Reg. Wt.	189	9.75	2.71	12.46

B1010 Floor Construction

Precast Beam and Plank with No Topping

Precast Beam and Plank with 2" Topping

B1010 236 — Precast Beam & Plank with No Topping

	BAY SIZE (FT.)	SUPERIMPOSED LOAD (P.S.F.)	PLANK THICKNESS (IN.)	TOTAL DEPTH (IN.)	TOTAL LOAD (P.S.F.)	COST PER S.F.		
						MAT.	INST.	TOTAL
6200	25x25	40	8	28	118	17.40	4.23	21.63
6400		75	8	36	158	18.15	4.23	22.38
6500		100	8	36	183	18.15	4.23	22.38
7000	25x30	40	8	36	110	16.50	4.23	20.73
7200		75	10	36	159	17.35	3.97	21.32
7400		100	12	36	188	18.75	3.76	22.51
7600	30x30	40	8	36	121	17.05	4.23	21.28
8000		75	10	44	140	17.95	4.23	22.18
8250		100	12	52	206	20.50	3.76	24.26
8500	30x35	40	12	44	135	17.40	3.76	21.16
8750		75	12	52	176	18.45	3.76	22.21
9000	35x35	40	12	52	141	18.50	3.76	22.26
9250		75	12	60	181	19.40	3.76	23.16
9500	35x40	40	12	52	137	17.70	4.70	22.40

B1010 238 — Precast Beam & Plank with 2" Topping

	BAY SIZE (FT.)	SUPERIMPOSED LOAD (P.S.F.)	PLANK THICKNESS (IN.)	TOTAL DEPTH (IN.)	TOTAL LOAD (P.S.F.)	COST PER S.F.		
						MAT.	INST.	TOTAL
4300	20x20	40	6	22	135	17.75	6.25	24
4400		75	6	24	173	18.85	6.25	25.10
4500		100	6	28	200	19.40	6.25	25.65
4600	20x25	40	6	26	134	16.70	6.25	22.95
5000		75	8	30	177	18.05	5.90	23.95
5200		100	8	30	202	18.05	5.90	23.95
5400	25x25	40	6	38	143	18.45	6.20	24.65
5600		75	8	38	183	18.45	6.20	24.65
6000		100	8	46	216	20.50	5.90	26.40
6200	25x30	40	8	38	144	17.50	5.85	23.35
6400		75	10	46	200	19.10	5.60	24.70
6600		100	10	46	225	19.10	5.60	24.70
7000	30x30	40	8	46	150	18.90	5.85	24.75
7200		75	10	54	181	20	5.85	25.85
7600		100	10	54	231	20	5.85	25.85
7800	30x35	40	10	54	166	18.80	5.55	24.35
8000		75	12	54	200	19.15	5.35	24.50

B1010 Floor Construction

General: Beams and double tees priced here are for plant produced prestressed members transported to the site and erected.

The 2″ structural topping is applied after the beams and double tees are in place and is reinforced with W.W.F.

Note: Deduct from prices 20% for Southern States. Add to prices 10% for Western States.

With Topping

B1010 239		Precast Double "T" & 2″ Topping on Precast Beams						
	BAY SIZE (FT.)	SUPERIMPOSED LOAD (P.S.F.)	DEPTH (IN.)		TOTAL LOAD (P.S.F.)	COST PER S.F.		
						MAT.	INST.	TOTAL
3000	25x30	40	38		130	17.30	5.05	22.35
3100		75	38		168	17.30	5.05	22.35
3300		100	46		196	18.05	5.05	23.10
3600	30x30	40	46		150	18.70	5.05	23.75
3750		75	46		174	18.70	5.05	23.75
4000		100	54		203	19.60	5.05	24.65
4100	30x40	40	46		136	15.25	4.77	20.02
4300		75	54		173	15.90	4.77	20.67
4400		100	62		204	16.70	4.77	21.47
4600	30x50	40	54		138	14.60	4.61	19.21
4800		75	54		181	15.20	4.61	19.81
5000		100	54		219	16.75	4.35	21.10
5200	30x60	40	62		151	15.40	4.36	19.76
5400		75	62		192	16.10	4.36	20.46
5600		100	62		215	16.10	4.36	20.46
5800	35x40	40	54		139	16.25	4.75	21
6000		75	62		179	17.60	4.62	22.22
6250		100	62		212	18	4.62	22.62
6500	35x50	40	62		142	15.50	4.61	20.11
6750		75	62		186	16.20	4.61	20.81
7300		100	62		231	18.60	4.35	22.95
7600	35x60	40	54		154	17.45	4.23	21.68
7750		75	54		179	17.90	4.23	22.13
8000		100	62		224	18.70	4.23	22.93
8250	40x40	40	62		145	17.85	5.70	23.55
8400		75	62		187	18.50	5.70	24.20
8750		100	62		223	19.70	5.70	25.40
9000	40x50	40	62		151	16.60	5.55	22.15
9300		75	62		193	17.50	4.75	22.25
9800	40x60	40	62		164	21.50	4.93	26.43

B1010 Floor Construction

General: The following table is based upon structural wide flange (WF) beam and girder framing. Non-composite action is assumed between WF framing and decking. Deck costs not included.

The deck spans the short direction. The steel beams and girders are fireproofed with sprayed fiber fireproofing.

No columns included in price.

B1010 241		W Shape Beams & Girders						
	BAY SIZE (FT.) BEAM X GIRD	SUPERIMPOSED LOAD (P.S.F.)	STEEL FRAMING DEPTH (IN.)	FIREPROOFING (S.F. PER S.F.)	TOTAL LOAD (P.S.F.)	COST PER S.F.		
						MAT.	INST.	TOTAL
1350	15x20	40	12	.535	50	5.70	1.76	7.46
1400		40	16	.65	90	7.45	2.27	9.72
1450		75	18	.694	125	9.80	2.86	12.66
1500		125	24	.796	175	13.55	3.98	17.53
2000	20x20	40	12	.55	50	6.35	1.94	8.29
2050		40	14	.579	90	8.75	2.52	11.27
2100		75	16	.672	125	10.45	3	13.45
2150		125	16	.714	175	12.50	3.67	16.17
2900	20x25	40	16	.53	50	7.05	2.08	9.13
2950		40	18	.621	96	11.10	3.12	14.22
3000		75	18	.651	131	12.80	3.55	16.35
3050		125	24	.77	200	16.90	4.80	21.70
5200	25x25	40	18	.486	50	7.65	2.19	9.84
5300		40	18	.592	96	11.45	3.17	14.62
5400		75	21	.668	131	13.85	3.80	17.65
5450		125	24	.738	191	18.25	5.10	23.35
6300	25x30	40	21	.568	50	8.75	2.51	11.26
6350		40	21	.694	90	11.80	3.34	15.14
6400		75	24	.776	125	15.25	4.22	19.47
6450		125	30	.904	175	18.35	5.25	23.60
7700	30x30	40	21	.52	50	9.35	2.63	11.98
7750		40	24	.629	103	14.45	3.93	18.38
7800		75	30	.715	138	17.20	4.64	21.84
7850		125	36	.822	206	22.50	6.30	28.80
8250	30x35	40	21	.508	50	10.70	2.94	13.64
8300		40	24	.651	109	15.85	4.27	20.12
8350		75	33	.732	150	19.25	5.15	24.40
8400		125	36	.802	225	24	6.60	30.60
9550	35x35	40	27	.560	50	11.05	3.06	14.11
9600		40	36	.706	109	20	5.35	25.35
9650		75	36	.750	150	21	5.55	26.55
9820		125	36	.797	225	28	7.65	35.65

B10 Superstructure

B1010 Floor Construction

Description: Table below lists costs for light gage CEE or PUNCHED DOUBLE joists to suit the span and loading with the minimum thickness subfloor required by the joist spacing.

	SPAN (FT.)	SUPERIMPOSED LOAD (P.S.F.)	FRAMING DEPTH (IN.)	FRAMING SPAC. (IN.)	TOTAL LOAD (P.S.F.)	COST PER S.F.		
B1010 244 Light Gauge Steel Floor Systems						MAT.	INST.	TOTAL
1500	15	40	8	16	54	3.23	1.94	5.17
1550			8	24	54	3.37	1.99	5.36
1600		65	10	16	80	4.33	2.38	6.71
1650			10	24	80	4.25	2.32	6.57
1700		75	10	16	90	4.33	2.38	6.71
1750			10	24	90	4.25	2.32	6.57
1800		100	10	16	116	5.75	3.02	8.77
1850			10	24	116	4.25	2.32	6.57
1900		125	10	16	141	5.75	3.02	8.77
1950			10	24	141	4.25	2.32	6.57
2500	20	40	8	16	55	3.75	2.20	5.95
2550			8	24	55	3.37	1.99	5.36
2600		65	8	16	80	4.40	2.51	6.91
2650			10	24	80	4.26	2.34	6.60
2700		75	10	16	90	4.07	2.26	6.33
2750			12	24	90	4.79	2.02	6.81
2800		100	10	16	115	4.07	2.26	6.33
2850			10	24	116	5.55	2.60	8.15
2900		125	12	16	142	6.55	2.56	9.11
2950			12	24	141	6.05	2.78	8.83
3500	25	40	10	16	55	4.33	2.38	6.71
3550			10	24	55	4.25	2.32	6.57
3600		65	10	16	81	5.75	3.02	8.77
3650			12	24	81	5.35	2.19	7.54
3700		75	12	16	92	6.50	2.55	9.05
3750			12	24	91	6.05	2.78	8.83
3800		100	12	16	117	7.40	2.82	10.22
3850		125	12	16	143	8.50	3.73	12.23
4500	30	40	12	16	57	6.50	2.55	9.05
4550			12	24	56	5.30	2.18	7.48
4600		65	12	16	82	7.35	2.81	10.16
4650		75	12	16	92	7.35	2.81	10.16

B10 Superstructure

B1010 Floor Construction

Description: The table below lists costs per S.F. for a floor system on bearing walls using open web steel joists, galvanized steel slab form and 2-1/2" concrete slab reinforced with welded wire fabric. Costs of the bearing walls are not included.

B1010 246		Deck & Joists on Bearing Walls						
	SPAN (FT.)	SUPERIMPOSED LOAD (P.S.F.)	JOIST SPACING FT. - IN.	DEPTH (IN.)	TOTAL LOAD (P.S.F.)	COST PER S.F.		
						MAT.	INST.	TOTAL
1050	20	40	2-0	14-1/2	83	6.75	3.47	10.22
1070		65	2-0	16-1/2	109	7.35	3.68	11.03
1100		75	2-0	16-1/2	119	7.35	3.68	11.03
1120		100	2-0	18-1/2	145	7.40	3.69	11.09
1150		125	1-9	18-1/2	170	9.10	4.40	13.50
1170	25	40	2-0	18-1/2	84	8.20	4.09	12.29
1200		65	2-0	20-1/2	109	8.50	4.20	12.70
1220		75	2-0	20-1/2	119	8.20	3.94	12.14
1250		100	2-0	22-1/2	145	8.35	3.98	12.33
1270		125	1-9	22-1/2	170	9.60	4.36	13.96
1300	30	40	2-0	22-1/2	84	8.35	3.58	11.93
1320		65	2-0	24-1/2	110	9.10	3.74	12.84
1350		75	2-0	26-1/2	121	9.45	3.82	13.27
1370		100	2-0	26-1/2	146	10.10	3.95	14.05
1400		125	2-0	24-1/2	172	11	5.20	16.20
1420	35	40	2-0	26-1/2	85	10.40	4.97	15.37
1450		65	2-0	28-1/2	111	10.75	5.10	15.85
1470		75	2-0	28-1/2	121	10.75	5.10	15.85
1500		100	1-11	28-1/2	147	11.70	5.45	17.15
1520		125	1-8	28-1/2	172	12.85	5.85	18.70
1550	5/8" gyp. fireproof.							
1560	On metal furring, add					.83	2.58	3.41

B10 Superstructure

B1010 Floor Construction

The table below lists costs for a floor system on exterior bearing walls and interior columns and beams using open web steel joists, galvanized steel slab form, 2-1/2″ concrete slab reinforced with welded wire fabric.

Columns are sized to accommodate one floor plus roof loading but costs for columns are from suspended floor to slab on grade only. Costs of the bearing walls are not included.

Steel Joists and Concrete Slab on Walls and Beams

B1010 248		Steel Joists on Beam & Wall						
	BAY SIZE (FT.)	SUPERIMPOSED LOAD (P.S.F.)	DEPTH (IN.)	TOTAL LOAD (P.S.F.)	COLUMN ADD	COST PER S.F.		
						MAT.	INST.	TOTAL
1720	25x25	40	23	84		10	4.41	14.41
1730					columns	.47	.12	.59
1750	25x25	65	29	110		10.55	4.55	15.10
1760					columns	.47	.12	.59
1770	25x25	75	26	120		10.90	4.46	15.36
1780					columns	.55	.13	.68
1800	25x25	100	29	145		12.35	4.88	17.23
1810					columns	.55	.13	.68
1820	25x25	125	29	170		12.95	5.05	18
1830					columns	.61	.15	.76
2020	30x30	40	29	84		10.70	4.01	14.71
2030					columns	.42	.11	.53
2050	30x30	65	29	110		12.15	4.34	16.49
2060					columns	.42	.11	.53
2070	30x30	75	32	120		12.50	4.42	16.92
2080					columns	.49	.12	.61
2100	30x30	100	35	145		13.90	4.73	18.63
2110					columns	.57	.13	.70
2120	30x30	125	35	172		15.20	6.10	21.30
2130					columns	.68	.16	.84
2170	30x35	40	29	85		12.15	4.33	16.48
2180					columns	.41	.09	.50
2200	30x35	65	29	111		13.70	5.70	19.40
2210					columns	.47	.11	.58
2220	30x35	75	32	121		13.65	5.70	19.35
2230					columns	.48	.12	.60
2250	30x35	100	35	148		14.80	4.92	19.72
2260					columns	.58	.15	.73
2270	30x35	125	38	173		16.65	5.35	22
2280					columns	.59	.15	.74
2320	35x35	40	32	85		12.50	4.42	16.92
2330					columns	.42	.11	.53
2350	35x35	65	35	111		14.45	5.90	20.35
2360					columns	.50	.12	.62
2370	35x35	75	35	121		14.75	6	20.75
2380					columns	.50	.12	.62
2400	35x35	100	38	148		15.40	6.20	21.60
2410					columns	.62	.15	.77

B10 Superstructure

B1010 Floor Construction

| B1010 248 | | | | | | Steel Joists on Beam & Wall | | |

	BAY SIZE (FT.)	SUPERIMPOSED LOAD (P.S.F.)	DEPTH (IN.)	TOTAL LOAD (P.S.F.)	COLUMN ADD	COST PER S.F.		
						MAT.	INST.	TOTAL
2460	5/8 gyp. fireproof.							
2475	On metal furring, add					.83	2.58	3.41

B10 Superstructure

B1010 Floor Construction

The table below lists costs per S.F. for a floor system on steel columns and beams using open web steel joists, galvanized steel slab form, 2-1/2″ concrete slab reinforced with welded wire fabric.

Columns are sized to accommodate one floor plus roof loading but costs for columns are from floor to grade only.

Steel Joists and Concrete Slab on Steel Columns and Beams

B1010 250		Steel Joists, Beams & Slab on Columns						
	BAY SIZE (FT.)	SUPERIMPOSED LOAD (P.S.F.)	DEPTH (IN.)	TOTAL LOAD (P.S.F.)	COLUMN ADD	COST PER S.F.		
						MAT.	INST.	TOTAL
2350	15x20	40	17	83		9.60	4.05	13.65
2400					column	1.56	.37	1.93
2450	15x20	65	19	108		10.65	4.31	14.96
2500					column	1.56	.37	1.93
2550	15x20	75	19	119		11.10	4.44	15.54
2600					column	1.70	.41	2.11
2650	15x20	100	19	144		11.80	4.63	16.43
2700					column	1.70	.41	2.11
2750	15x20	125	19	170		13.25	5.25	18.50
2800					column	2.27	.55	2.82
2850	20x20	40	19	83		10.45	4.24	14.69
2900					column	1.28	.31	1.59
2950	20x20	65	23	109		11.55	4.54	16.09
3000					column	1.70	.41	2.11
3100	20x20	75	26	119		12.20	4.70	16.90
3200					column	1.70	.41	2.11
3400	20x20	100	23	144		12.65	4.81	17.46
3450					column	1.70	.41	2.11
3500	20x20	125	23	170		14.10	5.20	19.30
3600					column	2.04	.49	2.53
3700	20x25	40	44	83		12	4.87	16.87
3800					column	1.36	.33	1.69
3900	20x25	65	26	110		13.10	5.15	18.25
4000					column	1.36	.33	1.69
4100	20x25	75	26	120		12.85	4.91	17.76
4200					column	1.63	.40	2.03
4300	20x25	100	26	145		13.65	5.10	18.75
4400					column	1.63	.40	2.03
4500	20x25	125	29	170		15.25	5.55	20.80
4600					column	1.91	.47	2.38
4700	25x25	40	23	84		12.90	5.05	17.95
4800					column	1.31	.32	1.63
4900	25x25	65	29	110		13.65	5.30	18.95
5000					column	1.31	.32	1.63
5100	25x25	75	26	120		14.25	5.25	19.50
5200					column	1.52	.37	1.89

B1010 Floor Construction

| B1010 250 | | Steel Joists, Beams & Slab on Columns | | | | | | |

	BAY SIZE (FT.)	SUPERIMPOSED LOAD (P.S.F.)	DEPTH (IN.)	TOTAL LOAD (P.S.F.)	COLUMN ADD	COST PER S.F.		
						MAT.	INST.	TOTAL
5300	25x25	100	29	145		15.90	5.70	21.60
5400					column	1.52	.37	1.89
5500	25x25	125	32	170		16.80	5.95	22.75
5600					column	1.69	.41	2.10
5700	25x30	40	29	84		13.40	5.30	18.70
5800					column	1.27	.31	1.58
5900	25x30	65	29	110		13.95	5.55	19.50
6000					column	1.27	.31	1.58
6050	25x30	75	29	120		15	5	20
6100					column	1.41	.35	1.76
6150	25x30	100	29	145		16.30	5.30	21.60
6200					column	1.41	.35	1.76
6250	25x30	125	32	170		17.55	6.55	24.10
6300					column	1.62	.39	2.01
6350	30x30	40	29	84		13.95	4.78	18.73
6400					column	1.17	.28	1.45
6500	30x30	65	29	110		15.90	5.25	21.15
6600					column	1.17	.28	1.45
6700	30x30	75	32	120		16.25	5.30	21.55
6800					column	1.35	.33	1.68
6900	30x30	100	35	145		18.05	5.75	23.80
7000					column	1.57	.39	1.96
7100	30x30	125	35	172		19.80	7.15	26.95
7200					column	1.75	.43	2.18
7300	30x35	40	29	85		15.80	5.20	21
7400					column	1	.24	1.24
7500	30x35	65	29	111		17.60	6.60	24.20
7600					column	1.30	.32	1.62
7700	30x35	75	32	121		17.60	6.60	24.20
7800					column	1.32	.32	1.64
7900	30x35	100	35	148		19.10	5.95	25.05
8000					column	1.62	.39	2.01
8100	30x35	125	38	173		21	6.45	27.45
8200					column	1.65	.40	2.05
8300	35x35	40	32	85		16.25	5.30	21.55
8400					column	1.16	.28	1.44
8500	35x35	65	35	111		18.55	6.90	25.45
8600					column	1.39	.33	1.72
9300	35x35	75	38	121		19.05	7	26.05
9400					column	1.39	.33	1.72
9500	35x35	100	38	148		20.50	7.40	27.90
9600					column	1.71	.41	2.12
9750	35x35	125	41	173		22	6.65	28.65
9800					column	1.75	.43	2.18
9810	5/8 gyp. fireproof.							
9815	On metal furring, add					.83	2.58	3.41

B1010 Floor Construction

General: Composite construction of wide flange beams and concrete slabs is most efficiently used when loads are heavy and spans are moderately long. It is stiffer with less deflection than non-composite construction of similar depth and spans.

Composite Beam & Cast in Place Slab

B1010 252	Composite Beam & Cast in Place Slab							
	BAY SIZE (FT.)	SUPERIMPOSED LOAD (P.S.F.)	SLAB THICKNESS (IN.)	TOTAL DEPTH (FT.-IN.)	TOTAL LOAD (P.S.F.)	COST PER S.F.		
						MAT.	INST.	TOTAL
3800	20x25	40	4	1 - 8	94	12.40	9.60	22
3900		75	4	1 - 8	130	14.35	10.20	24.55
4000		125	4	1 - 10	181	16.65	10.90	27.55
4100		200	5	2 - 2	272	21.50	12.40	33.90
4200	25x25	40	4-1/2	1 - 8-1/2	99	12.95	9.70	22.65
4300		75	4-1/2	1 - 10-1/2	136	15.55	10.45	26
4400		125	5-1/2	2 - 0-1/2	200	18.25	11.40	29.65
4500		200	5-1/2	2 - 2-1/2	278	22.50	12.60	35.10
4600	25x30	40	4-1/2	1 - 8-1/2	100	14.40	10.15	24.55
4700		75	4-1/2	1 - 10-1/2	136	16.80	10.85	27.65
4800		125	5-1/2	2 - 2-1/2	202	20.50	11.95	32.45
4900		200	5-1/2	2 - 5-1/2	279	24.50	13.10	37.60
5000	30x30	40	4	1 - 8	95	14.40	10.10	24.50
5200		75	4	2 - 1	131	16.60	10.70	27.30
5400		125	4	2 - 4	183	20.50	11.85	32.35
5600		200	5	2 - 10	274	26	13.45	39.45
5800	30x35	40	4	2 - 1	95	15.25	10.30	25.55
6000		75	4	2 - 4	139	18.25	11.15	29.40
6250		125	4	2 - 4	185	23.50	12.65	36.15
6500		200	5	2 - 11	276	28.50	14.15	42.65
8000	35x35	40	4	2 - 1	96	16.05	10.55	26.60
8250		75	4	2 - 4	133	19	11.40	30.40
8500		125	4	2 - 10	185	23	12.45	35.45
8750		200	5	3 - 5	276	29.50	14.35	43.85
9000	35x40	40	4	2 - 4	97	17.80	11	28.80
9250		75	4	2 - 4	134	21	11.75	32.75
9500		125	4	2 - 7	186	25	13	38
9750		200	5	3 - 5	278	32.50	15.05	47.55

B10 Superstructure

B1010   Floor Construction

The table below lists costs per S.F. for floors using steel beams and girders, composite steel deck, concrete slab reinforced with W.W.F. and sprayed fiber fireproofing (non- asbestos) on the steel beams and girders and on the steel deck.

Wide Flange, Composite Deck & Slab

B1010 254		W Shape, Composite Deck, & Slab						
	BAY SIZE (FT.) BEAM X GIRD	SUPERIMPOSED LOAD (P.S.F.)	SLAB THICKNESS (IN.)	TOTAL DEPTH (FT.-IN.)	TOTAL LOAD (P.S.F.)	COST PER S.F.		
						MAT.	INST.	TOTAL
0540	15x20	40	5	1-7	89	13.40	5.90	19.30
0560		75	5	1-9	125	14.75	6.30	21.05
0580		125	5	1-11	176	17.25	6.90	24.15
0600		200	5	2-2	254	21.50	7.85	29.35
0700	20x20	40	5	1-9	90	14.60	6.15	20.75
0720		75	5	1-9	126	16.50	6.70	23.20
0740		125	5	1-11	177	18.75	7.30	26.05
0760		200	5	2-5	255	22.50	8.15	30.65
0780	20x25	40	5	1-9	90	14.90	6.15	21.05
0800		75	5	1-9	127	18.60	7.05	25.65
0820		125	5	1-11	180	22.50	7.75	30.25
0840		200	5	2-5	256	24.50	8.80	33.30
0940	25x25	40	5	1-11	91	16.90	6.55	23.45
0960		75	5	2-5	178	19.50	7.25	26.75
0980		125	5	2-5	181	24	8.10	32.10
1000		200	5-1/2	2-8-1/2	263	27	9.45	36.45
1400	25x30	40	5	2-5	91	16.70	6.75	23.45
1500		75	5	2-5	128	20.50	7.55	28.05
1600		125	5	2-8	180	23	8.40	31.40
1700		200	5	2-11	259	29	9.70	38.70
2200	30x30	40	5	2-2	92	18.55	7.15	25.70
2300		75	5	2-5	129	22	8	30
2400		125	5	2-11	182	26	9.05	35.05
2500		200	5	3-2	263	35	11.40	46.40
2600	30x35	40	5	2-5	94	20.50	7.60	28.10
2700		75	5	2-11	131	23.50	8.45	31.95
2800		125	5	3-2	183	28	9.50	37.50
2900		200	5-1/2	3-5-1/2	268	34	10.95	44.95
3800	35x35	40	5	2-8	94	21	7.60	28.60
3900		75	5	2-11	131	24.50	8.45	32.95
4000		125	5	3-5	184	29	9.70	38.70
4100		200	5-1/2	3-5-1/2	270	38	11.55	49.55

B10 Superstructure

B1010 Floor Construction

Composite Beam, Deck & Slab

The table below lists costs per S.F. for floors using composite steel beams with welded shear studs, composite steel deck and light weight concrete slab reinforced with W.W.F.

	BAY SIZE (FT.)	SUPERIMPOSED LOAD (P.S.F.)	SLAB THICKNESS (IN.)	TOTAL DEPTH (FT.-IN.)	TOTAL LOAD (P.S.F.)	COST PER S.F.		
						MAT.	INST.	TOTAL
2400	20x25	40	5-1/2	1 - 5-1/2	80	13.85	5.50	19.35
2500		75	5-1/2	1 - 9-1/2	115	14.40	5.55	19.95
2750		125	5-1/2	1 - 9-1/2	167	17.70	6.50	24.20
2900		200	6-1/4	1 - 11-1/2	251	19.85	7	26.85
3000	25x25	40	5-1/2	1 - 9-1/2	82	13.70	5.25	18.95
3100		75	5-1/2	1 - 11-1/2	118	15.30	5.35	20.65
3200		125	5-1/2	2 - 2-1/2	169	15.95	5.75	21.70
3300		200	6-1/4	2 - 6-1/4	252	22	6.70	28.70
3400	25x30	40	5-1/2	1 - 11-1/2	83	14	5.20	19.20
3600		75	5-1/2	1 - 11-1/2	119	15.10	5.25	20.35
3900		125	5-1/2	1 - 11-1/2	170	17.60	5.95	23.55
4000		200	6-1/4	2 - 6-1/4	252	22	6.80	28.80
4200	30x30	40	5-1/2	1 - 11-1/2	81	13.95	5.40	19.35
4400		75	5-1/2	2 - 2-1/2	116	15.15	5.60	20.75
4500		125	5-1/2	2 - 5-1/2	168	18.40	6.30	24.70
4700		200	6-1/4	2 - 9-1/4	252	22	7.30	29.30
4900	30x35	40	5-1/2	2 - 2-1/2	82	14.65	5.55	20.20
5100		75	5-1/2	2 - 5-1/2	117	16.05	5.70	21.75
5300		125	5-1/2	2 - 5-1/2	169	19	6.45	25.45
5500		200	6-1/4	2 - 9-1/4	254	22	7.35	29.35
5750	35x35	40	5-1/2	2 - 5-1/2	84	15.75	5.55	21.30
6000		75	5-1/2	2 - 5-1/2	121	18	5.95	23.95
7000		125	5-1/2	2 - 8-1/2	170	21	6.85	27.85
7200		200	5-1/2	2 - 11-1/2	254	24.50	7.55	32.05
7400	35x40	40	5-1/2	2 - 5-1/2	85	17.40	6	23.40
7600		75	5-1/2	2 - 5-1/2	121	18.95	6.20	25.15
8000		125	5-1/2	2 - 5-1/2	171	21.50	6.95	28.45
9000		200	5-1/2	2 - 11-1/2	255	26.50	7.90	34.40

B1010 256 — **Composite Beams, Deck & Slab**

B10 Superstructure

B1010 Floor Construction

Description: Table B1010 258 lists S.F. costs for steel deck and concrete slabs for various spans.

Description: Table B1010 261 lists the S.F. costs for wood joists and a minimum thickness plywood subfloor.

Description: Table B1010 264 lists the S.F. costs, total load, and member sizes, for various bay sizes and loading conditions.

Metal Deck/Concrete Fill

Wood Joist

Wood Beam and Joist

B1010 258	Metal Deck/Concrete Fill							
	SUPERIMPOSED LOAD (P.S.F.)	DECK SPAN (FT.)	DECK GAGE DEPTH	SLAB THICKNESS (IN.)	TOTAL LOAD (P.S.F.)	COST PER S.F.		
						MAT.	INST.	TOTAL
0900	125	6	22 1-1/2	4	164	3.41	2.23	5.64
0920		7	20 1-1/2	4	164	3.81	2.35	6.16
0950		8	20 1-1/2	4	165	3.81	2.35	6.16
0970		9	18 1-1/2	4	165	4.55	2.35	6.90
1000		10	18 2	4	165	4.42	2.49	6.91
1020		11	18 3	5	169	5.50	2.69	8.19

B1010 261	Wood Joist	COST PER S.F.		
		MAT.	INST.	TOTAL
2902	Wood joist, 2" x 8", 12" O.C.	1.39	1.68	3.07
2950	16" O.C.	1.18	1.44	2.62
3000	24" O.C.	1.39	1.33	2.72
3300	2"x10", 12" O.C.	1.75	1.91	3.66
3350	16" O.C.	1.44	1.61	3.05
3400	24" O.C.	1.56	1.44	3
3700	2"x12", 12" O.C.	2.21	1.93	4.14
3750	16" O.C.	1.79	1.62	3.41
3800	24" O.C.	1.80	1.45	3.25
4100	2"x14", 12" O.C.	3.10	2.11	5.21
4150	16" O.C.	2.47	1.76	4.23
4200	24" O.C.	2.25	1.55	3.80
7101	Note: Subfloor cost is included in these prices			

B1010 264	Wood Beam & Joist						
	BAY SIZE (FT.)	SUPERIMPOSED LOAD (P.S.F.)	GIRDER BEAM (IN.)	JOISTS (IN.)	TOTAL LOAD (P.S.F.)	COST PER S.F.		
						MAT.	INST.	TOTAL
2000	15x15	40	8 x 12 4 x 12	2 x 6 @ 16	53	8.60	3.91	12.51
2050		75	8 x 16 4 x 16	2 x 8 @ 16	90	11.25	4.28	15.53
2100		125	12 x 16 6 x 16	2 x 8 @ 12	144	17.20	5.45	22.65
2150		200	14 x 22 12 x 16	2 x 10 @ 12	227	35	8.35	43.35
3000	20x20	40	10 x 14 10 x 12	2 x 8 @ 16	63	11	3.93	14.93
3050		75	12 x 16 8 x 16	2 x 10 @ 16	102	14.80	4.39	19.19
3100		125	14 x 22 12 x 16	2 x 10 @ 12	163	29.50	7.05	36.55

B10 Superstructure

B1010 Floor Construction

Steel Column Fireproofing

"T" Shaped Precast Beams

"L" Shaped Precast Beams

B1010 720			Steel Column Fireproofing					
	ENCASEMENT SYSTEM	COLUMN SIZE (IN.)	THICKNESS (IN.)	FIRE RATING (HRS.)	WEIGHT (P.L.F.)	COST PER V.L.F.		
						MAT.	INST.	TOTAL
3000	Concrete	8	1	1	110	6.70	31.50	38.20
3300		14	1	1	258	11.20	45.50	56.70
3400			2	3	325	13.15	52	65.15
3450	Gypsum board	8	1/2	2	8	3.18	19.25	22.43
3550	1 layer	14	1/2	2	18	3.44	20.50	23.94
3600	Gypsum board	8	1	3	14	4.43	24.50	28.93
3650	1/2" fire rated	10	1	3	17	4.76	26	30.76
3700	2 layers	14	1	3	22	4.93	27	31.93
3750	Gypsum board	8	1-1/2	3	23	5.90	31	36.90
3800	1/2" fire rated	10	1-1/2	3	27	6.65	34.50	41.15
3850	3 layers	14	1-1/2	3	35	7.40	37.50	44.90
3900	Sprayed fiber	8	1-1/2	2	6.3	3.93	6.10	10.03
3950	Direct application		2	3	8.3	5.40	8.40	13.80
4050		10	1-1/2	2	7.9	4.75	7.40	12.15
4200		14	1-1/2	2	10.8	5.90	9.15	15.05

B10 Superstructure

B1020 Roof Construction

The table below lists prices per S.F. for roof rafters and sheathing by nominal size and spacing. Sheathing is 5/16″ CDX for 12″ and 16″ spacing and 3/8″ CDX for 24″ spacing.

Factors for Converting Inclined to Horizontal

Roof Slope	Approx. Angle	Factor	Roof Slope	Approx. Angle	Factor
Flat	0°	1.000	12 in 12	45.0°	1.414
1 in 12	4.8°	1.003	13 in 12	47.3°	1.474
2 in 12	9.5°	1.014	14 in 12	49.4°	1.537
3 in 12	14.0°	1.031	15 in 12	51.3°	1.601
4 in 12	18.4°	1.054	16 in 12	53.1°	1.667
5 in 12	22.6°	1.083	17 in 12	54.8°	1.734
6 in 12	26.6°	1.118	18 in 12	56.3°	1.803
7 in 12	30.3°	1.158	19 in 12	57.7°	1.873
8 in 12	33.7°	1.202	20 in 12	59.0°	1.943
9 in 12	36.9°	1.250	21 in 12	60.3°	2.015
10 in 12	39.8°	1.302	22 in 12	61.4°	2.088
11 in 12	42.5°	1.357	23 in 12	62.4°	2.162

B1020 102	Wood/Flat or Pitched	COST PER S.F.		
		MAT.	INST.	TOTAL
2500	Flat rafter, 2″x4″, 12″ O.C.	.96	1.53	2.49
2550	16″ O.C.	.87	1.31	2.18
2600	24″ O.C.	.69	1.12	1.81
2900	2″x6″, 12″ O.C.	1.24	1.52	2.76
2950	16″ O.C.	1.09	1.31	2.40
3000	24″ O.C.	.83	1.12	1.95
3300	2″x8″, 12″ O.C.	1.46	1.64	3.10
3350	16″ O.C.	1.25	1.40	2.65
3400	24″ O.C.	.94	1.18	2.12
3700	2″x10″, 12″ O.C.	1.82	1.87	3.69
3750	16″ O.C.	1.51	1.57	3.08
3800	24″ O.C.	1.11	1.29	2.40
4100	2″x12″, 12″ O.C.	2.28	1.89	4.17
4150	16″ O.C.	1.86	1.58	3.44
4200	24″ O.C.	1.35	1.30	2.65
4500	2″x14″, 12″ O.C.	3.17	2.76	5.93
4550	16″ O.C.	2.54	2.24	4.78
4600	24″ O.C.	1.80	1.74	3.54
4900	3″x6″, 12″ O.C.	2.47	1.62	4.09
4950	16″ O.C.	2.01	1.38	3.39
5000	24″ O.C.	1.44	1.17	2.61
5300	3″x8″, 12″ O.C.	3.14	1.79	4.93
5350	16″ O.C.	2.51	1.51	4.02
5400	24″ O.C.	1.78	1.25	3.03
5700	3″x10″, 12″ O.C.	4.19	2.05	6.24
5750	16″ O.C.	3.30	1.71	5.01
5800	24″ O.C.	2.30	1.38	3.68
6100	3″x12″, 12″ O.C.	4.90	2.47	7.37
6150	16″ O.C.	3.83	2.01	5.84
6200	24″ O.C.	2.66	1.59	4.25
7001	Wood truss, 4 in 12 slope, 24″ O.C., 24′ to 29′ span	4.03	2.34	6.37
7100	30′ to 43′ span	4.27	2.34	6.61
7200	44′ to 60′ span	4.53	2.34	6.87

B1020 Roof Construction

**Steel Joists, Beams and Deck
on Bearing Walls**

The table below lists the cost per S.F. for a roof system with steel columns, beams and deck using open web steel joists and 1-1/2″ galvanized metal deck. Perimeter of system is supported on bearing walls.

Fireproofing is not included. Costs/S.F. are based on a building 4 bays long and 4 bays wide.

Column costs are additive. Costs for the bearing walls are not included.

B1020 108		Steel Joists, Beams & Deck on Columns & Walls						
	BAY SIZE (FT.)	SUPERIMPOSED LOAD (P.S.F.)	DEPTH (IN.)	TOTAL LOAD (P.S.F.)	COLUMN ADD	COST PER S.F.		
						MAT.	INST.	TOTAL
3000	25x25	20	18	40		5.15	1.31	6.46
3100					columns	.44	.08	.52
3200		30	22	50		5.85	1.55	7.40
3300					columns	.59	.12	.71
3400		40	20	60		6.15	1.55	7.70
3500					columns	.59	.12	.71
3600	25x30	20	22	40		5.50	1.27	6.77
3700					columns	.49	.09	.58
3800		30	20	50		5.95	1.64	7.59
3900					columns	.49	.09	.58
4000		40	25	60		6.45	1.45	7.90
4100					columns	.59	.12	.71
4200	30x30	20	25	42		6	1.37	7.37
4300					columns	.41	.08	.49
4400		30	22	52		6.55	1.48	8.03
4500					columns	.49	.09	.58
4600		40	28	62		6.80	1.52	8.32
4700					columns	.49	.09	.58
4800	30x35	20	22	42		6.25	1.41	7.66
4900					columns	.42	.08	.50
5000		30	28	52		6.55	1.47	8.02
5100					columns	.42	.08	.50
5200		40	25	62		7.10	1.60	8.70
5300					columns	.49	.09	.58
5400	35x35	20	28	42		6.25	1.42	7.67
5500					columns	.36	.07	.43

B10 Superstructure

B1020 Roof Construction

Steel Joists, Beams and Deck on Columns

Description: The table below lists the cost per S.F. for a roof system with steel columns, beams, and deck, using open web steel joists and 1-1/2″ galvanized metal deck.

Column costs are additive. Fireproofing is not included.

B1020 112	Steel Joists, Beams, & Deck on Columns						
	BAY SIZE (FT.)	SUPERIMPOSED LOAD (P.S.F.)	DEPTH (IN.)	TOTAL LOAD (P.S.F.)	COLUMN ADD	COST PER S.F.		
						MAT.	INST.	TOTAL
1100	15x20	20	16	40		5.10	1.31	6.41
1200					columns	2.55	.49	3.04
1500		40	18	60		5.70	1.45	7.15
1600					columns	2.55	.49	3.04
2300	20x25	20	18	40		5.80	1.46	7.26
2400					columns	1.53	.29	1.82
2700		40	20	60		6.50	1.63	8.13
2800					columns	2.04	.40	2.44
2900	25x25	20	18	40		6.75	1.64	8.39
3000					columns	1.22	.24	1.46
3100		30	22	50		7.30	1.86	9.16
3200					columns	1.63	.32	1.95
3300		40	20	60		7.80	1.89	9.69
3400					columns	1.63	.32	1.95
3500	25x30	20	22	40		6.65	1.49	8.14
3600					columns	1.36	.27	1.63
3900		40	25	60		8	1.76	9.76
4000					columns	1.63	.32	1.95
4100	30x30	20	25	42		7.50	1.65	9.15
4200					columns	1.14	.23	1.37
4300		30	22	52		8.25	1.81	10.06
4400					columns	1.36	.27	1.63
4500		40	28	62		8.60	1.88	10.48
4600					columns	1.36	.27	1.63
5300	35x35	20	28	42		8.15	1.79	9.94
5400					columns	1	.20	1.20
5500		30	25	52		9.10	1.98	11.08
5600					columns	1.17	.23	1.40
5700		40	28	62		9.80	2.12	11.92
5800					columns	1.29	.25	1.54

Reed Construction Data
The leader in construction information and BIM solutions

Reed Construction Data, Inc. is a leading provider of construction information and building information modeling (BIM) solutions. The company's portfolio of information products and services is designed specifically to help construction industry professionals advance their business with timely and accurate project, product, and cost data. Reed Construction Data is a division of Reed Business Information, a member of the Reed Elsevier PLC group of companies.

Cost Information

RSMeans, the undisputed market leader in construction costs, provides current cost and estimating information through its innovative MeansCostworks.com® web-based solution. In addition, RSMeans publishes annual cost books, estimating software, and a rich library of reference books. RSMeans also conducts a series of professional seminars and provides construction cost consulting for owners, manufacturers, designers, and contractors to sharpen personal skills and maximize the effective use of cost estimating and management tools.

Project Data

Reed Construction Data assembles one of the largest databases of public and private project data for use by contractors, distributors, and building product manufacturers in the U.S. and Canadian markets. In addition, Reed Construction Data is the North American construction community's premier resource for project leads and bid documents. Reed Bulletin and Reed CONNECT™ provide project leads and project data through all stages of construction for many of the country's largest public sector, commercial, industrial, and multi-family residential projects.

Research and Analytics

Reed Construction Data's forecasting tools cover most aspects of the construction business in the U.S. and Canada. With a vast network of resources, Reed Construction Data is uniquely qualified to give you the information you need to keep your business profitable.

SmartBIM Solutions

Reed Construction Data has emerged as the leader in the field of building information modeling (BIM) with products and services that have helped to advance the evolution of BIM. Through an in-depth SmartBIM Object Creation Program, BPMs can rely on Reed to create high-quality, real world objects embedded with superior cost data (RSMeans). In addition, Reed has made it easy for architects to manage these manufacturer-specific objects as well as generic objects in Revit with the SmartBIM Library.

SmartBuilding Index

The leading industry source for product research, product documentation, BIM objects, design ideas, and source locations. Search, select, and specify available building products with our online directories of manufacturer profiles, MANU-SPEC, SPEC-DATA, guide specs, manufacturer catalogs, building codes, historical project data, and BIM objects.

SmartBuilding Studio

A vibrant online resource library which brings together a comprehensive catalog of commercial interior finishes and products under a single standard for high-definition imagery, product data, and searchable attributes.

Associated Construction Publications (ACP)

Reed Construction Data's regional construction magazines cover the nation through a network of 14 regional magazines. Serving the construction market for more than 100 years, our magazines are a trusted source of news and information in the local and national construction communities.

For more information, please visit our website at www.reedconstructiondata.com

The leader in construction information and BIM solutions.

Reed Construction Data

B10 Superstructure

B1020 Roof Construction

Description: The table below lists the cost per S.F. for a roof system using open web steel joists and 1-1/2" galvanized metal deck. The system is assumed supported on bearing walls or other suitable support. Costs for the supports are not included.

B1020 116		**Steel Joists & Deck on Bearing Walls**				COST PER S.F.		
	BAY SIZE (FT.)	SUPERIMPOSED LOAD (P.S.F.)	DEPTH (IN.)	TOTAL LOAD (P.S.F.)		MAT.	INST.	TOTAL
1100	20	20	13-1/2	40		3.39	.95	4.34
1200		30	15-1/2	50		3.46	.97	4.43
1300		40	15-1/2	60		3.70	1.05	4.75
1400	25	20	17-1/2	40		3.71	1.05	4.76
1500		30	17-1/2	50		3.99	1.13	5.12
1600		40	19-1/2	60		4.04	1.15	5.19
1700	30	20	19-1/2	40		4.03	1.14	5.17
1800		30	21-1/2	50		4.12	1.17	5.29
1900		40	23-1/2	60		4.42	1.26	5.68
2000	35	20	23-1/2	40		4.38	1.06	5.44
2100		30	25-1/2	50		4.52	1.09	5.61
2200		40	25-1/2	60		4.79	1.14	5.93
2300	40	20	25-1/2	41		4.86	1.16	6.02
2400		30	25-1/2	51		5.15	1.22	6.37
2500		40	25-1/2	61		5.35	1.26	6.61
2600	45	20	27-1/2	41		5.40	1.75	7.15
2700		30	31-1/2	51		5.70	1.86	7.56
2800		40	31-1/2	61		6	1.96	7.96
2900	50	20	29-1/2	42		6	1.97	7.97
3000		30	31-1/2	52		6.55	2.16	8.71
3100		40	31-1/2	62		6.95	2.29	9.24
3200	60	20	37-1/2	42		7.70	2.04	9.74
3300		30	37-1/2	52		8.50	2.26	10.76
3400		40	37-1/2	62		8.50	2.26	10.76
3500	70	20	41-1/2	42		8.50	2.26	10.76
3600		30	41-1/2	52		9.05	2.41	11.46
3700		40	41-1/2	64		11	2.92	13.92
3800	80	20	45-1/2	44		12.30	3.19	15.49
3900		30	45-1/2	54		12.30	3.19	15.49
4000		40	45-1/2	64		13.60	3.53	17.13
4400	100	20	57-1/2	44		10.65	2.70	13.35
4500		30	57-1/2	54		12.70	3.21	15.91
4600		40	57-1/2	65		14	3.54	17.54
4700	125	20	69-1/2	44		14.90	3.60	18.50
4800		30	69-1/2	56		17.40	4.19	21.59
4900		40	69-1/2	67		19.85	4.78	24.63

B1020 Roof Construction

Description: The table below lists costs for a roof system supported on exterior bearing walls and interior columns. Costs include bracing, joist girders, open web steel joists and 1-1/2″ galvanized metal deck.

Column costs are additive. Fireproofing is not included.

Costs/S.F. are based on a building 4 bays long and 4 bays wide. Costs for bearing walls are not included.

B1020 120		Steel Joists, Joist Girders & Deck on Columns & Walls						
	BAY SIZE (FT.) GIRD X JOISTS	SUPERIMPOSED LOAD (P.S.F.)	DEPTH (IN.)	TOTAL LOAD (P.S.F.)	COLUMN ADD	COST PER S.F.		
						MAT.	INST.	TOTAL
2350	30x35	20	32-1/2	40		4.82	1.35	6.17
2400					columns	.42	.08	.50
2550		40	36-1/2	60		5.40	1.50	6.90
2600					columns	.49	.09	.58
3000	35x35	20	36-1/2	40		5.20	1.66	6.86
3050					columns	.36	.07	.43
3200		40	36-1/2	60		5.85	1.83	7.68
3250					columns	.46	.09	.55
3300	35x40	20	36-1/2	40		5.30	1.53	6.83
3350					columns	.37	.07	.44
3500		40	36-1/2	60		6	1.69	7.69
3550					columns	.41	.08	.49
3900	40x40	20	40-1/2	41		5.90	1.97	7.87
3950					columns	.36	.07	.43
4100		40	40-1/2	61		6.60	2.14	8.74
4150					columns	.36	.07	.43
5100	45x50	20	52-1/2	41		6.60	2.26	8.86
5150					columns	.25	.05	.30
5300		40	52-1/2	61		7.95	2.72	10.67
5350					columns	.37	.07	.44
5400	50x45	20	56-1/2	41		6.40	2.17	8.57
5450					columns	.25	.05	.30
5600		40	56-1/2	61		8.15	2.78	10.93
5650					columns	.35	.07	.42
5700	50x50	20	56-1/2	42		7.15	2.44	9.59
5750					columns	.26	.05	.31
5900		40	59	64		9.75	2.74	12.49
5950					columns	.36	.07	.43
6300	60x50	20	62-1/2	43		7.20	2.48	9.68
6350					columns	.41	.08	.49
6500		40	71	65		9.95	2.82	12.77
6550					columns	.54	.11	.65

B10 Superstructure

B1020 Roof Construction

Description: The table below lists the cost per S.F. for a roof system supported on columns. Costs include joist girders, open web steel joists and 1-1/2″ galvanized metal deck. Column costs are additive.

Fireproofing is not included. Costs/S.F. are based on a building 4 bays long and 4 bays wide.

Costs for wind columns are not included.

B1020 124		Steel Joists & Joist Girders on Columns						
	BAY SIZE (FT.) GIRD X JOISTS	SUPERIMPOSED LOAD (P.S.F.)	DEPTH (IN.)	TOTAL LOAD (P.S.F.)	COLUMN ADD	COST PER S.F.		
						MAT.	INST.	TOTAL
2000	30x30	20	17-1/2	40		4.94	1.57	6.51
2050					columns	1.36	.27	1.63
2100		30	17-1/2	50		5.35	1.69	7.04
2150					columns	1.36	.27	1.63
2200		40	21-1/2	60		5.50	1.74	7.24
2250					columns	1.36	.27	1.63
2300	30x35	20	32-1/2	40		5.15	1.52	6.67
2350					columns	1.17	.23	1.40
2500		40	36-1/2	60		5.80	1.69	7.49
2550					columns	1.36	.27	1.63
3200	35x40	20	36-1/2	40		6.40	2.30	8.70
3250					columns	1.02	.20	1.22
3400		40	36-1/2	60		7.20	2.55	9.75
3450					columns	1.13	.21	1.34
3800	40x40	20	40-1/2	41		6.50	2.30	8.80
3850					columns	.99	.19	1.18
4000		40	40-1/2	61		7.25	2.54	9.79
4050					columns	.99	.19	1.18
4100	40x45	20	40-1/2	41		6.95	2.58	9.53
4150					columns	.88	.17	1.05
4200		30	40-1/2	51		7.50	2.79	10.29
4250					columns	.88	.17	1.05
4300		40	40-1/2	61		8.35	3.06	11.41
4350					columns	.99	.19	1.18
5000	45x50	20	52-1/2	41		7.30	2.67	9.97
5050					columns	.70	.13	.83
5200		40	52-1/2	61		8.90	3.19	12.09
5250					columns	.91	.17	1.08
5300	50x45	20	56-1/2	41		7.15	2.59	9.74
5350					columns	.70	.13	.83
5500		40	56-1/2	61		9.15	3.26	12.41
5550					columns	.91	.17	1.08
5600	50x50	20	56-1/2	42		7.95	2.88	10.83
5650					columns	.71	.13	.84
5800		40	59	64		8.50	2.61	11.11
5850					columns	1	.19	1.19

B20 Exterior Enclosure

B2010 Exterior Walls

Cast in Place Concrete Wall

Flat Precast Concrete Wall

Table B2010 101 describes a concrete wall system for exterior closure. There are several types of wall finishes priced from plain finish to a finish with 3/4″ rustication strip.

In Table B2010 103, precast concrete wall panels are either solid or insulated. Prices below are based on delivery within 50 miles of a plant. Small jobs can double the prices below. For large, highly repetitive jobs, deduct up to 15%.

B2010 101	Cast In Place Concrete	COST PER S.F.		
		MAT.	INST.	TOTAL
2100	Conc wall reinforced, 8′ high, 6″ thick, plain finish, 3000 PSI	4.83	14.20	19.03
2700	Aged wood liner, 3000 PSI	6	16.15	22.15
3000	Sand blast light 1 side, 3000 PSI	5.60	16.10	21.70
3700	3/4″ bevel rustication strip, 3000 PSI	4.98	15.05	20.03
4000	8″ thick, plain finish, 3000 PSI	5.80	14.55	20.35
4100	4000 PSI	5.90	14.55	20.45
4300	Rub concrete 1 side, 3000 PSI	5.80	17.15	22.95
4550	8″ thick, aged wood liner, 3000 PSI	6.95	16.50	23.45
4750	Sand blast light 1 side, 3000 PSI	6.55	16.50	23.05
5300	3/4″ bevel rustication strip, 3000 PSI	5.95	15.40	21.35
5600	10″ thick, plain finish, 3000 PSI	6.70	14.90	21.60
5900	Rub concrete 1 side, 3000 PSI	6.70	17.50	24.20
6200	Aged wood liner, 3000 PSI	7.85	16.85	24.70
6500	Sand blast light 1 side, 3000 PSI	7.45	16.85	24.30
7100	3/4″ bevel rustication strip, 3000 PSI	6.85	15.75	22.60
7400	12″ thick, plain finish, 3000 PSI	7.75	15.35	23.10
7700	Rub concrete 1 side, 3000 PSI	7.75	17.90	25.65
8000	Aged wood liner, 3000 PSI	8.90	17.25	26.15
8300	Sand blast light 1 side, 3000 PSI	8.50	17.25	25.75
8900	3/4″ bevel rustication strip, 3000 PSI	7.90	16.20	24.10

B2010 102			Flat Precast Concrete			COST PER S.F.		
	THICKNESS (IN.)	PANEL SIZE (FT.)	FINISHES	RIGID INSULATION (IN)	TYPE	MAT.	INST.	TOTAL
3000	4	5x18	smooth gray	none	low rise	9.05	7.20	16.25
3050		6x18				7.55	6	13.55
3100		8x20				15.10	2.90	18
3150		12x20				14.30	2.74	17.04
3200	6	5x18	smooth gray	2	low rise	10.25	7.65	17.90
3250		6x18				8.75	6.50	15.25
3300		8x20				16.55	3.49	20.04
3350		12x20				15	3.20	18.20
3400	8	5x18	smooth gray	2	low rise	21.50	4.41	25.91
3450		6x18				20	4.20	24.20
3500		8x20				18.45	3.86	22.31
3550		12x20				16.80	3.55	20.35

B20 Exterior Enclosure

B2010 Exterior Walls

B2010 102	Flat Precast Concrete

	THICKNESS (IN.)	PANEL SIZE (FT.)	FINISHES	RIGID INSULATION (IN)	TYPE	COST PER S.F.		
						MAT.	INST.	TOTAL
3600	4	4x8	white face	none	low rise	36	4.15	40.15
3650		8x8				27	3.11	30.11
3700		10x10				23.50	2.74	26.24
3750		20x10				21.50	2.47	23.97
3800	5	4x8	white face	none	low rise	36.50	4.24	40.74
3850		8x8				27.50	3.21	30.71
3900		10x10				24.50	2.83	27.33
3950		20x20				22.50	2.60	25.10
4000	6	4x8	white face	none	low rise	38	4.39	42.39
4050		8x8				29	3.33	32.33
4100		10x10				25.50	2.94	28.44
4150		20x10				23.50	2.70	26.20
4200	6	4x8	white face	2	low rise	39.50	4.95	44.45
4250		8x8				30	3.89	33.89
4300		10x10				26.50	3.50	30
4350		20x10				23.50	2.70	26.20
4400	7	4x8	white face	none	low rise	39	4.49	43.49
4450		8x8				30	3.46	33.46
4500		10x10				27	3.10	30.10
4550		20x10				24.50	2.83	27.33
4600	7	4x8	white face	2	low rise	40	5.05	45.05
4650		8x8				31	4.02	35.02
4700		10x10				28	3.66	31.66
4750		20x10				26	3.39	29.39
4800	8	4x8	white face	none	low rise	40	4.60	44.60
4850		8x8				30.50	3.54	34.04
4900		10x10				27.50	3.19	30.69
4950		20x10				25.50	2.93	28.43
5000	8	4x8	white face	2	low rise	41	5.15	46.15
5050		8x8				32	4.10	36.10
5100		10x10				29	3.75	32.75
5150		20x10				26.50	3.49	29.99

B2010 103	Fluted Window or Mullion Precast Concrete

	THICKNESS (IN.)	PANEL SIZE (FT.)	FINISHES	RIGID INSULATION (IN)	TYPE	COST PER S.F.		
						MAT.	INST.	TOTAL
5200	4	4x8	smooth gray	none	high rise	20.50	16.45	36.95
5250		8x8				14.80	11.75	26.55
5300		10x10				29	5.55	34.55
5350		20x10				25.50	4.87	30.37
5400	5	4x8	smooth gray	none	high rise	21	16.70	37.70
5450		8x8				15.35	12.15	27.50
5500		10x10				30	5.80	35.80
5550		20x10				26.50	5.10	31.60
5600	6	4x8	smooth gray	none	high rise	21.50	17.20	38.70
5650		8x8				15.85	12.60	28.45
5700		10x10				31	5.95	36.95
5750		20x10				27.50	5.30	32.80
5800	6	4x8	smooth gray	2	high rise	23	17.75	40.75
5850		8x8				17.10	13.15	30.25
5900		10x10				32	6.50	38.50
5950		20x10				29	5.85	34.85

B20 Exterior Enclosure

B2010 Exterior Walls

B2010 103 — Fluted Window or Mullion Precast Concrete

	THICKNESS (IN.)	PANEL SIZE (FT.)	FINISHES	RIGID INSULATION (IN)	TYPE	COST PER S.F.		
						MAT.	INST.	TOTAL
6000	7	4x8	smooth gray	none	high rise	22	17.55	39.55
6050		8x8				16.30	12.90	29.20
6100		10x10				32.50	6.20	38.70
6150		20x10				28.50	5.45	33.95
6200	7	4x8	smooth gray	2	high rise	23.50	18.10	41.60
6250		8x8				17.55	13.45	31
6300		10x10				33.50	6.75	40.25
6350		20x10				29.50	6	35.50
6400	8	4x8	smooth gray	none	high rise	22.50	17.75	40.25
6450		8x8				16.70	13.25	29.95
6500		10x10				33.50	6.40	39.90
6550		20x10				30	5.70	35.70
6600	8	4x8	smooth gray	2	high rise	23.50	18.35	41.85
6650		8x8				17.95	13.80	31.75
6700		10x10				34.50	6.95	41.45
6750		20x10				31	6.30	37.30

B2010 104 — Ribbed Precast Concrete

	THICKNESS (IN.)	PANEL SIZE (FT.)	FINISHES	RIGID INSULATION (IN.)	TYPE	COST PER S.F.		
						MAT.	INST.	TOTAL
6800	4	4x8	aggregate	none	high rise	27	16.45	43.45
6850		8x8				19.75	12	31.75
6900		10x10				34	4.35	38.35
6950		20x10				30	3.85	33.85
7000	5	4x8	aggregate	none	high rise	27.50	16.70	44.20
7050		8x8				20.50	12.35	32.85
7100		10x10				35	4.51	39.51
7150		20x10				31.50	4.03	35.53
7200	6	4x8	aggregate	none	high rise	28	17.05	45.05
7250		8x8				21	12.70	33.70
7300		10x10				36	4.66	40.66
7350		20x10				32.50	4.18	36.68
7400	6	4x8	aggregate	2	high rise	29.50	17.60	47.10
7450		8x8				22	13.25	35.25
7500		10x10				37.50	5.20	42.70
7550		20x10				33.50	4.74	38.24
7600	7	4x8	aggregate	none	high rise	28.50	17.45	45.95
7650		8x8				21.50	13	34.50
7700		10x10				37.50	4.80	42.30
7750		20x10				33.50	4.29	37.79
7800	7	4x8	aggregate	2	high rise	30	18	48
7850		8x8				22.50	13.55	36.05
7900		10x10				38.50	5.35	43.85
7950		20x10				34.50	4.85	39.35
8000	8	4x8	aggregate	none	high rise	29	17.75	46.75
8050		8x8				22	13.40	35.40
8100		10x10				38.50	4.97	43.47
8150		20x10				35	4.50	39.50
8200	8	4x8	aggregate	2	high rise	30.50	18.30	48.80
8250		8x8				23.50	13.95	37.45
8300		10x10				40	5.55	45.55
8350		20x10				36	5.05	41.05

B20 Exterior Enclosure

B2010 Exterior Walls

B2010 105						Precast Concrete Specialties			

	TYPE	SIZE				COST PER L.F.		
						MAT.	INST.	TOTAL
8400	Coping, precast	6" wide				19.15	11.75	30.90
8450	Stock units	10" wide				19.15	12.60	31.75
8460		12" wide				22.50	13.55	36.05
8480		14" wide				18.80	14.70	33.50
8500	Window sills	6" wide				12.95	12.60	25.55
8550	Precast	10" wide				17.15	14.70	31.85
8600		14" wide				16.50	17.65	34.15
8610								

B2010 Exterior Walls

The advantage of tilt up construction is in the low cost of forms and placing of concrete and reinforcing. Tilt up has been used for several types of buildings, including warehouses, stores, offices, and schools. The panels are cast in forms on the ground, or floor slab. Most jobs use 5-1/2" thick solid reinforced concrete panels.

Tilt Up Concrete Panel

B2010 106	Tilt-Up Concrete Panel	COST PER S.F.		
		MAT.	INST.	TOTAL
3200	Tilt up conc panels, broom finish, 5-1/2" thick, 3000 PSI	6.50	5.45	11.95
3250	5000 PSI	6.55	5.35	11.90
3300	6" thick, 3000 PSI	7.25	5.60	12.85
3350	5000 PSI	7.30	5.50	12.80
3400	7-1/2" thick, 3000 PSI	9.55	5.80	15.35
3450	5000 PSI	9.65	5.70	15.35
3500	8" thick, 3000 PSI	10.30	5.95	16.25
3550	5000 PSI	10.45	5.85	16.30
3700	Steel trowel finish, 5-1/2" thick, 3000 PSI	6.50	5.50	12
3750	5000 PSI	6.55	5.40	11.95
3800	6" thick, 3000 PSI	7.25	5.65	12.90
3850	5000 PSI	7.30	5.55	12.85
3900	7-1/2" thick, 3000 PSI	9.55	5.85	15.40
3950	5000 PSI	9.65	5.75	15.40
4000	8" thick, 3000 PSI	10.30	6	16.30
4050	5000 PSI	10.45	5.90	16.35
4200	Exp. aggregate finish, 5-1/2" thick, 3000 PSI	6.90	5.60	12.50
4250	5000 PSI	6.95	5.50	12.45
4300	6" thick, 3000 PSI	7.65	5.70	13.35
4350	5000 PSI	7.70	5.60	13.30
4400	7-1/2" thick, 3000 PSI	10	5.90	15.90
4450	5000 PSI	10.05	5.80	15.85
4500	8" thick, 3000 PSI	10.75	6.05	16.80
4550	5000 PSI	10.85	5.95	16.80
4600	Exposed aggregate & vert. rustication 5-1/2" thick, 3000 PSI	9.10	6.85	15.95
4650	5000 PSI	9.15	6.75	15.90
4700	6" thick, 3000 PSI	9.85	7	16.85
4750	5000 PSI	9.90	6.90	16.80
4800	7-1/2" thick, 3000 PSI	12.20	7.20	19.40
4850	5000 PSI	12.25	7.10	19.35
4900	8" thick, 3000 PSI	12.95	7.35	20.30
4950	5000 PSI	13.05	7.25	20.30
5000	Vertical rib & light sandblast, 5-1/2" thick, 3000 PSI	8.40	9	17.40
5100	6" thick, 3000 PSI	9.15	9.15	18.30
5200	7-1/2" thick, 3000 PSI	11.50	9.35	20.85
5300	8" thick, 3000 PSI	12.20	9.50	21.70
6000	Broom finish w/2" polystyrene insulation, 6" thick, 3000 PSI	6.15	6.70	12.85
6100	Broom finish 2" fiberplank insulation, 6" thick, 3000 PSI	6.85	6.60	13.45
6200	Exposed aggregate w/2" polystyrene insulation, 6" thick, 3000 PSI	6.40	6.60	13
6300	Exposed aggregate 2" fiberplank insulation, 6" thick, 3000 PSI	7.10	6.55	13.65

B2010 Exterior Walls

Exterior concrete block walls are defined in the following terms; structural reinforcement, weight, percent solid, size, strength and insulation. Within each of these categories, two to four variations are shown. No costs are included for brick shelf or relieving angles.

B2010 109 — Concrete Block Wall - Regular Weight

	TYPE	SIZE (IN.)	STRENGTH (P.S.I.)	CORE FILL		COST PER S.F.		
						MAT.	INST.	TOTAL
1200	Hollow	4x8x16	2,000	none		1.75	5.45	7.20
1250			4,500	none		1.53	5.45	6.98
1400		8x8x16	2,000	perlite		3.05	6.60	9.65
1410				styrofoam		4.54	6.20	10.74
1440				none		2.55	6.20	8.75
1450			4,500	perlite		2.84	6.60	9.44
1460				styrofoam		4.33	6.20	10.53
1490				none		2.34	6.20	8.54
2000	75% solid	4x8x16	2,000	none		1.90	5.50	7.40
2100	75% solid	6x8x16	2,000	perlite		2.23	6.05	8.28
2500	Solid	4x8x16	2,000	none		2.15	5.65	7.80
2700		8x8x16	2,000	none		3.70	6.45	10.15

B2010 110 — Concrete Block Wall - Lightweight

	TYPE	SIZE (IN.)	WEIGHT (P.C.F.)	CORE FILL		COST PER S.F.		
						MAT.	INST.	TOTAL
3100	Hollow	8x4x16	105	perlite		2.03	6.25	8.28
3110				styrofoam		3.52	5.85	9.37
3200		4x8x16	105	none		2.06	5.30	7.36
3250			85	none		1.90	5.20	7.10
3300		6x8x16	105	perlite		3.08	6	9.08
3310				styrofoam		4.73	5.70	10.43
3340				none		2.74	5.70	8.44
3400		8x8x16	105	perlite		3.83	6.45	10.28
3410				styrofoam		5.30	6.05	11.35
3440				none		3.33	6.05	9.38
3450			85	perlite		3.20	6.30	9.50
4000	75% solid	4x8x16	105	none		2.08	5.40	7.48
4050			85	none		3.28	5.25	8.53
4100		6x8x16	105	perlite		2.48	5.90	8.38
4500	Solid	4x8x16	105	none		2.70	5.55	8.25
4700		8x8x16	105	none		4.11	6.40	10.51

B2010 Exterior Walls

B2010 111	Reinforced Concrete Block Wall - Regular Weight							
	TYPE	SIZE (IN.)	STRENGTH (P.S.I.)	VERT. REINF & GROUT SPACING		COST PER S.F.		
						MAT.	INST.	TOTAL
5200	Hollow	4x8x16	2,000	#4 @ 48"		1.95	6.05	8
5300		6x8x16	2,000	#4 @ 48"		2.72	6.45	9.17
5330				#5 @ 32"		3.02	6.70	9.72
5340				#5 @ 16"		3.59	7.55	11.14
5350			4,500	#4 @ 28"		2.24	6.45	8.69
5390				#5 @ 16"		3.11	7.55	10.66
5400	Hollow	8x8x16	2,000	#4 @ 48"		2.86	6.85	9.71
5430				#5 @ 32"		3.22	7.25	10.47
5440				#5 @ 16"		3.88	8.25	12.13
5450		8x8x16	4,500	#4 @ 48"		2.69	6.95	9.64
5490				#5 @ 16"		3.67	8.25	11.92
5500		12x8x16	2,000	#4 @ 48"		4.37	8.80	13.17
5540				#5 @ 16"		5.60	10.15	15.75
6100	75% solid	6x8x16	2,000	#4 @ 48"		2.25	6.40	8.65
6140				#5 @ 16"		2.90	7.30	10.20
6150			4,500	#4 @ 48"		2.74	6.40	9.14
6190				#5 @ 16"		3.39	7.30	10.69
6200		8x8x16	2,000	#4 @ 48"		3.27	6.90	10.17
6230				#5 @ 32"		3.52	7.15	10.67
6240				#5 @ 16"		3.95	8	11.95
6250			4,500	#4 @ 48"		3.66	6.90	10.56
6280				#5 @ 32"		3.91	7.15	11.06
6290				#5 @ 16"		4.34	8	12.34
6500	Solid-double	2-4x8x16	2,000	#4 @ 48" E.W.		5.30	12.75	18.05
6530	Wythe			#5 @ 16" E.W.		6.40	13.55	19.95
6550			4,500	#4 @ 48" E.W.		5.15	12.55	17.70
6580				#5 @ 16" E.W.		6.30	13.35	19.65

B2010 112	Reinforced Concrete Block Wall - Lightweight							
	TYPE	SIZE (IN.)	WEIGHT (P.C.F.)	VERT REINF. & GROUT SPACING		COST PER S.F.		
						MAT.	INST.	TOTAL
7100	Hollow	8x4x16	105	#4 @ 48"		1.88	6.60	8.48
7140				#5 @ 16"		2.86	7.90	10.76
7150			85	#4 @ 48"		4.73	6.45	11.18
7190				#5 @ 16"		5.70	7.75	13.45
7400		8x8x16	105	#4 @ 48"		3.68	6.80	10.48
7440				#5 @ 16"		4.66	8.10	12.76
7450		8x8x16	85	#4 @ 48"		3.05	6.65	9.70
7490				#5 @ 16"		4.03	7.95	11.98
7800		8x8x24	105	#4 @ 48"		2.84	7.30	10.14
7840				#5 @ 16"		3.82	8.60	12.42
7850			85	#4 @ 48"		5.15	6.25	11.40
7890				#5 @ 16"		6.15	7.55	13.70
8100	75% solid	6x8x16	105	#4 @ 48"		2.50	6.25	8.75
8130				#5 @ 32"		2.74	6.45	9.19
8150			85	#4 @ 48"		3.97	6.15	10.12
8180				#5 @ 32"		4.21	6.35	10.56
8200		8x8x16	105	#4 @ 48"		3.28	6.75	10.03
8230				#5 @ 32"		3.53	7	10.53
8250			85	#4 @ 48"		5.35	6.60	11.95
8280				#5 @ 32"		5.60	6.85	12.45

B20 Exterior Enclosure

B2010 Exterior Walls

B2010 112	Reinforced Concrete Block Wall - Lightweight							

	TYPE	SIZE (IN.)	WEIGHT (P.C.F.)	VERT REINF. & GROUT SPACING		COST PER S.F.		
						MAT.	INST.	TOTAL
8500	Solid-double	2-4x8x16	105	#4 @ 48"		6.40	12.55	18.95
8530	Wythe		105	#5 @ 16"		7.50	13.35	20.85
8600		2-6x8x16	105	#4 @ 48"		7.20	13.50	20.70
8630				#5 @ 16"		8.30	14.25	22.55
8650			85	#4 @ 48"		9.20	12.90	22.10

B2010 Exterior Walls

Exterior split ribbed block walls are defined in the following terms; structural reinforcement, weight, percent solid, size, number of ribs and insulation. Within each of these categories two to four variations are shown. No costs are included for brick shelf or relieving angles. Costs include control joints every 20' and horizontal reinforcing.

B2010 113				Split Ribbed Block Wall - Regular Weight				
	TYPE	SIZE (IN.)	RIBS	CORE FILL		COST PER S.F.		
						MAT.	INST.	TOTAL
1220	Hollow	4x8x16	4	none		3.15	6.75	9.90
1250			8	none		3.67	6.75	10.42
1280			16	none		2.99	6.85	9.84
1430		8x8x16	8	perlite		6.25	8	14.25
1440				styrofoam		7.75	7.60	15.35
1450				none		5.75	7.60	13.35
1530		12x8x16	8	perlite		7.75	10.60	18.35
1540				styrofoam		9.40	9.85	19.25
1550				none		6.95	9.85	16.80
2120	75% solid	4x8x16	4	none		3.94	6.85	10.79
2150			8	none		4.67	6.85	11.52
2180			16	none		3.79	6.95	10.74
2520	Solid	4x8x16	4	none		4.81	6.95	11.76
2550			8	none		5.60	6.95	12.55
2580			16	none		4.55	7.05	11.60

B2010 115				Reinforced Split Ribbed Block Wall - Regular Weight				
	TYPE	SIZE (IN.)	RIBS	VERT. REINF. & GROUT SPACING		COST PER S.F.		
						MAT.	INST.	TOTAL
5200	Hollow	4x8x16	4	#4 @ 48"		3.35	7.35	10.70
5230			8	#4 @ 48"		3.87	7.35	11.22
5260			16	#4 @ 48"		3.19	7.45	10.64
5430		8x8x16	8	#4 @ 48"		6.10	8.35	14.45
5440				#5 @ 32"		6.40	8.65	15.05
5450				#5 @ 16"		7.05	9.65	16.70
5530		12x8x16	8	#4 @ 48"		7.45	10.60	18.05
5540				#5 @ 32"		7.80	10.90	18.70
5550				#5 @ 16"		8.65	11.95	20.60
6230	75% solid	6x8x16	8	#4 @ 48"		6	7.75	13.75
6240				#5 @ 32"		6.25	7.95	14.20
6250				#5 @ 16"		6.65	8.65	15.30
6330		8x8x16	8	#4 @ 48"		7.55	8.30	15.85
6340				#5 @ 32"		7.80	8.55	16.35
6350				#5 @ 16"		8.20	9.40	17.60

B2010 Exterior Walls

Exterior split face block walls are defined in the following terms; structural reinforcement, weight, percent solid, size, scores and insulation. Within each of these categories two to four variations are shown. No costs are included for brick shelf or relieving angles. Costs include control joints every 20' and horizontal reinforcing.

B2010 117				Split Face Block Wall - Regular Weight				

	TYPE	SIZE (IN.)	SCORES	CORE FILL		COST PER S.F.		
						MAT.	INST.	TOTAL
1200	Hollow	8x4x16	0	perlite		9.20	8.80	18
1210				styrofoam		10.70	8.40	19.10
1240				none		8.70	8.40	17.10
1250			1	perlite		9.85	8.80	18.65
1260				styrofoam		11.35	8.40	19.75
1290				none		9.35	8.40	17.75
1300		12x4x16	0	perlite		11.90	10	21.90
1310				styrofoam		13.55	9.25	22.80
1340				none		11.10	9.25	20.35
1350			1	perlite		12.55	10	22.55
1360				styrofoam		14.20	9.25	23.45
1390				none		11.75	9.25	21
1400		4x8x16	0	none		3.15	6.65	9.80
1450			1	none		3.55	6.75	10.30
1500		6x8x16	0	perlite		4.41	7.65	12.06
1510				styrofoam		6.05	7.35	13.40
1540				none		4.07	7.35	11.42
1550			1	perlite		4.83	7.75	12.58
1560				styrofoam		6.50	7.45	13.95
1590				none		4.49	7.45	11.94
1600		8x8x16	0	perlite		5.35	8.25	13.60
1610				styrofoam		6.85	7.85	14.70
1640				none		4.85	7.85	12.70
1650		8x8x16	1	perlite		5.65	8.40	14.05
1660				styrofoam		7.15	8	15.15
1690				none		5.15	8	13.15
1700		12x8x16	0	perlite		6.80	10.80	17.60
1710				styrofoam		8.45	10.05	18.50
1740				none		6	10.05	16.05
1750			1	perlite		7.15	11	18.15
1760				styrofoam		8.80	10.25	19.05
1790				none		6.35	10.25	16.60

B20 Exterior Enclosure

B2010 Exterior Walls

B2010 117	Split Face Block Wall - Regular Weight							

	TYPE	SIZE (IN.)	SCORES	CORE FILL		COST PER S.F.		
						MAT.	INST.	TOTAL
1800	75% solid	8x4x16	0	perlite		11.20	8.75	19.95
1840				none		10.95	8.55	19.50
1852		8x4x16	1	perlite		11.90	8.75	20.65
1890				none		11.65	8.55	20.20
2000		4x8x16	0	none		3.94	6.75	10.69
2050			1	none		4.28	6.85	11.13
2400	Solid	8x4x16	0	none		13.40	8.70	22.10
2450			1	none		14.05	8.70	22.75
2900		12x8x16	0	none		9.15	10.40	19.55
2950			1	none		9.45	10.60	20.05

B2010 119	Reinforced Split Face Block Wall - Regular Weight							

	TYPE	SIZE (IN.)	SCORES	VERT. REINF. & GROUT SPACING		COST PER S.F.		
						MAT.	INST.	TOTAL
5200	Hollow	8x4x16	0	#4 @ 48"		9.05	9.15	18.20
5210				#5 @ 32"		9.35	9.45	18.80
5240				#5 @ 16"		10	10.45	20.45
5250			1	#4 @ 48"		9.70	9.15	18.85
5260				#5 @ 32"		10	9.45	19.45
5290				#5 @ 16"		10.65	10.45	21.10
5700		12x8x16	0	#4 @ 48"		6.50	10.80	17.30
5710				#5 @ 32"		6.85	11.10	17.95
5740				#5 @ 16"		7.70	12.15	19.85
5750			1	#4 @ 48"		6.85	11	17.85
5760				#5 @ 32"		7.20	11.30	18.50
5790				#5 @ 16"		8.05	12.35	20.40
6000	75% solid	8x4x16	0	#4 @ 48"		11.15	9.15	20.30
6010				#5 @ 32"		11.40	9.40	20.80
6040				#5 @ 16"		11.80	10.25	22.05
6050			1	#4 @ 48"		11.85	9.15	21
6060				#5 @ 32"		12.10	9.40	21.50
6090				#5 @ 16"		12.50	10.25	22.75
6100		12x4x16	0	#4 @ 48"		14.35	10.10	24.45
6110				#5 @ 32"		14.65	10.30	24.95
6140				#5 @ 16"		15.15	11	26.15
6150			1	#4 @ 48"		14.90	10.10	25
6160				#5 @ 32"		15.20	10.30	25.50
6190				#5 @ 16"		15.75	11.15	26.90
6700	Solid-double	2-4x8x16	0	#4 @ 48" E.W.		10.60	15.15	25.75
6710	Wythe			#5 @ 32" E.W.		11	15.30	26.30
6740				#5 @ 16" E.W.		11.75	15.95	27.70
6750			1	#4 @ 48" E.W.		11.25	15.35	26.60
6760				#5 @ 32" E.W.		11.65	15.50	27.15
6790				#5 @ 16" E.W.		12.40	16.15	28.55
6800		2-6x8x16	0	#4 @ 48" E.W.		13.85	16.70	30.55
6810				#5 @ 32" E.W.		14.25	16.85	31.10
6840				#5 @ 16" E.W.		15	17.45	32.45
6850			1	#4 @ 48" E.W.		14.45	16.90	31.35
6860				#5 @ 32" E.W.		14.85	17.05	31.90
6890				#5 @ 16" E.W.		15.60	17.65	33.25

B2010 Exterior Walls

Exterior solid brick walls are defined in the following terms; structural reinforcement, type of brick, thickness, and bond. Nine different types of face bricks are presented, with single wythes shown in four different bonds. Shelf angles are included in system components. These walls do not include ties and as such are not tied to a backup wall.

B2010 125	Solid Brick Walls - Single Wythe							
						COST PER S.F.		
	TYPE	THICKNESS (IN.)	BOND			MAT.	INST.	TOTAL
1000	Common	4	running			4.57	12.25	16.82
1010			common			5.45	14	19.45
1050			Flemish			6	17	23
1100			English			6.65	18	24.65
1150	Standard	4	running			7.60	12.70	20.30
1160			common			9	14.65	23.65
1200			Flemish			9.95	17.50	27.45
1250			English			11.15	18.55	29.70
1300	Glazed	4	running			15.65	13.20	28.85
1310			common			18.65	15.35	34
1350			Flemish			20.50	18.55	39.05
1400			English			23	19.80	42.80
1600	Economy	4	running			5.35	9.70	15.05
1610			common			6.25	11.10	17.35
1650			Flemish			6.85	13.20	20.05
1700			English			16.40	14	30.40
1900	Fire	4-1/2	running			11.45	11.10	22.55
1910			common			13.55	12.70	26.25
1950			Flemish			15	15.35	30.35
2000			English			16.85	16.10	32.95
2050	King	3-1/2	running			4.73	9.95	14.68
2060			common			5.50	11.45	16.95
2100			Flemish			6.55	13.70	20.25
2150			English			7.25	14.30	21.55
2200	Roman	4	running			7.50	11.45	18.95
2210			common			8.90	13.20	22.10
2250			Flemish			9.80	15.70	25.50
2300			English			11.05	17	28.05
2350	Norman	4	running			6.50	9.50	16
2360			common			7.65	10.80	18.45
2400			Flemish			16.95	12.95	29.90
2450			English			9.45	13.70	23.15

B20 Exterior Enclosure

B2010 Exterior Walls

B2010 125 — Solid Brick Walls - Single Wythe

	TYPE	THICKNESS (IN.)	BOND			COST PER S.F.		
						MAT.	INST.	TOTAL
2500	Norwegian	4	running			5.40	8.45	13.85
2510			common			6.30	9.60	15.90
2550			Flemish			6.90	11.45	18.35
2600			English			7.70	12.05	19.75

B2010 126 — Solid Brick Walls - Double Wythe

	TYPE	THICKNESS (IN.)	COLLAR JOINT THICKNESS (IN.)			COST PER S.F.		
						MAT.	INST.	TOTAL
4100	Common	8	3/4			8.45	20.50	28.95
4150	Standard	8	1/2			14.45	21.50	35.95
4200	Glazed	8	1/2			30.50	22	52.50
4250	Engineer	8	1/2			10.25	18.80	29.05
4300	Economy	8	3/4			9.90	16.15	26.05
4350	Double	8	3/4			12.50	13.05	25.55
4400	Fire	8	3/4			22	18.80	40.80
4450	King	7	3/4			8.65	16.50	25.15
4500	Roman	8	1			14.20	19.35	33.55
4550	Norman	8	3/4			12.25	15.80	28.05
4600	Norwegian	8	3/4			10	14	24
4650	Utility	8	3/4			9.30	12.10	21.40
4700	Triple	8	3/4			7.30	10.90	18.20
4750	SCR	12	3/4			13.70	16.20	29.90
4800	Norwegian	12	3/4			11.75	14.30	26.05
4850	Jumbo	12	3/4			12.80	12.50	25.30

B2010 127 — Reinforced Brick Walls

	TYPE	THICKNESS (IN.)	WYTHE	REINF. & SPACING		COST PER S.F.		
						MAT.	INST.	TOTAL
7200	Common	4"	1	#4 @ 48"vert		4.82	12.75	17.57
7220				#5 @ 32"vert		5.05	12.90	17.95
7230				#5 @ 16"vert		5.45	13.50	18.95
7700	King	9"	2	#4 @ 48" E.W.		9.15	17.30	26.45
7720				#5 @ 32" E.W.		9.60	17.40	27
7730				#5 @ 16" E.W.		10.30	17.95	28.25
7800	Common	10"	2	#4 @ 48" E.W.		8.95	21.50	30.45
7820				#5 @ 32" E.W.		9.40	21.50	30.90
7830				#5 @ 16" E.W.		10.10	22	32.10
7900	Standard	10"	2	#4 @ 48" E.W.		14.95	22	36.95
7920				#5 @ 32" E.W.		15.40	22	37.40
7930				#5 @ 16" E.W.		16.10	22.50	38.60
8100	Economy	10"	2	#4 @ 48" E.W.		10.40	16.95	27.35
8120				#5 @ 32" E.W.		10.85	17.05	27.90
8130				#5 @ 16" E.W.		11.55	17.60	29.15
8200	Double	10"	2	#4 @ 48" E.W.		13	13.85	26.85
8220				#5 @ 32" E.W.		13.45	13.95	27.40
8230				#5 @ 16" E.W.		14.15	14.50	28.65
8300	Fire	10"	2	#4 @ 48" E.W.		22.50	19.60	42.10
8320				#5 @ 32" E.W.		23	19.70	42.70
8330				#5 @ 16" E.W.		24	20	44
8400	Roman	10"	2	#4 @ 48" E.W.		14.70	20	34.70
8420				#5 @ 32" E.W.		15.15	20	35.15
8430				#5 @ 16" E.W.		15.85	20.50	36.35

B20 Exterior Enclosure

B2010 Exterior Walls

| B2010 127 | | | | | **Reinforced Brick Walls** | | | | |

	TYPE	THICKNESS (IN.)	WYTHE	REINF. & SPACING		COST PER S.F.		
						MAT.	INST.	TOTAL
8500	Norman	10"	2	#4 @ 48" E.W.		12.75	16.60	29.35
8520				#5 @ 32" E.W.		13.20	16.70	29.90
8530				#5 @ 16" E.W.		13.90	17.25	31.15
8600	Norwegian	10"	2	#4 @ 48" E.W.		10.50	14.80	25.30
8620				#5 @ 32" E.W.		10.95	14.90	25.85
8630				#5 @ 16" E.W.		11.65	15.45	27.10
8800	Triple	10"	2	#4 @ 48" E.W.		7.80	11.70	19.50
8820				#5 @ 32" E.W.		8.25	11.80	20.05
8830				#5 @ 16" E.W.		8.95	12.35	21.30

B2010 Exterior Walls

Stone Veneer

The table below lists costs per S.F. for stone veneer walls on various backup using different stone.

B2010 128	Stone Veneer	COST PER S.F.		
		MAT.	INST.	TOTAL
2000	Ashlar veneer, 4", 2"x4" stud backup, 16" O.C., 8' high, low priced stone	12.10	16.30	28.40
2100	Metal stud backup, 8' high, 16" O.C.	13.40	16.55	29.95
2150	24" O.C.	12.95	16.10	29.05
2200	Conc. block backup, 4" thick	13.40	19.95	33.35
2300	6" thick	14.05	20	34.05
2350	8" thick	14.20	20.50	34.70
2400	10" thick	15.05	22	37.05
2500	12" thick	15.50	23.50	39
3100	High priced stone, wood stud backup, 10' high, 16" O.C.	18.40	18.45	36.85
3200	Metal stud backup, 10' high, 16" O.C.	19.65	18.70	38.35
3250	24" O.C.	19.20	18.25	37.45
3300	Conc. block backup, 10' high, 4" thick	19.65	22	41.65
3350	6" thick	20.50	22	42.50
3400	8" thick	20.50	23	43.50
3450	10" thick	21.50	24	45.50
3500	12" thick	22	25.50	47.50
4000	Indiana limestone 2" thick, sawn finish, wood stud backup, 10' high, 16" O.C.	40.50	12.50	53
4100	Metal stud backup, 10' high, 16" O.C.	41.50	12.75	54.25
4150	24" O.C.	41	12.30	53.30
4200	Conc. block backup, 4" thick	41.50	16.15	57.65
4250	6" thick	42.50	16.35	58.85
4300	8" thick	42.50	16.90	59.40
4350	10" thick	43.50	18.10	61.60
4400	12" thick	44	19.80	63.80
4450	2" thick, smooth finish, wood stud backup, 8' high, 16" O.C.	40.50	12.50	53
4550	Metal stud backup, 8' high, 16" O.C.	41.50	12.55	54.05
4600	24" O.C.	41	12.15	53.15
4650	Conc. block backup, 4" thick	41.50	15.95	57.45
4700	6" thick	42.50	16.15	58.65
4750	8" thick	42.50	16.70	59.20
4800	10" thick	43.50	18.10	61.60
4850	12" thick	44	19.80	63.80
5350	4" thick, smooth finish, wood stud backup, 8' high, 16" O.C.	47	12.50	59.50
5450	Metal stud backup, 8' high, 16" O.C.	48	12.75	60.75
5500	24" O.C.	47.50	12.30	59.80
5550	Conc. block backup, 4" thick	48	16.15	64.15
5600	6" thick	49	16.35	65.35
5650	8" thick	49	16.90	65.90

B2010 Exterior Walls

B2010 128	Stone Veneer	COST PER S.F.		
		MAT.	INST.	TOTAL
5700	10" thick	50	18.10	68.10
5750	12" thick	50.50	19.80	70.30
6000	Granite, gray or pink, 2" thick, wood stud backup, 8' high, 16" O.C.	24	23.50	47.50
6100	Metal studs, 8' high, 16" O.C.	25	24	49
6150	24" O.C.	24.50	23.50	48
6200	Conc. block backup, 4" thick	25	27.50	52.50
6250	6" thick	26	27.50	53.50
6300	8" thick	26	28	54
6350	10" thick	27	29	56
6400	12" thick	27.50	31	58.50
6900	4" thick, wood stud backup, 8' high, 16" O.C.	38.50	27	65.50
7000	Metal studs, 8' high, 16" O.C.	39.50	27.50	67
7050	24" O.C.	39	27	66
7100	Conc. block backup, 4" thick	39.50	31	70.50
7150	6" thick	40.50	31	71.50
7200	8" thick	40.50	31.50	72
7250	10" thick	41.50	32.50	74
7300	12" thick	42	34.50	76.50

B2010 Exterior Walls

Exterior brick veneer/stud backup walls are defined in the following terms; type of brick and studs, stud spacing and bond. All systems include a brick shelf, ties to the backup and all necessary dampproofing and insulation.

B2010 129				Brick Veneer/Wood Stud Backup				
	FACE BRICK	STUD BACKUP	STUD SPACING (IN.)	BOND		COST PER S.F.		
						MAT.	INST.	TOTAL
1100	Standard	2x4-wood	16	running		9.15	15.05	24.20
1120				common		10.55	17	27.55
1140				Flemish		11.50	19.85	31.35
1160				English		12.70	21	33.70
1400		2x6-wood	16	running		9.45	15.15	24.60
1420				common		10.85	17.10	27.95
1440				Flemish		11.80	19.95	31.75
1460				English		13	21	34
1700	Glazed	2x4-wood	16	running		17.20	15.55	32.75
1720				common		20	17.70	37.70
1740				Flemish		22	21	43
1760				English		24.50	22	46.50
2300	Engineer	2x4-wood	16	running		7.10	13.45	20.55
2320				common		8	15.05	23.05
2340				Flemish		8.65	17.70	26.35
2360				English		9.50	18.45	27.95
2900	Roman	2x4-wood	16	running		9.05	13.80	22.85
2920				common		10.45	15.55	26
2940				Flemish		11.35	18.05	29.40
2960				English		12.60	19.35	31.95
4100	Norwegian	2x4-wood	16	running		6.95	10.80	17.75
4120				common		7.85	11.95	19.80
4140				Flemish		8.45	13.80	22.25
4160				English		9.25	14.40	23.65

B2010 130				Brick Veneer/Metal Stud Backup				
	FACE BRICK	STUD BACKUP	STUD SPACING (IN.)	BOND		COST PER S.F.		
						MAT.	INST.	TOTAL
5100	Standard	25ga.x6"NLB	24	running		9.15	14.75	23.90
5120				common		10.55	16.70	27.25
5140				Flemish		11	18.95	29.95
5160				English		12.70	20.50	33.20
5200		20ga.x3-5/8"NLB	16	running		9.30	15.35	24.65
5220				common		10.70	17.30	28
5240				Flemish		11.65	20	31.65
5260				English		12.85	21	33.85

B2010 Exterior Walls

| B2010 130 | | | | Brick Veneer/Metal Stud Backup | | | | |

	FACE BRICK	STUD BACKUP	STUD SPACING (IN.)	BOND		COST PER S.F.		
						MAT.	INST.	TOTAL
5400	Standard	16ga.x3-5/8"LB	16	running		10.25	15.55	25.80
5420				common		11.65	17.50	29.15
5440				Flemish		12.60	20.50	33.10
5460				English		13.80	21.50	35.30
5700	Glazed	25ga.x6"NLB	24	running		17.20	15.25	32.45
5720				common		20	17.40	37.40
5740				Flemish		22	20.50	42.50
5760				English		24.50	22	46.50
5800		20ga.x3-5/8"NLB	24	running		17.15	15.40	32.55
5820				common		20	17.55	37.55
5840				Flemish		22	21	43
5860				English		24.50	22	46.50
6000		16ga.x3-5/8"LB	16	running		18.30	16.05	34.35
6020				common		21.50	18.20	39.70
6040				Flemish		23.50	21.50	45
6060				English		25.50	22.50	48
6300	Engineer	25ga.x6"NLB	24	running		7.05	13.15	20.20
6320				common		8	14.75	22.75
6340				Flemish		8.65	17.40	26.05
6360				English		9.50	18.15	27.65
6400		20ga.x3-5/8"NLB	16	running		7.20	13.75	20.95
6420				common		8.15	15.35	23.50
6440				Flemish		8.80	18	26.80
6460				English		9.65	18.75	28.40
6900	Roman	25ga.x6"NLB	24	running		9.05	13.50	22.55
6920				common		10.45	15.20	25.65
6940				Flemish		11.35	17.75	29.10
6960				English		12.60	19.05	31.65
7000		20ga.x3-5/8"NLB	16	running		9.20	14.10	23.30
7020				common		10.60	15.85	26.45
7040				Flemish		11.50	18.35	29.85
7060				English		12.75	19.65	32.40
7500	Norman	25ga.x6"NLB	24	running		8.05	11.55	19.60
7520				common		9.20	12.85	22.05
7540				Flemish		18.50	15	33.50
7560				English		11	15.75	26.75
7600		20ga.x3-5/8"NLB	24	running		8	11.70	19.70
7620				common		9.15	13	22.15
7640				Flemish		18.45	15.15	33.60
7660				English		10.95	15.90	26.85
8100	Norwegian	25ga.x6"NLB	24	running		6.95	10.50	17.45
8120				common		7.85	11.65	19.50
8140				Flemish		8.45	13.50	21.95
8160				English		9.25	14.10	23.35
8400		16ga.x3-5/8"LB	16	running		8.05	11.30	19.35
8420				common		8.95	12.45	21.40
8440				Flemish		9.55	14.30	23.85
8460				English		10.35	14.90	25.25

B2010 Exterior Walls

Brick Face Composite Wall

B2010 132		Brick Face Composite Wall - Double Wythe						
	FACE BRICK	BACKUP MASONRY	BACKUP THICKNESS (IN.)	BACKUP CORE FILL		COST PER S.F.		
						MAT.	INST.	TOTAL
1000	Standard	common brick	4	none		11.25	23	34.25
1040		SCR brick	6	none		13.95	20.50	34.45
1080		conc. block	4	none		9.75	18.75	28.50
1120			6	perlite		10.75	19.40	30.15
1160				styrofoam		12.40	19.15	31.55
1200			8	perlite		11	19.85	30.85
1240				styrofoam		12.50	19.50	32
1520		glazed block	4	none		17.50	20	37.50
1560			6	perlite		18.55	20.50	39.05
1600				styrofoam		20	20.50	40.50
1640			8	perlite		18.80	21	39.80
1680				styrofoam		20.50	20.50	41
1720		clay tile	4	none		15.80	18	33.80
1760			6	none		17.70	18.50	36.20
1800			8	none		20.50	19.15	39.65
1840		glazed tile	4	none		20.50	23.50	44
1880								
2000	Glazed	common brick	4	none		19.30	23.50	42.80
2040		SCR brick	6	none		22	21	43
2080		conc. block	4	none		17.80	19.25	37.05
2200			8	perlite		19.05	20.50	39.55
2240				styrofoam		20.50	20	40.50
2280		L.W. block	4	none		18.10	19.10	37.20
2400			8	perlite		19.85	20	39.85
2520		glazed block	4	none		25.50	20.50	46
2560			6	perlite		26.50	21	47.50
2600				styrofoam		28	21	49
2640			8	perlite		27	21.50	48.50
2680				styrofoam		28.50	21	49.50

B2010 Exterior Walls

B2010 132	Brick Face Composite Wall - Double Wythe

	FACE BRICK	BACKUP MASONRY	BACKUP THICKNESS (IN.)	BACKUP CORE FILL		COST PER S.F.		
						MAT.	INST.	TOTAL
2720	Glazed	clay tile	4	none		24	18.50	42.50
2760			6	none		26	19	45
2800			8	none		28.50	19.65	48.15
2840		glazed tile	4	none		28.50	24	52.50
3000	Engineer	common brick	4	none		9.20	21.50	30.70
3040		SCR brick	6	none		11.85	19.15	31
3080		conc. block	4	none		7.65	17.15	24.80
3200			8	perlite		8.95	18.25	27.20
3280		L.W. block	4	none		6.15	12.15	18.30
3322		L.W. block	6	perlite		8.95	17.65	26.60
3520		glazed block	4	none		15.40	18.45	33.85
3560			6	perlite		16.45	19	35.45
3600				styrofoam		18.10	18.75	36.85
3640			8	perlite		16.70	19.50	36.20
3680				styrofoam		18.20	19.15	37.35
3720		clay tile	4	none		13.70	16.40	30.10
3800			8	none		18.25	17.55	35.80
3840		glazed tile	4	none		18.20	22	40.20
4000	Roman	common brick	4	none		11.15	22	33.15
4040		SCR brick	6	none		13.85	19.50	33.35
4080		conc. block	4	none		9.65	17.50	27.15
4120			6	perlite		10.65	18.15	28.80
4200			8	perlite		10.90	18.60	29.50
5000	Norman	common brick	4	none		10.15	20	30.15
5040		SCR brick	6	none		12.85	17.55	30.40
5280		L.W. block	4	none		8.95	15.40	24.35
5320			6	perlite		10	16.25	26.25
5720		clay tile	4	none		14.70	14.80	29.50
5760			6	none		16.60	15.30	31.90
5840		glazed tile	4	none		19.20	20.50	39.70
6000	Norwegian	common brick	4	none		9.05	19.05	28.10
6040		SCR brick	6	none		11.75	16.50	28.25
6080		conc. block	4	none		7.55	14.50	22.05
6120			6	perlite		8.55	15.15	23.70
6160				styrofoam		10.20	14.90	25.10
6200			8	perlite		8.80	15.60	24.40
6520		glazed block	4	none		15.30	15.80	31.10
6560			6	perlite		16.35	16.35	32.70
7000	Utility	common brick	4	none		8.70	18.05	26.75
7040		SCR brick	6	none		11.40	15.50	26.90
7080		conc. block	4	none		7.15	13.50	20.65
7120			6	perlite		8.15	14.15	22.30
7160				styrofoam		9.80	13.90	23.70
7200			8	perlite		8.45	14.60	23.05
7240				styrofoam		9.95	14.25	24.20
7280		L.W. block	4	none		7.50	13.35	20.85
7320			6	perlite		8.50	14	22.50
7520		glazed block	4	none		14.95	14.80	29.75
7560			6	perlite		16	15.35	31.35
7720		clay tile	4	none		13.25	12.75	26
7760			6	none		15.15	13.25	28.40
7840		glazed tile	4	none		17.75	18.50	36.25

B20 Exterior Enclosure

B2010 Exterior Walls

B2010 133	Brick Face Composite Wall - Triple Wythe

	FACE BRICK	MIDDLE WYTHE	INSIDE MASONRY	TOTAL THICKNESS (IN.)		COST PER S.F.		
						MAT.	INST.	TOTAL
8000	Standard	common brick	standard brick	12		17.90	33	50.90
8100		4" conc. brick	standard brick	12		17.30	34	51.30
8120		4" conc. brick	common brick	12		14.25	34	48.25
8200	Glazed	common brick	standard brick	12		26	33.50	59.50
8300		4" conc. brick	standard brick	12		25.50	34.50	60
8320		4" conc. brick	glazed brick	12		22.50	34.50	57

B2010 Exterior Walls

Brick Face Cavity Wall

B2010 134		Brick Face Cavity Wall							
	FACE BRICK	**BACKUP MASONRY**	**TOTAL THICKNESS (IN.)**	**CAVITY INSULATION**		**COST PER S.F.**			
							MAT.	**INST.**	**TOTAL**
1000	Standard	4" common brick	10	polystyrene			11.20	23.50	34.70
1020				none			10.90	22.50	33.40
1040		6" SCR brick	12	polystyrene			13.90	21	34.90
1060				none			13.65	20	33.65
1080		4" conc. block	10	polystyrene			9.65	18.80	28.45
1100				none			9.40	18.15	27.55
1120		6" conc. block	12	polystyrene			10.35	19.20	29.55
1140				none			10.05	18.55	28.60
1160		4" L.W. block	10	polystyrene			9.95	18.65	28.60
1180				none			9.70	18	27.70
1200		6" L.W. block	12	polystyrene			10.65	19.05	29.70
1220				none			10.40	18.40	28.80
1240		4" glazed block	10	polystyrene			17.40	20	37.40
1260				none			17.15	19.45	36.60
1280		6" glazed block	12	polystyrene			17.45	20	37.45
1300				none			17.20	19.45	36.65
1320		4" clay tile	10	polystyrene			15.70	18.05	33.75
1340				none			15.45	17.45	32.90
1360		4" glazed tile	10	polystyrene			20	24	44
1380				none			19.95	23	42.95
1500	Glazed	4" common brick	10	polystyrene			19.25	24	43.25
1520				none			18.95	23	41.95
1580		4" conc. block	10	polystyrene			17.70	19.30	37
1600				none			17.45	18.65	36.10
1660		4" L.W. block	10	polystyrene			18	19.15	37.15
1680				none			17.75	18.50	36.25
1740		4" glazed block	10	polystyrene			25.50	20.50	46
1760				none			25	19.95	44.95
1820		4" clay tile	10	polystyrene			24	18.55	42.55
1840				none			23.50	17.95	41.45
1860		4" glazed tile	10	polystyrene			25.50	19.05	44.55
1880				none			25.50	18.40	43.90
2000	Engineer	4" common brick	10	polystyrene			9.10	21.50	30.60
2020				none			8.85	21	29.85
2080		4" conc. block	10	polystyrene			7.55	17.20	24.75
2100				none			7.30	16.55	23.85

B2010 Exterior Walls

B2010 134	Brick Face Cavity Wall

	FACE BRICK	BACKUP MASONRY	TOTAL THICKNESS (IN.)	CAVITY INSULATION		COST PER S.F.		
						MAT.	INST.	TOTAL
2162	Engineer	4" L.W. block	10	polystyrene		7.90	17.05	24.95
2180				none		7.60	16.40	24
2240		4" glazed block	10	polystyrene		15.35	18.50	33.85
2260				none		15.10	17.85	32.95
2320		4" clay tile	10	polystyrene		13.65	16.45	30.10
2340				none		13.40	15.85	29.25
2360		4" glazed tile	10	polystyrene		18.15	22	40.15
2380				none		17.90	21.50	39.40
2500	Roman	4" common brick	10	polystyrene		11.10	22	33.10
2520				none		10.80	21.50	32.30
2580		4" conc. block	10	polystyrene		9.55	17.55	27.10
2600				none		9.30	16.90	26.20
2660		4" L.W. block	10	polystyrene		9.85	17.40	27.25
2680				none		9.60	16.75	26.35
2740		4" glazed block	10	polystyrene		17.30	18.85	36.15
2760				none		17.05	18.20	35.25
2820		4" clay tile	10	polystyrene		15.60	16.80	32.40
2840				none		15.35	16.20	31.55
2860		4" glazed tile	10	polystyrene		20	22.50	42.50
2880				none		19.85	22	41.85
3000	Norman	4" common brick	10	polystyrene		10.10	20	30.10
3020				none		9.80	19.50	29.30
3080		4" conc. block	10	polystyrene		8.55	15.60	24.15
3100				none		8.30	14.95	23.25
3160		4" L.W. block	10	polystyrene		8.85	15.45	24.30
3180				none		8.60	14.80	23.40
3240		4" glazed block	10	polystyrene		16.30	16.90	33.20
3260				none		16.05	16.25	32.30
3320		4" clay tile	10	polystyrene		14.60	14.85	29.45
3340				none		14.35	14.25	28.60
3360		4" glazed tile	10	polystyrene		19.10	20.50	39.60
3380				none		18.85	19.95	38.80
3500	Norwegian	4" common brick	10	polystyrene		9	19.10	28.10
3520				none		8.70	18.45	27.15
3580		4" conc. block	10	polystyrene		7.45	14.55	22
3600				none		7.20	13.90	21.10
3660		4" L.W. block	10	polystyrene		7.75	14.40	22.15
3680				none		7.50	13.75	21.25
3740		4" glazed block	10	polystyrene		15.20	15.85	31.05
3760				none		14.95	15.20	30.15
3820		4" clay tile	10	polystyrene		13.50	13.80	27.30
3840				none		13.25	13.20	26.45
3860		4" glazed tile	10	polystyrene		18	19.55	37.55
3880				none		17.75	18.90	36.65
4000	Utility	4" common brick	10	polystyrene		8.60	18.10	26.70
4020				none		8.35	17.45	25.80
4080		4" conc. block	10	polystyrene		7.10	13.55	20.65
4100				none		6.85	12.90	19.75
4160		4" L.W. block	10	polystyrene		7.40	13.40	20.80
4180				none		7.15	12.75	19.90
4240		4" glazed block	10	polystyrene		14.85	14.85	29.70
4260				none		14.60	14.20	28.80

B20 Exterior Enclosure

B2010 Exterior Walls

B2010 134 — Brick Face Cavity Wall

	FACE BRICK	BACKUP MASONRY	TOTAL THICKNESS (IN.)	CAVITY INSULATION		COST PER S.F. MAT.	COST PER S.F. INST.	COST PER S.F. TOTAL
4320	Utility	4" clay tile	10	polystyrene		13.15	12.80	25.95
4340				none		12.90	12.20	25.10
4360		4" glazed tile	10	polystyrene		17.65	18.55	36.20
4380				none		17.40	17.90	35.30

B2010 135 — Brick Face Cavity Wall - Insulated Backup

	FACE BRICK	BACKUP MASONRY	TOTAL THICKNESS (IN.)	BACKUP CORE FILL		COST PER S.F. MAT.	COST PER S.F. INST.	COST PER S.F. TOTAL
5100	Standard	6" conc. block	10	perlite		10.40	18.85	29.25
5120				styrofoam		12.05	18.55	30.60
5180		6" L.W. block	10	perlite		10.70	18.70	29.40
5200				styrofoam		12.35	18.40	30.75
5260		6" glazed block	10	perlite		18.20	20	38.20
5280				styrofoam		19.85	19.75	39.60
5340		6" clay tile	10	none		17.35	17.90	35.25
5360		8" clay tile	12	none		20	18.55	38.55
5600	Glazed	6" conc. block	10	perlite		18.45	19.35	37.80
5620				styrofoam		20	19.05	39.05
5680		6" L.W. block	10	perlite		18.75	19.20	37.95
5700				styrofoam		20.50	18.90	39.40
5760		6" glazed block	10	perlite		26.50	20.50	47
5780				styrofoam		28	20.50	48.50
5840		6" clay tile	10	none		25.50	18.40	43.90
5860		8" clay tile	8	none		28	19.05	47.05
6100	Engineer	6" conc. block	10	perlite		8.30	17.25	25.55
6120				styrofoam		9.95	16.95	26.90
6180		6" L.W. block	10	perlite		8.65	17.10	25.75
6200				styrofoam		10.30	16.80	27.10
6260		6" glazed block	10	perlite		16.10	18.45	34.55
6280				styrofoam		17.75	18.15	35.90
6340		6" clay tile	10	none		15.30	16.30	31.60
6360		8" clay tile	12	none		17.95	16.95	34.90
6600	Roman	6" conc. block	10	perlite		10.30	17.60	27.90
6620				styrofoam		11.95	17.30	29.25
6680		6" L.W. block	10	perlite		10.60	17.45	28.05
6700				styrofoam		12.25	17.15	29.40
6760		6" glazed block	10	perlite		18.10	18.80	36.90
6780				styrofoam		19.75	18.50	38.25
6840		6" clay tile	10	none		17.25	16.65	33.90
6860		8" clay tile	12	none		19.95	17.30	37.25
7100	Norman	6" conc. block	10	perlite		9.30	15.65	24.95
7120				styrofoam		10.95	15.35	26.30
7180		6" L.W. block	10	perlite		9.60	15.50	25.10
7200				styrofoam		11.25	15.20	26.45
7260		6" glazed block	10	perlite		17.10	16.85	33.95
7280				styrofoam		18.75	16.55	35.30
7340		6" clay tile	10	none		16.25	14.70	30.95
7360		8" clay tile	12	none		18.95	15.35	34.30
7600	Norwegian	6" conc. block	10	perlite		8.20	14.60	22.80
7620				styrofoam		9.85	14.30	24.15
7680		6" L.W. block	10	perlite		8.50	14.45	22.95
7700				styrofoam		10.15	14.15	24.30

B2010 Exterior Walls

| B2010 135 | | | | Brick Face Cavity Wall - Insulated Backup | | | | |

	FACE BRICK	BACKUP MASONRY	TOTAL THICKNESS (IN.)	BACKUP CORE FILL		COST PER S.F.		
						MAT.	INST.	TOTAL
7760	Norwegian	6" glazed block	10	perlite		16	15.80	31.80
7780				styrofoam		17.65	15.50	33.15
7840		6" clay tile	10	none		15.15	13.65	28.80
7860		8" clay tile	12	none		17.85	14.30	32.15
8100	Utility	6" conc. block	10	perlite		7.85	13.60	21.45
8120				styrofoam		9.50	13.30	22.80
8180		6" L.W. block	10	perlite		8.15	13.45	21.60
8200				styrofoam		9.80	13.15	22.95
8260		6" glazed block	10	perlite		15.65	14.80	30.45
8280				styrofoam		17.30	14.50	31.80
8340		6" clay tile	10	none		14.80	12.65	27.45
8360		8" clay tile	12	none		17.45	13.30	30.75

B2010 Exterior Walls

Glass Block

B2010 140	Glass Block	COST PER S.F.		
		MAT.	INST.	TOTAL
2300	Glass block 4" thick, 6"x6" plain, under 1,000 S.F.	22.50	22	44.50
2400	1,000 to 5,000 S.F.	22	18.95	40.95
2500	Over 5,000 S.F.	21.50	17.80	39.30
2600	Solar reflective, under 1,000 S.F.	31.50	29.50	61
2700	1,000 to 5,000 S.F.	31	25.50	56.50
2800	Over 5,000 S.F.	30	24	54
3500	8"x8" plain, under 1,000 S.F.	15.55	16.35	31.90
3600	1,000 to 5,000 S.F.	15.25	14.10	29.35
3700	Over 5,000 S.F.	14.75	12.75	27.50
3800	Solar reflective, under 1,000 S.F.	22	22	44
3900	1,000 to 5,000 S.F.	21.50	18.85	40.35
4000	Over 5,000 S.F.	20.50	16.95	37.45
5000	12"x12" plain, under 1,000 S.F.	17.70	15.15	32.85
5100	1,000 to 5,000 S.F.	17.35	12.75	30.10
5200	Over 5,000 S.F.	16.85	11.65	28.50
5300	Solar reflective, under 1,000 S.F.	25	20.50	45.50
5400	1,000 to 5,000 S.F.	24.50	16.95	41.45
5600	Over 5,000 S.F.	23.50	15.40	38.90
5800	3" thinline, 6"x6" plain, under 1,000 S.F.	16.65	22	38.65
5900	Over 5,000 S.F.	16.65	22	38.65
6000	Solar reflective, under 1,000 S.F.	23.50	29.50	53
6100	Over 5,000 S.F.	25	24	49
6200	8"x8" plain, under 1,000 S.F.	10.40	16.35	26.75
6300	Over 5,000 S.F.	10.05	12.75	22.80
6400	Solar reflective, under 1,000 S.F.	14.55	22	36.55
6500	Over 5,000 S.F.	14.05	16.95	31

B2010 Exterior Walls

Metal Siding

The table below lists costs for metal siding of various descriptions, not including the steel frame, or the structural steel, of a building. Costs are per S.F. including all accessories and insulation.

B2010 146	Metal Siding Panel	COST PER S.F.		
		MAT.	INST.	TOTAL
1400	Metal siding aluminum panel, corrugated, .024" thick, natural	2.71	3.56	6.27
1450	Painted	2.87	3.56	6.43
1500	.032" thick, natural	3.02	3.56	6.58
1550	Painted	3.55	3.56	7.11
1600	Ribbed 4" pitch, .032" thick, natural	2.96	3.56	6.52
1650	Painted	3.59	3.56	7.15
1700	.040" thick, natural	3.38	3.56	6.94
1750	Painted	3.88	3.56	7.44
1800	.050" thick, natural	3.83	3.56	7.39
1850	Painted	4.33	3.56	7.89
1900	8" pitch panel, .032" thick, natural	2.82	3.41	6.23
1950	Painted	3.44	3.43	6.87
2000	.040" thick, natural	3.25	3.43	6.68
2050	Painted	3.71	3.47	7.18
2100	.050" thick, natural	3.70	3.45	7.15
2150	Painted	4.22	3.48	7.70
3000	Steel, corrugated or ribbed, 29 Ga. .0135" thick, galvanized	2.11	3.25	5.36
3050	Colored	2.97	3.30	6.27
3100	26 Ga. .0179" thick, galvanized	2.18	3.27	5.45
3150	Colored	3.06	3.32	6.38
3200	24 Ga. .0239" thick, galvanized	2.74	3.28	6.02
3250	Colored	3.42	3.33	6.75
3300	22 Ga. .0299" thick, galvanized	2.75	3.30	6.05
3350	Colored	4.09	3.35	7.44
3400	20 Ga. .0359" thick, galvanized	2.75	3.30	6.05
3450	Colored	4.34	3.54	7.88
4100	Sandwich panels, factory fab., 1" polystyrene, steel core, 26 Ga., galv.	5.30	5.10	10.40
4200	Colored, 1 side	6.50	5.10	11.60
4300	2 sides	8.35	5.10	13.45
4400	2" polystyrene, steel core, 26 Ga., galvanized	6.25	5.10	11.35
4500	Colored, 1 side	7.45	5.10	12.55
4600	2 sides	9.30	5.10	14.40
4700	22 Ga., baked enamel exterior	12.30	5.40	17.70
4800	Polyvinyl chloride exterior	12.95	5.40	18.35
5100	Textured aluminum, 4' x 8' x 5/16" plywood backing, single face	4.10	3.10	7.20
5200	Double face	4.88	3.10	7.98

B2010 Exterior Walls

Exterior Wall Siding

The table below lists costs per S.F. for exterior walls with wood siding. A variety of systems are presented using both wood and metal studs at 16″ and 24″ O.C.

B2010 148	Panel, Shingle & Lap Siding	COST PER S.F.		
		MAT.	INST.	TOTAL
1400	Wood siding w/2″ x 4″ studs, 16″O.C., insul. wall, 5/8″ text 1-11 fir plywood	2.97	5.10	8.07
1450	5/8″ text 1-11 cedar plywood	4.43	4.91	9.34
1500	1″ x 4″ vert T.&G. redwood	5.15	5.85	11
1600	1″ x 8″ vert T.&G. redwood	5.30	5.10	10.40
1650	1″ x 5″ rabbetted cedar bev. siding	6.85	5.30	12.15
1700	1″ x 6″ cedar drop siding	6.95	5.30	12.25
1750	1″ x 12″ rough sawn cedar	3.10	5.10	8.20
1800	1″ x 12″ sawn cedar, 1″ x 4″ battens	6.05	4.79	10.84
1850	1″ x 10″ redwood shiplap siding	5.40	4.93	10.33
1900	18″ no. 1 red cedar shingles, 5-1/2″ exposed	5.45	6.70	12.15
1950	6″ exposed	5.20	6.45	11.65
2000	6-1/2″ exposed	4.92	6.25	11.17
2100	7″ exposed	4.66	6.05	10.71
2150	7-1/2″ exposed	4.39	5.80	10.19
3000	8″ wide aluminum siding	3.28	4.54	7.82
3150	8″ plain vinyl siding	2.61	4.62	7.23
3250	8″ insulated vinyl siding	2.91	5.10	8.01
3300				
3400	2″ x 6″ studs, 16″ O.C., insul. wall, w/ 5/8″ text 1-11 fir plywood	3.47	5.20	8.67
3500	5/8″ text 1-11 cedar plywood	4.93	5.20	10.13
3600	1″ x 4″ vert T.&G. redwood	5.65	6	11.65
3700	1″ x 8″ vert T.&G. redwood	5.80	5.25	11.05
3800	1″ x 5″ rabbetted cedar bev siding	7.35	5.45	12.80
3900	1″ x 6″ cedar drop siding	7.45	5.45	12.90
4000	1″ x 12″ rough sawn cedar	3.60	5.25	8.85
4200	1″ x 12″ sawn cedar, 1″ x 4″ battens	6.55	4.93	11.48
4500	1″ x 10″ redwood shiplap siding	5.90	5.05	10.95
4550	18″ no. 1 red cedar shingles, 5-1/2″ exposed	5.95	6.85	12.80
4600	6″ exposed	5.70	6.60	12.30
4650	6-1/2″ exposed	5.40	6.40	11.80
4700	7″ exposed	5.15	6.15	11.30
4750	7-1/2″ exposed	4.89	5.95	10.84
4800	8″ wide aluminum siding	3.78	4.68	8.46
4850	8″ plain vinyl siding	3.11	4.76	7.87
4900	8″ insulated vinyl siding	3.41	5.25	8.66
4910				
5000	2″ x 6″ studs, 24″ O.C., insul. wall, 5/8″ text 1-11, fir plywood	3.33	4.92	8.25
5050	5/8″ text 1-11 cedar plywood	4.79	4.92	9.71
5100	1″ x 4″ vert T.&G. redwood	5.50	5.70	11.20
5150	1″ x 8″ vert T.&G. redwood	5.65	4.95	10.60

B2010 Exterior Walls

B2010 148	Panel, Shingle & Lap Siding	COST PER S.F.		
		MAT.	INST.	TOTAL
5200	1" x 5" rabbetted cedar bev siding	7.20	5.15	12.35
5250	1" x 6" cedar drop siding	7.30	5.15	12.45
5300	1" x 12" rough sawn cedar	3.46	4.96	8.42
5400	1" x 12" sawn cedar, 1" x 4" battens	6.40	4.63	11.03
5450	1" x 10" redwood shiplap siding	5.75	4.77	10.52
5500	18" no. 1 red cedar shingles, 5-1/2" exposed	5.80	6.55	12.35
5550	6" exposed	5.55	6.30	11.85
5650	7" exposed	5	5.85	10.85
5700	7-1/2" exposed	4.75	5.65	10.40
5750	8" wide aluminum siding	3.64	4.38	8.02
5800	8" plain vinyl siding	2.97	4.46	7.43
5850	8" insulated vinyl siding	3.27	4.96	8.23
5900	3-5/8" metal studs, 16 Ga., 16" OC insul.wall, 5/8"text 1-11 fir plywood	4.27	5.35	9.62
5950	5/8" text 1-11 cedar plywood	5.75	5.35	11.10
6000	1" x 4" vert T.&G. redwood	6.45	6.10	12.55
6050	1" x 8" vert T.&G. redwood	6.60	5.35	11.95
6100	1" x 5" rabbetted cedar bev siding	8.15	5.55	13.70
6150	1" x 6" cedar drop siding	8.25	5.60	13.85
6200	1" x 12" rough sawn cedar	4.40	5.40	9.80
6250	1" x 12" sawn cedar, 1" x 4" battens	7.35	5.05	12.40
6300	1" x 10" redwood shiplap siding	6.70	5.20	11.90
6350	18" no. 1 red cedar shingles, 5-1/2" exposed	6.75	6.95	13.70
6500	6" exposed	6.50	6.75	13.25
6550	6-1/2" exposed	6.20	6.50	12.70
6600	7" exposed	5.95	6.30	12.25
6650	7-1/2" exposed	5.95	6.10	12.05
6700	8" wide aluminum siding	4.58	4.80	9.38
6750	8" plain vinyl siding	3.91	4.71	8.62
6800	8" insulated vinyl siding	4.21	5.20	9.41
7000	3-5/8" metal studs, 16 Ga. 24" OC insul wall, 5/8" text 1-11 fir plywood	3.81	4.75	8.56
7050	5/8" text 1-11 cedar plywood	5.25	4.75	10
7100	1" x 4" vert T.&G. redwood	6	5.70	11.70
7150	1" x 8" vert T.&G. redwood	6.15	4.95	11.10
7200	1" x 5" rabbetted cedar bev siding	7.65	5.15	12.80
7250	1" x 6" cedar drop siding	7.75	5.15	12.90
7300	1" x 12" rough sawn cedar	3.94	4.96	8.90
7350	1" x 12" sawn cedar 1" x 4" battens	6.90	4.63	11.53
7400	1" x 10" redwood shiplap siding	6.20	4.77	10.97
7450	18" no. 1 red cedar shingles, 5-1/2" exposed	6.55	6.75	13.30
7500	6" exposed	6.30	6.55	12.85
7550	6-1/2" exposed	6.05	6.30	12.35
7600	7" exposed	5.75	6.10	11.85
7650	7-1/2" exposed	5.25	5.65	10.90
7700	8" wide aluminum siding	4.12	4.38	8.50
7750	8" plain vinyl siding	3.45	4.46	7.91
7800	8" insul. vinyl siding	3.75	4.96	8.71

B2010 Exterior Walls

The table below lists costs for some typical stucco walls including all the components as demonstrated in the component block below. Prices are presented for backup walls using wood studs, metal studs and CMU.

Exterior Stucco Wall

B2010 151	Stucco Wall	COST PER S.F.		
		MAT.	INST.	TOTAL
2100	Cement stucco, 7/8" th., plywood sheathing, stud wall, 2" x 4", 16" O.C.	2.80	8.20	11
2200	24" O.C.	2.72	7.95	10.67
2300	2" x 6", 16" O.C.	3.30	8.35	11.65
2400	24" O.C.	3.16	8.05	11.21
2500	No sheathing, metal lath on stud wall, 2" x 4", 16" O.C.	2.01	7.15	9.16
2600	24" O.C.	1.93	6.90	8.83
2700	2" x 6", 16" O.C.	2.51	7.30	9.81
2800	24" O.C.	2.37	7	9.37
2900	1/2" gypsum sheathing, 3-5/8" metal studs, 16" O.C.	3.78	7.45	11.23
2950	24" O.C.	3.32	7	10.32
3000	Cement stucco, 5/8" th., 2 coats on std. CMU block, 8"x 16", 8" thick	2.66	10.05	12.71
3100	10" thick	3.81	10.35	14.16
3200	12" thick	3.91	11.90	15.81
3300	Std. light Wt. block 8" x 16", 8" thick	3.70	9.90	13.60
3400	10" thick	4.66	10.15	14.81
3500	12" thick	5.25	11.60	16.85
3600	3 coat stucco, self furring metal lath 3.4 Lb/SY, on 8" x 16", 8" thick	2.70	10.65	13.35
3700	10" thick	3.85	10.95	14.80
3800	12" thick	3.95	12.50	16.45
3900	Lt. Wt. block, 8" thick	3.74	10.50	14.24
4000	10" thick	4.70	10.75	15.45
4100	12" thick	5.30	12.20	17.50

B2010 152	E.I.F.S.	COST PER S.F.		
		MAT.	INST.	TOTAL
5100	E.I.F.S., plywood sheathing, stud wall, 2" x 4", 16" O.C., 1" EPS	4.13	10.10	14.23
5110	2" EPS	4.54	10.10	14.64
5120	3" EPS	4.72	10.10	14.82
5130	4" EPS	5.20	10.10	15.30
5140	2" x 6", 16" O.C., 1" EPS	4.63	10.25	14.88
5150	2" EPS	5.05	10.25	15.30
5160	3" EPS	5.20	10.25	15.45
5170	4" EPS	5.70	10.25	15.95
5180	Cement board sheathing, 3-5/8" metal studs, 16" O.C., 1" EPS	5.50	12	17.50
5190	2" EPS	5.90	12	17.90
5200	3" EPS	6.10	12	18.10
5210	4" EPS	7.35	12	19.35
5220	6" metal studs, 16" O.C., 1" EPS	6.25	12.05	18.30
5230	2" EPS	6.65	12.05	18.70

B20 Exterior Enclosure

B2010 Exterior Walls

B2010 152	E.I.F.S.	COST PER S.F.		
		MAT.	INST.	TOTAL
5240	3" EPS	6.85	12.05	18.90
5250	4" EPS	8.10	12.05	20.15
5260	CMU block, 8" x 8" x 16", 1" EPS	4.61	13.65	18.26
5270	2" EPS	5	13.65	18.65
5280	3" EPS	5.20	13.65	18.85
5290	4" EPS	5.70	13.65	19.35
5300	8" x 10" x 16", 1" EPS	5.75	13.95	19.70
5310	2" EPS	6.15	13.95	20.10
5320	3" EPS	6.35	13.95	20.30
5330	4" EPS	6.85	13.95	20.80
5340	8" x 12" x 16", 1" EPS	5.85	15.50	21.35
5350	2" EPS	6.25	15.50	21.75
5360	3" EPS	6.45	15.50	21.95
5370	4" EPS	6.95	15.50	22.45

B2010 Exterior Walls

Description: The table below lists costs, $/S.F., for channel girts with sag rods and connector angles top and bottom for various column spacings, building heights and wind loads. Additive costs are shown for wind columns.

B2010 154			**Metal Siding Support**					
	BLDG. HEIGHT (FT.)	WIND LOAD (P.S.F.)	COL. SPACING (FT.)		INTERMEDIATE COLUMNS	COST PER S.F.		
						MAT.	INST.	TOTAL
3000	18	20	20			2.74	3.59	6.33
3100					wind cols.	1.41	.28	1.69
3200		20	25			2.99	3.65	6.64
3300					wind cols.	1.13	.21	1.34
3400		20	30			3.30	3.70	7
3500					wind cols.	.94	.19	1.13
3600		20	35			3.64	3.77	7.41
3700					wind cols.	.81	.16	.97
3800		30	20			3.03	3.65	6.68
3900					wind cols.	1.41	.28	1.69
4000		30	25			3.30	3.70	7
4100					wind cols.	1.13	.21	1.34
4200		30	30			3.65	3.77	7.42
4300					wind cols.	1.28	.25	1.53
4600	30	20	20			2.65	2.55	5.20
4700					wind cols.	2	.39	2.39
4800		20	25			2.98	2.61	5.59
4900					wind cols.	1.60	.31	1.91
5000		20	30			3.36	2.68	6.04
5100					wind cols.	1.57	.31	1.88
5200		20	35			3.75	2.76	6.51
5300					wind cols.	1.56	.31	1.87
5400		30	20			3.02	2.61	5.63
5500					wind cols.	2.36	.45	2.81
5600		30	25			3.35	2.68	6.03
5700					wind cols.	2.25	.44	2.69
5800		30	30			3.77	2.76	6.53
5900					wind cols.	2.12	.41	2.53

B20 Exterior Enclosure

B2020 Exterior Windows

The table below lists window systems by material, type and size. Prices between sizes listed can be interpolated with reasonable accuracy. Prices include frame, hardware, and casing as illustrated in the component block below.

	MATERIAL	TYPE	GLAZING	SIZE	DETAIL	COST PER UNIT		
B2020 102				Wood Windows				
						MAT.	INST.	TOTAL
3000	Wood	double hung	std. glass	2'-8" x 4'-6"		238	216	454
3050				3'-0" x 5'-6"		315	248	563
3100			insul. glass	2'-8" x 4'-6"		255	216	471
3150				3'-0" x 5'-6"		340	248	588
3200		sliding	std. glass	3'-4" x 2'-7"		288	180	468
3250				4'-4" x 3'-3"		330	197	527
3300				5'-4" x 6'-0"		420	237	657
3350			insul. glass	3'-4" x 2'-7"		350	210	560
3400				4'-4" x 3'-3"		400	229	629
3450				5'-4" x 6'-0"		505	270	775
3500		awning	std. glass	2'-10" x 1'-9"		215	108	323
3600				4'-4" x 2'-8"		350	105	455
3700			insul. glass	2'-10" x 1'-9"		263	125	388
3800				4'-4" x 2'-8"		430	115	545
3900		casement	std. glass	1'-10" x 3'-2"	1 lite	320	142	462
3950				4'-2" x 4'-2"	2 lite	575	189	764
4000				5'-11" x 5'-2"	3 lite	870	244	1,114
4050				7'-11" x 6'-3"	4 lite	1,225	293	1,518
4100				9'-11" x 6'-3"	5 lite	1,600	340	1,940
4150			insul. glass	1'-10" x 3'-2"	1 lite	320	142	462
4200				4'-2" x 4'-2"	2 lite	590	189	779
4250				5'-11" x 5'-2"	3 lite	920	244	1,164
4300				7'-11" x 6'-3"	4 lite	1,300	293	1,593
4350				9'-11" x 6'-3"	5 lite	1,675	340	2,015
4400		picture	std. glass	4'-6" x 4'-6"		455	244	699
4450				5'-8" x 4'-6"		515	271	786
4500			insul. glass	4'-6" x 4'-6"		560	284	844
4550				5'-8" x 4'-6"		630	315	945
4600		fixed bay	std. glass	8' x 5'		1,600	445	2,045
4650				9'-9" x 5'-4"		1,125	630	1,755
4700			insul. glass	8' x 5'		2,150	445	2,595
4750				9'-9" x 5'-4"		1,225	630	1,855
4800		casement bay	std. glass	8' x 5'		1,600	510	2,110
4850			insul. glass	8' x 5'		1,800	510	2,310
4900		vert. bay	std. glass	8' x 5'		1,800	510	2,310
4950			insul. glass	8' x 5'		1,875	510	2,385

B20 Exterior Enclosure

B2020 Exterior Windows

B2020 104 — Steel Windows

	MATERIAL	TYPE	GLAZING	SIZE	DETAIL	COST PER UNIT		
						MAT.	INST.	TOTAL
5000	Steel	double hung	1/4" tempered	2'-8" x 4'-6"		835	170	1,005
5050				3'-4" x 5'-6"		1,275	259	1,534
5100			insul. glass	2'-8" x 4'-6"		855	194	1,049
5150				3'-4" x 5'-6"		1,300	296	1,596
5202		horiz. pivoted	std. glass	2' x 2'		251	56.50	307.50
5250				3' x 3'		565	127	692
5300				4' x 4'		1,000	226	1,226
5350				6' x 4'		1,500	340	1,840
5400			insul. glass	2' x 2'		257	65	322
5450				3' x 3'		580	146	726
5500				4' x 4'		1,025	259	1,284
5550				6' x 4'		1,550	390	1,940
5600		picture window	std. glass	3' x 3'		355	127	482
5650				6' x 4'		940	340	1,280
5700			insul. glass	3' x 3'		370	146	516
5750				6' x 4'		980	390	1,370
5800		industrial security	std. glass	2'-9" x 4'-1"		755	159	914
5850				4'-1" x 5'-5"		1,475	315	1,790
5900			insul. glass	2'-9" x 4'-1"		775	182	957
5950				4'-1" x 5'-5"		1,525	360	1,885
6000		comm. projected	std. glass	3'-9" x 5'-5"		1,250	288	1,538
6050				6'-9" x 4'-1"		1,675	390	2,065
6100			insul. glass	3'-9" x 5'-5"		1,275	330	1,605
6150				6'-9" x 4'-1"		1,725	445	2,170
6200		casement	std. glass	4'-2" x 4'-2"	2 lite	985	246	1,231
6250			insul. glass	4'-2" x 4'-2"		1,000	281	1,281
6300			std. glass	5'-11" x 5'-2"	3 lite	1,850	435	2,285
6350			insul. glass	5'-11" x 5'-2"		1,900	495	2,395

B2020 106 — Aluminum Windows

	MATERIAL	TYPE	GLAZING	SIZE	DETAIL	COST PER UNIT		
						MAT.	INST.	TOTAL
6400	Aluminum	awning	std. glass	3'-1" x 3'-2"		370	127	497
6450				4'-5" x 5'-3"		420	159	579
6500			insul. glass	3'-1" x 3'-2"		445	152	597
6550				4'-5" x 5'-3"		505	191	696
6600		sliding	std. glass	3' x 2'		230	127	357
6650				5' x 3'		350	142	492
6700				8' x 4'		370	212	582
6750				9' x 5'		560	320	880
6800			insul. glass	3' x 2'		246	127	373
6850				5' x 3'		410	142	552
6900				8' x 4'		595	212	807
6950				9' x 5'		900	320	1,220
7000		single hung	std. glass	2' x 3'		218	127	345
7050				2'-8" x 6'-8"		385	159	544
7100				3'-4" x 5'-0"		315	142	455
7150			insul. glass	2' x 3'		264	127	391
7200				2'-8" x 6'-8"		495	159	654
7250				3'-4" x 5'		350	142	492
7300		double hung	std. glass	2' x 3'		289	85	374
7350				2'-8" x 6'-8"		860	252	1,112

B2020 Exterior Windows

B2020 106			Aluminum Windows			COST PER UNIT		
	MATERIAL	TYPE	GLAZING	SIZE	DETAIL	MAT.	INST.	TOTAL
7400	Aluminum	double hung		3'-4" x 5'		805	236	1,041
7450			insul. glass	2' x 3'		299	97	396
7500				2'-8" x 6'-8"		885	288	1,173
7550				3'-4" x 5'-0"		830	270	1,100
7600		casement	std. glass	3'-1" x 3'-2"		248	138	386
7650				4'-5" x 5'-3"		600	335	935
7700			insul. glass	3'-1" x 3'-2"		264	158	422
7750				4'-5" x 5'-3"		640	385	1,025
7800		hinged swing	std. glass	3' x 4'		530	170	700
7850				4' x 5'		885	283	1,168
7900			insul. glass	3' x 4'		550	194	744
7950				4' x 5'		915	325	1,240
8200		picture unit	std. glass	2'-0" x 3'-0"		159	85	244
8250				2'-8" x 6'-8"		470	252	722
8300				3'-4" x 5'-0"		440	236	676
8350			insul. glass	2'-0" x 3'-0"		169	97	266
8400				2'-8" x 6'-8"		500	288	788
8450				3'-4" x 5'-0"		470	270	740
8500		awning type	std. glass	3'-0" x 3'-0"	2 lite	445	91	536
8550				3'-0" x 4'-0"	3 lite	520	127	647
8600				3'-0" x 5'-4"	4 lite	625	127	752
8650				4'-0" x 5'-4"	4 lite	685	142	827
8700			insul. glass	3'-0" x 3'-0"	2 lite	475	91	566
8750				3'-0" x 4'-0"	3 lite	595	127	722
8800				3'-0" x 5'-4"	4 lite	735	127	862
8850				4'-0" x 5'-4"	4 lite	825	142	967
9051								
9052								

B2020 210	Tubular Aluminum Framing	COST/S.F. OPNG.		
		MAT.	INST.	TOTAL
1100	Alum flush tube frame, for 1/4" glass, 1-3/4"x4", 5'x6' opng, no inter horizontals	12.55	11.45	24
1150	One intermediate horizontal	17.10	13.45	30.55
1200	Two intermediate horizontals	21.50	15.45	36.95
1250	5' x 20' opening, three intermediate horizontals	11.50	9.50	21
1400	1-3/4" x 4-1/2", 5' x 6' opening, no intermediate horizontals	14.60	11.45	26.05
1450	One intermediate horizontal	19.35	13.45	32.80
1500	Two intermediate horizontals	24	15.45	39.45
1550	5' x 20' opening, three intermediate horizontals	13.05	9.50	22.55
1700	For insulating glass, 2"x4-1/2", 5'x6' opening, no intermediate horizontals	15.05	12.05	27.10
1750	One intermediate horizontal	19.50	14.15	33.65
1800	Two intermediate horizontals	24	16.25	40.25
1850	5' x 20' opening, three intermediate horizontals	13.10	10	23.10
2000	Thermal break frame, 2-1/4"x4-1/2", 5'x6'opng, no intermediate horizontals	15.70	12.20	27.90
2050	One intermediate horizontal	21	14.65	35.65
2100	Two intermediate horizontals	26	17.15	43.15
2150	5' x 20' opening, three intermediate horizontals	14.20	10.40	24.60

B2020 Exterior Windows

The table below lists costs of curtain wall and spandrel panels per S.F. Costs do not include structural framing used to hang the panels.

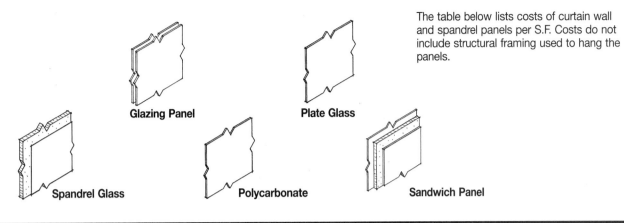

Glazing Panel

Plate Glass

Spandrel Glass

Polycarbonate

Sandwich Panel

B2020 220	Curtain Wall Panels	COST PER S.F.		
		MAT.	INST.	TOTAL
1000	Glazing panel, insulating, 1/2" thick, 2 lites 1/8" float, clear	9.85	9.85	19.70
1100	Tinted	13.80	9.85	23.65
1200	5/8" thick units, 2 lites 3/16" float, clear	14.25	10.35	24.60
1300	Tinted	14.65	10.35	25
1400	1" thick units, 2 lites, 1/4" float, clear	16.70	12.45	29.15
1500	Tinted	24.50	12.45	36.95
1600	Heat reflective film inside	28.50	11	39.50
1700	Light and heat reflective glass, tinted	32.50	11	43.50
2000	Plate glass, 1/4" thick, clear	5.80	7.80	13.60
2050	Tempered	8.20	7.80	16
2100	Tinted	8.15	7.80	15.95
2200	3/8" thick, clear	10.05	12.45	22.50
2250	Tempered	16.60	12.45	29.05
2300	Tinted	16.30	12.45	28.75
2400	1/2" thick, clear	19.95	16.95	36.90
2450	Tempered	24.50	16.95	41.45
2500	Tinted	28.50	16.95	45.45
2600	3/4" thick, clear	36.50	26.50	63
2650	Tempered	43	26.50	69.50
3000	Spandrel glass, panels, 1/4" plate glass insul w/fiberglass, 1" thick	17.40	7.80	25.20
3100	2" thick	20.50	7.80	28.30
3200	Galvanized steel backing, add	6.05		6.05
3300	3/8" plate glass, 1" thick	29	7.80	36.80
3400	2" thick	32	7.80	39.80
4000	Polycarbonate, masked, clear or colored, 1/8" thick	10.15	5.50	15.65
4100	3/16" thick	11.60	5.65	17.25
4200	1/4" thick	13.05	6	19.05
4300	3/8" thick	23	6.20	29.20
5000	Facing panel, textured al, 4' x 8' x 5/16" plywood backing, sgl face	4.10	3.10	7.20
5100	Double face	4.88	3.10	7.98
5200	4' x 10' x 5/16" plywood backing, single face	4.58	3.10	7.68
5300	Double face	6.45	3.10	9.55
5400	4' x 12' x 5/16" plywood backing, single face	6.50	3.10	9.60
5500	Sandwich panel, 22 Ga. galv., both sides 2" insulation, enamel exterior	12.30	5.40	17.70
5600	Polyvinylidene fluoride exterior finish	12.95	5.40	18.35
5700	26 Ga., galv. both sides, 1" insulation, colored 1 side	6.50	5.10	11.60
5800	Colored 2 sides	8.35	5.10	13.45

B2030 Exterior Doors

Costs are listed for exterior door systems by material, type and size. Prices between sizes listed can be interpolated with reasonable accuracy. Prices are per opening for a complete door system including frame.

B2030 110				Glazed Doors, Steel or Aluminum				
	MATERIAL	TYPE	DOORS	SPECIFICATION	OPENING	COST PER OPNG.		
						MAT.	INST.	TOTAL
5600	St. Stl. & glass	revolving	stock unit	manual oper.	6'-0" x 7'-0"	41,500	8,450	49,950
5650				auto Cntrls.	6'-10" x 7'-0"	57,000	9,000	66,000
5700	Bronze		stock unit	manual oper.	6'-10" x 7'-0"	48,400	17,000	65,400
5750				auto Cntrls.	6'-10" x 7'-0"	64,000	17,600	81,600
5800	St. Stl. & glass	balanced	standard	economy	3'-0" x 7'-0"	8,900	1,425	10,325
5850				premium	3'-0" x 7'-0"	15,500	1,875	17,375
6300	Alum. & glass	w/o transom	narrow stile	w/panic Hrdwre.	3'-0" x 7'-0"	1,500	925	2,425
6350				dbl. door, Hrdwre.	6'-0" x 7'-0"	2,600	1,525	4,125
6400			wide stile	hdwre.	3'-0" x 7'-0"	1,825	910	2,735
6450				dbl. door, Hdwre.	6'-0" x 7'-0"	3,550	1,825	5,375
6500			full vision	hdwre.	3'-0" x 7'-0"	2,225	1,450	3,675
6550				dbl. door, Hdwre.	6'-0" x 7'-0"	3,100	2,075	5,175
6600			non-standard	hdwre.	3'-0" x 7'-0"	2,000	910	2,910
6650				dbl. door, Hdwre.	6'-0" x 7'-0"	4,000	1,825	5,825
6700			bronze fin.	hdwre.	3'-0" x 7'-0"	1,475	910	2,385
6750				dbl. door, Hrdwre.	6'-0" x 7'-0"	2,925	1,825	4,750
6800			black fin.	hdwre.	3'-0" x 7'-0"	2,125	910	3,035
6850				dbl. door, Hdwre.	6'-0" x 7'-0"	4,250	1,825	6,075
6900		w/transom	narrow stile	hdwre.	3'-0" x 10'-0"	2,125	1,050	3,175
6950				dbl. door, Hdwre.	6'-0" x 10'-0"	3,025	1,825	4,850
7000			wide stile	hdwre.	3'-0" x 10'-0"	2,450	1,275	3,725
7050				dbl. door, Hdwre.	6'-0" x 10'-0"	3,275	2,175	5,450
7100			full vision	hdwre.	3'-0" x 10'-0"	2,700	1,400	4,100
7150				dbl. door, Hdwre.	6'-0" x 10'-0"	3,550	2,400	5,950

B2030 Exterior Doors

B2030 110	Glazed Doors, Steel or Aluminum

	MATERIAL	TYPE	DOORS	SPECIFICATION	OPENING	COST PER OPNG.		
						MAT.	INST.	TOTAL
7200	Alum. & glass	w/transom	non-standard	hdwre.	3'-0" x 10'-0"	2,075	980	3,055
7250				dbl. door, Hdwre.	6'-0" x 10'-0"	4,150	1,950	6,100
7300			bronze fin.	hdwre.	3'-0" x 10'-0"	1,550	980	2,530
7350				dbl. door, Hdwre.	6'-0" x 10'-0"	3,075	1,950	5,025
7400			black fin.	hdwre.	3'-0" x 10'-0"	2,200	980	3,180
7450				dbl. door, Hdwre.	6'-0" x 10'-0"	4,400	1,950	6,350
7500		revolving	stock design	minimum	6'-10" x 7'-0"	20,600	3,400	24,000
7550				average	6'-0" x 7'-0"	24,900	4,250	29,150
7600				maximum	6'-10" x 7'-0"	43,000	5,675	48,675
7650				min., automatic	6'-10" x 7'-0"	36,300	3,950	40,250
7700				avg., automatic	6'-10" x 7'-0"	40,600	4,800	45,400
7750				max., automatic	6'-10" x 7'-0"	58,500	6,225	64,725
7800		balanced	standard	economy	3'-0" x 7'-0"	6,375	1,425	7,800
7850				premium	3'-0" x 7'-0"	8,000	1,825	9,825
7900		mall front	sliding panels	alum. fin.	16'-0" x 9'-0"	3,325	720	4,045
7950					24'-0" x 9'-0"	4,825	1,325	6,150
8000				bronze fin.	16'-0" x 9'-0"	3,875	840	4,715
8050					24'-0" x 9'-0"	5,625	1,550	7,175
8100			fixed panels	alum. fin.	48'-0" x 9'-0"	9,000	1,025	10,025
8150				bronze fin.	48'-0" x 9'-0"	10,500	1,200	11,700
8200		sliding entrance	5' x 7' door	electric oper.	12'-0" x 7'-6"	8,475	1,325	9,800
8250		sliding patio	temp. glass	economy	6'-0" x 7'-0"	1,475	248	1,723
8300			temp. glass	economy	12'-0" x 7'-0"	3,500	330	3,830
8350				premium	6'-0" x 7'-0"	2,225	370	2,595
8400					12'-0" x 7'-0"	5,250	495	5,745

B20 Exterior Enclosure

B2030 Exterior Doors

Exterior Door System

The table below lists exterior door systems by material, type and size. Prices between sizes listed can be interpolated with reasonable accuracy. Prices are per opening for a complete door system including frame and required hardware.

Wood doors in this table are designed with wood frames and metal doors with hollow metal frames. Depending upon quality the total material and installation cost of a wood door frame is about the same as a hollow metal frame.

B2030 210 — Wood Doors

	MATERIAL	TYPE	DOORS	SPECIFICATION	OPENING	COST PER OPNG.		
						MAT.	INST.	TOTAL
2350	Birch	solid core	single door	hinged	2'-6" x 6'-8"	1,175	269	1,444
2400					2'-6" x 7'-0"	1,175	271	1,446
2450					2'-8" x 7'-0"	1,175	271	1,446
2500					3'-0" x 7'-0"	1,200	276	1,476
2550			double door	hinged	2'-6" x 6'-8"	2,150	485	2,635
2600					2'-6" x 7'-0"	2,150	495	2,645
2650					2'-8" x 7'-0"	2,150	495	2,645
2700					3'-0" x 7'-0"	2,200	505	2,705
2750	Wood	combination	storm & screen	hinged	3'-0" x 6'-8"	325	66	391
2800					3'-0" x 7'-0"	360	72.50	432.50
2850		overhead	panels, H.D.	manual oper.	8'-0" x 8'-0"	830	495	1,325
2900					10'-0" x 10'-0"	1,250	550	1,800
2950					12'-0" x 12'-0"	1,800	660	2,460
3000					14'-0" x 14'-0"	2,800	760	3,560
3050					20'-0" x 16'-0"	6,550	1,525	8,075
3100				electric oper.	8'-0" x 8'-0"	1,850	745	2,595
3150					10'-0" x 10'-0"	2,275	800	3,075
3200					12'-0" x 12'-0"	2,825	910	3,735
3250					14'-0" x 14'-0"	3,825	1,000	4,825
3300					20'-0" x 16'-0"	7,775	2,025	9,800

B2030 220 — Steel Doors

	MATERIAL	TYPE	DOORS	SPECIFICATION	OPENING	COST PER OPNG.		
						MAT.	INST.	TOTAL
3350	Steel 18 Ga.	hollow metal	1 door w/frame	no label	2'-6" x 7'-0"	1,300	280	1,580
3400					2'-8" x 7'-0"	1,300	280	1,580
3450					3'-0" x 7'-0"	1,300	280	1,580
3500					3'-6" x 7'-0"	1,400	294	1,694
3550			1 door w/frame	no label	4'-0" x 8'-0"	1,625	295	1,920
3600			2 doors w/frame	no label	5'-0" x 7'-0"	2,500	520	3,020
3650					5'-4" x 7'-0"	2,500	520	3,020
3700					6'-0" x 7'-0"	2,500	520	3,020
3750					7'-0" x 7'-0"	2,700	545	3,245
3800					8'-0" x 8'-0"	3,175	550	3,725
3850			1 door w/frame	"A" label	2'-6" x 7'-0"	1,575	335	1,910

B2030 Exterior Doors

B2030 220 — Steel Doors

	MATERIAL	TYPE	DOORS	SPECIFICATION	OPENING	COST PER OPNG.		
						MAT.	INST.	TOTAL
3900	Steel 18 Ga.	hollow metal	1 door w/frame	"A" label	2'-8" x 7'-0"	1,550	340	1,890
3950					3'-0" x 7'-0"	1,550	340	1,890
4000					3'-6" x 7'-0"	1,675	350	2,025
4050					4'-0" x 8'-0"	1,800	365	2,165
4100			2 doors w/frame	"A" label	5'-0" x 7'-0"	3,000	625	3,625
4150					5'-4" x 7'-0"	3,050	630	3,680
4200					6'-0" x 7'-0"	3,050	630	3,680
4250					7'-0" x 7'-0"	3,275	645	3,920
4300					8'-0" x 8'-0"	3,300	650	3,950
4350	Steel 24 Ga.	overhead	sectional	manual oper.	8'-0" x 8'-0"	805	495	1,300
4400					10'-0" x 10'-0"	1,025	550	1,575
4450					12'-0" x 12'-0"	1,300	660	1,960
4500					20'-0" x 14'-0"	3,725	1,425	5,150
4550				electric oper.	8'-0" x 8'-0"	1,825	745	2,570
4600					10'-0" x 10'-0"	2,050	800	2,850
4650					12'-0" x 12'-0"	2,325	910	3,235
4700					20'-0" x 14'-0"	4,950	1,925	6,875
4750	Steel	overhead	rolling	manual oper.	8'-0" x 8'-0"	1,125	795	1,920
4800					10'-0" x 10'-0"	1,925	910	2,835
4850					12'-0" x 12'-0"	2,025	1,050	3,075
4900					14'-0" x 14'-0"	3,100	1,600	4,700
4950					20'-0" x 12'-0"	2,175	1,425	3,600
5000					20'-0" x 16'-0"	3,550	2,125	5,675
5050				electric oper.	8'-0" x 8'-0"	2,300	1,050	3,350
5100					10'-0" x 10'-0"	3,100	1,175	4,275
5150					12'-0" x 12'-0"	3,200	1,300	4,500
5200					14'-0" x 14'-0"	4,275	1,850	6,125
5250					20'-0" x 12'-0"	3,325	1,675	5,000
5300					20'-0" x 16'-0"	4,700	2,375	7,075
5350				fire rated	10'-0" x 10'-0"	2,075	1,150	3,225
5400			rolling grille	manual oper.	10'-0" x 10'-0"	2,700	1,275	3,975
5450					15'-0" x 8'-0"	3,125	1,600	4,725
5500		vertical lift	1 door w/frame	motor operator	16'-0" x 16'-0"	23,100	5,325	28,425
5550					32'-0" x 24'-0"	49,200	3,550	52,750

B2030 230 — Aluminum Doors

	MATERIAL	TYPE	DOORS	SPECIFICATION	OPENING	COST PER OPNG.		
						MAT.	INST.	TOTAL
6000	Aluminum	combination	storm & screen	hinged	3'-0" x 6'-8"	335	71	406
6050					3'-0" x 7'-0"	370	78	448
6100		overhead	rolling grille	manual oper.	12'-0" x 12'-0"	4,325	2,250	6,575
6150				motor oper.	12'-0" x 12'-0"	5,700	2,500	8,200
6200	Alum. & Fbrgls.	overhead	heavy duty	manual oper.	12'-0" x 12'-0"	2,800	660	3,460
6250				electric oper.	12'-0" x 12'-0"	3,825	910	4,735

B30 Roofing

B3010 Roof Coverings

Built Up Ply

Multiple ply roofing is the most popular covering for minimum pitch roofs.

B3010 105	Built-Up	COST PER S.F.		
		MAT.	INST.	TOTAL
1200	Asphalt flood coat w/gravel; not incl. insul, flash., nailers			
1300				
1400	Asphalt base sheets & 3 plies #15 asphalt felt, mopped	.93	1.59	2.52
1500	On nailable deck	.98	1.67	2.65
1600	4 plies #15 asphalt felt, mopped	1.28	1.75	3.03
1700	On nailable deck	1.14	1.85	2.99
1800	Coated glass base sheet, 2 plies glass (type IV), mopped	.94	1.59	2.53
1900	For 3 plies	1.12	1.75	2.87
2000	On nailable deck	1.05	1.85	2.90
2300	4 plies glass fiber felt (type IV), mopped	1.37	1.75	3.12
2400	On nailable deck	1.23	1.85	3.08
2500	Organic base sheet & 3 plies #15 organic felt, mopped	.99	1.77	2.76
2600	On nailable deck	.96	1.85	2.81
2700	4 plies #15 organic felt, mopped	1.20	1.59	2.79
2750				
2800	Asphalt flood coat, smooth surface, not incl. insul, flash., nailers			
2900	Asphalt base sheet & 3 plies #15 asphalt felt, mopped	.99	1.46	2.45
3000	On nailable deck	.92	1.52	2.44
3100	Coated glass fiber base sheet & 2 plies glass fiber felt, mopped	.88	1.40	2.28
3200	On nailable deck	.83	1.46	2.29
3300	For 3 plies, mopped	1.06	1.52	2.58
3400	On nailable deck	.99	1.59	2.58
3700	4 plies glass fiber felt (type IV), mopped	1.24	1.52	2.76
3800	On nailable deck	1.17	1.59	2.76
3900	Organic base sheet & 3 plies #15 organic felt, mopped	.97	1.46	2.43
4000	On nailable decks	.90	1.52	2.42
4100	4 plies #15 organic felt, mopped	1.14	1.59	2.73
4200	Coal tar pitch with gravel surfacing			
4300	4 plies #15 tarred felt, mopped	1.72	1.67	3.39
4400	3 plies glass fiber felt (type IV), mopped	1.40	1.85	3.25
4500	Coated glass fiber base sheets 2 plies glass fiber felt, mopped	1.40	1.85	3.25
4600	On nailable decks	1.24	1.95	3.19
4800	3 plies glass fiber felt (type IV), mopped	1.95	1.67	3.62
4900	On nailable decks	1.79	1.75	3.54

B3010 Roof Coverings

Full Adhered Single Ply Membrane

Preformed Metal

Flat Seam

Batten Seam

Standing Seam

B3010 120	Single Ply Membrane	COST PER S.F.		
		MAT.	INST.	TOTAL
1000	CSPE (Chlorosulfonated polyethylene), 35 mils, fully adhered	2.04	.88	2.92
2000	EPDM (Ethylene propylene diene monomer), 45 mils, fully adhered	1.02	.88	1.90
4000	Modified bit., SBS modified, granule surface cap sheet, mopped, 150 mils	.58	1.75	2.33
4500	APP modified, granule surface cap sheet, torched, 180 mils	.65	1.15	1.80
6000	Reinforced PVC, 48 mils, loose laid and ballasted with stone	1.28	.45	1.73
6200	Fully adhered with adhesive	1.71	.88	2.59

B3010 130	Preformed Metal Roofing	COST PER S.F.		
		MAT.	INST.	TOTAL
0200	Corrugated roofing, aluminum, mill finish, .0175" thick, .272 P.S.F.	.96	1.62	2.58
0250	.0215" thick, .334 P.S.F.	1.27	1.62	2.89

B3010 135	Formed Metal	COST PER S.F.		
		MAT.	INST.	TOTAL
1000	Batten seam, formed copper roofing, 3" min slope, 16 oz., 1.2 P.S.F.	11.85	5.40	17.25
1100	18 oz., 1.35 P.S.F.	13.10	5.90	19
2000	Zinc copper alloy, 3" min slope, .020" thick, .88 P.S.F.	11.85	4.93	16.78
3000	Flat seam, copper, 1/4" min. slope, 16 oz., 1.2 P.S.F.	11.85	4.93	16.78
3100	18 oz., 1.35 P.S.F.	13.10	5.15	18.25
5000	Standing seam, copper, 2-1/2" min. slope, 16 oz., 1.25 P.S.F.	11.85	4.53	16.38
5100	18 oz., 1.40 P.S.F.	13.10	4.93	18.03
6000	Zinc copper alloy, 2-1/2" min. slope, .020" thick, .87 P.S.F.	11.85	4.93	16.78
6100	.032" thick, 1.39 P.S.F.	19.85	5.40	25.25

B30 Roofing

B3010 Roof Coverings

Shingle and Tile

Shingles and tiles are practical in applications where the roof slope is more than 3-1/2″ per foot of rise. Lines 1100 through 6000 list the various materials and the weight per square foot.

Roof Edges

The table below lists the costs for various types of roof perimeter edge treatments.

Roof edge systems include the cost per L.F. for a 2″ x 8″ treated wood nailer fastened at 4′-0″ O.C. and a diagonally cut 4″ x 6″ treated wood cant.

Roof edge and base flashing are assumed to be made from the same material.

B3010 140	Shingle & Tile	COST PER S.F.		
		MAT.	INST.	TOTAL
1095	Asphalt roofing			
1100	Strip shingles, 4″ slope, inorganic class A 210-235 lb/Sq.	.60	.97	1.57
1150	Organic, class C, 235-240 lb./sq.	.61	1.06	1.67
1200	Premium laminated multi-layered, class A, 260-300 lb./Sq.	.93	1.45	2.38
1545	Metal roofing			
1550	Alum., shingles, colors, 3″ min slope, .019″ thick, 0.4 PSF	2.64	1.07	3.71
1850	Steel, colors, 3″ min slope, 26 gauge, 1.0 PSF	3.28	2.17	5.45
2795	Slate roofing			
2800	4″ min. slope, shingles, 3/16″ thick, 8.0 PSF	5.35	2.71	8.06
3495	Wood roofing			
3500	4″ min slope, cedar shingles, 16″ x 5″, 5″ exposure 1.6 PSF	3.76	2.14	5.90
4000	Shakes, 18″, 8-1/2″ exposure, 2.8 PSF	2.74	2.64	5.38
5095	Tile roofing			
5100	Aluminum, mission, 3″ min slope, .019″ thick, 0.65 PSF	9.05	2.06	11.11
6000	Clay, Americana, 3″ minimum slope, 8 PSF	7.40	2.87	10.27

B3010 Roof Coverings

B3010 320	Roof Deck Rigid Insulation	COST PER S.F.		
		MAT.	INST.	TOTAL
0100	Fiberboard low density, 1/2" thick, R1.39			
0150	1" thick R2.78	.46	.58	1.04
0300	1 1/2" thick R4.17	.70	.58	1.28
0350	2" thick R5.56	.94	.58	1.52
0370	Fiberboard high density, 1/2" thick R1.3	.24	.46	.70
0380	1" thick R2.5	.48	.58	1.06
0390	1 1/2" thick R3.8	.73	.58	1.31
0410	Fiberglass, 3/4" thick R2.78	.62	.46	1.08
0450	15/16" thick R3.70	.81	.46	1.27
0550	1-5/16" thick R5.26	1.39	.46	1.85
0650	2 7/16" thick R10	1.71	.58	2.29
1510	Polyisocyanurate 2#/CF density, 1" thick R7.14	.55	.33	.88
1550	1 1/2" thick R10.87	.69	.37	1.06
1600	2" thick R14.29	.89	.42	1.31
1650	2 1/2" thick R16.67	1.11	.44	1.55
1700	3" thick R21.74	1.39	.46	1.85
1750	3 1/2" thick R25	2.12	.46	2.58
1800	Tapered for drainage	2.12	.33	2.45
1810	Expanded polystyrene, 1#/CF density, 3/4" thick R2.89	.34	.31	.65
1820	2" thick R7.69	.68	.37	1.05
1830	Extruded polystyrene, 15 PSI compressive strength, 1" thick, R5	.54	.31	.85
1835	2" thick R10	.68	.37	1.05
1840	3" thick R15	1.40	.46	1.86
2550	40 PSI compressive strength, 1" thick R5	.56	.31	.87
2600	2" thick R10	1.06	.37	1.43
2650	3" thick R15	1.54	.46	2
2700	4" thick R20	2.07	.46	2.53
2750	Tapered for drainage	.78	.33	1.11
2810	60 PSI compressive strength, 1" thick R5	.65	.32	.97
2850	2" thick R10	1.16	.38	1.54
2900	Tapered for drainage	.94	.33	1.27
3050	Composites with 2" EPS			
3060	1" Fiberboard	1.18	.49	1.67
3070	7/16" oriented strand board	1.39	.58	1.97
3080	1/2" plywood	1.51	.58	2.09
3090	1" perlite	1.23	.58	1.81
4000	Composites with 1-1/2" polyisocyanurate			
4010	1" fiberboard	1.61	.58	2.19
4020	1" perlite	1.68	.54	2.22
4030	7/16" oriented strand board	1.95	.58	2.53

B3010 420	Roof Edges					COST PER L.F.		
	EDGE TYPE	DESCRIPTION	SPECIFICATION	FACE HEIGHT		MAT.	INST.	TOTAL
1000	Aluminum	mill finish	.050" thick	4"		12.10	9.60	21.70
1100				6"		12.55	9.90	22.45
1300		duranodic	.050" thick	4"		11.85	9.60	21.45
1400				6"		13	9.90	22.90
1600		painted	.050" thick	4"		12.85	9.60	22.45
1700				6"		14.20	9.90	24.10
2000	Copper	plain	16 oz.	4"		8.10	9.60	17.70
2100				6"		9.45	9.90	19.35
2300			20 oz.	4"		9	10.95	19.95
2400				6"		10.05	10.55	20.60
2700	Sheet Metal	galvanized	20 Ga.	4"		9.25	11.35	20.60
2800				6"		11.10	11.35	22.45

B30 Roofing

B3010 Roof Coverings

B3010 420		Roof Edges						

	EDGE TYPE	DESCRIPTION	SPECIFICATION	FACE HEIGHT		COST PER L.F.		
						MAT.	INST.	TOTAL
3000	Sheet metal	galvanized	24 Ga.	4"		8.10	9.60	17.70
3100				6"		9.45	9.60	19.05

B3010 430		Flashing						

	MATERIAL	BACKING	SIDES	SPECIFICATION	QUANTITY	COST PER S.F.		
						MAT.	INST.	TOTAL
0040	Aluminum	none		.019"		1.12	3.18	4.30
0050				.032"		1.63	3.18	4.81
0300		fabric	2	.004"		1.55	1.40	2.95
0400		mastic		.004"		1.55	1.40	2.95
0700	Copper	none		16 oz.	<500 lbs.	7.45	4.01	11.46
0800				24 oz.	<500 lbs.	10.60	4.40	15
2000	Copper lead	fabric	1	2 oz.		2.66	1.40	4.06
3500	PVC black	none		.010"		.29	1.62	1.91
3700				.030"		.52	1.62	2.14
4200	Neoprene			1/16"		2.23	1.62	3.85
4500	Stainless steel	none		.015"	<500 lbs.	5.75	4.01	9.76
4600	Copper clad				>2000 lbs.	5.55	2.98	8.53
5000	Plain			32 ga.		4.17	2.98	7.15

B30 Roofing

B3010 Roof Coverings

B3010 610 — Gutters

	SECTION	MATERIAL	THICKNESS	SIZE	FINISH	COST PER L.F. MAT.	COST PER L.F. INST.	COST PER L.F. TOTAL
0050	Box	aluminum	.027"	5"	enameled	2.70	4.84	7.54
0100					mill	2.55	4.84	7.39
0200			.032"	5"	enameled	3.21	4.84	8.05
0500		copper	16 Oz.	4"	lead coated	12.50	4.84	17.34
0600					mill	6.75	4.84	11.59
1000		steel galv.	28 Ga.	5"	enameled	1.66	4.84	6.50
1200			26 Ga.	5"	mill	1.75	4.84	6.59
1800		vinyl		4"	colors	1.25	4.51	5.76
1900				5"	colors	1.49	4.51	6
2300		hemlock or fir		4"x5"	treated	8.80	4.96	13.76
3000	Half round	copper	16 Oz.	4"	lead coated	11.55	4.84	16.39
3100					mill	6.90	4.84	11.74
3600		steel galv.	28 Ga.	5"	enameled	1.66	4.84	6.50
4102		stainless steel		5"	mill	6.95	4.84	11.79
5000		vinyl		4"	white	1	4.51	5.51

B3010 620 — Downspouts

	MATERIALS	SECTION	SIZE	FINISH	THICKNESS	COST PER V.L.F. MAT.	COST PER V.L.F. INST.	COST PER V.L.F. TOTAL
0100	Aluminum	rectangular	2"x3"	embossed mill	.020"	1.14	3.06	4.20
0150				enameled	.020"	1.75	3.06	4.81
0250			3"x4"	enameled	.024"	2.81	4.15	6.96
0300		round corrugated	3"	enameled	.020"	1.56	3.06	4.62
0350			4"	enameled	.025"	2.63	4.15	6.78
0500	Copper	rectangular corr.	2"x3"	mill	16 Oz.	7.90	3.06	10.96
0600		smooth		mill	16 Oz.	7.70	3.06	10.76
0700		rectangular corr.	3"x4"	mill	16 Oz.	10.20	4.01	14.21
1300	Steel	rectangular corr.	2"x3"	galvanized	28 Ga.	1.76	3.06	4.82
1350				epoxy coated	24 Ga.	2.09	3.06	5.15
1400		smooth		galvanized	28 Ga.	1.91	3.06	4.97
1450		rectangular corr.	3"x4"	galvanized	28 Ga.	1.76	4.01	5.77
1500				epoxy coated	24 Ga.	2.33	4.01	6.34
1550		smooth		galvanized	28 Ga.	2.76	4.01	6.77
1652		round corrugated	3"	galvanized	28 Ga.	1.91	3.06	4.97
1700			4"	galvanized	28 Ga.	2.73	4.01	6.74
1750			5"	galvanized	28 Ga.	4.05	4.47	8.52
2552	S.S. tubing sch.5	rectangular	3"x4"	mill		69	4.01	73.01

B3010 630 — Gravel Stop

	MATERIALS	SECTION	SIZE	FINISH	THICKNESS	COST PER L.F. MAT.	COST PER L.F. INST.	COST PER L.F. TOTAL
5100	Aluminum	extruded	4"	mill	.050"	6.55	4.01	10.56
5200			4"	duranodic	.050"	6.30	4.01	10.31
5300			8"	mill	.050"	8.80	4.65	13.45
5400			8"	duranodic	.050"	8.55	4.65	13.20
6000			12"-2 pc.	duranodic	.050"	10.75	5.80	16.55
6100	Stainless	formed	6"	mill	24 Ga.	12.25	4.31	16.56

B30 Roofing

B3020 Roof Openings

Roof Hatch

Smoke Hatch

Skylight

B3020 110	Skylights	COST PER S.F.		
		MAT.	INST.	TOTAL
5100	Skylights, plastic domes, insul curbs, nom. size to 10 S.F., single glaze	25.50	12.15	37.65
5200	Double glazing	29	14.95	43.95
5300	10 S.F. to 20 S.F., single glazing	25	4.93	29.93
5400	Double glazing	24.50	6.20	30.70
5500	20 S.F. to 30 S.F., single glazing	21	4.19	25.19
5600	Double glazing	23	4.93	27.93
5700	30 S.F. to 65 S.F., single glazing	18.15	3.19	21.34
5800	Double glazing	23	4.19	27.19
6000	Sandwich panels fiberglass, 1-9/16" thick, 2 S.F. to 10 S.F.	17.60	9.75	27.35
6100	10 S.F. to 18 S.F.	15.75	7.35	23.10
6200	2-3/4" thick, 25 S.F. to 40 S.F.	25.50	6.60	32.10
6300	40 S.F. to 70 S.F.	21	5.90	26.90

B3020 210	Hatches	COST PER OPNG.		
		MAT.	INST.	TOTAL
0200	Roof hatches, with curb, and 1" fiberglass insulation, 2'-6"x3'-0", aluminum	1,075	195	1,270
0300	Galvanized steel 165 lbs.	655	195	850
0400	Primed steel 164 lbs.	550	195	745
0500	2'-6"x4'-6" aluminum curb and cover, 150 lbs.	935	216	1,151
0600	Galvanized steel 220 lbs.	660	216	876
0650	Primed steel 218 lbs.	910	216	1,126
0800	2'x6"x8'-0" aluminum curb and cover, 260 lbs.	1,825	295	2,120
0900	Galvanized steel, 360 lbs.	1,750	295	2,045
0950	Primed steel 358 lbs.	1,325	295	1,620
1200	For plexiglass panels, add to the above	440		440
2100	Smoke hatches, unlabeled not incl. hand winch operator, 2'-6"x3', galv	795	236	1,031
2200	Plain steel, 160 lbs.	665	236	901
2400	2'-6"x8'-0",galvanized steel, 360 lbs.	1,925	325	2,250
2500	Plain steel, 350 lbs.	1,450	325	1,775
3000	4'-0"x8'-0", double leaf low profile, aluminum cover, 359 lb.	2,850	243	3,093
3100	Galvanized steel 475 lbs.	2,425	243	2,668
3200	High profile, aluminum cover, galvanized curb, 361 lbs.	2,750	243	2,993

C1010 Partitions

The Concrete Block Partition Systems are defined by weight and type of block, thickness, type of finish and number of sides finished. System components include joint reinforcing on alternate courses and vertical control joints.

C1010 102	Concrete Block Partitions - Regular Weight						
	TYPE	THICKNESS (IN.)	TYPE FINISH	SIDES FINISHED		COST PER S.F.	
						MAT.	INST.	TOTAL
1000	Hollow	4	none	0		1.75	5.45	7.20
1010			gyp. plaster 2 coat	1		2.13	7.90	10.03
1020				2		2.51	10.40	12.91
1200			portland - 3 coat	1		2.02	8.30	10.32
1400			5/8" drywall	1		2.38	7.15	9.53
1500		6	none	0		2.42	5.85	8.27
1510			gyp. plaster 2 coat	1		2.80	8.30	11.10
1520				2		3.18	10.80	13.98
1700			portland - 3 coat	1		2.69	8.70	11.39
1900			5/8" drywall	1		3.05	7.55	10.60
1910				2		3.68	9.25	12.93
2000		8	none	0		2.55	6.20	8.75
2010			gyp. plaster 2 coat	1		2.93	8.70	11.63
2020			gyp. plaster 2 coat	2		3.31	11.20	14.51
2200			portland - 3 coat	1		2.82	9.10	11.92
2400			5/8" drywall	1		3.18	7.90	11.08
2410				2		3.81	9.60	13.41
2500		10	none	0		3.43	6.50	9.93
2510			gyp. plaster 2 coat	1		3.81	9	12.81
2520				2		4.19	11.45	15.64
2700			portland - 3 coat	1		3.70	9.35	13.05
2900			5/8" drywall	1		4.06	8.20	12.26
2910				2		4.69	9.90	14.59
3000	Solid	2	none	0		1.61	5.40	7.01
3010			gyp. plaster	1		1.99	7.85	9.84
3020				2		2.37	10.35	12.72
3200			portland - 3 coat	1		1.88	8.25	10.13
3400			5/8" drywall	1		2.24	7.10	9.34
3410				2		2.87	8.80	11.67
3500		4	none	0		2.15	5.65	7.80
3510			gyp. plaster	1		2.62	8.15	10.77
3520				2		2.91	10.60	13.51
3700			portland - 3 coat	1		2.42	8.50	10.92
3900			5/8" drywall	1		2.78	7.35	10.13
3910				2		3.41	9.05	12.46

C10 Interior Construction

C1010 Partitions

C1010 102 — Concrete Block Partitions - Regular Weight

	TYPE	THICKNESS (IN.)	TYPE FINISH	SIDES FINISHED		COST PER S.F.		
						MAT.	INST.	TOTAL
4002	Solid	6	none	0		2.40	6.05	8.45
4010			gyp. plaster	1		2.78	8.50	11.28
4020				2		3.16	11	14.16
4200			portland - 3 coat	1		2.67	8.90	11.57
4400			5/8" drywall	1		3.03	7.75	10.78
4410				2		3.66	9.45	13.11

C1010 104 — Concrete Block Partitions - Lightweight

	TYPE	THICKNESS (IN.)	TYPE FINISH	SIDES FINISHED		COST PER S.F.		
						MAT.	INST.	TOTAL
5000	Hollow	4	none	0		2.06	5.30	7.36
5010			gyp. plaster	1		2.44	7.75	10.19
5020				2		2.82	10.25	13.07
5200			portland - 3 coat	1		2.33	8.15	10.48
5400			5/8" drywall	1		2.69	7	9.69
5410				2		3.32	8.70	12.02
5500		6	none	0		2.74	5.70	8.44
5520			gyp. plaster	2		3.50	10.65	14.15
5700			portland - 3 coat	1		3.01	8.55	11.56
5900			5/8" drywall	1		3.37	7.40	10.77
5910				2		4	9.10	13.10
6000		8	none	0		3.33	6.05	9.38
6010			gyp. plaster	1		3.71	8.55	12.26
6020				2		4.09	11.05	15.14
6200			portland - 3 coat	1		3.60	8.95	12.55
6400			5/8" drywall	1		3.96	7.75	11.71
6410				2		4.59	9.45	14.04
6500		10	none	0		4.29	6.35	10.64
6510			gyp. plaster	1		4.67	8.85	13.52
6520				2		5.05	11.30	16.35
6700			portland - 3 coat	1		4.56	9.20	13.76
6900			5/8" drywall	1		4.92	8.05	12.97
6910				2		5.55	9.75	15.30
7000	Solid	4	none	0		2.70	5.55	8.25
7010			gyp. plaster	1		3.08	8	11.08
7020				2		3.46	10.50	13.96
7200			portland - 3 coat	1		2.97	8.40	11.37
7400			5/8" drywall	1		3.33	7.25	10.58
7410				2		3.96	8.95	12.91
7500		6	none	0		3.10	6	9.10
7510			gyp. plaster	1		3.57	8.55	12.12
7520				2		3.86	10.95	14.81
7700			portland - 3 coat	1		3.37	8.85	12.22
7900			5/8" drywall	1		3.73	7.70	11.43
7910				2		4.36	9.40	13.76
8000		8	none	0		4.11	6.40	10.51
8010			gyp. plaster	1		4.49	8.90	13.39
8020				2		4.87	11.40	16.27
8200			portland - 3 coat	1		4.38	9.30	13.68
8400			5/8" drywall	1		4.74	8.10	12.84
8410				2		5.35	9.80	15.15

C1010 Partitions

Single Wythe

Structural facing tile
8W series
8″ x 16″

Double Wythe

Single Wythe

Structural facing tile
6T series
5-1/3″ x 12″

Double Wythe

C1010 120	Tile Partitions	COST PER S.F.		
		MAT.	INST.	TOTAL
1000	8W series 8″x16″, 4″ thick wall, reinf every 2 courses, glazed 1 side	13.25	6.55	19.80
1100	Glazed 2 sides	14.80	6.95	21.75
1200	Glazed 2 sides, using 2 wythes of 2″ thick tile	20.50	12.50	33
1300	6″ thick wall, horizontal reinf every 2 courses, glazed 1 side	23.50	6.85	30.35
1400	Glazed 2 sides, each face different color, 2″ and 4″ tile	23.50	12.80	36.30
1500	8″ thick wall, glazed 2 sides using 2 wythes of 4″ thick tile	26.50	13.10	39.60
1600	10″ thick wall, glazed 2 sides using 1 wythe of 4″ tile & 1 wythe of 6″ tile	37	13.40	50.40
1700	Glazed 2 sides cavity wall, using 2 wythes of 4″ thick tile	26.50	13.10	39.60
1800	12″ thick wall, glazed 2 sides using 2 wythes of 6″ thick tile	47	13.70	60.70
1900	Glazed 2 sides cavity wall, using 2 wythes of 4″ thick tile	26.50	13.10	39.60
2100	6T series 5-1/3″x12″ tile, 4″ thick, non load bearing glazed one side,	12.15	10.25	22.40
2200	Glazed two sides	15.60	11.55	27.15
2300	Glazed two sides, using two wythes of 2″ thick tile	18.60	20	38.60
2400	6″ thick, glazed one side	18.25	10.75	29
2500	Glazed two sides	21.50	12.20	33.70
2600	Glazed two sides using 2″ thick tile and 4″ thick tile	21.50	20.50	42
2700	8″ thick, glazed one side	24	12.55	36.55
2800	Glazed two sides using two wythes of 4″ thick tile	24.50	20.50	45
2900	Glazed two sides using 6″ thick tile and 2″ thick tile	27.50	21	48.50
3000	10″ thick cavity wall, glazed two sides using two wythes of 4″ tile	24.50	20.50	45
3100	12″ thick, glazed two sides using 4″ thick tile and 8″ thick tile	36	23	59
3200	2″ thick facing tile, glazed one side, on 6″ concrete block	12.50	11.90	24.40
3300	On 8″ concrete block	12.60	12.25	24.85
3400	On 10″ concrete block	13.50	12.50	26

C10 Interior Construction

C1010 Partitions

Gypsum board, single layer each side on wood studs.

Gypsum board, sound deadening board each side, with 1-1/2" insulation on wood studs.

Gypsum board, two layers each side on wood studs.

Gypsum board, single layer each side on metal studs.

Gypsum board, sound deadening board each side on metal studs.

Gypsum board two layers one side, single layer opposite side, with 3-1/2" insulation on metal studs.

C1010 124			Drywall Partitions/Wood Stud Framing					
						COST PER S.F.		
	FACE LAYER	BASE LAYER	FRAMING	OPPOSITE FACE	INSULATION	MAT.	INST.	TOTAL
1200	5/8" FR drywall	none	2 x 4, @ 16" O.C.	same	0	1.24	2.99	4.23
1250				5/8" reg. drywall	0	1.21	2.99	4.20
1300				nothing	0	.79	1.99	2.78
1400		1/4" SD gypsum	2 x 4 @ 16" O.C.	same	1-1/2" fiberglass	2.76	4.59	7.35
1450				5/8" FR drywall	1-1/2" fiberglass	2.35	4.04	6.39
1500				nothing	1-1/2" fiberglass	1.90	3.04	4.94
1600		resil. channels	2 x 4 @ 16", O.C.	same	1-1/2" fiberglass	2.29	5.80	8.09
1650				5/8" FR drywall	1-1/2" fiberglass	2.12	4.65	6.77
1700				nothing	1-1/2" fiberglass	1.67	3.65	5.32
1800		5/8" FR drywall	2 x 4 @ 24" O.C.	same	0	1.96	3.80	5.76
1850				5/8" FR drywall	0	1.56	3.30	4.86
1900				nothing	0	1.11	2.30	3.41
2200		5/8" FR drywall	2 rows-2 x 4	same	2" fiberglass	3.13	5.50	8.63
2250			16"O.C.	5/8" FR drywall	2" fiberglass	2.73	4.98	7.71
2300				nothing	2" fiberglass	2.28	3.98	6.26
2400	5/8" WR drywall	none	2 x 4, @ 16" O.C.	same	0	1.48	2.99	4.47
2450				5/8" FR drywall	0	1.36	2.99	4.35
2500				nothing	0	.91	1.99	2.90
2600		5/8" FR drywall	2 x 4, @ 24" O.C.	same	0	2.20	3.80	6
2650				5/8" FR drywall	0	1.68	3.30	4.98
2700				nothing	0	1.23	2.30	3.53
2800	5/8 VF drywall	none	2 x 4, @ 16" O.C.	same	0	2.58	3.19	5.77
2850				5/8" FR drywall	0	1.91	3.09	5
2900				nothing	0	1.46	2.09	3.55

C1010 Partitions

C1010 124	Drywall Partitions/Wood Stud Framing

	FACE LAYER	BASE LAYER	FRAMING	OPPOSITE FACE	INSULATION	COST PER S.F.		
						MAT.	INST.	TOTAL
3000	5/8 VF drywall	5/8″ FR drywall	2 x 4 , 24″ O.C.	same	0	3.30	4	7.30
3050				5/8″ FR drywall	0	2.23	3.40	5.63
3100				nothing	0	1.78	2.40	4.18
3200	1/2″ reg drywall	3/8″ reg drywall	2 x 4, @ 16″ O.C.	same	0	1.80	3.99	5.79
3252				5/8″ FR drywall	0	1.52	3.49	5.01
3300				nothing	0	1.07	2.49	3.56

C1010 126	Drywall Partitions/Metal Stud Framing

	FACE LAYER	BASE LAYER	FRAMING	OPPOSITE FACE	INSULATION	COST PER S.F.		
						MAT.	INST.	TOTAL
5200	5/8″ FR drywall	none	1-5/8″ @ 24″ O.C.	same	0	1.15	2.65	3.80
5250				5/8″ reg. drywall	0	1.12	2.65	3.77
5300				nothing	0	.70	1.65	2.35
5400			3-5/8″ @ 24″ O.C.	same	0	1.26	2.67	3.93
5450				5/8″ reg. drywall	0	1.23	2.67	3.90
5500				nothing	0	.81	1.67	2.48
5600		1/4″ SD gypsum	1-5/8″ @ 24″ O.C.	same	0	1.97	3.75	5.72
5650				5/8″ FR drywall	0	1.56	3.20	4.76
5700				nothing	0	1.11	2.20	3.31
5800			2-1/2″ @ 24″ O.C.	same	0	2.04	3.76	5.80
5850				5/8″ FR drywall	0	1.63	3.21	4.84
5900				nothing	0	1.18	2.21	3.39
6000		5/8″ FR drywall	2-1/2″ @ 16″ O.C.	same	0	2.35	4.27	6.62
6050				5/8″ FR drywall	0	1.95	3.77	5.72
6100				nothing	0	1.50	2.77	4.27
6200			3-5/8″ @ 24″ O.C.	same	0	2.06	3.67	5.73
6250				5/8″ FR drywall	3-1/2″ fiberglass	2.29	3.54	5.83
6300				nothing	0	1.21	2.17	3.38
6400	5/8″ WR drywall	none	1-5/8″ @ 24″ O.C.	same	0	1.39	2.65	4.04
6450				5/8″ FR drywall	0	1.27	2.65	3.92
6500				nothing	0	.82	1.65	2.47
6600			3-5/8″ @ 24″ O.C.	same	0	1.50	2.67	4.17
6650				5/8″ FR drywall	0	1.38	2.67	4.05
6700				nothing	0	.93	1.67	2.60
6800		5/8″ FR drywall	2-1/2″ @ 16″ O.C.	same	0	2.59	4.27	6.86
6850				5/8″ FR drywall	0	2.07	3.77	5.84
6900				nothing	0	1.62	2.77	4.39
7000			3-5/8″ @ 24″ O.C.	same	0	2.30	3.67	5.97
7050				5/8″ FR drywall	3-1/2″ fiberglass	2.41	3.54	5.95
7100				nothing	0	1.33	2.17	3.50
7200	5/8″ VF drywall	none	1-5/8″ @ 24″ O.C.	same	0	2.49	2.85	5.34
7250				5/8″ FR drywall	0	1.82	2.75	4.57
7300				nothing	0	1.37	1.75	3.12
7400			3-5/8″ @ 24″ O.C.	same	0	2.60	2.87	5.47
7450				5/8″ FR drywall	0	1.93	2.77	4.70
7500				nothing	0	1.48	1.77	3.25
7600		5/8″ FR drywall	2-1/2″ @ 16″ O.C.	same	0	3.69	4.47	8.16
7650				5/8″ FR drywall	0	2.62	3.87	6.49
7700				nothing	0	2.17	2.87	5.04
7800			3-5/8″ @ 24″ O.C.	same	0	3.40	3.87	7.27
7850				5/8″ FR drywall	3-1/2″ fiberglass	2.96	3.64	6.60
7900				nothing	0	1.88	2.27	4.15

C1010 Partitions

C1010 128	Drywall Components	COST PER S.F.		
		MAT.	INST.	TOTAL
0060	Metal studs, 24" O.C. including track, load bearing, 20 gage, 2-1/2"	.87	.93	1.80
0080	3-5/8"	1.03	.95	1.98
0100	4"	.93	.96	1.89
0120	6"	1.38	.98	2.36
0140	Metal studs, 24" O.C. including track, load bearing, 18 gage, 2-1/2"	.87	.93	1.80
0160	3-5/8"	1.03	.95	1.98
0180	4"	.93	.96	1.89
0200	6"	1.38	.98	2.36
0220	16 gage, 2-1/2"	1	1.06	2.06
0240	3-5/8"	1.20	1.08	2.28
0260	4"	1.27	1.10	2.37
0280	6"	1.60	1.13	2.73
0300	Non load bearing, 25 gage, 1-5/8"	.25	.65	.90
0340	3-5/8"	.36	.67	1.03
0360	4"	.39	.67	1.06
0380	6"	.55	.68	1.23
0400	20 gage, 2-1/2"	.48	.83	1.31
0420	3-5/8"	.52	.84	1.36
0440	4"	.61	.84	1.45
0460	6"	.72	.85	1.57
0540	Wood studs including blocking, shoe and double top plate, 2"x4", 12"O.C.	.42	1.24	1.66
0560	16" O.C.	.34	.99	1.33
0580	24" O.C.	.26	.80	1.06
0600	2"x6", 12" O.C.	.75	1.42	2.17
0620	16" O.C.	.60	1.10	1.70
0640	24" O.C.	.46	.86	1.32
0642	Furring one side only, steel channels, 3/4", 12" O.C.	.38	1.79	2.17
0644	16" O.C.	.34	1.59	1.93
0646	24" O.C.	.23	1.20	1.43
0647	1-1/2" , 12" O.C.	.51	2	2.51
0648	16" O.C.	.46	1.75	2.21
0649	24" O.C.	.30	1.38	1.68
0650	Wood strips, 1" x 3", on wood, 12" O.C.	.38	.90	1.28
0651	16" O.C.	.29	.68	.97
0652	On masonry, 12" O.C.	.38	1	1.38
0653	16" O.C.	.29	.75	1.04
0654	On concrete, 12" O.C.	.38	1.91	2.29
0655	16" O.C.	.29	1.43	1.72
0665	Gypsum board, one face only, exterior sheathing, 1/2"	.67	.88	1.55
0680	Interior, fire resistant, 1/2"	.35	.50	.85
0700	5/8"	.40	.50	.90
0720	Sound deadening board 1/4"	.41	.55	.96
0740	Standard drywall 3/8"	.34	.50	.84
0760	1/2"	.34	.50	.84
0780	5/8"	.37	.50	.87
0800	Tongue & groove coreboard 1"	.97	2.06	3.03
0820	Water resistant, 1/2"	.53	.50	1.03
0840	5/8"	.52	.50	1.02
0860	Add for the following:, foil backing	.15		.15
0880	Fiberglass insulation, 3-1/2"	.63	.37	1
0900	6"	.85	.37	1.22
0920	Rigid insulation 1"	.54	.50	1.04
0940	Resilient furring @ 16" O.C.	.24	1.55	1.79
0960	Taping and finishing	.05	.50	.55
0980	Texture spray	.09	.53	.62
1000	Thin coat plaster	.09	.62	.71
1040	2"x4" staggered studs 2"x6" plates & blocking	.48	1.08	1.56

C10 Interior Construction

C1010 Partitions

Gypsum plaster and gypsum lath on metal studs.

Gypsum plaster and diamond metal lath on metal studs.

Gypsum plaster and gypsum lath on wood studs.

Gypsum plaster and diamond metal lath on wood studs.

C1010 140	Plaster Partitions/Metal Stud Framing							
	TYPE	FRAMING	LATH	OPPOSITE FACE		COST PER S.F.		
						MAT.	INST.	TOTAL
1000	2 coat gypsum	2-1/2" @ 16"O.C.	3/8" gypsum	same		2.93	7.15	10.08
1010				nothing		1.74	4.15	5.89
1100		3-1/4" @ 24"O.C.	1/2" gypsum	same		2.95	6.90	9.85
1110				nothing		1.72	3.86	5.58
1500	2 coat vermiculite	2-1/2" @ 16"O.C.	3/8" gypsum	same		2.88	7.75	10.63
1510				nothing		1.72	4.46	6.18
1600		3-1/4" @ 24"O.C.	1/2" gypsum	same		2.90	7.50	10.40
1610				nothing		1.70	4.17	5.87
2000	3 coat gypsum	2-1/2" @ 16"O.C.	3/8" gypsum	same		2.81	8.10	10.91
2010				nothing		1.68	4.63	6.31
2020			3.4 lb. diamond	same		2.59	8.10	10.69
2030				nothing		1.57	4.63	6.20
2040			2.75 lb. ribbed	same		2.35	8.10	10.45
2050				nothing		1.45	4.63	6.08
2100		3-1/4" @ 24"O.C.	1/2" gypsum	same		2.83	7.85	10.68
2110				nothing		1.66	4.34	6
2120			3.4 lb. ribbed	same		2.77	7.85	10.62
2130				nothing		1.62	4.34	5.96
3500	3 coat gypsum	2-1/2" @ 16"O.C.	3/8" gypsum	same		3.42	10.40	13.82
3510	W/med. Keenes			nothing		1.99	5.80	7.79
3520			3.4 lb. diamond	same		3.20	10.40	13.60
3530				nothing		1.88	5.80	7.68
3540			2.75 lb. ribbed	same		2.96	10.40	13.36
3550				nothing		1.76	5.80	7.56
3600		3-1/4" @ 24"O.C.	1/2" gypsum	same		3.44	10.15	13.59
3610				nothing		1.97	5.50	7.47
3620			3.4 lb. ribbed	same		3.38	10.15	13.53
3630				nothing		1.93	5.50	7.43

C1010 Partitions

| C1010 140 | | Plaster Partitions/Metal Stud Framing | | | | | | |

	TYPE	FRAMING	LATH	OPPOSITE FACE		COST PER S.F.		
						MAT.	INST.	TOTAL
4000	3 coat gypsum	2-1/2" @ 16"O.C.	3/8" gypsum	same		3.43	11.55	14.98
4010	W/hard Keenes			nothing		1.99	6.35	8.34
4022			3.4 lb. diamond	same		3.21	11.55	14.76
4032				nothing		1.88	6.35	8.23
4040			2.75 lb. ribbed	same		2.97	11.55	14.52
4050				nothing		1.76	6.35	8.11
4100		3-1/4" @ 24"O.C.	1/2" gypsum	same		3.45	11.30	14.75
4110				nothing		1.97	6.05	8.02
4120			3.4 lb. ribbed	same		3.39	11.30	14.69
4130				nothing		1.93	6.05	7.98

| C1010 142 | | Plaster Partitions/Wood Stud Framing | | | | | | |

	TYPE	FRAMING	LATH	OPPOSITE FACE		COST PER S.F.		
						MAT.	INST.	TOTAL
5000	2 coat gypsum	2"x4" @ 16"O.C.	3/8" gypsum	same		2.84	6.95	9.79
5010				nothing		1.65	4.06	5.71
5100		2"x4" @ 24"O.C.	1/2" gypsum	same		2.85	6.85	9.70
5110				nothing		1.62	3.90	5.52
5500	2 coat vermiculite	2"x4" @ 16"O.C.	3/8" gypsum	same		2.79	7.60	10.39
5510				nothing		1.63	4.37	6
5600		2"x4" @ 24"O.C.	1/2" gypsum	same		2.80	7.50	10.30
5610				nothing		1.60	4.21	5.81
6000	3 coat gypsum	2"x4" @ 16"O.C.	3/8" gypsum	same		2.72	7.95	10.67
6010				nothing		1.59	4.54	6.13
6020			3.4 lb. diamond	same		2.50	8	10.50
6030				nothing		1.48	4.57	6.05
6040			2.75 lb. ribbed	same		2.24	8.05	10.29
6050				nothing		1.35	4.59	5.94
6100		2"x4" @ 24"O.C.	1/2" gypsum	same		2.73	7.80	10.53
6110				nothing		1.56	4.38	5.94
6120			3.4 lb. ribbed	same		2.18	7.90	10.08
6130				nothing		1.28	4.42	5.70
7500	3 coat gypsum	2"x4" @ 16"O.C.	3/8" gypsum	same		3.33	10.25	13.58
7510	W/med Keenes			nothing		1.90	5.70	7.60
7520			3.4 lb. diamond	same		3.11	10.30	13.41
7530				nothing		1.79	5.70	7.49
7540			2.75 lb. ribbed	same		2.85	10.35	13.20
7550				nothing		1.66	5.75	7.41
7600		2"x4" @ 24"O.C.	1/2" gypsum	same		3.34	10.10	13.44
7610				nothing		1.87	5.55	7.42
7620			3.4 lb. ribbed	same		3.28	10.25	13.53
7630				nothing		1.83	5.60	7.43
8000	3 coat gypsum	2"x4" @ 16"O.C.	3/8" gypsum	same		3.34	11.40	14.74
8010	W/hard Keenes			nothing		1.90	6.25	8.15
8020			3.4 lb. diamond	same		3.12	11.45	14.57
8030				nothing		1.79	6.30	8.09
8040			2.75 lb. ribbed	same		2.86	11.50	14.36
8050				nothing		1.66	6.30	7.96
8100		2"x4" @ 24"O.C.	1/2" gypsum	same		3.35	11.30	14.65
8110				nothing		1.87	6.10	7.97
8120			3.4 lb. ribbed	same		3.29	11.45	14.74
8130				nothing		1.83	6.20	8.03

C1010 Partitions

C1010 144	Plaster Partition Components	COST PER S.F.		
		MAT.	INST.	TOTAL
0060	Metal studs, 16" O.C., including track, non load bearing, 25 gage, 1-5/8"	.43	1.01	1.44
0080	2-1/2"	.43	1.01	1.44
0100	3-1/4"	.50	1.03	1.53
0120	3-5/8"	.50	1.03	1.53
0140	4"	.53	1.04	1.57
0160	6"	.75	1.05	1.80
0180	Load bearing, 20 gage, 2-1/2"	1.18	1.29	2.47
0200	3-5/8"	1.40	1.31	2.71
0220	4"	1.47	1.34	2.81
0240	6"	1.87	1.36	3.23
0260	16 gage 2-1/2"	1.38	1.46	2.84
0280	3-5/8"	1.66	1.50	3.16
0300	4"	1.75	1.53	3.28
0320	6"	2.20	1.55	3.75
0340	Wood studs, including blocking, shoe and double plate, 2"x4", 12" O.C.	.42	1.24	1.66
0360	16" O.C.	.34	.99	1.33
0380	24" O.C.	.26	.80	1.06
0400	2"x6", 12" O.C.	.75	1.42	2.17
0420	16" O.C.	.60	1.10	1.70
0440	24" O.C.	.46	.86	1.32
0460	Furring one face only, steel channels, 3/4", 12" O.C.	.38	1.79	2.17
0480	16" O.C.	.34	1.59	1.93
0500	24" O.C.	.23	1.20	1.43
0520	1-1/2", 12" O.C.	.51	2	2.51
0540	16" O.C.	.46	1.75	2.21
0560	24" O.C.	.30	1.38	1.68
0580	Wood strips 1"x3", on wood., 12" O.C.	.38	.90	1.28
0600	16" O.C.	.29	.68	.97
0620	On masonry, 12" O.C.	.38	1	1.38
0640	16" O.C.	.29	.75	1.04
0660	On concrete, 12" O.C.	.38	1.91	2.29
0680	16" O.C.	.29	1.43	1.72
0700	Gypsum lath. plain or perforated, nailed to studs, 3/8" thick	.67	.55	1.22
0720	1/2" thick	.72	.58	1.30
0740	Clipped to studs, 3/8" thick	.67	.62	1.29
0760	1/2" thick	.72	.67	1.39
0780	Metal lath, diamond painted, nailed to wood studs, 2.5 lb.	.41	.55	.96
0800	3.4 lb.	.56	.58	1.14
0820	Screwed to steel studs, 2.5 lb.	.41	.58	.99
0840	3.4 lb.	.49	.62	1.11
0860	Rib painted, wired to steel, 2.75 lb	.44	.62	1.06
0880	3.4 lb	.68	.67	1.35
0900	4.0 lb	.70	.72	1.42
0910				
0920	Gypsum plaster, 2 coats	.46	2.36	2.82
0940	3 coats	.66	2.84	3.50
0960	Perlite or vermiculite plaster, 2 coats	.50	2.67	3.17
0980	3 coats	.77	3.32	4.09
1000	Stucco, 3 coats, 1" thick, on wood framing	.82	4.69	5.51
1020	On masonry	.29	3.67	3.96
1100	Metal base galvanized and painted 2-1/2" high	.76	1.75	2.51

C10 Interior Construction

C1010 Partitions

Folding Accordion

Folding Leaf

Movable and Borrow Lites

C1010 205	Partitions	COST PER S.F.		
		MAT.	INST.	TOTAL
0360	Folding accordion, vinyl covered, acoustical, 3 lb. S.F., 17 ft max. hgt	28	9.90	37.90
0380	5 lb. per S.F. 27 ft max height	39	10.45	49.45
0400	5.5 lb. per S.F., 17 ft. max height	46	11	57
0420	Commercial, 1.75 lb per S.F., 8 ft. max height	24.50	4.41	28.91
0440	2.0 Lb per S.F., 17 ft. max height	25	6.60	31.60
0460	Industrial, 4.0 lb. per S.F. 27 ft max height	36	13.20	49.20
0480	Vinyl clad wood or steel, electric operation 6 psf	53	6.20	59.20
0500	Wood, non acoustic, birch or mahogany	28.50	3.30	31.80
0560	Folding leaf, aluminum framed acoustical 12 ft.high.,5.5 lb per S.F.,min.	39	16.50	55.50
0580	Maximum	47	33	80
0600	6.5 lb. per S.F., minimum	40.50	16.50	57
0620	Maximum	50.50	33	83.50
0640	Steel acoustical, 7.5 per S.F., vinyl faced, minimum	57.50	16.50	74
0660	Maximum	70	33	103
0680	Wood acoustic type, vinyl faced to 18' high 6 psf, minimum	54	16.50	70.50
0700	Average	64.50	22	86.50
0720	Maximum	83.50	33	116.50
0740	Formica or hardwood faced, minimum	55.50	16.50	72
0760	Maximum	59.50	33	92.50
0780	Wood, low acoustical type to 12 ft. high 4.5 psf	40.50	19.80	60.30
0840	Demountable, trackless wall, cork finish, semi acous, 1-5/8"th, min	42.50	3.05	45.55
0860	Maximum	40	5.20	45.20
0880	Acoustic, 2" thick, minimum	34	3.25	37.25
0900	Maximum	59	4.41	63.41
0920	In-plant modular office system, w/prehung steel door			
0940	3" thick honeycomb core panels			
0960	12' x 12', 2 wall	10.35	.62	10.97
0970	4 wall	9.65	.83	10.48
0980	16' x 16', 2 wall	10.10	.44	10.54
0990	4 wall	6.80	.44	7.24
1000	Gypsum, demountable, 3" to 3-3/4" thick x 9' high, vinyl clad	6.95	2.28	9.23
1020	Fabric clad	17.20	2.50	19.70
1040	1.75 system, vinyl clad hardboard, paper honeycomb core panel			
1060	1-3/4" to 2-1/2" thick x 9' high	11.65	2.28	13.93
1080	Unitized gypsum panel system, 2" to 2-1/2" thick x 9' high			
1100	Vinyl clad gypsum	15	2.28	17.28
1120	Fabric clad gypsum	24.50	2.50	27
1140	Movable steel walls, modular system			
1160	Unitized panels, 48" wide x 9' high			
1180	Baked enamel, pre-finished	16.85	1.83	18.68
1200	Fabric clad	24.50	1.96	26.46

C10 Interior Construction

C1020 Interior Doors

Single Leaf　　　　　**Sliding Entrance**　　　　　**Rolling Overhead**

C1020 102	Special Doors	MAT.	INST.	TOTAL
			COST PER OPNG.	
2500	Single leaf, wood, 3'-0"x7'-0"x1 3/8", birch, solid core	380	185	565
2510	Hollow core	310	178	488
2530	Hollow core, lauan	293	178	471
2540	Louvered pine	425	178	603
2550	Paneled pine	435	178	613
2600	Hollow metal, comm. quality, flush, 3'-0"x7'-0"x1-3/8"	680	195	875
2650	3'-0"x10'-0" openings with panel	1,125	253	1,378
2700	Metal fire, comm. quality, 3'-0"x7'-0"x1-3/8"	815	203	1,018
2800	Kalamein fire, comm. quality, 3'-0"x7'-0"x1-3/4"	575	370	945
3200	Double leaf, wood, hollow core, 2 - 3'-0"x7'-0"x1-3/8"	540	340	880
3300	Hollow metal, comm. quality, B label, 2'-3'-0"x7'-0"x1-3/8"	1,575	425	2,000
3400	6'-0"x10'-0" opening, with panel	2,200	525	2,725
3500	Double swing door system, 12'-0"x7'-0", mill finish	8,225	2,800	11,025
3700	Black finish	8,350	2,875	11,225
3800	Sliding entrance door and system mill finish	9,625	2,350	11,975
3900	Bronze finish	10,500	2,550	13,050
4000	Black finish	11,000	2,650	13,650
4100	Sliding panel mall front, 16'x9' opening, mill finish	3,325	720	4,045
4200	Bronze finish	4,325	935	5,260
4300	Black finish	5,325	1,150	6,475
4400	24'x9' opening mill finish	4,825	1,325	6,150
4500	Bronze finish	6,275	1,725	8,000
4600	Black finish	7,725	2,125	9,850
4700	48'x9' opening mill finish	9,000	1,025	10,025
4800	Bronze finish	11,700	1,325	13,025
4900	Black finish	14,400	1,650	16,050
5000	Rolling overhead steel door, manual, 8' x 8' high	1,125	795	1,920
5100	10' x 10' high	1,925	910	2,835
5200	20' x 10' high	3,225	1,275	4,500
5300	12' x 12' high	2,025	1,050	3,075
5400	Motor operated, 8' x 8' high	2,300	1,050	3,350
5500	10' x 10' high	3,100	1,175	4,275
5600	20' x 10' high	4,400	1,525	5,925
5700	12' x 12' high	3,200	1,300	4,500
5800	Roll up grille, aluminum, manual, 10' x 10' high, mill finish	3,000	1,550	4,550
5900	Bronze anodized	4,800	1,550	6,350
6000	Motor operated, 10' x 10' high, mill finish	4,375	1,800	6,175
6100	Bronze anodized	6,175	1,800	7,975
6200	Steel, manual, 10' x 10' high	2,700	1,275	3,975
6300	15' x 8' high	3,125	1,600	4,725
6400	Motor operated, 10' x 10' high	4,075	1,525	5,600
6500	15' x 8' high	4,500	1,850	6,350

C10 Interior Construction

C1030 Fittings

Toilet Units

Entrance Screens

Urinal Screens

C1030 110	Toilet Partitions	COST PER UNIT		
		MAT.	INST.	TOTAL
0380	Toilet partitions, cubicles, ceiling hung, marble	1,825	470	2,295
0400	Painted metal	515	248	763
0420	Plastic laminate	665	248	913
0460	Stainless steel	1,950	248	2,198
0480	Handicap addition	430		430
0520	Floor and ceiling anchored, marble	2,000	375	2,375
0540	Painted metal	535	198	733
0560	Plastic laminate	675	198	873
0600	Stainless steel	1,850	198	2,048
0620	Handicap addition	305		305
0660	Floor mounted marble	1,175	315	1,490
0680	Painted metal	595	142	737
0700	Plastic laminate	600	142	742
0740	Stainless steel	1,675	142	1,817
0760	Handicap addition	300		300
0780	Juvenile deduction	43.50		43.50
0820	Floor mounted with handrail marble	1,125	315	1,440
0840	Painted metal	485	165	650
0860	Plastic laminate	635	165	800
0900	Stainless steel	1,250	165	1,415
0920	Handicap addition	365		365
0960	Wall hung, painted metal	655	142	797
1020	Stainless steel	1,725	142	1,867
1040	Handicap addition	365		365
1080	Entrance screens, floor mounted, 54" high, marble	745	104	849
1100	Painted metal	256	66	322
1140	Stainless steel	935	66	1,001
1300	Urinal screens, floor mounted, 24" wide, laminated plastic	400	124	524
1320	Marble	630	147	777
1340	Painted metal	219	124	343
1380	Stainless steel	615	124	739
1428	Wall mounted wedge type, painted metal	141	99	240
1460	Stainless steel	595	99	694
1500	Partitions, shower stall, single wall, painted steel, 2'-8" x 2'-8"	1,000	232	1,232
1510	Fiberglass, 2'-8" x 2'-8"	775	258	1,033
1520	Double wall, enameled steel, 2'-8" x 2'-8"	1,075	232	1,307
1530	Stainless steel, 2'-8" x 2'-8"	1,875	232	2,107
1560	Tub enclosure, sliding panels, tempered glass, aluminum frame	400	291	691
1570	Chrome/brass frame-deluxe	1,175	385	1,560

C1030 Fittings

C1030 310	Storage Specialties, EACH	COST EACH		
		MAT.	INST.	TOTAL
0200	Lockers, steel, single tier, 5' to 6' high, per opening, min.	149	41.50	190.50
0210	Maximum	261	48.50	309.50
0600	Shelving, metal industrial, braced, 3' wide, 1' deep	21.50	10.90	32.40
0610	2' deep	31	11.60	42.60

C1030 510	Identifying/Visual Aid Specialties, EACH	COST EACH		
		MAT.	INST.	TOTAL
0100	Control boards, magnetic, porcelain finish, framed, 24" x 18"	219	124	343
0110	96" x 48"	930	198	1,128
0120	Directory boards, outdoor, black plastic, 36" x 24"	810	495	1,305
0130	36" x 36"	935	660	1,595
0140	Indoor, economy, open faced, 18" x 24"	146	142	288
0500	Signs, interior electric exit sign, wall mounted, 6"	117	70	187
0510	Street, reflective alum., dbl. face, 4 way, w/bracket	184	33	217
0520	Letters, cast aluminum, 1/2" deep, 4" high	24.50	27.50	52
0530	1" deep, 10" high	54.50	27.50	82
0540	Plaques, cast aluminum, 20" x 30"	1,150	248	1,398
0550	Cast bronze, 36" x 48"	4,450	495	4,945

C1030 520	Identifying/Visual Aid Specialties, S.F.	COST PER S.F.		
		MAT.	INST.	TOTAL
0100	Bulletin board, cork sheets, no frame, 1/4" thick	3.96	3.42	7.38
0120	Aluminum frame, 1/4" thick, 3' x 5'	8.05	4.15	12.20
0200	Chalkboards, wall hung, alum, frame & chalktrough	11.80	2.20	14
0210	Wood frame & chalktrough	11.45	2.37	13.82
0220	Sliding board, one board with back panel	45	2.11	47.11
0230	Two boards with back panel	69	2.11	71.11
0240	Liquid chalk type, alum. frame & chalktrough	10.70	2.20	12.90
0250	Wood frame & chalktrough	21.50	2.20	23.70

C1030 710	Bath and Toilet Accessories, EACH	COST EACH		
		MAT.	INST.	TOTAL
0100	Specialties, bathroom accessories, st. steel, curtain rod, 5' long, 1" diam.	45.50	38	83.50
0120	Dispenser, towel, surface mounted	46	31	77
0140	Grab bar, 1-1/4" diam., 12" long	32.50	20.50	53
0160	Mirror, framed with shelf, 18" x 24"	249	25	274
0170	72" x 24"	415	82.50	497.50
0180	Toilet tissue dispenser, surface mounted, single roll	21.50	16.50	38
0200	Towel bar, 18" long	43.50	21.50	65
0300	Medicine cabinets, sliding mirror doors, 20" x 16" x 4-3/4", unlighted	111	71	182
0310	24" x 19" x 8-1/2", lighted	175	99	274

C1030 730	Bath and Toilet Accessories, L.F.	COST PER L.F.		
		MAT.	INST.	TOTAL
0100	Partitions, hospital curtain, ceiling hung, polyester oxford cloth	19.10	4.84	23.94
0110	Designer oxford cloth	35	6.15	41.15

C1030 830	Fabricated Cabinets & Counters, L.F.	COST PER L.F.		
		MAT.	INST.	TOTAL
0110	Household, base, hardwood, one top drawer & one door below x 12" wide	250	40	290
0120	Four drawer x 24" wide	450	44.50	494.50
0130	Wall, hardwood, 30" high with one door x 12" wide	211	45	256
0140	Two doors x 48" wide	440	54	494
0150	Counter top-laminated plastic, stock, economy	9.55	16.50	26.05
0160	Custom-square edge, 7/8" thick	13.80	37	50.80

C10 Interior Construction

C1030 Fittings

C1030 830	Fabricated Cabinets & Counters, L.F.	COST PER L.F.		
		MAT.	INST.	TOTAL
0170	School, counter, wood, 32" high	225	49.50	274.50
0180	Metal, 84" high	405	66	471

C1030 910	Other Fittings, EACH	COST EACH		
		MAT.	INST.	TOTAL
0500	Mail boxes, horizontal, rear loaded, aluminum, 5" x 6" x 15" deep	38.50	14.60	53.10
0510	Front loaded, aluminum, 10" x 12" x 15" deep	119	25	144
0520	Vertical, front loaded, aluminum, 15" x 5" x 6" deep	33	14.60	47.60
0530	Bronze, duranodic finish	44	14.60	58.60
0540	Letter slot, post office	121	62	183
0550	Mail counter, window, post office, with grille	605	248	853
0700	Turnstiles, one way, 4' arm, 46" diam., manual	855	198	1,053
0710	Electric	1,400	825	2,225
0720	3 arm, 5'-5" diam. & 7' high, manual	4,250	990	5,240
0730	Electric	4,800	1,650	6,450

C2010 Stair Construction

The table below lists the cost per flight for 4'-0" wide stairs. Side walls are not included. Railings are included in the prices.

C2010 110	Stairs	COST PER FLIGHT		
		MAT.	INST.	TOTAL
0470	Stairs, C.I.P. concrete, w/o landing, 12 risers, w/o nosing	1,275	2,075	3,350
0480	With nosing	2,025	2,375	4,400
0550	W/landing, 12 risers, w/o nosing	1,450	2,575	4,025
0560	With nosing	2,200	2,875	5,075
0570	16 risers, w/o nosing	1,775	3,225	5,000
0580	With nosing	2,775	3,625	6,400
0590	20 risers, w/o nosing	2,100	3,875	5,975
0600	With nosing	3,350	4,375	7,725
0610	24 risers, w/o nosing	2,425	4,550	6,975
0620	With nosing	3,925	5,150	9,075
0630	Steel, grate type w/nosing & rails, 12 risers, w/o landing	6,425	1,075	7,500
0640	With landing	8,400	1,500	9,900
0660	16 risers, with landing	10,500	1,850	12,350
0680	20 risers, with landing	12,700	2,225	14,925
0700	24 risers, with landing	14,800	2,600	17,400
0701				
0710	Cement fill metal pan & picket rail, 12 risers, w/o landing	8,400	1,075	9,475
0720	With landing	11,000	1,650	12,650
0740	16 risers, with landing	13,800	2,000	15,800
0760	20 risers, with landing	16,600	2,350	18,950
0780	24 risers, with landing	19,400	2,725	22,125
0790	Cast iron tread & pipe rail, 12 risers, w/o landing	8,400	1,075	9,475
0800	With landing	11,000	1,650	12,650
1120	Wood, prefab box type, oak treads, wood rails 3'-6" wide, 14 risers	2,100	415	2,515
1150	Prefab basement type, oak treads, wood rails 3'-0" wide, 14 risers	1,050	102	1,152

C30 Interior Finishes

C3010 Wall Finishes

C3010 230	Paint & Covering	COST PER S.F.		
		MAT.	INST.	TOTAL
0060	Painting, interior on plaster and drywall, brushwork, primer & 1 coat	.11	.65	.76
0080	Primer & 2 coats	.17	.86	1.03
0100	Primer & 3 coats	.24	1.06	1.30
0120	Walls & ceilings, roller work, primer & 1 coat	.11	.43	.54
0140	Primer & 2 coats	.17	.56	.73
0160	Woodwork incl. puttying, brushwork, primer & 1 coat	.11	.94	1.05
0180	Primer & 2 coats	.17	1.24	1.41
0200	Primer & 3 coats	.24	1.69	1.93
0260	Cabinets and casework, enamel, primer & 1 coat	.12	1.06	1.18
0280	Primer & 2 coats	.18	1.30	1.48
0300	Masonry or concrete, latex, brushwork, primer & 1 coat	.23	.88	1.11
0320	Primer & 2 coats	.30	1.26	1.56
0340	Addition for block filler	.15	1.08	1.23
0380	Fireproof paints, intumescent, 1/8" thick 3/4 hour	2.14	.86	3
0400	3/16" thick 1 hour	4.78	1.30	6.08
0420	7/16" thick 2 hour	6.15	3.01	9.16
0440	1-1/16" thick 3 hour	10.05	6.05	16.10
0500	Gratings, primer & 1 coat	.26	1.32	1.58
0600	Pipes over 12" diameter	.49	4.22	4.71
0700	Structural steel, brushwork, light framing 300-500 S.F./Ton	.08	1.63	1.71
0720	Heavy framing 50-100 S.F./Ton	.08	.82	.90
0740	Spraywork, light framing 300-500 S.F./Ton	.09	.36	.45
0760	Heavy framing 50-100 S.F./Ton	.09	.40	.49
0800	Varnish, interior wood trim, no sanding sealer & 1 coat	.08	1.06	1.14
0820	Hardwood floor, no sanding 2 coats	.17	.22	.39
0840	Wall coatings, acrylic glazed coatings, minimum	.31	.80	1.11
0860	Maximum	.66	1.38	2.04
0880	Epoxy coatings, minimum	.41	.80	1.21
0900	Maximum	1.23	2.48	3.71
0940	Exposed epoxy aggregate, troweled on, 1/16" to 1/4" aggregate, minimum	.62	1.80	2.42
0960	Maximum	1.32	3.25	4.57
0980	1/2" to 5/8" aggregate, minimum	1.22	3.25	4.47
1000	Maximum	2.09	5.30	7.39
1020	1" aggregate, minimum	2.12	4.69	6.81
1040	Maximum	3.25	7.65	10.90
1060	Sprayed on, minimum	.56	1.43	1.99
1080	Maximum	1.06	2.91	3.97
1100	High build epoxy 50 mil, minimum	.68	1.08	1.76
1120	Maximum	1.17	4.44	5.61
1140	Laminated epoxy with fiberglass minimum	.74	1.43	2.17
1160	Maximum	1.31	2.91	4.22
1180	Sprayed perlite or vermiculite 1/16" thick, minimum	.26	.14	.40
1200	Maximum	.75	.66	1.41
1260	Wall coatings, vinyl plastic, minimum	.33	.57	.90
1280	Maximum	.83	1.76	2.59
1300	Urethane on smooth surface, 2 coats, minimum	.26	.37	.63
1320	Maximum	.58	.63	1.21
1340	3 coats, minimum	.35	.50	.85
1360	Maximum	.79	.90	1.69
1380	Ceramic-like glazed coating, cementitious, minimum	.48	.96	1.44
1420	Resin base, minimum	.33	.66	.99
1440	Maximum	.54	1.28	1.82
1460	Wall coverings, aluminum foil	1.07	1.54	2.61
1500	Vinyl backing	5.70	1.77	7.47
1520	Cork tiles, 12"x12", light or dark, 3/16" thick	4.68	1.77	6.45
1540	5/16" thick	3.98	1.81	5.79
1560	Basketweave, 1/4" thick	6.15	1.77	7.92
1580	Natural, non-directional, 1/2" thick	7.35	1.77	9.12

C3010 Wall Finishes

C3010 230	Paint & Covering	COST PER S.F.		
		MAT.	INST.	TOTAL
1600	12"x36", granular, 3/16" thick	1.32	1.10	2.42
1620	1" thick	1.71	1.15	2.86
1640	12"x12", polyurethane coated, 3/16" thick	4.14	1.77	5.91
1660	5/16" thick	5.90	1.81	7.71
1661	Paneling, prefinished plywood, birch	1.49	2.36	3.85
1662	Mahogany, African	2.76	2.48	5.24
1663	Philippine (lauan)	1.19	1.98	3.17
1664	Oak or cherry	3.54	2.48	6.02
1665	Rosewood	5.05	3.10	8.15
1666	Teak	3.54	2.48	6.02
1667	Chestnut	5.25	2.64	7.89
1668	Pecan	2.27	2.48	4.75
1669	Walnut	5.75	2.48	8.23
1670	Wood board, knotty pine, finished	1.89	4.10	5.99
1671	Rough sawn cedar	2.35	4.10	6.45
1672	Redwood	5.05	4.10	9.15
1673	Aromatic cedar	3.96	4.40	8.36
1680	Cork wallpaper, paper backed, natural	2.09	.89	2.98
1700	Color	2.92	.89	3.81
1720	Gypsum based, fabric backed, minimum	.88	.53	1.41
1740	Average	1.29	.59	1.88
1760	Maximum	1.43	.66	2.09
1780	Vinyl wall covering, fabric back, light weight	.72	.66	1.38
1800	Medium weight	.85	.89	1.74
1820	Heavy weight	1.75	.98	2.73
1840	Wall paper, double roll, solid pattern, avg. workmanship	.39	.66	1.05
1860	Basic pattern, avg. workmanship	.69	.79	1.48
1880	Basic pattern, quality workmanship	2.59	.98	3.57
1900	Grass cloths with lining paper, minimum	.81	1.06	1.87
1920	Maximum	2.61	1.21	3.82
1940	Ceramic tile, thin set, 4-1/4" x 4-1/4"	2.74	4.20	6.94
1942				
1960	12" x 12"	3.98	2.66	6.64

C3010 235	Paint Trim	COST PER L.F.		
		MAT.	INST.	TOTAL
2040	Painting, wood trim, to 6" wide, enamel, primer & 1 coat	.12	.53	.65
2060	Primer & 2 coats	.18	.67	.85
2080	Misc. metal brushwork, ladders	.52	5.30	5.82
2100	Pipes, to 4" dia.	.06	1.11	1.17
2120	6" to 8" dia.	.13	2.22	2.35
2140	10" to 12" dia.	.38	3.32	3.70
2160	Railings, 2" pipe	.18	2.64	2.82
2180	Handrail, single	.10	1.06	1.16
2185	Caulking & Sealants, Polyureathane, In place, 1 or 2 component, 1/2" X 1/4"	.39	3.39	3.78

C30 Interior Finishes

C3020 Floor Finishes

C3020 410	Tile & Covering	COST PER S.F.		
		MAT.	INST.	TOTAL
0060	Carpet tile, nylon, fusion bonded, 18" x 18" or 24" x 24", 24 oz.	3.55	.62	4.17
0080	35 oz.	4.16	.62	4.78
0100	42 oz.	5.55	.62	6.17
0140	Carpet, tufted, nylon, roll goods, 12' wide, 26 oz.	5.10	.71	5.81
0160	36 oz.	7.85	.71	8.56
0180	Woven, wool, 36 oz.	13.30	.71	14.01
0200	42 oz.	13.65	.71	14.36
0220	Padding, add to above, minimum	.56	.33	.89
0240	Maximum	.95	.33	1.28
0260	Composition flooring, acrylic, 1/4" thick	1.60	4.79	6.39
0280	3/8" thick	2.04	5.55	7.59
0300	Epoxy, minimum	2.74	3.68	6.42
0320	Maximum	3.32	5.10	8.42
0340	Epoxy terrazzo, minimum	5.80	5.40	11.20
0360	Maximum	8.50	7.20	15.70
0380	Mastic, hot laid, 1-1/2" thick, minimum	4.02	3.61	7.63
0400	Maximum	5.15	4.79	9.94
0420	Neoprene 1/4" thick, minimum	3.95	4.56	8.51
0440	Maximum	5.45	5.80	11.25
0460	Polyacrylate with ground granite 1/4", minimum	3.28	3.39	6.67
0480	Maximum	5.65	5.15	10.80
0500	Polyester with colored quart 2 chips 1/16", minimum	2.94	2.33	5.27
0520	Maximum	4.64	3.68	8.32
0540	Polyurethane with vinyl chips, minimum	7.70	2.33	10.03
0560	Maximum	11.10	2.90	14
0600	Concrete topping, granolithic concrete, 1/2" thick	.32	3.96	4.28
0620	1" thick	.63	4.07	4.70
0640	2" thick	1.26	4.68	5.94
0660	Heavy duty 3/4" thick, minimum	.43	6.15	6.58
0680	Maximum	.80	7.35	8.15
0700	For colors, add to above, minimum	.47	1.42	1.89
0720	Maximum	.78	1.56	2.34
0740	Exposed aggregate finish, minimum	.25	.72	.97
0760	Maximum	.73	.96	1.69
0780	Abrasives, .25 P.S.F. add to above, minimum	.48	.53	1.01
0800	Maximum	.67	.53	1.20
0820	Dust on coloring, add, minimum	.47	.34	.81
0840	Maximum	.78	.72	1.50
0860	Floor coloring using 0.6 psf powdered color, 1/2" integral, minimum	5.35	3.96	9.31
0880	Maximum	5.65	3.96	9.61
0900	Dustproofing, add, minimum	.19	.24	.43
0920	Maximum	.68	.34	1.02
0930	Paint	.30	1.26	1.56
0940	Hardeners, metallic add, minimum	.56	.53	1.09
0960	Maximum	1.67	.78	2.45
0980	Non-metallic, minimum	.25	.53	.78
1000	Maximum	.75	.78	1.53
1020	Integral topping and finish, 1:1:2 mix, 3/16" thick	.11	2.33	2.44
1040	1/2" thick	.28	2.46	2.74
1060	3/4" thick	.43	2.75	3.18
1080	1" thick	.57	3.11	3.68
1100	Terrazzo, minimum	3.16	14.40	17.56
1120	Maximum	5.90	18.05	23.95
1280	Resilient, asphalt tile, 1/8" thick on concrete, minimum	1.24	1.12	2.36
1300	Maximum	1.36	1.12	2.48
1320	On wood, add for felt underlay	.22		.22
1340	Cork tile, minimum	6.35	1.42	7.77
1360	Maximum	13.65	1.42	15.07

C30 Interior Finishes

C3020 Floor Finishes

C3020 410	Tile & Covering	COST PER S.F.		
		MAT.	INST.	TOTAL
1380	Polyethylene, in rolls, minimum	3.36	1.62	4.98
1400	Maximum	6.35	1.62	7.97
1420	Polyurethane, thermoset, minimum	5.05	4.46	9.51
1440	Maximum	6.05	8.95	15
1460	Rubber, sheet goods, minimum	6.90	3.72	10.62
1480	Maximum	11	4.96	15.96
1500	Tile, minimum	6.80	1.12	7.92
1520	Maximum	12.65	1.62	14.27
1580	Vinyl, composition tile, minimum	.96	.89	1.85
1600	Maximum	2.35	.89	3.24
1620	Vinyl tile, minimum	3.25	.89	4.14
1640	Maximum	6.45	.89	7.34
1660	Sheet goods, minimum	3.85	1.79	5.64
1680	Maximum	6.55	2.23	8.78
1720	Tile, ceramic natural clay	5.15	4.36	9.51
1730	Marble, synthetic 12"x12"x5/8"	6.65	13.30	19.95
1740	Porcelain type, minimum	5.20	4.36	9.56
1760	Maximum	7.15	5.15	12.30
1800	Quarry tile, mud set, minimum	7.75	5.70	13.45
1820	Maximum	10	7.25	17.25
1840	Thin set, deduct		1.14	1.14
1860	Terrazzo precast, minimum	4.57	5.95	10.52
1880	Maximum	9.85	5.95	15.80
1900	Non-slip, minimum	19.45	14.70	34.15
1920	Maximum	21.50	19.60	41.10
2020	Wood block, end grain factory type, natural finish, 2" thick	5.10	3.96	9.06
2040	Fir, vertical grain, 1"x4", no finish, minimum	2.84	1.94	4.78
2060	Maximum	3.01	1.94	4.95
2080	Prefinished white oak, prime grade, 2-1/4" wide	7.05	2.92	9.97
2100	3-1/4" wide	9	2.68	11.68
2120	Maple strip, sanded and finished, minimum	4.51	4.25	8.76
2140	Maximum	5.45	4.25	9.70
2160	Oak strip, sanded and finished, minimum	4.36	4.25	8.61
2180	Maximum	5.25	4.25	9.50
2200	Parquetry, sanded and finished, minimum	5.70	4.43	10.13
2220	Maximum	6.30	6.30	12.60
2260	Add for sleepers on concrete, treated, 24" O.C., 1"x2"	2.50	2.55	5.05
2280	1"x3"	2.28	1.98	4.26
2300	2"x4"	.73	1	1.73
2340	Underlayment, plywood, 3/8" thick	1.13	.66	1.79
2350	1/2" thick	1.10	.68	1.78
2360	5/8" thick	1.42	.71	2.13
2370	3/4" thick	1.46	.76	2.22
2380	Particle board, 3/8" thick	.45	.66	1.11
2390	1/2" thick	.50	.68	1.18
2400	5/8" thick	.59	.71	1.30
2410	3/4" thick	.72	.76	1.48
2420	Hardboard, 4' x 4', .215" thick	.55	.66	1.21

C30 Interior Finishes

C3030 Ceiling Finishes

2 Coats of Plaster on Gypsum
Lath on Wood Furring

Fiberglass Board on
Exposed Suspended Grid System

Plaster and Metal Lath
on Metal Furring

C3030 105 — Plaster Ceilings

	TYPE	LATH	FURRING	SUPPORT		COST PER S.F. MAT.	INST.	TOTAL
2400	2 coat gypsum	3/8" gypsum	1"x3" wood, 16" O.C.	wood		1.53	4.94	6.47
2500	Painted			masonry		1.53	5.05	6.58
2600				concrete		1.53	5.60	7.13
2700	3 coat gypsum	3.4# metal	1"x3" wood, 16" O.C.	wood		1.62	5.35	6.97
2800	Painted			masonry		1.62	5.45	7.07
2900				concrete		1.62	6.05	7.67
3000	2 coat perlite	3/8" gypsum	1"x3" wood, 16" O.C.	wood		1.57	5.20	6.77
3100	Painted			masonry		1.57	5.30	6.87
3200				concrete		1.57	5.90	7.47
3300	3 coat perlite	3.4# metal	1"x3" wood, 16" O.C.	wood		1.46	5.30	6.76
3400	Painted			masonry		1.46	5.40	6.86
3500				concrete		1.46	6	7.46
3600	2 coat gypsum	3/8" gypsum	3/4" CRC, 12" O.C.	1-1/2" CRC, 48"O.C.		1.62	5.85	7.47
3700	Painted		3/4" CRC, 16" O.C.	1-1/2" CRC, 48"O.C.		1.58	5.30	6.88
3800			3/4" CRC, 24" O.C.	1-1/2" CRC, 48"O.C.		1.47	4.87	6.34
3900	2 coat perlite	3/8" gypsum	3/4" CRC, 12" O.C.	1-1/2" CRC, 48"O.C		1.66	6.35	8.01
4000	Painted		3/4" CRC, 16" O.C.	1-1/2" CRC, 48"O.C.		1.62	5.80	7.42
4100			3/4" CRC, 24" O.C.	1-1/2" CRC, 48"O.C.		1.51	5.35	6.86
4200	3 coat gypsum	3.4# metal	3/4" CRC, 12" O.C.	1-1/2" CRC, 36" O.C.		2.02	7.80	9.82
4300	Painted		3/4" CRC, 16" O.C.	1-1/2" CRC, 36" O.C.		1.98	7.25	9.23
4400			3/4" CRC, 24" O.C.	1-1/2" CRC, 36" O.C.		1.87	6.80	8.67
4500	3 coat perlite	3.4# metal	3/4" CRC, 12" O.C.	1-1/2" CRC, 36" O.C.		2.13	8.50	10.63
4600	Painted		3/4" CRC, 16" O.C.	1-1/2" CRC, 36" O.C.		2.09	7.95	10.04
4700			3/4" CRC, 24" O.C.	1-1/2" CRC, 36" O.C.		1.98	7.50	9.48

C3030 110 — Drywall Ceilings

	TYPE	FINISH	FURRING	SUPPORT		COST PER S.F. MAT.	INST.	TOTAL
4800	1/2" F.R. drywall	painted and textured	1"x3" wood, 16" O.C.	wood		.89	3.06	3.95
4900				masonry		.89	3.15	4.04
5000				concrete		.89	3.76	4.65
5100	5/8" F.R. drywall	painted and textured	1"x3" wood, 16" O.C.	wood		.93	3.06	3.99
5200				masonry		.93	3.15	4.08
5300				concrete		.93	3.76	4.69
5400	1/2" F.R. drywall	painted and textured	7/8" resil. channels	24" O.C.		.75	2.96	3.71

C30 Interior Finishes

C3030 Ceiling Finishes

C3030 110 — Drywall Ceilings

	TYPE	FINISH	FURRING	SUPPORT		COST PER S.F.		
						MAT.	INST.	TOTAL
5500	1/2" F.R. drywall	painted and textured	1"x2" wood	stud clips		.93	2.85	3.78
5602		painted	1-5/8" metal studs	24" O.C.		.85	2.64	3.49
5700	5/8" F.R. drywall	painted and textured	1-5/8" metal studs	24" O.C.		.89	2.64	3.53
5702								

C3030 140 — Plaster Ceiling Components

		COST PER S.F.		
		MAT.	INST.	TOTAL
0060	Plaster, gypsum incl. finish, 2 coats			
0080	3 coats	.66	3.20	3.86
0100	Perlite, incl. finish, 2 coats	.50	3.14	3.64
0120	3 coats	.77	3.91	4.68
0140	Thin coat on drywall	.09	.62	.71
0200	Lath, gypsum, 3/8" thick	.67	.77	1.44
0220	1/2" thick	.72	.80	1.52
0240	5/8" thick	.40	.94	1.34
0260	Metal, diamond, 2.5 lb.	.41	.62	1.03
0280	3.4 lb.	.56	.67	1.23
0300	Flat rib, 2.75 lb.	.44	.62	1.06
0320	3.4 lb.	.68	.67	1.35
0440	Furring, steel channels, 3/4" galvanized, 12" O.C.	.38	2	2.38
0460	16" O.C.	.34	1.45	1.79
0480	24" O.C.	.23	1	1.23
0500	1-1/2" galvanized, 12" O.C.	.51	2.22	2.73
0520	16" O.C.	.46	1.62	2.08
0540	24" O.C.	.30	1.08	1.38
0560	Wood strips, 1"x3", on wood, 12" O.C.	.38	1.42	1.80
0580	16" O.C.	.29	1.07	1.36
0600	24" O.C.	.19	.71	.90
0620	On masonry, 12" O.C.	.38	1.55	1.93
0640	16" O.C.	.29	1.16	1.45
0660	24" O.C.	.19	.78	.97
0680	On concrete, 12" O.C.	.38	2.36	2.74
0700	16" O.C.	.29	1.77	2.06
0720	24" O.C.	.19	1.18	1.37
0940	Paint on plaster or drywall, roller work, primer + 1 coat	.11	.43	.54
0960	Primer + 2 coats	.17	.56	.73

C3030 210 — Acoustical Ceilings

	TYPE	TILE	GRID	SUPPORT		COST PER S.F.		
						MAT.	INST.	TOTAL
5800	5/8" fiberglass board	24" x 48"	tee	suspended		1.72	1.48	3.20
5900		24" x 24"	tee	suspended		1.89	1.62	3.51
6000	3/4" fiberglass board	24" x 48"	tee	suspended		2.78	1.52	4.30
6100		24" x 24"	tee	suspended		2.95	1.66	4.61
6500	5/8" mineral fiber	12" x 12"	1"x3" wood, 12" O.C.	wood		2.42	3.07	5.49
6600				masonry		2.42	3.20	5.62
6700				concrete		2.42	4.01	6.43
6800	3/4" mineral fiber	12" x 12"	1"x3" wood, 12" O.C.	wood		2.44	3.07	5.51
6900				masonry		2.44	3.07	5.51
7000				concrete		2.44	3.07	5.51
7100	3/4" mineral fiber on	12" x 12"	25 ga. channels	runners		2.98	3.77	6.75
7102	5/8" F.R. drywall							

C30 Interior Finishes

C3030 Ceiling Finishes

C3030 210				Acoustical Ceilings				
						COST PER S.F.		
	TYPE	TILE	GRID	SUPPORT		MAT.	INST.	TOTAL
7200 7201 7202	5/8" plastic coated Mineral fiber	12" x 12"	25 ga. channels	adhesive backed		2.45	1.65	4.10
7300 7301 7302	3/4" plastic coated Mineral fiber	12" x 12"		adhesive backed		2.47	1.65	4.12
7400 7401 7402	3/4" mineral fiber	12" x 12"	conceal 2" bar & channels	suspended		2.66	3.72	6.38

C3030 240	Acoustical Ceiling Components	COST PER S.F.		
		MAT.	INST.	TOTAL
2480	Ceiling boards, eggcrate, acrylic, 1/2" x 1/2" x 1/2" cubes	2	.99	2.99
2500	Polystyrene, 3/8" x 3/8" x 1/2" cubes	1.68	.97	2.65
2520	1/2" x 1/2" x 1/2" cubes	2.24	.99	3.23
2540	Fiberglass boards, plain, 5/8" thick	.80	.79	1.59
2560	3/4" thick	1.86	.83	2.69
2580	Grass cloth faced, 3/4" thick	3.34	.99	4.33
2600	1" thick	2.38	1.02	3.40
2620	Luminous panels, prismatic, acrylic	2.44	1.24	3.68
2640	Polystyrene	1.25	1.24	2.49
2660	Flat or ribbed, acrylic	4.25	1.24	5.49
2680	Polystyrene	2.92	1.24	4.16
2700	Drop pan, white, acrylic	6.25	1.24	7.49
2720	Polystyrene	5.20	1.24	6.44
2740	Mineral fiber boards, 5/8" thick, standard	.86	.73	1.59
2760	Plastic faced	1.35	1.24	2.59
2780	2 hour rating	1.29	.73	2.02
2800	Perforated aluminum sheets, .024 thick, corrugated painted	2.48	1.01	3.49
2820	Plain	4.14	.99	5.13
3080	Mineral fiber, plastic coated, 12" x 12" or 12" x 24", 5/8" thick	2.04	1.65	3.69
3100	3/4" thick	2.06	1.65	3.71
3120	Fire rated, 3/4" thick, plain faced	1.54	1.65	3.19
3140	Mylar faced	1.30	1.65	2.95
3160	Add for ceiling primer	.14		.14
3180	Add for ceiling cement	.41		.41
3240	Suspension system, furring, 1" x 3" wood 12" O.C.	.38	1.42	1.80
3260	T bar suspension system, 2' x 4' grid	.69	.62	1.31
3280	2' x 2' grid	.86	.76	1.62
3300	Concealed Z bar suspension system 12" module	.72	.95	1.67
3320	Add to above for 1-1/2" carrier channels 4' O.C.	.17	1.05	1.22
3340	Add to above for carrier channels for recessed lighting	.30	1.08	1.38

D10 Conveying

D1010 Elevators and Lifts

Hydraulic

Traction Geared

Traction Gearless

D1010 110	Hydraulic	MAT.	INST.	TOTAL
		COST EACH		
1300	Pass. elev., 1500 lb., 2 Floors, 100 FPM	49,000	13,800	62,800
1400	5 Floors, 100 FPM	99,000	36,800	135,800
1600	2000 lb., 2 Floors, 100 FPM	51,000	13,800	64,800
1700	5 floors, 100 FPM	101,500	36,800	138,300
1900	2500 lb., 2 Floors, 100 FPM	52,500	13,800	66,300
2000	5 floors, 100 FPM	102,500	36,800	139,300
2200	3000 lb., 2 Floors, 100 FPM	53,500	13,800	67,300
2300	5 floors, 100 FPM	104,000	36,800	140,800
2500	3500 lb., 2 Floors, 100 FPM	56,000	13,800	69,800
2600	5 floors, 100 FPM	106,000	36,800	142,800
2800	4000 lb., 2 Floors, 100 FPM	57,500	13,800	71,300
2900	5 floors, 100 FPM	107,500	36,800	144,300
3100	4500 lb., 2 Floors, 100 FPM	59,000	13,800	72,800
3200	5 floors, 100 FPM	109,000	36,800	145,800
4000	Hospital elevators, 3500 lb., 2 Floors, 100 FPM	82,500	13,800	96,300
4100	5 floors, 100 FPM	143,000	36,800	179,800
4300	4000 lb., 2 Floors, 100 FPM	82,500	13,800	96,300
4400	5 floors, 100 FPM	143,000	36,800	179,800
4600	4500 lb., 2 Floors, 100 FPM	90,000	13,800	103,800
4800	5 floors, 100 FPM	150,500	36,800	187,300
4900	5000 lb., 2 Floors, 100 FPM	92,500	13,800	106,300
5000	5 floors, 100 FPM	153,000	36,800	189,800
6700	Freight elevators (Class "B"), 3000 lb., 2 Floors, 50 FPM	132,000	17,600	149,600
6800	5 floors, 100 FPM	257,000	46,200	303,200
7000	4000 lb., 2 Floors, 50 FPM	141,000	17,600	158,600
7100	5 floors, 100 FPM	266,500	46,200	312,700
7500	10,000 lb., 2 Floors, 50 FPM	180,000	17,600	197,600
7600	5 floors, 100 FPM	305,000	46,200	351,200
8100	20,000 lb., 2 Floors, 50 FPM	248,000	17,600	265,600
8200	5 Floors, 100 FPM	373,000	46,200	419,200

D1010 140	Traction Geared Elevators	MAT.	INST.	TOTAL
		COST EACH		
1300	Passenger, 2000 Lb., 5 floors, 200 FPM	124,000	34,700	158,700
1500	15 floors, 350 FPM	300,500	105,000	405,500
1600	2500 Lb., 5 floors, 200 FPM	128,000	34,700	162,700
1800	15 floors, 350 FPM	305,000	105,000	410,000
1900	3000 Lb., 5 floors, 200 FPM	129,000	34,700	163,700
2100	15 floors, 350 FPM	305,500	105,000	410,500
2200	3500 Lb., 5 floors, 200 FPM	132,500	34,700	167,200
2400	15 floors, 350 FPM	309,000	105,000	414,000
2500	4000 Lb., 5 floors, 200 FPM	131,500	34,700	166,200
2700	15 floors, 350 FPM	308,500	105,000	413,500

D10 Conveying

D1010 Elevators and Lifts

D1010 140	Traction Geared Elevators	COST EACH		
		MAT.	INST.	TOTAL
2800	4500 Lb., 5 floors, 200 FPM	134,000	34,700	168,700
3000	15 floors, 350 FPM	310,500	105,000	415,500
3100	5000 Lb., 5 floors, 200 FPM	136,000	34,700	170,700
3300	15 floors, 350 FPM	312,500	105,000	417,500
4000	Hospital, 3500 Lb., 5 floors, 200 FPM	127,000	34,700	161,700
4200	15 floors, 350 FPM	301,500	105,000	406,500
4300	4000 Lb., 5 floors, 200 FPM	127,000	34,700	161,700
4500	15 floors, 350 FPM	301,500	105,000	406,500
4600	4500 Lb., 5 floors, 200 FPM	133,000	34,700	167,700
4800	15 floors, 350 FPM	307,500	105,000	412,500
4900	5000 Lb., 5 floors, 200 FPM	134,500	34,700	169,200
5100	15 floors, 350 FPM	309,000	105,000	414,000
6000	Freight, 4000 Lb., 5 floors, 50 FPM class 'B'	142,000	36,000	178,000
6200	15 floors, 200 FPM class 'B'	357,500	125,000	482,500
6300	8000 Lb., 5 floors, 50 FPM class 'B'	174,500	36,000	210,500
6500	15 floors, 200 FPM class 'B'	390,000	125,000	515,000
7000	10,000 Lb., 5 floors, 50 FPM class 'B'	192,500	36,000	228,500
7200	15 floors, 200 FPM class 'B'	556,000	125,000	681,000
8000	20,000 Lb., 5 floors, 50 FPM class 'B'	222,500	36,000	258,500
8200	15 floors, 200 FPM class 'B'	585,500	125,000	710,500

D1010 150	Traction Gearless Elevators	COST EACH		
		MAT.	INST.	TOTAL
1700	Passenger, 2500 Lb., 10 floors, 200 FPM	339,000	90,500	429,500
1900	30 floors, 600 FPM	705,500	231,000	936,500
2000	3000 Lb., 10 floors, 200 FPM	340,000	90,500	430,500
2200	30 floors, 600 FPM	706,500	231,000	937,500
2300	3500 Lb., 10 floors, 200 FPM	343,000	90,500	433,500
2500	30 floors, 600 FPM	710,000	231,000	941,000
2700	50 floors, 800 FPM	1,048,500	371,500	1,420,000
2800	4000 lb., 10 floors, 200 FPM	342,500	90,500	433,000
3000	30 floors, 600 FPM	709,000	231,000	940,000
3200	50 floors, 800 FPM	1,048,000	371,500	1,419,500
3300	4500 lb., 10 floors, 200 FPM	344,500	90,500	435,000
3500	30 floors, 600 FPM	711,500	231,000	942,500
3700	50 floors, 800 FPM	1,050,000	371,500	1,421,500
3800	5000 lb., 10 floors, 200 FPM	346,500	90,500	437,000
4000	30 floors, 600 FPM	713,500	231,000	944,500
4200	50 floors, 800 FPM	1,052,000	371,500	1,423,500
6000	Hospital, 3500 Lb., 10 floors, 200 FPM	336,500	90,500	427,000
6200	30 floors, 600 FPM	699,000	231,000	930,000
6400	4000 Lb., 10 floors, 200 FPM	336,500	90,500	427,000
6600	30 floors, 600 FPM	699,000	231,000	930,000
6800	4500 Lb., 10 floors, 200 FPM	343,000	90,500	433,500
7000	30 floors, 600 FPM	705,000	231,000	936,000
7200	5000 Lb., 10 floors, 200 FPM	344,000	90,500	434,500
7400	30 floors, 600 FPM	706,000	231,000	937,000

D10 Conveying

D1090 Other Conveying Systems

D1090 510	Chutes	COST EACH		
		MAT.	INST.	TOTAL
0100	Chutes, linen or refuse incl. sprinklers, galv. steel, 18" diam., per floor	1,225	330	1,555
0110	Aluminized steel, 36" diam., per floor	1,650	415	2,065
0120	Mail, 8-3/4" x 3-1/2", aluminum & glass, per floor	715	232	947
0130	Bronze or stainless, per floor	1,100	258	1,358

Minimum Plumbing Fixture Requirements

Type of Building or Occupancy (2)	Water Closets (14) (Fixtures per Person)		Urinals (5,10) (Fixtures per Person)		Lavatories (Fixtures per Person)		Bathtubs or Showers (Fixtures per Person)	Drinking Fountains (Fixtures per Person) (3, 13)
	Male	Female	Male	Female	Male	Female		
Assembly Places-Theatres, Auditoriums, Convention Halls, etc.-for permanent employee use	1: 1 - 15 2: 16 - 35 3: 36 - 55 Over 55, add 1 fixture for each additional 40 persons	1: 1 - 15 2: 16 - 35 3: 36 - 55	0: 1 - 9 1: 10 - 50 Add one fixture for each additional 50 males		1 per 40	1 per 40		
Assembly Places-Theatres, Auditoriums, Convention Halls, etc. - for public use	1: 1 - 100 2: 101 - 200 3: 201 - 400 Over 400, add 1 fixture for each additional 500 males and 1 for each additional 125 females	3: 1 - 50 4: 51 - 100 8: 101 - 200 11: 201 - 400	1: 1 - 100 2: 101 - 200 3: 201 - 400 4: 401 - 600 Over 600, add 1 fixture for each additional 300 males		1: 1 - 200 2: 201 - 400 3: 401 - 750 Over 750, add 1 fixture for each additional 500 persons	1: 1 - 200 2: 201 - 400 3: 401 - 750		1: 1 - 150 2: 151 - 400 3: 401 - 750 Over 750, add one fixture for each additional 500 persons
Dormitories (9) School or Labor	1 per 10 Add 1 fixture for each additional 25 males (over 10) and 1 for each additional 20 females (over 8)	1 per 8	1 per 25 Over 150, add 1 fixture for each additional 50 males		1 per 12 Over 12 add 1 fixture for each additional 20 males and 1 for each 15 additional females	1 per 12	1 per 8 For females add 1 bathtub per 30. Over 150, add 1 per 20	1 per 150 (12)
Dormitories- for Staff Use	1: 1 - 15 2: 16 - 35 3: 36 - 55 Over 55, add 1 fixture for each additional 40 persons	1: 1 - 15 3: 16 - 35 4: 36 - 55	1 per 50		1 per 40	1 per 40	1 per 8	

Dwellings:
  Single Dwelling
  Multiple Dwelling
  or Apartment House
	1 per dwelling
	1 per dwelling or apartment unit
					1 per dwelling
					1 per dwelling or apartment unit
							1 per dwelling
							1 per dwelling or apartment unit	

Hospital Waiting rooms	1 per room				1 per room			1 per 150 (12)
Hospitals- for employee use	1: 1 - 15 2: 16 - 35 3: 36 - 55 Over 55, add 1 fixture for each additional 40 persons	1: 1 - 15 3: 16 - 35 4: 36 - 55	0: 1 - 9 1: 10 - 50 Add 1 fixture for each additional 50 males		1 per 40	1 per 40		

Hospitals:
  Individual Room
  Ward Room
	1 per room
	1 per 8 patients
					1 per room
					1 per 10 patients
							1 per room
							1 per 20 patients
								1 per 150 (12)

Industrial (6) Warehouses Workshops, Foundries and similar establishments- for employee use	1: 1 -10 2: 11 - 25 3: 26 - 50 4: 51 - 75 5: 76 - 100 Over 100, add 1 fixture for each additional 30 persons	1: 1 -10 2: 11 - 25 3: 26 - 50 4: 51 - 75 5: 76 - 100			Up to 100, per 10 persons Over 100, 1 per 15 persons (7, 8)		1 shower for each 15 persons exposed to excessive heat or to skin contamination with poisonous, infectious or irritating material	1 per 150 (12)
Institutional - Other than Hospitals or Penal Institutions (on each occupied floor)	1 per 25	1 per 20	0: 1 - 9 1: 10 - 50 Add 1 fixture for each additional 50 males		1 per 10	1 per 10	1 per 8	1 per 150 (12)
Institutional - Other than Hospitals or Penal Institutions (on each occupied floor)- for employee use	1: 1 - 15 2: 16 - 35 3: 36 - 55 Over 55, add 1 fixture for each additional 40 persons	1: 1 - 15 3: 16 - 35 4: 36 - 55	0: 1 - 9 1: 10 - 50 Add 1 fixture for each additional 50 males		1 per 40	1 per 40	1 per 8	1 per 150 (12)
Office or Public Buildings	1: 1 - 100 2: 101 - 200 3: 201 - 400 Over 400, add 1 fixture for each additional 500 males and 1 for each additional 150 females	3: 1 - 50 4: 51 - 100 8: 101 - 200 11: 201 - 400	1: 1 - 100 2: 101 - 200 3: 201 - 400 4: 401 - 600 Over 600, add 1 fixture for each additional 300 males		1: 1 - 200 2: 201 - 400 3: 401 - 750 Over 750, add 1 fixture for each additional 500 persons	1: 1 - 200 2: 201 - 400 3: 401 - 750		1 per 150 (12)
Office or Public Buildings - for employee use	1: 1 - 15 2: 16 - 35 3: 36 - 55 Over 55, add 1 fixture for each additional 40 persons	1: 1 - 15 3: 16 - 35 4: 36 - 55	0: 1 - 9 1: 10 - 50 Add 1 fixture for each additional 50 males		1 per 40	1 per 40		

D2010 Plumbing Fixtures

Minimum Plumbing Fixture Requirements

Type of Building or Occupancy	Water Closets (14) (Fixtures per Person)		Urinals (5, 10) (Fixtures per Person)		Lavatories (Fixtures per Person)		Bathtubs or Showers (Fixtures per Person)	Drinking Fountains (Fixtures per Person) (3, 13)
	Male	Female	Male	Female	Male	Female		
Penal Institutions - for employee use	1: 1 - 15 2: 16 - 35 3: 36 - 55 Over 55, add 1 fixture for each additional 40 persons	1: 1 - 15 3: 16 - 35 4: 36 - 55	0: 1 - 9 1: 10 - 50 Add 1 fixture for each additional 50 males		1 per 40	1 per 40		1 per 150 (12)
Penal Institutions - for prison use Cell	1 per cell				1 per cell			1 per cellblock floor
Exercise room	1 per exercise room		1 per exercise room		1 per exercise room			1 per exercise room
Restaurants, Pubs and Lounges (11)	1: 1 - 50 2: 51 - 150 3: 151 - 300 Over 300, add 1 fixture for each additional 200 persons	1: 1 - 50 2: 51 - 150 4: 151 - 300	1: 1 - 150 Over 150, add 1 fixture for each additional 150 males		1: 1 - 150 2: 151 - 200 3: 201 - 400 Over 400, add 1 fixture for each additional 400 persons	1: 1 - 150 2: 151 - 200 3: 201 - 400		
Schools - for staff use All Schools	1: 1 - 15 2: 16 - 35 3: 36 - 55 Over 55, add 1 fixture for each additional 40 persons	1: 1 - 15 3: 16 - 35 4: 36 - 55	1 per 50		1 per 40	1 per 40		
Schools - for student use: Nursery	1: 1 - 20 2: 21 - 50 Over 50, add 1 fixture for each additional 50 persons	1: 1 - 20 2: 21 - 50			1: 1 - 25 2: 26 - 50 Over 50, add 1 fixture for each additional 50 persons	1: 1 - 25 2: 26 - 50		1 per 150 (12)
Elementary	1 per 30	1 per 25	1 per 75		1 per 35	1 per 35		1 per 150 (12)
Secondary	1 per 40	1 per 30	1 per 35		1 per 40	1 per 40		1 per 150 (12)
Others (Colleges, Universities, Adult Centers, etc.	1 per 40	1 per 30	1 per 35		1 per 40	1 per 40		1 per 150 (12)
Worship Places Educational and Activities Unit	1 per 150	1 per 75	1 per 150		1 per 2 water closets			1 per 150 (12)
Worship Places Principal Assembly Place	1 per 150	1 per 75	1 per 150		1 per 2 water closets			1 per 150 (12)

Notes:
1. The figures shown are based upon one (1) fixture being the minimum required for the number of persons indicated or any fraction thereof.
2. Building categories not shown on this table shall be considered separately by the Administrative Authority.
3. Drinking fountains shall not be installed in toilet rooms.
4. Laundry trays. One (1) laundry tray or one (1) automatic washer standpipe for each dwelling unit or one (1) laundry trays or one (1) automatic washer standpipes, or combination thereof, for each twelve (12) apartments. Kitchen sinks, one (1) for each dwelling or apartment unit.
5. For each urinal added in excess of the minimum required, one water closet may be deducted. The number of water closets shall not be reduced to less than two-thirds (2/3) of the minimum requirement.
6. As required by ANSI Z4.1-1968, Sanitation in Places of Employment.
7. Where there is exposure to skin contamination with poisonous, infectious, or irritating materials, provide one (1) lavatory for each five (5) persons.
8. Twenty-four (24) lineal inches of wash sink or eighteen (18) inches of a circular basin, when provided with water outlets for such space shall be considered equivalent to one (1) lavatory.
9. Laundry trays, one (1) for each fifty (50) persons. Service sinks, one (1) for each hundred (100) persons.
10. General. In applying this schedule of facilities, consideration shall be given to the accessibility of the fixtures. Conformity purely on a numerical basis may not result in an installation suited to the need of the individual establishment. For example, schools should be provided with toilet facilities on each floor having classrooms.
    a. Surrounding materials, wall and floor space to a point two (2) feet in front of urinal lip and four (4) feet above the floor, and at least two (2) feet to each side of the urinal shall be lined with non-absorbent materials.
    b. Trough urinals shall be prohibited.
11. A restaurant is defined as a business which sells food to be consumed on the premises.
    a. The number of occupants for a drive-in restaurant shall be considered as equal to the number of parking stalls.
    b. Employee toilet facilities shall not to be included in the above restaurant requirements. Hand washing facilities shall be available in the kitchen for employees.
12. Where food is consumed indoors, water stations may be substituted for drinking fountains. Offices, or public buildings for use by more than six (6) persons shall have one (1) drinking fountain for the first one hundred fifty (150) persons and one additional fountain for each three hundred (300) persons thereafter.
13. There shall be a minimum of one (1) drinking fountain per occupied floor in schools, theaters, auditoriums, dormitories, offices of public building.
14. The total number of water closets for females shall be at least equal to the total number of water closets and urinals required for males.

D2010 Plumbing Fixtures

One Piece Wall Hung Water Closet

Wall Hung Urinal

Side By Side Water Closet Group

Floor Mount Water Closet

Stall Type Urinal

Back to Back Water Closet Group

D2010 110	Water Closet Systems	COST EACH		
		MAT.	INST.	TOTAL
1800	Water closet, vitreous china, elongated			
1840	Tank type, wall hung			
1880	Close coupled two piece	1,300	625	1,925
1920	Floor mount, one piece	915	660	1,575
1960	One piece low profile	775	660	1,435
2000	Two piece close coupled	535	660	1,195
2040	Bowl only with flush valve			
2080	Wall hung	1,175	705	1,880
2120	Floor mount	720	670	1,390
2122				
2160	Floor mount, ADA compliant with 18" high bowl	685	690	1,375

D2010 120	Water Closets, Group	COST EACH		
		MAT.	INST.	TOTAL
1760	Water closets, battery mount, wall hung, side by side, first closet	1,250	725	1,975
1800	Each additional water closet, add	1,150	685	1,835
3000	Back to back, first pair of closets	2,075	965	3,040
3100	Each additional pair of closets, back to back	2,025	950	2,975

D2010 210	Urinal Systems	COST EACH		
		MAT.	INST.	TOTAL
2000	Urinal, vitreous china, wall hung	515	695	1,210
2040	Stall type	1,025	830	1,855

D2010 Plumbing Fixtures

Systems are complete with trim and rough-in (supply, waste and vent) to connect to supply branches and waste mains.

Vanity Top

Wall Hung

Counter Top Single Bowl

Counter Top Double Bowl

D2010 310	Lavatory Systems	COST EACH		
		MAT.	INST.	TOTAL
1560	Lavatory w/trim, vanity top, PE on CI, 20" x 18", Vanity top by others.	500	625	1,125
1600	19" x 16" oval	420	625	1,045
1640	18" round	485	625	1,110
1680	Cultured marble, 19" x 17"	435	625	1,060
1720	25" x 19"	445	625	1,070
1760	Stainless, self-rimming, 25" x 22"	605	625	1,230
1800	17" x 22"	595	625	1,220
1840	Steel enameled, 20" x 17"	420	640	1,060
1880	19" round	420	640	1,060
1920	Vitreous china, 20" x 16"	530	655	1,185
1960	19" x 16"	535	655	1,190
2000	22" x 13"	540	655	1,195
2040	Wall hung, PE on CI, 18" x 15"	760	690	1,450
2080	19" x 17"	775	690	1,465
2120	20" x 18"	725	690	1,415
2160	Vitreous china, 18" x 15"	655	705	1,360
2200	19" x 17"	590	705	1,295
2240	24" x 20"	925	705	1,630
2241				
2243				

D2010 410	Kitchen Sink Systems	COST EACH		
		MAT.	INST.	TOTAL
1720	Kitchen sink w/trim, countertop, PE on CI, 24"x21", single bowl	595	685	1,280
1760	30" x 21" single bowl	600	685	1,285
1800	32" x 21" double bowl	735	735	1,470
1840	42" x 21" double bowl	890	745	1,635
1880	Stainless steel, 19" x 18" single bowl	855	685	1,540
1920	25" x 22" single bowl	920	685	1,605
1960	33" x 22" double bowl	1,225	735	1,960
2000	43" x 22" double bowl	1,400	745	2,145
2040	44" x 22" triple bowl	1,425	775	2,200
2080	44" x 24" corner double bowl	1,075	745	1,820
2120	Steel, enameled, 24" x 21" single bowl	455	685	1,140
2160	32" x 21" double bowl	505	735	1,240
2240	Raised deck, PE on CI, 32" x 21", dual level, double bowl	660	930	1,590
2280	42" x 21" dual level, triple bowl	1,250	1,025	2,275

D2010 Plumbing Fixtures

Laboratory Sink

Service Sink

Systems are complete with trim and rough-in (supply, waste and vent) to connect to supply branches and waste mains.

Single Compartment Sink

Double Compartment Sink

D2010 420	Laundry Sink Systems	COST EACH		
		MAT.	INST.	TOTAL
1740	Laundry sink w/trim, PE on CI, black iron frame			
1760	24" x 20", single compartment	750	670	1,420
1840	48" x 21" double compartment	1,225	725	1,950
1920	Molded stone, on wall, 22" x 21" single compartment	450	670	1,120
1960	45"x 21" double compartment	560	725	1,285
2040	Plastic, on wall or legs, 18" x 23" single compartment	405	655	1,060
2080	20" x 24" single compartment	440	655	1,095
2120	36" x 23" double compartment	500	710	1,210
2160	40" x 24" double compartment	585	710	1,295

D2010 430	Laboratory Sink Systems	COST EACH		
		MAT.	INST.	TOTAL
1580	Laboratory sink w/trim, polyethylene, single bowl,			
1600	Double drainboard, 54" x 24" O.D.	1,325	875	2,200
1640	Single drainboard, 47" x 24"O.D.	1,525	875	2,400
1680	70" x 24" O.D.	1,400	875	2,275
1760	Flanged, 14-1/2" x 14-1/2" O.D.	400	785	1,185
1800	18-1/2" x 18-1/2" O.D.	490	785	1,275
1840	23-1/2" x 20-1/2" O.D.	530	785	1,315
1920	Polypropylene, cup sink, oval, 7" x 4" O.D.	282	695	977
1960	10" x 4-1/2" O.D.	266	695	961
1961				

D2010   Plumbing Fixtures

D2010 440	Service Sink Systems	COST EACH		
		MAT.	INST.	TOTAL
4260	Service sink w/trim, PE on CI, corner floor, 28″ x 28″, w/rim guard	1,475	880	2,355
4300	Wall hung w/rim guard, 22″ x 18″	1,750	1,025	2,775
4340	24″ x 20″	1,800	1,025	2,825
4380	Vitreous china, wall hung 22″ x 20″	1,625	1,025	2,650
4383				

D2010 Plumbing Fixtures

Recessed Bathtub

Corner Bathtub

Circular Wash Fountain

Semi-Circular Wash Fountain

Systems are complete with trim and rough-in (supply, waste and vent) to connect to supply branches and waste mains.

D2010 510	Bathtub Systems	COST EACH		
		MAT.	INST.	TOTAL
2000	Bathtub, recessed, P.E. on Cl., 48" x 42"	2,350	770	3,120
2040	72" x 36"	2,425	860	3,285
2080	Mat bottom, 5' long	1,250	750	2,000
2120	5'-6" long	1,750	770	2,520
2160	Corner, 48" x 42"	2,350	750	3,100
2200	Formed steel, enameled, 4'-6" long	770	690	1,460

D2010 610	Group Wash Fountain Systems	COST EACH		
		MAT.	INST.	TOTAL
1740	Group wash fountain, precast terrazzo			
1760	Circular, 36" diameter	4,925	1,125	6,050
1800	54" diameter	6,025	1,225	7,250
1840	Semi-circular, 36" diameter	4,575	1,125	5,700
1880	54" diameter	5,425	1,225	6,650
1960	Stainless steel, circular, 36" diameter	4,825	1,050	5,875
2000	54" diameter	6,125	1,175	7,300
2040	Semi-circular, 36" diameter	4,275	1,050	5,325
2080	54" diameter	5,350	1,175	6,525
2160	Thermoplastic, circular, 36" diameter	3,725	850	4,575
2200	54" diameter	4,275	990	5,265
2240	Semi-circular, 36" diameter	3,475	850	4,325
2280	54" diameter	4,150	990	5,140

D2010 Plumbing Fixtures

Square Shower Stall

Corner Angle Shower Stall

Systems are complete with trim and rough-in (supply, waste and vent) to connect to supply branches and waste mains.

Wall Mounted, Low Back

Wall Mounted, No Back

D2010 710	Shower Systems	COST EACH		
		MAT.	INST.	TOTAL
1560	Shower, stall, baked enamel, molded stone receptor, 30" square	1,075	720	1,795
1600	32" square	680	725	1,405
1640	Terrazzo receptor, 32" square	1,100	725	1,825
1680	36" square	1,550	735	2,285
1720	36" corner angle	1,675	735	2,410
1800	Fiberglass one piece, three walls, 32" square	790	710	1,500
1840	36" square	855	710	1,565
1880	Polypropylene, molded stone receptor, 30" square	830	1,050	1,880
1920	32" square	845	1,050	1,895
1960	Built-in head, arm, bypass, stops and handles	254	271	525

D2010 810	Drinking Fountain Systems	COST EACH		
		MAT.	INST.	TOTAL
1740	Drinking fountain, one bubbler, wall mounted			
1760	Non recessed			
1800	Bronze, no back	1,225	410	1,635
1840	Cast iron, enameled, low back	830	410	1,240
1880	Fiberglass, 12" back	1,475	410	1,885
1920	Stainless steel, no back	1,525	410	1,935
1960	Semi-recessed, poly marble	1,050	410	1,460
2040	Stainless steel	1,175	410	1,585
2080	Vitreous china	860	410	1,270
2120	Full recessed, poly marble	1,175	410	1,585
2200	Stainless steel	1,150	410	1,560
2240	Floor mounted, pedestal type, aluminum	770	555	1,325
2320	Bronze	1,700	555	2,255
2360	Stainless steel	1,700	555	2,255

D20 Plumbing

D2010 Plumbing Fixtures

Wall Hung Water Cooler

Floor Mounted Water Cooler

Systems are complete with trim and rough-in (supply, waste and vent) to connect to supply branches and waste mains.

D2010 820	Water Cooler Systems	COST EACH		
		MAT.	INST.	TOTAL
1840	Water cooler, electric, wall hung, 8.2 GPH	900	525	1,425
1880	Dual height, 14.3 GPH	1,275	540	1,815
1920	Wheelchair type, 7.5 G.P.H.	2,000	525	2,525
1960	Semi recessed, 8.1 G.P.H.	1,250	525	1,775
2000	Full recessed, 8 G.P.H.	1,875	565	2,440
2040	Floor mounted, 14.3 G.P.H.	975	460	1,435
2080	Dual height, 14.3 G.P.H.	1,350	555	1,905
2120	Refrigerated compartment type, 1.5 G.P.H.	1,600	460	2,060

D20 Plumbing

D2010 Plumbing Fixtures

Two Fixture   Three Fixture

*Common wall is with adjacent bathrom

D2010 920	Two Fixture Bathroom, Two Wall Plumbing	COST EACH		
		MAT.	INST.	TOTAL
1180	Bathroom, lavatory & water closet, 2 wall plumbing, stand alone	1,500	1,725	3,225
1200	Share common plumbing wall*	1,350	1,475	2,825

D2010 922	Two Fixture Bathroom, One Wall Plumbing	COST EACH		
		MAT.	INST.	TOTAL
2220	Bathroom, lavatory & water closet, one wall plumbing, stand alone	1,375	1,550	2,925
2240	Share common plumbing wall*	1,225	1,300	2,525

D2010 924	Three Fixture Bathroom, One Wall Plumbing	COST EACH		
		MAT.	INST.	TOTAL
1150	Bathroom, three fixture, one wall plumbing			
1160	Lavatory, water closet & bathtub			
1170	Stand alone	2,425	2,000	4,425
1180	Share common plumbing wall *	2,075	1,450	3,525

D2010 926	Three Fixture Bathroom, Two Wall Plumbing	COST EACH		
		MAT.	INST.	TOTAL
2130	Bathroom, three fixture, two wall plumbing			
2140	Lavatory, water closet & bathtub			
2160	Stand alone	2,425	2,025	4,450
2180	Long plumbing wall common *	2,175	1,625	3,800
3610	Lavatory, bathtub & water closet			
3620	Stand alone	2,650	2,325	4,975
3640	Long plumbing wall common *	2,475	2,100	4,575
4660	Water closet, corner bathtub & lavatory			
4680	Stand alone	3,525	2,050	5,575
4700	Long plumbing wall common *	3,200	1,550	4,750
6100	Water closet, stall shower & lavatory			
6120	Stand alone	2,275	2,300	4,575
6140	Long plumbing wall common *	2,125	2,125	4,250
7060	Lavatory, corner stall shower & water closet			
7080	Stand alone	3,075	2,050	5,125
7100	Short plumbing wall common *	2,525	1,375	3,900

D20 Plumbing

D2010 Plumbing Fixtures

Four Fixture

Five Fixture

*Common wall is with adjacent bathroom

D2010 928	Four Fixture Bathroom, Two Wall Plumbing	COST EACH		
		MAT.	INST.	TOTAL
1140	Bathroom, four fixture, two wall plumbing			
1150	Bathtub, water closet, stall shower & lavatory			
1160	Stand alone	2,925	2,225	5,150
1180	Long plumbing wall common *	2,600	1,700	4,300
2260	Bathtub, lavatory, corner stall shower & water closet			
2280	Stand alone	3,925	2,225	6,150
2320	Long plumbing wall common *	3,600	1,725	5,325
3620	Bathtub, stall shower, lavatory & water closet			
3640	Stand alone	3,275	2,775	6,050
3660	Long plumbing wall (opp. door) common *	2,950	2,275	5,225
4680	Bathroom, four fixture, three wall plumbing			
4700	Bathtub, stall shower, lavatory & water closet			
4720	Stand alone	4,450	3,075	7,525
4761	Long plumbing wall (opposite door) common*	4,300	2,825	7,125

D2010 932	Five Fixture Bathroom, Two Wall Plumbing	COST EACH		
		MAT.	INST.	TOTAL
1320	Bathroom, five fixture, two wall plumbing			
1340	Bathtub, water closet, stall shower & two lavatories			
1360	Stand alone	4,000	3,550	7,550
1400	One short plumbing wall common *	3,675	3,050	6,725
2360	Bathroom, five fixture, three wall plumbing			
2380	Water closet, bathtub, two lavatories & stall shower			
2400	Stand alone	5,000	3,550	8,550
2440	One short plumbing wall common *	4,675	3,050	7,725
4080	Bathroom, five fixture, one wall plumbing			
4100	Bathtub, two lavatories, corner stall shower & water closet			
4120	Stand alone	4,775	3,175	7,950
4160	Share common wall *	4,000	2,050	6,050

D2020 Domestic Water Distribution

Installation includes piping and fittings
within 10' of heater. Gas heaters require
vent piping (not included with these units).

Gas Fired

Oil Fired

D2020 220	Gas Fired Water Heaters - Residential Systems	COST EACH		
		MAT.	INST.	TOTAL
2200	Gas fired water heater, residential, 100° F rise			
2260	30 gallon tank, 32 GPH	1,225	1,225	2,450
2300	40 gallon tank, 32 GPH	1,750	1,350	3,100
2340	50 gallon tank, 63 GPH	1,800	1,375	3,175
2380	75 gallon tank, 63 GPH	2,375	1,525	3,900
2420	100 gallon tank, 63 GPH	2,725	1,600	4,325
2422				

D2020 230	Oil Fired Water Heaters - Residential Systems	COST EACH		
		MAT.	INST.	TOTAL
2200	Oil fired water heater, residential, 100° F rise			
2220	30 gallon tank, 103 GPH	1,275	1,125	2,400
2260	50 gallon tank, 145 GPH	2,025	1,250	3,275
2300	70 gallon tank, 164 GPH	2,925	1,425	4,350
2340	85 gallon tank, 181 GPH	6,250	1,450	7,700

D2020 Domestic Water Distribution

Systems below include piping and fittings within 10' of heater. Electric water heaters do not require venting. Gas fired heaters require vent piping (not included in these prices).

Electric

Gas Fired

D2020 240	Electric Water Heaters - Commercial Systems	COST EACH		
		MAT.	INST.	TOTAL
1800	Electric water heater, commercial, 100°F rise			
1820	50 gallon tank, 9 KW 37 GPH	4,975	1,025	6,000
1860	80 gal, 12 KW 49 GPH	6,475	1,250	7,725
1900	36 KW 147 GPH	8,925	1,350	10,275
1940	120 gal, 36 KW 147 GPH	9,575	1,475	11,050
1980	150 gal, 120 KW 490 GPH	29,000	1,575	30,575
2020	200 gal, 120 KW 490 GPH	30,700	1,600	32,300
2060	250 gal, 150 KW 615 GPH	34,300	1,875	36,175
2100	300 gal, 180 KW 738 GPH	38,500	1,975	40,475
2140	350 gal, 30 KW 123 GPH	28,300	2,150	30,450
2180	180 KW 738 GPH	37,800	2,150	39,950
2220	500 gal, 30 KW 123 GPH	36,200	2,525	38,725
2260	240 KW 984 GPH	54,000	2,525	56,525
2300	700 gal, 30 KW 123 GPH	45,100	2,875	47,975
2340	300 KW 1230 GPH	63,000	2,875	65,875
2380	1000 gal, 60 KW 245 GPH	52,500	4,025	56,525
2420	480 KW 1970 GPH	85,000	4,025	89,025
2460	1500 gal, 60 KW 245 GPH	76,500	4,950	81,450
2500	480 KW 1970 GPH	121,000	4,950	125,950

D2020 250	Gas Fired Water Heaters - Commercial Systems	COST EACH		
		MAT.	INST.	TOTAL
1760	Gas fired water heater, commercial, 100°F rise			
1780	75.5 MBH input, 63 GPH	2,950	1,575	4,525
1820	95 MBH input, 86 GPH	5,425	1,575	7,000
1860	100 MBH input, 91 GPH	6,075	1,650	7,725
1900	115 MBH input, 110 GPH	6,300	1,700	8,000
1980	155 MBH input, 150 GPH	7,525	1,900	9,425
2020	175 MBH input, 168 GPH	8,050	2,025	10,075
2060	200 MBH input, 192 GPH	8,325	2,300	10,625
2100	240 MBH input, 230 GPH	8,775	2,500	11,275
2140	300 MBH input, 278 GPH	9,875	2,850	12,725
2180	390 MBH input, 374 GPH	11,500	2,875	14,375
2220	500 MBH input, 480 GPH	15,300	3,100	18,400
2260	600 MBH input, 576 GPH	21,700	3,350	25,050

D2020 Domestic Water Distribution

Oil Fired

Oil fired water heater systems include piping and fittings within 10' of heater. Oil fired heaters require vent piping (not included in these systems).

D2020 260	Oil Fired Water Heaters - Commercial Systems	COST EACH		
		MAT.	INST.	TOTAL
1800	Oil fired water heater, commercial, 100°F rise			
1820	140 gal., 140 MBH input, 134 GPH	16,200	1,325	17,525
1900	140 gal., 255 MBH input, 247 GPH	17,500	1,675	19,175
1940	140 gal., 270 MBH input, 259 GPH	21,400	1,900	23,300
1980	140 gal., 400 MBH input, 384 GPH	22,200	2,200	24,400
2060	140 gal., 720 MBH input, 691 GPH	23,600	2,300	25,900
2100	221 gal., 300 MBH input, 288 GPH	31,400	2,525	33,925
2140	221 gal., 600 MBH input, 576 GPH	34,800	2,550	37,350
2180	221 gal., 800 MBH input, 768 GPH	35,300	2,650	37,950
2220	201 gal., 1000 MBH input, 960 GPH	36,100	2,675	38,775
2260	201 gal., 1250 MBH input, 1200 GPH	36,800	2,750	39,550
2300	201 gal., 1500 MBH input, 1441 GPH	39,800	2,825	42,625
2340	411 gal., 600 MBH input, 576 GPH	40,200	2,875	43,075
2380	411 gal., 800 MBH input, 768 GPH	40,500	2,950	43,450
2420	411 gal., 1000 MBH input, 960 GPH	43,900	3,400	47,300
2460	411 gal., 1250 MBH input, 1200 GPH	44,800	3,500	48,300
2500	397 gal., 1500 MBH input, 1441 GPH	45,200	3,600	48,800
2540	397 gal., 1750 MBH input, 1681 GPH	48,600	3,725	52,325

D20 Plumbing

D2040 Rain Water Drainage

Design Assumptions: Vertical conductor size is based on a maximum rate of rainfall of 4" per hour. To convert roof area to other rates multiply "Max. S.F. Roof Area" shown by four and divide the result by desired local rate. The answer is the local roof area that may be handled by the indicated pipe diameter.

Basic cost is for roof drain, 10' of vertical leader and 10' of horizontal, plus connection to the main.

Pipe Dia.	Max. S.F. Roof Area	Gallons per Min.
2"	544	23
3"	1610	67
4"	3460	144
5"	6280	261
6"	10,200	424
8"	22,000	913

D2040 210	Roof Drain Systems	COST EACH		
		MAT.	INST.	TOTAL
1880	Roof drain, DWV PVC, 2" diam., piping, 10' high	212	585	797
1920	For each additional foot add	2.42	17.85	20.27
1960	3" diam., 10' high	277	685	962
2000	For each additional foot add	4.63	19.90	24.53
2040	4" diam., 10' high	335	770	1,105
2080	For each additional foot add	6.35	22	28.35
2120	5" diam., 10' high	985	890	1,875
2160	For each additional foot add	14.15	24.50	38.65
2200	6" diam., 10' high	970	985	1,955
2240	For each additional foot add	11.60	27	38.60
2280	8" diam., 10' high	2,775	1,675	4,450
2320	For each additional foot add	27	34	61
3940	C.I., soil, single hub, service wt., 2" diam. piping, 10' high	410	640	1,050
3980	For each additional foot add	7	16.70	23.70
4120	3" diam., 10' high	580	695	1,275
4160	For each additional foot add	9.80	17.55	27.35
4200	4" diam., 10' high	660	760	1,420
4240	For each additional foot add	12.65	19.15	31.80
4280	5" diam., 10' high	1,050	835	1,885
4320	For each additional foot add	17.40	21.50	38.90
4360	6" diam., 10' high	1,175	895	2,070
4400	For each additional foot add	21.50	22.50	44
4440	8" diam., 10' high	2,975	1,825	4,800
4480	For each additional foot add	33.50	38	71.50
6040	Steel galv. sch 40 threaded, 2" diam. piping, 10' high	550	620	1,170
6080	For each additional foot add	11.40	16.45	27.85
6120	3" diam., 10' high	1,075	895	1,970
6160	For each additional foot add	24	24.50	48.50
6200	4" diam., 10' high	1,525	1,150	2,675
6240	For each additional foot add	35.50	29.50	65
6280	5" diam., 10' high	1,650	955	2,605
6320	For each additional foot add	38	28.50	66.50
6360	6" diam, 10' high	1,875	1,225	3,100
6400	For each additional foot add	44.50	39	83.50
6440	8" diam., 10' high	3,675	1,800	5,475
6480	For each additional foot add	53	44.50	97.50

D2090 Other Plumbing Systems

D2090 810	Piping - Installed - Unit Costs	COST PER L.F.		
		MAT.	INST.	TOTAL
0840	Cast iron, soil, B & S, service weight, 2″ diameter	7	16.70	23.70
0860	3″ diameter	9.80	17.55	27.35
0880	4″ diameter	12.65	19.15	31.80
0900	5″ diameter	17.40	21.50	38.90
0920	6″ diameter	21.50	22.50	44
0940	8″ diameter	33.50	38	71.50
0960	10″ diameter	55.50	41.50	97
0980	12″ diameter	79	46.50	125.50
1040	No hub, 1-1/2″ diameter	7.55	14.85	22.40
1060	2″ diameter	7.85	15.75	23.60
1080	3″ diameter	10.80	16.45	27.25
1100	4″ diameter	13.95	18.15	32.10
1120	5″ diameter	19.85	19.75	39.60
1140	6″ diameter	24	20.50	44.50
1160	8″ diameter	44	32.50	76.50
1180	10″ diameter	73.50	36.50	110
1220	Copper tubing, hard temper, solder, type K, 1/2″ diameter	5.50	7.50	13
1260	3/4″ diameter	10.10	7.90	18
1280	1″ diameter	13.30	8.85	22.15
1300	1-1/4″ diameter	16.60	10.45	27.05
1320	1-1/2″ diameter	21.50	11.70	33.20
1340	2″ diameter	33.50	14.65	48.15
1360	2-1/2″ diameter	50	17.55	67.55
1380	3″ diameter	70	19.50	89.50
1400	4″ diameter	118	27.50	145.50
1480	5″ diameter	262	33	295
1500	6″ diameter	380	43	423
1520	8″ diameter	675	48	723
1560	Type L, 1/2″ diameter	4.21	7.25	11.46
1600	3/4″ diameter	6.60	7.70	14.30
1620	1″ diameter	9.95	8.60	18.55
1640	1-1/4″ diameter	13.75	10.10	23.85
1660	1-1/2″ diameter	17.70	11.25	28.95
1680	2″ diameter	27.50	13.95	41.45
1700	2-1/2″ diameter	42	17	59
1720	3″ diameter	57	18.80	75.80
1740	4″ diameter	96	27	123
1760	5″ diameter	204	31	235
1780	6″ diameter	282	41	323
1800	8″ diameter	495	45.50	540.50
1840	Type M, 1/2″ diameter	3.07	6.95	10.02
1880	3/4″ diameter	4.85	7.50	12.35
1900	1″ diameter	7.60	8.35	15.95
1920	1-1/4″ diameter	11.25	9.75	21
1940	1-1/2″ diameter	15.40	10.85	26.25
1960	2″ diameter	24	13.30	37.30
1980	2-1/2″ diameter	36.50	16.45	52.95
2000	3″ diameter	48.50	18.15	66.65
2020	4″ diameter	87	26.50	113.50
2060	6″ diameter	282	39	321
2080	8″ diameter	495	43	538
2120	Type DWV, 1-1/4″ diameter	12	9.75	21.75
2160	1-1/2″ diameter	15.10	10.85	25.95
2180	2″ diameter	20.50	13.30	33.80
2200	3″ diameter	36	18.15	54.15
2220	4″ diameter	63.50	26.50	90
2240	5″ diameter	181	29.50	210.50
2260	6″ diameter	259	39	298

D2090 Other Plumbing Systems

D2090 810	Piping - Installed - Unit Costs	COST PER L.F.		
		MAT.	INST.	TOTAL
2280	8" diameter	630	43	673
2800	Plastic,PVC, DWV, schedule 40, 1-1/4" diameter	2.03	13.95	15.98
2820	1-1/2" diameter	2	16.25	18.25
2830	2" diameter	2.42	17.85	20.27
2840	3" diameter	4.63	19.90	24.53
2850	4" diameter	6.35	22	28.35
2890	6" diameter	11.60	27	38.60
3010	Pressure pipe 200 PSI, 1/2" diameter	1.33	10.85	12.18
3030	3/4" diameter	1.63	11.45	13.08
3040	1" diameter	1.80	12.70	14.50
3050	1-1/4" diameter	2.36	13.95	16.31
3060	1-1/2" diameter	2.62	16.25	18.87
3070	2" diameter	3.19	17.85	21.04
3080	2-1/2" diameter	5.65	18.80	24.45
3090	3" diameter	6.60	19.90	26.50
3100	4" diameter	10.90	22	32.90
3110	6" diameter	19.60	27	46.60
3120	8" diameter	30.50	34	64.50
4000	Steel, schedule 40, black, threaded, 1/2" diameter	2.46	9.30	11.76
4020	3/4" diameter	2.90	9.60	12.50
4030	1" diameter	4.26	11.05	15.31
4040	1-1/4" diameter	5.40	11.85	17.25
4050	1-1/2" diameter	6.35	13.15	19.50
4060	2" diameter	8.45	16.45	24.90
4070	2-1/2" diameter	13.25	21	34.25
4080	3" diameter	17.05	24.50	41.55
4090	4" diameter	25	29.50	54.50
4100	Grooved, 5" diameter	29	28.50	57.50
4110	6" diameter	37	39	76
4120	8" diameter	57	44.50	102
4130	10" diameter	85.50	53	139
4140	12" diameter	91	60.50	152
4200	Galvanized, threaded, 1/2" diameter	3.42	9.30	12.70
4220	3/4" diameter	4.14	9.60	13.75
4230	1" diameter	5.65	11.05	16.70
4240	1-1/4" diameter	7.20	11.85	19.05
4250	1-1/2" diameter	8.55	13.15	21.70
4260	2" diameter	11.40	16.45	27.85
4270	2-1/2" diameter	19.35	21	40.35
4280	3" diameter	24	24.50	48.50
4290	4" diameter	35.50	29.50	65
4300	Grooved, 5" diameter	38	28.50	66.50
4310	6" diameter	44.50	39	83.50
4320	8" diameter	53	44.50	97.50
4330	10" diameter	79	53	132
4340	12" diameter	102	60.50	163
5010	Flanged, black, 1" diameter	9.30	15.95	25.25
5020	1-1/4" diameter	10.70	17.40	28.10
5030	1-1/2" diameter	11.45	19.20	30.65
5040	2" diameter	13.90	25	38.90
5050	2-1/2" diameter	16.40	31	47.40
5060	3" diameter	23	35	58
5070	4" diameter	28.50	43	71.50
5080	5" diameter	39.50	53	92.50
5090	6" diameter	48.50	68	116.50
5100	8" diameter	73.50	89.50	163
5110	10" diameter	120	106	226
5120	12" diameter	142	121	263

D2090 Other Plumbing Systems

D2090 810	Piping - Installed - Unit Costs	COST PER L.F.		
		MAT.	INST.	TOTAL
5720	Grooved joints, black, 3/4" diameter	3.89	8.25	12.14
5730	1" diameter	4.65	9.30	13.95
5740	1-1/4" diameter	6	10.10	16.10
5750	1-1/2" diameter	6.95	11.45	18.40
5760	2" diameter	8.85	14.65	23.50
5770	2-1/2" diameter	11.10	18.50	29.60
5900	3" diameter	17.70	21	38.70
5910	4" diameter	21	23.50	44.50
5920	5" diameter	29	28.50	57.50
5930	6" diameter	37	39	76
5940	8" diameter	57	44.50	101.50
5950	10" diameter	85.50	53	138.50
5960	12" diameter	91	60.50	151.50

D3010   Energy Supply

Forced Hot Water Heating System

Fin Tube Radiation

Terminal Unit Heater

D3010 510	Apartment Building Heating - Fin Tube Radiation	COST PER S.F.		
		MAT.	INST.	TOTAL
1740	Heating systems, fin tube radiation, forced hot water			
1760	1,000 S.F. area, 10,000 C.F. volume	6.08	5.65	11.73
1800	10,000 S.F. area, 100,000 C.F. volume	2.84	3.42	6.26
1840	20,000 S.F. area, 200,000 C.F. volume	3.31	3.83	7.14
1880	30,000 S.F. area, 300,000 C.F. volume	3.19	3.72	6.91

D3010 520	Commercial Building Heating - Fin Tube Radiation	COST PER S.F.		
		MAT.	INST.	TOTAL
1940	Heating systems, fin tube radiation, forced hot water			
1960	1,000 S.F. bldg, one floor	13.80	12.80	26.60
2000	10,000 S.F., 100,000 C.F., total two floors	3.87	4.86	8.73
2040	100,000 S.F., 1,000,000 C.F., total three floors	1.61	2.16	3.77
2080	1,000,000 S.F., 10,000,000 C.F., total five floors	.82	1.15	1.97

D3010 530	Commercial Bldg. Heating - Terminal Unit Heaters	COST PER S.F.		
		MAT.	INST.	TOTAL
1860	Heating systems, terminal unit heaters, forced hot water			
1880	1,000 S.F. bldg., one floor	13.30	11.70	25
1920	10,000 S.F. bldg., 100,000 C.F. total two floors	3.44	4.02	7.46
1960	100,000 S.F. bldg., 1,000,000 C.F. total three floors	1.60	1.98	3.58
2000	1,000,000 S.F. bldg., 10,000,000 C.F. total five floors	1.03	1.28	2.31

D3020 Heat Generating Systems

Small Electric Boiler Systems Considerations:

1. Terminal units are fin tube baseboard radiation rated at 720 BTU/hr with 200° water temperature or 820 BTU/hr steam.
2. Primary use being for residential or smaller supplementary areas, the floor levels are based on 7-1/2″ ceiling heights.
3. All distribution piping is copper for boilers through 205 MBH. All piping for larger systems is steel pipe.

Large Electric Boiler System Considerations:

1. Terminal units are all unit heaters of the same size. Quantities are varied to accommodate total requirements.
2. All air is circulated through the heaters a minimum of three times per hour.
3. As the capacities are adequate for commercial use, gross output rating by floor levels are based on 10′ ceiling height.
4. All distribution piping is black steel pipe.

D3020 102	Small Heating Systems, Hydronic, Electric Boilers	COST PER S.F.		
		MAT.	INST.	TOTAL
1100	Small heating systems, hydronic, electric boilers			
1120	Steam, 1 floor, 1480 S.F., 61 M.B.H.	6.76	7.30	14.06
1160	3,000 S.F., 123 M.B.H.	5.25	6.40	11.65
1200	5,000 S.F., 205 M.B.H.	4.60	5.90	10.50
1240	2 floors, 12,400 S.F., 512 M.B.H.	3.38	5.85	9.23
1280	3 floors, 24,800 S.F., 1023 M.B.H.	3.53	5.80	9.33
1320	34,750 S.F., 1,433 M.B.H.	3.25	5.65	8.90
1360	Hot water, 1 floor, 1,000 S.F., 41 M.B.H.	9.75	4.07	13.82
1400	2,500 S.F., 103 M.B.H.	7.70	7.30	15
1440	2 floors, 4,850 S.F., 205 M.B.H.	8.10	8.75	16.85
1480	3 floors, 9,700 S.F., 410 M.B.H.	9.15	9.05	18.20

D3020 104	Large Heating Systems, Hydronic, Electric Boilers	COST PER S.F.		
		MAT.	INST.	TOTAL
1230	Large heating systems, hydronic, electric boilers			
1240	9,280 S.F., 135 K.W., 461 M.B.H., 1 floor	3.28	2.73	6.01
1280	14,900 S.F., 240 K.W., 820 M.B.H., 2 floors	4.11	4.43	8.54
1320	18,600 S.F., 296 K.W., 1,010 M.B.H., 3 floors	4.07	4.81	8.88
1360	26,100 S.F., 420 K.W., 1,432 M.B.H., 4 floors	4.02	4.74	8.76
1400	39,100 S.F., 666 K.W., 2,273 M.B.H., 4 floors	3.45	3.95	7.40
1440	57,700 S.F., 900 K.W., 3,071 M.B.H., 5 floors	3.58	3.87	7.45
1480	111,700 S.F., 1,800 K.W., 6,148 M.B.H., 6 floors	3.23	3.33	6.56
1520	149,000 S.F., 2,400 K.W., 8,191 M.B.H., 8 floors	3.22	3.33	6.55
1560	223,300 S.F., 3,600 K.W., 12,283 M.B.H., 14 floors	3.46	3.78	7.24

D3020 Heat Generating Systems

Boiler Selection: The maximum allowable working pressures are limited by ASME "Code for Heating Boilers" to 15 PSI for steam and 160 PSI for hot water heating boilers, with a maximum temperature limitation of 250° F. Hot water boilers are generally rated for a working pressure of 30 PSI. High pressure boilers are governed by the ASME "Code for Power Boilers" which is used almost universally for boilers operating over 15 PSIG. High pressure boilers used for a combination of heating/process loads are usually designed for 150 PSIG.

Boiler ratings are usually indicated as either Gross or Net Output. The Gross Load is equal to the Net Load plus a piping and pickup allowance. When this allowance cannot be determined, divide the gross output rating by 1.25 for a value equal to or greater than the next heat loss requirement of the building.

Table below lists installed cost per boiler and includes insulating jacket, standard controls, burner and safety controls. Costs do not include piping or boiler base pad. Outputs are Gross.

D3020 106	Boilers, Hot Water & Steam	COST EACH		
		MAT.	INST.	TOTAL
0600	Boiler, electric, steel, hot water, 12 K.W., 41 M.B.H.	3,950	1,250	5,200
0620	30 K.W., 103 M.B.H.	4,700	1,350	6,050
0640	60 K.W., 205 M.B.H.	5,750	1,475	7,225
0660	120 K.W., 410 M.B.H.	6,450	1,800	8,250
0680	210 K.W., 716 M.B.H.	7,550	2,700	10,250
0700	510 K.W., 1,739 M.B.H.	18,900	5,050	23,950
0720	720 K.W., 2,452 M.B.H.	22,800	5,700	28,500
0740	1,200 K.W., 4,095 M.B.H.	29,600	6,525	36,125
0760	2,100 K.W., 7,167 M.B.H.	56,500	8,225	64,725
0780	3,600 K.W., 12,283 M.B.H.	87,000	13,900	100,900
0820	Steam, 6 K.W., 20.5 M.B.H.	3,725	1,350	5,075
0840	24 K.W., 81.8 M.B.H.	4,625	1,475	6,100
0860	60 K.W., 205 M.B.H.	6,375	1,625	8,000
0880	150 K.W., 512 M.B.H.	9,225	2,500	11,725
0900	510 K.W., 1,740 M.B.H.	24,300	6,175	30,475
0920	1,080 K.W., 3,685 M.B.H.	34,100	8,875	42,975
0940	2,340 K.W., 7,984 M.B.H.	69,500	13,900	83,400
0980	Gas, cast iron, hot water, 80 M.B.H.	1,875	1,550	3,425
1000	100 M.B.H.	2,150	1,675	3,825
1020	163 M.B.H.	2,800	2,250	5,050
1040	280 M.B.H.	4,075	2,500	6,575
1060	544 M.B.H.	8,650	4,450	13,100
1080	1,088 M.B.H.	13,100	5,650	18,750
1100	2,000 M.B.H.	23,100	8,825	31,925
1120	2,856 M.B.H.	25,500	11,300	36,800
1140	4,720 M.B.H.	68,500	15,600	84,100
1160	6,970 M.B.H.	98,500	25,400	123,900
1180	For steam systems under 2,856 M.B.H., add 8%			
1520	Oil, cast iron, hot water, 109 M.B.H.	2,075	1,875	3,950
1540	173 M.B.H.	2,600	2,250	4,850
1560	236 M.B.H.	3,375	2,650	6,025
1580	1,084 M.B.H.	9,600	6,000	15,600
1600	1,600 M.B.H.	12,600	8,625	21,225
1620	2,480 M.B.H.	19,500	11,000	30,500
1640	3,550 M.B.H.	25,000	13,200	38,200
1660	Steam systems same price as hot water			

D3020 Heat Generating Systems

Unit Heater

Fossil Fuel Boiler System Considerations:

1. Terminal units are horizontal unit heaters. Quantities are varied to accommodate total heat loss per building.
2. Unit heater selection was determined by their capacity to circulate the building volume a minimum of three times per hour in addition to the BTU output.
3. Systems shown are forced hot water. Steam boilers cost slightly more than hot water boilers. However, this is compensated for by the smaller size or fewer terminal units required with steam.
4. Floor levels are based on 10′ story heights.
5. MBH requirements are gross boiler output.

D3020 108	Heating Systems, Unit Heaters	COST PER S.F.		
		MAT.	INST.	TOTAL
1260	Heating systems, hydronic, fossil fuel, terminal unit heaters,			
1280	Cast iron boiler, gas, 80 M.B.H., 1,070 S.F. bldg.	8.74	8.69	17.43
1320	163 M.B.H., 2,140 S.F. bldg.	5.80	5.90	11.70
1360	544 M.B.H., 7,250 S.F. bldg.	4.23	4.18	8.41
1400	1,088 M.B.H., 14,500 S.F. bldg.	3.59	3.96	7.55
1440	3,264 M.B.H., 43,500 S.F. bldg.	3	3.01	6.01
1480	5,032 M.B.H., 67,100 S.F. bldg.	3.45	3.06	6.51
1520	Oil, 109 M.B.H., 1,420 S.F. bldg.	9	7.75	16.75
1560	235 M.B.H., 3,150 S.F. bldg.	6	5.50	11.50
1600	940 M.B.H., 12,500 S.F. bldg.	4.36	3.62	7.98
1640	1,600 M.B.H., 21,300 S.F. bldg.	3.97	3.48	7.45
1680	2,480 M.B.H., 33,100 S.F. bldg.	3.95	3.13	7.08
1720	3,350 M.B.H., 44,500 S.F. bldg.	3.63	3.20	6.83
1800	300 M.B.H., 4,000 S.F. bldg.	6.10	3.94	10.04
1840	2,360 M.B.H., 31,500 S.F. bldg.	4.45	3.19	7.64

D3020 Heat Generating Systems

Fin Tube Radiator

Fossil Fuel Boiler System Considerations:

1. Terminal units are commercial steel fin tube radiation. Quantities are varied to accommodate total heat loss per building.
2. Systems shown are forced hot water. Steam boilers cost slightly more than hot water boilers. However, this is compensated for by the smaller size or fewer terminal units required with steam.
3. Floor levels are based on 10′ story heights.
4. MBH requirements are gross boiler output.

D3020 110	Heating System, Fin Tube Radiation	COST PER S.F.		
		MAT.	INST.	TOTAL
3230	Heating systems, hydronic, fossil fuel, fin tube radiation			
3240	Cast iron boiler, gas, 80 MBH, 1,070 S.F. bldg.	11.03	13.08	24.11
3280	169 M.B.H., 2,140 S.F. bldg.	6.90	8.25	15.15
3320	544 M.B.H., 7,250 S.F. bldg.	5.85	7.10	12.95
3360	1,088 M.B.H., 14,500 S.F. bldg.	5.30	6.95	12.25
3400	3,264 M.B.H., 43,500 S.F. bldg.	4.82	6.15	10.97
3440	5,032 M.B.H., 67,100 S.F. bldg.	5.30	6.20	11.50
3480	Oil, 109 M.B.H., 1,420 S.F. bldg.	12.80	14.25	27.05
3520	235 M.B.H., 3,150 S.F. bldg.	7.60	8.35	15.95
3560	940 M.B.H., 12,500 S.F. bldg.	6.10	6.65	12.75
3600	1,600 M.B.H., 21,300 S.F. bldg.	5.80	6.60	12.40
3640	2,480 M.B.H., 33,100 S.F. bldg.	5.80	6.30	12.10
3680	3,350 M.B.H., 44,500 S.F. bldg.	5.45	6.35	11.80
3720	Coal, 148 M.B.H., 1,975 S.F. bldg.	9.65	7.90	17.55
3760	300 M.B.H., 4,000 S.F. bldg.	7.65	6.75	14.40
3800	2,360 M.B.H., 31,500 S.F. bldg.	6.25	6.25	12.50
4080	Steel boiler, oil, 97 M.B.H., 1,300 S.F. bldg.	11.55	11.75	23.30
4120	315 M.B.H., 4,550 S.F. bldg.	5.20	5.95	11.15
4160	525 M.B.H., 7,000 S.F. bldg.	7.15	6.85	14
4200	1,050 M.B.H., 14,000 S.F. bldg.	6.25	6.85	13.10
4240	2,310 M.B.H., 30,800 S.F. bldg.	5.75	6.30	12.05
4280	3,150 M.B.H., 42,000 S.F. bldg.	5.75	6.40	12.15

D3030 Cooling Generating Systems

Chilled Water Supply & Return Piping

Air Cooled Water Chiller Unit

Roof

Insulate

Return

Fan Coil Unit

Supply

Finish Ceiling

*Cooling requirements would lead to a choice of multiple chillers.

D3030 110	Chilled Water, Air Cooled Condenser Systems	COST PER S.F.		
		MAT.	INST.	TOTAL
1180	Packaged chiller, air cooled, with fan coil unit			
1200	Apartment corridors, 3,000 S.F., 5.50 ton	7.07	7.23	14.30
1240	6,000 S.F., 11.00 ton	5.85	5.80	11.65
1280	10,000 S.F., 18.33 ton	5	4.40	9.40
1320	20,000 S.F., 36.66 ton	3.73	3.18	6.91
1360	40,000 S.F., 73.33 ton	4.16	3.20	7.36
1440	Banks and libraries, 3,000 S.F., 12.50 ton	10.60	8.50	19.10
1480	6,000 S.F., 25.00 ton	9.60	6.95	16.55
1520	10,000 S.F., 41.66 ton	7.15	4.83	11.98
1560	20,000 S.F., 83.33 ton	7.30	4.39	11.69
1600	40,000 S.F., 167 ton*			
1680	Bars and taverns, 3,000 S.F., 33.25 ton	17.30	10.10	27.40
1720	6,000 S.F., 66.50 ton	15.95	7.80	23.75
1760	10,000 S.F., 110.83 ton	12.75	2.50	15.25
1800	20,000 S.F., 220 ton*			
1840	40,000 S.F., 440 ton*			
1920	Bowling alleys, 3,000 S.F., 17.00 ton	13.15	9.55	22.70
1960	6,000 S.F., 34.00 ton	9.70	6.75	16.45
2000	10,000 S.F., 56.66 ton	8.60	4.88	13.48
2040	20,000 S.F., 113.33 ton	8.40	4.44	12.84
2080	40,000 S.F., 227 ton*			
2160	Department stores, 3,000 S.F., 8.75 ton	9.55	8.05	17.60
2200	6,000 S.F., 17.50 ton	7.50	6.25	13.75
2240	10,000 S.F., 29.17 ton	5.75	4.50	10.25
2280	20,000 S.F., 58.33 ton	4.95	3.33	8.28
2320	40,000 S.F., 116.66 ton	5.45	3.31	8.76
2400	Drug stores, 3,000 S.F., 20.00 ton	14.85	9.90	24.75
2440	6,000 S.F., 40.00 ton	11.35	7.35	18.70
2480	10,000 S.F., 66.66 ton	11.25	6.30	17.55
2520	20,000 S.F., 133.33 ton	10.40	4.68	15.08
2560	40,000 S.F., 267 ton*			
2640	Factories, 2,000 S.F., 10.00 ton	9.30	8.15	17.45
2680	6,000 S.F., 20.00 ton	8.35	6.60	14.95
2720	10,000 S.F., 33.33 ton	6.20	4.61	10.81
2760	20,000 S.F., 66.66 ton	6.35	4.17	10.52
2800	40,000 S.F., 133.33 ton	5.95	3.40	9.35
2880	Food supermarkets, 3,000 S.F., 8.50 ton	9.40	8	17.40
2920	6,000 S.F., 17.00 ton	7.35	6.20	13.55
2960	10,000 S.F., 28.33 ton	5.50	4.38	9.88
3000	20,000 S.F., 56.66 ton	4.78	3.26	8.04
3040	40,000 S.F., 113.33 ton	5.20	3.32	8.52
3120	Medical centers, 3,000 S.F., 7.00 ton	8.50	7.80	16.30

D30 HVAC

D3030 Cooling Generating Systems

D3030 110	Chilled Water, Air Cooled Condenser Systems	COST PER S.F.		
		MAT.	INST.	TOTAL
3160	6,000 S.F., 14.00 ton	6.60	6	12.60
3200	10,000 S.F., 23.33 ton	5.80	4.58	10.38
3240	20,000 S.F., 46.66 ton	4.39	3.36	7.75
3280	40,000 S.F., 93.33 ton	4.68	3.25	7.93
3360	Offices, 3,000 S.F., 9.50 ton	8.65	7.95	16.60
3400	6,000 S.F., 19.00 ton	8.10	6.50	14.60
3440	10,000 S.F., 31.66 ton	6	4.56	10.56
3480	20,000 S.F., 63.33 ton	6.30	4.24	10.54
3520	40,000 S.F., 126.66 ton	5.85	3.43	9.28
3600	Restaurants, 3,000 S.F., 15.00 ton	11.85	8.85	20.70
3640	6,000 S.F., 30.00 ton	9.20	6.75	15.95
3680	10,000 S.F., 50.00 ton	8.20	4.96	13.16
3720	20,000 S.F., 100.00 ton	8.45	4.65	13.10
3760	40,000 S.F., 200 ton*			
3840	Schools and colleges, 3,000 S.F., 11.50 ton	10.10	8.35	18.45
3880	6,000 S.F., 23.00 ton	9.10	6.80	15.90
3920	10,000 S.F., 38.33 ton	6.75	4.74	11.49
3960	20,000 S.F., 76.66 ton	7.10	4.43	11.53
4000	40,000 S.F., 153 ton*			

D3030 Cooling Generating Systems

*Cooling requirements would lead to a
choice of multiple chillers

D3030 115	Chilled Water, Cooling Tower Systems	COST PER S.F.		
		MAT.	INST.	TOTAL
1300	Packaged chiller, water cooled, with fan coil unit			
1320	Apartment corridors, 4,000 S.F., 7.33 ton	6.43	7.08	13.51
1360	6,000 S.F., 11.00 ton	5.40	6.10	11.50
1400	10,000 S.F., 18.33 ton	4.88	4.51	9.39
1440	20,000 S.F., 26.66 ton	3.86	3.37	7.23
1480	40,000 S.F., 73.33 ton	3.85	3.44	7.29
1520	60,000 S.F., 110.00 ton	4.06	3.54	7.60
1600	Banks and libraries, 4,000 S.F., 16.66 ton	10.05	7.75	17.80
1640	6,000 S.F., 25.00 ton	8.65	6.85	15.50
1680	10,000 S.F., 41.66 ton	7	5.10	12.10
1720	20,000 S.F., 83.33 ton	6.50	4.89	11.39
1760	40,000 S.F., 166.66 ton	7.20	5.90	13.10
1800	60,000 S.F., 250.00 ton	7.30	6.30	13.60
1880	Bars and taverns, 4,000 S.F., 44.33 ton	16.25	9.80	26.05
1920	6,000 S.F., 66.50 ton	18.60	10.15	28.75
1960	10,000 S.F., 110.83 ton	14.75	8.05	22.80
2000	20,000 S.F., 221.66 ton	17.80	8.15	25.95
2040	40,000 S.F., 440 ton*			
2080	60,000 S.F., 660 ton*			
2160	Bowling alleys, 4,000 S.F., 22.66 ton	11.75	8.55	20.30
2200	6,000 S.F., 34.00 ton	10.30	7.55	17.85
2240	10,000 S.F., 56.66 ton	9.55	5.50	15.05
2280	20,000 S.F., 113.33 ton	8.25	5.15	13.40
2320	40,000 S.F., 226.66 ton	10.05	5.90	15.95
2360	60,000 S.F., 340 ton			
2440	Department stores, 4,000 S.F., 11.66 ton	7.25	7.65	14.90
2480	6,000 S.F., 17.50 ton	7.55	6.30	13.85
2520	10,000 S.F., 29.17 ton	5.75	4.72	10.47
2560	20,000 S.F., 58.33 ton	5.10	3.51	8.61
2600	40,000 S.F., 116.66 ton	4.94	3.63	8.57
2640	60,000 S.F., 175.00 ton	6.55	5.75	12.30
2720	Drug stores, 4,000 S.F., 26.66 ton	12.55	8.80	21.35
2760	6,000 S.F., 40.00 ton	11	7.55	18.55
2800	10,000 S.F., 66.66 ton	11.80	6.85	18.65
2840	20,000 S.F., 133.33 ton	9.50	5.40	14.90
2880	40,000 S.F., 266.67 ton	9.90	6.70	16.60
2920	60,000 S.F., 400 ton*			
3000	Factories, 4,000 S.F., 13.33 ton	8.80	7.40	16.20
3040	6,000 S.F., 20.00 ton	7.70	6.40	14.10
3080	10,000 S.F., 33.33 ton	6.20	4.87	11.07
3120	20,000 S.F., 66.66 ton	6.55	4.38	10.93
3160	40,000 S.F., 133.33 ton	5.55	3.75	9.30

D3030 Cooling Generating Systems

D3030 115	Chilled Water, Cooling Tower Systems	COST PER S.F.		
		MAT.	INST.	TOTAL
3200	60,000 S.F., 200.00 ton	6.60	6	12.60
3280	Food supermarkets, 4,000 S.F., 11.33 ton	7.10	7.60	14.70
3320	6,000 S.F., 17.00 ton	6.90	6.20	13.10
3360	10,000 S.F., 28.33 ton	5.65	4.69	10.34
3400	20,000 S.F., 56.66 ton	5.15	3.52	8.67
3440	40,000 S.F., 113.33 ton	4.89	3.64	8.53
3480	60,000 S.F., 170.00 ton	6.50	5.70	12.20
3560	Medical centers, 4.000 S.F., 9.33 ton	6.25	7	13.25
3600	6,000 S.F., 14.00 ton	6.65	6.05	12.70
3640	10,000 S.F., 23.33 ton	5.25	4.53	9.78
3680	20,000 S.F., 46.66 ton	4.63	3.43	8.06
3720	40,000 S.F., 93.33 ton	4.41	3.56	7.97
3760	60,000 S.F., 140.00 ton	5.40	5.80	11.20
3840	Offices, 4,000 S.F., 12.66 ton	8.55	7.30	15.85
3880	6,000 S.F., 19.00 ton	7.65	6.55	14.20
3920	10,000 S.F., 31.66 ton	6.20	4.92	11.12
3960	20,000 S.F., 63.33 ton	6.40	4.40	10.80
4000	40,000 S.F., 126.66 ton	5.90	5.65	11.55
4040	60,000 S.F., 190.00 ton	6.40	5.95	12.35
4120	Restaurants, 4,000 S.F., 20.00 ton	10.60	7.95	18.55
4160	6,000 S.F., 30.00 ton	9.35	7.05	16.40
4200	10,000 S.F., 50.00 ton	8.80	5.30	14.10
4240	20,000 S.F., 100.00 ton	7.70	5.15	12.85
4280	40,000 S.F., 200.00 ton	7.75	5.70	13.45
4320	60,000 S.F., 300.00 ton	8.20	6.50	14.70
4400	Schools and colleges, 4,000 S.F., 15.33 ton	9.55	7.60	17.15
4440	6,000 S.F., 23.00 ton	8.20	6.70	14.90
4480	10,000 S.F., 38.33 ton	6.65	4.98	11.63
4520	20,000 S.F., 76.66 ton	6.30	4.82	11.12
4560	40,000 S.F., 153.33 ton	6.80	5.75	12.55
4600	60,000 S.F., 230.00 ton	6.80	6.05	12.85

D3050 Terminal & Package Units

Roof

Roof Top Unit

Return Duct

Fan Coil Unit

Return Grille

Supply Diff.

Insulated Supply Duct

Finish Ceiling

*Above normal capacity

Rooftop, Single Zone System

D3050 150	Rooftop Single Zone Unit Systems	COST PER S.F.		
		MAT.	INST.	TOTAL
1260	Rooftop, single zone, air conditioner			
1280	Apartment corridors, 500 S.F., .92 ton	3.80	3.45	7.25
1320	1,000 S.F., 1.83 ton	3.81	3.41	7.22
1360	1500 S.F., 2.75 ton	2.47	2.82	5.29
1400	3,000 S.F., 5.50 ton	2.26	2.72	4.98
1440	5,000 S.F., 9.17 ton	2.25	2.47	4.72
1480	10,000 S.F., 18.33 ton	2.61	2.33	4.94
1560	Banks or libraries, 500 S.F., 2.08 ton	8.65	7.75	16.40
1600	1,000 S.F., 4.17 ton	5.60	6.40	12
1640	1,500 S.F., 6.25 ton	5.10	6.15	11.25
1680	3,000 S.F., 12.50 ton	5.10	5.60	10.70
1720	5,000 S.F., 20.80 ton	5.90	5.30	11.20
1760	10,000 S.F., 41.67 ton	5.10	5.25	10.35
1840	Bars and taverns, 500 S.F. 5.54 ton	11.90	10.30	22.20
1880	1,000 S.F., 11.08 ton	11.85	8.85	20.70
1920	1,500 S.F., 16.62 ton	11.95	8.25	20.20
1960	3,000 S.F., 33.25 ton	12.20	7.90	20.10
2000	5,000 S.F., 55.42 ton	11.30	7.90	19.20
2040	10,000 S.F., 110.83 ton*			
2080	Bowling alleys, 500 S.F., 2.83 ton	7.60	8.70	16.30
2120	1,000 S.F., 5.67 ton	6.95	8.40	15.35
2160	1,500 S.F., 8.50 ton	6.95	7.65	14.60
2200	3,000 S.F., 17.00 ton	7	7.35	14.35
2240	5,000 S.F., 28.33 ton	7.10	7.15	14.25
2280	10,000 S.F., 56.67 ton	6.70	7.15	13.85
2360	Department stores, 500 S.F., 1.46 ton	6.05	5.45	11.50
2400	1,000 S.F., 2.92 ton	3.93	4.49	8.42
2480	3,000 S.F., 8.75 ton	3.58	3.93	7.51
2520	5,000 S.F., 14.58 ton	3.60	3.77	7.37
2560	10,000 S.F., 29.17 ton	3.67	3.68	7.35
2640	Drug stores, 500 S.F., 3.33 ton	8.95	10.25	19.20
2680	1,000 S.F., 6.67 ton	8.20	9.85	18.05
2720	1,500 S.F., 10.00 ton	8.20	9	17.20
2760	3,000 S.F., 20.00 ton	9.45	8.45	17.90
2800	5,000 S.F., 33.33 ton	8.40	8.40	16.80
2840	10,000 S.F., 66.67 ton	7.85	8.40	16.25
2920	Factories, 500 S.F., 1.67 ton	6.90	6.20	13.10
3000	1,500 S.F., 5.00 ton	4.10	4.94	9.04
3040	3,000 S.F., 10.00 ton	4.09	4.50	8.59
3080	5,000 S.F., 16.67 ton	4.12	4.31	8.43
3120	10,000 S.F., 33.33 ton	4.19	4.21	8.40
3200	Food supermarkets, 500 S.F., 1.42 ton	5.90	5.30	11.20

D30 HVAC

D3050 Terminal & Package Units

D3050 150	Rooftop Single Zone Unit Systems	COST PER S.F.		
		MAT.	INST.	TOTAL
3240	1,000 S.F., 2.83 ton	3.78	4.34	8.12
3280	1,500 S.F., 4.25 ton	3.48	4.20	7.68
3320	3,000 S.F., 8.50 ton	3.48	3.82	7.30
3360	5,000 S.F., 14.17 ton	3.50	3.67	7.17
3400	10,000 S.F., 28.33 ton	3.56	3.58	7.14
3480	Medical centers, 500 S.F., 1.17 ton	4.85	4.35	9.20
3520	1,000 S.F., 2.33 ton	4.84	4.34	9.18
3560	1,500 S.F., 3.50 ton	3.15	3.59	6.74
3640	5,000 S.F., 11.67 ton	2.87	3.15	6.02
3680	10,000 S.F., 23.33 ton	3.31	2.96	6.27
3760	Offices, 500 S.F., 1.58 ton	6.60	5.90	12.50
3800	1,000 S.F., 3.17 ton	4.27	4.88	9.15
3840	1,500 S.F., 4.75 ton	3.90	4.68	8.58
3880	3,000 S.F., 9.50 ton	3.89	4.26	8.15
3920	5,000 S.F., 15.83 ton	3.91	4.09	8
3960	10,000 S.F., 31.67 ton	3.99	3.99	7.98
4000	Restaurants, 500 S.F., 2.50 ton	10.40	9.30	19.70
4040	1,000 S.F., 5.00 ton	6.15	7.40	13.55
4080	1,500 S.F., 7.50 ton	6.15	6.75	12.90
4120	3,000 S.F., 15.00 ton	6.20	6.45	12.65
4160	5,000 S.F., 25.00 ton	7.10	6.35	13.45
4200	10,000 S.F., 50.00 ton	5.90	6.30	12.20
4240	Schools and colleges, 500 S.F., 1.92 ton	7.95	7.15	15.10
4280	1,000 S.F., 3.83 ton	5.15	5.90	11.05
4360	3,000 S.F., 11.50 ton	4.71	5.15	9.86
4400	5,000 S.F., 19.17 ton	5.45	4.87	10.32

D3050 Terminal & Package Units

*Note A: Small single zone unit recommended.

*Note B: A combination of multizone units recommended.

D3050 155	Rooftop Multizone Unit Systems	COST PER S.F.		
		MAT.	INST.	TOTAL
1240	Rooftop, multizone, air conditioner			
1260	Apartment corridors, 1,500 S.F., 2.75 ton. See Note A.			
1280	3,000 S.F., 5.50 ton	9.50	4.40	13.90
1320	10,000 S.F., 18.30 ton	6.65	4.23	10.88
1360	15,000 S.F., 27.50 ton	6.50	4.21	10.71
1400	20,000 S.F., 36.70 ton	6.80	4.22	11.02
1440	25,000 S.F., 45.80 ton	6.35	4.25	10.60
1520	Banks or libraries, 1,500 S.F., 6.25 ton	21.50	10	31.50
1560	3,000 S.F., 12.50 ton	18.30	9.80	28.10
1600	10,000 S.F., 41.67 ton	14.40	9.65	24.05
1640	15,000 S.F., 62.50 ton	10.80	9.60	20.40
1680	20,000 S.F., 83.33 ton	10.85	9.60	20.45
1720	25,000 S.F., 104.00 ton	9.75	9.55	19.30
1800	Bars and taverns, 1,500 S.F., 16.62 ton	46	14.35	60.35
1840	3,000 S.F., 33.24 ton	38.50	13.85	52.35
1880	10,000 S.F., 110.83 ton	23.50	13.80	37.30
1920	15,000 S.F., 165 ton, See Note B			
1960	20,000 S.F., 220 ton, See Note B			
2000	25,000 S.F., 275 ton, See Note B			
2080	Bowling alleys, 1,500 S.F., 8.50 ton	29.50	13.60	43.10
2120	3,000 S.F., 17.00 ton	25	13.30	38.30
2160	10,000 S.F., 56.70 ton	19.60	13.15	32.75
2200	15,000 S.F., 85.00 ton	14.75	13.05	27.80
2240	20,000 S.F., 113.00 ton	13.25	12.95	26.20
2280	25,000 S.F., 140.00 ton see Note B			
2360	Department stores, 1,500 S.F., 4.37 ton, See Note A.			
2400	3,000 S.F., 8.75 ton	15.10	7	22.10
2440	10,000 S.F., 29.17 ton	10.80	6.70	17.50
2520	20,000 S.F., 58.33 ton	7.55	6.70	14.25
2560	25,000 S.F., 72.92 ton	7.55	6.70	14.25
2640	Drug stores, 1,500 S.F., 10.00 ton	34.50	15.95	50.45
2680	3,000 S.F., 20.00 ton	24	15.40	39.40
2720	10,000 S.F., 66.66 ton	17.30	15.35	32.65
2760	15,000 S.F., 100.00 ton	15.60	15.30	30.90
2800	20,000 S.F., 135 ton, See Note B			
2840	25,000 S.F., 165 ton, See Note B			
2920	Factories, 1,500 S.F., 5 ton, See Note A			
3000	10,000 S.F., 33.33 ton	12.35	7.65	20
3040	15,000 S.F., 50.00 ton	11.55	7.70	19.25
3080	20,000 S.F., 66.66 ton	8.65	7.65	16.30

D3050 Terminal & Package Units

D3050 155	Rooftop Multizone Unit Systems	COST PER S.F.		
		MAT.	INST.	TOTAL
3120	25,000 S.F., 83.33 ton	8.65	7.65	16.30
3200	Food supermarkets, 1,500 S.F., 4.25 ton, See Note A			
3240	3,000 S.F., 8.50 ton	14.65	6.80	21.45
3280	10,000 S.F., 28.33 ton	10.05	6.50	16.55
3320	15,000 S.F., 42.50 ton	9.80	6.55	16.35
3360	20,000 S.F., 56.67 ton	7.35	6.50	13.85
3400	25,000 S.F., 70.83 ton	7.35	6.50	13.85
3480	Medical centers, 1,500 S.F., 3.5 ton, See Note A			
3520	3,000 S.F., 7.00 ton	12.10	5.60	17.70
3560	10,000 S.F., 23.33 ton	8.15	5.35	13.50
3600	15,000 S.F., 35.00 ton	8.65	5.35	14
3640	20,000 S.F., 46.66 ton	8.05	5.40	13.45
3680	25,000 S.F., 58.33 ton	6.05	5.35	11.40
3760	Offices, 1,500 S.F., 4.75 ton, See Note A			
3800	3,000 S.F., 9.50 ton	16.40	7.60	24
3840	10,000 S.F., 31.66 ton	11.75	7.30	19.05
3920	20,000 S.F., 63.33 ton	8.20	7.30	15.50
3960	25,000 S.F., 79.16 ton	8.20	7.30	15.50
4000	Restaurants, 1,500 S.F., 7.50 ton	26	12	38
4040	3,000 S.F., 15.00 ton	22	11.75	33.75
4080	10,000 S.F., 50.00 ton	17.30	11.60	28.90
4120	15,000 S.F., 75.00 ton	12.95	11.50	24.45
4160	20,000 S.F., 100.00 ton	11.70	11.50	23.20
4200	25,000 S.F., 125 ton, See Note B			
4240	Schools and colleges, 1,500 S.F., 5.75 ton	19.85	9.20	29.05
4320	10,000 S.F., 38.33 ton	13.25	8.90	22.15
4360	15,000 S.F., 57.50 ton	9.95	8.80	18.75
4400	20,000 S.F.,76.66 ton	9.95	8.80	18.75
4440	25,000 S.F., 95.83 ton	9.40	8.80	18.20
2280	25,000 S.F., 140.00 ton see Note B			
4200	25,000 S.F., 125 ton, See Note B			

D3050 Terminal & Package Units

Self-Contained Water Cooled System

D3050 160	Self-contained, Water Cooled Unit Systems	COST PER S.F.		
		MAT.	INST.	TOTAL
1280	Self-contained, water cooled unit	2.92	2.75	5.67
1300	Apartment corridors, 500 S.F., .92 ton	2.90	2.75	5.65
1320	1,000 S.F., 1.83 ton	2.92	2.74	5.66
1360	3,000 S.F., 5.50 ton	2.42	2.30	4.72
1400	5,000 S.F., 9.17 ton	2.40	2.13	4.53
1440	10,000 S.F., 18.33 ton	3.50	1.92	5.42
1520	Banks or libraries, 500 S.F., 2.08 ton	6.10	2.77	8.87
1560	1,000 S.F., 4.17 ton	5.50	5.25	10.75
1600	3,000 S.F., 12.50 ton	5.40	4.87	10.27
1640	5,000 S.F., 20.80 ton	7.90	4.36	12.26
1680	10,000 S.F., 41.66 ton	6.90	4.37	11.27
1760	Bars and taverns, 500 S.F., 5.54 ton	13.20	4.67	17.87
1800	1,000 S.F., 11.08 ton	13.70	8.30	22
1840	3,000 S.F., 33.25 ton	18.35	6.60	24.95
1880	5,000 S.F., 55.42 ton	17.20	6.85	24.05
1920	10,000 S.F., 110.00 ton	16.95	6.80	23.75
2000	Bowling alleys, 500 S.F., 2.83 ton	8.35	3.78	12.13
2040	1,000 S.F., 5.66 ton	7.45	7.10	14.55
2080	3,000 S.F., 17.00 ton	10.80	5.95	16.75
2120	5,000 S.F., 28.33 ton	9.75	5.70	15.45
2160	10,000 S.F., 56.66 ton	9.10	5.85	14.95
2200	Department stores, 500 S.F., 1.46 ton	4.30	1.95	6.25
2240	1,000 S.F., 2.92 ton	3.84	3.66	7.50
2280	3,000 S.F., 8.75 ton	3.79	3.40	7.19
2320	5,000 S.F., 14.58 ton	5.55	3.05	8.60
2360	10,000 S.F., 29.17 ton	5	2.94	7.94
2440	Drug stores, 500 S.F., 3.33 ton	9.80	4.44	14.24
2480	1,000 S.F., 6.66 ton	8.80	8.35	17.15
2520	3,000 S.F., 20.00 ton	12.70	7	19.70
2560	5,000 S.F., 33.33 ton	13.30	6.90	20.20
2600	10,000 S.F., 66.66 ton	10.75	6.85	17.60
2680	Factories, 500 S.F., 1.66 ton	4.90	2.21	7.11
2720	1,000 S.F. 3.37 ton	4.44	4.23	8.67
2760	3,000 S.F., 10.00 ton	4.33	3.90	8.23
2800	5,000 S.F., 16.66 ton	6.35	3.49	9.84
2840	10,000 S.F., 33.33 ton	5.75	3.36	9.11
2920	Food supermarkets, 500 S.F., 1.42 ton	4.17	1.88	6.05
2960	1,000 S.F., 2.83 ton	4.50	4.23	8.73

D30 HVAC

D3050 Terminal & Package Units

D3050 160	Self-contained, Water Cooled Unit Systems	COST PER S.F.		
		MAT.	INST.	TOTAL
3000	3,000 S.F., 8.50 ton	3.73	3.55	7.28
3040	5,000 S.F., 14.17 ton	3.68	3.30	6.98
3080	10,000 S.F., 28.33 ton	4.88	2.86	7.74
3160	Medical centers, 500 S.F., 1.17 ton	3.43	1.55	4.98
3200	1,000 S.F., 2.33 ton	3.72	3.49	7.21
3240	3,000 S.F., 7.00 ton	3.08	2.93	6.01
3280	5,000 S.F., 11.66 ton	3.04	2.72	5.76
3320	10,000 S.F., 23.33 ton	4.44	2.44	6.88
3400	Offices, 500 S.F., 1.58 ton	4.66	2.11	6.77
3440	1,000 S.F., 3.17 ton	5.05	4.74	9.79
3480	3,000 S.F., 9.50 ton	4.12	3.70	7.82
3520	5,000 S.F., 15.83 ton	6	3.32	9.32
3560	10,000 S.F., 31.67 ton	5.45	3.20	8.65
3640	Restaurants, 500 S.F., 2.50 ton	7.35	3.33	10.68
3680	1,000 S.F., 5.00 ton	6.60	6.30	12.90
3720	3,000 S.F., 15.00 ton	6.50	5.85	12.35
3760	5,000 S.F., 25.00 ton	9.50	5.25	14.75
3800	10,000 S.F., 50.00 ton	5.85	4.81	10.66
3880	Schools and colleges, 500 S.F., 1.92 ton	5.65	2.56	8.21
3920	1,000 S.F., 3.83 ton	5.05	4.81	9.86
3960	3,000 S.F., 11.50 ton	4.99	4.47	9.46
4000	5,000 S.F., 19.17 ton	7.30	4.02	11.32
4040	10,000 S.F., 38.33 ton	6.35	4.02	10.37

D3050 Terminal & Package Units

Self-Contained Air Cooled System

D3050 165	Self-contained, Air Cooled Unit Systems	COST PER S.F.		
		MAT.	INST.	TOTAL
1300	Self-contained, air cooled unit			
1320	Apartment corridors, 500 S.F., .92 ton	4.30	3.50	7.80
1360	1,000 S.F., 1.83 ton	4.28	3.49	7.77
1400	3,000 S.F., 5.50 ton	3.68	3.25	6.93
1440	5,000 S.F., 9.17 ton	3.28	3.09	6.37
1480	10,000 S.F., 18.33 ton	3.15	2.84	5.99
1560	Banks or libraries, 500 S.F., 2.08 ton	9.20	4.47	13.67
1600	1,000 S.F., 4.17 ton	8.30	7.40	15.70
1640	3,000 S.F., 12.50 ton	7.45	7.05	14.50
1680	5,000 S.F., 20.80 ton	7.20	6.45	13.65
1720	10,000 S.F., 41.66 ton	8.30	6.35	14.65
1800	Bars and taverns, 500 S.F., 5.54 ton	19.30	9.75	29.05
1840	1,000 S.F., 11.08 ton	19.15	14.10	33.25
1880	3,000 S.F., 33.25 ton	18.90	12.25	31.15
1920	5,000 S.F., 55.42 ton	20.50	12.25	32.75
1960	10,000 S.F., 110.00 ton	20.50	12.20	32.70
2040	Bowling alleys, 500 S.F., 2.83 ton	12.55	6.10	18.65
2080	1,000 S.F., 5.66 ton	11.35	10.05	21.40
2120	3,000 S.F., 17.00 ton	9.75	8.80	18.55
2160	5,000 S.F., 28.33 ton	10	8.60	18.60
2200	10,000 S.F., 56.66 ton	10.70	8.60	19.30
2240	Department stores, 500 S.F., 1.46 ton	6.45	3.14	9.59
2280	1,000 S.F., 2.92 ton	5.85	5.15	11
2320	3,000 S.F., 8.75 ton	5.20	4.92	10.12
2360	5,000 S.F., 14.58 ton	5.20	4.92	10.12
2400	10,000 S.F., 29.17 ton	5.10	4.43	9.53
2480	Drug stores, 500 S.F., 3.33 ton	14.75	7.20	21.95
2520	1,000 S.F., 6.66 ton	13.35	11.80	25.15
2560	3,000 S.F., 20.00 ton	11.55	10.40	21.95
2600	5,000 S.F., 33.33 ton	11.75	10.15	21.90
2640	10,000 S.F., 66.66 ton	12.65	10.15	22.80
2720	Factories, 500 S.F., 1.66 ton	7.45	3.61	11.06
2760	1,000 S.F., 3.33 ton	6.70	5.95	12.65
2800	3,000 S.F., 10.00 ton	5.95	5.65	11.60
2840	5,000 S.F., 16.66 ton	5.70	5.20	10.90
2880	10,000 S.F., 33.33 ton	5.85	5.05	10.90
2960	Food supermarkets, 500 S.F., 1.42 ton	6.30	3.06	9.36
3000	1,000 S.F., 2.83 ton	6.60	5.40	12
3040	3,000 S.F., 8.50 ton	5.70	5.05	10.75
3080	5,000 S.F., 14.17 ton	5.05	4.79	9.84
3120	10,000 S.F., 28.33 ton	4.99	4.30	9.29
3200	Medical centers, 500 S.F., 1.17 ton	5.20	2.52	7.72

D30 HVAC

D3050 Terminal & Package Units

D3050 165	Self-contained, Air Cooled Unit Systems	COST PER S.F.		
		MAT.	INST.	TOTAL
3240	1,000 S.F., 2.33 ton	5.45	4.46	9.91
3280	3,000 S.F., 7.00 ton	4.69	4.15	8.84
3320	5,000 S.F., 16.66 ton	4.18	3.95	8.13
3360	10,000 S.F., 23.33 ton	4.02	3.63	7.65
3440	Offices, 500 S.F., 1.58 ton	7	3.41	10.41
3480	1,000 S.F., 3.16 ton	7.40	6.05	13.45
3520	3,000 S.F., 9.50 ton	5.65	5.35	11
3560	5,000 S.F., 15.83 ton	5.45	4.93	10.38
3600	10,000 S.F., 31.66 ton	5.60	4.81	10.41
3680	Restaurants, 500 S.F., 2.50 ton	11.05	5.40	16.45
3720	1,000 S.F., 5.00 ton	10.05	8.90	18.95
3760	3,000 S.F., 15.00 ton	8.95	8.45	17.40
3800	5,000 S.F., 25.00 ton	8.75	7.60	16.35
3840	10,000 S.F., 50.00 ton	10.15	7.55	17.70
3920	Schools and colleges, 500 S.F., 1.92 ton	8.50	4.13	12.63
3960	1,000 S.F., 3.83 ton	7.75	6.80	14.55
4000	3,000 S.F., 11.50 ton	6.85	6.45	13.30
4040	5,000 S.F., 19.17 ton	6.60	5.95	12.55
4080	10,000 S.F., 38.33 ton	7.65	5.85	13.50

D3050 Terminal & Package Units

Air Cooled Condensing Unit
Refrigerant Piping
Roof
Supply Duct
Finish Ceiling
Return Grille
DX Air Handling Unit
Supply Diffuser

*Cooling requirements would lead to more than one system.

D3050 170	Split Systems With Air Cooled Condensing Units	COST PER S.F.		
		MAT.	INST.	TOTAL
1260	Split system, air cooled condensing unit			
1280	Apartment corridors, 1,000 S.F., 1.83 ton	2.45	2.18	4.63
1320	2,000 S.F., 3.66 ton	1.77	2.20	3.97
1360	5,000 S.F., 9.17 ton	2.04	2.78	4.82
1400	10,000 S.F., 18.33 ton	2.32	3.03	5.35
1440	20,000 S.F., 36.66 ton	2.58	3.08	5.66
1520	Banks and libraries, 1,000 S.F., 4.17 ton	4.02	5	9.02
1560	2,000 S.F., 8.33 ton	4.63	6.30	10.93
1600	5,000 S.F., 20.80 ton	5.25	6.85	12.10
1640	10,000 S.F., 41.66 ton	5.85	6.95	12.80
1680	20,000 S.F., 83.32 ton	7.05	7.25	14.30
1760	Bars and taverns, 1,000 S.F., 11.08 ton	10.50	9.90	20.40
1800	2,000 S.F., 22.16 ton	13.80	11.85	25.65
1840	5,000 S.F., 55.42 ton	12.45	11.10	23.55
1880	10,000 S.F., 110.84 ton	16.05	11.55	27.60
1920	20,000 S.F., 220 ton*			
2000	Bowling alleys, 1,000 S.F., 5.66 ton	5.70	9.40	15.10
2040	2,000 S.F., 11.33 ton	6.30	8.55	14.85
2080	5,000 S.F., 28.33 ton	7.20	9.35	16.55
2120	10,000 S.F., 56.66 ton	8	9.50	17.50
2160	20,000 S.F., 113.32 ton	11.10	10.30	21.40
2320	Department stores, 1,000 S.F., 2.92 ton	2.86	3.46	6.32
2360	2,000 S.F., 5.83 ton	2.94	4.85	7.79
2400	5,000 S.F., 14.58 ton	3.24	4.41	7.65
2440	10,000 S.F., 29.17 ton	3.69	4.82	8.51
2480	20,000 S.F., 58.33 ton	4.10	4.89	8.99
2560	Drug stores, 1,000 S.F., 6.66 ton	6.70	11.05	17.75
2600	2,000 S.F., 13.32 ton	7.40	10.05	17.45
2640	5,000 S.F., 33.33 ton	9.40	11.15	20.55
2680	10,000 S.F., 66.66 ton	10.20	11.70	21.90
2720	20,000 S.F., 133.32 ton*			
2800	Factories, 1,000 S.F., 3.33 ton	3.25	3.94	7.19
2840	2,000 S.F., 6.66 ton	3.36	5.55	8.91
2880	5,000 S.F., 16.66 ton	4.22	5.50	9.72
2920	10,000 S.F., 33.33 ton	4.69	5.60	10.29
2960	20,000 S.F., 66.66 ton	5.10	5.85	10.95
3040	Food supermarkets, 1,000 S.F., 2.83 ton	2.78	3.35	6.13
3080	2,000 S.F., 5.66 ton	2.85	4.71	7.56
3120	5,000 S.F., 14.66 ton	3.15	4.28	7.43
3160	10,000 S.F., 28.33 ton	3.58	4.69	8.27
3200	20,000 S.F., 56.66 ton	3.98	4.76	8.74
3280	Medical centers, 1,000 S.F., 2.33 ton	2.46	2.72	5.18

D30 HVAC

D3050 Terminal & Package Units

D3050 170	Split Systems With Air Cooled Condensing Units	COST PER S.F.		
		MAT.	INST.	TOTAL
3320	2,000 S.F., 4.66 ton	2.35	3.88	6.23
3360	5,000 S.F., 11.66 ton	2.60	3.53	6.13
3400	10,000 S.F., 23.33 ton	2.95	3.85	6.80
3440	20,000 S.F., 46.66 ton	3.28	3.91	7.19
3520	Offices, 1,000 S.F., 3.17 ton	3.09	3.75	6.84
3560	2,000 S.F., 6.33 ton	3.18	5.25	8.43
3600	5,000 S.F., 15.83 ton	3.52	4.78	8.30
3640	10,000 S.F., 31.66 ton	4.01	5.25	9.26
3680	20,000 S.F., 63.32 ton	4.85	5.55	10.40
3760	Restaurants, 1,000 S.F., 5.00 ton	5.05	8.30	13.35
3800	2,000 S.F., 10.00 ton	5.55	7.55	13.10
3840	5,000 S.F., 25.00 ton	6.35	8.25	14.60
3880	10,000 S.F., 50.00 ton	7.05	8.40	15.45
3920	20,000 S.F., 100.00 ton	9.80	9.10	18.90
4000	Schools and colleges, 1,000 S.F., 3.83 ton	3.69	4.62	8.31
4040	2,000 S.F., 7.66 ton	4.27	5.80	10.07
4080	5,000 S.F., 19.17 ton	4.86	6.35	11.21
4120	10,000 S.F., 38.33 ton	5.40	6.45	11.85

D3090 Other HVAC Systems/Equip

Vitrified Clay Garage Exhaust System

Dual Exhaust System

D3090 320	Garage Exhaust Systems	COST PER BAY		
		MAT.	**INST.**	**TOTAL**
1040	Garage, single exhaust, 3" outlet, cars & light trucks, one bay	2,700	1,150	3,850
1060	Additional bays up to seven bays	420	160	580
1500	4" outlet, trucks, one bay	2,725	1,150	3,875
1520	Additional bays up to six bays	435	160	595
1600	5" outlet, diesel trucks, one bay	3,050	1,150	4,200
1650	Additional single bays up to six	770	184	954
1700	Two adjoining bays	3,050	1,150	4,200
2000	Dual exhaust, 3" outlets, pair of adjoining bays	3,125	1,250	4,375
2100	Additional pairs of adjoining bays	795	184	979

Reference - System Classification

System Classification
Rules for installation of sprinkler systems vary depending on the classification of occupancy falling into one of three categories as follows:

Light Hazard Occupancy
The protection area allotted per sprinkler should not exceed 225 S.F., with the maximum distance between lines and sprinklers on lines being 15'. The sprinklers do not need to be staggered. Branch lines should not exceed eight sprinklers on either side of a cross main. Each large area requiring more than 100 sprinklers and without a sub-dividing partition should be supplied by feed mains or risers sized for ordinary hazard occupancy.
Maximum system area = 52,000 S.F.

Included in this group are:

Churches	Nursing Homes
Clubs	Offices
Educational	Residential
Hospitals	Restaurants
Institutional	Theaters and Auditoriums
Libraries	(except stages and prosceniums)
(except large stack rooms)	Unused Attics
Museums	

Ordinary Hazard Occupancy
The protection area allotted per sprinkler shall not exceed 130 S.F. of noncombustible ceiling and 130 S.F. of combustible ceiling. The maximum allowable distance between sprinkler lines and sprinklers on line is 15'. Sprinklers shall be staggered if the distance between heads exceeds 12'. Branch lines should not exceed eight sprinklers on either side of a cross main.
Maximum system area = 52,000 S.F.

Included in this group are:

Group 1	Group 2
Automotive Parking and Showrooms	Cereal Mills
Bakeries	Chemical Plants—Ordinary
Beverage manufacturing	Confectionery Products
Canneries	Distilleries
Dairy Products Manufacturing/Processing	Dry Cleaners
Electronic Plans	Feed Mills
Glass and Glass Products Manufacturing	Horse Stables
Laundries	Leather Goods Manufacturing
Restaurant Service Areas	Libraries—Large Stack Room Areas
	Machine Shops
	Metal Working
	Mercantile
	Paper and Pulp Mills
	Paper Process Plants
	Piers and Wharves
	Post Offices
	Printing and Publishing
	Repair Garages
	Stages
	Textile Manufacturing
	Tire Manufacturing
	Tobacco Products Manufacturing
	Wood Machining
	Wood Product Assembly

Extra Hazard Occupancy
The protection area allotted per sprinkler shall not exceed 100 S.F. of noncombustible ceiling and 100 S.F. of combustible ceiling. The maximum allowable distance between lines and between sprinklers on lines is 12'. Sprinklers on alternate lines shall be staggered if the distance between sprinklers on lines exceeds 8'. Branch lines should not exceed six sprinklers on either side of a cross main.
Maximum system area:
Design by pipe schedule = 25,000 S.F.
Design by hydraulic calculation = 40,000 S.F.

Included in this group are:

Group 1	Group 2
Aircraft hangars	Asphalt Saturating
Combustible Hydraulic Fluid Use Area	Flammable Liquids Spraying
Die Casting	Flow Coating
Metal Extruding	Manufactured/Modular Home
Plywood/Particle Board Manufacturing	Building Assemblies (where
Printing (inkls with flash points < 100 degrees F	finished enclosure is present and has combustible interiors)
Rubber Reclaiming, Compounding, Drying, Milling, Vulcanizing	Open Oil Quenching
Saw Mills	Plastics Processing
Textile Picking, Opening, Blending, Garnetting, Carding, Combing of Cotton, Synthetics, Wood Shoddy, or Burlap	Solvent Cleaning
	Varnish and Paint Dipping
Upholstering with Plastic Foams	

Reference - Sprinkler Systems (Automatic)

Sprinkler systems may be classified by type as follows:

1. **Wet Pipe System.** A system employing automatic sprinklers attached to a piping system containing water and connected to a water supply so that water discharges immediately from sprinklers opened by a fire.

2. **Dry Pipe System.** A system employing automatic sprinklers attached to a piping system containing air under pressure, the release of which as from the opening of sprinklers permits the water pressure to open a valve known as a "dry pipe valve". The water then flows into the piping system and out the opened sprinklers.

3. **Pre-Action System.** A system employing automatic sprinklers attached to a piping system containing air that may or may not be under pressure, with a supplemental heat responsive system of generally more sensitive characteristics than the automatic sprinklers themselves, installed in the same areas as the sprinklers; actuation of the heat responsive system, as from a fire, opens a valve which permits water to flow into the sprinkler piping system and to be discharged from any sprinklers which may be open.

4. **Deluge System.** A system employing open sprinklers attached to a piping system connected to a water supply through a valve which is opened by the operation of a heat responsive system installed in the same areas as the sprinklers. When this valve opens, water flows into the piping system and discharges from all sprinklers attached thereto.

5. **Combined Dry Pipe and Pre-Action Sprinkler System.** A system employing automatic sprinklers attached to a piping system containing air under pressure with a supplemental heat responsive system of generally more sensitive characteristics than the automatic sprinklers themselves, installed in the same areas as the sprinklers; operation of the heat responsive system, as from a fire, actuates tripping devices which open dry pipe valves simultaneously and without loss of air pressure in the system. Operation of the heat responsive system also opens approved air exhaust valves at the end of the feed main which facilitates the filling of the system with water which usually precedes the opening of sprinklers. The heat responsive system also serves as an automatic fire alarm system.

6. **Limited Water Supply System.** A system employing automatic sprinklers and conforming to these standards but supplied by a pressure tank of limited capacity.

7. **Chemical Systems.** Systems using halon, carbon dioxide, dry chemical or high expansion foam as selected for special requirements. Agent may extinguish flames by chemically inhibiting flame propagation, suffocate flames by excluding oxygen, interrupting chemical action of oxygen uniting with fuel or sealing and cooling the combustion center.

8. **Firecycle System.** Firecycle is a fixed fire protection sprinkler system utilizing water as its extinguishing agent. It is a time delayed, recycling, preaction type which automatically shuts the water off when heat is reduced below the detector operating temperature and turns the water back on when that temperature is exceeded. The system senses a fire condition through a closed circuit electrical detector system which controls water flow to the fire automatically. Batteries supply up to 90 hour emergency power supply for system operation. The piping system is dry (until water is required) and is monitored with pressurized air. Should any leak in the system piping occur, an alarm will sound, but water will not enter the system until heat is sensed by a firecycle detector.

Area coverage sprinkler systems may be laid out and fed from the supply in any one of several patterns as shown below. It is desirable, if possible, to utilize a central feed and achieve a shorter flow path from the riser to the furthest sprinkler. This permits use of the smallest sizes of pipe possible with resulting savings.

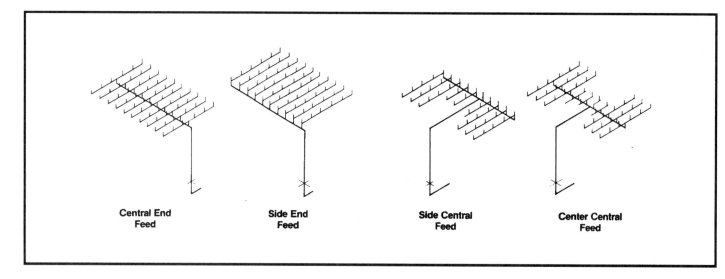

| Central End Feed | Side End Feed | Side Central Feed | Center Central Feed |

D4010 Sprinklers

Dry Pipe System: A system employing automatic sprinklers attached to a piping system containing air under pressure, the release of which as from the opening of sprinklers permits the water pressure to open a valve known as a "dry pipe valve". The water then flows into the piping system and out the opened sprinklers.

All areas are assumed to be open.

D4010 310	Dry Pipe Sprinkler Systems	COST PER S.F.		
		MAT.	INST.	TOTAL
0520	Dry pipe sprinkler systems, steel, black, sch. 40 pipe			
0530	Light hazard, one floor, 500 S.F.	7.25	5.25	12.50
0560	1000 S.F.	4.06	3.03	7.09
0580	2000 S.F.	3.88	3.06	6.94
0600	5000 S.F.	2.05	2.08	4.13
0620	10,000 S.F.	1.47	1.74	3.21
0640	50,000 S.F.	1.17	1.52	2.69
0660	Each additional floor, 500 S.F.	1.64	2.54	4.18
0680	1000 S.F.	1.35	2.08	3.43
0700	2000 S.F.	1.32	1.95	3.27
0720	5000 S.F.	1.10	1.65	2.75
0740	10,000 S.F.	1.01	1.52	2.53
0760	50,000 S.F.	.94	1.34	2.28
1000	Ordinary hazard, one floor, 500 S.F.	7.40	5.30	12.70
1020	1000 S.F.	4.18	3.07	7.25
1040	2000 S.F.	3.98	3.21	7.19
1060	5000 S.F.	2.34	2.22	4.56
1080	10,000 S.F.	1.93	2.29	4.22
1100	50,000 S.F.	1.70	2.15	3.85
1140	Each additional floor, 500 S.F.	1.79	2.60	4.39
1160	1000 S.F.	1.58	2.34	3.92
1180	2000 S.F.	1.56	2.13	3.69
1200	5000 S.F.	1.44	1.82	3.26
1220	10,000 S.F.	1.33	1.80	3.13
1240	50,000 S.F.	1.24	1.55	2.79
1500	Extra hazard, one floor, 500 S.F.	10.15	6.60	16.75
1520	1000 S.F.	6.20	4.73	10.93
1540	2000 S.F.	4.47	4.12	8.59
1560	5000 S.F.	2.77	3.10	5.87
1580	10,000 S.F.	2.83	2.95	5.78
1600	50,000 S.F.	2.99	2.83	5.82
1660	Each additional floor, 500 S.F.	2.37	3.23	5.60
1680	1000 S.F.	2.32	3.06	5.38
1700	2000 S.F.	2.17	3.07	5.24
1720	5000 S.F.	1.87	2.67	4.54
1740	10,000 S.F.	2.14	2.44	4.58
1760	50,000 S.F.	2.15	2.33	4.48
2020	Grooved steel, black, sch. 40 pipe, light hazard, one floor, 2000 S.F.	3.73	2.65	6.38
2060	10,000 S.F.	1.49	1.52	3.01
2100	Each additional floor, 2000 S.F.	1.31	1.57	2.88

D4010 Sprinklers

D4010 310	Dry Pipe Sprinkler Systems	COST PER S.F.		
		MAT.	INST.	TOTAL
2150	10,000 S.F.	1.03	1.30	2.33
2200	Ordinary hazard, one floor, 2000 S.F.	3.95	2.79	6.74
2250	10,000 S.F.	1.89	1.92	3.81
2300	Each additional floor, 2000 S.F.	1.53	1.71	3.24
2350	10,000 S.F.	1.44	1.71	3.15
2400	Extra hazard, one floor, 2000 S.F.	4.47	3.54	8.01
2450	10,000 S.F.	2.56	2.48	5.04
2500	Each additional floor, 2000 S.F.	2.18	2.54	4.72
2550	10,000 S.F.	1.98	2.18	4.16
3050	Grooved steel black sch. 10 pipe, light hazard, one floor, 2000 S.F.	3.66	2.62	6.28
3100	10,000 S.F.	1.43	1.49	2.92
3150	Each additional floor, 2000 S.F.	1.24	1.54	2.78
3200	10,000 S.F.	.97	1.27	2.24
3250	Ordinary hazard, one floor, 2000 S.F.	3.88	2.78	6.66
3300	10,000 S.F.	1.75	1.88	3.63
3350	Each additional floor, 2000 S.F.	1.46	1.70	3.16
3400	10,000 S.F.	1.30	1.67	2.97
3450	Extra hazard, one floor, 2000 S.F.	4.41	3.51	7.92
3500	10,000 S.F.	2.42	2.44	4.86
3550	Each additional floor, 2000 S.F.	2.12	2.51	4.63
3600	10,000 S.F.	1.90	2.15	4.05
4050	Copper tubing, type M, light hazard, one floor, 2000 S.F.	4.56	2.61	7.17
4100	10,000 S.F.	2.26	1.52	3.78
4150	Each additional floor, 2000 S.F.	2.16	1.56	3.72
4200	10,000 S.F.	1.80	1.31	3.11
4250	Ordinary hazard, one floor, 2000 S.F.	4.94	2.92	7.86
4300	10,000 S.F.	2.82	1.78	4.60
4350	Each additional floor, 2000 S.F.	2.96	1.82	4.78
4400	10,000 S.F.	2.30	1.54	3.84
4450	Extra hazard, one floor, 2000 S.F.	5.75	3.55	9.30
4500	10,000 S.F.	5.50	2.66	8.16
4550	Each additional floor, 2000 S.F.	3.47	2.55	6.02
4600	10,000 S.F.	3.96	2.34	6.30
5050	Copper tubing, type M, T-drill system, light hazard, one floor			
5060	2000 S.F.	4.58	2.45	7.03
5100	10,000 S.F.	2.14	1.27	3.41
5150	Each additional floor, 2000 S.F.	2.18	1.40	3.58
5200	10,000 S.F.	1.68	1.06	2.74
5250	Ordinary hazard, one floor, 2000 S.F.	4.75	2.51	7.26
5300	10,000 S.F.	2.73	1.60	4.33
5350	Each additional floor, 2000 S.F.	2.33	1.43	3.76
5400	10,000 S.F.	2.15	1.32	3.47
5450	Extra hazard, one floor, 2000 S.F.	5.20	2.94	8.14
5500	10,000 S.F.	4.55	1.95	6.50
5550	Each additional floor, 2000 S.F.	2.91	1.94	4.85
5600	10,000 S.F.	3.03	1.63	4.66

D4010 Sprinklers

Preaction System: A system employing automatic sprinklers attached to a piping system containing air that may or may not be under pressure, with a supplemental heat responsive system of generally more sensitive characteristics than the automatic sprinklers themselves, installed in the same areas as the sprinklers. Actuation of the heat responsive system, as from a fire, opens a valve which permits water to flow into the sprinkler piping system and to be discharged from those sprinklers which were opened by heat from the fire.

All areas are assumed to be opened.

D4010 350	**Preaction Sprinkler Systems**	COST PER S.F.		
		MAT.	INST.	TOTAL
0520	Preaction sprinkler systems, steel, black, sch. 40 pipe			
0530	Light hazard, one floor, 500 S.F.	7.10	4.17	11.27
0560	1000 S.F.	4.10	3.10	7.20
0580	2000 S.F.	3.76	3.05	6.81
0600	5000 S.F.	1.96	2.07	4.03
0620	10,000 S.F.	1.39	1.74	3.13
0640	50,000 S.F.	1.10	1.51	2.61
0660	Each additional floor, 500 S.F.	1.81	2.25	4.06
0680	1000 S.F.	1.37	2.07	3.44
0700	2000 S.F.	1.34	1.94	3.28
0720	5000 S.F.	1.01	1.64	2.65
0740	10,000 S.F.	.93	1.52	2.45
0760	50,000 S.F.	.94	1.34	2.28
1000	Ordinary hazard, one floor, 500 S.F.	7.40	4.53	11.93
1020	1000 S.F.	4.07	3.06	7.13
1040	2000 S.F.	4.08	3.21	7.29
1060	5000 S.F.	2.15	2.21	4.36
1080	10,000 S.F.	1.70	2.28	3.98
1100	50,000 S.F.	1.45	2.13	3.58
1140	Each additional floor, 500 S.F.	2.09	2.61	4.70
1160	1000 S.F.	1.36	2.11	3.47
1180	2000 S.F.	1.25	2.11	3.36
1200	5000 S.F.	1.36	1.97	3.33
1220	10,000 S.F.	1.25	2.07	3.32
1240	50,000 S.F.	1.13	1.82	2.95
1500	Extra hazard, one floor, 500 S.F.	9.85	5.80	15.65
1520	1000 S.F.	5.65	4.30	9.95
1540	2000 S.F.	4.09	4.09	8.18

D4010 Sprinklers

D4010 350	Preaction Sprinkler Systems	COST PER S.F.		
		MAT.	INST.	TOTAL
1560	5000 S.F.	2.66	3.35	6.01
1580	10,000 S.F.	2.50	3.27	5.77
1600	50,000 S.F.	2.59	3.15	5.74
1660	Each additional floor, 500 S.F.	2.40	3.22	5.62
1680	1000 S.F.	1.94	3.03	4.97
1700	2000 S.F.	1.79	3.04	4.83
1720	5000 S.F.	1.52	2.69	4.21
1740	10,000 S.F.	1.66	2.46	4.12
1760	50,000 S.F.	1.59	2.29	3.88
2020	Grooved steel, black, sch. 40 pipe, light hazard, one floor, 2000 S.F.	3.75	2.64	6.39
2060	10,000 S.F.	1.41	1.52	2.93
2100	Each additional floor of 2000 S.F.	1.33	1.56	2.89
2150	10,000 S.F.	.95	1.30	2.25
2200	Ordinary hazard, one floor, 2000 S.F.	3.84	2.78	6.62
2250	10,000 S.F.	1.66	1.91	3.57
2300	Each additional floor, 2000 S.F.	1.42	1.70	3.12
2350	10,000 S.F.	1.21	1.70	2.91
2400	Extra hazard, one floor, 2000 S.F.	4.09	3.51	7.60
2450	10,000 S.F.	2.09	2.45	4.54
2500	Each additional floor, 2000 S.F.	1.80	2.51	4.31
2550	10,000 S.F.	1.48	2.15	3.63
3050	Grooved steel, black, sch. 10 pipe light hazard, one floor, 2000 S.F.	3.68	2.61	6.29
3100	10,000 S.F.	1.35	1.49	2.84
3150	Each additional floor, 2000 S.F.	1.26	1.53	2.79
3200	10,000 S.F.	.89	1.27	2.16
3250	Ordinary hazard, one floor, 2000 S.F.	3.71	2.59	6.30
3300	10,000 S.F.	1.31	1.85	3.16
3350	Each additional floor, 2000 S.F.	1.35	1.69	3.04
3400	10,000 S.F.	1.07	1.66	2.73
3450	Extra hazard, one floor, 2000 S.F.	4.03	3.48	7.51
3500	10,000 S.F.	1.92	2.41	4.33
3550	Each additional floor, 2000 S.F.	1.74	2.48	4.22
3600	10,000 S.F.	1.40	2.12	3.52
4050	Copper tubing, type M, light hazard, one floor, 2000 S.F.	4.59	2.60	7.19
4100	10,000 S.F.	2.19	1.52	3.71
4150	Each additional floor, 2000 S.F.	2.19	1.55	3.74
4200	10,000 S.F.	1.51	1.29	2.80
4250	Ordinary hazard, one floor, 2000 S.F.	4.83	2.91	7.74
4300	10,000 S.F.	2.59	1.77	4.36
4350	Each additional floor, 2000 S.F.	2.20	1.61	3.81
4400	10,000 S.F.	1.85	1.41	3.26
4450	Extra hazard, one floor, 2000 S.F.	5.40	3.52	8.92
4500	10,000 S.F.	4.95	2.63	7.58
4550	Each additional floor, 2000 S.F.	3.09	2.52	5.61
4600	10,000 S.F.	3.46	2.31	5.77
5050	Copper tubing, type M, T-drill system, light hazard, one floor			
5060	2000 S.F.	4.60	2.44	7.04
5100	10,000 S.F.	2.06	1.27	3.33
5150	Each additional floor, 2000 S.F.	2.20	1.39	3.59
5200	10,000 S.F.	1.60	1.06	2.66
5250	Ordinary hazard, one floor, 2000 S.F.	4.64	2.50	7.14
5300	10,000 S.F.	2.50	1.59	4.09
5350	Each additional floor, 2000 S.F.	2.22	1.42	3.64
5400	10,000 S.F.	2.05	1.38	3.43
5450	Extra hazard, one floor, 2000 S.F.	4.82	2.91	7.73
5500	10,000 S.F.	4.02	1.92	5.94
5550	Each additional floor, 2000 S.F.	2.53	1.91	4.44
5600	10,000 S.F.	2.53	1.60	4.13

D4010 Sprinklers

Deluge System: A system employing open sprinklers attached to a piping system connected to a water supply through a valve which is opened by the operation of a heat responsive system installed in the same areas as the sprinklers. When this valve opens, water flows into the piping system and discharges from all sprinklers attached thereto.

D4010 370	**Deluge Sprinkler Systems**	COST PER S.F.		
		MAT.	INST.	TOTAL
0520	Deluge sprinkler systems, steel, black, sch. 40 pipe			
0530	Light hazard, one floor, 500 S.F.	21	4.22	25.22
0560	1000 S.F.	11.05	2.97	14.02
0580	2000 S.F.	7.50	3.06	10.56
0600	5000 S.F.	3.45	2.07	5.52
0620	10,000 S.F.	2.14	1.74	3.88
0640	50,000 S.F.	1.26	1.51	2.77
0660	Each additional floor, 500 S.F.	1.81	2.25	4.06
0680	1000 S.F.	1.37	2.07	3.44
0700	2000 S.F.	1.34	1.94	3.28
0720	5000 S.F.	1.01	1.64	2.65
0740	10,000 S.F.	.93	1.52	2.45
0760	50,000 S.F.	.95	1.38	2.33
1000	Ordinary hazard, one floor, 500 S.F.	22	4.84	26.84
1020	1000 S.F.	11.05	3.08	14.13
1040	2000 S.F.	7.80	3.22	11.02
1060	5000 S.F.	3.64	2.21	5.85
1080	10,000 S.F.	2.45	2.28	4.73
1100	50,000 S.F.	1.64	2.16	3.80
1140	Each additional floor, 500 S.F.	2.09	2.61	4.70
1160	1000 S.F.	1.36	2.11	3.47
1180	2000 S.F.	1.25	2.11	3.36
1200	5000 S.F.	1.25	1.81	3.06
1220	10,000 S.F.	1.19	1.84	3.03
1240	50,000 S.F.	1.06	1.66	2.72
1500	Extra hazard, one floor, 500 S.F.	24	5.85	29.85
1520	1000 S.F.	12.95	4.47	17.42
1540	2000 S.F.	7.80	4.10	11.90
1560	5000 S.F.	3.87	3.08	6.95
1580	10,000 S.F.	3.13	2.99	6.12
1600	50,000 S.F.	3.07	2.88	5.95
1660	Each additional floor, 500 S.F.	2.40	3.22	5.62

D4010 Sprinklers

D4010 370	Deluge Sprinkler Systems	COST PER S.F.		
		MAT.	INST.	TOTAL
1680	1000 S.F.	1.94	3.03	4.97
1700	2000 S.F.	1.79	3.04	4.83
1720	5000 S.F.	1.52	2.69	4.21
1740	10,000 S.F.	1.71	2.56	4.27
1760	50,000 S.F.	1.69	2.44	4.13
2000	Grooved steel, black, sch. 40 pipe, light hazard, one floor			
2020	2000 S.F.	7.50	2.65	10.15
2060	10,000 S.F.	2.18	1.54	3.72
2100	Each additional floor, 2,000 S.F.	1.33	1.56	2.89
2150	10,000 S.F.	.95	1.30	2.25
2200	Ordinary hazard, one floor, 2000 S.F.	3.84	2.78	6.62
2250	10,000 S.F.	2.41	1.91	4.32
2300	Each additional floor, 2000 S.F.	1.42	1.70	3.12
2350	10,000 S.F.	1.21	1.70	2.91
2400	Extra hazard, one floor, 2000 S.F.	7.80	3.52	11.32
2450	10,000 S.F.	2.86	2.46	5.32
2500	Each additional floor, 2000 S.F.	1.80	2.51	4.31
2550	10,000 S.F.	1.48	2.15	3.63
3000	Grooved steel, black, sch. 10 pipe, light hazard, one floor			
3050	2000 S.F.	6.95	2.50	9.45
3100	10,000 S.F.	2.10	1.49	3.59
3150	Each additional floor, 2000 S.F.	1.26	1.53	2.79
3200	10,000 S.F.	.89	1.27	2.16
3250	Ordinary hazard, one floor, 2000 S.F.	7.50	2.78	10.28
3300	10,000 S.F.	2.06	1.85	3.91
3350	Each additional floor, 2000 S.F.	1.35	1.69	3.04
3400	10,000 S.F.	1.07	1.66	2.73
3450	Extra hazard, one floor, 2000 S.F.	7.75	3.49	11.24
3500	10,000 S.F.	2.67	2.41	5.08
3550	Each additional floor, 2000 S.F.	1.74	2.48	4.22
3600	10,000 S.F.	1.40	2.12	3.52
4000	Copper tubing, type M, light hazard, one floor			
4050	2000 S.F.	8.30	2.61	10.91
4100	10,000 S.F.	2.93	1.52	4.45
4150	Each additional floor, 2000 S.F.	2.18	1.55	3.73
4200	10,000 S.F.	1.51	1.29	2.80
4250	Ordinary hazard, one floor, 2000 S.F.	8.55	2.92	11.47
4300	10,000 S.F.	3.34	1.77	5.11
4350	Each additional floor, 2000 S.F.	2.20	1.61	3.81
4400	10,000 S.F.	1.85	1.41	3.26
4450	Extra hazard, one floor, 2000 S.F.	9.10	3.53	12.63
4500	10,000 S.F.	5.75	2.64	8.39
4550	Each additional floor, 2000 S.F.	3.09	2.52	5.61
4600	10,000 S.F.	3.46	2.31	5.77
5000	Copper tubing, type M, T-drill system, light hazard, one floor			
5050	2000 S.F.	8.35	2.45	10.80
5100	10,000 S.F.	2.81	1.27	4.08
5150	Each additional floor, 2000 S.F.	2.11	1.41	3.52
5200	10,000 S.F.	1.60	1.06	2.66
5250	Ordinary hazard, one floor, 2000 S.F.	8.35	2.51	10.86
5300	10,000 S.F.	3.25	1.59	4.84
5350	Each additional floor, 2000 S.F.	2.22	1.42	3.64
5400	10,000 S.F.	2.05	1.38	3.43
5450	Extra hazard, one floor, 2000 S.F.	8.55	2.92	11.47
5500	10,000 S.F.	4.77	1.92	6.69
5550	Each additional floor, 2000 S.F.	2.53	1.91	4.44
5600	10,000 S.F.	2.53	1.60	4.13

D4010 Sprinklers

Firecycle is a fixed fire protection sprinkler system utilizing water as its extinguishing agent. It is a time delayed, recycling, preaction type which automatically shuts the water off when heat is reduced below the detector operating temperature and turns the water back on when that temperature is exceeded.

The system senses a fire condition through a closed circuit electrical detector system which controls water flow to the fire automatically. Batteries supply up to 90 hours emergency power supply for system operation. The piping system is dry (until water is required) and is monitored with pressurized air. Should any leak in the system piping occur, an alarm will sound, but water will not enter the system until heat is sensed by a Firecycle detector.

D4010 390	Firecycle Sprinkler Systems	COST PER S.F.		
		MAT.	INST.	TOTAL
0520	Firecycle sprinkler systems, steel black sch. 40 pipe			
0530	Light hazard, one floor, 500 S.F.	30	8.20	38.20
0560	1000 S.F.	15.45	5.10	20.55
0580	2000 S.F.	9.35	3.75	13.10
0600	5000 S.F.	4.20	2.35	6.55
0620	10,000 S.F.	2.56	1.88	4.44
0640	50,000 S.F.	1.37	1.54	2.91
0660	Each additional floor of 500 S.F.	1.85	2.26	4.11
0680	1000 S.F.	1.39	2.08	3.47
0700	2000 S.F.	1.15	1.93	3.08
0720	5000 S.F.	1.02	1.64	2.66
0740	10,000 S.F.	.99	1.52	2.51
0760	50,000 S.F.	.99	1.38	2.37
1000	Ordinary hazard, one floor, 500 S.F.	30	8.55	38.55
1020	1000 S.F.	15.40	5.10	20.50
1040	2000 S.F.	9.45	3.90	13.35
1060	5000 S.F.	4.39	2.49	6.88
1080	10,000 S.F.	2.87	2.42	5.29
1100	50,000 S.F.	1.91	2.41	4.32
1140	Each additional floor, 500 S.F.	2.13	2.62	4.75
1160	1000 S.F.	1.38	2.12	3.50
1180	2000 S.F.	1.40	1.95	3.35
1200	5000 S.F.	1.26	1.81	3.07
1220	10,000 S.F.	1.16	1.79	2.95
1240	50,000 S.F.	1.07	1.59	2.66
1500	Extra hazard, one floor, 500 S.F.	32.50	9.80	42.30
1520	1000 S.F.	16.75	6.30	23.05
1540	2000 S.F.	9.70	4.79	14.49
1560	5000 S.F.	4.62	3.36	7.98
1580	10,000 S.F.	3.62	3.40	7.02

D4010 Sprinklers

D4010 390	Firecycle Sprinkler Systems	COST PER S.F.		
		MAT.	INST.	TOTAL
1600	50,000 S.F.	3.10	3.74	6.84
1660	Each additional floor, 500 S.F.	2.44	3.23	5.67
1680	1000 S.F.	1.96	3.04	5
1700	2000 S.F.	1.81	3.05	4.86
1720	5000 S.F.	1.53	2.69	4.22
1740	10,000 S.F.	1.72	2.46	4.18
1760	50,000 S.F.	1.70	2.36	4.06
2020	Grooved steel, black, sch. 40 pipe, light hazard, one floor			
2030	2000 S.F.	9.35	3.34	12.69
2060	10,000 S.F.	2.81	2.30	5.11
2100	Each additional floor, 2000 S.F.	1.35	1.57	2.92
2150	10,000 S.F.	1.01	1.30	2.31
2200	Ordinary hazard, one floor, 2000 S.F.	9.45	3.48	12.93
2250	10,000 S.F.	3.01	2.17	5.18
2300	Each additional floor, 2000 S.F.	1.44	1.71	3.15
2350	10,000 S.F.	1.27	1.70	2.97
2400	Extra hazard, one floor, 2000 S.F.	9.70	4.21	13.91
2450	10,000 S.F.	3.21	2.58	5.79
2500	Each additional floor, 2000 S.F.	1.82	2.52	4.34
2550	10,000 S.F.	1.54	2.15	3.69
3050	Grooved steel, black, sch. 10 pipe light hazard, one floor,			
3060	2000 S.F.	9.25	3.31	12.56
3100	10,000 S.F.	2.52	1.63	4.15
3150	Each additional floor, 2000 S.F.	1.28	1.54	2.82
3200	10,000 S.F.	.95	1.27	2.22
3250	Ordinary hazard, one floor, 2000 S.F.	9.35	3.47	12.82
3300	10,000 S.F.	2.69	2.01	4.70
3350	Each additional floor, 2000 S.F.	1.37	1.70	3.07
3400	10,000 S.F.	1.13	1.66	2.79
3450	Extra hazard, one floor, 2000 S.F.	9.60	4.18	13.78
3500	10,000 S.F.	3.07	2.54	5.61
3550	Each additional floor, 2000 S.F.	1.76	2.49	4.25
3600	10,000 S.F.	1.46	2.12	3.58
4060	Copper tubing, type M, light hazard, one floor, 2000 S.F.	10.15	3.30	13.45
4100	10,000 S.F.	3.35	1.66	5.01
4150	Each additional floor, 2000 S.F.	2.20	1.56	3.76
4200	10,000 S.F.	1.78	1.31	3.09
4250	Ordinary hazard, one floor, 2000 S.F.	10.40	3.61	14.01
4300	10,000 S.F.	3.76	1.91	5.67
4350	Each additional floor, 2000 S.F.	2.22	1.62	3.84
4400	10,000 S.F.	1.88	1.39	3.27
4450	Extra hazard, one floor, 2000 S.F.	10.95	4.22	15.17
4500	10,000 S.F.	6.20	2.80	9
4550	Each additional floor, 2000 S.F.	3.11	2.53	5.64
4600	10,000 S.F.	3.52	2.31	5.83
5060	Copper tubing, type M, T-drill system, light hazard, one floor 2000 S.F.	10.20	3.14	13.34
5100	10,000 S.F.	3.23	1.41	4.64
5150	Each additional floor, 2000 S.F.	2.35	1.48	3.83
5200	10,000 S.F.	1.66	1.06	2.72
5250	Ordinary hazard, one floor, 2000 S.F.	10.25	3.20	13.45
5300	10,000 S.F.	3.67	1.73	5.40
5350	Each additional floor, 2000 S.F.	2.24	1.43	3.67
5400	10,000 S.F.	2.11	1.38	3.49
5450	Extra hazard, one floor, 2000 S.F.	10.40	3.61	14.01
5500	10,000 S.F.	5.15	2.05	7.20
5550	Each additional floor, 2000 S.F.	2.55	1.92	4.47
5600	10,000 S.F.	2.59	1.60	4.19

D4010 Sprinklers

Wet Pipe System. A system employing automatic sprinklers attached to a piping system containing water and connected to a water supply so that water discharges immediately from sprinklers opened by heat from a fire.

All areas are assumed to be open.

D4010 410	Wet Pipe Sprinkler Systems	COST PER S.F.		
		MAT.	INST.	TOTAL
0520	Wet pipe sprinkler systems, steel, black, sch. 40 pipe			
0530	Light hazard, one floor, 500 S.F.	2.10	2.68	4.78
0560	1000 S.F.	3.27	2.75	6.02
0580	2000 S.F.	3.40	2.76	6.16
0600	5000 S.F.	1.65	1.94	3.59
0620	10,000 S.F.	1.10	1.67	2.77
0640	50,000 S.F.	.85	1.48	2.33
0660	Each additional floor, 500 S.F.	1.01	2.27	3.28
0680	1000 S.F.	.98	2.11	3.09
0700	2000 S.F.	.93	1.92	2.85
0720	5000 S.F.	.68	1.62	2.30
0740	10,000 S.F.	.64	1.50	2.14
0760	50,000 S.F.	.57	1.16	1.73
1000	Ordinary hazard, one floor, 500 S.F.	2.32	2.88	5.20
1020	1000 S.F.	3.24	2.71	5.95
1040	2000 S.F.	3.50	2.91	6.41
1060	5000 S.F.	1.84	2.08	3.92
1080	10,000 S.F.	1.41	2.21	3.62
1100	50,000 S.F.	1.17	2.07	3.24
1140	Each additional floor, 500 S.F.	1.27	2.57	3.84
1160	1000 S.F.	.95	2.09	3.04
1180	2000 S.F.	1.04	2.10	3.14
1200	5000 S.F.	1.05	1.98	3.03
1220	10,000 S.F.	.96	2.05	3.01
1240	50,000 S.F.	.87	1.81	2.68
1500	Extra hazard, one floor, 500 S.F.	6.15	4.44	10.59
1520	1000 S.F.	4.09	3.82	7.91
1540	2000 S.F.	3.73	3.91	7.64
1560	5000 S.F.	2.55	3.41	5.96
1580	10,000 S.F.	2.19	3.24	5.43
1600	50,000 S.F.	2.37	3.11	5.48
1660	Each additional floor, 500 S.F.	1.58	3.18	4.76
1680	1000 S.F.	1.53	3.01	4.54
1700	2000 S.F.	1.38	3.02	4.40
1720	5000 S.F.	1.19	2.67	3.86
1740	10,000 S.F.	1.37	2.44	3.81
1760	50,000 S.F.	1.36	2.32	3.68
2020	Grooved steel, black sch. 40 pipe, light hazard, one floor, 2000 S.F.	3.38	2.35	5.73
2060	10,000 S.F.	1.36	1.51	2.87
2100	Each additional floor, 2000 S.F.	.92	1.54	2.46

D4010 Sprinklers

D4010 410	Wet Pipe Sprinkler Systems	COST PER S.F.		
		MAT.	INST.	TOTAL
2150	10,000 S.F.	.66	1.28	1.94
2200	Ordinary hazard, one floor, 2000 S.F.	3.47	2.49	5.96
2250	10,000 S.F.	1.37	1.84	3.21
2300	Each additional floor, 2000 S.F.	1.01	1.68	2.69
2350	10,000 S.F.	.92	1.68	2.60
2400	Extra hazard, one floor, 2000 S.F.	3.72	3.22	6.94
2450	10,000 S.F.	1.77	2.38	4.15
2500	Each additional floor, 2000 S.F.	1.39	2.49	3.88
2550	10,000 S.F.	1.19	2.13	3.32
3050	Grooved steel black sch. 10 pipe, light hazard, one floor, 2000 S.F.	3.31	2.32	5.63
3100	10,000 S.F.	1.06	1.42	2.48
3150	Each additional floor, 2000 S.F.	.85	1.51	2.36
3200	10,000 S.F.	.60	1.25	1.85
3250	Ordinary hazard, one floor, 2000 S.F.	3.40	2.48	5.88
3300	10,000 S.F.	1.23	1.80	3.03
3350	Each additional floor, 2000 S.F.	.94	1.67	2.61
3400	10,000 S.F.	.78	1.64	2.42
3450	Extra hazard, one floor, 2000 S.F.	3.66	3.19	6.85
3500	10,000 S.F.	1.63	2.34	3.97
3550	Each additional floor, 2000 S.F.	1.33	2.46	3.79
3600	10,000 S.F.	1.11	2.10	3.21
4050	Copper tubing, type M, light hazard, one floor, 2000 S.F.	4.21	2.31	6.52
4100	10,000 S.F.	1.89	1.45	3.34
4150	Each additional floor, 2000 S.F.	1.77	1.53	3.30
4200	10,000 S.F.	1.43	1.29	2.72
4250	Ordinary hazard, one floor, 2000 S.F.	4.46	2.62	7.08
4300	10,000 S.F.	2.30	1.70	4
4350	Each additional floor, 2000 S.F.	2.09	1.73	3.82
4400	10,000 S.F.	1.78	1.51	3.29
4450	Extra hazard, one floor, 2000 S.F.	5	3.23	8.23
4500	10,000 S.F.	4.66	2.56	7.22
4550	Each additional floor, 2000 S.F.	2.68	2.50	5.18
4600	10,000 S.F.	3.17	2.29	5.46
5050	Copper tubing, type M, T-drill system, light hazard, one floor			
5060	2000 S.F.	4.23	2.15	6.38
5100	10,000 S.F.	1.77	1.20	2.97
5150	Each additional floor, 2000 S.F.	1.79	1.37	3.16
5200	10,000 S.F.	1.31	1.04	2.35
5250	Ordinary hazard, one floor, 2000 S.F.	4.27	2.21	6.48
5300	10,000 S.F.	2.21	1.52	3.73
5350	Each additional floor, 2000 S.F.	1.81	1.40	3.21
5400	10,000 S.F.	1.76	1.36	3.12
5450	Extra hazard, one floor, 2000 S.F.	4.45	2.62	7.07
5500	10,000 S.F.	3.73	1.85	5.58
5550	Each additional floor, 2000 S.F.	2.19	1.93	4.12
5600	10,000 S.F.	2.24	1.58	3.82

D40 Fire Protection

D4020 Standpipes

Roof — Roof connections with hose gate valves (for combustible roof)

Hose connections on each floor (size based on class of service)

Check Valve — Siamese inlet connections (for fire department use)

D4020 310	Wet Standpipe Risers, Class I	COST PER FLOOR		
		MAT.	INST.	TOTAL
0550	Wet standpipe risers, Class I, steel black sch. 40, 10' height			
0560	4" diameter pipe, one floor	3,600	2,900	6,500
0580	Additional floors	855	905	1,760
0600	6" diameter pipe, one floor	5,925	5,100	11,025
0620	Additional floors	1,450	1,425	2,875
0640	8" diameter pipe, one floor	8,850	6,175	15,025
0660	Additional floors	2,075	1,725	3,800

D4020 310	Wet Standpipe Risers, Class II	COST PER FLOOR		
		MAT.	INST.	TOTAL
1030	Wet standpipe risers, Class II, steel black sch. 40, 10' height			
1040	2" diameter pipe, one floor	1,425	1,050	2,475
1060	Additional floors	465	405	870
1080	2-1/2" diameter pipe, one floor	2,075	1,525	3,600
1100	Additional floors	535	470	1,005

D4020 310	Wet Standpipe Risers, Class III	COST PER FLOOR		
		MAT.	INST.	TOTAL
1530	Wet standpipe risers, Class III, steel black sch. 40, 10' height			
1540	4" diameter pipe, one floor	3,700	2,900	6,600
1560	Additional floors	725	755	1,480
1580	6" diameter pipe, one floor	6,025	5,100	11,125
1600	Additional floors	1,500	1,425	2,925
1620	8" diameter pipe, one floor	8,950	6,175	15,125
1640	Additional floors	2,125	1,725	3,850

D4020 Standpipes

Roof

Roof connections with hose gate valves (for combustible roof)

Hose connections on each floor (size based on class of service)

Check valve

Siamese inlet connections (for fire department use)

D4020 330	Dry Standpipe Risers, Class I	COST PER FLOOR		
		MAT.	INST.	TOTAL
0530	Dry standpipe riser, Class I, steel black sch. 40, 10' height			
0540	4" diameter pipe, one floor	2,250	2,350	4,600
0560	Additional floors	790	855	1,645
0580	6" diameter pipe, one floor	4,325	4,050	8,375
0600	Additional floors	1,375	1,375	2,750
0620	8" diameter pipe, one floor	6,500	4,925	11,425
0640	Additional floors	2,000	1,675	3,675

D4020 330	Dry Standpipe Risers, Class II	COST PER FLOOR		
		MAT.	INST.	TOTAL
1030	Dry standpipe risers, Class II, steel black sch. 40, 10' height			
1040	2" diameter pipe, one floor	1,225	1,100	2,325
1060	Additional floor	400	355	755
1080	2-1/2" diameter pipe, one floor	1,650	1,275	2,925
1100	Additional floors	470	425	895

D4020 330	Dry Standpipe Risers, Class III	COST PER FLOOR		
		MAT.	INST.	TOTAL
1530	Dry standpipe risers, Class III, steel black sch. 40, 10' height			
1540	4" diameter pipe, one floor	2,300	2,325	4,625
1560	Additional floors	670	775	1,445
1580	6" diameter pipe, one floor	4,375	4,050	8,425
1600	Additional floors	1,425	1,375	2,800
1620	8" diameter pipe, one floor	6,550	4,925	11,475
1640	Additional floor	2,050	1,675	3,725

D4020 Standpipes

D4020 410	Fire Hose Equipment	COST EACH		
		MAT.	INST.	TOTAL
0100	Adapters, reducing, 1 piece, FxM, hexagon, cast brass, 2-1/2" x 1-1/2"	50		50
0200	Pin lug, 1-1/2" x 1"	43.50		43.50
0250	3" x 2-1/2"	113		113
0300	For polished chrome, add 75% mat.			
0400	Cabinets, D.S. glass in door, recessed, steel box, not equipped			
0500	Single extinguisher, steel door & frame	95.50	130	225.50
0550	Stainless steel door & frame	204	130	334
0600	Valve, 2-1/2" angle, steel door & frame	123	86.50	209.50
0650	Aluminum door & frame	149	86.50	235.50
0700	Stainless steel door & frame	200	86.50	286.50
0750	Hose rack assy, 2-1/2" x 1-1/2" valve & 100' hose, steel door & frame	229	173	402
0800	Aluminum door & frame	335	173	508
0850	Stainless steel door & frame	450	173	623
0900	Hose rack assy, & extinguisher, 2-1/2" x 1-1/2" valve & hose, steel door & fram	228	208	436
0950	Aluminum	430	208	638
1000	Stainless steel	785	208	993
1550	Compressor, air, dry pipe system, automatic, 200 gal., 3/4 H.P.	765	445	1,210
1600	520 gal., 1 H.P.	805	445	1,250
1650	Alarm, electric pressure switch (circuit closer)	265	22	287
2500	Couplings, hose, rocker lug, cast brass, 1-1/2"	48		48
2550	2-1/2"	63		63
3000	Escutcheon plate, for angle valves, polished brass, 1-1/2"	12.15		12.15
3050	2-1/2"	29		29
3500	Fire pump, electric, w/controller, fittings, relief valve			
3550	4" pump, 30 H.P., 500 G.P.M.	21,800	3,250	25,050
3600	5" pump, 40 H.P., 1000 G.P.M.	31,800	3,675	35,475
3650	5" pump, 100 H.P., 1000 G.P.M.	35,700	4,075	39,775
3700	For jockey pump system, add	3,900	520	4,420
5000	Hose, per linear foot, synthetic jacket, lined,			
5100	300 lb. test, 1-1/2" diameter	2.63	.40	3.03
5150	2-1/2" diameter	4.38	.47	4.85
5200	500 lb. test, 1-1/2" diameter	2.72	.40	3.12
5250	2-1/2" diameter	4.72	.47	5.19
5500	Nozzle, plain stream, polished brass, 1-1/2" x 10"	43.50		43.50
5550	2-1/2" x 15" x 13/16" or 1-1/2"	78.50		78.50
5600	Heavy duty combination adjustable fog and straight stream w/handle 1-1/2"	405		405
5650	2-1/2" direct connection	505		505
6000	Rack, for 1-1/2" diameter hose 100 ft. long, steel	54	52	106
6050	Brass	66.50	52	118.50
6500	Reel, steel, for 50 ft. long 1-1/2" diameter hose	118	74.50	192.50
6550	For 75 ft. long 2-1/2" diameter hose	192	74.50	266.50
7050	Siamese, w/plugs & chains, polished brass, sidewalk, 4" x 2-1/2" x 2-1/2"	520	415	935
7100	6" x 2-1/2" x 2-1/2"	710	520	1,230
7200	Wall type, flush, 4" x 2-1/2" x 2-1/2"	470	208	678
7250	6" x 2-1/2" x 2-1/2"	655	226	881
7300	Projecting, 4" x 2-1/2" x 2-1/2"	430	208	638
7350	6" x 2-1/2" x 2-1/2"	725	226	951
7400	For chrome plate, add 15% mat.			
8000	Valves, angle, wheel handle, 300 Lb., rough brass, 1-1/2"	63.50	48	111.50
8050	2-1/2"	66	82.50	148.50
8100	Combination pressure restricting, 1-1/2"	77	48	125
8150	2-1/2"	153	82.50	235.50
8200	Pressure restricting, adjustable, satin brass, 1-1/2"	109	48	157
8250	2-1/2"	152	82.50	234.50
8300	Hydrolator, vent and drain, rough brass, 1-1/2"	64.50	48	112.50
8350	2-1/2"	64.50	48	112.50
8400	Cabinet assy, incls. adapter, rack, hose, and nozzle	675	315	990

D4090 Other Fire Protection Systems

General: Automatic fire protection (suppression) systems other than water sprinklers may be desired for special environments, high risk areas, isolated locations or unusual hazards. Some typical applications would include:

1. Paint dip tanks
2. Securities vaults
3. Electronic data processing
4. Tape and data storage
5. Transformer rooms
6. Spray booths
7. Petroleum storage
8. High rack storage

Piping and wiring costs are highly variable and are not included.

D4090 910	FM200 Fire Suppression	COST EACH		
		MAT.	INST.	TOTAL
0020	Detectors with brackets			
0040	Fixed temperature heat detector	36	70	106
0060	Rate of temperature rise detector	42.50	70	112.50
0080	Ion detector (smoke) detector	96	90.50	186.50
0200	Extinguisher agent			
0240	200 lb FM200, container	6,700	263	6,963
0280	75 lb carbon dioxide cylinder	1,100	176	1,276
0320	Dispersion nozzle			
0340	FM200 1-1/2" dispersion nozzle	58	42	100
0380	Carbon dioxide 3" x 5" dispersion nozzle	58	32.50	90.50
0420	Control station			
0440	Single zone control station with batteries	1,475	560	2,035
0470	Multizone (4) control station with batteries	2,825	1,125	3,950
0500	Electric mechanical release	144	286	430
0550	Manual pull station	52	97.50	149.50
0640	Battery standby power 10" x 10" x 17"	885	140	1,025
0740	Bell signalling device	63.50	70	133.50
0900	Standard low-rise sprinkler accessory package, 3 story	2,725	1,550	4,275
1000	Standard mid-rise sprinkler accessory package, 8 story	5,900	2,975	8,875
1100	Standard high-rise sprinkler accessory package, 16 story	12,400	5,825	18,225

D4090 920	FM200 Systems	COST PER C.F.		
		MAT.	INST.	TOTAL
0820	Average FM200 system, minimum			1.52
0840	Maximum			3.03

D5010 Electrical Service/Distribution

Electric Service

Feeder Installation

Switchgear

D5010 120	Electric Service, 3 Phase - 4 Wire	COST EACH		
		MAT.	INST.	TOTAL
0200	Service installation, includes breakers, metering, 20' conduit & wire			
0220	3 phase, 4 wire, 120/208 volts, 60 A	995	865	1,860
0240	100 A	1,225	1,050	2,275
0280	200 A	1,875	1,600	3,475
0320	400 A	4,450	2,925	7,375
0360	600 A	8,400	3,975	12,375
0400	800 A	10,800	4,800	15,600
0440	1000 A	13,300	5,500	18,800
0480	1200 A	16,700	5,625	22,325
0520	1600 A	30,200	8,075	38,275
0560	2000 A	33,300	9,200	42,500
0570	Add 25% for 277/480 volt			

D5010 230	Feeder Installation	COST PER L.F.		
		MAT.	INST.	TOTAL
0200	Feeder installation 600 V, including RGS conduit and XHHW wire, 60 A	8.45	10.45	18.90
0240	100 A	12.65	13.85	26.50
0280	200 A	28	21.50	49.50
0320	400 A	56.50	43	99.50
0360	600 A	122	70	192
0400	800 A	177	83.50	260.50
0440	1000 A	200	107	307
0480	1200 A	263	110	373
0520	1600 A	355	167	522
0560	2000 A	400	214	614
1200	Branch installation 600 V, including EMT conduit and THW wire, 15 A	1.44	5.35	6.79
1240	20 A	1.44	5.35	6.79
1280	30 A	2.26	6.55	8.81
1320	50 A	3.74	7.55	11.29
1360	65 A	4.54	8	12.54
1400	85 A	7.25	9.50	16.75
1440	100 A	9.30	10	19.30
1480	130 A	11.95	11.25	23.20
1520	150 A	14.65	12.95	27.60
1560	200 A	19.60	14.60	34.20

D5010 Electrical Service/Distribution

D5010 240	Switchgear	COST EACH		
		MAT.	INST.	TOTAL
0200	Switchgear inst., incl. swbd., panels & circ bkr, 400 A, 120/208volt	4,300	3,525	7,825
0240	600 A	13,400	4,775	18,175
0280	800 A	17,000	6,800	23,800
0320	1200 A	20,300	10,400	30,700
0360	1600 A	27,400	14,600	42,000
0400	2000 A	34,600	18,600	53,200
0410	Add 20% for 277/480 volt			

D50 Electrical

D5020 Lighting and Branch Wiring

Duplex Wall Receptacle

Undercarpet Receptacle System

D5020 110	Receptacle (by Wattage)	COST PER S.F.		
		MAT.	INST.	TOTAL
0190	Receptacles include plate, box, conduit, wire & transformer when required			
0200	2.5 per 1000 S.F., .3 watts per S.F.	.45	1.26	1.71
0240	With transformer	.52	1.32	1.84
0280	4 per 1000 S.F., .5 watts per S.F.	.50	1.47	1.97
0320	With transformer	.60	1.56	2.16
0360	5 per 1000 S.F., .6 watts per S.F.	.60	1.72	2.32
0400	With transformer	.73	1.84	2.57
0440	8 per 1000 S.F., .9 watts per S.F.	.62	1.90	2.52
0480	With transformer	.80	2.07	2.87
0520	10 per 1000 S.F., 1.2 watts per S.F.	.65	2.07	2.72
0560	With transformer	.94	2.35	3.29
0600	16.5 per 1000 S.F., 2.0 watts per S.F.	.76	2.59	3.35
0640	With transformer	1.26	3.07	4.33
0680	20 per 1000 S.F.,2.4 watts per S.F.	.80	2.82	3.62
0720	With transformer	1.39	3.38	4.77

D5020 115	Receptacles, Floor	COST PER S.F.		
		MAT.	INST.	TOTAL
0200	Receptacle systems, underfloor duct, 5' on center, low density	6.45	2.75	9.20
0240	High density	7	3.54	10.54
0280	7' on center, low density	5.10	2.36	7.46
0320	High density	5.65	3.15	8.80
0400	Poke thru fittings, low density	1.15	1.34	2.49
0440	High density	2.26	2.65	4.91
0520	Telepoles, using Romex, low density	1.14	.82	1.96
0560	High density	2.27	1.66	3.93
0600	Using EMT, low density	1.21	1.08	2.29
0640	High density	2.40	2.19	4.59
0720	Conduit system with floor boxes, low density	1.09	.93	2.02
0760	High density	2.19	1.90	4.09
0840	Undercarpet power system, 3 conductor with 5 conductor feeder, low density	1.46	.33	1.79
0880	High density	2.81	.70	3.51

D50 Electrical

D5020 Lighting and Branch Wiring

Description: Table 9.2-542 includes the cost for switch, plate, box, conduit in slab or EMT exposed and copper wire. Add 20% for exposed conduit.

No power required for switches.

Federal energy guidelines recommend the maximum lighting area controlled per switch shall not exceed 1000 S.F. and that areas over 500 S.F. shall be so controlled that total illumination can be reduced by at least 50%.

D5020 130	Wall Switch by Sq. Ft.	COST PER S.F.		
		MAT.	INST.	TOTAL
0200	Wall switches, 1.0 per 1000 S.F.	.07	.20	.27
0240	1.2 per 1000 S.F.	.07	.22	.29
0280	2.0 per 1000 S.F.	.11	.32	.43
0320	2.5 per 1000 S.F.	.13	.41	.54
0360	5.0 per 1000 S.F.	.29	.87	1.16
0400	10.0 per 1000 S.F.	.58	1.76	2.34

D5020 135	Miscellaneous Power	COST PER S.F.		
		MAT.	INST.	TOTAL
0200	Miscellaneous power, to .5 watts	.04	.10	.14
0240	.8 watts	.06	.14	.20
0280	1 watt	.07	.19	.26
0320	1.2 watts	.08	.22	.30
0360	1.5 watts	.10	.26	.36
0400	1.8 watts	.12	.30	.42
0440	2 watts	.14	.36	.50
0480	2.5 watts	.17	.45	.62
0520	3 watts	.21	.52	.73

D5020 140	Central A. C. Power (by Wattage)	COST PER S.F.		
		MAT.	INST.	TOTAL
0200	Central air conditioning power, 1 watt	.09	.22	.31
0220	2 watts	.10	.25	.35
0240	3 watts	.14	.28	.42
0280	4 watts	.20	.38	.58
0320	6 watts	.49	.52	1.01
0360	8 watts	.46	.54	1
0400	10 watts	.64	.63	1.27

D5020 Lighting and Branch Wiring

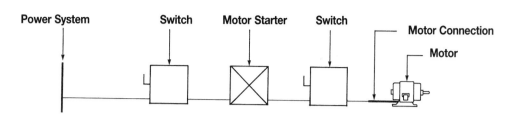

Motor Installation

D5020 145	Motor Installation	COST EACH		
		MAT.	INST.	TOTAL
0200	Motor installation, single phase, 115V, 1/3 HP motor size	675	850	1,525
0240	1 HP motor size	700	850	1,550
0280	2 HP motor size	750	905	1,655
0320	3 HP motor size	835	925	1,760
0360	230V, 1 HP motor size	680	865	1,545
0400	2 HP motor size	710	865	1,575
0440	3 HP motor size	790	930	1,720
0520	Three phase, 200V, 1-1/2 HP motor size	795	950	1,745
0560	3 HP motor size	845	1,025	1,870
0600	5 HP motor size	905	1,150	2,055
0640	7-1/2 HP motor size	940	1,175	2,115
0680	10 HP motor size	1,475	1,475	2,950
0720	15 HP motor size	1,900	1,625	3,525
0760	20 HP motor size	2,475	1,875	4,350
0800	25 HP motor size	2,525	1,900	4,425
0840	30 HP motor size	3,925	2,225	6,150
0880	40 HP motor size	5,050	2,625	7,675
0920	50 HP motor size	8,375	3,075	11,450
0960	60 HP motor size	8,825	3,250	12,075
1000	75 HP motor size	11,500	3,700	15,200
1040	100 HP motor size	27,100	4,375	31,475
1080	125 HP motor size	28,300	4,800	33,100
1120	150 HP motor size	33,600	5,625	39,225
1160	200 HP motor size	34,100	6,875	40,975
1240	230V, 1-1/2 HP motor size	760	940	1,700
1280	3 HP motor size	810	1,025	1,835
1320	5 HP motor size	865	1,125	1,990
1360	7-1/2 HP motor size	865	1,125	1,990
1400	10 HP motor size	1,325	1,400	2,725
1440	15 HP motor size	1,550	1,525	3,075
1480	20 HP motor size	2,275	1,825	4,100
1520	25 HP motor size	2,475	1,875	4,350
1560	30 HP motor size	2,525	1,900	4,425
1600	40 HP motor size	4,625	2,575	7,200
1640	50 HP motor size	4,900	2,725	7,625
1680	60 HP motor size	8,375	3,075	11,450
1720	75 HP motor size	10,500	3,500	14,000
1760	100 HP motor size	11,800	3,900	15,700
1800	125 HP motor size	27,500	4,500	32,000
1840	150 HP motor size	30,200	5,125	35,325
1880	200 HP motor size	31,800	5,700	37,500
1960	460V, 2 HP motor size	930	945	1,875
2000	5 HP motor size	975	1,025	2,000
2040	10 HP motor size	1,000	1,125	2,125
2080	15 HP motor size	1,300	1,300	2,600
2120	20 HP motor size	1,350	1,400	2,750

D5020 Lighting and Branch Wiring

D5020 145	Motor Installation	COST EACH		
		MAT.	INST.	TOTAL
2160	25 HP motor size	1,500	1,475	2,975
2200	30 HP motor size	1,875	1,575	3,450
2240	40 HP motor size	2,425	1,700	4,125
2280	50 HP motor size	2,750	1,875	4,625
2320	60 HP motor size	3,925	2,200	6,125
2360	75 HP motor size	4,650	2,450	7,100
2400	100 HP motor size	5,200	2,725	7,925
2440	125 HP motor size	8,675	3,100	11,775
2480	150 HP motor size	10,500	3,450	13,950
2520	200 HP motor size	12,600	3,900	16,500
2600	575V, 2 HP motor size	930	945	1,875
2640	5 HP motor size	975	1,025	2,000
2680	10 HP motor size	1,000	1,125	2,125
2720	20 HP motor size	1,300	1,300	2,600
2760	25 HP motor size	1,350	1,400	2,750
2800	30 HP motor size	1,875	1,575	3,450
2840	50 HP motor size	2,025	1,625	3,650
2880	60 HP motor size	3,875	2,200	6,075
2920	75 HP motor size	3,925	2,200	6,125
2960	100 HP motor size	4,650	2,450	7,100
3000	125 HP motor size	8,000	3,050	11,050
3040	150 HP motor size	8,675	3,100	11,775
3080	200 HP motor size	11,200	3,500	14,700

D5020 155	Motor Feeder	COST PER L.F.		
		MAT.	INST.	TOTAL
0200	Motor feeder systems, single phase, feed up to 115V 1HP or 230V 2 HP	2.59	6.60	9.19
0240	115V 2HP, 230V 3HP	2.79	6.70	9.49
0280	115V 3HP	3.20	7	10.20
0360	Three phase, feed to 200V 3HP, 230V 5HP, 460V 10HP, 575V 10HP	2.77	7.15	9.92
0440	200V 5HP, 230V 7.5HP, 460V 15HP, 575V 20HP	3.07	7.30	10.37
0520	200V 10HP, 230V 10HP, 460V 30HP, 575V 30HP	3.68	7.70	11.38
0600	200V 15HP, 230V 15HP, 460V 40HP, 575V 50HP	4.92	8.80	13.72
0680	200V 20HP, 230V 25HP, 460V 50HP, 575V 60HP	7.25	11.20	18.45
0760	200V 25HP, 230V 30HP, 460V 60HP, 575V 75HP	8.20	11.35	19.55
0840	200V 30HP	12.60	11.70	24.30
0920	230V 40HP, 460V 75HP, 575V 100HP	11.95	12.80	24.75
1000	200V 40HP	19	13.70	32.70
1080	230V 50HP, 460V 100HP, 575V 125HP	16.75	15.15	31.90
1160	200V 50HP, 230V 60HP, 460V 125HP, 575V 150HP	28	16.05	44.05
1240	200V 60HP, 460V 150HP	24.50	18.80	43.30
1320	230V 75HP, 575V 200HP	36.50	19.60	56.10
1400	200V 75HP	37.50	20	57.50
1480	230V 100HP, 460V 200HP	43	23.50	66.50
1560	200V 100HP	58.50	29	87.50
1640	230V 125HP	66	31.50	97.50
1720	200V 125HP, 230V 150HP	79.50	36.50	116
1800	200V 150HP	81	39	120
1880	200V 200HP	117	58.50	175.50
1960	230V 200HP	106	43.50	149.50

D50 Electrical

D5020 Lighting and Branch Wiring

Fluorescent Fixture

Incandescent Fixture

D5020 210	Fluorescent Fixtures (by Wattage)	COST PER S.F.		
		MAT.	INST.	TOTAL
0190	Fluorescent fixtures recess mounted in ceiling			
0195	T-12, standard 40 watt lamps			
0200	1 watt per S.F., 20 FC, 5 fixtures @40 watts per 1000 S.F.	.80	1.65	2.45
0240	2 watt per S.F., 40 FC, 10 fixtures @40 watt per 1000 S.F.	1.60	3.23	4.83
0280	3 watt per S.F., 60 FC, 15 fixtures @40 watt per 1000 S.F	2.40	4.88	7.28
0320	4 watt per S.F., 80 FC, 20 fixtures @40 watt per 1000 S.F.	3.20	6.45	9.65
0400	5 watt per S.F., 100 FC, 25 fixtures @40 watt per 1000 S.F.	4	8.10	12.10
0450	T-8, energy saver 32 watt lamps			
0500	0.8 watt per S.F., 20 FC, 5 fixtures @32 watt per 1000 S.F.	.88	1.65	2.53
0520	1.6 watt per S.F., 40 FC, 10 fixtures @32 watt per 1000 S.F.	1.76	3.23	4.99
0540	2.4 watt per S.F., 60 FC, 15 fixtures @ 32 watt per 1000 S.F	2.64	4.88	7.52
0560	3.2 watt per S.F., 80 FC, 20 fixtures @32 watt per 1000 S.F.	3.52	6.45	9.97
0580	4 watt per S.F., 100 FC, 25 fixtures @32 watt per 1000 S.F.	4.40	8.10	12.50

D5020 216	Incandescent Fixture (by Wattage)	COST PER S.F.		
		MAT.	INST.	TOTAL
0190	Incandescent fixture recess mounted, type A			
0200	1 watt per S.F., 8 FC, 6 fixtures per 1000 S.F.	1.01	1.32	2.33
0240	2 watt per S.F., 16 FC, 12 fixtures per 1000 S.F.	2.02	2.64	4.66
0280	3 watt per S.F., 24 FC, 18 fixtures, per 1000 S.F.	3	3.92	6.92
0320	4 watt per S.F., 32 FC, 24 fixtures per 1000 S.F.	4.01	5.25	9.26
0400	5 watt per S.F., 40 FC, 30 fixtures per 1000 S.F.	5	6.55	11.55

D50 Electrical

D5020 Lighting and Branch Wiring

High Bay Fixture

Low Bay Fixture

D5020 226	H.I.D. Fixture, High Bay, 16′ (by Wattage)	COST PER S.F.		
		MAT.	INST.	TOTAL
0190	High intensity discharge fixture, 16′ above work plane			
0240	1 watt/S.F., type E, 42 FC, 1 fixture/1000 S.F.	1.11	1.39	2.50
0280	Type G, 52 FC, 1 fixture/1000 S.F.	1.11	1.39	2.50
0320	Type C, 54 FC, 2 fixture/1000 S.F.	1.27	1.55	2.82
0440	2 watt/S.F., type E, 84 FC, 2 fixture/1000 S.F.	2.23	2.81	5.04
0480	Type G, 105 FC, 2 fixture/1000 S.F.	2.23	2.81	5.04
0520	Type C, 108 FC, 4 fixture/1000 S.F.	2.56	3.09	5.65
0640	3 watt/S.F., type E, 126 FC, 3 fixture/1000 S.F.	3.35	4.19	7.54
0680	Type G, 157 FC, 3 fixture/1000 S.F.	3.35	4.19	7.54
0720	Type C, 162 FC, 6 fixture/1000 S.F.	3.83	4.62	8.45
0840	4 watt/S.F., type E, 168 FC, 4 fixture/1000 S.F.	4.48	5.60	10.08
0880	Type G, 210 FC, 4 fixture/1000 S.F.	4.48	5.60	10.08
0920	Type C, 243 FC, 9 fixture/1000 S.F.	5.55	6.45	12
1040	5 watt/S.F., type E, 210 FC, 5 fixture/1000 S.F.	5.55	7	12.55
1080	Type G, 262 FC, 5 fixture/1000 S.F.	5.55	7	12.55
1120	Type C, 297 FC, 11 fixture/1000 S.F.	6.80	8	14.80

D5020 234	H.I.D. Fixture, High Bay, 30′ (by Wattage)	COST PER S.F.		
		MAT.	INST.	TOTAL
0190	High intensity discharge fixture, 30′ above work plane			
0240	1 watt/S.F., type E, 37 FC, 1 fixture/1000 S.F.	1.28	1.76	3.04
0280	Type G, 45 FC., 1 fixture/1000 S.F.	1.28	1.76	3.04
0320	Type F, 50 FC, 1 fixture/1000 S.F.	1.06	1.37	2.43
0440	2 watt/S.F., type E, 74 FC, 2 fixtures/1000 S.F.	2.53	3.50	6.03
0480	Type G, 92 FC, 2 fixtures/1000 S.F.	2.53	3.50	6.03
0520	Type F, 100 FC, 2 fixtures/1000 S.F.	2.13	2.79	4.92
0640	3 watt/S.F., type E, 110 FC, 3 fixtures/1000 S.F.	3.83	5.30	9.13
0680	Type G, 138FC, 3 fixtures/1000 S.F.	3.83	5.30	9.13
0720	Type F, 150 FC, 3 fixtures/1000 S.F.	3.19	4.16	7.35
0840	4 watt/S.F., type E, 148 FC, 4 fixtures/1000 S.F.	5.10	7.05	12.15
0880	Type G, 185 FC, 4 fixtures/1000 S.F.	5.10	7.05	12.15
0920	Type F, 200 FC, 4 fixtures/1000 S.F.	4.29	5.60	9.89
1040	5 watt/S.F., type E, 185 FC, 5 fixtures/1000 S.F.	6.35	8.80	15.15
1080	Type G, 230 FC, 5 fixtures/1000 S.F.	6.35	8.80	15.15
1120	Type F, 250 FC, 5 fixtures/1000 S.F.	5.35	6.95	12.30

D5020 238	H.I.D. Fixture, Low Bay, 8′-10′ (by Wattage)	COST PER S.F.		
		MAT.	INST.	TOTAL
0190	High intensity discharge fixture, 8′-10′ above work plane			
0240	1 watt/S.F., type J, 30 FC, 4 fixtures/1000 S.F.	2.51	2.86	5.37
0280	Type K, 29 FC, 5 fixtures/1000 S.F.	2.53	2.58	5.11
0400	2 watt/S.F., type J, 52 FC, 7 fixtures/1000 S.F.	4.48	5.25	9.73
0440	Type K, 63 FC, 11 fixtures/1000 S.F.	5.45	5.40	10.85
0560	3 watt/S.F., type J, 81 FC, 11 fixtures/1000 S.F.	6.90	7.90	14.80

D50 Electrical

D5020 Lighting and Branch Wiring

D5020 238	H.I.D. Fixture, Low Bay, 8'-10' (by Wattage)	COST PER S.F.		
		MAT.	INST.	TOTAL
0600	Type K, 92 FC, 16 fixtures/1000 S.F.	8	7.95	15.95
0720	4 watt/S.F., type J, 103 FC, 14 fixtures/1000 S.F.	8.90	10.40	19.30
0760	Type K, 127 FC, 22 fixtures/1000 S.F.	10.90	10.80	21.70
0880	5 watt/S.F., type J, 133 FC, 18 fixtures/1000 S.F.	11.35	13.15	24.50
0920	Type K, 155 FC, 27 fixtures/1000 S.F.	13.45	13.35	26.80

D5020 242	H.I.D. Fixture, Low Bay, 16' (by Wattage)	COST PER S.F.		
		MAT.	INST.	TOTAL
0190	High intensity discharge fixture, mounted 16' above work plane			
0240	1 watt/S.F., type J, 28 FC, 4 fixt./1000 S.F.	2.67	3.24	5.91
0280	Type K, 27 FC, 5 fixt./1000 S.F.	2.97	3.64	6.61
0400	2 watt/S.F., type J, 48 FC, 7 fixt/1000 S.F.	4.91	6.25	11.16
0440	Type K, 58 FC, 11 fixt/1000 S.F.	6.30	7.40	13.70
0560	3 watt/S.F., type J, 75 FC, 11 fixt/1000 S.F.	7.55	9.50	17.05
0600	Type K, 85 FC, 16 fixt/1000 S.F.	9.30	11.05	20.35
0720	4 watt/S.F., type J, 95 FC, 14 fixt/1000 S.F.	9.80	12.55	22.35
0760	Type K, 117 FC, 22 fixt/1000 S.F.	12.60	14.80	27.40
0880	5 watt/S.F., type J, 122 FC, 18 fixt/1000 S.F.	12.50	15.75	28.25
0920	Type K, 143 FC, 27 fixt/1000 S.F.	15.60	18.45	34.05

D50 Electrical

D5030 Communications and Security

D5030 910	Communication & Alarm Systems	COST EACH		
		MAT.	**INST.**	**TOTAL**
0200	Communication & alarm systems, includes outlets, boxes, conduit & wire			
0210	Sound system, 6 outlets	6,475	7,075	13,550
0220	12 outlets	9,050	11,300	20,350
0240	30 outlets	15,700	21,400	37,100
0280	100 outlets	47,800	71,500	119,300
0320	Fire detection systems, non-addressable, 12 detectors	3,125	5,875	9,000
0360	25 detectors	5,450	9,925	15,375
0400	50 detectors	10,700	19,700	30,400
0440	100 detectors	19,900	35,800	55,700
0450	Addressable type, 12 detectors	4,225	5,925	10,150
0452	25 detectors	7,675	10,000	17,675
0454	50 detectors	14,900	19,700	34,600
0456	100 detectors	28,600	36,100	64,700
0458	Fire alarm control panel, 8 zone	740	1,125	1,865
0459	12 zone	2,525	1,675	4,200
0460	Fire alarm command center, addressable without voice	2,600	980	3,580
0480	Intercom systems, 6 stations	3,625	4,850	8,475
0560	25 stations	12,700	18,600	31,300
0640	100 stations	48,300	68,000	116,300
0680	Master clock systems, 6 rooms	4,800	7,900	12,700
0720	12 rooms	7,425	13,500	20,925
0760	20 rooms	10,100	19,100	29,200
0800	30 rooms	17,100	35,100	52,200
0840	50 rooms	27,800	59,000	86,800
0920	Master TV antenna systems, 6 outlets	3,250	5,000	8,250
0960	12 outlets	6,050	9,300	15,350
1000	30 outlets	12,400	21,600	34,000
1040	100 outlets	41,100	70,500	111,600

D5090 Other Electrical Systems

Generator System

A: Engine
B: Battery
C: Charger
D: Transfer Switch
E: Muffler

D5090 210	Generators (by kW)	COST PER kW		
		MAT.	INST.	TOTAL
0190	Generator sets, include battery, charger, muffler & transfer switch			
0200	Gas/gasoline operated, 3 phase, 4 wire, 277/480 volt, 7.5 kW	1,125	246	1,371
0240	11.5 kW	1,050	187	1,237
0280	20 kW	705	121	826
0320	35 kW	480	79	559
0360	80 kW	345	47.50	392.50
0400	100 kW	305	46.50	351.50
0440	125 kW	495	43.50	538.50
0480	185 kW	445	33	478
0560	Diesel engine with fuel tank, 30 kW	720	92	812
0600	50 kW	530	73	603
0640	75 kW	460	58	518
0680	100 kW	385	49	434
0760	150 kW	310	39	349
0840	200 kW	260	32	292
0880	250 kW	246	26.50	272.50
0960	350 kW	214	21.50	235.50
1040	500 kW	233	16.95	249.95

E10 Equipment

E1010 Commercial Equipment

E1010 110	Security/Vault, EACH	COST EACH		
		MAT.	INST.	TOTAL
0100	Bank equipment, drive up window, drawer & mike, no glazing, economy	6,150	1,275	7,425
0110	Deluxe	9,775	2,550	12,325
0120	Night depository, economy	7,975	1,275	9,250
0130	Deluxe	11,300	2,550	13,850
0140	Pneumatic tube systems, 2 station, standard	27,100	4,275	31,375
0150	Teller, automated, 24 hour, single unit	47,000	4,275	51,275
0160	Teller window, bullet proof glazing, 44" x 60"	3,550	780	4,330
0170	Pass through, painted steel, 72" x 40"	3,775	1,600	5,375
0300	Safe, office type, 1 hr. rating, 34" x 20" x 20"	2,400		2,400
0310	4 hr. rating, 62" x 33" x 20"	9,375		9,375
0320	Data storage, 4 hr. rating, 63" x 44" x 16"	14,400		14,400
0330	Jewelers, 63" x 44" x 16"	24,700		24,700
0340	Money, "B" label, 9" x 14" x 14"	580		580
0350	Tool and torch resistive, 24" x 24" x 20"	10,100	196	10,296
0500	Security gates-scissors type, painted steel, single, 6' high, 5-1/2' wide	198	320	518
0510	Double gate, 7-1/2' high, 14' wide	595	635	1,230

E1010 510	Mercantile Equipment, EACH	COST EACH		
		MAT.	INST.	TOTAL
0015	Barber equipment, chair, hydraulic, economy	445	20.50	465.50
0020	Deluxe	3,025	31	3,056
0100	Checkout counter, single belt	2,725	78.50	2,803.50
0110	Double belt, power take-away	4,600	87	4,687
0200	Display cases, freestanding, glass and aluminum, 3'-6" x 3' x 1'-0" deep	1,325	124	1,449
0220	Wall mounted, glass and aluminum, 3' x 4' x 1'-4" deep	2,350	198	2,548
0320	Frozen food, chest type, 12 ft. long	7,375	325	7,700

E1010 610	Laundry/Dry Cleaning, EACH	COST EACH		
		MAT.	INST.	TOTAL
0100	Laundry equipment, dryers, gas fired, residential, 16 lb. capacity	690	195	885
0110	Commercial, 30 lb. capacity, single	3,400	195	3,595
0120	Dry cleaners, electric, 20 lb. capacity	36,600	5,725	42,325
0130	30 lb. capacity	52,500	7,625	60,125
0140	Ironers, commercial, 120" with canopy, 8 roll	178,500	16,400	194,900
0150	Institutional, 110", single roll	32,700	2,800	35,500
0160	Washers, residential, 4 cycle	880	195	1,075
0170	Commercial, coin operated, deluxe	3,325	195	3,520

E10 Equipment

E1020   Institutional Equipment

E1020 110	Ecclesiastical Equipment, EACH	COST EACH		
		MAT.	INST.	TOTAL
0090	Church equipment, altar, wood, custom, plain	2,100	355	2,455
0100	Granite, custom, deluxe	33,400	4,700	38,100
0110	Baptistry, fiberglass, economy	4,200	1,275	5,475
0120	Bells & carillons, keyboard operation	18,000	8,100	26,100
0130	Confessional, wood, single, economy	2,675	825	3,500
0140	Double, deluxe	16,700	2,475	19,175
0150	Steeples, translucent fiberglas, 30" square, 15' high	6,000	1,675	7,675

E1020 130	Ecclesiastical Equipment, L.F.	COST PER L.F.		
		MAT.	INST.	TOTAL
0100	Arch. equip., church equip. pews, bench type, hardwood, economy	81.50	25	106.50
0110	Deluxe	161	33	194

E1020 210	Library Equipment, EACH	COST EACH		
		MAT.	INST.	TOTAL
0110	Library equipment, carrels, metal, economy	265	99	364
0120	Hardwood, deluxe	865	124	989

E1020 230	Library Equipment, L.F.	COST PER L.F.		
		MAT.	INST.	TOTAL
0100	Library equipment, book shelf, metal, single face, 90" high x 10" shelf	124	41.50	165.50
0110	Double face, 90" high x 10" shelf	270	89.50	359.50
0120	Charging desk, built-in, with counter, plastic laminate	286	71	357

E1020 310	Theater and Stage Equipment, EACH	COST EACH		
		MAT.	INST.	TOTAL
0200	Movie equipment, changeover, economy	510		510
0210	Film transport, incl. platters and autowind, economy	5,525		5,525
0220	Lamphouses, incl. rectifiers, xenon, 1000W	7,300	280	7,580
0230	4000W	10,400	375	10,775
0240	Projector mechanisms, 35 mm, economy	12,200		12,200
0250	Deluxe	16,700		16,700
0260	Sound systems, incl. amplifier, single, economy	3,625	620	4,245
0270	Dual, Dolby/super sound	18,200	1,400	19,600
0280	Projection screens, wall hung, manual operation, 50 S.F., economy	325	99	424
0290	Electric operation, 100 S.F., deluxe	2,175	495	2,670

E1020 320	Theater and Stage Equipment, S.F.	COST PER S.F.		
		MAT.	INST.	TOTAL
0090	Movie equipment, projection screens, rigid in wall, acrylic, 1/4" thick	46	4.79	50.79
0100	1/2" thick	53.50	7.20	60.70
0110	Stage equipment, curtains, velour, medium weight	8.70	1.65	10.35
0120	Silica based yarn, fireproof	16.70	19.80	36.50
0130	Stages, portable with steps, folding legs, 8" high	28.50		28.50
0140	Telescoping platforms, aluminum, deluxe	41.50	26	67.50

E1020 330	Theater and Stage Equipment, L.F.	COST PER L.F.		
		MAT.	INST.	TOTAL
0100	Stage equipment, curtain track, heavy duty	65	55	120
0110	Lights, border, quartz, colored	190	28	218

E1020 610	Detention Equipment, EACH	COST EACH		
		MAT.	INST.	TOTAL
0110	Detention equipment, cell front rolling door, 7/8" bars, 5' x 7' high	6,425	1,375	7,800
0120	Cells, prefab., including front, 5' x 7' x 7' deep	10,600	1,825	12,425

E1020 Institutional Equipment

E1020 610	Detention Equipment, EACH	COST EACH		
		MAT.	INST.	TOTAL
0130	Doors and frames, 3' x 7', single plate	4,625	680	5,305
0140	Double plate	5,700	680	6,380
0150	Toilet apparatus, incl wash basin	3,700	855	4,555
0160	Visitor cubicle, vision panel, no intercom	3,225	1,375	4,600

E1020 710	Laboratory Equipment, EACH	COST EACH		
		MAT.	INST.	TOTAL
0110	Laboratory equipment, glassware washer, distilled water, economy	6,000	635	6,635
0120	Deluxe	13,400	1,150	14,550
0140	Radio isotope	18,600		18,600

E1020 720	Laboratory Equipment, S.F.	COST PER S.F.		
		MAT.	INST.	TOTAL
0100	Arch. equip., lab equip., counter tops, acid proof, economy	34.50	12.10	46.60
0110	Stainless steel	105	12.10	117.10

E1020 730	Laboratory Equipment, L.F.	COST PER L.F.		
		MAT.	INST.	TOTAL
0110	Laboratory equipment, cabinets, wall, open	142	49.50	191.50
0120	Base, drawer units	450	55	505
0130	Fume hoods, not incl. HVAC, economy	570	184	754
0140	Deluxe incl. fixtures	1,850	415	2,265

E1020 810	Medical Equipment, EACH	COST EACH		
		MAT.	INST.	TOTAL
0100	Dental equipment, central suction system, economy	1,550	490	2,040
0110	Compressor-air, deluxe	7,550	1,000	8,550
0120	Chair, hydraulic, economy	2,925	1,000	3,925
0130	Deluxe	15,100	2,025	17,125
0140	Drill console with accessories, economy	1,650	315	1,965
0150	Deluxe	5,225	315	5,540
0160	X-ray unit, portable	2,400	127	2,527
0170	Panoramic unit	15,300	845	16,145
0300	Medical equipment, autopsy table, standard	9,275	585	9,860
0310	Deluxe	18,400	975	19,375
0320	Incubators, economy	3,575		3,575
0330	Deluxe	21,100		21,100
0700	Station, scrub-surgical, single, economy	4,325	195	4,520
0710	Dietary, medium, with ice	16,800		16,800
0720	Sterilizers, general purpose, single door, 20" x 20" x 28"	16,000		16,000
0730	Floor loading, double door, 28" x 67" x 52"	213,500		213,500
0740	Surgery tables, standard	14,500	910	15,410
0750	Deluxe	32,900	1,275	34,175
0770	Tables, standard, with base cabinets, economy	1,375	330	1,705
0780	Deluxe	4,050	495	4,545
0790	X-ray, mobile, economy	14,000		14,000
0800	Stationary, deluxe	228,000		228,000

E10 Equipment

E1030 Vehicular Equipment

E1030 110	Vehicular Service Equipment, EACH	COST EACH		
		MAT.	INST.	TOTAL
0110	Automotive equipment, compressors, electric, 1-1/2 H.P., std. controls	410	925	1,335
0120	5 H.P., dual controls	2,225	1,375	3,600
0130	Hoists, single post, 4 ton capacity, swivel arms	5,475	3,475	8,950
0140	Dual post, 12 ton capacity, adjustable frame	9,400	9,250	18,650
0150	Lube equipment, 3 reel type, with pumps	8,700	2,775	11,475
0160	Product dispenser, 6 nozzles, w/vapor recovery, not incl. piping, installed	22,900		22,900
0800	Scales, dial type, built in floor, 5 ton capacity, 8' x 6' platform	8,000	2,975	10,975
0810	10 ton capacity, 9' x 7' platform	11,800	4,250	16,050
0820	Truck (including weigh bridge), 20 ton capacity, 24' x 10'	12,300	4,950	17,250

E1030 210	Parking Control Equipment, EACH	COST EACH		
		MAT.	INST.	TOTAL
0110	Parking equipment, automatic gates, 8 ft. arm, one way	4,050	1,025	5,075
0120	Traffic detectors, single treadle	2,600	465	3,065
0130	Booth for attendant, economy	6,400		6,400
0140	Deluxe	27,400		27,400
0150	Ticket printer/dispenser, rate computing	8,875	800	9,675
0160	Key station on pedestal	800	273	1,073

E1030 310	Loading Dock Equipment, EACH	COST EACH		
		MAT.	INST.	TOTAL
0110	Dock bumpers, rubber blocks, 4-1/2" thick, 10" high, 14" long	50	19.05	69.05
0120	6" thick, 20" high, 11" long	147	38	185
0130	Dock boards, H.D., 5' x 5', aluminum, 5000 lb. capacity	1,250		1,250
0140	16,000 lb. capacity	1,650		1,650
0150	Dock levelers, hydraulic, 7' x 8', 10 ton capacity	9,325	1,325	10,650
0160	Dock lifters, platform, 6' x 6', portable, 3000 lb. capacity	9,150		9,150
0170	Dock shelters, truck, scissor arms, economy	1,625	495	2,120
0180	Deluxe	2,100	990	3,090

E10 Equipment

E1090 Other Equipment

E1090 210	Solid Waste Handling Equipment, EACH	COST EACH		
		MAT.	INST.	TOTAL
0110	Waste handling, compactors, single bag, 250 lbs./hr., hand fed	8,775	580	9,355
0120	Heavy duty industrial, 5 C.Y. capacity	20,100	2,775	22,875
0130	Incinerator, electric, 100 lbs./hr., economy	12,800	2,800	15,600
0140	Gas, 2000 lbs./hr., deluxe	554,500	22,300	576,800
0150	Shredder, no baling, 35 tons/hr.	288,500		288,500
0160	Incl. baling, 50 tons/day	576,000		576,000

E1090 350	Food Service Equipment, EACH	COST EACH		
		MAT.	INST.	TOTAL
0110	Kitchen equipment, bake oven, single deck	4,750	132	4,882
0120	Broiler, without oven	3,900	132	4,032
0130	Commercial dish washer, semiautomatic, 50 racks/hr.	9,175	810	9,985
0140	Automatic, 275 racks/hr.	30,200	3,450	33,650
0150	Cooler, beverage, reach-in, 6 ft. long	4,750	176	4,926
0160	Food warmer, counter, 1.65 kw	735		735
0170	Fryers, with submerger, single	1,600	151	1,751
0180	Double	2,625	211	2,836
0185	Ice maker, 1000 lb. per day, with bin	5,250	1,050	6,300
0190	Kettles, steam jacketed, 20 gallons	7,350	238	7,588
0200	Range, restaurant type, burners, 2 ovens and 24" griddle	8,200	176	8,376
0210	Range hood, incl. carbon dioxide system, economy	4,500	350	4,850
0220	Deluxe	34,800	1,050	35,850

E1090 360	Food Service Equipment, S.F.	COST PER S.F.		
		MAT.	INST.	TOTAL
0110	Refrigerators, prefab, walk-in, 7'-6" high, 6' x 6'	146	18.10	164.10
0120	12' x 20'	92	9.05	101.05

E1090 410	Residential Equipment, EACH	COST EACH		
		MAT.	INST.	TOTAL
0110	Arch. equip., appliances, range, cook top, 4 burner, economy	268	93.50	361.50
0120	Built in, single oven 30" wide, economy	760	93.50	853.50
0130	Standing, single oven-21" wide, economy	345	78.50	423.50
0140	Double oven-30" wide, deluxe	1,800	78.50	1,878.50
0150	Compactor, residential, economy	530	99	629
0160	Deluxe	565	165	730
0170	Dish washer, built-in, 2 cycles, economy	227	286	513
0180	4 or more cycles, deluxe	1,125	570	1,695
0190	Garbage disposer, sink type, economy	73.50	114	187.50
0200	Deluxe	211	114	325
0210	Refrigerator, no frost, 10 to 12 C.F., economy	490	78.50	568.50
0220	21 to 29 C.F., deluxe	3,250	261	3,511
0300	Washing machine, automatic	1,200	585	1,785

E1090 610	School Equipment, EACH	COST EACH		
		MAT.	INST.	TOTAL
0110	School equipment, basketball backstops, wall mounted, wood, fixed	1,250	870	2,120
0120	Suspended type, electrically operated	5,350	1,675	7,025
0130	Bleachers-telescoping, manual operation, 15 tier, economy (per seat)	83.50	31	114.50
0140	Power operation, 30 tier, deluxe (per seat)	320	54.50	374.50
0150	Weight lifting gym, universal, economy	435	785	1,220
0160	Deluxe	16,200	1,575	17,775
0170	Scoreboards, basketball, 1 side, economy	2,450	735	3,185
0180	4 sides, deluxe	6,175	10,100	16,275
0800	Vocational shop equipment, benches, metal	405	198	603
0810	Wood	555	198	753

E10 Equipment

E1090 Other Equipment

E1090 610	School Equipment, EACH	COST EACH		
		MAT.	INST.	TOTAL
0820	Dust collector, not incl. ductwork, 6' diam.	3,300	530	3,830
0830	Planer, 13" x 6"	1,375	248	1,623

E1090 620	School Equipment, S.F.	COST PER S.F.		
		MAT.	INST.	TOTAL
0110	School equipment, gym mats, naugahyde cover, 2" thick	3.95		3.95
0120	Wrestling, 1" thick, heavy duty	6.05		6.05

E1090 810	Athletic, Recreational, and Therapeutic Equipment, EACH	COST EACH		
		MAT.	INST.	TOTAL
0050	Bolwling alley with gutters	583,000	125,000	708,000
0060	Combo. table and ball rack	7,500		7,500
0070	Bowling alley automatic scorer	131,000		131,000
0110	Sauna, prefabricated, incl. heater and controls, 7' high, 6' x 4'	5,100	755	5,855
0120	10' x 12'	12,300	1,675	13,975
0130	Heaters, wall mounted, to 200 C.F.	665		665
0140	Floor standing, to 1000 C.F., 12500 W	2,775	187	2,962
0610	Shooting range incl. bullet traps, controls, separators, ceilings, economy	13,200	6,550	19,750
0620	Deluxe	40,700	11,100	51,800
0650	Sport court, squash, regulation, in existing building, economy			38,500
0660	Deluxe			42,900
0670	Racketball, regulation, in existing building, economy	41,700	9,400	51,100
0680	Deluxe	45,100	18,800	63,900
0700	Swimming pool equipment, diving stand, stainless steel, 1 meter	6,025	365	6,390
0710	3 meter	10,300	2,475	12,775
0720	Diving boards, 16 ft. long, aluminum	3,375	365	3,740
0730	Fiberglass	2,675	365	3,040
0740	Filter system, sand, incl. pump, 6000 gal./hr.	1,600	650	2,250
0750	Lights, underwater, 12 volt with transformer, 300W	268	560	828
0760	Slides, fiberglass with aluminum handrails & ladder, 6' high, straight	4,000	620	4,620
0780	12' high, straight with platform	5,625	825	6,450

E1090 820	Athletic, Recreational, and Therapeutic Equipment, S.F.	COST PER S.F.		
		MAT.	INST.	TOTAL
0110	Swimming pools, residential, vinyl liner, metal sides	11.60	6.80	18.40
0120	Concrete sides	13.90	12.40	26.30
0130	Gunite shell, plaster finish, 350 S.F.	26	26	52
0140	800 S.F.	21	14.90	35.90
0150	Motel, gunite shell, plaster finish	32	32	64
0160	Municipal, gunite shell, tile finish, formed gutters	122	56	178

E20 Furnishings

E2010 Fixed Furnishings

E2010 310	Window Treatment, EACH	COST EACH		
		MAT.	INST.	TOTAL
0110	Furnishings, blinds, exterior, aluminum, louvered, 1'-4" wide x 3'-0" long	61	49.50	110.50
0120	1'-4" wide x 6'-8" long	123	55	178
0130	Hemlock, solid raised, 1'-4" wide x 3'-0" long	44.50	49.50	94
0140	1'-4" wide x 6'-9" long	72.50	55	127.50
0150	Polystyrene, louvered, 1'-3" wide x 3'-3" long	30.50	49.50	80
0160	1'-3" wide x 6'-8" long	51	55	106
0200	Interior, wood folding panels, louvered, 7" x 20" (per pair)	50.50	29	79.50
0210	18" x 40" (per pair)	185	29	214

E2010 320	Window Treatment, S.F.	COST PER S.F.		
		MAT.	INST.	TOTAL
0110	Furnishings, blinds-interior, venetian-aluminum, stock, 2" slats, economy	2.02	.84	2.86
0120	Custom, 1" slats, deluxe	7.85	1.13	8.98
0130	Vertical, PVC or cloth, T&B track, economy	5.95	1.08	7.03
0140	Deluxe	12.90	1.24	14.14
0150	Draperies, unlined, economy	3.05		3.05
0160	Lightproof, deluxe	9.15		9.15
0510	Shades, mylar, wood roller, single layer, non-reflective	6.70	.72	7.42
0520	Metal roller, triple layer, heat reflective	11.95	.72	12.67
0530	Vinyl, light weight, 4 ga.	.61	.72	1.33
0540	Heavyweight, 6 ga.	1.84	.72	2.56
0550	Vinyl coated cotton, lightproof decorator shades	2.31	.72	3.03
0560	Woven aluminum, 3/8" thick, light and fireproof	5.90	1.42	7.32

E2010 420	Fixed Floor Grilles and Mats, S.F.	COST PER S.F.		
		MAT.	INST.	TOTAL
0110	Floor mats, recessed, inlaid black rubber, 3/8" thick, solid	21.50	2.53	24.03
0120	Colors, 1/2" thick, perforated	25.50	2.53	28.03
0130	Link-including nosings, steel-galvanized, 3/8" thick	18.60	2.53	21.13
0140	Vinyl, in colors	22.50	2.53	25.03

E2010 510	Fixed Multiple Seating, EACH	COST EACH		
		MAT.	INST.	TOTAL
0110	Seating, painted steel, upholstered, economy	145	28.50	173.50
0120	Deluxe	465	35.50	500.50
0400	Seating, lecture hall, pedestal type, economy	182	45	227
0410	Deluxe	610	68.50	678.50
0500	Auditorium chair, veneer construction	193	45	238
0510	Fully upholstered, spring seat	232	45	277

E20 Furnishings

E2020 Moveable Furnishings

E2020 210	Furnishings/EACH	COST EACH		
		MAT.	INST.	TOTAL
0200	Hospital furniture, beds, manual, economy	800		800
0210	Deluxe	2,750		2,750
0220	All electric, economy	1,450		1,450
0230	Deluxe	4,525		4,525
0240	Patient wall systems, no utilities, economy, per room	1,125		1,125
0250	Deluxe, per room	2,100		2,100
0300	Hotel furnishings, standard room set, economy, per room	2,550		2,550
0310	Deluxe, per room	9,125		9,125
0500	Office furniture, standard employee set, economy, per person	505		505
0510	Deluxe, per person	2,250		2,250
0550	Posts, portable, pedestrian traffic control, economy	114		114
0560	Deluxe	340		340
0700	Restaurant furniture, booth, molded plastic, stub wall and 2 seats, economy	355	248	603
0710	Deluxe	1,225	330	1,555
0720	Upholstered seats, foursome, single-economy	650	99	749
0730	Foursome, double-deluxe	1,500	165	1,665

E2020 220	Furniture and Accessories, L.F.	COST PER L.F.		
		MAT.	INST.	TOTAL
0210	Dormitory furniture, desk top (built-in),laminated plastc, 24"deep, economy	40.50	19.80	60.30
0220	30" deep, deluxe	228	25	253
0230	Dressing unit, built-in, economy	223	82.50	305.50
0240	Deluxe	670	124	794
0310	Furnishings, cabinets, hospital, base, laminated plastic	258	99	357
0320	Stainless steel	475	99	574
0330	Counter top, laminated plastic, no backsplash	44.50	25	69.50
0340	Stainless steel	144	25	169
0350	Nurses station, door type, laminated plastic	299	99	398
0360	Stainless steel	570	99	669
0710	Restaurant furniture, bars, built-in, back bar	190	99	289
0720	Front bar	262	99	361
0910	Wardrobes & coatracks, standing, steel, single pedestal, 30" x 18" x 63"	103		103
0920	Double face rack, 39" x 26" x 70"	105		105
0930	Wall mounted rack, steel frame & shelves, 12" x 15" x 26"	76.50	7.05	83.55
0940	12" x 15" x 50"	47	3.69	50.69

F10 Special Construction

F1010 Special Structures

F1010 120	Air-Supported Structures, S.F.	COST PER S.F.		
		MAT.	INST.	TOTAL
0110	Air supported struc., polyester vinyl fabric, 24 oz., warehouse, 5000 S.F.	26.50	.31	26.81
0120	50,000 S.F.	12.35	.25	12.60
0130	Tennis, 7,200 S.F.	24.50	.26	24.76
0140	24,000 S.F.	17.30	.26	17.56
0150	Woven polyethylene, 6 oz., shelter, 3,000 S.F.	7.25	.52	7.77
0160	24,000 S.F.	6.15	.26	6.41
0170	Teflon coated fiberglass, stadium cover, economy	60.50	.14	60.64
0180	Deluxe	71.50	.19	71.69
0190	Air supported storage tank covers, reinf. vinyl fabric, 12 oz., 400 S.F.	24.50	.44	24.94
0200	18,000 S.F.	7.75	.40	8.15

F1010 210	Pre-Engineered Structures, EACH	COST EACH		
		MAT.	INST.	TOTAL
0600	Radio towers, guyed, 40 lb. section, 50' high, 70 MPH basic wind speed	3,600	1,275	4,875
0610	90 lb. section, 400' high, wind load 70 MPH basic wind speed	46,700	13,900	60,600
0620	Self supporting, 60' high, 70 MPH basic wind speed	7,500	2,425	9,925
0630	190' high, wind load 90 MPH basic wind speed	44,600	9,750	54,350
0700	Shelters, aluminum frame, acrylic glazing, 8' high, 3' x 9'	3,475	1,125	4,600
0710	9' x 12'	8,050	1,750	9,800

F1010 320	Other Special Structures, S.F.	COST PER S.F.		
		MAT.	INST.	TOTAL
0110	Swimming pool enclosure, transluscent, freestanding, economy	16.25	4.96	21.21
0120	Deluxe	247	14.15	261.15
0510	Tension structures, steel frame, polyester vinyl fabric, 12,000 S.F.	12.70	2.28	14.98
0520	20,800 S.F.	12.90	2.06	14.96

F1010 330	Special Structures, EACH	COST EACH		
		MAT.	INST.	TOTAL
0110	Kiosks, round, 5' diam., 1/4" fiberglass wall, 8' high	6,375		6,375
0120	Rectangular, 5' x 9', 1" insulated dbl. wall fiberglass, 7'-6" high	11,000		11,000
0200	Silos, concrete stave, industrial, 12' diam., 35' high	46,600	20,500	67,100
0210	25' diam., 75' high	175,000	45,100	220,100
0220	Steel prefab, 30,000 gal., painted, economy	20,900	5,250	26,150
0230	Epoxy-lined, deluxe	43,300	10,500	53,800

F1010 340	Special Structures, S.F.	COST PER S.F.		
		MAT.	INST.	TOTAL
0110	Comfort stations, prefab, mobile on steel frame, economy	51		51
0120	Permanent on concrete slab, deluxe	315	45.50	360.50
0210	Domes, bulk storage, wood framing, wood decking, 50' diam.	42.50	1.67	44.17
0220	116' diam.	25.50	1.93	27.43
0230	Steel framing, metal decking, 150' diam.	30.50	11.10	41.60
0240	400' diam.	25	8.45	33.45
0250	Geodesic, wood framing, wood panels, 30' diam.	30.50	1.79	32.29
0260	60' diam.	23	1.18	24.18
0270	Aluminum framing, acrylic panels, 40' diam.			116
0280	Aluminum panels, 400' diam.			26
0310	Garden house, prefab, wood, shell only, 48 S.F.	36	4.96	40.96
0320	200 S.F.	41	20.50	61.50
0410	Greenhouse, shell-stock, residential, lean-to, 8'-6" long x 3'-10" wide	42.50	29	71.50
0420	Freestanding, 8'-6" long x 13'-6" wide	38	9.20	47.20
0430	Commercial-truss frame, under 2000 S.F., deluxe			34.50
0440	Over 5,000 S.F., economy			23.50
0450	Institutional-rigid frame, under 500 S.F., deluxe			84
0460	Over 2,000 S.F., economy			33.50
0510	Hangar, prefab, galv. roof and walls, bottom rolling doors, economy	14.65	4.39	19.04
0520	Electric bifolding doors, deluxe	16.75	5.70	22.45

F10 Special Construction

F1020 Integrated Construction

F1020 110 — Integrated Construction, EACH

		COST EACH		
		MAT.	INST.	TOTAL
0110	Integrated ceilings, radiant electric, 2' x 4' panel, manila finish	99	22.50	121.50
0120	ABS plastic finish	112	43	155

F1020 120 — Integrated Construction, S.F.

		COST PER S.F.		
		MAT.	INST.	TOTAL
0110	Integrated ceilings, Luminaire, suspended, 5' x 5' modules, 50% lighted	4.32	11.85	16.17
0120	100% lighted	6.45	21.50	27.95
0130	Dimensionaire, 2' x 4' module tile system, no air bar	2.61	1.98	4.59
0140	With air bar, deluxe	3.83	1.98	5.81
0220	Pedestal access floors, stl pnls, no stringers, vinyl covering, >6000 S.F.	11.45	1.55	13
0230	With stringers, high pressure laminate covering, under 6000 S.F.	10.65	1.65	12.30
0240	Carpet covering, under 6000 S.F.	14.05	1.65	15.70
0250	Aluminum panels, no stringers, no covering	34.50	1.98	36.48

F1020 250 — Special Purpose Room, EACH

		COST EACH		
		MAT.	INST.	TOTAL
0110	Portable booth, acoustical, 27 db 1000 hz., 15 S.F. floor	3,975		3,975
0120	55 S.F. flr.	8,175		8,175

F1020 260 — Special Purpose Room, S.F.

		COST PER S.F.		
		MAT.	INST.	TOTAL
0110	Anechoic chambers, 7' high, 100 cps cutoff, 25 S.F.			2,550
0120	200 cps cutoff, 100 S.F.			1,325
0130	Audiometric rooms, under 500 S.F.	56.50	20	76.50
0140	Over 500 S.F.	53.50	16.50	70
0300	Darkrooms, shell, not including door, 240 S.F., 8' high	29.50	8.25	37.75
0310	64 S.F., 12' high	75	15.50	90.50
0510	Music practice room, modular, perforated steel, under 500 S.F.	34.50	14.15	48.65
0520	Over 500 S.F.	29.50	12.40	41.90

F1020 330 — Special Construction, L.F.

		COST PER L.F.		
		MAT.	INST.	TOTAL
0110	Spec. const., air curtains, shipping & receiving, 8' high x 5' wide, economy	385	116	501
0120	20' high x 8' wide, heated, deluxe	1,375	291	1,666
0130	Customer entrance, 10' high x 5' wide, economy	385	116	501
0140	12' high x 4' wide, heated, deluxe	720	145	865

F10 Special Construction

F1030 Special Construction Systems

F1030 120	Sound, Vibration, and Seismic Construction, S.F.	COST PER S.F.		
		MAT.	INST.	TOTAL
0020	Special construction, acoustical, enclosure, 4" thick, 8 psf panels	35.50	20.50	56
0030	Reverb chamber, 4" thick, parallel walls	50	25	75
0110	Sound absorbing panels, 2'-6" x 8', painted metal	13.30	6.90	20.20
0120	Vinyl faced	10.35	6.20	16.55
0130	Flexible transparent curtain, clear	8	8.10	16.10
0140	With absorbing foam, 75% coverage	11.20	8.10	19.30
0150	Strip entrance, 2/3 overlap	8.25	12.90	21.15
0160	Full overlap	6.95	15.15	22.10
0200	Audio masking system, plenum mounted, over 10,000 S.F.	.79	.25	1.04
0210	Ceiling mounted, under 5,000 S.F.	1.38	.47	1.85

F1030 210	Radiation Protection, EACH	COST EACH		
		MAT.	INST.	TOTAL
0110	Shielding, lead x-ray protection, radiography room, 1/16" lead, economy	8,900	3,375	12,275
0120	Deluxe	10,700	5,600	16,300
0210	Deep therapy x-ray, 1/4" lead, economy	24,800	10,500	35,300
0220	Deluxe	30,600	14,000	44,600

F1030 220	Radiation Protection, S.F.	COST PER S.F.		
		MAT.	INST.	TOTAL
0110	Shielding, lead, gypsum board, 5/8" thick, 1/16" lead	8	5.85	13.85
0120	1/8" lead	15.50	6.65	22.15
0130	Lath, 1/16" thick	7.90	6.25	14.15
0140	1/8" thick	15.90	7	22.90
0150	Radio frequency, copper, prefab screen type, economy	34.50	5.50	40
0160	Deluxe	39	6.85	45.85

F1030 910	Other Special Construction Systems, EACH	COST EACH		
		MAT.	INST.	TOTAL
0110	Disappearing stairways, folding, pine, 8'-6" ceiling	215	124	339
0200	Fire escape, galvanized steel, 8'-0" to 10'-6" ceiling	1,675	990	2,665
0220	Automatic electric, wood, 8' to 9' ceiling	9,075	990	10,065
0300	Fireplace prefabricated, freestanding or wall hung, economy	1,275	380	1,655
0310	Deluxe	3,425	550	3,975
0320	Woodburning stoves, cast iron, economy	1,250	760	2,010
0330	Deluxe	3,025	1,250	4,275

F10 Special Construction

F1040 Special Facilities

F1040 210	Ice Rinks, EACH	COAT EACH		
		MAT.	INST.	TOTAL
0100	Ice skating rink, 85' x 200', 55° system, 5 mos., 100 ton			655,500
0110	90° system, 12 mos., 135 ton			715,000
0120	Dash boards, acrylic screens, polyethylene coated plywood	143,000	33,500	176,500
0130	Fiberglass and aluminum construction	154,000	33,500	187,500

F1040 510	Liquid & Gas Storage Tanks, EACH	COST EACH		
		MAT.	INST.	TOTAL
0100	Tanks, steel, ground level, 100,000 gal.			228,000
0110	10,000,000 gal.			4,124,500
0120	Elevated water, 50,000 gal.		177,500	388,000
0130	1,000,000 gal.		828,500	1,811,000
0150	Cypress wood, ground level, 3,000 gal.	9,675	9,900	19,575
0160	Redwood, ground level, 45,000 gal.	73,500	26,800	100,300

F10 Special Construction

F1040 Special Facilities

F1040 910	Special Construction, EACH	COST EACH		
		MAT.	INST.	TOTAL
0110	Special construction, bowling alley incl. pinsetter, scorer etc., economy	44,300	9,900	54,200
0120	Deluxe	56,000	11,000	67,000
0130	For automatic scorer, economy, add	9,050		9,050
0140	Deluxe, add	10,900		10,900
0200	Chimney, metal, high temp. steel jacket, factory lining, 24" diameter	11,600	4,675	16,275
0210	60" diameter	84,000	20,300	104,300
0220	Poured concrete, brick lining, 10' diam., 200' high			1,600,000
0230	20' diam., 500' high			7,000,000
0240	Radial brick, 3'-6" I.D., 75' high			268,000
0250	7' I.D., 175' high			800,500
0300	Control tower, modular, 12' x 10', incl. instrumentation, economy			737,000
0310	Deluxe			1,028,500
0400	Garage costs, residential, prefab, wood, single car economy	5,600	990	6,590
0410	Two car deluxe	11,800	1,975	13,775
0500	Hangars, prefab, galv. steel, bottom rolling doors, economy (per plane)	14,200	4,850	19,050
0510	Electrical bi-folding doors, deluxe (per plane)	22,700	6,675	29,375

G10 Site Preparation

G1030 Site Earthwork

G1030 805	Trenching Common Earth	COST PER L.F.		
		MAT.	INST.	TOTAL
1310	Trenching, common earth, no slope, 2' wide, 2' deep, 3/8 C.Y. bucket		2.17	2.17
1330	4' deep, 3/8 C.Y. bucket		4.29	4.29
1360	10' deep, 1 C.Y. bucket		9.72	9.72
1400	4' wide, 2' deep, 3/8 C.Y. bucket		4.45	4.45
1420	4' deep, 1/2 C.Y. bucket		7.55	7.55
1450	10' deep, 1 C.Y. bucket		20.70	20.70
1480	18' deep, 2-1/2 C.Y. bucket		32.60	32.60
1520	6' wide, 6' deep, 5/8 C.Y. bucket w/trench box		20.35	20.35
1540	10' deep, 1 C.Y. bucket		27.10	27.10
1570	20' deep, 3-1/2 C.Y. bucket		46.85	46.85
1640	8' wide, 12' deep, 1-1/2 C.Y. bucket w/trench box		38.10	38.10
1680	24' deep, 3-1/2 C.Y. bucket		73.50	73.50
1730	10' wide, 20' deep, 3-1/2 C.Y. bucket w/trench box		65.50	65.50
1740	24' deep, 3-1/2 C.Y. bucket		86.50	86.50
3500	1 to 1 slope, 2' wide, 2' deep, 3/8 C.Y. bucket		4.30	4.30
3540	4' deep, 3/8 C.Y. bucket		12.73	12.73
3600	10' deep, 1 C.Y. bucket		58	58
3800	4' wide, 2' deep, 3/8 C.Y. bucket		6.56	6.56
3840	4' deep, 1/2 C.Y. bucket		14.93	14.93
3900	10' deep, 1 C.Y. bucket		73	73
4030	6' wide, 6' deep, 5/8 C.Y. bucket w/trench box		40.35	40.35
4050	10' deep, 1 C.Y. bucket		65	65
4080	20' deep, 3-1/2 C.Y. bucket		198	198
4500	8' wide, 12' deep, 1-1/2 C.Y. bucket w/trench box		97	97
4650	24' deep, 3-1/2 C.Y. bucket		300	300
4800	10' wide, 20' deep, 3-1/2 C.Y. bucket w/trench box		201	201
4850	24' deep, 3-1/2 C.Y. bucket		318	318

G1030 806	Trenching Loam & Sandy Clay	COST PER L.F.		
		MAT.	INST.	TOTAL
1310	Trenching, loam & sandy clay, no slope, 2' wide, 2' deep, 3/8 C.Y. bucket		2.14	2.14

G1030 807	Trenching Sand & Gravel	COST PER L.F.		
		MAT.	INST.	TOTAL
1310	Trenching, sand & gravel, no slope, 2' wide, 2' deep, 3/8 C.Y. bucket		2.01	2.01
1320	3' deep, 3/8 C.Y. bucket		3.33	3.33
1330	4' deep, 3/8 C.Y. bucket		3.97	3.97
1340	6' deep, 3/8 C.Y. bucket		5.05	5.05
1350	8' deep, 1/2 C.Y. bucket		6.70	6.70
1360	10' deep, 1 C.Y. bucket		7.05	7.05
1400	4' wide, 2' deep, 3/8 C.Y. bucket		4.10	4.10
1410	3' deep, 3/8 C.Y. bucket		6.05	6.05
1420	4' deep, 1/2 C.Y. bucket		6.95	6.95
1430	6' deep, 1/2 C.Y. bucket		10.85	10.85
1440	8' deep, 1/2 C.Y. bucket		15.60	15.60
1450	10' deep, 1 C.Y. bucket		15.50	15.50
1460	12' deep, 1 C.Y. bucket		19.35	19.35
1470	15' deep, 1-1/2 C.Y. bucket		23	23
1480	18' deep, 2-1/2 C.Y. bucket		25.50	25.50
1520	6' wide, 6' deep, 5/8 C.Y. bucket w/trench box		18.80	18.80
1530	8' deep, 3/4 C.Y. bucket		25	25
1540	10' deep, 1 C.Y. bucket		24	24
1550	12' deep, 1-1/2 C.Y. bucket		27	27
1560	16' deep, 2 C.Y. bucket		35.50	35.50
1570	20' deep, 3-1/2 C.Y. bucket		42.50	42.50
1580	24' deep, 3-1/2 C.Y. bucket		52.50	52.50

G1030 Site Earthwork

G1030 807	Trenching Sand & Gravel	COST PER L.F.		
		MAT.	INST.	TOTAL
1640	8' wide, 12' deep, 1-1/2 C.Y. bucket w/trench box		36	36
1650	15' deep, 1-1/2 C.Y. bucket		45	45
1660	18' deep, 2-1/2 C.Y. bucket		51.50	51.50
1680	24' deep, 3-1/2 C.Y. bucket		69	69
1730	10' wide, 20' deep, 3-1/2 C.Y. bucket w/trench box		69	69
1740	24' deep, 3-1/2 C.Y. bucket		86	86
1780	12' wide, 20' deep, 3-1/2 C.Y. bucket w/trench box		82	82
1790	25' deep, bucket		107	107
1800	1/2:1 slope, 2' wide, 2' deep, 3/8 CY bk		3	3
1810	3' deep, 3/8 C.Y. bucket		5.20	5.20
1820	4' deep, 3/8 C.Y. bucket		7.90	7.90
1840	6' deep, 3/8 C.Y. bucket		12.50	12.50
1860	8' deep, 1/2 C.Y. bucket		20	20
1880	10' deep, 1 C.Y. bucket		25	25
2300	4' wide, 2' deep, 3/8 C.Y. bucket		5.10	5.10
2310	3' deep, 3/8 C.Y. bucket		8.25	8.25
2320	4' deep, 1/2 C.Y. bucket		10.45	10.45
2340	6' deep, 1/2 C.Y. bucket		19	19
2360	8' deep, 1/2 C.Y. bucket		31.50	31.50
2380	10' deep, 1 C.Y. bucket		35	35
2400	12' deep, 1 C.Y. bucket		59	59
2430	15' deep, 1-1/2 C.Y. bucket		66	66
2460	18' deep, 2-1/2 C.Y. bucket		104	104
2840	6' wide, 6' deep, 5/8 C.Y. bucket w/trench box		28	28
2860	8' deep, 3/4 C.Y. bucket		41.50	41.50
2880	10' deep, 1 C.Y. bucket		44.50	44.50
2900	12' deep, 1-1/2 C.Y. bucket		55	55
2940	16' deep, 2 C.Y. bucket		83.50	83.50
2980	20' deep, 3-1/2 C.Y. bucket		114	114
3020	24' deep, 3-1/2 C.Y. bucket		159	159
3100	8' wide, 12' deep, 1-1/4 C.Y. bucket w/trench box		64	64
3120	15' deep, 1-1/2 C.Y. bucket		88.50	88.50
3140	18' deep, 2-1/2 C.Y. bucket		112	112
3180	24' deep, 3-1/2 C.Y. bucket		176	176
3270	10' wide, 20' deep, 3-1/2 C.Y. bucket w/trench box		142	142
3280	24' deep, 3-1/2 C.Y. bucket		193	193
3370	12' wide, 20' deep, 3-1/2 C.Y. bucket w/trench box		157	157
3380	25' deep, bucket		222	222
3500	1:1 slope, 2' wide, 2' deep, 3/8 CY bk		5.30	5.30
3520	3' deep, 3/8 C.Y. bucket		7.40	7.40
3540	4' deep, 3/8 C.Y. bucket		11.80	11.80
3560	6' deep, 3/8 C.Y. bucket		12.50	12.50
3580	8' deep, 1/2 C.Y. bucket		33.50	33.50
3600	10' deep, 1 C.Y. bucket		42.50	42.50
3800	4' wide, 2' deep, 3/8 C.Y. bucket		6.05	6.05
3820	3' deep, 3/8 C.Y. bucket		10.45	10.45
3840	4' deep, 1/2 C.Y. bucket		13.85	13.85
3860	6' deep, 1/2 C.Y. bucket		27	27
3880	8' deep, 1/2 C.Y. bucket		46.50	46.50
3900	10' deep, 1 C.Y. bucket		55	55
3920	12' deep, 1 C.Y. bucket		78.50	78.50
3940	15' deep, 1-1/2 C.Y. bucket		110	110
3960	18' deep, 2-1/2 C.Y. bucket		143	143
4030	6' wide, 6' deep, 5/8 C.Y. bucket w/trench box		37.50	37.50
4040	8' deep, 3/4 C.Y. bucket		58.50	58.50
4050	10' deep, 1 C.Y. bucket		65	65
4060	12' deep, 1-1/2 C.Y. bucket		82.50	82.50
4070	16' deep, 2 C.Y. bucket		132	132

G10 Site Preparation

G1030 Site Earthwork

G1030 807	Trenching Sand & Gravel	COST PER L.F.		
		MAT.	INST.	TOTAL
4080	20' deep, 3-1/2 C.Y. bucket		186	186
4090	24' deep, 3-1/2 C.Y. bucket		265	265
4500	8' wide, 12' deep, 1-1/2 C.Y. bucket w/trench box		91.50	91.50
4550	15' deep, 1-1/2 C.Y. bucket		131	131
4600	18' deep, 2-1/2 C.Y. bucket		173	173
4650	24' deep, 3-1/2 C.Y. bucket		281	281
4800	10' wide, 20' deep, 3-1/2 C.Y. bucket w/trench box		213	213
4850	24' deep, 3-1/2 C.Y. bucket		299	299
4950	12' wide, 20' deep, 3-1/2 C.Y. bucket w/trench box		227	227
4980	25' deep, bucket		340	340

G1030 815	Pipe Bedding	COST PER L.F.		
		MAT.	INST.	TOTAL
1440	Pipe bedding, side slope 0 to 1, 1' wide, pipe size 6" diameter	.77	.57	1.34
1460	2' wide, pipe size 8" diameter	1.71	1.27	2.98
1500	Pipe size 12" diameter	1.80	1.34	3.14
1600	4' wide, pipe size 20" diameter	4.48	3.32	7.80
1660	Pipe size 30" diameter	4.75	3.52	8.27
1680	6' wide, pipe size 32" diameter	8.30	6.15	14.45
1740	8' wide, pipe size 60" diameter	13.85	10.20	24.05
1780	12' wide, pipe size 84" diameter	27	20	47

G20 Site Improvements

G2010 Roadways

G2010 210	Roadway Pavement	COST PER L.F.		
		MAT.	INST.	TOTAL
1500	Roadways, bituminous conc. paving, 2-1/2" thick, 20' wide	62.50	54.50	117
1580	30' wide	80.50	61.50	142
1800	3" thick, 20' wide	66.50	53.50	120
1880	30' wide	86.50	62.50	149
2100	4" thick, 20' wide	73	53.50	126.50
2180	30' wide	96	62.50	158.50
2400	5" thick, 20' wide	87	54	141
2480	30' wide	117	63	180
3000	6" thick, 20' wide	95	55.50	150.50
3080	30' wide	129	66	195
3300	8" thick 20' wide	122	56	178
3380	30' wide	169	66.50	235.50
3600	12" thick 20' wide	153	59.50	212.50
3700	32' wide	228	74.50	302.50

G2020 210	Parking Lots	COST PER CAR		
		MAT.	INST.	TOTAL
1500	Parking lot, 90° angle parking, 3" bituminous paving, 6" gravel base	835	355	1,190
1540	10" gravel base	980	415	1,395
1560	4" bituminous paving, 6" gravel base	960	355	1,315
1600	10" gravel base	1,100	415	1,515
1620	6" bituminous paving, 6" gravel base	1,225	370	1,595
1660	10" gravel base	1,375	430	1,805
1800	60° angle parking, 3" bituminous paving, 6" gravel base	835	355	1,190
1840	10" gravel base	980	415	1,395
1860	4" bituminous paving, 6" gravel base	960	355	1,315
1900	10" gravel base	1,100	415	1,515
1920	6" bituminous paving, 6" gravel base	1,225	370	1,595
1960	10" gravel base	1,375	430	1,805
2200	45° angle parking, 3" bituminous paving, 6" gravel base	855	365	1,220
2240	10" gravel base	1,000	425	1,425
2260	4" bituminous paving, 6" gravel base	980	365	1,345
2300	10" gravel base	1,125	425	1,550
2320	6" bituminous paving, 6" gravel base	1,275	380	1,655
2360	10" gravel base	1,425	440	1,865

G2040 810	Flagpoles	COST EACH		
		MAT.	INST.	TOTAL
0110	Flagpoles, on grade, aluminum, tapered, 20' high	1,075	575	1,650
0120	70' high	8,675	1,425	10,100
0130	Fiberglass, tapered, 23' high	1,200	575	1,775
0140	59' high	6,950	1,275	8,225
0150	Concrete, internal halyard, 20' high	1,725	455	2,180
0160	100' high	17,100	1,150	18,250

G2040 950	Other Site Development, EACH	COST EACH		
		MAT.	INST.	TOTAL
0110	Grandstands, permanent, closed deck, steel, economy (per seat)			60.50
0120	Deluxe (per seat)			121
0130	Composite design, economy (per seat)			143
0140	Deluxe (per seat)			286

G30 Site Mechanical Utilities

G3030 Storm Sewer

G3030 210	Manholes & Catch Basins	COST PER EACH		
		MAT.	INST.	TOTAL
1920	Manhole/catch basin, brick, 4' I.D. riser, 4' deep	1,050	1,550	2,600
1980	10' deep	2,325	3,625	5,950
3200	Block, 4' I.D. riser, 4' deep	995	1,250	2,245
3260	10' deep	1,950	2,975	4,925
4620	Concrete, cast-in-place, 4' I.D. riser, 4' deep	1,225	2,125	3,350
4680	10' deep	2,750	5,100	7,850
5820	Concrete, precast, 4' I.D. riser, 4' deep	1,575	1,125	2,700
5880	10' deep	3,025	2,575	5,600
6200	6' I.D. riser, 4' deep	2,425	1,650	4,075
6260	10' deep	4,975	3,875	8,850

G4020 Site Lighting

Light Pole

G4020 210	Light Pole (Installed)	COST EACH		
		MAT.	INST.	TOTAL
0200	Light pole, aluminum, 20' high, 1 arm bracket	1,150	985	2,135
0240	2 arm brackets	1,275	985	2,260
0280	3 arm brackets	1,400	1,025	2,425
0320	4 arm brackets	1,525	1,025	2,550
0360	30' high, 1 arm bracket	2,075	1,225	3,300
0400	2 arm brackets	2,200	1,225	3,425
0440	3 arm brackets	2,325	1,275	3,600
0480	4 arm brackets	2,450	1,275	3,725
0680	40' high, 1 arm bracket	2,550	1,650	4,200
0720	2 arm brackets	2,675	1,650	4,325
0760	3 arm brackets	2,800	1,700	4,500
0800	4 arm brackets	2,925	1,700	4,625
0840	Steel, 20' high, 1 arm bracket	1,400	1,050	2,450
0880	2 arm brackets	1,500	1,050	2,550
0920	3 arm brackets	1,525	1,075	2,600
0960	4 arm brackets	1,650	1,075	2,725
1000	30' high, 1 arm bracket	1,600	1,325	2,925
1040	2 arm brackets	1,725	1,325	3,050
1080	3 arm brackets	1,725	1,375	3,100
1120	4 arm brackets	1,850	1,375	3,225
1320	40' high, 1 arm bracket	2,100	1,775	3,875
1360	2 arm brackets	2,200	1,775	3,975
1400	3 arm brackets	2,225	1,825	4,050
1440	4 arm brackets	2,350	1,825	4,175

Reference Section

All the reference information is in one section making it easy to find what you need to know . . . and easy to use the book on a daily basis. This section is visually identified by a vertical gray bar on the edge of pages.

In this Reference Section, we've included General Conditions; Location Factors for adjusting costs to the region you are in; Historical Cost Indexes for cost comparisons over time; a Glossary; an explanation of all Abbreviations used in the book; and sample Estimating Forms.

Table of Contents

451

General Conditions, Overhead & Profit

The total building costs in the Commercial/Industrial/Institutional section include a 10% allowance for general conditions and a 15% allowance for the general contractor's overhead and profit and contingencies.

General contractor overhead includes indirect costs such as permits, Workers' Compensation, insurances, supervision and bonding fees. Overhead will vary with the size of project, the contractor's operating procedures and location. Profits will vary with economic activity and local conditions.

Contingencies provide for unforeseen construction difficulties which include material shortages and weather. In all situations, the appraiser should give consideration to possible adjustment of the 25% factor used in developing the Commercial/Industrial/Institutional models.

Architectural Fees

Tabulated below are typical percentage fees by project size, for good professional architectural service. Fees may vary from those listed depending upon degree of design difficulty and economic conditions in any particular area.

Rates can be interpolated horizontally and vertically. Various portions of the same project requiring different rates should be adjusted proportionately. For alterations, add 50% to the fee for the first $500,000 of project cost and add 25% to the fee for project cost over $500,000.

Architectural fees tabulated below include Engineering fees.

Insurance Exclusions

Many insurance companies exclude from coverage such items as architect's fees, excavation, foundations below grade, underground piping and site preparation. Since exclusions vary among insurance companies, it is recommended that for greatest accuracy each exclusion be priced separately using the unit-in-place section.

As a rule of thumb, exclusions can be calculated at 9% of total building cost plus the appropriate allowance for architect's fees.

Building Types	Total Project Size in Thousands of Dollars						
	100	250	500	1,000	5,000	10,000	50,000
Factories, garages, warehouses, repetitive housing	9.0%	8.0%	7.0%	6.2%	5.3%	4.9%	4.5%
Apartments, banks, schools, libraries, offices, municipal buildings	12.2	12.3	9.2	8.0	7.0	6.6	6.2
Churches, hospitals, homes, laboratories, museums, research	15.0	13.6	12.7	11.9	9.5	8.8	8.0
Memorials, monumental work, decorative furnishings	—	16.0	14.5	13.1	10.0	9.0	8.3

Location Factors

Costs shown in *RSMeans Square Foot Costs* are based on national averages for materials and installation. To adjust these costs to a specific location, simply multiply the base cost by the factor for that city.

The data is arranged alphabetically by state and postal zip code numbers. For a city not listed, use the factor for a nearby city with similar economic characteristics.

STATE/ZIP	CITY	Residential	Commercial
ALABAMA			
350-352	Birmingham	.87	.89
354	Tuscaloosa	.78	.82
355	Jasper	.72	.79
356	Decatur	.78	.81
357-358	Huntsville	.84	.86
359	Gadsden	.75	.81
360-361	Montgomery	.77	.82
362	Anniston	.73	.78
363	Dothan	.76	.78
364	Evergreen	.74	.80
365-366	Mobile	.82	.85
367	Selma	.74	.79
368	Phenix City	.75	.81
369	Butler	.75	.80
ALASKA			
995-996	Anchorage	1.25	1.21
997	Fairbanks	1.28	1.21
998	Juneau	1.24	1.19
999	Ketchikan	1.28	1.26
ARIZONA			
850,853	Phoenix	.85	.89
852	Mesa/Tempe	.82	.86
855	Globe	.78	.84
856-857	Tucson	.83	.87
859	Show Low	.80	.85
860	Flagstaff	.85	.89
863	Prescott	.79	.84
864	Kingman	.83	.86
865	Chambers	.79	.84
ARKANSAS			
716	Pine Bluff	.80	.84
717	Camden	.68	.74
718	Texarkana	.73	.76
719	Hot Springs	.69	.75
720-722	Little Rock	.84	.86
723	West Memphis	.79	.81
724	Jonesboro	.77	.82
725	Batesville	.74	.78
726	Harrison	.76	.79
727	Fayetteville	.71	.77
728	Russellville	.76	.79
729	Fort Smith	.77	.82
CALIFORNIA			
900-902	Los Angeles	1.08	1.08
903-905	Inglewood	1.03	1.02
906-908	Long Beach	1.02	1.03
910-912	Pasadena	1.02	1.02
913-916	Van Nuys	1.05	1.04
917-918	Alhambra	1.06	1.03
919-921	San Diego	1.04	1.05
922	Palm Springs	1.02	1.02
923-924	San Bernardino	1.03	1.01
925	Riverside	1.07	1.06
926-927	Santa Ana	1.04	1.03
928	Anaheim	1.07	1.07
930	Oxnard	1.08	1.07
931	Santa Barbara	1.07	1.07
932-933	Bakersfield	1.06	1.06
934	San Luis Obispo	1.05	1.04
935	Mojave	1.03	1.02
936-938	Fresno	1.09	1.07
939	Salinas	1.11	1.09
940-941	San Francisco	1.26	1.24
942,956-958	Sacramento	1.12	1.09
943	Palo Alto	1.16	1.12
944	San Mateo	1.23	1.16
945	Vallejo	1.15	1.11
946	Oakland	1.22	1.18
947	Berkeley	1.22	1.15
948	Richmond	1.24	1.14
949	San Rafael	1.22	1.15
950	Santa Cruz	1.14	1.12
951	San Jose	1.21	1.17
952	Stockton	1.08	1.07
953	Modesto	1.08	1.07

STATE/ZIP	CITY	Residential	Commercial
CALIFORNIA (CONT'D)			
954	Santa Rosa	1.16	1.13
955	Eureka	1.11	1.07
959	Marysville	1.09	1.07
960	Redding	1.09	1.08
961	Susanville	1.09	1.07
COLORADO			
800-802	Denver	.93	.95
803	Boulder	.92	.92
804	Golden	.90	.92
805	Fort Collins	.88	.92
806	Greeley	.78	.86
807	Fort Morgan	.91	.91
808-809	Colorado Springs	.89	.93
810	Pueblo	.90	.92
811	Alamosa	.86	.91
812	Salida	.89	.91
813	Durango	.89	.92
814	Montrose	.86	.90
815	Grand Junction	.90	.91
816	Glenwood Springs	.88	.92
CONNECTICUT			
060	New Britain	1.09	1.07
061	Hartford	1.09	1.08
062	Willimantic	1.10	1.07
063	New London	1.09	1.05
064	Meriden	1.09	1.07
065	New Haven	1.10	1.08
066	Bridgeport	1.10	1.08
067	Waterbury	1.10	1.08
068	Norwalk	1.10	1.08
069	Stamford	1.11	1.11
D.C.			
200-205	Washington	.96	1.00
DELAWARE			
197	Newark	1.02	1.03
198	Wilmington	1.03	1.03
199	Dover	1.02	1.03
FLORIDA			
320,322	Jacksonville	.80	.84
321	Daytona Beach	.89	.90
323	Tallahassee	.77	.80
324	Panama City	.74	.79
325	Pensacola	.81	.85
326,344	Gainesville	.80	.86
327-328,347	Orlando	.89	.90
329	Melbourne	.90	.93
330-332,340	Miami	.86	.90
333	Fort Lauderdale	.84	.88
334,349	West Palm Beach	.84	.86
335-336,346	Tampa	.91	.91
337	St. Petersburg	.78	.84
338	Lakeland	.88	.90
339,341	Fort Myers	.86	.87
342	Sarasota	.89	.88
GEORGIA			
300-303,399	Atlanta	.90	.90
304	Statesboro	.71	.77
305	Gainesville	.78	.83
306	Athens	.78	.84
307	Dalton	.74	.79
308-309	Augusta	.80	.83
310-312	Macon	.81	.84
313-314	Savannah	.81	.83
315	Waycross	.74	.81
316	Valdosta	.72	.78
317,398	Albany	.77	.83
318-319	Columbus	.82	.85
HAWAII			
967	Hilo	1.19	1.16
968	Honolulu	1.21	1.19

Location Factors

STATE/ZIP	CITY	Residential	Commercial
STATES & POSS.			
969	Guam	.97	1.07
IDAHO			
832	Pocatello	.86	.90
833	Twin Falls	.72	.82
834	Idaho Falls	.74	.83
835	Lewiston	.96	.97
836-837	Boise	.86	.90
838	Coeur d'Alene	.93	.95
ILLINOIS			
600-603	North Suburban	1.11	1.09
604	Joliet	1.14	1.10
605	South Suburban	1.11	1.09
606-608	Chicago	1.20	1.15
609	Kankakee	1.00	.99
610-611	Rockford	1.06	1.05
612	Rock Island	.97	.97
613	La Salle	1.05	1.00
614	Galesburg	.99	.97
615-616	Peoria	1.03	1.01
617	Bloomington	1.01	1.00
618-619	Champaign	1.03	1.01
620-622	East St. Louis	1.01	.99
623	Quincy	.99	.95
624	Effingham	.98	.95
625	Decatur	1.01	.99
626-627	Springfield	1.01	1.00
628	Centralia	.99	.96
629	Carbondale	.95	.93
INDIANA			
460	Anderson	.90	.90
461-462	Indianapolis	.93	.93
463-464	Gary	1.01	.99
465-466	South Bend	.90	.90
467-468	Fort Wayne	.89	.88
469	Kokomo	.91	.88
470	Lawrenceburg	.85	.85
471	New Albany	.85	.85
472	Columbus	.90	.88
473	Muncie	.91	.89
474	Bloomington	.92	.89
475	Washington	.89	.88
476-477	Evansville	.90	.91
478	Terre Haute	.90	.92
479	Lafayette	.91	.89
IOWA			
500-503,509	Des Moines	.89	.89
504	Mason City	.76	.81
505	Fort Dodge	.75	.80
506-507	Waterloo	.78	.81
508	Creston	.79	.82
510-511	Sioux City	.84	.86
512	Sibley	.72	.76
513	Spencer	.73	.77
514	Carroll	.73	.77
515	Council Bluffs	.81	.89
516	Shenandoah	.73	.77
520	Dubuque	.84	.89
521	Decorah	.74	.77
522-524	Cedar Rapids	.92	.91
525	Ottumwa	.82	.85
526	Burlington	.85	.85
527-528	Davenport	.95	.95
KANSAS			
660-662	Kansas City	.98	.96
664-666	Topeka	.79	.85
667	Fort Scott	.87	.86
668	Emporia	.74	.81
669	Belleville	.78	.83
670-672	Wichita	.79	.84
673	Independence	.84	.84
674	Salina	.77	.83
675	Hutchinson	.78	.80
676	Hays	.81	.83
677	Colby	.82	.83
678	Dodge City	.81	.85
679	Liberal	.79	.83
KENTUCKY			
400-402	Louisville	.91	.92
403-405	Lexington	.88	.88

STATE/ZIP	CITY	Residential	Commercial
KENTUCKY (CONT'D)			
406	Frankfort	.85	.89
407-409	Corbin	.75	.81
410	Covington	.97	.96
411-412	Ashland	.91	.94
413-414	Campton	.76	.82
415-416	Pikeville	.83	.89
417-418	Hazard	.72	.78
420	Paducah	.89	.89
421-422	Bowling Green	.89	.90
423	Owensboro	.86	.89
424	Henderson	.90	.89
425-426	Somerset	.76	.82
427	Elizabethtown	.87	.87
LOUISIANA			
700-701	New Orleans	.86	.89
703	Thibodaux	.82	.85
704	Hammond	.77	.81
705	Lafayette	.80	.83
706	Lake Charles	.82	.84
707-708	Baton Rouge	.84	.86
710-711	Shreveport	.78	.81
712	Monroe	.73	.80
713-714	Alexandria	.74	.80
MAINE			
039	Kittery	.86	.85
040-041	Portland	.88	.88
042	Lewiston	.87	.87
043	Augusta	.88	.87
044	Bangor	.86	.87
045	Bath	.86	.86
046	Machias	.87	.85
047	Houlton	.88	.86
048	Rockland	.87	.85
049	Waterville	.86	.86
MARYLAND			
206	Waldorf	.85	.88
207-208	College Park	.88	.92
209	Silver Spring	.86	.90
210-212	Baltimore	.90	.93
214	Annapolis	.84	.91
215	Cumberland	.86	.88
216	Easton	.67	.73
217	Hagerstown	.86	.89
218	Salisbury	.73	.77
219	Elkton	.79	.80
MASSACHUSETTS			
010-011	Springfield	1.04	1.01
012	Pittsfield	1.02	.99
013	Greenfield	1.00	.98
014	Fitchburg	1.11	1.04
015-016	Worcester	1.12	1.07
017	Framingham	1.13	1.07
018	Lowell	1.13	1.10
019	Lawrence	1.13	1.09
020-022, 024	Boston	1.20	1.15
023	Brockton	1.12	1.08
025	Buzzards Bay	1.10	1.04
026	Hyannis	1.10	1.06
027	New Bedford	1.12	1.07
MICHIGAN			
480,483	Royal Oak	1.00	.97
481	Ann Arbor	1.01	.98
482	Detroit	1.06	1.03
484-485	Flint	.97	.97
486	Saginaw	.91	.92
487	Bay City	.92	.92
488-489	Lansing	.96	.96
490	Battle Creek	.92	.92
491	Kalamazoo	.91	.91
492	Jackson	.92	.92
493,495	Grand Rapids	.80	.83
494	Muskegon	.87	.88
496	Traverse City	.78	.83
497	Gaylord	.81	.84
498-499	Iron Mountain	.87	.90
MINNESOTA			
550-551	Saint Paul	1.11	1.07
553-555	Minneapolis	1.15	1.10
556-558	Duluth	1.07	1.02

Location Factors

STATE/ZIP	CITY	Residential	Commercial
MINNESOTA (CONT'd)			
559	Rochester	1.03	1.01
560	Mankato	1.01	.99
561	Windom	.82	.88
562	Willmar	.83	.90
563	St. Cloud	1.06	1.05
564	Brainerd	.96	.97
565	Detroit Lakes	.95	.96
566	Bemidji	.94	.97
567	Thief River Falls	.94	.95
MISSISSIPPI			
386	Clarksdale	.78	.81
387	Greenville	.84	.88
388	Tupelo	.79	.83
389	Greenwood	.80	.82
390-392	Jackson	.85	.87
393	Meridian	.83	.86
394	Laurel	.80	.84
395	Biloxi	.82	.83
396	Mccomb	.77	.81
397	Columbus	.78	.82
MISSOURI			
630-631	St. Louis	1.03	1.03
633	Bowling Green	.95	.94
634	Hannibal	.86	.89
635	Kirksville	.80	.88
636	Flat River	.94	.95
637	Cape Girardeau	.88	.94
638	Sikeston	.82	.88
639	Poplar Bluff	.83	.88
640-641	Kansas City	1.03	1.02
644-645	St. Joseph	.93	.95
646	Chillicothe	.87	.84
647	Harrisonville	.96	.96
648	Joplin	.83	.85
650-651	Jefferson City	.87	.92
652	Columbia	.87	.93
653	Sedalia	.85	.90
654-655	Rolla	.87	.85
656-658	Springfield	.87	.89
MONTANA			
590-591	Billings	.88	.90
592	Wolf Point	.84	.89
593	Miles City	.86	.88
594	Great Falls	.89	.91
595	Havre	.82	.89
596	Helena	.88	.90
597	Butte	.87	.90
598	Missoula	.85	.88
599	Kalispell	.83	.87
NEBRASKA			
680-681	Omaha	.91	.91
683-685	Lincoln	.87	.89
686	Columbus	.87	.88
687	Norfolk	.91	.90
688	Grand Island	.92	.91
689	Hastings	.93	.92
690	Mccook	.85	.88
691	North Platte	.92	.92
692	Valentine	.85	.88
693	Alliance	.85	.87
NEVADA			
889-891	Las Vegas	1.03	1.06
893	Ely	.85	.88
894-895	Reno	.93	.97
897	Carson City	.94	.97
898	Elko	.91	.90
NEW HAMPSHIRE			
030	Nashua	.94	.94
031	Manchester	.94	.94
032-033	Concord	.92	.92
034	Keene	.75	.78
035	Littleton	.81	.81
036	Charleston	.74	.76
037	Claremont	.75	.76
038	Portsmouth	.93	.94

STATE/ZIP	CITY	Residential	Commercial
NEW JERSEY			
070-071	Newark	1.12	1.10
072	Elizabeth	1.14	1.08
073	Jersey City	1.10	1.08
074-075	Paterson	1.11	1.09
076	Hackensack	1.10	1.08
077	Long Branch	1.11	1.07
078	Dover	1.11	1.08
079	Summit	1.11	1.08
080,083	Vineland	1.08	1.05
081	Camden	1.09	1.06
082,084	Atlantic City	1.11	1.05
085-086	Trenton	1.10	1.07
087	Point Pleasant	1.09	1.07
088-089	New Brunswick	1.11	1.08
NEW MEXICO			
870-872	Albuquerque	.85	.90
873	Gallup	.85	.90
874	Farmington	.85	.90
875	Santa Fe	.86	.91
877	Las Vegas	.85	.89
878	Socorro	.85	.89
879	Truth/Consequences	.84	.87
880	Las Cruces	.83	.85
881	Clovis	.85	.88
882	Roswell	.85	.89
883	Carrizozo	.85	.90
884	Tucumcari	.86	.89
NEW YORK			
100-102	New York	1.37	1.31
103	Staten Island	1.31	1.27
104	Bronx	1.33	1.26
105	Mount Vernon	1.14	1.14
106	White Plains	1.17	1.14
107	Yonkers	1.18	1.17
108	New Rochelle	1.18	1.14
109	Suffern	1.13	1.09
110	Queens	1.31	1.27
111	Long Island City	1.34	1.28
112	Brooklyn	1.35	1.28
113	Flushing	1.33	1.28
114	Jamaica	1.33	1.27
115,117,118	Hicksville	1.20	1.20
116	Far Rockaway	1.32	1.28
119	Riverhead	1.21	1.21
120-122	Albany	.94	.96
123	Schenectady	.95	.97
124	Kingston	1.02	1.06
125-126	Poughkeepsie	1.19	1.12
127	Monticello	1.04	1.06
128	Glens Falls	.88	.92
129	Plattsburgh	.92	.92
130-132	Syracuse	.96	.96
133-135	Utica	.94	.94
136	Watertown	.93	.96
137-139	Binghamton	.93	.93
140-142	Buffalo	1.04	1.02
143	Niagara Falls	1.00	.99
144-146	Rochester	.96	.97
147	Jamestown	.87	.90
148-149	Elmira	.85	.91
NORTH CAROLINA			
270,272-274	Greensboro	.83	.79
271	Winston-Salem	.83	.79
275-276	Raleigh	.84	.80
277	Durham	.83	.80
278	Rocky Mount	.73	.74
279	Elizabeth City	.75	.75
280	Gastonia	.84	.78
281-282	Charlotte	.85	.80
283	Fayetteville	.82	.81
284	Wilmington	.81	.77
285	Kinston	.74	.73
286	Hickory	.78	.75
287-288	Asheville	.81	.78
289	Murphy	.73	.71
NORTH DAKOTA			
580-581	Fargo	.78	.85
582	Grand Forks	.75	.82
583	Devils Lake	.78	.82
584	Jamestown	.73	.79
585	Bismarck	.78	.85

455

Location Factors

STATE/ZIP	CITY	Residential	Commercial
NORTH DAKOTA (CONT'D)			
586	Dickinson	.76	.84
587	Minot	.81	.87
588	Williston	.76	.83
OHIO			
430-432	Columbus	.93	.93
433	Marion	.89	.89
434-436	Toledo	1.00	.98
437-438	Zanesville	.88	.89
439	Steubenville	.93	.93
440	Lorain	.98	.96
441	Cleveland	1.01	1.00
442-443	Akron	.98	.96
444-445	Youngstown	.95	.94
446-447	Canton	.93	.92
448-449	Mansfield	.93	.92
450	Hamilton	.92	.91
451-452	Cincinnati	.92	.92
453-454	Dayton	.91	.91
455	Springfield	.92	.91
456	Chillicothe	.94	.93
457	Athens	.87	.88
458	Lima	.90	.92
OKLAHOMA			
730-731	Oklahoma City	.79	.83
734	Ardmore	.78	.81
735	Lawton	.80	.83
736	Clinton	.76	.81
737	Enid	.76	.82
738	Woodward	.76	.80
739	Guymon	.67	.69
740-741	Tulsa	.77	.80
743	Miami	.81	.82
744	Muskogee	.71	.74
745	Mcalester	.73	.77
746	Ponca City	.77	.80
747	Durant	.77	.80
748	Shawnee	.75	.80
749	Poteau	.77	.81
OREGON			
970-972	Portland	1.00	1.01
973	Salem	.98	1.00
974	Eugene	.99	1.00
975	Medford	.98	1.00
976	Klamath Falls	.98	1.00
977	Bend	1.00	1.00
978	Pendleton	.98	.97
979	Vale	.97	.92
PENNSYLVANIA			
150-152	Pittsburgh	.96	.98
153	Washington	.93	.96
154	Uniontown	.90	.95
155	Bedford	.87	.93
156	Greensburg	.93	.96
157	Indiana	.90	.95
158	Dubois	.89	.95
159	Johnstown	.89	.94
160	Butler	.91	.94
161	New Castle	.91	.93
162	Kittanning	.93	.95
163	Oil City	.89	.92
164-165	Erie	.93	.93
166	Altoona	.87	.92
167	Bradford	.89	.93
168	State College	.90	.93
169	Wellsboro	.90	.94
170-171	Harrisburg	.94	.96
172	Chambersburg	.89	.93
173-174	York	.91	.95
175-176	Lancaster	.91	.92
177	Williamsport	.85	.88
178	Sunbury	.91	.94
179	Pottsville	.91	.93
180	Lehigh Valley	1.01	1.02
181	Allentown	1.03	1.01
182	Hazleton	.90	.94
183	Stroudsburg	.91	.97
184-185	Scranton	.95	.97
186-187	Wilkes-Barre	.92	.94
188	Montrose	.90	.94
189	Doylestown	1.05	1.05

STATE/ZIP	CITY	Residential	Commercial
PENNSYLVANIA (CONT'D)			
190-191	Philadelphia	1.16	1.13
193	Westchester	1.10	1.07
194	Norristown	1.09	1.09
195-196	Reading	.97	.98
PUERTO RICO			
009	San Juan	.75	.80
RHODE ISLAND			
028	Newport	1.06	1.03
029	Providence	1.06	1.03
SOUTH CAROLINA			
290-292	Columbia	.84	.80
293	Spartanburg	.84	.78
294	Charleston	.87	.83
295	Florence	.80	.78
296	Greenville	.83	.78
297	Rock Hill	.82	.77
298	Aiken	.97	.86
299	Beaufort	.82	.76
SOUTH DAKOTA			
570-571	Sioux Falls	.79	.83
572	Watertown	.75	.80
573	Mitchell	.77	.80
574	Aberdeen	.77	.82
575	Pierre	.77	.81
576	Mobridge	.75	.80
577	Rapid City	.78	.82
TENNESSEE			
370-372	Nashville	.84	.88
373-374	Chattanooga	.75	.81
375,380-381	Memphis	.81	.86
376	Johnson City	.70	.80
377-379	Knoxville	.72	.79
382	McKenzie	.72	.80
383	Jackson	.70	.78
384	Columbia	.71	.79
385	Cookeville	.71	.81
TEXAS			
750	McKinney	.73	.79
751	Waxahachie	.74	.80
752-753	Dallas	.83	.85
754	Greenville	.68	.73
755	Texarkana	.72	.78
756	Longview	.67	.74
757	Tyler	.73	.80
758	Palestine	.66	.72
759	Lufkin	.70	.74
760-761	Fort Worth	.81	.82
762	Denton	.75	.77
763	Wichita Falls	.78	.80
764	Eastland	.71	.73
765	Temple	.74	.76
766-767	Waco	.76	.81
768	Brownwood	.68	.73
769	San Angelo	.71	.76
770-772	Houston	.85	.88
773	Huntsville	.68	.73
774	Wharton	.69	.76
775	Galveston	.83	.86
776-777	Beaumont	.80	.82
778	Bryan	.73	.82
779	Victoria	.73	.77
780	Laredo	.72	.77
781-782	San Antonio	.80	.83
783-784	Corpus Christi	.77	.78
785	McAllen	.75	.76
786-787	Austin	.79	.81
788	Del Rio	.66	.70
789	Giddings	.69	.72
790-791	Amarillo	.76	.81
792	Childress	.74	.77
793-794	Lubbock	.74	.80
795-796	Abilene	.74	.78
797	Midland	.75	.78
798-799,885	El Paso	.73	.78
UTAH			
840-841	Salt Lake City	.81	.88
842,844	Ogden	.78	.85
843	Logan	.79	.86

Location Factors

STATE/ZIP	CITY	Residential	Commercial
UTAH (CONT'd)			
845	Price	.70	.78
846-847	Provo	.80	.87
VERMONT			
050	White River Jct.	.76	.80
051	Bellows Falls	.78	.82
052	Bennington	.80	.83
053	Brattleboro	.80	.84
054	Burlington	.81	.86
056	Montpelier	.82	.84
057	Rutland	.81	.85
058	St. Johnsbury	.78	.80
059	Guildhall	.77	.79
VIRGINIA			
220-221	Fairfax	1.02	.93
222	Arlington	1.03	.93
223	Alexandria	1.07	.95
224-225	Fredericksburg	.94	.88
226	Winchester	.91	.86
227	Culpeper	.99	.88
228	Harrisonburg	.89	.86
229	Charlottesville	.90	.86
230-232	Richmond	.98	.88
233-235	Norfolk	1.00	.89
236	Newport News	.99	.88
237	Portsmouth	.92	.86
238	Petersburg	.96	.87
239	Farmville	.88	.81
240-241	Roanoke	.97	.85
242	Bristol	.85	.81
243	Pulaski	.83	.80
244	Staunton	.90	.84
245	Lynchburg	.95	.86
246	Grundy	.83	.80
WASHINGTON			
980-981,987	Seattle	1.02	1.04
982	Everett	1.04	1.02
983-984	Tacoma	1.02	1.03
985	Olympia	1.01	1.02
986	Vancouver	.97	1.01
988	Wenatchee	.92	.95
989	Yakima	.96	.98
990-992	Spokane	.99	.95
993	Richland	.97	.96
994	Clarkston	.96	.94
WEST VIRGINIA			
247-248	Bluefield	.88	.89
249	Lewisburg	.90	.92
250-253	Charleston	.95	.95
254	Martinsburg	.86	.90
255-257	Huntington	.96	.96
258-259	Beckley	.90	.93
260	Wheeling	.92	.96
261	Parkersburg	.91	.95
262	Buckhannon	.91	.95
263-264	Clarksburg	.91	.95
265	Morgantown	.92	.95
266	Gassaway	.91	.95
267	Romney	.89	.92
268	Petersburg	.91	.93
WISCONSIN			
530,532	Milwaukee	1.07	1.03
531	Kenosha	1.03	1.00
534	Racine	1.02	1.00
535	Beloit	.98	.97
537	Madison	.98	.98
538	Lancaster	.97	.94
539	Portage	.96	.95
540	New Richmond	.99	.95
541-543	Green Bay	1.00	.96
544	Wausau	.94	.92
545	Rhinelander	.94	.94
546	La Crosse	.94	.94
547	Eau Claire	.97	.95
548	Superior	.98	.96
549	Oshkosh	.94	.93
WYOMING			
820	Cheyenne	.82	.86
821	Yellowstone Nat. Pk.	.74	.81
822	Wheatland	.74	.82

STATE/ZIP	CITY	Residential	Commercial
WYOMING (CONT'd)			
823	Rawlins	.75	.83
824	Worland	.74	.81
825	Riverton	.73	.81
826	Casper	.76	.83
827	Newcastle	.74	.81
828	Sheridan	.79	.84
829-831	Rock Springs	.78	.83
CANADIAN FACTORS (reflect Canadian currency)			
ALBERTA			
	Calgary	1.14	1.14
	Edmonton	1.13	1.14
	Fort McMurray	1.14	1.13
	Lethbridge	1.11	1.09
	Lloydminster	1.06	1.05
	Medicine Hat	1.07	1.05
	Red Deer	1.07	1.05
BRITISH COLUMBIA			
	Kamloops	1.05	1.06
	Prince George	1.05	1.07
	Vancouver	1.06	1.11
	Victoria	.99	1.02
MANITOBA			
	Brandon	1.02	1.00
	Portage la Prairie	1.02	.99
	Winnipeg	1.02	1.04
NEW BRUNSWICK			
	Bathurst	.94	.95
	Dalhousie	.94	.95
	Fredericton	1.01	.98
	Moncton	.95	.96
	Newcastle	.94	.95
	St. John	1.01	.98
NEWFOUNDLAND			
	Corner Brook	.96	.98
	St. Johns	.98	.99
NORTHWEST TERRITORIES			
	Yellowknife	1.07	1.06
NOVA SCOTIA			
	Bridgewater	.97	.99
	Dartmouth	.98	1.00
	Halifax	1.00	1.02
	New Glasgow	.97	.99
	Sydney	.96	.97
	Truro	.97	.99
	Yarmouth	.97	.99
ONTARIO			
	Barrie	1.13	1.08
	Brantford	1.14	1.09
	Cornwall	1.14	1.08
	Hamilton	1.16	1.12
	Kingston	1.14	1.09
	Kitchener	1.09	1.05
	London	1.14	1.10
	North Bay	1.11	1.07
	Oshawa	1.13	1.08
	Ottawa	1.16	1.11
	Owen Sound	1.11	1.08
	Peterborough	1.12	1.08
	Sarnia	1.14	1.09
	Sault Ste Marie	1.07	1.04
	St. Catharines	1.10	1.05
	Sudbury	1.07	1.04
	Thunder Bay	1.12	1.05
	Timmins	1.11	1.07
	Toronto	1.17	1.14
	Windsor	1.11	1.05
PRINCE EDWARD ISLAND			
	Charlottetown	.92	.95
	Summerside	.92	.95
QUEBEC			
	Cap-de-la-Madeleine	1.13	1.04
	Charlesbourg	1.13	1.04
	Chicoutimi	1.16	1.05
	Gatineau	1.12	1.03

Location Factors

STATE/ZIP	CITY	Residential	Commercial
QUEBEC (CONT'D)			
	Granby	1.12	
	Hull	1.12	1.04
	Joliette	1.13	1.04
	Laval	1.12	1.03
	Montreal	1.18	1.10
	Quebec	1.18	1.09
	Rimouski	1.16	1.06
	Rouyn-Noranda	1.12	1.03
	Saint Hyacinthe	1.12	1.03
	Sherbrooke	1.12	1.04
	Sorel	1.13	1.04
	St. Jerome	1.12	1.03
	Trois Rivieres	1.13	1.04
SASKATCHEWAN			
	Moose Jaw	.94	.95
	Prince Albert	.93	.94
	Regina	.96	.96
	Saskatoon	.95	.95
YUKON			
	Whitehorse	.93	.95

Historical Cost Indexes

The following tables are the estimated Historical Cost Indexes based on a 30-city national average with a base of 100 on January 1, 1993. The indexes may be used to:

1. Estimate and compare construction costs for different years in the same city.
2. Estimate and compare construction costs in different cities for the same year.
3. Estimate and compare construction costs in different cities for different years.
4. Compare construction trends in any city with the national average.

EXAMPLES

1. Estimate and compare construction costs for different years in the same city.

   A. To estimate the construction cost of a building in Lexington, KY in 1970, knowing that it cost $915,000 in 2009.

   Index Lexington, KY in 1970 = 26.9
   Index Lexington, KY in 2009 = 162.2

   $$\frac{Index\ 1970}{Index\ 2009} \quad x \quad Cost\ 2009 \quad = \quad Cost\ 1970$$

   $$\frac{26.9}{162.2} \quad x \quad \$915,000 \quad = \quad \$151,500$$

   Construction Cost in Lexington, KY in 1970 = $151,500

   B. To estimate the current construction cost of a building in Boston, MA that was built in 1980 for $900,000.

   Index Boston, MA in 1980 = 64.0
   Index Boston, MA in 2009 = 212.6

   $$\frac{Index\ 2009}{Index\ 1980} \quad x \quad Cost\ 1980 \quad = \quad Cost\ 2009$$

   $$\frac{212.6}{64.0} \quad x \quad \$900,000 \quad = \quad \$2,989,500$$

   Construction Cost in Boston in 2009 = $2,989,500

2. Estimate and compare construction costs in different cities for the same year.

   To compare the construction cost of a building in Topeka, KS in 2009 with the known cost of $800,000 in Baltimore, MD in 2009

   Index Topeka, KS in 2009 = 156.4
   Index Baltimore, MD in 2009 = 170.9

   $$\frac{Index\ Topeka}{Index\ Baltimore} \quad x \quad Cost\ Baltimore \quad = \quad Cost\ Topeka$$

   $$\frac{156.4}{170.9} \quad x \quad \$800,000 \quad = \quad \$732,000$$

   Construction Cost in Topeka in 2009 = $732,000

3. Estimate and compare construction costs in different cities for different years.

   To compare the construction cost of a building in Detroit, MI in 2009 with the known construction cost of $5,000,000 for the same building in San Francisco, CA in 1980.

   Index Detroit, MI in 2009 = 190.0
   Index San Francisco, CA in 1980 = 75.2

   $$\frac{Index\ Detroit\ 2009}{Index\ San\ Francisco\ 1980} \quad x\ Cost\ San\ Francisco\ 1980 = Cost\ Detroit\ 2009$$

   $$\frac{190.0}{75.2} \quad x \quad \$5,000,000 \quad = \quad \$12,633,000$$

   Construction Cost in Detroit in 2009 = $12,633,000

4. Compare construction trends in any city with the national average.

   To compare the construction cost in Las Vegas, NV from 1975 to 2009 with the increase in the National Average during the same time period.

   Index Las Vegas, NV for 1975 = 42.8   For 2009 = 190.8
   Index 30 City Average for 1975 = 43.7   For 2009 = 185.9

   A. National Average escalation   = $\dfrac{Index - 30\ City\ 2009}{Index - 30\ City\ 1975}$
   From 1975 to 2009

   $$= \quad \frac{185.9}{43.7}$$

   National Average escalation
   From 1975 to 2009   =   4.25   or increased by 425%

   B. Escalation for Las Vegas, NV   = $\dfrac{Index\ Las\ Vegas,\ NV\ 2009}{Index\ Las\ Vegas,\ NV\ 1975}$
   From 1975 to 2009

   $$= \quad \frac{190.8}{42.8}$$

   Las Vegas escalation 1975 — 2009 = 4.46 – 1.00 = 3.46 x 100 = 346%

   Conclusion: Construction costs in Las Vegas are higher than National average costs and increased at a greater rate from 1975 to 2009 than the National Average.

Historical Cost Indexes

	National	Alabama					Alaska	Arizona		Arkansas		California				
Year	30 City Average	Birming- ham	Hunts- ville	Mobile	Mont- gomery	Tusca- loosa	Anchor- age	Phoenix	Tuscon	Fort Smith	Little Rock	Anaheim	Bakers- field	Fresno	Los Angeles	Oxnard
Jan 2009	185.9E	162.3E	158.6E	155.2E	149.4E	149.9E	225.0E	163.6E	158.8E	149.7E	157.1E	197.3E	194.7E	197.2E	199.2E	196.8E
2008	171.0	150.3	146.9	143.7	138.4	138.8	210.8	152.2	148.4	138.6	145.4	182.7	179.8	183.2	184.7	182.6
2007	165.0	146.9	143.3	140.3	134.8	135.4	206.8	147.7	143.1	134.8	141.7	175.1	173.7	176.7	177.5	176.2
2006	156.2	135.7	133.6	126.7	124.0	122.2	196.4	137.9	134.8	123.2	127.2	166.8	164.9	169.8	167.3	167.4
2005	146.7	127.9	125.4	119.3	116.6	114.6	185.6	128.5	124.1	115.4	119.4	156.5	153.0	157.9	157.1	156.4
2004	132.8	115.9	112.9	107.0	104.9	102.9	167.0	116.5	113.6	103.8	108.6	142.1	139.5	143.8	142.0	140.9
2003	129.7	113.1	110.7	104.8	102.6	100.8	163.5	113.9	110.6	101.8	105.8	139.4	137.2	142.1	139.6	138.7
2002	126.7	110.0	103.4	103.6	101.7	99.7	159.5	113.3	110.2	100.4	102.1	136.5	133.4	136.0	136.4	136.3
2001	122.2	106.0	100.5	100.6	98.3	95.9	152.6	109.0	106.4	97.3	98.7	132.5	129.3	132.8	132.4	132.4
2000	118.9	104.1	98.9	99.1	94.2	94.6	148.3	106.9	104.9	94.4	95.7	129.4	125.2	129.4	129.9	129.7
1999	116.6	101.2	97.4	97.7	92.5	92.8	145.9	105.5	103.3	93.1	94.4	127.9	123.6	126.9	128.7	128.2
1998	113.6	96.2	94.0	94.8	90.3	89.8	143.8	102.1	101.0	90.6	91.4	125.2	120.3	123.9	125.8	124.7
1997	111.5	94.6	92.4	93.3	88.8	88.3	142.0	101.8	100.7	89.3	90.1	124.0	119.1	122.3	124.6	123.5
1996	108.9	90.9	91.4	92.3	87.5	86.5	140.4	98.3	97.4	86.5	86.7	121.7	117.2	120.2	122.4	121.5
1995	105.6	87.8	88.0	88.8	84.5	83.5	138.0	96.1	95.5	85.0	85.6	120.1	115.7	117.5	120.9	120.0
1994	103.0	85.9	86.1	86.8	82.6	81.7	134.0	93.7	93.1	82.4	83.9	118.1	113.6	114.5	119.0	117.5
1993	100.0	82.8	82.7	86.1	82.1	78.7	132.0	90.9	90.9	80.9	82.4	115.0	111.3	112.9	115.6	115.4
1992	97.9	81.7	81.5	84.9	80.9	77.6	128.6	88.8	89.4	79.7	81.2	113.5	108.1	110.3	113.7	113.7
1991	95.7	80.5	78.9	83.9	79.8	76.6	127.4	88.1	88.5	78.3	80.0	111.0	105.8	108.2	110.9	111.4
1990	93.2	79.4	77.6	82.7	78.4	75.2	125.8	86.4	87.0	77.1	77.9	107.7	102.7	103.3	107.5	107.4
1989	91.0	77.5	76.1	81.1	76.8	73.7	123.4	85.3	85.6	75.6	76.4	105.6	101.0	101.7	105.3	105.4
1988	88.5	75.7	74.7	79.7	75.4	72.3	121.4	84.4	84.3	74.2	75.0	102.7	97.6	99.2	102.6	102.3
1985	81.8	71.1	70.5	72.7	70.9	67.9	116.0	78.1	77.7	69.6	71.0	95.4	92.2	92.6	94.6	96.6
1980	60.7	55.2	54.1	56.8	56.8	54.0	91.4	63.7	62.4	53.1	55.8	68.7	69.4	68.7	67.4	69.9
1975	43.7	40.0	40.9	41.8	40.1	37.8	57.3	44.5	45.2	39.0	38.7	47.6	46.6	47.7	48.3	47.0
1970	27.8	24.1	25.1	25.8	25.4	24.2	43.0	27.2	28.5	24.3	22.3	31.0	30.8	31.1	29.0	31.0
1965	21.5	19.6	19.3	19.6	19.5	18.7	34.9	21.8	22.0	18.7	18.5	23.9	23.7	24.0	22.7	23.9
1960	19.5	17.9	17.5	17.8	17.7	16.9	31.7	19.9	20.0	17.0	16.8	21.7	21.5	21.8	20.6	21.7
1955	16.3	14.8	14.7	14.9	14.9	14.2	26.6	16.7	16.7	14.3	14.5	18.2	18.1	18.3	17.3	18.2
1950	13.5	12.2	12.1	12.3	12.3	11.7	21.9	13.8	13.8	11.8	11.6	15.1	15.0	15.1	14.3	15.1
1945	8.6	7.8	7.7	7.8	7.8	7.5	14.0	8.8	8.8	7.5	7.4	9.6	9.5	9.6	9.1	9.6
1940	6.6	6.0	6.0	6.1	6.0	5.8	10.8	6.8	6.8	5.8	5.7	7.4	7.4	7.4	7.0	7.4

	National	California							Colorado			Connecticut				
Year	30 City Average	River- side	Sacra- mento	San Diego	San Francisco	Santa Barbara	Stockton	Vallejo	Colorado Springs	Denver	Pueblo	Bridge- Port	Bristol	Hartford	New Britain	New Haven
Jan 2009	185.9E	195.5E	200.5E	193.3E	228.1E	195.9E	196.9E	205.2E	170.9E	174.8E	169.1E	198.9E	197.7E	198.7E	197.2E	199.2E
2008	171.0	181.3	186.0	179.9	210.6	181.5	183.2	191.6	158.8	161.9	157.9	185.8	184.8	185.7	184.5	186.1
2007	165.0	174.7	179.1	173.6	201.1	175.2	178.2	185.9	152.6	155.9	152.3	179.6	178.3	179.6	178.0	179.5
2006	156.2	166.0	172.2	164.0	191.2	166.4	171.0	177.9	146.1	149.2	145.0	169.5	168.0	169.5	167.7	169.8
2005	146.7	155.4	161.1	153.8	179.7	155.7	160.0	167.2	138.3	141.1	136.4	160.8	159.4	159.6	159.1	160.8
2004	132.8	141.0	147.2	139.0	163.6	141.4	144.2	148.6	125.6	127.2	123.5	143.6	142.9	142.4	142.7	144.3
2003	129.7	138.7	144.9	136.6	162.1	139.2	141.8	146.4	123.1	123.7	120.7	141.3	140.2	140.3	140.0	141.6
2002	126.7	135.3	138.4	134.2	157.9	135.9	136.9	143.8	118.7	121.3	116.7	133.8	133.6	133.1	133.3	134.8
2001	122.2	131.4	135.5	129.7	151.8	131.5	134.1	140.2	113.3	117.2	113.1	128.6	128.5	128.5	128.3	128.6
2000	118.9	128.0	131.5	127.1	146.9	128.7	130.7	137.1	109.8	111.8	108.8	122.7	122.6	122.9	122.4	122.7
1999	116.6	126.5	129.4	124.9	145.1	127.2	127.9	135.3	107.0	109.1	107.1	121.1	121.3	121.2	121.1	121.3
1998	113.6	123.7	125.9	121.3	141.9	123.7	124.7	132.5	103.3	106.5	103.7	119.1	119.4	120.0	119.7	120.0
1997	111.5	122.6	124.7	120.3	139.2	122.4	123.3	130.6	101.1	104.4	102.0	119.2	119.5	119.9	119.7	120.0
1996	108.9	120.6	122.4	118.4	136.8	120.5	121.4	128.1	98.5	101.4	99.7	117.4	117.6	117.9	117.8	118.1
1995	105.6	119.2	119.5	115.4	133.8	119.0	119.0	122.5	96.1	98.9	96.8	116.0	116.5	116.9	116.3	116.5
1994	103.0	116.6	114.8	113.4	131.4	116.4	115.5	119.7	94.2	95.9	94.6	114.3	115.0	115.3	114.7	114.8
1993	100.0	114.3	112.5	111.3	129.6	114.3	115.2	117.5	92.1	93.8	92.6	108.7	107.8	108.4	106.3	106.1
1992	97.9	111.8	110.8	109.5	127.9	112.2	112.7	115.2	90.6	91.5	90.9	106.7	106.0	106.6	104.5	104.4
1991	95.7	109.4	108.5	107.7	125.8	110.0	111.0	113.4	88.9	90.4	89.4	97.6	97.3	98.0	97.3	97.9
1990	93.2	107.0	104.9	105.6	121.8	106.4	105.4	111.4	86.9	88.8	88.8	96.3	95.9	96.6	95.9	96.5
1989	91.0	105.2	103.1	103.6	119.2	104.6	103.7	109.8	85.6	88.7	87.4	94.5	94.2	94.9	94.2	94.6
1988	88.5	102.6	100.5	101.0	115.2	101.4	101.2	107.2	83.8	87.0	85.6	92.8	92.6	93.3	92.6	93.0
1985	81.8	94.8	92.2	94.2	106.2	93.2	93.7	97.0	79.6	81.0	80.4	86.1	86.2	87.2	86.1	86.3
1980	60.7	68.7	71.3	68.1	75.2	71.1	71.2	71.9	60.7	60.9	59.5	61.5	60.7	61.9	60.7	61.4
1975	43.7	47.3	49.1	47.7	49.8	46.1	47.6	46.5	43.1	42.7	42.5	45.2	45.5	46.0	45.2	46.0
1970	27.8	31.0	32.1	30.7	31.6	31.3	31.8	31.8	27.6	26.1	27.3	29.2	28.2	29.6	28.2	29.3
1965	21.5	23.9	24.8	24.0	23.7	24.1	24.5	24.5	21.3	20.9	21.0	22.4	21.7	22.6	21.7	23.2
1960	19.5	21.7	22.5	21.7	21.5	21.9	22.3	22.2	19.3	19.0	19.1	20.0	19.8	20.0	19.8	20.0
1955	16.3	18.2	18.9	18.2	18.0	18.4	18.7	18.6	16.2	15.9	16.0	16.8	16.6	16.8	16.6	16.8
1950	13.5	15.0	15.6	15.0	14.9	15.2	15.4	15.4	13.4	13.2	13.2	13.9	13.7	13.8	13.7	13.9
1945	8.6	9.6	10.0	9.6	9.5	9.7	9.9	9.8	8.5	8.4	8.4	8.8	8.7	8.8	8.7	8.9
1940	6.6	7.4	7.7	7.4	7.3	7.5	7.6	7.6	6.6	6.5	6.5	6.8	6.8	6.8	6.8	6.8

Year	National 30 City Average	Connecticut Norwalk	Stamford	Waterbury	Delaware Wilmington	D.C. Washington	Florida Fort Lauderdale	Jacksonville	Miami	Orlando	Tallahassee	Tampa	Georgia Albany	Atlanta	Columbus	Macon
Jan 2009	185.9E	198.7E	203.8E	198.4E	189.4E	183.3E	161.5E	154.0E	165.8E	165.4E	146.1E	167.4E	152.0E	165.8E	155.5E	153.9E
2008	171.0	185.2	189.2	185.3	177.5	170.4	149.6	142.5	152.9	153.0	135.3	155.6	140.4	153.2	143.9	142.5
2007	165.0	178.6	183.3	179.1	173.4	163.1	145.6	139.0	148.9	148.1	131.7	152.0	135.8	148.0	140.1	138.1
2006	156.2	169.7	173.7	169.1	158.3	153.0	136.0	126.5	136.7	134.3	118.4	136.6	123.6	139.9	126.3	124.3
2005	146.7	160.9	164.1	160.5	149.6	142.7	126.0	119.2	127.1	126.0	110.7	127.7	115.7	131.7	118.9	116.1
2004	132.8	143.6	146.5	143.2	136.1	126.8	114.8	107.8	115.6	113.2	100.6	116.6	102.8	119.6	102.0	105.0
2003	129.7	140.6	145.1	140.8	133.0	124.6	109.0	105.6	110.5	107.8	98.0	103.5	100.5	115.8	99.1	102.7
2002	126.7	133.5	136.2	133.9	129.4	120.3	107.1	104.6	107.4	106.7	97.3	102.8	99.7	113.0	98.3	101.9
2001	122.2	128.2	131.5	128.8	124.8	115.9	104.3	100.8	104.6	103.8	94.8	100.3	96.4	109.1	95.5	99.0
2000	118.9	121.5	126.4	123.2	117.4	113.8	102.4	99.0	101.8	101.2	93.6	98.9	94.9	106.1	94.1	97.0
1999	116.6	120.0	122.0	121.2	116.6	111.5	101.4	98.0	100.8	100.1	92.5	97.9	93.5	102.9	92.8	95.8
1998	113.6	118.8	120.8	120.1	112.3	109.6	99.4	96.2	99.0	98.4	90.9	96.3	91.4	100.8	90.4	93.3
1997	111.5	118.9	120.9	120.2	110.7	106.4	98.6	95.4	98.2	97.2	89.9	95.5	89.9	98.5	88.7	91.6
1996	108.9	117.0	119.0	118.4	108.6	105.4	95.9	92.6	95.9	95.1	88.0	93.6	86.6	94.3	83.9	88.3
1995	105.6	115.5	117.9	117.3	106.1	102.3	94.0	90.9	93.7	93.5	86.5	92.2	85.0	92.0	82.6	86.8
1994	103.0	113.9	116.4	115.8	105.0	99.6	92.2	88.9	91.8	91.5	84.5	90.2	82.3	89.6	80.6	83.7
1993	100.0	108.8	110.6	104.8	101.5	96.3	87.4	86.1	87.1	88.5	82.1	87.7	79.5	85.7	77.8	80.9
1992	97.9	107.2	109.0	103.1	100.3	94.7	85.7	84.0	85.3	87.1	80.8	86.2	78.2	84.3	76.5	79.6
1991	95.7	100.6	103.2	96.5	94.5	92.9	85.1	82.8	85.2	85.5	79.7	86.3	76.3	82.6	75.4	78.4
1990	93.2	96.3	98.9	95.1	92.5	90.4	83.9	81.1	84.0	82.9	78.4	85.0	75.0	80.4	74.0	76.9
1989	91.0	94.4	97.0	93.4	89.5	87.5	82.4	79.7	82.5	81.6	76.9	83.5	73.4	78.6	72.4	75.3
1988	88.5	92.3	94.9	91.8	87.7	85.0	80.9	78.0	81.0	80.1	75.4	81.9	71.8	76.9	70.8	73.6
1985	81.8	85.3	86.8	85.6	81.1	78.8	76.7	73.4	78.3	73.9	70.6	77.3	66.9	70.3	66.1	68.1
1980	60.7	60.7	60.9	62.3	58.5	59.6	55.3	55.8	56.5	56.7	53.5	57.2	51.7	54.0	51.1	51.3
1975	43.7	44.7	45.0	46.3	42.9	43.7	42.1	40.3	43.2	41.5	38.1	41.3	37.5	38.4	36.2	36.5
1970	27.8	28.1	28.2	28.9	27.0	26.3	25.7	22.8	27.0	26.2	24.5	24.2	23.8	25.2	22.8	23.4
1965	21.5	21.6	21.7	22.2	20.9	21.8	19.8	17.4	19.3	20.2	18.9	18.6	18.3	19.8	17.6	18.0
1960	19.5	19.7	19.7	20.2	18.9	19.4	18.0	15.8	17.6	18.3	17.2	16.9	16.7	17.1	16.0	16.4
1955	16.3	16.5	16.5	17.0	15.9	16.3	15.1	13.2	14.7	15.4	14.4	14.1	14.0	14.4	13.4	13.7
1950	13.5	13.6	13.7	14.0	13.1	13.4	12.5	11.0	12.2	12.7	11.9	11.7	11.5	11.9	11.0	11.3
1945	8.6	8.7	8.7	8.9	8.4	8.6	7.9	7.0	7.8	8.1	7.6	7.5	7.4	7.6	7.0	7.2
1940	6.6	6.7	6.7	6.9	6.4	6.6	6.1	5.4	6.0	6.2	5.8	5.7	5.7	5.8	5.5	5.6

Year	National 30 City Average	Georgia Savannah	Hawaii Honolulu	Idaho Boise	Pocatello	Illinois Chicago	Decatur	Joliet	Peoria	Rockford	Springfield	Indiana Anderson	Evansville	Fort Wayne	Gary	Indianapolis
Jan 2009	185.9E	152.8E	219.8E	165.1E	165.0E	211.3E	181.9E	200.3E	185.9E	192.8E	182.0E	164.7E	168.0E	162.1E	181.1E	171.4E
2008	171.0	141.1	205.5	153.2	152.8	195.7	168.7	185.5	172.2	180.2	168.5	152.4	155.8	151.0	169.0	159.2
2007	165.0	136.9	200.4	148.0	148.1	186.5	160.5	177.7	166.0	175.0	159.2	149.0	152.6	147.7	165.4	154.1
2006	156.2	124.0	191.9	141.6	141.2	177.8	153.8	168.7	155.2	163.3	152.8	139.8	144.4	139.1	154.5	144.6
2005	146.7	116.9	181.2	133.8	132.5	164.4	145.0	158.3	146.5	151.2	145.4	131.4	136.2	131.0	146.4	137.4
2004	132.8	105.6	161.2	122.0	120.5	149.3	129.3	145.1	133.9	138.1	130.3	120.9	123.5	120.1	132.2	125.0
2003	129.7	103.0	159.4	120.2	118.7	146.2	127.4	143.5	132.4	137.3	128.4	119.7	121.4	118.2	131.4	122.5
2002	126.7	102.0	157.2	118.3	117.1	141.2	123.8	138.5	129.6	132.9	125.3	117.5	119.0	116.4	129.1	120.5
2001	122.2	99.0	150.0	114.3	113.4	135.8	120.1	133.7	124.3	127.8	119.8	113.4	115.6	112.1	123.4	116.4
2000	118.9	97.5	144.8	112.9	112.1	131.2	115.1	124.6	119.0	122.2	116.2	109.8	111.5	108.4	117.8	113.2
1999	116.6	96.0	143.0	110.2	109.6	129.6	113.0	122.6	116.4	120.7	113.8	107.1	109.3	106.8	112.8	110.6
1998	113.6	93.7	140.4	107.4	107.0	125.2	110.1	119.8	113.8	115.5	111.1	105.0	107.2	104.5	111.1	108.0
1997	111.5	92.0	139.8	104.6	104.7	121.3	107.8	117.5	111.5	113.1	108.9	101.9	104.4	101.8	110.3	105.2
1996	108.9	88.6	134.5	102.2	102.1	118.8	106.6	116.2	109.3	111.5	106.5	100.0	102.1	99.9	107.5	102.7
1995	105.6	87.4	130.3	99.5	98.2	114.2	98.5	110.5	102.3	103.6	98.1	96.4	97.2	95.0	100.7	100.1
1994	103.0	85.3	124.0	94.8	95.0	111.3	97.3	108.9	100.9	102.2	97.0	93.6	95.8	93.6	99.1	97.1
1993	100.0	82.0	122.0	92.2	92.1	107.6	95.6	106.8	98.9	99.6	95.2	91.2	94.3	91.5	96.7	93.9
1992	97.9	80.8	120.0	91.0	91.0	104.3	94.4	104.2	97.3	98.2	94.0	89.5	92.9	89.9	95.0	91.5
1991	95.7	79.5	106.1	89.5	89.4	100.9	92.3	100.0	95.9	95.8	91.5	87.8	91.4	88.3	93.3	89.1
1990	93.2	77.9	104.7	88.2	88.1	98.4	90.9	98.4	93.7	94.0	90.1	84.6	89.3	83.4	88.4	87.1
1989	91.0	76.0	102.8	86.6	86.5	93.7	89.4	92.8	91.6	92.1	88.6	82.3	87.8	81.7	86.5	85.1
1988	88.5	74.5	101.1	83.9	83.7	90.6	87.5	89.8	88.5	87.9	86.8	80.8	85.5	80.0	84.6	83.1
1985	81.8	68.9	94.7	78.0	78.0	82.4	81.9	83.4	83.7	83.0	81.5	75.2	79.8	75.0	77.8	77.1
1980	60.7	52.2	68.9	60.3	59.5	62.8	62.3	63.4	64.5	61.6	61.1	56.5	59.0	56.7	59.8	57.9
1975	43.7	36.9	44.6	40.8	40.5	45.7	43.1	44.5	44.7	42.7	42.4	39.5	41.7	39.9	41.9	40.6
1970	27.8	21.0	30.4	26.7	26.6	29.1	28.0	28.6	29.0	27.5	27.6	25.4	26.4	25.5	27.1	26.2
1965	21.5	16.4	21.8	20.6	20.5	22.7	21.5	22.1	22.4	21.2	21.3	19.5	20.4	19.7	20.8	20.7
1960	19.5	14.9	19.8	18.7	18.6	20.2	19.6	20.0	20.3	19.2	19.3	17.7	18.7	17.9	18.9	18.4
1955	16.3	12.5	16.6	15.7	15.6	16.9	16.4	16.8	17.0	16.1	16.2	14.9	15.7	15.0	15.9	15.5
1950	13.5	10.3	13.7	13.0	12.9	14.0	13.6	13.9	14.0	13.3	13.4	12.3	12.9	12.4	13.1	12.8
1945	8.6	6.6	8.8	8.3	8.2	8.9	8.6	8.9	9.0	8.5	8.6	7.8	8.3	7.9	8.4	8.1
1940	6.6	5.1	6.8	6.4	6.3	6.9	6.7	6.8	6.9	6.5	6.6	6.0	6.4	6.1	6.5	6.3

461

Historical Cost Indexes

Year	National 30 City Average	Indiana Muncie	Indiana South Bend	Indiana Terre Haute	Iowa Cedar Rapids	Iowa Daven-port	Iowa Des Moines	Iowa Sioux City	Iowa Water-loo	Kansas Topeka	Kansas Wichita	Kentucky Lexing-ton	Kentucky Louis-ville	Louisiana Baton Rouge	Louisiana Lake Charles	Louisiana New Orleans
Jan 2009	185.9E	164.1E	165.4E	168.7E	167.8E	173.9E	164.0E	158.0E	149.2E	156.4E	154.2E	162.2E	168.6E	156.5E	152.8E	162.0E
2008	171.0	153.0	153.1	155.9	157.3	163.4	152.4	147.8	139.2	144.9	143.1	152.0	156.8	144.0	141.5	149.3
2007	165.0	149.7	149.5	152.7	152.4	157.6	148.9	144.0	135.0	141.0	138.7	147.6	150.9	139.4	138.0	145.2
2006	156.2	141.2	141.3	144.0	145.8	151.1	142.9	137.3	128.4	134.7	132.9	129.0	143.0	129.4	129.8	135.8
2005	146.7	131.0	133.8	135.2	135.7	142.1	133.5	129.4	120.0	125.9	125.6	121.5	135.1	120.8	117.7	126.2
2004	132.8	120.0	120.6	123.0	122.7	128.6	121.6	116.2	108.3	112.6	113.4	110.1	120.3	105.6	109.0	115.3
2003	129.7	118.7	119.2	121.7	120.9	126.4	120.2	114.6	105.9	109.4	111.3	107.8	118.7	103.3	106.1	112.4
2002	126.7	116.7	117.1	119.9	116.0	122.1	116.7	111.4	103.7	107.8	109.6	106.2	116.4	102.5	105.0	110.6
2001	122.2	112.8	111.6	115.5	112.5	117.5	113.2	107.7	100.6	104.3	105.2	103.3	112.7	99.4	101.3	104.6
2000	118.9	109.1	106.9	110.8	108.8	112.6	108.9	99.0	98.1	101.0	101.1	101.4	109.3	97.8	99.7	102.1
1999	116.6	105.7	103.7	107.9	104.1	109.2	107.6	96.7	96.4	98.7	99.3	99.7	106.6	96.1	97.6	99.5
1998	113.6	103.6	102.3	106.0	102.5	106.8	103.8	95.1	94.8	97.4	97.4	97.4	101.5	94.7	96.1	97.7
1997	111.5	101.3	101.2	104.6	101.2	105.6	102.5	93.8	93.5	96.3	96.2	96.1	99.9	93.3	96.8	96.2
1996	108.9	99.4	99.1	102.1	99.1	102.0	99.1	90.9	90.7	93.6	93.8	94.4	98.3	91.5	94.9	94.3
1995	105.6	95.6	94.6	97.2	95.4	96.6	94.4	88.2	88.9	91.1	91.0	91.6	94.6	89.6	92.7	91.6
1994	103.0	93.2	93.0	95.8	93.0	92.3	92.4	85.9	86.8	89.4	88.1	89.6	92.2	87.7	89.9	89.5
1993	100.0	91.0	90.7	94.4	91.1	90.5	90.7	84.1	85.2	87.4	86.2	87.3	89.4	86.4	88.5	87.8
1992	97.9	89.4	89.3	93.1	89.8	89.3	89.4	82.9	83.8	86.2	85.0	85.8	88.0	85.2	87.3	86.6
1991	95.7	87.7	87.5	91.7	88.5	88.0	88.4	81.8	81.6	84.4	83.9	84.1	84.4	83.2	85.3	85.8
1990	93.2	83.9	85.1	89.1	87.1	86.5	86.9	80.4	80.3	83.2	82.6	82.8	82.6	82.0	84.1	84.5
1989	91.0	81.9	83.5	86.7	85.3	84.9	85.3	79.0	78.8	81.6	81.2	81.3	80.1	80.6	82.7	83.4
1988	88.5	80.4	82.1	84.9	83.4	83.1	83.6	76.8	77.2	80.1	79.5	79.4	78.2	79.0	81.1	82.0
1985	81.8	75.2	76.0	79.2	75.9	77.6	77.1	72.1	72.3	75.0	74.7	75.5	74.5	74.9	78.5	78.2
1980	60.7	56.1	58.3	59.7	62.7	59.6	61.8	59.3	57.7	58.9	58.0	59.3	59.8	59.1	60.0	57.2
1975	43.7	39.2	40.3	41.9	43.1	41.0	43.1	41.0	40.0	42.0	43.1	42.7	42.5	40.4	40.6	41.5
1970	27.8	25.3	26.2	27.0	28.2	26.9	27.6	26.6	25.9	27.0	25.5	26.9	25.9	25.0	26.7	27.2
1965	21.5	19.5	20.2	20.9	21.7	20.8	21.7	20.5	20.0	20.8	19.6	20.7	20.3	19.4	20.6	20.4
1960	19.5	17.8	18.3	18.9	19.7	18.8	19.5	18.6	18.2	18.9	17.8	18.8	18.4	17.6	18.7	18.5
1955	16.3	14.9	15.4	15.9	16.6	15.8	16.3	15.6	15.2	15.8	15.0	15.8	15.4	14.8	15.7	15.6
1950	13.5	12.3	12.7	13.1	13.7	13.1	13.5	12.9	12.6	13.1	12.4	13.0	12.8	12.2	13.0	12.8
1945	8.6	7.8	8.1	8.4	8.7	8.3	8.6	8.2	8.0	8.3	7.9	8.3	8.1	7.8	8.3	8.2
1940	6.6	6.0	6.3	6.5	6.7	6.4	6.6	6.3	6.2	6.4	6.1	6.4	6.3	6.0	6.4	6.3

Year	National 30 City Average	Louisiana Shreve-port	Maine Lewis-ton	Maine Portland	Maryland Balti-more	Mass. Boston	Mass. Brockton	Mass. Fall River	Mass. Law-rence	Mass. Lowell	Mass. New Bedford	Mass. Pitts-field	Mass. Spring-field	Mass. Wor-cester	Michigan Ann Arbor	Michigan Dear-born
Jan 2009	185.9E	147.1E	159.6E	162.0E	170.9E	212.6E	199.3E	197.7E	201.2E	201.7E	197.1E	182.4E	185.8E	196.9E	180.6E	189.0E
2008	171.0	135.9	149.1	150.9	157.7	198.6	186.4	185.1	188.7	189.1	184.5	171.3	173.7	184.5	169.6	176.4
2007	165.0	132.4	146.4	148.2	152.5	191.8	180.4	179.7	182.0	181.7	179.1	166.8	169.1	180.4	166.8	172.5
2006	156.2	125.2	140.2	139.8	144.9	180.4	171.6	170.4	172.3	173.5	170.4	157.3	159.7	171.6	161.1	165.5
2005	146.7	117.4	131.7	131.4	135.8	169.6	161.3	159.6	162.2	162.3	159.6	147.6	150.7	158.8	147.5	155.8
2004	132.8	105.7	119.8	119.4	121.4	154.1	144.4	143.6	146.4	146.7	143.6	131.9	136.5	143.2	135.9	142.1
2003	129.7	103.9	117.7	117.3	118.1	150.2	139.6	140.6	143.3	143.6	140.5	128.9	133.8	139.0	134.1	139.0
2002	126.7	102.1	117.5	117.1	115.6	145.6	136.0	135.0	136.9	137.3	134.9	124.7	128.3	134.9	131.4	134.3
2001	122.2	98.3	114.6	114.3	111.7	140.9	132.1	131.4	133.1	132.9	131.3	120.3	124.5	130.1	126.8	129.8
2000	118.9	96.2	105.7	105.4	107.7	138.9	129.4	128.4	129.7	130.3	128.3	116.7	121.0	127.3	124.3	125.8
1999	116.6	94.8	105.0	104.7	106.4	136.2	126.7	126.3	127.3	127.4	126.2	115.0	118.8	123.8	117.5	122.7
1998	113.6	92.0	102.8	102.5	104.1	132.8	125.0	124.7	125.0	125.3	124.6	114.2	117.1	122.6	116.0	119.7
1997	111.5	90.1	101.7	101.4	102.2	132.1	124.4	124.7	125.1	125.6	124.7	114.6	117.4	122.9	113.5	117.9
1996	108.9	88.5	99.5	99.2	99.6	128.7	121.6	122.2	121.7	122.1	122.2	112.7	115.0	120.2	112.9	116.2
1995	105.6	86.7	96.8	96.5	96.1	128.6	119.6	117.7	120.5	119.6	117.2	110.7	112.7	114.8	106.1	110.5
1994	103.0	85.0	95.2	94.9	94.4	124.9	114.2	113.7	116.8	116.9	111.6	109.2	111.1	112.9	105.0	109.0
1993	100.0	83.6	93.0	93.0	93.1	121.1	111.7	111.5	114.9	114.3	109.4	107.1	108.9	110.3	102.8	105.8
1992	97.9	82.3	91.7	91.7	90.9	118.0	110.0	109.8	110.9	110.2	107.7	105.7	107.2	108.6	101.5	103.2
1991	95.7	81.6	89.8	89.9	89.1	115.2	105.9	104.7	107.3	105.5	105.1	100.6	103.2	105.7	95.1	98.3
1990	93.2	80.4	88.5	88.6	85.6	110.9	103.7	102.9	105.2	102.8	102.6	98.7	101.4	103.2	93.2	96.5
1989	91.0	78.8	86.7	86.7	83.5	107.2	100.4	99.7	102.9	99.7	99.8	94.7	96.4	99.1	89.6	94.6
1988	88.5	77.0	84.0	84.1	81.2	102.5	96.7	96.4	97.7	96.6	96.4	91.9	92.7	94.4	85.8	91.4
1985	81.8	73.6	76.7	77.0	72.7	92.8	88.8	88.7	89.4	88.2	88.6	85.0	85.6	86.7	80.1	82.7
1980	60.7	58.7	57.3	58.5	53.6	64.0	63.7	64.1	63.4	62.7	63.1	61.8	62.0	62.3	62.9	64.0
1975	43.7	40.5	41.9	42.1	39.8	46.6	45.8	45.7	46.2	45.7	46.1	45.5	45.8	46.0	44.1	44.9
1970	27.8	26.4	26.5	25.8	25.1	29.2	29.1	29.2	29.1	28.9	29.0	28.6	28.5	28.7	28.5	28.9
1965	21.5	20.3	20.4	19.4	20.2	23.0	22.5	22.5	22.5	22.2	22.4	22.0	22.4	22.1	22.0	22.3
1960	19.5	18.5	18.6	17.6	17.5	20.5	20.4	20.4	20.4	20.2	20.3	20.0	20.1	20.1	20.0	20.2
1955	16.3	15.5	15.6	14.7	14.7	17.2	17.1	17.1	17.1	16.9	17.1	16.8	16.9	16.8	16.7	16.9
1950	13.5	12.8	12.9	12.2	12.1	14.2	14.1	14.1	14.1	14.0	14.1	13.8	14.0	13.9	13.8	14.0
1945	8.6	8.1	8.2	7.8	7.7	9.1	9.0	9.0	9.0	8.9	9.0	8.9	8.9	8.9	8.8	8.9
1940	6.6	6.3	6.3	6.0	6.0	7.0	6.9	7.0	7.0	6.9	6.9	6.8	6.9	6.8	6.8	6.9

Historical Cost Indexes

Year	National 30 City Average	Michigan Detroit	Flint	Grand Rapids	Kala-mazoo	Lansing	Sagi-naw	Minnesota Duluth	Minne-apolis	Roches-ter	Mississippi Biloxi	Jackson	Missouri Kansas City	St. Joseph	St. Louis	Spring-field
Jan 2009	185.9E	190.0E	177.5E	151.5E	167.5E	174.5E	168.6E	188.5E	202.3E	185.6E	153.2E	158.8E	188.2E	174.9E	189.3E	163.8E
2008	171.0	177.0	164.4	140.2	155.7	161.6	158.7	177.4	190.6	175.0	141.8	147.1	175.5	164.2	176.2	152.9
2007	165.0	172.9	161.7	137.2	152.6	158.7	156.1	174.2	184.5	171.4	137.7	131.2	169.0	157.8	170.6	145.3
2006	156.2	165.8	153.8	131.1	146.0	152.5	150.6	166.2	173.9	161.9	123.3	117.1	162.0	152.4	159.3	139.4
2005	146.7	156.2	145.1	123.1	134.9	142.0	140.6	157.3	164.6	152.8	116.0	109.9	151.3	143.1	149.4	131.7
2004	132.8	141.9	129.6	112.8	122.5	129.7	127.6	139.1	150.1	137.0	105.3	99.3	135.1	126.9	135.5	116.3
2003	129.7	138.7	127.9	110.9	120.8	127.7	126.0	136.7	146.5	134.5	101.4	96.7	131.9	124.8	133.3	113.9
2002	126.7	134.2	126.9	107.4	119.9	125.3	124.4	131.9	139.5	130.0	101.0	96.3	128.4	122.7	129.6	112.5
2001	122.2	129.4	122.4	104.3	115.2	120.1	119.7	131.4	136.1	124.9	98.7	93.9	121.8	114.7	125.5	106.7
2000	118.9	125.3	119.8	102.7	111.6	117.6	115.9	124.1	131.1	120.1	97.2	92.9	118.2	111.9	122.4	104.3
1999	116.6	122.6	113.7	100.9	106.1	111.5	110.1	120.3	126.5	117.1	95.7	91.8	114.9	106.1	119.8	101.1
1998	113.6	119.5	112.3	99.7	104.9	110.1	108.7	117.7	124.6	115.5	92.1	89.5	108.2	104.2	115.9	98.9
1997	111.5	117.6	110.8	98.9	104.2	108.7	107.5	115.9	121.9	114.1	90.9	88.3	106.4	102.4	113.2	97.3
1996	108.9	116.0	109.7	93.9	103.5	107.8	106.9	115.0	120.4	113.5	87.1	85.4	103.2	99.9	110.1	94.6
1995	105.6	110.1	104.1	91.1	96.4	98.3	102.3	100.3	111.9	102.5	84.2	83.5	99.7	96.1	106.3	89.5
1994	103.0	108.5	102.9	89.7	95.0	97.1	101.1	99.1	109.3	101.1	82.4	81.7	97.3	92.8	103.2	88.2
1993	100.0	105.4	100.7	87.7	92.7	95.3	98.8	99.8	106.7	99.9	80.0	79.4	94.5	90.9	99.9	86.2
1992	97.9	102.9	99.4	86.4	91.4	94.0	97.4	98.5	105.3	98.6	78.8	78.2	92.9	89.5	98.6	84.9
1991	95.7	97.6	93.6	85.3	90.0	92.1	90.3	96.0	102.9	97.3	77.6	76.9	90.9	88.1	96.4	83.9
1990	93.2	96.0	91.7	84.0	87.4	90.2	88.9	94.3	100.5	95.6	76.4	75.6	89.5	86.8	94.0	82.6
1989	91.0	94.0	90.2	82.4	85.8	88.6	87.4	92.5	97.4	93.7	75.0	74.1	87.3	85.1	91.8	80.9
1988	88.5	90.8	87.8	80.7	84.1	86.3	85.6	90.5	94.6	91.9	73.5	72.9	84.9	82.9	89.0	79.2
1985	81.8	81.6	80.7	75.8	79.4	80.0	80.5	85.2	87.9	86.5	69.4	68.5	78.2	77.3	80.8	73.2
1980	60.7	62.5	63.1	59.8	61.7	60.3	63.3	63.4	64.1	64.8	54.3	54.3	59.1	60.4	59.7	56.7
1975	43.7	45.8	44.9	41.8	43.7	44.1	44.8	45.2	46.0	45.3	37.9	37.7	42.7	45.4	44.8	41.1
1970	27.8	29.7	28.5	26.7	28.1	27.8	28.7	28.9	29.5	28.9	24.4	21.3	25.3	28.3	28.4	26.0
1965	21.5	22.1	21.9	20.6	21.7	21.5	22.2	22.3	23.4	22.3	18.8	16.4	20.6	21.8	21.8	20.0
1960	19.5	20.1	20.0	18.7	19.7	19.5	20.1	20.3	20.5	20.2	17.1	14.9	19.0	19.8	19.5	18.2
1955	16.3	16.8	16.7	15.7	16.5	16.3	16.9	17.0	17.2	17.0	14.3	12.5	16.0	16.6	16.3	15.2
1950	13.5	13.9	13.8	13.0	13.6	13.5	13.9	14.0	14.2	14.0	11.8	10.3	13.2	13.7	13.5	12.6
1945	8.6	8.9	8.8	8.3	8.7	8.6	8.9	8.9	9.0	8.9	7.6	6.6	8.4	8.8	8.6	8.0
1940	6.6	6.8	6.8	6.4	6.7	6.6	6.9	6.9	7.0	6.9	5.8	5.1	6.5	6.7	6.7	6.2

Year	National 30 City Average	Montana Billings	Great Falls	Nebraska Lincoln	Omaha	Nevada Las Vegas	Reno	New Hampshire Man-chester	Nashua	New Jersey Camden	Jersey City	Newark	Pater-son	Trenton	NM Albu-querque	NY Albany
Jan 2009	185.9E	165.3E	167.0E	163.7E	167.0E	190.8E	179.0E	173.5E	173.0E	195.4E	198.9E	201.7E	200.2E	197.5E	165.2E	177.2E
2008	171.0	153.1	154.6	152.0	154.8	176.2	167.0	162.5	161.9	184.0	187.3	189.6	188.7	185.6	152.6	166.0
2007	165.0	147.7	147.8	147.7	150.3	166.7	161.4	159.3	158.7	179.3	183.3	185.2	184.6	181.7	146.7	159.4
2006	156.2	140.5	140.8	132.4	140.7	160.0	154.5	146.7	146.4	167.7	171.0	173.6	172.2	169.9	140.1	151.6
2005	146.7	131.9	132.4	124.1	132.8	149.5	145.4	136.8	136.5	159.0	162.4	163.9	163.0	161.5	130.4	142.2
2004	132.8	118.2	117.9	112.1	119.4	137.1	130.5	124.2	123.9	143.4	145.4	147.2	146.6	146.0	118.3	128.6
2003	129.7	115.8	115.8	110.2	117.3	133.8	128.3	122.4	122.2	142.0	144.3	145.7	145.0	143.0	116.4	126.8
2002	126.7	114.6	114.5	108.2	115.0	131.9	126.4	119.9	119.6	135.9	138.5	140.2	140.1	137.6	114.6	122.6
2001	122.2	117.5	117.7	101.3	111.6	127.8	122.5	116.2	116.2	133.4	136.7	136.9	136.8	136.0	111.4	119.2
2000	118.9	113.7	113.9	98.8	107.0	125.8	118.2	111.9	111.9	128.4	130.5	132.6	132.4	130.6	109.0	116.5
1999	116.6	112.1	112.7	96.6	104.8	121.9	114.4	109.6	109.6	125.3	128.8	131.6	129.5	129.6	106.7	114.6
1998	113.6	109.7	109.1	94.9	101.2	118.1	111.4	110.1	110.0	124.3	127.7	128.9	128.5	127.9	103.8	113.0
1997	111.5	107.4	107.6	93.3	99.5	114.6	109.8	108.6	108.6	121.9	125.1	126.4	126.4	125.2	100.8	110.0
1996	108.9	108.0	107.5	91.4	97.4	111.8	108.9	106.5	106.4	119.1	122.5	122.5	123.8	121.8	98.6	108.3
1995	105.6	104.7	104.9	85.6	93.4	108.5	105.0	100.9	100.8	107.0	112.2	111.9	112.1	111.2	96.3	103.6
1994	103.0	100.2	100.7	84.3	91.0	105.3	102.2	97.8	97.7	105.4	110.6	110.1	110.6	108.2	93.4	102.4
1993	100.0	97.9	97.8	82.3	88.7	102.8	99.9	95.4	95.4	103.7	109.0	108.2	109.0	106.4	90.1	99.8
1992	97.9	95.5	96.4	81.1	87.4	99.4	98.4	90.4	90.4	102.0	107.5	107.0	107.8	103.9	87.5	98.5
1991	95.7	94.2	95.1	80.1	86.4	97.5	95.8	87.8	87.8	95.0	98.5	95.6	100.0	96.6	86.2	96.1
1990	93.2	92.9	93.8	78.8	85.0	96.3	94.5	86.3	86.3	93.0	93.5	93.8	97.2	94.5	84.9	93.2
1989	91.0	91.3	92.2	77.3	83.6	94.8	92.3	84.8	84.8	90.4	91.4	91.8	95.5	91.9	83.3	88.4
1988	88.5	89.5	90.4	75.8	82.0	93.0	90.5	83.2	83.2	87.7	89.5	89.7	92.5	89.5	81.7	86.8
1985	81.8	83.9	84.3	71.5	77.5	87.6	85.0	78.1	78.1	81.3	83.3	83.5	84.5	82.8	76.7	79.5
1980	60.7	63.9	64.6	58.5	63.5	64.6	63.0	56.6	56.0	58.6	60.6	60.1	60.0	58.9	59.0	59.5
1975	43.7	43.1	43.8	40.9	43.2	42.8	41.9	41.3	40.8	42.3	43.4	44.2	43.9	43.8	40.3	43.9
1970	27.8	28.5	28.9	26.4	26.8	29.4	28.0	26.2	25.6	27.2	27.8	29.0	27.8	27.4	26.4	28.3
1965	21.5	22.0	22.3	20.3	20.6	22.4	21.6	20.6	19.7	20.9	21.4	23.8	21.4	21.3	20.6	22.3
1960	19.5	20.0	20.3	18.5	18.7	20.2	19.6	18.0	17.9	19.0	19.4	19.4	19.4	19.2	18.5	19.3
1955	16.3	16.7	17.0	15.5	15.7	16.9	16.4	15.1	15.0	16.0	16.3	16.3	16.3	16.1	15.6	16.2
1950	13.5	13.9	14.0	12.8	13.0	14.0	13.6	12.5	12.4	13.2	13.5	13.5	13.5	13.3	12.9	13.4
1945	8.6	8.8	9.0	8.1	8.3	8.9	8.7	8.0	7.9	8.4	8.6	8.6	8.6	8.5	8.2	8.5
1940	6.6	6.8	6.9	6.3	6.4	6.9	6.7	6.2	6.1	6.5	6.6	6.6	6.6	6.6	6.3	6.6

463

Historical Cost Indexes

Year	National 30 City Average	New York Bing-hamton	Buffalo	New York	Roches-ter	Schen-ectady	Syracuse	Utica	Yonkers	North Carolina Charlotte	Durham	Greens-boro	Raleigh	Winston-Salem	N. Dakota Fargo	Ohio Akron
Jan 2009	185.9E	171.4E	186.4E	241.5E	178.1E	178.0E	177.4E	172.3E	215.0E	146.6E	147.4E	145.1E	146.6E	144.3E	155.9E	176.8E
2008	171.0	160.7	173.9	226.8	168.0	167.2	165.9	161.6	202.1	135.8	136.6	134.6	135.3	133.8	145.0	165.4
2007	165.0	155.7	168.6	215.2	163.5	159.8	159.8	155.0	196.5	132.8	133.7	131.7	132.1	131.0	139.8	159.0
2006	156.2	147.5	159.2	204.5	155.2	151.1	150.9	146.4	186.0	125.1	124.6	123.8	124.2	123.0	133.2	152.8
2005	146.7	137.5	149.4	194.0	147.3	142.2	141.9	136.9	176.1	110.5	112.1	112.1	112.1	111.1	125.3	144.4
2004	132.8	123.5	136.1	177.7	132.0	128.4	127.3	124.4	161.8	98.9	100.0	100.1	100.3	99.4	113.0	131.9
2003	129.7	121.8	132.8	173.4	130.3	126.7	124.7	121.7	160.0	96.2	97.5	97.5	97.8	96.8	110.6	129.6
2002	126.7	119.0	128.5	170.1	127.1	123.0	121.8	118.4	154.8	94.8	96.1	96.1	96.3	95.5	106.3	127.2
2001	122.2	116.0	125.2	164.4	123.1	120.1	118.6	115.3	151.4	91.5	92.9	92.9	93.3	92.3	103.1	123.5
2000	118.9	112.4	122.3	159.2	120.0	117.5	115.1	112.4	144.8	90.3	91.6	91.6	91.9	91.0	97.9	117.8
1999	116.6	108.6	120.2	155.9	116.8	114.7	113.7	108.5	140.6	89.3	90.2	90.3	90.5	90.1	96.8	116.0
1998	113.6	109.0	119.1	154.4	117.2	114.2	113.6	108.6	141.4	88.2	89.1	89.2	89.4	89.0	95.2	113.1
1997	111.5	107.2	115.7	150.3	115.2	111.2	110.6	107.0	138.8	86.8	87.7	87.8	87.9	87.6	93.5	110.6
1996	108.9	105.5	114.0	148.0	113.3	109.5	108.2	105.2	137.2	85.0	85.9	86.0	86.1	85.8	91.7	107.9
1995	105.6	99.3	110.1	140.7	106.6	104.6	104.0	97.7	129.4	81.8	82.6	82.6	82.7	82.6	88.4	103.4
1994	103.0	98.1	107.2	137.1	105.2	103.5	102.3	96.5	128.1	80.3	81.1	81.0	81.2	81.1	87.0	102.2
1993	100.0	95.7	102.2	133.3	102.0	100.8	99.6	94.1	126.3	78.2	78.9	78.9	78.9	78.8	85.8	100.3
1992	97.9	93.7	100.1	128.2	99.3	99.3	98.1	90.5	123.9	77.1	77.7	77.8	77.8	77.6	83.1	98.0
1991	95.7	89.5	96.8	124.4	96.0	96.7	95.5	88.7	121.5	76.0	76.6	76.8	76.7	76.6	82.6	96.8
1990	93.2	87.0	94.1	118.1	94.6	93.9	91.0	85.6	111.4	74.8	75.4	75.5	75.5	75.3	81.3	94.6
1989	91.0	85.4	91.8	114.8	92.6	89.6	88.4	84.1	102.1	73.3	74.0	74.0	74.0	73.8	79.7	92.9
1988	88.5	83.6	89.4	106.5	87.9	87.7	86.5	82.4	99.9	71.5	72.2	72.2	72.2	72.0	78.2	91.1
1985	81.8	77.5	83.2	94.9	81.6	80.1	81.0	77.0	92.8	66.9	67.6	67.7	67.6	67.5	73.4	86.8
1980	60.7	58.0	60.6	66.0	60.5	60.3	61.6	58.5	65.9	51.1	52.2	52.5	51.7	50.8	57.4	62.3
1975	43.7	42.5	45.1	49.5	44.8	43.7	44.8	42.7	47.1	36.1	37.0	37.0	37.3	36.1	39.3	44.7
1970	27.8	27.0	28.9	32.8	29.2	27.8	28.5	26.9	30.0	20.9	23.7	23.6	23.4	23.0	25.8	28.3
1965	21.5	20.8	22.2	25.5	22.8	21.4	21.9	20.7	23.1	16.0	18.2	18.2	18.0	17.7	19.9	21.8
1960	19.5	18.9	19.9	21.5	19.7	19.5	19.9	18.8	21.0	14.4	16.6	16.6	16.4	16.1	18.1	19.9
1955	16.3	15.8	16.7	18.1	16.5	16.3	16.7	15.8	17.6	12.1	13.9	13.9	13.7	13.5	15.2	16.6
1950	13.5	13.1	13.8	14.9	13.6	13.5	13.8	13.0	14.5	10.0	11.5	11.5	11.4	11.2	12.5	13.7
1945	8.6	8.3	8.8	9.5	8.7	8.6	8.8	8.3	9.3	6.4	7.3	7.3	7.2	7.1	8.0	8.8
1940	6.6	6.4	6.8	7.4	6.7	6.7	6.8	6.4	7.2	4.9	5.6	5.6	5.6	5.5	6.2	6.8

Year	National 30 City Average	Ohio Canton	Cincin-nati	Cleve-land	Colum-bus	Dayton	Lorain	Spring-field	Toledo	Youngs-town	Oklahoma Lawton	Oklahoma City	Tulsa	Oregon Eugene	Port-land	PA Allen-town
Jan 2009	185.9E	169.6E	169.9E	183.1E	171.1E	167.7E	175.8E	167.8E	180.1E	173.2E	152.4E	151.2E	147.3E	184.0E	185.7E	186.7E
2008	171.0	158.4	159.0	170.6	160.2	157.5	163.9	157.5	168.2	161.5	141.3	140.6	136.2	172.3	174.7	175.7
2007	165.0	153.2	152.4	164.9	155.1	149.4	158.1	150.3	161.6	156.7	136.2	135.6	132.5	168.5	169.5	169.4
2006	156.2	147.0	144.9	156.9	146.9	142.9	151.8	143.4	154.5	150.3	129.5	129.5	125.7	160.5	161.7	160.1
2005	146.7	137.7	136.8	147.9	138.0	134.5	143.3	134.8	145.5	141.1	121.6	121.6	117.8	150.9	152.2	150.2
2004	132.8	125.6	124.3	135.6	126.2	121.1	131.5	122.0	132.8	129.3	109.3	109.4	107.3	136.8	137.8	133.9
2003	129.7	123.7	121.3	132.7	123.7	119.4	126.1	120.0	130.4	126.2	107.5	107.8	105.0	134.2	135.9	130.3
2002	126.7	120.9	119.2	130.0	120.9	117.5	124.3	118.0	128.4	123.6	106.3	106.0	104.1	132.2	133.9	128.1
2001	122.2	117.7	115.9	125.9	117.1	114.0	120.4	114.7	123.8	119.9	102.0	102.1	99.7	129.8	131.2	123.5
2000	118.9	112.7	110.1	121.3	112.5	109.0	115.0	109.2	115.7	114.1	98.7	98.9	97.5	126.3	127.4	119.9
1999	116.6	110.6	107.9	118.7	109.5	107.1	112.0	106.4	113.6	111.9	97.5	97.4	96.2	120.9	124.3	117.8
1998	113.6	108.5	105.0	114.8	106.8	104.5	109.4	104.1	111.2	109.0	94.3	94.5	94.5	120.0	122.2	115.7
1997	111.5	106.4	102.9	112.9	104.7	102.6	107.4	102.2	108.5	107.2	92.9	93.2	93.0	118.1	119.6	113.9
1996	108.9	103.2	100.2	110.1	101.1	98.8	103.8	98.4	105.1	104.1	90.9	91.3	91.2	114.4	116.1	112.0
1995	105.6	98.8	97.1	106.4	99.1	94.6	97.1	92.0	100.6	100.1	85.3	88.0	89.0	112.2	114.3	108.3
1994	103.0	97.7	95.0	104.8	95.4	93.0	96.0	90.1	99.3	98.9	84.0	86.5	87.6	107.6	109.2	106.4
1993	100.0	95.9	92.3	101.9	93.9	90.6	93.9	87.7	98.3	97.2	81.3	83.9	84.8	107.2	108.8	103.6
1992	97.9	94.5	90.6	98.7	92.6	89.2	92.2	86.4	97.1	96.0	80.1	82.7	83.4	101.1	102.6	101.6
1991	95.7	93.4	88.6	97.2	90.6	87.7	91.1	84.9	94.2	92.7	78.3	80.7	81.4	99.5	101.1	98.9
1990	93.2	92.1	86.7	95.1	88.1	85.9	89.4	83.4	92.9	91.8	77.1	79.9	80.0	98.0	99.4	95.0
1989	91.0	90.3	84.6	93.6	85.8	82.6	87.9	81.0	91.4	88.4	75.8	78.5	78.4	95.2	97.0	92.4
1988	88.5	89.1	82.9	92.1	83.9	81.0	85.2	79.5	89.6	86.9	74.3	77.0	76.8	93.5	95.3	89.8
1985	81.8	83.7	78.2	86.2	77.7	76.3	80.7	74.3	84.7	82.7	71.2	73.9	73.8	88.7	90.1	81.9
1980	60.7	60.6	59.8	61.0	58.6	56.6	60.8	56.2	64.0	61.5	52.2	55.5	57.2	68.9	68.4	58.7
1975	43.7	44.0	43.7	44.6	42.2	40.5	42.7	40.0	44.9	44.5	38.6	38.6	39.2	44.6	44.9	42.3
1970	27.8	27.8	28.2	29.6	26.7	27.0	27.5	25.4	29.1	29.6	24.1	23.4	24.1	30.4	30.0	27.3
1965	21.5	21.5	20.7	21.1	20.2	20.0	21.1	19.6	21.3	21.0	18.5	17.9	20.6	23.4	22.6	21.0
1960	19.5	19.5	19.3	19.6	18.7	18.1	19.2	17.8	19.4	19.1	16.9	16.1	18.1	21.2	21.1	19.1
1955	16.3	16.3	16.2	16.5	15.7	15.2	16.1	14.9	16.2	16.0	14.1	13.5	15.1	17.8	17.7	16.0
1950	13.5	13.5	13.3	13.6	13.0	12.5	13.3	12.3	13.4	13.2	11.7	11.2	12.5	14.7	14.6	13.2
1945	8.6	8.6	8.5	8.7	8.3	8.0	8.5	7.8	8.6	8.4	7.4	7.1	8.0	9.4	9.3	8.4
1940	6.6	6.7	6.5	6.7	6.4	6.2	6.5	6.1	6.6	6.5	5.7	5.5	6.1	7.2	7.2	6.5

464

Year	National 30 City Average	Pennsylvania Erie	Harris-burg	Phila-delphia	Pitts-burgh	Reading	Scranton	RI Provi-dence	South Carolina Charles-ton	Colum-bia	South Dakota Rapid City	Sioux Falls	Tennessee Chatta-nooga	Knox-ville	Memphis	Nash-ville
Jan 2009	185.9E	171.5E	176.5E	208.9E	180.0E	180.7E	178.7E	189.8E	152.6E	145.9E	149.8E	152.9E	147.0E	144.1E	157.8E	160.0E
2008	171.0	161.0	165.7	196.2	169.2	169.3	167.9	178.2	142.1	134.8	138.4	141.3	136.7	133.4	146.0	147.5
2007	165.0	156.0	159.4	188.4	163.2	164.9	163.1	174.4	139.3	131.7	129.5	133.0	133.7	130.6	143.1	144.4
2006	156.2	148.7	150.7	177.8	155.5	155.0	153.9	163.6	122.4	118.7	122.8	126.6	127.0	123.9	136.5	135.5
2005	146.7	140.6	141.5	166.4	146.6	145.5	142.9	155.6	110.2	109.6	115.2	118.5	116.0	115.1	128.6	127.9
2004	132.8	127.0	126.2	148.4	133.1	128.7	129.0	138.8	98.9	98.4	103.9	107.3	105.0	104.5	115.6	115.8
2003	129.7	124.5	123.8	145.5	130.5	126.3	126.0	135.9	96.1	95.7	101.3	104.0	102.8	102.3	111.1	110.9
2002	126.7	122.1	121.1	142.0	127.8	123.4	124.2	131.1	94.8	93.9	100.2	103.2	102.2	101.2	107.6	109.2
2001	122.2	118.9	118.4	136.9	124.1	120.4	121.1	127.8	92.1	91.3	96.8	99.7	98.2	96.5	102.6	104.3
2000	118.9	114.6	114.0	132.1	120.9	115.9	117.7	122.8	89.9	89.1	94.2	98.0	96.9	95.0	100.8	100.8
1999	116.6	113.8	112.8	129.8	119.6	114.8	116.3	121.6	89.0	88.1	92.4	95.8	95.9	93.4	99.7	98.6
1998	113.6	109.8	110.5	126.6	117.1	112.5	113.7	120.3	88.0	87.2	90.7	93.6	94.9	92.4	97.2	96.8
1997	111.5	108.5	108.8	123.3	113.9	110.8	112.3	118.9	86.5	85.6	89.3	92.1	93.6	90.8	96.1	94.9
1996	108.9	106.8	105.7	120.3	110.8	108.7	109.9	117.2	84.9	84.1	87.2	90.0	91.6	88.8	94.0	91.9
1995	105.6	99.8	100.6	117.1	106.3	103.6	103.8	111.1	82.7	82.2	84.0	84.7	89.2	86.0	91.2	87.6
1994	103.0	98.2	99.4	115.2	103.7	102.2	102.6	109.6	81.2	80.6	82.7	83.1	87.4	84.1	89.0	84.8
1993	100.0	94.9	96.6	107.4	99.0	98.6	99.8	108.2	78.4	77.9	81.2	81.9	85.1	82.3	86.8	81.9
1992	97.9	92.1	94.8	105.3	96.5	96.7	97.8	106.9	77.1	76.8	80.0	80.7	83.6	77.7	85.4	80.7
1991	95.7	90.5	92.5	101.7	93.2	94.1	94.8	96.1	75.0	75.8	78.7	79.7	81.9	76.6	83.0	79.1
1990	93.2	88.5	89.5	98.5	91.2	90.8	91.3	94.1	73.7	74.5	77.2	78.4	79.9	75.1	81.3	77.1
1989	91.0	86.6	86.5	94.2	89.4	87.9	88.8	92.4	72.1	72.8	75.7	77.0	78.5	73.7	81.2	74.8
1988	88.5	84.9	84.0	89.8	87.7	85.3	86.4	89.1	70.5	71.3	74.2	75.5	77.0	72.1	79.7	73.1
1985	81.8	79.6	77.2	82.2	81.5	77.7	80.0	83.0	65.9	66.9	69.7	71.2	72.5	67.7	74.3	66.7
1980	60.7	59.4	56.9	58.7	61.3	57.7	58.4	59.2	50.6	51.1	54.8	57.1	55.0	51.6	55.9	53.1
1975	43.7	43.4	42.2	44.5	44.5	43.6	41.9	42.7	35.0	36.4	37.7	39.6	39.1	37.3	40.7	37.0
1970	27.8	27.5	25.8	27.5	28.7	27.1	27.0	27.3	22.8	22.9	24.8	25.8	24.9	22.0	23.0	22.8
1965	21.5	21.1	20.1	21.7	22.4	20.9	20.8	21.9	17.6	17.6	19.1	19.9	19.2	17.1	18.3	17.4
1960	19.5	19.1	18.6	19.4	19.7	19.0	18.9	19.1	16.0	16.0	17.3	18.1	17.4	15.5	16.6	15.8
1955	16.3	16.1	15.6	16.3	16.5	15.9	15.9	16.0	13.4	13.4	14.5	15.1	14.6	13.0	13.9	13.3
1950	13.5	13.3	12.9	13.5	13.6	13.2	13.1	13.2	11.1	11.1	12.0	12.5	12.1	10.7	11.5	10.9
1945	8.6	8.5	8.2	8.6	8.7	8.4	8.3	8.4	7.1	7.1	7.7	8.0	7.7	6.8	7.3	7.0
1940	6.6	6.5	6.3	6.6	6.7	6.4	6.5	6.5	5.4	5.5	5.9	6.2	6.0	5.3	5.6	5.4

Year	National 30 City Average	Texas Abi-lene	Ama-rillo	Austin	Beau-mont	Corpus Christi	Dallas	El Paso	Fort Worth	Houston	Lubbock	Odessa	San Antonio	Waco	Wichita Falls	Utah Ogden
Jan 2009	185.9E	142.1E	147.1E	147.6E	150.6E	143.2E	155.9E	141.3E	149.5E	161.1E	145.5E	139.4E	152.3E	147.0E	146.6E	155.9E
2008	171.0	132.4	137.0	137.5	140.3	133.2	144.1	130.7	138.3	149.1	134.7	128.9	141.0	136.1	136.1	144.8
2007	165.0	128.5	132.2	131.6	137.1	128.8	138.6	126.2	134.8	146.0	130.2	125.0	136.4	132.1	132.1	140.0
2006	156.2	121.6	125.7	125.5	130.1	122.0	131.1	120.3	127.6	138.2	123.2	118.3	129.8	125.2	125.3	133.4
2005	146.7	113.8	117.3	117.9	121.3	114.4	123.7	112.5	119.4	129.0	115.5	110.5	121.3	116.1	117.3	126.0
2004	132.8	102.9	106.3	105.7	108.5	102.9	112.0	101.5	108.4	115.9	104.7	99.9	108.4	104.8	105.1	115.2
2003	129.7	100.5	104.4	103.9	106.1	100.4	109.3	99.5	105.5	113.4	102.3	97.6	104.8	102.7	103.0	112.7
2002	126.7	99.9	101.9	102.4	104.5	98.9	107.9	98.7	104.6	111.5	100.5	95.9	103.7	100.5	101.4	110.8
2001	122.2	93.4	98.4	99.8	102.3	96.6	103.8	95.5	100.9	107.8	97.6	93.4	100.5	97.8	97.3	107.7
2000	118.9	93.4	98.1	99.1	101.6	96.9	102.7	92.4	99.9	106.0	97.6	93.4	99.4	97.3	96.9	104.6
1999	116.6	91.8	94.5	96.0	99.7	94.0	101.0	90.7	97.6	104.6	96.0	92.1	98.0	94.8	95.5	103.3
1998	113.6	89.8	92.7	94.2	97.9	91.8	97.9	88.4	94.5	101.3	93.3	90.1	94.8	92.7	92.9	98.5
1997	111.5	88.4	91.3	92.8	96.8	90.3	96.1	87.0	93.3	100.1	91.9	88.8	93.4	91.4	91.5	96.1
1996	108.9	86.8	89.6	90.9	95.3	88.5	94.1	86.7	91.5	97.9	90.2	87.1	92.3	89.7	89.9	94.1
1995	105.6	85.2	87.4	89.3	93.7	87.4	91.4	85.2	89.5	95.9	88.4	85.6	88.9	86.4	86.8	92.2
1994	103.0	83.3	85.5	87.0	91.8	84.6	89.6	82.2	87.4	93.4	87.0	83.9	87.0	84.9	85.4	89.4
1993	100.0	81.4	83.4	84.9	90.0	82.8	87.8	80.2	85.5	91.1	84.8	81.9	85.0	83.0	83.5	87.1
1992	97.9	80.3	82.3	83.8	88.9	81.6	86.2	79.0	84.1	89.8	83.6	80.7	83.9	81.8	82.4	85.1
1991	95.7	79.2	81.1	82.8	87.8	80.6	85.9	78.0	83.3	87.9	82.7	79.8	83.3	80.6	81.5	84.1
1990	93.2	78.0	80.1	81.3	86.5	79.3	84.5	76.7	82.1	85.4	81.5	78.6	80.7	79.6	80.3	83.4
1989	91.0	76.6	78.7	79.7	85.1	77.8	82.5	75.2	80.6	84.0	80.0	77.0	78.7	78.1	78.7	81.9
1988	88.5	75.0	77.2	78.3	83.9	76.3	81.1	73.6	79.8	82.5	78.4	75.5	77.1	76.5	77.4	80.4
1985	81.8	71.1	72.5	74.5	79.3	72.3	77.6	69.4	75.1	79.6	74.0	71.2	73.9	71.7	73.3	75.2
1980	60.7	53.4	55.2	54.5	57.6	54.5	57.9	53.1	57.0	59.4	55.6	57.2	55.0	54.9	55.4	62.2
1975	43.7	37.6	39.0	39.0	39.6	38.1	40.7	38.0	40.4	41.2	38.9	37.9	39.0	38.6	38.0	40.0
1970	27.8	24.5	24.9	24.9	25.7	24.5	25.5	23.7	25.9	25.4	25.1	24.6	23.3	24.8	24.5	26.8
1965	21.5	18.9	19.2	19.2	19.9	18.9	19.9	19.0	19.9	20.0	19.4	19.0	18.5	19.2	18.9	20.6
1960	19.5	17.1	17.4	17.4	18.1	17.1	18.2	17.0	18.1	18.2	17.6	17.3	16.8	17.4	17.2	18.8
1955	16.3	14.4	14.6	14.6	15.1	14.4	15.3	14.3	15.2	15.2	14.8	14.5	14.1	14.6	14.4	15.7
1950	13.5	11.9	12.1	12.1	12.5	11.9	12.6	11.8	12.5	12.6	12.2	12.0	11.6	12.1	11.9	13.0
1945	8.6	7.6	7.7	7.7	8.0	7.6	8.0	7.5	8.0	8.0	7.8	7.6	7.4	7.7	7.6	8.3
1940	6.6	5.9	5.9	5.9	6.1	5.8	6.2	5.8	6.2	6.2	6.0	5.9	5.7	5.9	5.8	6.4

Historical Cost Indexes

Year	National 30 City Average	Utah Salt Lake City	Vermont Burlington	Vermont Rutland	Virginia Alexandria	Virginia Newport News	Virginia Norfolk	Virginia Richmond	Virginia Roanoke	Washington Seattle	Washington Spokane	Washington Tacoma	West Virginia Charleston	West Virginia Huntington	Wisconsin Green Bay	Wisconsin Kenosha
Jan 2009	185.9E	161.3E	158.6E	156.6E	173.7E	161.8E	162.0E	162.1E	156.7E	191.2E	173.9E	189.0E	174.4E	177.1E	176.8E	183.2
2008	171.0	149.7	147.2	145.8	161.8	150.8	150.8	150.9	145.9	176.9	162.4	174.9	162.8	165.7	165.6	172.2
2007	165.0	144.6	144.4	143.2	155.0	146.2	146.1	147.3	142.9	171.4	156.7	168.7	158.7	160.1	159.4	164.8
2006	156.2	137.7	131.7	130.7	146.2	133.7	134.6	134.8	129.7	162.9	150.1	160.7	149.4	151.4	152.7	157.1
2005	146.7	129.4	124.6	123.8	136.5	123.5	124.4	125.4	112.2	153.9	141.9	151.5	140.8	141.0	144.4	148.3
2004	132.8	117.8	113.0	112.3	121.5	108.9	110.2	110.9	99.7	138.0	127.6	134.5	124.8	125.6	128.9	132.4
2003	129.7	116.0	110.7	110.1	119.5	104.8	106.2	108.6	97.0	134.9	125.9	133.0	123.3	123.6	127.4	130.9
2002	126.7	113.7	109.0	108.4	115.1	102.9	104.1	106.6	95.2	132.7	123.9	131.4	121.2	120.9	123.0	127.3
2001	122.2	109.1	105.7	105.2	110.8	99.8	100.3	102.9	92.1	127.9	120.3	125.7	114.6	117.5	119.1	123.6
2000	118.9	106.5	98.9	98.3	108.1	96.5	97.6	100.2	90.7	124.6	118.3	122.9	111.5	114.4	114.6	119.1
1999	116.6	104.5	98.2	97.7	106.1	95.6	96.5	98.8	89.8	123.3	116.7	121.6	110.6	113.4	112.1	115.8
1998	113.6	99.5	97.8	97.3	104.1	93.7	93.9	97.0	88.3	119.4	114.3	118.3	106.7	109.0	109.5	112.9
1997	111.5	97.2	96.6	96.3	101.2	91.6	91.7	92.9	86.9	118.1	111.7	117.2	105.3	107.7	105.6	109.1
1996	108.9	94.9	95.1	94.8	99.7	90.2	90.4	91.6	85.5	115.2	109.2	114.3	103.1	104.8	103.8	106.4
1995	105.6	93.1	91.1	90.8	96.3	86.0	86.4	87.8	82.8	113.7	107.4	112.8	95.8	97.2	97.6	97.9
1994	103.0	90.2	89.5	89.3	93.9	84.6	84.8	86.3	81.4	109.9	104.0	108.3	94.3	95.3	96.3	96.2
1993	100.0	87.9	87.6	87.6	91.6	82.9	83.0	84.3	79.5	107.3	103.9	106.7	92.6	93.5	94.0	94.3
1992	97.9	86.0	86.1	86.1	90.1	81.0	81.6	82.0	78.3	105.1	101.4	103.7	91.4	92.3	92.0	92.1
1991	95.7	84.9	84.2	84.2	88.2	77.6	77.9	79.8	77.3	102.2	100.0	102.2	89.7	88.6	88.6	89.8
1990	93.2	84.3	83.0	82.9	86.1	76.3	76.7	77.6	76.1	100.1	98.5	100.5	86.1	86.8	86.7	87.8
1989	91.0	82.8	81.4	81.3	83.4	74.9	75.2	76.0	74.2	96.1	96.8	98.4	84.5	84.8	84.3	85.6
1988	88.5	81.3	79.7	79.7	81.0	73.4	73.7	74.1	72.3	94.2	95.0	96.6	82.8	83.1	82.0	83.0
1985	81.8	75.9	74.8	74.9	75.1	68.7	68.8	69.5	67.2	88.3	89.0	91.2	77.7	77.7	76.7	77.4
1980	60.7	57.0	55.3	58.3	57.3	52.5	52.4	54.3	51.3	67.9	66.3	66.7	57.7	58.3	58.6	58.3
1975	43.7	40.1	41.8	43.9	41.7	37.2	36.9	37.1	37.1	44.9	44.4	44.5	41.0	40.0	40.9	40.5
1970	27.8	26.1	25.4	26.8	26.2	23.9	21.5	22.0	23.7	28.8	29.3	29.6	26.1	25.8	26.4	26.5
1965	21.5	20.0	19.8	20.6	20.2	18.4	17.1	17.2	18.3	22.4	22.5	22.8	20.1	19.9	20.3	20.4
1960	19.5	18.4	18.0	18.8	18.4	16.7	15.4	15.6	16.6	20.4	20.8	20.8	18.3	18.1	18.4	18.6
1955	16.3	15.4	15.1	15.7	15.4	14.0	12.9	13.1	13.9	17.1	17.4	17.4	15.4	15.2	15.5	15.6
1950	13.5	12.7	12.4	13.0	12.7	11.6	10.7	10.8	11.5	14.1	14.4	14.4	12.7	12.5	12.8	12.9
1945	8.6	8.1	7.9	8.3	8.1	7.4	6.8	6.9	7.3	9.0	9.2	9.2	8.1	8.0	8.1	8.2
1940	6.6	6.3	6.1	6.4	6.2	5.7	5.3	5.3	5.7	7.0	7.1	7.1	6.2	6.2	6.3	6.3

Year	National 30 City Average	Wisconsin Madison	Wisconsin Milwaukee	Wisconsin Racine	Wyoming Cheyenne	Canada Calgary	Canada Edmonton	Canada Hamilton	Canada London	Canada Montreal	Canada Ottawa	Canada Quebec	Canada Toronto	Canada Vancouver	Canada Winnipeg
Jan 2009	185.9E	179.2E	189.3E	184.1E	157.2E	206.6E	207.4E	206.7E	202.3E	201.6E	203.0E	200.3E	209.1E	200.9E	189.5E
2008	171.0	168.4	176.3	173.0	146.5	190.2	191.0	194.3	188.6	186.7	188.0	186.8	194.6	184.9	174.4
2007	165.0	160.7	168.9	164.7	141.3	183.1	184.4	186.8	182.4	181.7	182.4	182.0	187.7	180.6	170.1
2006	156.2	152.8	158.8	157.2	128.0	163.6	164.8	169.0	164.9	158.6	163.7	159.0	170.6	169.3	155.9
2005	146.7	145.2	148.4	147.9	118.9	154.4	155.7	160.0	156.2	149.2	155.1	150.1	162.6	159.4	146.9
2004	132.8	129.7	134.2	132.7	104.7	138.8	139.4	142.6	140.4	134.5	141.0	135.7	146.0	141.7	129.7
2003	129.7	128.4	131.1	131.5	102.6	133.6	134.2	139.1	137.0	130.2	137.4	131.4	142.3	137.0	124.4
2002	126.7	124.5	128.2	126.5	101.7	122.5	122.2	136.3	134.1	127.2	134.9	128.4	139.8	134.6	121.4
2001	122.2	120.6	123.9	123.3	99.0	117.5	117.4	131.5	129.1	124.4	130.2	125.5	134.7	130.2	117.2
2000	118.9	116.6	120.5	118.7	98.1	115.9	115.8	130.0	127.5	122.8	128.8	124.1	133.1	128.4	115.6
1999	116.6	115.9	117.4	115.5	96.9	115.3	115.2	128.1	125.6	120.8	126.8	121.9	131.2	127.1	115.2
1998	113.6	110.8	113.4	112.7	95.4	112.5	112.4	126.3	123.9	119.0	124.7	119.6	128.5	123.8	113.7
1997	111.5	106.1	110.1	109.3	93.2	110.7	110.6	124.4	121.9	114.6	122.8	115.4	125.7	121.9	111.4
1996	108.9	104.4	107.1	106.5	91.1	109.1	109.0	122.6	120.2	112.9	121.0	113.6	123.9	119.0	109.7
1995	105.6	96.5	103.9	97.8	87.6	107.4	107.4	119.9	117.5	110.8	118.2	111.5	121.6	116.2	107.6
1994	103.0	94.5	100.6	96.1	85.4	106.7	106.6	116.5	114.2	109.5	115.0	110.2	117.8	115.1	105.5
1993	100.0	91.3	96.7	93.8	82.9	104.7	104.5	113.7	111.9	106.9	112.0	107.0	114.9	109.2	102.8
1992	97.9	89.2	93.9	91.7	81.7	103.4	103.2	112.4	110.7	104.2	110.8	103.6	113.7	108.0	101.6
1991	95.7	86.2	91.6	89.3	80.3	102.1	102.0	108.2	106.7	101.8	106.9	100.4	109.0	106.8	98.6
1990	93.2	84.3	88.9	87.3	79.1	98.0	97.1	103.8	101.4	99.0	102.7	96.8	104.6	103.2	95.1
1989	91.0	81.8	86.4	85.0	77.8	95.6	94.7	98.2	96.5	94.9	98.5	92.5	98.5	97.6	92.9
1988	88.5	79.9	84.1	82.4	76.3	93.9	93.0	95.4	93.2	90.6	94.0	88.3	94.8	95.9	90.0
1985	81.8	74.3	77.4	77.0	72.3	90.2	89.1	85.8	84.9	82.4	83.9	79.9	86.5	89.8	83.4
1980	60.7	56.8	58.8	58.1	56.9	64.9	63.3	63.7	61.9	59.2	60.9	59.3	60.9	65.0	61.7
1975	43.7	40.7	43.3	40.7	40.6	42.2	41.6	42.9	41.6	39.7	41.5	39.0	42.2	42.4	39.2
1970	27.8	26.5	29.4	26.5	26.0	28.9	28.6	28.5	27.8	25.6	27.6	26.0	25.6	26.0	23.1
1965	21.5	20.6	21.8	20.4	20.0	22.3	22.0	22.0	21.4	18.7	21.2	20.1	19.4	20.5	17.5
1960	19.5	18.1	19.0	18.6	18.2	20.2	20.0	20.0	19.5	17.0	19.3	18.2	17.6	18.6	15.8
1955	16.3	15.2	15.9	15.6	15.2	17.0	16.8	16.7	16.3	14.3	16.2	15.3	14.8	15.5	13.3
1950	13.5	12.5	13.2	12.9	12.6	14.0	13.9	13.8	13.5	11.8	13.4	12.6	12.2	12.8	10.9
1945	8.6	8.0	8.4	8.2	8.0	9.0	8.8	8.8	8.6	7.5	8.5	8.0	7.8	8.2	7.0
1940	6.6	6.2	6.5	6.3	6.2	6.9	6.8	6.8	6.6	5.8	6.6	6.2	6.0	6.3	5.4

Glossary

Accent lighting–Fixtures or directional beams of light arranged to bring attention to an object or area.

Acoustical material–A material fabricated for the sole purpose of absorbing sound.

Addition–An expansion to an existing structure generally in the form of a room, floor(s) or wing(s). An increase in the floor area or volume of a structure.

Aggregate–Materials such as sand, gravel, stone, vermiculite, perlite and fly ash (slag) which are essential components in the production of concrete, mortar or plaster.

Air conditioning system–An air treatment to control the temperature, humidity and cleanliness of air and to provide for its distribution throughout the structure.

Air curtain–A stream of air directed downward to prevent the loss of hot or cool air from the structure and inhibit the entrance of dust and insects. Generally installed at loading platforms.

Anodized aluminum–Aluminum treated by an electrolytic process to produce an oxide film that is corrosion resistant.

Arcade–A covered passageway between buildings often with shops and offices on one or both sides.

Ashlar–A square-cut building stone or a wall constructed of cut stone.

Asphalt paper–A sheet paper material either coated or saturated with asphalt to increase its strength and resistance to water.

Asphalt shingles–Shingles manufactured from saturated roofing felts, coated with asphalt and topped with mineral granules to prevent weathering.

Assessment ratio–The ratio between the market value and assessed valuation of a property.

Back-up–That portion of a masonry wall behind the exterior facing-usually load bearing.

Balcony–A platform projecting from a building either supported from below or cantilevered. Usually protected by railing.

Bay–Structural component consisting of beams and columns occurring consistently throughout the structure.

Beam–A horizontal structural framing member that transfers loads to vertical members (columns) or bearing walls.

Bid–An offer to perform work described in a contract at a specified price.

Board foot–A unit or measure equal in volume to a board one foot long, one foot wide and one inch thick.

Booster pump–A supplemental pump installed to increase or maintain adequate pressure in the system.

Bowstring roof–A roof supported by trusses fabricated in the shape of a bow and tied together by a straight member.

Brick veneer–A facing of brick laid against, but not bonded to, a wall.

Bridging–A method of bracing joists for stiffness, stability and load distribution.

Broom finish–A method of finishing concrete by lightly dragging a broom over freshly placed concrete.

Built-up roofing–Installed on flat or extremely low-pitched roofs, a roof covering composed of plies or laminations of saturated roofing felts alternated with layers of coal tar pitch or asphalt and surfaced with a layer of gravel or slag in a thick coat of asphalt and finished with a capping sheet.

Caisson–A watertight chamber to expedite work on foundations or structure below water level.

Cantilever–A structural member supported only at one end.

Carport–An automobile shelter having one or more sides open to the weather.

Casement window–A vertical opening window having one side fixed.

Caulking–A resilient compound generally having a silicone or rubber base used to prevent infiltration of water or outside air.

Cavity wall–An exterior wall usually of masonry having an inner and outer wall separated by a continuous air space for thermal insulation.

Cellular concrete–A lightweight concrete consisting of cement mixed with gas-producing materials which, in turn, forms bubbles resulting in good insulating qualities.

Chattel mortgage–A secured interest in a property as collateral for payment of a note.

Clapboard–A wood siding used as exterior covering in frame construction. It is applied horizontally with grain running lengthwise. The thickest section of the board is on the bottom.

Clean room–An environmentally controlled room usually found in medical facilities or precision manufacturing spaces where it is essential to eliminate dust, lint or pathogens.

Close studding–A method of construction whereby studs are spaced close together and the intervening spaces are plastered.

Cluster housing–A closely grouped series of houses resulting in a high density land use.

Cofferdam–A watertight enclosure used for foundation construction in waterfront areas. Water is pumped out of the cofferdam allowing free access to the work area.

Collar joint–Joint between the collar beam and roof rafters.

Column–A vertical structural member supporting horizontal members (beams) along the direction of its longitudinal axis.

Combination door (windows)–Door (windows) having interchangeable screens and glass for seasonal use.

Common area–Spaces either inside or outside the building designated for use by the occupant of the building but not the general public.

Common wall–A wall used jointly by two dwelling units.

Compound wall–A wall constructed of more than one material.

Concrete–A composite material consisting of sand, coarse aggregate (gravel, stone or slag), cement and water that when mixed and allowed to harden forms a hard stone-like material.

Concrete floor hardener–A mixture of chemicals applied to the surface of concrete to produce a dense, wear-resistant bearing surface.

Conduit–A tube or pipe used to protect electric wiring.

Coping–The top cover or capping on a wall.

Craneway–Steel or concrete column and beam supports and rails on which a crane travels.

Curtain wall–A non-bearing exterior wall not supported by beams or girders of the steel frame.

Dampproofing–Coating of a surface to prevent the passage of water or moisture.

Dead load–Total weight of all structural components plus permanently attached fixtures and equipment.

Decibel–Unit of acoustical measurement.

Decorative block–A concrete masonry unit having a special treatment on its face.

Deed restriction–A restriction on the use of a property as set forth in the deed.

Depreciation–A loss in property value caused by physical aging, functional or economic obsolescence.

Direct heating–Heating of spaces by means of exposed heated surfaces (stove, fire, radiators, etc.).

Distributed load–A load that acts evenly over a structural member.

Dock bumper–A resilient material attached to a loading dock to absorb the impact of trucks backing in.

Dome–A curved roof shape spanning an area.

Dormer–A structure projecting from a sloping roof.

Double-hung window–A window having two vertical sliding sashes-one covering the upper section and one covering the lower.

Downspout–A vertical pipe for carrying rain water from the roof to the ground.

Drain tile–A hollow tile used to drain water-soaked soil.

Dressed size–In lumber about 3/8″ - 1/2″ less than nominal size after planing and sawing.

Drop panel–The depressed surface on the bottom side of a flat concrete slab which surrounds a column.

Dry pipe sprinkler system–A sprinkler system that is activated with water only in case of fire. Used in areas susceptible to freezing or to avoid the hazards of leaking pipes.

Drywall–An interior wall constructed of gypsum board, plywood or wood paneling. No water is required for application.

Duct–Pipe for transmitting warm or cold air. A pipe containing electrical cable or wires.

Dwelling–A structure designed as living quarters for one or more families.

Easement–A right of way or free access over land owned by another.

Eave–The lower edge of a sloping roof that overhangs the side wall of the structure.

Economic rent–That rent on a property sufficient to pay all operating costs exclusive of services and utilities.

Economic life–The term during which a structure is expected to be profitable. Generally shorter than the physical life of the structure.

Economic obsolescence–Loss in value due to unfavorable economic influences occurring from outside the structure itself.

Effective age–The age a structure appears to be based on observed physical condition determined by degree of maintenance and repair.

Elevator–A mechanism for vertical transport of personnel or freight equipped with car or platform.

Ell–A secondary wing or addition to a structure at right angles to the main structure.

Eminent domain–The right of the state to take private property for public use.

Envelope–The shape of a building indicating volume.

Equity–Value of an owner's interest calculated by subtracting outstanding mortgages and expenses from the value of the property.

Escheat–The assumption of ownership by the state, of property whose owner cannot be determined.

Facade–The exterior face of a building sometimes decorated with elaborate detail.

Face brick–Brick manufactured to present an attractive appearance.

Facing block–A concrete masonry unit having a decorative exterior finish.

Feasibility study–A detailed evaluation of a proposed project to determine its financial potential.

Felt paper–Paper sheathing used on walls as an insulator and to prevent infiltration of moisture.

Fenestration–Pertaining to the density and arrangement of windows in a structure.

Fiberboard–A building material composed of wood fiber compressed with a binding agent, produced in sheet form.

Fiberglass–Fine spun filaments of glass processed into various densities to produce thermal and acoustical insulation.

Field house–A long structure used for indoor athletic events.

Finish floor–The top or wearing surface of the floor system.

Fireproofing–The use of fire-resistant materials for the protection of structural members to ensure structural integrity in the event of fire.

Fire stop–A material or member used to seal an opening to prevent the spread of fire.

Flashing–A thin, impervious material such as copper or sheet metal used to prevent air or water penetration.

Float finish–A concrete finish accomplished by using a flat tool with a handle on the back.

Floor drain–An opening installed in a floor for removing excess water into a plumbing system.

Floor load–The live load the floor system has been designed for and which may be applied safely.

Flue–A heat-resistant enclosed passage in a chimney to remove gaseous products of combustion from a fireplace or boiler.

Footing–The lowest portion of the foundation wall that transmits the load directly to the soil.

Foundation–Below grade wall system that supports the structure.

Foyer–A portion of the structure that serves as a transitional space between the interior and exterior.

Frontage–That portion of a lot that is placed facing a street, public way or body of water.

Frost action–The effects of freezing and thawing on materials and the resultant structural damage.

Functional obsolescence–An inadequacy caused by outmoded design, dated construction materials or over or undersized areas, all of which cause excessive operating costs.

Furring strips–Wood or metal channels used for attaching gypsum or metal lath to masonry walls as a finish or for leveling purposes.

Gambrel roof (mansard)–A roof system having two pitches on each side.

Garden apartment–Two or three story apartment building with common outside areas.

General contractor–The prime contractor responsible for work on the construction site.

Girder–A principal horizontal supporting member.

Girt–A horizontal framing member for adding rigid support to columns which also acts as a support for sheathing or siding.

Grade beam–That portion of the foundation system that directly supports the exterior walls of the building.

Gross area–The total enclosed floor area of a building.

Ground area–The area computed by the exterior dimensions of the structure.

Ground floor–The floor of a building in closest proximity to the ground.

Grout–A mortar containing a high water content used to fill joints and cavities in masonry work.

Gutter–A wood or sheet metal conduit set along the building eaves used to channel rainwater to leaders or downspouts.

Gunite–A concrete mix placed pneumatically.

Gypsum–Hydrated calcium sulphate used as both a retarder in Portland cement and as a major ingredient in plaster of Paris. Used in sheets as a substitute for plaster.

Hall–A passageway providing access to various parts of a building–a large room for entertainment or assembly.

Hangar–A structure for the storage or repair of aircraft.

Head room–The distance between the top of the finished floor to the bottom of the finished ceiling.

Hearth–The floor of a fireplace and the adjacent area of fireproof material.

Heat pump–A mechanical device for providing either heating or air conditioning.

Hip roof–A roof whose four sides meet at a common point with no gabled ends.

I-beam–A structural member having a cross-section resembling the letter "I".

Insulation–Material used to reduce the effects of heat, cold or sound.

Jack rafter–An unusually short rafter generally found in hip roofs.

Jalousie–Adjustable glass louvers which pivot simultaneously in a common frame.

Jamb–The vertical member on either side of a door frame or window frame.

Joist–Parallel beams of timber, concrete or steel used to support floor and ceiling.

Junction box–A box that protects splices or joints in electrical wiring.

Kalamein door–Solid core wood doors clad with galvanized sheet metal.

Kip–A unit of weight equal to 1,000 pounds.

Lally column–A concrete-filled steel pipe used as a vertical support.

Laminated beam–A beam built up by gluing together several pieces of timber.

Landing–A platform between flights of stairs.

Lath–Strips of wood or metal used as a base for plaster.

Lead-lined door, sheetrock–Doors or sheetrock internally lined with sheet lead to provide protection from X-ray radiation.

Lean-to–A small shed or building addition with a single pitched roof attached to the exterior wall of the main building.

Lintel–A horizontal framing member used to carry a load over a wall opening.

Live load–The moving or movable load on a structure composed of furnishings, equipment or personnel weight.

Load bearing partition–A partition that supports a load in addition to its own weight.

Loading dock leveler–An adjustable platform for handling off-on loading of trucks.

Loft building–A commercial/industrial type building containing large, open unpartitioned floor areas.

Louver window–A window composed of a series of sloping, overlapping blades or slats that may be adjusted to admit varying degrees of air or light.

Main beam–The principal load bearing beam used to transmit loads directly to columns.

Mansard roof–A roof having a double pitch on all four sides, the lower level having the steeper pitch.

Mansion–An extremely large and imposing residence.

Masonry–The utilization of brick, stone, concrete or block for walls and other building components.

Mastic–A sealant or adhesive compound generally used as either a binding agent or floor finish.

Membrane fireproofing–A coating of metal lath and plaster to provide resistance to fire and heat.

Mesh reinforcement–An arrangement of wire tied or welded at their intersection to provide strength and resistance to cracking.

Metal lath–A diamond-shaped metallic base for plaster.

Mezzanine–A low story situated between two main floors.

Mill construction–A heavy timber construction that achieves fire resistance by using large wood structural members, noncombustible bearing and non-bearing walls and omitting concealed spaces under floors and roof.

Mixed occupancy–Two or more classes of occupancy in a single structure.

Molded brick–A specially shaped brick used for decorative purposes.

Monolithic concrete–Concrete poured in a continuous process so there are no joints.

Movable partition–A non-load bearing demountable partition that can be relocated and can be either ceiling height or partial height.

Moving walk–A continually moving horizontal passenger carrying device.

Multi-zone system–An air conditioning system that is capable of handling several individual zones simultaneously.

Net floor area–The usable, occupied area of a structure excluding stairwells, elevator shafts and wall thicknesses.

Non-combustible construction–Construction in which the walls, partitions and the structural members are of non-combustible materials.

Nurses call system–An electrically operated system for use by patients or personnel for summoning a nurse.

Occupancy rate–The number of persons per room, per dwelling unit, etc.

One-way joist connection–A framing system for floors and roofs in a concrete building consisting of a series of parallel joists supported by girders between columns.

Open web joist–A lightweight prefabricated metal chord truss.

Parapet–A low guarding wall at the point of a drop. That portion of the exterior wall that extends above the roof line.

Parging–A thin coat of mortar applied to a masonry wall used primarily for waterproofing purposes.

Parquet–Inlaid wood flooring set in a simple design.

Peaked roof–A roof of two or more slopes that rises to a peak.

Penthouse–A structure occupying usually half the roof area and used to house elevator, HVAC equipment, or other mechanical or electrical systems.

Pier–A column designed to support a concentrated load.

Pilaster–A column usually formed of the same material and integral with, but projecting from, a wall.

Pile–A concrete, wood or steel column usually less than 2′ in diameter which is driven or otherwise introduced into the soil to carry a vertical load or provide vertical support.

Pitched roof–A roof having one or more surfaces with a pitch greater than 10°.

Plenum–The space between the suspended ceiling and the floor above.

Plywood–Structural wood made of three or more layers of veneer and bonded together with glue.

Porch–A structure attached to a building to provide shelter for an entranceway or to serve as a semi-enclosed space.

Post and beam framing–A structural framing system where beams rest on posts rather than bearing walls.

Post-tensioned concrete–Concrete that has the reinforcing tendons tensioned after the concrete has set.

Pre-cast concrete–Concrete structural components fabricated at a location other than in-place.

Pre-stressed concrete–Concrete that has the reinforcing tendons tensioned prior to the concrete setting.

Purlins–A horizontal structural member supporting the roof deck and resting on the trusses, girders, beams or rafters.

Rafters–Structural members supporting the roof deck and covering.

Resilient flooring–A manufactured interior floor covering material in either sheet or tile form, that is resilient.

Ridge–The horizontal line at the junction of two sloping roof edges.

Rigid frame–A structural framing system in which all columns and beams are rigidly connected; there are no hinged joints.

Roof drain–A drain designed to accept rainwater and discharge it into a leader or downspout.

Rough floor–A layer of boards or plywood nailed to the floor joists that serves as a base for the finished floor.

Sanitary sewer–A sewer line designed to carry only liquid or waterborne waste from the structure to a central treatment plant.

Sawtooth roof–Roof shape found primarily in industrial roofs that creates the appearance of the teeth of a saw.

Semi-rigid frame–A structural system wherein the columns and beams are so attached that there is some flexibility at the joints.

Septic tank–A covered tank in which waste matter is decomposed by natural bacterial action.

Service elevator–A combination passenger and freight elevator.

Sheathing–The first covering of exterior studs by boards, plywood or particle board.

Shoring–Temporary bracing for structural components during construction.

Span–The horizontal distance between supports.

Specification–A detailed list of materials and requirements for construction of a building.

Storm sewer–A system of pipes used to carry rainwater or surface waters.

Stucco–A cement plaster used to cover exterior wall surfaces usually applied over a wood or metal lath base.

Stud–A vertical wooden structural component.

Subfloor–A system of boards, plywood or particle board laid over the floor joists to form a base.

Superstructure–That portion of the structure above the foundation or ground level.

Terrazzo–A durable floor finish made of small chips of colored stone or marble, embedded in concrete, then polished to a high sheen.

Tile–A thin piece of fired clay, stone or concrete used for floor, roof or wall finishes.

Unit cost–The cost per unit of measurement.

Vault–A room especially designed for storage.

Veneer–A thin surface layer covering a base of common material.

Vent–An opening serving as an outlet for air.

Wainscot–The lower portion of an interior wall whose surface differs from that of the upper wall.

Wall bearing construction–A structural system where the weight of the floors and roof are carried directly by the masonry walls rather than the structural framing system.

Waterproofing–Any of a number of materials applied to various surfaces to prevent the infiltration of water.

Weatherstrip–A thin strip of metal, wood or felt used to cover the joint between the door or window sash and the jamb, casing or sill to keep out air, dust, rain, etc.

Wing–A building section or addition projecting from the main structure.

X-ray protection–Lead encased in sheetrock or plaster to prevent the escape of radiation.

Abbreviations

A	Area Square Feet; Ampere	BTUH	BTU per Hour	Cwt.	100 Pounds
ABS	Acrylonitrile Butadiene Stryrene; Asbestos Bonded Steel	B.U.R.	Built-up Roofing	C.W.X.	Cool White Deluxe
		BX	Interlocked Armored Cable	C.Y.	Cubic Yard (27 cubic feet)
A.C.	Alternating Current; Air-Conditioning; Asbestos Cement; Plywood Grade A & C	°C	degree centegrade	C.Y./Hr.	Cubic Yard per Hour
		c	Conductivity, Copper Sweat	Cyl.	Cylinder
		C	Hundred; Centigrade	d	Penny (nail size)
		C/C	Center to Center, Cedar on Cedar	D	Deep; Depth; Discharge
A.C.I.	American Concrete Institute	C-C	Center to Center	Dis., Disch.	Discharge
AD	Plywood, Grade A & D	Cab.	Cabinet	Db.	Decibel
Addit.	Additional	Cair.	Air Tool Laborer	Dbl.	Double
Adj.	Adjustable	Calc	Calculated	DC	Direct Current
af	Audio-frequency	Cap.	Capacity	DDC	Direct Digital Control
A.G.A.	American Gas Association	Carp.	Carpenter	Demob.	Demobilization
Agg.	Aggregate	C.B.	Circuit Breaker	d.f.u.	Drainage Fixture Units
A.H.	Ampere Hours	C.C.A.	Chromate Copper Arsenate	D.H.	Double Hung
A hr.	Ampere-hour	C.C.F.	Hundred Cubic Feet	DHW	Domestic Hot Water
A.H.U.	Air Handling Unit	cd	Candela	DI	Ductile Iron
A.I.A.	American Institute of Architects	cd/sf	Candela per Square Foot	Diag.	Diagonal
AIC	Ampere Interrupting Capacity	CD	Grade of Plywood Face & Back	Diam., Dia	Diameter
Allow.	Allowance	CDX	Plywood, Grade C & D, exterior glue	Distrib.	Distribution
alt.	Altitude			Div.	Division
Alum.	Aluminum	Cefi.	Cement Finisher	Dk.	Deck
a.m.	Ante Meridiem	Cem.	Cement	D.L.	Dead Load; Diesel
Amp.	Ampere	CF	Hundred Feet	DLH	Deep Long Span Bar Joist
Anod.	Anodized	C.F.	Cubic Feet	Do.	Ditto
Approx.	Approximate	CFM	Cubic Feet per Minute	Dp.	Depth
Apt.	Apartment	c.g.	Center of Gravity	D.P.S.T.	Double Pole, Single Throw
Asb.	Asbestos	CHW	Chilled Water; Commercial Hot Water	Dr.	Drive
A.S.B.C.	American Standard Building Code			Drink.	Drinking
Asbe.	Asbestos Worker	C.I.	Cast Iron	D.S.	Double Strength
ASCE.	American Society of Civil Engineers	C.I.P.	Cast in Place	D.S.A.	Double Strength A Grade
A.S.H.R.A.E.	American Society of Heating, Refrig. & AC Engineers	Circ.	Circuit	D.S.B.	Double Strength B Grade
		C.L.	Carload Lot	Dty.	Duty
A.S.M.E.	American Society of Mechanical Engineers	Clab.	Common Laborer	DWV	Drain Waste Vent
		Clam	Common maintenance laborer	DX	Deluxe White, Direct Expansion
A.S.T.M.	American Society for Testing and Materials	C.L.F.	Hundred Linear Feet	dyn	Dyne
		CLF	Current Limiting Fuse	e	Eccentricity
Attchmt.	Attachment	CLP	Cross Linked Polyethylene	E	Equipment Only; East
Avg., Ave.	Average	cm	Centimeter	Ea.	Each
A.W.G.	American Wire Gauge	CMP	Corr. Metal Pipe	E.B.	Encased Burial
AWWA	American Water Works Assoc.	C.M.U.	Concrete Masonry Unit	Econ.	Economy
Bbl.	Barrel	CN	Change Notice	E.C.Y	Embankment Cubic Yards
B&B	Grade B and Better; Balled & Burlapped	Col.	Column	EDP	Electronic Data Processing
		CO₂	Carbon Dioxide	EIFS	Exterior Insulation Finish System
B.&S.	Bell and Spigot	Comb.	Combination	E.D.R.	Equiv. Direct Radiation
B.&W.	Black and White	Compr.	Compressor	Eq.	Equation
b.c.c.	Body-centered Cubic	Conc.	Concrete	EL	elevation
B.C.Y.	Bank Cubic Yards	Cont.	Continuous; Continued, Container	Elec.	Electrician; Electrical
BE	Bevel End	Corr.	Corrugated	Elev.	Elevator; Elevating
B.F.	Board Feet	Cos	Cosine	EMT	Electrical Metallic Conduit; Thin Wall Conduit
Bg. cem.	Bag of Cement	Cot	Cotangent		
BHP	Boiler Horsepower; Brake Horsepower	Cov.	Cover	Eng.	Engine, Engineered
		C/P	Cedar on Paneling	EPDM	Ethylene Propylene Diene Monomer
B.I.	Black Iron	CPA	Control Point Adjustment		
Bit., Bitum.	Bituminous	Cplg.	Coupling	EPS	Expanded Polystyrene
Bit., Conc.	Bituminous Concrete	C.P.M.	Critical Path Method	Eqhv.	Equip. Oper., Heavy
Bk.	Backed	CPVC	Chlorinated Polyvinyl Chloride	Eqlt.	Equip. Oper., Light
Bkrs.	Breakers	C.Pr.	Hundred Pair	Eqmd.	Equip. Oper., Medium
Bldg.	Building	CRC	Cold Rolled Channel	Eqmm.	Equip. Oper., Master Mechanic
Blk.	Block	Creos.	Creosote	Eqol.	Equip. Oper., Oilers
Bm.	Beam	Crpt.	Carpet & Linoleum Layer	Equip.	Equipment
Boil.	Boilermaker	CRT	Cathode-ray Tube	ERW	Electric Resistance Welded
B.P.M.	Blows per Minute	CS	Carbon Steel, Constant Shear Bar Joist	E.S.	Energy Saver
BR	Bedroom			Est.	Estimated
Brg.	Bearing	Csc	Cosecant	esu	Electrostatic Units
Brhe.	Bricklayer Helper	C.S.F.	Hundred Square Feet	E.W.	Each Way
Bric.	Bricklayer	CSI	Construction Specifications Institute	EWT	Entering Water Temperature
Brk.	Brick			Excav.	Excavation
Brng.	Bearing	C.T.	Current Transformer	Exp.	Expansion, Exposure
Brs.	Brass	CTS	Copper Tube Size	Ext.	Exterior
Brz.	Bronze	Cu	Copper, Cubic	Extru.	Extrusion
Bsn.	Basin	Cu. Ft.	Cubic Foot	f.	Fiber stress
Btr.	Better	cw	Continuous Wave	F	Fahrenheit; Female; Fill
Btu	British Thermal Unit	C.W.	Cool White; Cold Water	Fab.	Fabricated

471

FBGS	Fiberglass	H.P.	Horsepower; High Pressure	LE	Lead Equivalent
F.C.	Footcandles	H.P.F.	High Power Factor	LED	Light Emitting Diode
f.c.c.	Face-centered Cubic	Hr.	Hour	L.F.	Linear Foot
f'c.	Compressive Stress in Concrete;	Hrs./Day	Hours per Day	L.F. Nose	Linear Foot of Stair Nosing
	Extreme Compressive Stress	HSC	High Short Circuit	L.F. Rsr	Linear Foot of Stair Riser
F.E.	Front End	Ht.	Height	Lg.	Long; Length; Large
FEP	Fluorinated Ethylene Propylene	Htg.	Heating	L & H	Light and Heat
	(Teflon)	Htrs.	Heaters	LH	Long Span Bar Joist
F.G.	Flat Grain	HVAC	Heating, Ventilation & Air-	L.H.	Labor Hours
F.H.A.	Federal Housing Administration		Conditioning	L.L.	Live Load
Fig.	Figure	Hvy.	Heavy	L.L.D.	Lamp Lumen Depreciation
Fin.	Finished	HW	Hot Water	lm	Lumen
Fixt.	Fixture	Hyd.;Hydr.	Hydraulic	lm/sf	Lumen per Square Foot
Fl. Oz.	Fluid Ounces	Hz.	Hertz (cycles)	lm/W	Lumen per Watt
Flr.	Floor	I.	Moment of Inertia	L.O.A.	Length Over All
F.M.	Frequency Modulation;	IBC	International Building Code	log	Logarithm
	Factory Mutual	I.C.	Interrupting Capacity	L-O-L	Lateralolet
Fmg.	Framing	ID	Inside Diameter	long.	longitude
Fdn.	Foundation	I.D.	Inside Dimension; Identification	L.P.	Liquefied Petroleum; Low Pressure
Fori.	Foreman, Inside	I.F.	Inside Frosted	L.P.F.	Low Power Factor
Foro.	Foreman, Outside	I.M.C.	Intermediate Metal Conduit	LR	Long Radius
Fount.	Fountain	In.	Inch	L.S.	Lump Sum
fpm	Feet per Minute	Incan.	Incandescent	Lt.	Light
FPT	Female Pipe Thread	Incl.	Included; Including	Lt. Ga.	Light Gauge
Fr.	Frame	Int.	Interior	L.T.L.	Less than Truckload Lot
F.R.	Fire Rating	Inst.	Installation	Lt. Wt.	Lightweight
FRK	Foil Reinforced Kraft	Insul.	Insulation/Insulated	L.V.	Low Voltage
FRP	Fiberglass Reinforced Plastic	I.P.	Iron Pipe	M	Thousand; Material; Male;
FS	Forged Steel	I.P.S.	Iron Pipe Size		Light Wall Copper Tubing
FSC	Cast Body; Cast Switch Box	I.P.T.	Iron Pipe Threaded	M²CA	Meters Squared Contact Area
Ft.	Foot; Feet	I.W.	Indirect Waste	m/hr.; M.H.	Man-hour
Ftng.	Fitting	J	Joule	mA	Milliampere
Ftg.	Footing	J.I.C.	Joint Industrial Council	Mach.	Machine
Ft lb.	Foot Pound	K	Thousand; Thousand Pounds;	Mag. Str.	Magnetic Starter
Furn.	Furniture		Heavy Wall Copper Tubing, Kelvin	Maint.	Maintenance
FVNR	Full Voltage Non-Reversing	K.A.H.	Thousand Amp. Hours	Marb.	Marble Setter
FXM	Female by Male	KCMIL	Thousand Circular Mils	Mat; Mat'l.	Material
Fy.	Minimum Yield Stress of Steel	KD	Knock Down	Max.	Maximum
g	Gram	K.D.A.T.	Kiln Dried After Treatment	MBF	Thousand Board Feet
G	Gauss	kg	Kilogram	MBH	Thousand BTU's per hr.
Ga.	Gauge	kG	Kilogauss	MC	Metal Clad Cable
Gal., gal.	Gallon	kgf	Kilogram Force	M.C.F.	Thousand Cubic Feet
gpm, GPM	Gallon per Minute	kHz	Kilohertz	M.C.F.M.	Thousand Cubic Feet per Minute
Galv.	Galvanized	Kip.	1000 Pounds	M.C.M.	Thousand Circular Mils
Gen.	General	KJ	Kiljoule	M.C.P.	Motor Circuit Protector
G.F.I.	Ground Fault Interrupter	K.L.	Effective Length Factor	MD	Medium Duty
Glaz.	Glazier	K.L.F.	Kips per Linear Foot	M.D.O.	Medium Density Overlaid
GPD	Gallons per Day	Km	Kilometer	Med.	Medium
GPH	Gallons per Hour	K.S.F.	Kips per Square Foot	MF	Thousand Feet
GPM	Gallons per Minute	K.S.I.	Kips per Square Inch	M.F.B.M.	Thousand Feet Board Measure
GR	Grade	kV	Kilovolt	Mfg.	Manufacturing
Gran.	Granular	kVA	Kilovolt Ampere	Mfrs.	Manufacturers
Grnd.	Ground	K.V.A.R.	Kilovar (Reactance)	mg	Milligram
H	High Henry	KW	Kilowatt	MGD	Million Gallons per Day
H.C.	High Capacity	KWh	Kilowatt-hour	MGPH	Thousand Gallons per Hour
H.D.	Heavy Duty; High Density	L	Labor Only; Length; Long;	MH, M.H.	Manhole; Metal Halide; Man-Hour
H.D.O.	High Density Overlaid		Medium Wall Copper Tubing	MHz	Megahertz
H.D.P.E.	high density polyethelene	Lab.	Labor	Mi.	Mile
Hdr.	Header	lat	Latitude	MI	Malleable Iron; Mineral Insulated
Hdwe.	Hardware	Lath.	Lather	mm	Millimeter
Help.	Helper Average	Lav.	Lavatory	Mill.	Millwright
HEPA	High Efficiency Particulate Air	lb.; #	Pound	Min., min.	Minimum, minute
	Filter	L.B.	Load Bearing; L Conduit Body	Misc.	Miscellaneous
Hg	Mercury	L. & E.	Labor & Equipment	ml	Milliliter, Mainline
HIC	High Interrupting Capacity	lb./hr.	Pounds per Hour	M.L.F.	Thousand Linear Feet
HM	Hollow Metal	lb./L.F.	Pounds per Linear Foot	Mo.	Month
HMWPE	high molecular weight	lbf/sq.in.	Pound-force per Square Inch	Mobil.	Mobilization
	polyethylene	L.C.L.	Less than Carload Lot	Mog.	Mogul Base
H.O.	High Output	L.C.Y.	Loose Cubic Yard	MPH	Miles per Hour
Horiz.	Horizontal	Ld.	Load	MPT	Male Pipe Thread

Abbreviations

MRT	Mile Round Trip	Pl.	Plate	S.F.C.A.	Square Foot Contact Area
ms	Millisecond	Plah.	Plasterer Helper	S.F. Flr.	Square Foot of Floor
M.S.F.	Thousand Square Feet	Plas.	Plasterer	S.F.G.	Square Foot of Ground
Mstz.	Mosaic & Terrazzo Worker	Pluh.	Plumbers Helper	S.F. Hor.	Square Foot Horizontal
M.S.Y.	Thousand Square Yards	Plum.	Plumber	S.F.R.	Square Feet of Radiation
Mtd., mtd.	Mounted	Ply.	Plywood	S.F. Shlf.	Square Foot of Shelf
Mthe.	Mosaic & Terrazzo Helper	p.m.	Post Meridiem	S4S	Surface 4 Sides
Mtng.	Mounting	Pntd.	Painted	Shee.	Sheet Metal Worker
Mult.	Multi; Multiply	Pord.	Painter, Ordinary	Sin.	Sine
M.V.A.	Million Volt Amperes	pp	Pages	Skwk.	Skilled Worker
M.V.A.R.	Million Volt Amperes Reactance	PP, PPL	Polypropylene	SL	Saran Lined
MV	Megavolt	P.P.M.	Parts per Million	S.L.	Slimline
MW	Megawatt	Pr.	Pair	Sldr.	Solder
MXM	Male by Male	P.E.S.B.	Pre-engineered Steel Building	SLH	Super Long Span Bar Joist
MYD	Thousand Yards	Prefab.	Prefabricated	S.N.	Solid Neutral
N	Natural; North	Prefin.	Prefinished	S-O-L	Socketolet
nA	Nanoampere	Prop.	Propelled	sp	Standpipe
NA	Not Available; Not Applicable	PSF, psf	Pounds per Square Foot	S.P.	Static Pressure; Single Pole; Self-Propelled
N.B.C.	National Building Code	PSI, psi	Pounds per Square Inch		
NC	Normally Closed	PSIG	Pounds per Square Inch Gauge	Spri.	Sprinkler Installer
N.E.M.A.	National Electrical Manufacturers Assoc.	PSP	Plastic Sewer Pipe	spwg	Static Pressure Water Gauge
		Pspr.	Painter, Spray	S.P.D.T.	Single Pole, Double Throw
NEHB	Bolted Circuit Breaker to 600V.	Psst.	Painter, Structural Steel	SPF	Spruce Pine Fir
N.L.B.	Non-Load-Bearing	P.T.	Potential Transformer	S.P.S.T.	Single Pole, Single Throw
NM	Non-Metallic Cable	P. & T.	Pressure & Temperature	SPT	Standard Pipe Thread
nm	Nanometer	Ptd.	Painted	Sq.	Square; 100 Square Feet
No.	Number	Ptns.	Partitions	Sq. Hd.	Square Head
NO	Normally Open	Pu	Ultimate Load	Sq. In.	Square Inch
N.O.C.	Not Otherwise Classified	PVC	Polyvinyl Chloride	S.S.	Single Strength; Stainless Steel
Nose.	Nosing	Pvmt.	Pavement	S.S.B.	Single Strength B Grade
N.P.T.	National Pipe Thread	Pwr.	Power	sst, ss	Stainless Steel
NQOD	Combination Plug-on/Bolt on Circuit Breaker to 240V.	Q	Quantity Heat Flow	Sswk.	Structural Steel Worker
		Qt.	Quart	Sswl.	Structural Steel Welder
N.R.C.	Noise Reduction Coefficient/ Nuclear Regulator Commission	Quan., Qty.	Quantity	St.;Stl.	Steel
		Q.C.	Quick Coupling	S.T.C.	Sound Transmission Coefficient
N.R.S.	Non Rising Stem	r	Radius of Gyration	Std.	Standard
ns	Nanosecond	R	Resistance	Stg.	Staging
nW	Nanowatt	R.C.P.	Reinforced Concrete Pipe	STK	Select Tight Knot
OB	Opposing Blade	Rect.	Rectangle	STP	Standard Temperature & Pressure
OC	On Center	Reg.	Regular	Stpi.	Steamfitter, Pipefitter
OD	Outside Diameter	Reinf.	Reinforced	Str.	Strength; Starter; Straight
O.D.	Outside Dimension	Req'd.	Required	Strd.	Stranded
ODS	Overhead Distribution System	Res.	Resistant	Struct.	Structural
O.G.	Ogee	Resi.	Residential	Sty.	Story
O.H.	Overhead	Rgh.	Rough	Subj.	Subject
O&P	Overhead and Profit	RGS	Rigid Galvanized Steel	Subs.	Subcontractors
Oper.	Operator	R.H.W.	Rubber, Heat & Water Resistant; Residential Hot Water	Surf.	Surface
Opng.	Opening			Sw.	Switch
Orna.	Ornamental	rms	Root Mean Square	Swbd.	Switchboard
OSB	Oriented Strand Board	Rnd.	Round	S.Y.	Square Yard
O.S.&Y.	Outside Screw and Yoke	Rodm.	Rodman	Syn.	Synthetic
Ovhd.	Overhead	Rofc.	Roofer, Composition	S.Y.P.	Southern Yellow Pine
OWG	Oil, Water or Gas	Rofp.	Roofer, Precast	Sys.	System
Oz.	Ounce	Rohe.	Roofer Helpers (Composition)	t.	Thickness
P.	Pole; Applied Load; Projection	Rots.	Roofer, Tile & Slate	T	Temperature; Ton
p.	Page	R.O.W.	Right of Way	Tan	Tangent
Pape.	Paperhanger	RPM	Revolutions per Minute	T.C.	Terra Cotta
P.A.P.R.	Powered Air Purifying Respirator	R.S.	Rapid Start	T & C	Threaded and Coupled
PAR	Parabolic Reflector	Rsr	Riser	T.D.	Temperature Difference
Pc., Pcs.	Piece, Pieces	RT	Round Trip	Tdd	Telecommunications Device for the Deaf
P.C.	Portland Cement; Power Connector	S.	Suction; Single Entrance; South		
P.C.F.	Pounds per Cubic Foot	SC	Screw Cover	T.E.M.	Transmission Electron Microscopy
P.C.M.	Phase Contrast Microscopy	SCFM	Standard Cubic Feet per Minute	TFE	Tetrafluoroethylene (Teflon)
P.E.	Professional Engineer; Porcelain Enamel; Polyethylene; Plain End	Scaf.	Scaffold	T. & G.	Tongue & Groove; Tar & Gravel
		Sch., Sched.	Schedule		
		S.C.R.	Modular Brick	Th., Thk.	Thick
Perf.	Perforated	S.D.	Sound Deadening	Thn.	Thin
PEX	Cross linked polyethylene	S.D.R.	Standard Dimension Ratio	Thrded	Threaded
Ph.	Phase	S.E.	Surfaced Edge	Tilf.	Tile Layer, Floor
P.I.	Pressure Injected	Sel.	Select	Tilh.	Tile Layer, Helper
Pile.	Pile Driver	S.E.R., S.E.U.	Service Entrance Cable	THHN	Nylon Jacketed Wire
Pkg.	Package	S.F.	Square Foot	THW.	Insulated Strand Wire

473

Abbreviations

THWN	Nylon Jacketed Wire	USP	United States Primed	Wrck.	Wrecker
T.L.	Truckload	UTP	Unshielded Twisted Pair	W.S.P.	Water, Steam, Petroleum
T.M.	Track Mounted	V	Volt	WT., Wt.	Weight
Tot.	Total	V.A.	Volt Amperes	WWF	Welded Wire Fabric
T-O-L	Threadolet	V.C.T.	Vinyl Composition Tile	XFER	Transfer
T.S.	Trigger Start	VAV	Variable Air Volume	XFMR	Transformer
Tr.	Trade	VC	Veneer Core	XHD	Extra Heavy Duty
Transf.	Transformer	Vent.	Ventilation	XHHW, XLPE	Cross-Linked Polyethylene Wire
Trhv.	Truck Driver, Heavy	Vert.	Vertical		Insulation
Trlr	Trailer	V.F.	Vinyl Faced	XLP	Cross-linked Polyethylene
Trlt.	Truck Driver, Light	V.G.	Vertical Grain	Y	Wye
TTY	Teletypewriter	V.H.F.	Very High Frequency	yd	Yard
TV	Television	VHO	Very High Output	yr	Year
T.W.	Thermoplastic Water Resistant	Vib.	Vibrating	Δ	Delta
	Wire	V.L.F.	Vertical Linear Foot	%	Percent
UCI	Uniform Construction Index	Vol.	Volume	~	Approximately
UF	Underground Feeder	VRP	Vinyl Reinforced Polyester	Ø	Phase; diameter
UGND	Underground Feeder	W	Wire; Watt; Wide; West	@	At
U.H.F.	Ultra High Frequency	w/	With	#	Pound; Number
U.I.	United Inch	W.C.	Water Column; Water Closet	<	Less Than
U.L.	Underwriters Laboratory	W.F.	Wide Flange	>	Greater Than
Uld.	unloading	W.G.	Water Gauge	Z	zone
Unfin.	Unfinished	Wldg.	Welding		
URD	Underground Residential	W. Mile	Wire Mile		
	Distribution	W-O-L	Weldolet		
US	United States	W.R.	Water Resistant		

FORMS

RESIDENTIAL COST ESTIMATE

OWNER'S NAME: _____ APPRAISER: _____

RESIDENCE ADDRESS: _____ PROJECT: _____

CITY, STATE, ZIP CODE: _____ DATE: _____

CLASS OF CONSTRUCTION
- ☐ ECONOMY
- ☐ AVERAGE
- ☐ CUSTOM
- ☐ LUXURY

RESIDENCE TYPE
- ☐ 1 STORY
- ☐ 1-1/2 STORY
- ☐ 2 STORY
- ☐ 2-1/2 STORY
- ☐ 3 STORY
- ☐ BI-LEVEL
- ☐ TRI-LEVEL

CONFIGURATION
- ☐ DETACHED
- ☐ TOWN/ROW HOUSE
- ☐ SEMI-DETACHED

OCCUPANCY
- ☐ ONE STORY
- ☐ TWO FAMILY
- ☐ THREE FAMILY
- ☐ OTHER _____

EXTERIOR WALL SYSTEM
- ☐ WOOD SIDING—WOOD FRAME
- ☐ BRICK VENEER—WOOD FRAME
- ☐ STUCCO ON WOOD FRAME
- ☐ PAINTED CONCRETE BLOCK
- ☐ SOLID MASONRY (AVERAGE & CUSTOM)
- ☐ STONE VENEER—WOOD FRAME
- ☐ SOLID BRICK (LUXURY)
- ☐ SOLID STONE (LUXURY)

*LIVING AREA (Main Building)

First Level	_____	S.F.
Second Level	_____	S.F.
Third Level	_____	S.F.
Total	_____	S.F.

*LIVING AREA (Wing or Ell) (   )

First Level	_____	S.F.
Second Level	_____	S.F.
Third Level	_____	S.F.
Total	_____	S.F.

*LIVING AREA (Wing or Ell) (   )

First Level	_____	S.F.
Second Level	_____	S.F.
Third Level	_____	S.F.
Total	_____	S.F.

*Basement Area is not part of living area.

MAIN BUILDING	COSTS PER S.F. LIVING AREA
Cost per Square Foot of Living Area, from Page _____	$
Basement Addition: _____ % Finished, _____ % Unfinished	+
Roof Cover Adjustment: _____ Type, Page _____ (Add or Deduct)	( )
Central Air Conditioning: ☐ Separate Ducts ☐ Heating Ducts, Page _____	+
Heating System Adjustment: _____ Type, Page _____ (Add or Deduct)	( )
Main Building: Adjusted Cost per S.F. of Living Area	$

MAIN BUILDING TOTAL COST

$ _____ /S.F.	x	_____ S.F.	x	_____	=	$ _____
Cost per S.F. Living Area		Living Area		Town/Row House Multiplier (Use 1 for Detached)		TOTAL COST

WING OR ELL ( ) _____ STORY	COSTS PER S.F. LIVING AREA
Cost per Square Foot of Living Area, from Page _____	$
Basement Addition: _____ % Finished, _____ % Unfinished	+
Roof Cover Adjustment: _____ Type, Page _____ (Add or Deduct)	( )
Central Air Conditioning: ☐ Separate Ducts ☐ Heating Ducts, Page _____	+
Heating System Adjustment: _____ Type, Page _____ (Add or Deduct)	( )
Wing or Ell ( ): Adjusted Cost per S.F. of Living Area	$

WING OR ELL (   ) TOTAL COST

$ _____ /S.F.	x	_____ S.F.	=	$ _____
Cost per S.F. Living Area		Living Area		TOTAL COST

WING OR ELL ( ) _____ STORY	COSTS PER S.F. LIVING AREA
Cost per Square Foot of Living Area, from Page _____	$
Basement Addition: _____ % Finished, _____ % Unfinished	+
Roof Cover Adjustment: _____ Type, Page _____ (Add or Deduct)	( )
Central Air Conditioning: ☐ Separate Ducts ☐ Heating Ducts, Page _____	+
Heating System Adjustment: _____ Type, Page _____ (Add or Deduct)	( )
Wing or Ell ( ): Adjusted Cost per S.F. of Living Area	$

WING OR ELL (   ) TOTAL COST

$ _____ /S.F.	x	_____ S.F.	=	$ _____
Cost per S.F. Living Area		Living Area		TOTAL COST

TOTAL THIS PAGE [_____]

Page 1 of 2

475

RESIDENTIAL
COST ESTIMATE

Total Page 1			$
	QUANTITY	UNIT COST	
Additional Bathrooms: _____ Full, _____ Half			
Finished Attic: _____ Ft. x _____ Ft.	S.F.		+
Breezeway: ☐ Open ☐ Enclosed _____ Ft. x _____ Ft.	S.F.		+
Covered Porch: ☐ Open ☐ Enclosed _____ Ft. x _____ Ft.	S.F.		+
Fireplace: ☐ Interior Chimney ☐ Exterior Chimney ☐ No. of Flues ☐ Additional Fireplaces			+
Appliances:			+
Kitchen Cabinets Adjustment: (+/−)			
☐ Garage ☐ Carport: _____ Car(s) Description _____ (+/−)			
Miscellaneous:			+

ADJUSTED TOTAL BUILDING COST $ [   ]

REPLACEMENT COST	
ADJUSTED TOTAL BUILDING COST	$ _____
Site Improvements	
(A) Paving & Sidewalks	$ _____
(B) Landscaping	$ _____
(C) Fences	$ _____
(D) Swimming Pool	$ _____
(E) Miscellaneous	$ _____
TOTAL	$ _____
Location Factor	x _____
Location Replacement Cost	$ _____
Depreciation	−$ _____
LOCAL DEPRECIATED COST	$ _____

INSURANCE COST	
ADJUSTED TOTAL BUILDING COST	$ _____
Insurance Exclusions	
(A) Footings, Site Work, Underground Piping	−$ _____
(B) Architects' Fees	−$ _____
Total Building Cost Less Exclusion	$ _____
Location Factor	x _____
LOCAL INSURABLE REPLACEMENT COST	$ _____

SKETCH AND ADDITIONAL CALCULATIONS

CII APPRAISAL

5. DATE: _____

1. SUBJECT PROPERTY: _____
6. APPRAISER: _____

2. BUILDING: _____

3. ADDRESS: _____

4. BUILDING USE: _____
7. YEAR BUILT: _____

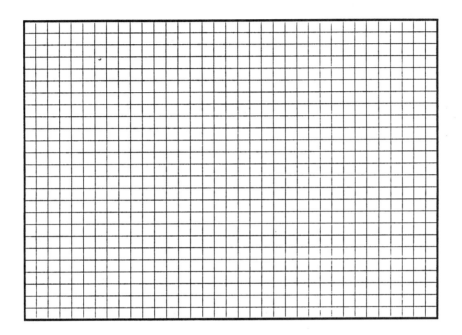

8. EXTERIOR WALL CONSTRUCTION: _____

9. FRAME: _____

10. GROUND FLOOR AREA: _____ S.F. 11. GROSS FLOOR AREA (EXCL. BASEMENT): _____ S.F.

12. NUMBER OF STORIES: _____ 13. STORY HEIGHT: _____

14. PERIMETER: _____ L.F. 15. BASEMENT AREA: _____ S.F.

16. GENERAL COMMENTS: _____

_____
_____
_____
_____
_____
_____

	NO.	DESCRIPTION	UNIT	UNIT COST	NEW SF COST	MODEL SF COST	+/- CHANGE
A SUBSTRUCTURE							
A	1010	Standard Foundations Bay size:	S.F. Gnd.				
A	1030	Slab on Grade Material: Thickness:	S.F. Slab				
A	2010	Basement Excavation Depth: Area:	S.F. Gnd.				
A	2020	Basement Walls	L.F. Walls				
B SHELL							
	B10 Superstructure						
B	1010	Floor Construction Elevated floors:	S.F. Floor				
			S.F. Floor				
B	1020	Roof Construction	S.F. Roof				
	B20 Exterior Enclosure						
B	2010	Exterior walls Material: Thickness: *% of wall*	S.F. Walls				
		Material: Thickness: *% of wall*	S.F. Walls				
B	2020	Exterior Windows Type: *% of wall*	S.F. Wind.				
		Type: *% of wall* Each	S.F. Wind.				
B	2030	Exterior Doors Type: Number:	Each				
		Type: Number:	Each				
	B30 Roofing						
B	3010	Roof Coverings Material:	S.F. Roof				
		Material:	S.F. Roof				
B	3020	Roof Openings	S.F. Opng.				
C INTERIORS							
C	1010	Partitions: Material: Density:	S.F. Part.				
		Material: Density:					
C	1020	Interior Doors Type: Number:	Each				
C	1030	Fittings	Each				
C	2010	Stair Construction	Flight				
C	3010	Wall Finishes Material: % of Wall	S.F. Walls				
		Material:	S.F. Walls				
C	3020	Floor Finishes Material:	S.F. Floor				
		Material:	S.F. Floor				
		Material:	S.F. Floor				
C	3030	Ceiling Finishes Material:	S.F. Ceil.				

NO.	SYSTEM/COMPONENT	DESCRIPTION	UNIT	UNIT COST	NEW S.F. COST	MODEL S.F. COST	+/- CHANGE
D SERVICES							
D10 Conveying							
D 1010	Elevators & Lifts	Type: Capacity: Stops:	Each				
D 1020	Escalators & Moving Walks	Type:	Each				
D20 Plumbing							
D 2010	Plumbing		Each				
D 2020	Domestic Water Distribution		S.F. Floor				
D 2040	Rain Water Drainage		S.F. Roof				
D30 HVAC							
D 3010	Energy Supply		Each				
D 3020	Heat Generating Systems	Type:	S.F. Floor				
D 3030	Cooling Generating Systems	Type:	S.F. Floor				
D 3090	Other HVAC Sys. & Equipment		Each				
D40 Fire Protection							
D 4010	Sprinklers		S.F. Floor				
D 4020	Standpipes		S.F. Floor				
D50 Electrical							
D 5010	Electrical Service/Distribution		S.F. Floor				
D 5020	Lighting & Branch Wiring		S.F. Floor				
D 5030	Communications & Security		S.F. Floor				
D 5090	Other Electrical Systems		S.F. Floor				
E EQUIPMENT & FURNISHINGS							
E 1010	Commercial Equipment		Each				
E 1020	Institutional Equipment		Each				
E 1030	Vehicular Equipment		Each				
E 1090	Other Equipment		Each				
F SPECIAL CONSTRUCTION							
F 1020	Integrated Construction		S.F.				
F 1040	Special Facilities		S.F.				
G BUILDING SITEWORK							

ITEM		Total Change
17	Model 020 total _____ $ _____	
18	Adjusted S.F. cost $ _____ item 17 +/- changes	
19	Building area - from item 11 _____ S.F. x adjusted S.F. cost..................................	$..
20	Basement area - from item 15 _____ S.F. x S.F. cost $	$..
21	Base building sub-total - item 20 + item 19	$.....
22	Miscellaneous addition (quality, etc.)	$......
23	Sub-total - item 22 + 21	$......
24	General conditions -25 % of item 23	$.......
25	Sub-total - item 24 + item 23..	$......
26	Architects fees _____ % of item 25	$
27	Sub-total - item 26 + item 27	$......
28	Location modifier	...x....
29	Local replacement cost - item 28 x item 27	$.....
30	Depreciation _____ % of item 29	$
31	Depreciated local replacement cost - item 29 less item 30	$.....
32	Exclusions	$.......
33	Net depreciated replacement cost - item 31 less item 32	$....

Base Cost per square foot floor area: *(from square foot table)*

Specify source:

Page: _____ , Model # _____ , Area _____ S.F.,
Exterior wall _____ , Frame _____

Adjustments for exterior wall variation:

Size Adjustment:
   (Interpolate)

Height Adjustment: _____ + _____ = _____

Adjusted Base Cost per square foot:

Building Cost $ _____ x _____ = _____
                        Adjusted Base Cost   *Floor Area*
                        per square foot

Basement Cost $ _____ x _____ = _____
                        Basement Cost   *Basement Area*

Lump Sum Additions

TOTAL BUILDING COST *(Sum of above costs)*   _____

Modifications: *(complexity, workmanship, size)*   +/- _____ %   _____

Location Modifier: City _____ Date _____   x _____

        Local cost of replacement   _____

Less depreciation: _____   - _____

        Local cost of replacement less depreciation $ _____

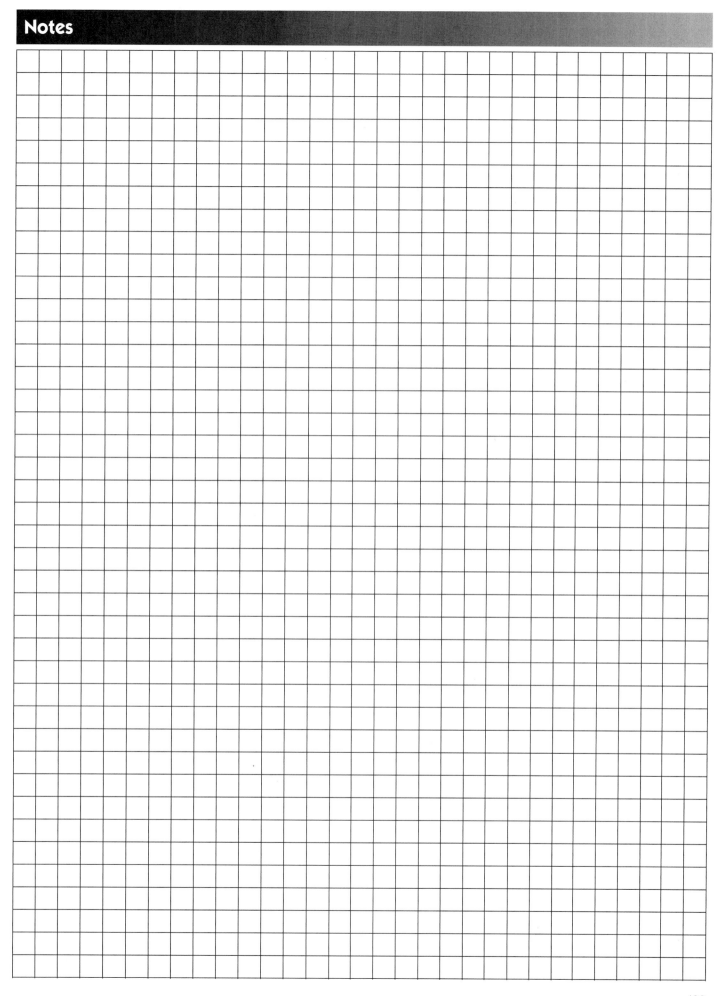

Reed Construction Data/RSMeans—
a tradition of excellence in construction cost
information and services since 1942

**For more information
visit the RSMeans website
at www.rsmeans.com**

Book Selection Guide

The following table provides definitive information on the content of each cost data publication. The number of lines of data provided in each unit price or assemblies division, as well as the number of reference tables and crews, is listed for each book. The presence of other elements such as equipment rental costs, historical cost indexes, city cost indexes, square foot models, or cross-referenced indexes is also indicated. You can use the table to help select the RSMeans book that has the quantity and type of information you most need in your work.

Unit Cost Divisions	Building Construction Costs	Mechanical	Electrical	Repair & Remodel	Square Foot	Site Work Landsc.	Assemblies	Interior	Concrete Masonry	Open Shop	Heavy Construction	Light Commercial	Facilities Construction	Plumbing	Residential
1	553	372	385	474		489		301	437	552	483	220	967	386	157
2	555	239	60	544		782		333	189	553	680	409	1022	262	217
3	1471	193	179	830		1321		266	1803	1467	1459	304	1386	162	309
4	878	24	0	677		683		557	1082	854	576	469	1125	0	385
5	1847	197	152	951		796		1023	751	1816	1045	809	1837	275	724
6	1870	78	80	1458		461		1191	310	1849	395	1562	1484	22	2076
7	1395	164	71	1418		506		513	439	1393	0	1091	1451	175	852
8	1941	61	10	1969		298		1677	673	1906	0	1389	2134	0	1319
9	1800	72	26	1606		292		1899	392	1761	0	1513	2035	54	1357
10	978	17	10	582		226		800	156	978	29	465	1072	233	212
11	998	209	160	476		126		848	28	983	0	223	1012	170	110
12	547	0	2	303		248		1691	12	536	19	349	1736	23	317
13	730	119	111	247		341		255	69	702	253	80	740	74	79
14	281	36	0	230		28		260	0	280	0	12	300	21	6
21	78	0	16	31		0		252	0	78	0	60	370	375	0
22	1090	6955	154	1049		1409		596	20	1069	1634	751	6824	8656	572
23	1189	7103	609	776		159		682	38	1103	110	649	5143	1860	425
26	1246	493	9825	991		743		1053	55	1234	527	1060	9761	438	557
27	75	0	200	34		9		67	0	75	35	48	198	0	4
28	87	59	94	64		0		69	0	72	0	42	109	45	27
31	1904	1162	1031	1225		2731		7	1635	1848	2804	1027	1964	1087	1033
32	737	46	0	603		3849		360	271	710	1366	353	1370	158	412
33	514	1020	429	199		1957		0	225	512	1883	114	1547	1189	112
34	99	0	20	4		134		0	31	55	113	0	119	0	0
35	18	0	0	0		166		0	0	18	281	0	83	0	0
41	52	0	0	24		7		22	0	52	30	0	59	14	0
44	98	90	0	0		32		0	0	23	32	0	98	105	0
Totals	23031	18709	13264	16765		17793		14722	8616	22479	13754	12999	45946	15784	11262

Assembly Divisions	Building Construction Costs	Mechanical	Electrical	Repair & Remodel	Square Foot	Site Work Landscape	Assemblies	Interior	Concrete Masonry	Open Shop	Heavy Construction	Light Commercial	Facilities Construction	Plumbing	Asm Div	Residential
A		15	0	182	150	530	598	0	536		570	154	24	0	1	374
B		0	0	808	2471	0	5577	329	1914		368	2014	143	0	2	217
C		0	0	630	864	0	1229	1567	146		0	765	242	0	3	588
D		1066	810	694	1779	0	2408	765	0		0	1270	1011	1068	4	861
E		0	0	85	260	0	297	5	0		0	257	5	0	5	392
F		0	0	0	123	0	124	0	0		0	123	3	0	6	358
G		528	294	311	202	2206	668	0	535		1194	200	293	685	7	301
															8	760
															9	80
															10	0
															11	0
															12	0
Totals		1609	1104	2710	5849	2736	10901	2666	3131		2132	4783	1721	1753		3931

Reference Section	Building Construction Costs	Mechanical	Electrical	Repair & Remodel	Square Foot	Site Work Landscape	Assemblies	Interior	Concrete Masonry	Open Shop	Heavy Construction	Light Commercial	Facilities Construction	Plumbing	Residential
Reference Tables	yes	yes	yes	yes	no	yes	yes	yes	yes	yes	yes	yes	yes	yes	yes
Models					105							46			32
Crews	482	482	482	462		482		482	482	460	482	460	462	482	460
Equipment Rental Costs	yes	yes	yes	yes		yes		yes	yes	yes	yes	yes	yes	yes	yes
Historical Cost Indexes	yes	yes	yes	yes	yes	yes	yes	yes	yes	yes	yes	yes	yes	yes	no
City Cost Indexes	yes	yes	yes	yes	yes	yes	yes	yes	yes	yes	yes	yes	yes	yes	yes

491

For more information
visit the RSMeans website
at www.rsmeans.com

Annual Cost Guides

RSMeans Square Foot Costs 2009

It's Accurate and Easy To Use!

- **Updated price information** based on nationwide figures from suppliers, estimators, labor experts, and contractors

- "How-to-Use" sections, with **clear examples** of commercial, residential, industrial, and institutional structures

- Realistic graphics, offering true-to-life illustrations of building projects

- Extensive information on using square foot cost data, including sample estimates and alternate pricing methods

$168.95 per copy
Available Oct. 2008
Catalog no. 60059

RSMeans Repair & Remodeling Cost Data 2009

Commercial/Residential

Use this valuable tool to estimate commercial and residential renovation and remodeling.

Includes: New costs for hundreds of unique methods, materials, and conditions that only come up in repair and remodeling, PLUS:

- Unit costs for over 15,000 construction components
- Installed costs for over 90 assemblies
- Over 930 "on-site" localization factors for the U.S. and Canada.

Unit prices organized according to MasterFormat 2004!

$132.95 per copy
Available Nov. 2008
Catalog no. 60049

RSMeans Electrical Cost Data 2009

Pricing information for every part of electrical cost planning. More than 13,000 unit and systems costs with design tables; clear specifications and drawings; engineering guides; illustrated estimating procedures; complete labor-hour and materials costs for better scheduling and procurement; and the latest electrical products and construction methods.

- A variety of special electrical systems including cathodic protection

- Costs for maintenance, demolition, HVAC/mechanical, specialties, equipment, and more

Unit prices organized according to MasterFormat 2004!

$154.95 per copy
Available Oct. 2008
Catalog no. 60039

RSMeans Electrical Change Order Cost Data 2009

RSMeans Electrical Change Order Cost Data provides you with electrical unit prices exclusively for pricing change orders—based on the recent, direct experience of contractors and suppliers. Analyze and check your own change order estimates against the experience others have had doing the same work. It also covers productivity analysis and change order cost justifications. With useful information for calculating the effects of change orders and dealing with their administration.

$154.95 per copy
Available Dec. 2008
Catalog no. 60239

Annual Cost Guides

For more information
visit the RSMeans website
at www.rsmeans.com

RSMeans Assemblies Cost Data 2009

RSMeans Assemblies Cost Data takes the guesswork out of preliminary or conceptual estimates. Now you don't have to try to calculate the assembled cost by working up individual component costs. We've done all the work for you.

Presents detailed illustrations, descriptions, specifications, and costs for every conceivable building assembly—240 types in all—arranged in the easy-to-use UNIFORMAT II system. Each illustrated "assembled" cost includes a complete grouping of materials and associated installation costs, including the installing contractor's overhead and profit.

$245.95 per copy
Available Oct. 2008
Catalog no. 60069

RSMeans Open Shop Building Construction Cost Data 2009

The latest costs for accurate budgeting and estimating of new commercial and residential construction. . . renovation work. . . change orders. . . cost engineering.

RSMeans Open Shop "BCCD" will assist you to:
• Develop benchmark prices for change orders
• Plug gaps in preliminary estimates and budgets
• Estimate complex projects
• Substantiate invoices on contracts
• Price ADA-related renovations

Unit prices organized according to MasterFormat 2004!

$154.95 per copy
Available Dec. 2008
Catalog no. 60159

RSMeans Residential Cost Data 2009

Contains square foot costs for 30 basic home models with the look of today, plus hundreds of custom additions and modifications you can quote right off the page. With costs for the 100 residential systems you're most likely to use in the year ahead. Complete with blank estimating forms, sample estimates, and step-by-step instructions.

Now contains line items for cultured stone and brick, PVC trim lumber, and TPO roofing.

$132.95 per copy
Available Oct. 2008
Catalog no. 60179

Unit prices organized according to MasterFormat 2004!

RSMeans Site Work & Landscape Cost Data 2009

Includes unit and assemblies costs for earthwork, sewerage, piped utilities, site improvements, drainage, paving, trees & shrubs, street openings/repairs, underground tanks, and more. Contains 57 tables of assemblies costs for accurate conceptual estimates.

Includes:
• Estimating for infrastructure improvements
• Environmentally-oriented construction
• ADA-mandated handicapped access
• Hazardous waste line items

$154.95 per copy
Available Nov. 2008
Catalog no. 60289

RSMeans Facilities Maintenance & Repair Cost Data 2009

RSMeans Facilities Maintenance & Repair Cost Data gives you a complete system to manage and plan your facility repair and maintenance costs and budget efficiently. Guidelines for auditing a facility and developing an annual maintenance plan. Budgeting is included, along with reference tables on cost and management, and information on frequency and productivity of maintenance operations.

The only nationally recognized source of maintenance and repair costs. Developed in cooperation with the Civil Engineering Research Laboratory (CERL) of the Army Corps of Engineers.

$336.95 per copy
Available Dec. 2008
Catalog no. 60309

RSMeans Light Commercial Cost Data 2009

Specifically addresses the light commercial market, which is a specialized niche in the construction industry. Aids you, the owner/designer/contractor, in preparing all types of estimates—from budgets to detailed bids. Includes new advances in methods and materials.

Assemblies Section allows you to evaluate alternatives in the early stages of design/planning.

Over 11,000 unit costs ensure that you have the prices you need. . . when you need them.

Unit prices organized according to MasterFormat 2004!

$132.95 per copy
Available Nov. 2008
Catalog no. 60189

For more information
visit the RSMeans website
at www.rsmeans.com

Annual Cost Guides

RSMeans Concrete & Masonry Cost Data 2009

Provides you with cost facts for virtually all concrete/masonry estimating needs, from complicated formwork to various sizes and face finishes of brick and block—all in great detail. The comprehensive Unit Price Section contains more than 7,500 selected entries. Also contains an Assemblies [Cost] Section, and a detailed Reference Section that supplements the cost data.

Unit prices organized according to MasterFormat 2004!

$139.95 per copy
Available Dec. 2008
Catalog no. 60119

RSMeans Labor Rates for the Construction Industry 2009

Complete information for estimating labor costs, making comparisons, and negotiating wage rates by trade for over 300 U.S. and Canadian cities. With 46 construction trades listed by local union number in each city, and historical wage rates included for comparison. Each city chart lists the county and is alphabetically arranged with handy visual flip tabs for quick reference.

$336.95 per copy
Available Dec. 2008
Catalog no. 60129

RSMeans Construction Cost Indexes 2009

What materials and labor costs will change unexpectedly this year? By how much?

• Breakdowns for 316 major cities
• National averages for 30 key cities
• Expanded five major city indexes
• Historical construction cost indexes

$335.00 per year (subscription)
Catalog no. 50149

$83.25 individual quarters
Catalog no. 60149 A,B,C,D

RSMeans Interior Cost Data 2009

Provides you with prices and guidance needed to make accurate interior work estimates. Contains costs on materials, equipment, hardware, custom installations, furnishings, and labor costs. . . for new and remodel commercial and industrial interior construction, including updated information on office furnishings, and reference information.

Unit prices organized according to MasterFormat 2004!

$154.95 per copy
Available Nov. 2008
Catalog no. 60099

RSMeans Heavy Construction Cost Data 2009

A comprehensive guide to heavy construction costs. Includes costs for highly specialized projects such as tunnels, dams, highways, airports, and waterways. Information on labor rates, equipment, and materials costs is included. Features unit price costs, systems costs, and numerous reference tables for costs and design.

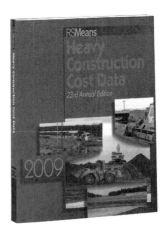

$154.95 per copy
Available Dec. 2008
Catalog no. 60169

Unit prices organized according to MasterFormat 2004!

Reference Books

For more information
visit the RSMeans website
at www.rsmeans.com

Value Engineering: Practical Applications

for Design, Construction, Maintenance & Operations

by Alphonse Dell'Isola, PE

A tool for immediate application—for engineers, architects, facility managers, owners, and contractors. Includes making the case for VE—the management briefing; integrating VE into planning, budgeting, and design; conducting life cycle costing; using VE methodology in design review and consultant selection; case studies; VE workbook; and a life cycle costing program on disk.

$79.95 per copy
Over 450 pages, illustrated, softcover
Catalog no. 67319A

The Building Professional's Guide to Contract Documents

3rd Edition

by Waller S. Poage, AIA, CSI, CVS

A comprehensive reference for owners, design professionals, contractors, and students

Structure your documents for maximum efficiency.
Effectively communicate construction requirements.
Understand the roles and responsibilities of construction professionals.
Improve methods of project delivery.

$64.95 per copy, 400 pages
Diagrams and construction forms, hardcover
Catalog no. 67261A

Cost Planning & Estimating for Facilities Maintenance

In this unique book, a team of facilities management authorities shares their expertise on:

Evaluating and budgeting maintenance operations
Maintaining and repairing key building components
Applying *RSMeans Facilities Maintenance & Repair Cost Data* to your estimating

Covers special maintenance requirements of the ten major building types.

$89.95 per copy
Over 475 pages, hardcover
Catalog no. 67314

Planning & Managing Interior Projects 2nd Edition

by Carol E. Farren, CFM

Expert guidance on managing renovation & relocation projects

This book guides you through every step in relocating to a new space or renovating an old one. From initial meeting through design and construction, to post-project administration, it helps you get the most for your company or client. Includes sample forms, spec lists, agreements, drawings, and much more!

$34.98 per copy
200 pages, softcover
Catalog no. 67245A

Facilities Operations & Engineering Reference

by the Association for Facilities Engineering and RSMeans

An all-in-one technical reference for planning and managing facility projects and solving day-to-day operations problems. Selected as the official certified plant engineer reference, this handbook covers financial analysis, maintenance, HVAC and energy efficiency, and more.

$54.98 per copy
Over 700 pages, illustrated, hardcover
Catalog no. 67318

Complete Book of Framing

by Scot Simpson

This straightforward, easy-to-learn method will help framers, carpenters, and handy homeowners build their skills in rough carpentry and framing. Shows how to frame all the parts of a house: floors, walls, roofs, door & window openings, and stairs—with hundreds of color photographs and drawings that show every detail.

The book gives beginners all the basics they need to go from zero framing knowledge to a journeyman level.

$29.95 per copy
352 pages, softcover
Catalog no. 67353

Life Cycle Costing for Facilities

by Alphonse Dell'Isola and Dr. Steven Kirk

Guidance for achieving higher quality design and construction projects at lower costs! Cost-cutting efforts often sacrifice quality to yield the cheapest product. Life cycle costing enables building designers and owners to achieve both. The authors of this book show how LCC can work for a variety of projects — from roads to HVAC upgrades to different types of buildings.

$99.95 per copy
450 pages, hardcover
Catalog no. 67341

Construction Business Management

by Nick Ganaway

Only 43% of construction firms stay in business after four years. Make sure your company thrives with valuable guidance from a pro with 25 years of success as a commercial contractor. Find out what it takes to build all aspects of a business that is profitable, enjoyable, and enduring. With a bonus chapter on retail construction.

$49.95 per copy
200 pages, softcover
Catalog no. 67352

For more information
visit the RSMeans website
at www.rsmeans.com

Reference Books

Interior Home Improvement Costs 9th Edition

Updated estimates for the most popular remodeling and repair projects—from small, do-it-yourself jobs to major renovations and new construction. Includes: kitchens & baths; new living space from your attic, basement, or garage; new floors, paint, and wallpaper; tearing out or building new walls; closets, stairs, and fireplaces; new energy-saving improvements, home theaters, and more!

$24.95 per copy
250 pages, illustrated, softcover
Catalog no. 67308E

Exterior Home Improvement Costs 9th Edition

Updated estimates for the most popular remodeling and repair projects—from small, do-it-yourself jobs, to major renovations and new construction. Includes: curb appeal projects—landscaping, patios, porches, driveways, and walkways; new windows and doors; decks, greenhouses, and sunrooms; room additions and garages; roofing, siding, and painting; "green" improvements to save energy & water.

$24.95 per copy
Over 275 pages, illustrated, softcover
Catalog no. 67309E

Builder's Essentials: Plan Reading & Material Takeoff

For Residential and Light Commercial Construction

by Wayne J. DelPico

A valuable tool for understanding plans and specs, and accurately calculating material quantities. Step-by-step instructions and takeoff procedures based on a full set of working drawings.

$35.95 per copy
Over 420 pages, softcover
Catalog no. 67307

Means Unit Price Estimating Methods

New 4th Edition

This new edition includes up-to-date cost data and estimating examples, updated to reflect changes to the CSI numbering system and new features of RSMeans cost data. It describes the most productive, universally accepted ways to estimate, and uses checklists and forms to illustrate shortcuts and timesavers. A model estimate demonstrates procedures. A new chapter explores computer estimating alternatives.

$59.95 per copy
Over 350 pages, illustrated, softcover
Catalog no. 67303B

Total Productive Facilities Management

by Richard W. Sievert, Jr.

Today, facilities are viewed as strategic resources. . . elevating the facility manager to the role of asset manager supporting the organization's overall business goals. Now, Richard Sievert Jr., in this well-articulated guidebook, sets forth a new operational standard for the facility manager's emerging role. . . a comprehensive program for managing facilities as a true profit center.

$29.98 per copy
275 pages, softcover
Catalog no. 67321

Green Building: Project Planning & Cost Estimating 2nd Edition

This new edition has been completely updated with the latest in green building technologies, design concepts, standards, and costs. Now includes a 2009 Green Building *CostWorks* CD with more than 300 green building assemblies and over 5,000 unit price line items for sustainable building. The new edition is also full-color with all new case studies—plus a new chapter on deconstruction, a key aspect of green building.

$129.95 per copy
350 pages, softcover
Catalog no. 67338A

Concrete Repair and Maintenance Illustrated

by Peter Emmons

Hundreds of illustrations show users how to analyze, repair, clean, and maintain concrete structures for optimal performance and cost effectiveness. From parking garages to roads and bridges to structural concrete, this comprehensive book describes the causes, effects, and remedies for concrete wear and failure. Invaluable for planning jobs, selecting materials, and training employees, this book is a must-have for concrete specialists, general contractors, facility managers, civil and structural engineers, and architects.

$69.95 per copy
300 pages, illustrated, softcover
Catalog no. 67146

Reference Books

**For more information
visit the RSMeans website
at www.rsmeans.com**

Means Illustrated Construction Dictionary Condensed, 2nd Edition

The best portable dictionary for office or field use—an essential tool for contractors, architects, insurance and real estate personnel, facility managers, homeowners, and anyone who needs quick, clear definitions for construction terms. The second edition has been further enhanced with updates and hundreds of new terms and illustrations . . . in keeping with the most recent developments in the construction industry.

Now with a quick-reference Spanish section. Includes tools and equipment, materials, tasks, and more!

$59.95 per copy
Over 500 pages, softcover
Catalog no. 67282A

Means Repair & Remodeling Estimating 4th Edition

By Edward B. Wetherill and RSMeans

This important reference focuses on the unique problems of estimating renovations of existing structures, and helps you determine the true costs of remodeling through careful evaluation of architectural details and a site visit.

New section on disaster restoration costs.

$69.95 per copy
Over 450 pages, illustrated, hardcover
Catalog no. 67265B

Facilities Planning & Relocation

by David D. Owen

A complete system for planning space needs and managing relocations. Includes step-by-step manual, over 50 forms, and extensive reference section on materials and furnishings.

New lower price and user-friendly format.

$49.98 per copy
Over 450 pages, softcover
Catalog no. 67301

Means Square Foot & UNIFORMAT Assemblies Estimating Methods 3rd Edition

Develop realistic square foot and assemblies costs for budgeting and construction funding. The new edition features updated guidance on square foot and assemblies estimating using UNIFORMAT II. An essential reference for anyone who performs conceptual estimates.

$69.95 per copy
Over 300 pages, illustrated, softcover
Catalog no. 67145B

Means Electrical Estimating Methods 3rd Edition

Expanded edition includes sample estimates and cost information in keeping with the latest version of the CSI MasterFormat and UNIFORMAT II. Complete coverage of fiber optic and uninterruptible power supply electrical systems, broken down by components, and explained in detail. Includes a new chapter on computerized estimating methods. A practical companion to *RSMeans Electrical Cost Data.*

$64.95 per copy
Over 325 pages, hardcover
Catalog no. 67230B

Means Mechanical Estimating Methods 4th Edition

Completely updated, this guide assists you in making a review of plans, specs, and bid packages, with suggestions for takeoff procedures, listings, substitutions, and pre-bid scheduling for all components of HVAC. Includes suggestions for budgeting labor and equipment usage. Compares materials and construction methods to allow you to select the best option.

$64.95 per copy
Over 350 pages, illustrated, softcover
Catalog no. 67294B

Means ADA Compliance Pricing Guide 2nd Edition

by Adaptive Environments and RSMeans

Completely updated and revised to the new 2004 *Americans with Disabilities Act Accessibility Guidelines,* this book features more than 70 of the most commonly needed modifications for ADA compliance. Projects range from installing ramps and walkways, widening doorways and entryways, and installing and refitting elevators, to relocating light switches and signage.

$79.95 per copy
Over 350 pages, illustrated, softcover
Catalog no. 67310A

Project Scheduling & Management for Construction 3rd Edition

by David R. Pierce, Jr.

A comprehensive yet easy-to-follow guide to construction project scheduling and control—from vital project management principles through the latest scheduling, tracking, and controlling techniques. The author is a leading authority on scheduling, with years of field and teaching experience at leading academic institutions. Spend a few hours with this book and come away with a solid understanding of this essential management topic.

$64.95 per copy
Over 300 pages, illustrated, hardcover
Catalog no. 67247B

For more information
visit the RSMeans website
at www.rsmeans.com

Reference Books

The Practice of Cost Segregation Analysis

by Bruce A. Desrosiers and Wayne J. DelPico

This expert guide walks you through the practice of cost segregation analysis, which enables property owners to defer taxes and benefit from "accelerated cost recovery" through depreciation deductions on assets that are properly identified and classified.

With a glossary of terms, sample cost segregation estimates for various building types, key information resources, and updates via a dedicated website, this book is a critical resource for anyone involved in cost segregation analysis.

$99.95 per copy
Over 225 pages
Catalog no. 67345

Preventive Maintenance for Multi-Family Housing

by John C. Maciha

Prepared by one of the nation's leading experts on multi-family housing.

This complete PM system for apartment and condominium communities features expert guidance, checklists for buildings and grounds maintenance tasks and their frequencies, a reusable wall chart to track maintenance, and a dedicated website featuring customizable electronic forms. A must-have for anyone involved with multi-family housing maintenance and upkeep.

$89.95 per copy
225 pages
Catalog no. 67346

How to Estimate with Means Data & CostWorks

New 3rd Edition

by RSMeans and Saleh A. Mubarak, Ph.D.

New 3rd Edition—fully updated with new chapters, plus new CD with updated *CostWorks* cost data and MasterFormat organization. Includes all major construction items—with more than 300 exercises and two sets of plans that show how to estimate for a broad range of construction items and systems—including general conditions and equipment costs.

$59.95 per copy
272 pages, softcover
Includes CostWorks CD
Catalog no. 67324B

Job Order Contracting

Expediting Construction Project Delivery

by Allen Henderson

Expert guidance to help you implement JOC—fast becoming the preferred project delivery method for repair and renovation, minor new construction, and maintenance projects in the public sector and in many states and municipalities. The author, a leading JOC expert and practitioner, shows how to:

- Establish a JOC program
- Evaluate proposals and award contracts
- Handle general requirements and estimating
- Partner for maximum benefits

$89.95 per copy
192 pages, illustrated, hardcover
Catalog no. 67348

Builder's Essentials: Estimating Building Costs

For the Residential & Light Commercial Contractor

by Wayne J. DelPico

Step-by-step estimating methods for residential and light commercial contractors. Includes a detailed look at every construction specialty—explaining all the components, takeoff units, and labor needed for well-organized, complete estimates. Covers correctly interpreting plans and specifications, and developing accurate and complete labor and material costs.

$29.95 per copy
Over 400 pages, illustrated, softcover
Catalog no. 67343

Building & Renovating Schools

This all-inclusive guide covers every step of the school construction process—from initial planning, needs assessment, and design, right through moving into the new facility. A must-have resource for anyone concerned with new school construction or renovation. With square foot cost models for elementary, middle, and high school facilities, and real-life case studies of recently completed school projects.

The contributors to this book—architects, construction project managers, contractors, and estimators who specialize in school construction—provide start-to-finish, expert guidance on the process.

$69.98 per copy
Over 425 pages, hardcover
Catalog no. 67342

The Homeowner's Guide to Mold By Michael Pugliese

Expert guidance to protect your health and your home.

Mold, whether caused by leaks, humidity or flooding, is a real health and financial issue—for homeowners and contractors. This full-color book explains:

- Construction and maintenance practices to prevent mold
- How to inspect for and remove mold
- Mold remediation procedures and costs
- What to do after a flood
- How to deal with insurance companies

$21.95 per copy
144 pages, softcover
Catalog no. 67344

Means Illustrated Construction Dictionary
Unabridged 3rd Edition, with CD-ROM

Long regarded as the industry's finest, *Means Illustrated Construction Dictionary* is now even better. With the addition of over 1,000 new terms and hundreds of new illustrations, it is the clear choice for the most comprehensive and current information. The companion CD-ROM that comes with this new edition adds many extra features: larger graphics, expanded definitions, and links to both CSI MasterFormat numbers and product information.

$99.95 per copy
Over 790 pages, illustrated, hardcover
Catalog no. 67292A

Means Landscape Estimating Methods

New 5th Edition

Answers questions about preparing competitive landscape construction estimates, with up-to-date cost estimates and the new MasterFormat classification system. Expanded and revised to address the latest materials and methods, including new coverage on approaches to green building. Includes:

- Step-by-step explanation of the estimating process
- Sample forms and worksheets that save time and prevent errors

$64.95 per copy
Over 350 pages, softcover
Catalog no. 67295C

Means Plumbing Estimating Methods 3rd Edition
by Joseph Galeno and Sheldon Greene

Updated and revised! This practical guide walks you through a plumbing estimate, from basic materials and installation methods through change order analysis. *Plumbing Estimating Methods* covers residential, commercial, industrial, and medical systems, and features sample takeoff and estimate forms and detailed illustrations of systems and components.

$29.98 per copy
330+ pages, softcover
Catalog no. 67283B

Understanding & Negotiating Construction Contracts
by Kit Werremeyer

Take advantage of the author's 30 years' experience in small-to-large (including international) construction projects. Learn how to identify, understand, and evaluate high risk terms and conditions typically found in all construction contracts—then negotiate to lower or eliminate the risk, improve terms of payment, and reduce exposure to claims and disputes. The author avoids "legalese" and gives real-life examples from actual projects.

$69.95 per copy
300 pages, softcover
Catalog no. 67350

Means Estimating Handbook
2nd Edition

Updated Second Edition answers virtually any estimating technical question—all organized by CSI MasterFormat. This comprehensive reference covers the full spectrum of technical data required to estimate construction costs. The book includes information on sizing, productivity, equipment requirements, code-mandated specifications, design standards, and engineering factors.

$99.95 per copy
Over 900 pages, hardcover
Catalog no. 67276A

Means Spanish/English Construction Dictionary 2nd Edition
by RSMeans and the International Code Council

This expanded edition features thousands of the most common words and useful phrases in the construction industry with easy-to-follow pronunciations in both Spanish and English. Over 800 new terms, phrases, and illustrations have been added. It also features a new stand-alone "Safety & Emergencies" section, with colored pages for quick access.

Unique to this dictionary are the systems illustrations showing the relationship of components in the most common building systems for all major trades.

$23.95 per copy
Over 400 pages
Catalog no. 67327A

How Your House Works
by Charlie Wing

A must-have reference for every homeowner, handyman, and contractor—for repair, remodeling, and new construction. This book uncovers the mysteries behind just about every major appliance and building element in your house—from electrical, heating and AC, to plumbing, framing, foundations and appliances. Clear, "exploded" drawings show exactly how things should be put together and how they function—what to check if they don't work, and what you can do that might save you having to call in a professional.

$21.95 per copy
160 pages, softcover
Catalog no. 67351

Means CostWorks® Training

This one-day course helps users become more familiar with the functionality of *Means CostWorks* program. Each menu, icon, screen, and function found in the program is explained in depth. Time is devoted to hands-on estimating exercises.

Some of what you'll learn:
- Search the database utilizing all navigation methods
- Export RSMeans Data to your preferred spreadsheet format
- View crews, assembly components, and much more!
- Automatically regionalize the database

You are required to bring your own laptop computer for this course.

When you register for this course you will receive an outline for your laptop requirements.

Unit Price Estimating

This interactive *two-day* seminar teaches attendees how to interpret project information and process it into final, detailed estimates with the greatest accuracy level.

The most important credential an estimator can take to the job is the ability to visualize construction, and estimate accurately.

Some of what you'll learn:
- Interpreting the design in terms of cost
- The most detailed, time-tested methodology for accurate pricing
- Key cost drivers—material, labor, equipment, staging, and subcontracts
- Understanding direct and indirect costs for accurate job cost accounting and change order management

Who should attend: Corporate and government estimators and purchasers, architects, engineers… and others needing to produce accurate project estimates.

Square Foot & Assemblies Estimating

This *two-day* course teaches attendees how to quickly deliver accurate square foot estimates using limited budget and design information.

Some of what you'll learn:
- How square foot costing gets the estimate done faster
- Taking advantage of a "systems" or "assemblies" format
- The RSMeans "building assemblies/square foot cost approach"
- How to create a reliable preliminary systems estimate using bare-bones design information

Who should attend: Facilities managers, facilities engineers, estimators, planners, developers, construction finance professionals… and others needing to make quick, accurate construction cost estimates at commercial, government, educational, and medical facilities.

Repair & Remodeling Estimating

This *two-day* seminar emphasizes all the underlying considerations unique to repair/remodeling estimating and presents the correct methods for generating accurate, reliable R&R project costs using the unit price and assemblies methods.

Some of what you'll learn:
- Estimating considerations—like labor-hours, building code compliance, working within existing structures, purchasing materials in smaller quantities, unforeseen deficiencies
- Identifying problems and providing solutions to estimating building alterations
- Rules for factoring in minimum labor costs, accurate productivity estimates, and allowances for project contingencies
- R&R estimating examples calculated using unit price and assemblies data

Who should attend: Facilities managers, plant engineers, architects, contractors, estimators, builders… and others who are concerned with the proper preparation and/or evaluation of repair and remodeling estimates.

Mechanical & Electrical Estimating

This *two-day* course teaches attendees how to prepare more accurate and complete mechanical/electrical estimates, avoiding the pitfalls of omission and double-counting, while understanding the composition and rationale within the RSMeans Mechanical/Electrical database.

Some of what you'll learn:
- The unique way mechanical and electrical systems are interrelated
- M&E estimates–conceptual, planning, budgeting, and bidding stages
- Order of magnitude, square foot, assemblies, and unit price estimating
- Comparative cost analysis of equipment and design alternatives

Who should attend: Architects, engineers, facilities managers, mechanical and electrical contractors… and others needing a highly reliable method for developing, understanding, and evaluating mechanical and electrical contracts.

Green Building Planning & Construction

In this *two-day* course, learn about tailoring a building and its placement on the site to the local climate, site conditions, culture, and community. Includes information to reduce resource consumption and augment resource supply.

Some of what you'll learn:
- Green technologies, materials, systems, and standards
- Energy efficiencies with energy modeling tools
- Cost vs. value of green products over their life cycle
- Low-cost strategies and economic incentives and funding
- Health, comfort, and productivity goals and techniques

Who should attend: Contractors, project managers, building owners, building officials, healthcare and insurance professionals.

Facilities Maintenance & Repair Estimating

This *two-day* course teaches attendees how to plan, budget, and estimate the cost of ongoing and preventive maintenance and repair for existing buildings and grounds.

Some of what you'll learn:
- The most financially favorable maintenance, repair, and replacement scheduling and estimating
- Auditing and value engineering facilities
- Preventive planning and facilities upgrading
- Determining both in-house and contract-out service costs
- Annual, asset-protecting M&R plan

Who should attend: Facility managers, maintenance supervisors, buildings and grounds superintendents, plant managers, planners, estimators… and others involved in facilities planning and budgeting.

Practical Project Management for Construction Professionals

In this *two-day* course, acquire the essential knowledge and develop the skills to effectively and efficiently execute the day-to-day responsibilities of the construction project manager.

Covers:
- General conditions of the construction contract
- Contract modifications: change orders and construction change directives
- Negotiations with subcontractors and vendors
- Effective writing: notification and communications
- Dispute resolution: claims and liens

Who should attend: Architects, engineers, owner's representatives, project managers.

Facilities Repair & Remodeling Estimating

In this *two-day* course, professionals working in facilities management can get help with their daily challenges to establish budgets for all phase of a project

Some of what you'll learn:
- Determine the full scope of a project
- Identify the scope of risks & opportunities
- Creative solutions to estimating issues
- Organizing estimates for presentation & discussion
- Special techniques for repair/remodel and maintenance projects
- Negotiating project change orders

Who should attend: Facility managers, engineers, contractors, facility trades-people, planners & project managers

Scheduling and Project Management

This *two-day* course teaches attendees the most current and proven scheduling and management techniques needed to bring projects in on time and on budget.

Some of what you'll learn:
- Crucial phases of planning and scheduling
- How to establish project priorities and develop realistic schedules and management techniques
- Critical Path and Precedence Methods
- Special emphasis on cost control

Who should attend: Construction project managers, supervisors, engineers, estimators, contractors... and others who want to improve their project planning, scheduling, and management skills.

Understanding & Negotiating Construction Contracts

In this *two-day* course, learn how to protect the assets of your company by justifying or eliminating commercial risk through negotiation of a contract's terms and conditions.

Some of what you'll learn:
- Myths and paradigms of contracts
- Scope of work, terms of payment, and scheduling
- Dispute resolution
- Why clients love CLAIMS!
- Negotiating issues and much more

Who should attend: Contractors & subcontractors, material suppliers, project managers, risk & insurance managers, procurement managers, owners and facility managers, corporate executives.

Site Work & Heavy Construction Estimating

This *two-day* course teaches attendees how to estimate earthwork, site utilities, foundations, and site improvements, using the assemblies and the unit price methods.

Some of what you'll learn:
- Basic site work and heavy construction estimating skills
- Estimating foundations, utilities, earthwork, and site improvements
- Correct equipment usage, quality control, and site investigation for estimating purposes

Who should attend: Project managers, design engineers, estimators, and contractors doing site work and heavy construction.

Assessing Scope of Work for Facility Construction Estimating

This *two-day* course is a practical training program that addresses the vital importance of understanding the SCOPE of projects in order to produce accurate cost estimates in a facility repair and remodeling environment.

Some of what you'll learn:
- Discussions of site visits, plans/specs, record drawings of facilities, and site-specific lists
- Review of CSI divisions, including means, methods, materials, and the challenges of scoping each topic
- Exercises in SCOPE identification and SCOPE writing for accurate estimating of projects
- Hands-on exercises that require SCOPE, take-off, and pricing

Who should attend: Corporate and government estimators, planners, facility managers, and others who need to produce accurate project estimates.

2009 RSMeans Seminar Schedule

Note: call for exact dates and details.

Location	Dates
Las Vegas, NV	March
Washington, DC	April
Denver, CO	May
San Francisco, CA	June
Washington, DC	September
Dallas, TX	September
Las Vegas, NV	October
Atlantic City, NJ	October
Orlando, FL	November
San Diego, CA	December

1-800-334-3509, ext. 5115

501

Professional Development

Registration Information

Register early... Save up to $100! Register 30 days before the start date of a seminar and save $100 off your total fee. *Note: This discount can be applied only once per order. It cannot be applied to team discount registrations or any other special offer.*

How to register Register by phone today! The RSMeans toll-free number for making reservations is **1-800-334-3509, ext. 5115.**

Individual seminar registration fee - $935. *Means CostWorks*® training registration fee - $375. To register by mail, complete the registration form and return, with your full fee, to: RSMeans Seminars, 63 Smiths Lane, Kingston, MA 02364.

Government pricing All federal government employees save off the regular seminar price. Other promotional discounts cannot be combined with the government discount.

Team discount program for two to four seminar registrations. Call for pricing: 1-800-334-3509, ext. 5115

Multiple course discounts When signing up for two or more courses, call for pricing.

Refund policy Cancellations will be accepted up to ten business days prior to the seminar start. There are no refunds for cancellations received later than ten working days prior to the first day of the seminar. A $150 processing fee will be applied for all cancellations. Written notice of cancellation is required. Substitutions can be made at any time before the session starts. **No-shows are subject to the full seminar fee.**

AACE approved courses Many seminars described and offered here have been approved for 14 hours (1.4 recertification credits) of credit by the AACE International Certification Board toward meeting the continuing education requirements for recertification as a Certified Cost Engineer/Certified Cost Consultant.

AIA Continuing Education We are registered with the AIA Continuing Education System (AIA/CES) and are committed to developing quality learning activities in accordance with the CES criteria. Many seminars meet the AIA/CES criteria for Quality Level 2. AIA members may receive (14) learning units (LUs) for each two-day RSMeans course.

NASBA CPE sponsor credits We are part of the National Registry of CPE sponsors. Attendees may be eligible for (16) CPE credits.

Daily course schedule The first day of each seminar session begins at 8:30 a.m. and ends at 4:30 p.m. The second day begins at 8:00 a.m. and ends at 4:00 p.m. Participants are urged to bring a hand-held calculator, since many actual problems will be worked out in each session.

Continental breakfast Your registration includes the cost of a continental breakfast, and a morning and afternoon refreshment break. These informal segments allow you to discuss topics of mutual interest with other seminar attendees. (You are free to make your own lunch and dinner arrangements.)

Hotel/transportation arrangements RSMeans arranges to hold a block of rooms at most host hotels. To take advantage of special group rates when making your reservation, be sure to mention that you are attending the RSMeans seminar. You are, of course, free to stay at the lodging place of your choice. (**Hotel reservations and transportation arrangements should be made directly by seminar attendees.**)

Important Class sizes are limited, so please register as soon as possible.

Note: Pricing subject to change.

Registration Form

ADDS-1000

Call 1-800-334-3509, ext. 5115 to register or FAX this form 1-800-632-6732. Visit our Web site: www.rsmeans.com

Please register the following people for the RSMeans construction seminars as shown here. We understand that we must make our own hotel reservations if overnight stays are necessary.

☐ Full payment of $_____enclosed.

☐ Bill me

Please print name of registrant(s)

(To appear on certificate of completion)

_____

_____

_____

_____

P.O. #: _____
GOVERNMENT AGENCIES MUST SUPPLY PURCHASE ORDER NUMBER OR TRAINING FORM.

Firm name_____

Address_____

City/State/Zip_____

Telephone no._____ Fax no._____

E-mail address_____

Charge registration(s) to: ☐ MasterCard ☐ VISA ☐ American Express

Account no._____Exp. date_____

Cardholder's signature_____

Seminar name_____

Seminar City _____

Please mail check to: RSMeans Seminars, 63 Smiths Lane, P.O. Box 800, Kingston, MA 02364 USA

503

**For more information
visit the RSMeans website
at www.rsmeans.com**

New Titles

The Gypsum Construction Handbook by USG

An invaluable reference of construction procedures for gypsum drywall, cement board, veneer plaster, and conventional plaster. This book includes the newest product developments, installation methods, fire- and sound-rated construction information, illustrated framing-to-finish application instructions, estimating and planning information, and more. Great for architects and engineers; contractors, builders, and dealers; apprentices and training programs; building inspectors and code officials; and anyone interested in gypsum construction. Features information on tools and safety practices, a glossary of construction terms, and a list of important agencies and associations. Also available in Spanish.

$29.95
Over 500 pages, illustrated, softcover
Catalog no. 67357

BIM Handbook
A Guide to Building Information Modeling for Owners, Managers, Designers, Engineers and Contractors
by Chuck Eastman, Paul Teicholz, Rafael Sacks, Kathleen Liston

Building Information Modeling (BIM) is a new approach to design, construction, and facility management where a digital depiction of the building process is used to facilitate the exchange of information in digital format.

The BIM Handbook provides an in-depth understanding of BIM technologies, the business and organizational issues associated with its implementation, and the profound advantages that effective use of BIM can provide to all members of a project team. Also includes a rich set of informative BIM case studies.

$85.00
504 pages
Catalog no. 67356

Universal Design Ideas for Style, Comfort & Safety
by RSMeans and Lexicon Consulting, Inc.

Incorporating universal design when building or remodeling helps people of any age and physical ability more fully and safely enjoy their living spaces. This book shows how universal design can be artfully blended into the most attractive homes. It discusses specialized products like adjustable countertops and chair lifts, as well as simple ways to enhance a home's safety and comfort. With color photos and expert guidance, every area of the home is covered. Includes budget estimates that give an idea how much projects will cost.

$21.95
160 pages, illustrated, softcover
Catalog no. 67354

Green Home Improvement
by Daniel D. Chiras, PhD

With energy costs rising and environmental awareness increasing, many people are looking to make their homes greener. This book makes it easy, with 65 projects and actual costs and projected savings that will help homeowners prioritize their green improvements.

Projects range from simple water savers that cost only a few dollars, to bigger-ticket items such as efficient appliances and HVAC systems, green roofing, flooring, landscaping, and much more. With color photos and cost estimates, each project compares options and describes the work involved, the benefits, and the savings.

$24.95
320 pages, illustrated, softcover
Catalog no. 67355

Residential & Light Commercial Construction Standards, 3rd Edition, Updated
by RSMeans and contributing authors

This book provides authoritative requirements and recommendations compiled from leading professional associations, industry publications, and building code organizations. This all-in-one reference helps establish a standard for workmanship, quickly resolve disputes, and avoid defect claims. Updated third edition includes new coverage of green building, seismic, hurricane, and mold-resistant construction, plus updated national building code requirements.

Covers all areas of building construction, including paving; concrete; framing; finish carpentry; insulation; roofing and siding; doors and windows; drywall; ceilings and floors; wall coverings; plumbing; HVAC; electrical; and more!

$59.95
Over 550 pages, illustrated, softcover
Catalog no. 67322B

2009 Order Form

Qty.	Book no.	COST ESTIMATING BOOKS	Unit Price	Total
	60069	Assemblies Cost Data 2009	$245.95	
	60019	Building Construction Cost Data 2009	154.95	
	60119	Concrete & Masonry Cost Data 2009	136.95	
	50149	Construction Cost Indexes 2009 (subscription)	335.00	
	60149A	Construction Cost Index–January 2009	83.75	
	60149B	Construction Cost Index–April 2009	83.75	
	60149C	Construction Cost Index–July 2009	83.75	
	60149D	Construction Cost Index–October 2009	83.75	
	60349	Contr. Pricing Guide: Resid. R & R Costs 2009	39.95	
	60339	Contr. Pricing Guide: Resid. Detailed 2009	39.95	
	60329	Contr. Pricing Guide: Resid. Sq. Ft. 2009	39.95	
	60239	Electrical Change Order Cost Data 2009	154.95	
	60039	Electrical Cost Data 2009	154.95	
	60209	Facilities Construction Cost Data 2009	368.95	
	60309	Facilities Maintenance & Repair Cost Data 2009	336.95	
	60169	Heavy Construction Cost Data 2009	154.95	
	60099	Interior Cost Data 2009	154.95	
	60129	Labor Rates for the Const. Industry 2009	336.95	
	60189	Light Commercial Cost Data 2009	132.95	
	60029	Mechanical Cost Data 2009	154.95	
	60159	Open Shop Building Const. Cost Data 2009	154.95	
	60219	Plumbing Cost Data 2009	154.95	
	60049	Repair & Remodeling Cost Data 2009	132.95	
	60179	Residential Cost Data 2009	132.95	
	60289	Site Work & Landscape Cost Data 2009	154.95	
	60059	Square Foot Costs 2009	168.95	
	62018	Yardsticks for Costing (2008)	149.95	
	62019	Yardsticks for Costing (2009)	154.95	
		REFERENCE BOOKS		
	67310A	ADA Compliance Pricing Guide, 2nd Ed.	79.95	
	67330	Bldrs Essentials: Adv. Framing Methods	12.98	
	67329	Bldrs Essentials: Best Bus. Practices for Bldrs	29.95	
	67298A	Bldrs Essentials: Framing/Carpentry 2nd Ed.	15.98	
	67298AS	Bldrs Essentials: Framing/Carpentry Spanish	15.98	
	67307	Bldrs Essentials: Plan Reading & Takeoff	35.95	
	67342	Building & Renovating Schools	69.98	
	67339	Building Security: Strategies & Costs	44.98	
	67353	Complete Book of Framing	29.95	
	67146	Concrete Repair & Maintenance Illustrated	69.95	
	67352	Construction Business Management	49.95	
	67314	Cost Planning & Est. for Facil. Maint.	89.95	
	67328A	Designing & Building with the IBC, 2nd Ed.	99.95	
	67230B	Electrical Estimating Methods, 3rd Ed.	64.95	
	64777A	Environmental Remediation Est. Methods, 2nd Ed.	49.98	
	67343	Estimating Bldg. Costs for Resi. & Lt. Comm.	29.95	
	67276A	Estimating Handbook, 2nd Ed.	$99.95	
	67318	Facilities Operations & Engineering Reference	54.98	
	67301	Facilities Planning & Relocation	49.98	
	67338A	Green Building: Proj. Planning & Cost Est., 2nd Ed.	129.95	
	67355	Green Home Improvement	24.95	

Qty.	Book no.	REFERENCE BOOKS (Cont.)	Unit Price	Total
	67357	The Gypsum Construction Handbook	29.95	
	67357S	The Gypsum Construction Handbook (Spanish)	29.95	
	67323	Historic Preservation: Proj. Planning & Est.	49.98	
	67349	Home Addition & Renovation Project Costs	19.98	
	67308E	Home Improvement Costs–Int. Projects, 9th Ed.	24.95	
	67309E	Home Improvement Costs–Ext. Projects, 9th Ed.	24.95	
	67344	Homeowner's Guide to Mold	21.95	
	67324B	How to Est.w/Means Data & CostWorks, 3rd Ed.	59.95	
	67351	How Your House Works	21.95	
	67282A	Illustrated Const. Dictionary, Condensed, 2nd Ed.	59.95	
	67292A	Illustrated Const. Dictionary, w/CD-ROM, 3rd Ed.	99.95	
	67348	Job Order Contracting	89.95	
	67347	Kitchen & Bath Project Costs	19.98	
	67295C	Landscape Estimating Methods, 5th Ed.	64.95	
	67341	Life Cycle Costing for Facilities	99.95	
	67294B	Mechanical Estimating Methods, 4th Ed.	64.95	
	67245A	Planning & Managing Interior Projects, 2nd Ed.	34.98	
	67283B	Plumbing Estimating Methods, 3rd Ed.	29.98	
	67345	Practice of Cost Segregation Analysis	99.95	
	67337	Preventive Maint. for Higher Education Facilities	149.95	
	67346	Preventive Maint. for Multi-Family Housing	89.95	
	67326	Preventive Maint. Guidelines for School Facil.	149.95	
	67247B	Project Scheduling & Management for Constr. 3rd Ed.	64.95	
	67265B	Repair & Remodeling Estimating Methods, 4th Ed.	69.95	
	67322B	Resi. & Light Commercial Const. Stds., 3rd Ed.	59.95	
	67327A	Spanish/English Construction Dictionary, 2nd Ed.	23.95	
	67145B	Sq. Ft. & Assem. Estimating Methods, 3rd Ed.	69.95	
	67321	Total Productive Facilities Management	29.98	
	67350	Understanding and Negotiating Const. Contracts	69.95	
	67303B	Unit Price Estimating Methods, 4th Ed.	59.95	
	67354	Universal Design	21.95	
	67319A	Value Engineering: Practical Applications	79.95	

MA residents add 5% state sales tax

Shipping & Handling**

Total (U.S. Funds)*

Prices are subject to change and are for U.S. delivery only. *Canadian customers may call for current prices. **Shipping & handling charges: Add 7% of total order for check and credit card payments. Add 9% of total order for invoiced orders.

Send order to: **ADDV-1000**

Name (please print) _____

Company _____

☐ **Company**

☐ **Home   Address** _____

_____

City/State/Zip _____

Phone # _____ P.O. # _____

(Must accompany all orders being billed)

RSMeans Project Cost Report

By filling out this report, your project data will contribute to the database that supports the RSMeans Project Cost Square Foot Data. When you fill out this form, RSMeans will provide a $30 discount off one of the RSMeans products advertised in the preceding pages. Please complete the form including all items where you have cost data, and all the items marked (✔).

$30.00 Discount per product for each report you submit.

Project Description (No remodeling projects, please.)

✔ Building Use (Office, School...) _____

✔ Address (City, State) _____

_____

✔ Building Area (SF) _____

✔ Frame (Wood, Steel...) _____

✔ Exterior Wall (Brick, Tilt-up...) _____

✔ Basement: (check one) ☐ Full ☐ Partial ☐ None

✔ Number Stories _____

✔ Floor-to-Floor Height _____

✔ Volume (C.F.) _____

% Air Conditioned _____ Tons _____

Total Project Cost $ _____

Owner _____

Architect _____

General Contractor _____

✔ Bid Date _____

✔ Typical Bay Size _____

✔ Capacity _____

✔ Labor Force: _____ % Union _____ % Non-Union

✔ Project Description (Circle one number in each line)

1. Economy 2. Average 3. Custom 4. Luxury
1. Square 2. Rectangular 3. Irregular 4. Very Irregular

Comments _____

_____

A	✔ General Conditions	$	
B	✔ Site Work	$	
C	✔ Concrete	$	
D	✔ Masonry	$	
E	✔ Metals	$	
F	✔ Wood & Plastics	$	
G	✔ Thermal & Moisture Protection	$	
GR	Roofing & Flashing	$	
H	✔ Doors and Windows	$	
J	✔ Finishes	$	
JP	Painting & Wall Covering	$	

K	✔ Specialties	$	
L	✔ Equipment	$	
M	✔ Furnishings	$	
N	✔ Special Construction	$	
P	✔ Conveying Systems	$	
Q	✔ Mechanical	$	
QP	Plumbing	$	
QB	HVAC	$	
R	✔ Electrical	$	
S	✔ Mech./Elec. Combined	$	

Please specify the RSMeans product you wish to receive. Complete the address information.

Product Name _____

Product Number _____

Your Name _____

Title _____

Company _____

☐ Company
☐ Home Street Address _____

City, State, Zip _____

Method of Payment:

Credit Card # _____

Expiration Date _____

Check _____

Purchase Order _____

R.S. Means Company, Inc.
Square Foot Costs Department
P.O. Box 800
Kingston, MA 02364-9988

Return by mail or Fax 888-492-6770.